NON-CONVENTIONAL SOURCES OF ENERGY

[A Textbook for Engineering Students]

G.D. Rai
B.Sc., B.E. (Mech.) M.E. (Hons)
MSESI, MISTE.
Formerly,
Department of Mechanical Engineering
Samrat Ashok Technological Institute
Vidisha (M.P.)

KHANNA PUBLISHERS®
Investing in Learning®
Operational Office :
4575/15, Onkar House, Ground Floor,
Opp. Happy School, Daryaganj, New Delhi 110 002
Phones : 011-45033819 • *Mob*. 09811541460, 9650669626
email : contactus@khannapublishers.in
website : khannapublishers.in

Published by :

Romesh Chander Khanna & Vineet Khanna
for **KHANNA PUBLISHERS**
2-B, Nath Market, Nai Sarak,
Delhi-110006 (India)

Visit us at : www.khannapublishers.in

Hologram & Description

To all readers of our books, to prevent yourself from being defrauded by piraters, please make sure that there is an Hologram on the cover of our books with the below specifications. If you find any book without Hologram and Description, please mail us at **contactus@khannapublishers.in**

Thanking you

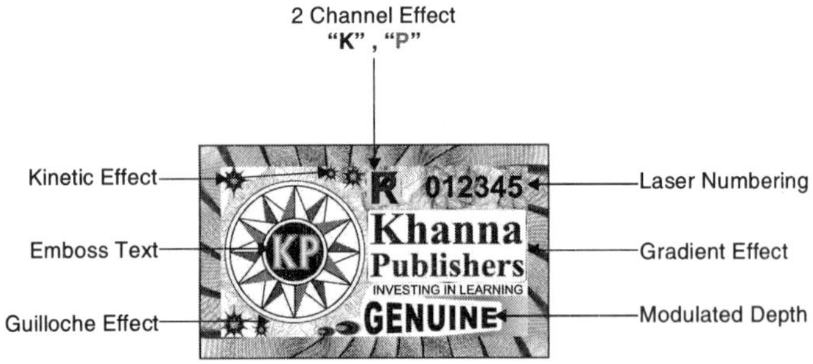

ISBN No. : 978-81-7409-073-8

First Edition : **1988**
Sixth Edition : **2017**
Sixth Reprint : **2026**

***Price* : ₹ 639.00**

Compiled by

Smt. C.K. RAI

To

My loving children

Atul and Rachna

Preface to the Sixth Edition

The author is highly thankful to the readers for the warm welcome given to the previous editions of this book and sending their appreciation and suggestions. As far as possible suggestions received from readers have been considered and relevant additions have been incorporated. The present edition has been made comprehensive for increasing its utility to the students.

Most of the chapters have been reinforced with additional materials wherever considered necessary. As per the suggestions by readers, the chapter on "Energy Conservation" is added.

Some more solved and unsolved problems are added after the relevant articles, for better understanding and use of the articles.

Suggestions for the further improvement in the book will be gratefully acknowledged.

April 20, 2017 G.D. RAI

Preface to the First Edition

It is a well known fact the known sources of fossil fuels in the world are depleting very fast and by the turn of the century, man will have to increasingly depend upon renewable resources of energy. For developing countries like India, large scale demand of heat energy for meeting day to day domestic, institutional and industrial requirements can be met by utilizing solar thermal systems, biogas, PV cells, wind energy, Geothermal, MHD etc. *i.e.* from non-conventional sources of energy. Non-conventional sources of energy is one of the areas of emerging technologies which have higher priority with reference to national needs. Energy is an important input in all sectors of any country's economy. The standard of living of a given country can be directly related to per capita energy consumption. Energy needs of a developing country like India is going to increase significantly in foreseable future, is a well known fact. Hence it is absolutely necessary to plan for this critical input, so that the country's development is not hampered.

Keeping the above consideration in view, a number of universities, colleges and polytechnics are now offering this course for the students. It appeared to author that there is a need for a conscise and practical book on non-conventional energy sources. The treatment of the book begins with fundamental and develops in a way that allows the practical man as well as the student to expand his knowledge progressively. Throughout the text attempt has been made to present the subject matter in a simple, lucid and precise manner. Sufficient solved examples are given to illustrate the use of the equations developed in the text. The book is written in 15 chapters. In dealing with the different non-conventional energy systems, the book has been divided into 7 *units, i.e.* :

1. Solar energy
2. Wind energy
3. Energy from biomass and biogas
4. Geothermal energy
5. Energy from oceans *i.e.* tidal energy, ocean thermal electric conversion (OTEC) and wave energy.
6. Chemical energy sources
7. MHD (Magneto hydrodynamics), Thermo nuclear fusion and small scale power generation sources such as thermoelectric and thermionic converters.

Unit 1. Solar energy source has been described under the chapters, solar radiation and its measurement, solar energy collectors, solar energy storage systems and solar energy applications.

Unit 2. Covers *Wind energy* conversion systems.

Unit 3. energy from Biomass is discussed under two types *i.e.* biogas generation and biomass gasification.

Unit 4. discusses the geothermal energy source.

Unit 5. different energy forms from the oceans are described in the order, ocean thermal electric conversion, tidal energy, wave energy and small scale hydroelectric system.

Unit 6. discusses the two types of chemical energy sources. Fuel cells and Batteries, and Hydrogen energy.

Unit 7. MHD, thermonuclear fusion and other small scale energy sources such as thermo electric and thermionic conversion systems are discussed.

It is regretted, however, that despite careful scrutiny of the proofs, several printing mistakes have nevertheless remained in the book. If such mistakes are pointed out by readers, the same will be gratefully received. The author sincerely hopes that readers of the text will be benefitted from approach presented and will oblige with useful suggestions.

In the last, the author feels highly indebted to all those who have been helpful directly or indirectly, in bringing the book to the present shape.

March 7, 1988 G.D. RAI

Contents

Chapter			Page
1.	**An Introduction to Energy Sources**		**1—33**
	1.1.	General	... 1
	1.2.	Energy Consumption as a Measure of Prosperity	... 1
	1.3.	World Energy Futures	... 2
	1.4.	Energy Sources and their Availability	... 3
		1.4.1. Introduction	... 3
		1.4.2. Commercial or Conventional Energy Sources	... 3
		1.4.3. New Energy Technologies	... 9
	1.5.	Renewable Energy Sources	... 9
		1.5.1. Advantages of Fluidised Bed Combustion Boilers	... 11
	1.6.	Non Conventional Sources	... 11
		1.6.1. Solar Energy	... 13
		1.6.2. Electricity from Solar Energy	... 16
		1.6.3. Wind Energy	... 17
		1.6.4. Energy from Biomass and Biogas	... 19
	1.7.	Energy plantation	... 20
		1.7.1. Biogas	... 20
		1.7.2. Ocean Thermal Energy Conversion	... 20
		1.7.3. Tidal Energy	... 21
	1.8.	Geothermal Energy	... 22
	1.9.	Hydrogen Energy	... 23
	1.10.	Fuel Cells	... 24
	1.11.	Magneto Hydro-Dynamics Generator	... 25
	1.12.	Thermionic Converter	... 26
	1.13.	Thermo Electric Power	... 26
	1.14.	Renewable Energy Resources	... 27
		1.14.1. Advantages of Renewable Energy	... 28
	1.15.	Prospects of Renewable Energy Sources	... 29
		Questions	... 33

UNIT 1—SOLAR-ENERGY

2.	**Solar Radiation and its Measurement**		**37—56**
	2.1.	Introduction	... 37
	2.2.	Solar Constant	... 37
	2.3.	Solar Radiation at the Earth's Surface	... 39
		2.3.1. Beam and Diffuse Solar Radiation	... 39
		2.3.2. Sun at Zenith	... 40
		2.3.3. Air mass	... 40
		2.3.4. Alternation of Beam Radiation	... 40
	2.4.	Solar Radiation Geometry	... 42

Chapter *Page*

2.4.1. Altitude angle α ... 43
2.4.2. Zenith angle θ_z ... 44
2.4.3. Solar Azimuth angles ... 44
2.4.4. The Slope (s) ... 44
2.4.5. Surface Azimuth ... 44
2.4.6. Incident angle ... 44
2.4.7. Day length ... 45
2.5. Local Solar Time ... 45
2.6. Solar Radiation Measurements ... 47
2.6.1. Pyrheliometers ... 47
2.6.2. Pyranometers ... 49
2.7. Sunshine Recoder ... 50
2.8. Solar Radiation Data ... 50
2.9. Estimation of Average Solar Radiation ... 51
2.10. Solar Radiation on Tilted Surfaces ... 54
2.10.1. Beam Radiation ... 54
2.10.2. Total Radiation ... 55
Questions ... 56

3. Solar Energy Collectors **57—95**
3.1. Introduction ... 57
3.2. Physical Principles of the Conversion of Solar Radiation into Heat ... 57
3.3. Flat-Plate Collectors ... 59
3.4. Transmissivity of Cover System ... 67
3.5. Energy Balance Equation and Collector Efficiency ... 71
3.6. Thermal Analysis of Flat-plate Collector and Useful Heat Gained by the Fluid ... 74
3.7. Concentrating Collector: Focusing Type ... 80
3.8. Advantages and Disadvantages of Concentrating Collectors Over Flat-Plate Type Collectors ... 86
3.9. Performance Analysis of a Cylindrical Parabolic Concentrating Collector ... 87
3.10. Selective Absorber Coatings ... 94
Questions ... 95

4. Solar Energy Storage **96—112**
4.1. Introduction ... 96
4.2. Solar Energy Storage Systems ... 97
4.3. Solar Pond ... 106
4.3.1. Introduction ... 106
4.3.2. Principle of Operation and Description of Non-convective Solar Pond ... 107
4.3.3. Extraction of Thermal Energy ... 109
4.3.4 Applications of Solar Ponds ... 110
Questions ... 111

Chapter			Page
5.	**Applications of Solar Energy**		**113—171**
5.1.	Introduction	...	113
5.2.	Solar Water Heating	...	114
5.3.	Space-Heating (or Solar Heating of Buildings)	...	117
5.4.	Space Cooling (or Solar Cooling of Buildings)	...	126
5.5.	Solar Thermal Electric Conversion	...	130
5.6.	Solar Electric Power Generation : Solar Photo-Voltaics	...	138
5.7.	Agriculture and Industrial Process Heat	...	149
5.8.	Solar Distillation	...	150
5.9.	Solar Pumping	...	152
5.10.	Solar Furnace	...	154
5.11.	Solar Cooking	...	156
5.12.	Solar Green Houses	...	160
5.13.	Solar Production of Hydrogen	...	169
	Questions	...	171

UNIT 2—WIND-ENERGY

6.	**Wind Energy**		**175—238**
6.1.	Introduction	...	175
6.2.	Basic Principles of Wind Energy Conversion	...	177
	6.2.1. The Nature of the Wind	...	177
	6.2.2. The Power in the Wind	...	178
	6.2.3. Forces on the Blades	...	184
	6.2.4. Wind Energy Conversion	...	185
6.3.	Wind Data and Energy Estimation	...	189
6.4.	Site Selection Considerations	...	193
6.5.	Basic Components of a WECS (*Wind Energy Conversion System*)	...	196
6.6.	Classification of WEC Systems	...	199
6.7.	Advantages and Disadvantages of WECS	...	200
6.8.	Wind Energy Collectors	...	201
	6.8.1. Introduction	...	201
	6.8.2. Horizontal-Axial Machines	...	201
	6.8.3. Design Consideration of Horizontal-Axis Machines	...	204
	6.8.4. Vertical-Axis Machines	...	212
6.9.	Analysis of Aerodynamic Forces Acting on the Blade	...	218
6.10.	Performance of Wind-machines	...	220
6.11.	Generating Systems	...	223
	6.11.1. Introduction	...	223
	6.11.2. Schemes for Electric Generation	...	224
	6.11.3. Generator Control	...	227
	6.11.4. Load Control	...	228
6.12.	Energy Storage	...	228
6.13.	Applications of Wind Energy	...	229

Chapter			Page
6.14.	Interconnected System	...	235
6.15.	Safety Systems	...	236
6.16.	Environmental Aspects	...	237
	Questions	...	237

UNIT 3—ENERGY FROM BIOMASS (Bio-Energy)

7.	**Energy from Biomass**		**241—329**
7.1.	Introduction	...	241
7.2.	Biomass Conversion Technologies	...	246
	7.2.1. Wet Processes	...	247
	7.2.2. Dry Processes	...	248
7.3.	Photosynthesis	...	249
7.4.	Biogas Generation	...	251
7.5.	Factors Affecting Biodigestion or Generation of Gas	...	255
7.6.	Classification of Biogas Plants	...	259
7.7.	Advantages and Disadvantages of Floating Drum Plant Advantages	...	261
7.8.	Advantages and Disadvantages of Fixed Dome Type Plant	...	262
7.9.	Types of Biogas Plants	...	262
7.10.	Constructional Details of some Main Digesters	...	270
7.11.	Bio-gas from Plant Wastes	...	272
7.12.	Community Biogas Plants	...	277
7.13.	Materials Used for Bio-gas Generation	...	278
7.14.	Selection of Site for a Biogas Plant	...	279
7.15.	Digester Design Considerations	...	280
7.16.	Methods for Maintaining Biogas Production	...	287
7.17.	Problems Related to Bio-gas Plants	...	289
7.18.	Starting a Biogas Plant	...	290
7.19.	Filling a Digester for Starting	...	290
7.20.	Fuel Properties of Bio-gas	...	290
7.21.	Utilization of Bio-gas	...	291
7.22.	Biomass as a Source of Energy	...	294
	7.22.1. Introduction	...	294
	7.22.2. Energy Plantation	...	295
	7.22.3. Advantages of Energy Plantation	...	296
	7.22.4. Plants Proposed for Energy Plantation	...	297
7.23.	Methods for Obtaining Energy from Biomass	...	299
7.24.	Thermal Gasification of Biomass	...	301
	7.24.1. Introduction	...	301
	7.24.2. Classification of Biomass Gasifiers	...	303
	7.24.3. Chemistry of the Gasification Process	...	308
	7.24.4. Applications of the Gasifier	...	310
	7.24.5. Problems in Developments of Gasifiers	...	312

Chapter			Page
7.25.	Advantages and Disadvantages of Bio-logical Conversion of Solar Energy	...	314
7.26.	Pyrolysis (destructive distillation)	...	314
7.27.	Further Thermochemical Processes	...	317
7.28.	Alternative Liquid Fuels (Alcohol Fuels)	...	318
7.29.	Conclusion	...	328
	Questions	...	328

UNIT 4—GEOTHERMAL ENERGY

8.	**Geothermal Energy**		**333—372**
8.1.	Introduction	...	333
8.2.	Estimates of Geothermal Power	...	336
8.3.	Nature of Geothermal Fields	...	336
8.4.	Geothermal Sources	...	336
8.5.	Hydrothermal (Convective) Resources	...	338
	8.5.1 Vapour Dominated Systems	...	339
	8.5.2 Liquid-Dominated Systems	...	343
8.6.	Geopressured Resources	...	350
8.7.	Hot Dry Rock Resources of Petrothermal Systems	...	351
8.8.	Magma Resources (Molten rock-chamber systems)	...	352
8.9.	Comparison of Flashed Steam and Total Flow Concept	...	353
8.10.	Interconnection of Geothermal Fossil Systems (*Hybrid Systems*)	...	354
8.11.	Prime Movers for Geothermal Energy Conversion	...	356
	8.11.1. Impulse Reaction Machines	...	356
	8.11.2. Positive Displacement Machines	...	358
	8.11.3. Impulse Machine	...	361
8.12.	Advantages and Disadvantages of Geothermal Energy over other Energy Forms	...	361
8.13.	Applications of Geothermal Energy	...	362
8.14.	Material Selection for Geothermal Power Plants	...	365
8.15.	Geothermal Exploration	...	366
8.16.	Geothermal Well Drilling	...	367
8.17.	Operational and Environmental Problems	...	368
8.18.	Geothermal Energy in India : Prospects	...	370
	Questions	...	372

UNIT 5—ENERGY FROM THE OCEANS

9.	**Energy from the Oceans**		**375—423**
9.1.	Introduction	...	375
9.2.	Ocean Thermal Electric Conversion (OTEC)	...	376
	9.2.1. Introduction	...	376
	9.2.2. Methods of Ocean Thermal Electric Power Generation	...	379

Chapter			Page
	9.2.3.	Open Cycle OTEC System ...	379
	9.2.4.	The Closed or Anderon, OTEC Cycle ...	381
	9.2.5.	Heat Exchangers (Evaporators) ...	383
	9.2.6.	Bio-fouling ...	384
	9.2.7.	Site-Selection ...	385
	9.2.8.	Energy Utilization ...	385
	9.2.9.	Hybrid Cycle ...	385
	9.2.10.	Conclusions ...	386
	9.2.11.	Prospects of Ocean Thermal Energy Conversion in India ...	386
9.3.	Energy from Tides ...		387
	9.3.1.	Introduction ...	387
	9.3.2.	Basic Principle of Tidal Power ...	389
	9.3.3.	Components of Tidal Power Plants ...	391
	9.3.4.	Operation Methods of Utilization of Tidal Energy ...	394
	9.3.5.	Estimate of Energy and Power in Simple Single Basin Tidal System ...	398
	9.3.6.	Estimation of Energy and Power in a Double Cycle System ...	400
	9.3.7.	Site Requirements ...	401
	9.3.8.	Storage ...	402
	9.3.9.	Advantages and Limitation of Tidal Power Generation ...	403
	9.3.10.	Prospects of Tidal Energy in India ...	404
9.4.	Ocean Waves ...		404
	9.4.1.	Introduction ...	404
	9.4.2.	Advantages and Disadvantages of Wave Energy ...	405
	9.4.3.	Energy and Power from the Waves ...	406
	9.4.4.	Wave-Energy Conversion Devices ...	407
9.5.	Small Scale Hydroelectric ...		411
	9.5.1.	Introduction ...	411
	9.5.2.	What is Small Hydel Development ? ...	413
	9.5.3.	Nature of Small Hydro Development ...	414
	9.5.4.	Classification of Small Hydro Power Stations ...	414
	9.5.5.	Components of a Hydroelectric Scheme ...	415
	9.5.6.	Civil Works Design Considerations for Mini and Micro Hydel Projects ...	416
	9.5.7.	Turbines and Generators for Small Scale Hydro Electric ...	419
	9.5.8.	Protection, Control and Management of Equipment ...	421
	9.5.9.	Advantages and Limitations of Small Scale Hydro-electric ...	421
	9.5.10.	Hybrid Systems ...	422
	9.5.11.	Conclusion ...	422
	Questions ...		423

Chapter *Page*

UNIT 6—CHEMICAL ENERGY SOURCES

10. Chemical Energy Sources **427—463**

10.1. General Introduction ... 427
10.2. Fuel Cells ... 427
10.2.1. Introduction ... 427
10.2.2. Design and Principle of Operation of a Fuel Cell (with, special reference to H_2, O_2, cell) ... 428
10.2.3. Classification of Fuel Cells ... 431
10.2.4. Types of Fuel Cells ... 431
10.2.5. Advantages and Disadvantages of Fuel Cell ... 437
10.2.6. Conversion Efficiency of Fuel Cells ... 437
10.2.7. Types of Electrodes ... 441
10.2.8. Work Output and EMF of Fuel Cell ... 441
10.2.9. Applications of Fuel Cells 448
10.3. Batteries ... 450
10.3.1. Introduction ... 450
10.3.2. Basic Battery Theory ... 450
10.3.3. Definitions of Fundamental Quantities ... 451
10.3.4. Battery Fundamental Characteristics ... 452
10.3.5. Different Types of Battery Arrangement ... 453
10.3.6. Classification of Batteries ... 454
10.3.7. Advantages of Batteries for Bulk Energy Storage ... 462
Questions ... 463

11. Hydrogen Energy **464—499**

11.1. Introduction ... 464
11.2. Hydrogen Production ... 466
11.2.1. Introduction ... 466
11.2.2. Electrolysis ... 467
11.2.3. Thermochemical Methods ... 471
11.2.4. Some Thermochemical Cyclic Processes ... 473
11.2.5. Fossil Fuel Methods ... 477
11.2.6. Solar Energy Methods ... 482
11.3. Hydrogen-Storage ... 484
11.4. Hydrogen Transpotation ... 488
11.5. Utilization of Hydrogen Gas ... 489
11.6. Hydrogen as an Altnernative Fuel for Motor Vehicles ... 493
11.7. Safety and Management ... 495
11.8. Hydrogen Technology Development in India ... 498
Review Questions ... 499

UNIT 7—ADDITIONAL ALTERNATE ENERGY SOURCES AND IMPROVED ENERGY UTILIZATION

12. Magneto Hydro Dynamic (MHD) Power Generation **503—530**

12.1 Introduction ... 503
12.2. Principles of MHD Power Generation ... 503

Chapter		Page
12.3.	MHD Systems	... 507
	12.3.1. Introduction	... 507
	12.3.2. Open-cycle Systems	... 508
	12.3.3. Closed-cycle Systems	... 510
12.4.	MHD Design Problems and Developments	... 513
12.5.	Advantages of MHD Systems	... 514
12.6.	Electrical Conditions : Voltage and Power Output of MHD Generator	... 515
12.7.	Gas Conductivity	... 519
12.8.	Materials for MHD Generators	... 523
12.9.	Magnetic Field	... 526
12.10.	Superconductivity	... 526
12.11.	International Status of MHD Power Generation and its Future Prospects	... 528
	Review Questions	... 530
13.	**Thermo Electric Power**	**531—551**
13.1.	Introduction	... 531
13.2.	Basic Principles of Thermoelectric Power Generation	... 531
13.3.	Thermoelectric Power Generator	... 538
13.4.	Performance Analysis of Thermo-electric Power Generator	... 539
13.5.	Thermoelectric Materials	... 548
13.6.	Selection of Materials	... 548
	Review Questions	... 550
14.	**Thermionic Generation**	**552—563**
14.1.	Introduction	... 552
14.2.	Thermionic Emission and Work Function	... 553
14.3.	Basic Thermionic Generator (the Basic Diode)	... 557
14.4.	Analysis of Thermionic Generator	... 557
	Review Questions	... 559
15.	**Thermonuclear Fusion Energy**	**564—592**
15.1.	Introduction	... 564
15.2.	The Basic : Nuclear Fusion and Reactions	... 567
15.3.	Requirements for Nuclear Fusion	... 571
15.4.	Plasma Confinement	... 573
15.5.	Magnetic-Confinement Fusion	... 574
15.6.	Inertial-confinement Fusion	... 581
15.7.	Muon Catalysed Fusion	... 586
15.8.	Characteristics of D-T Reaction	... 589
15.9.	Advantages of Nuclear Fusion	... 589
15.10.	Fusion Hydrid	... 589
15.11.	Cold Fusion (Current Status)	... 590
	Review Questions	... 592

Chapter		Page
16.	**Energy Storage and Distribution**	**593—627**
16.1.	Introduction	... 593
16.2.	Energy Storage Systems	... 594
16.3.	Mechanical Energy Storage	... 594
	16.3.1. Pumped Hydroelectric Storage	... 595
	16.3.2. Compressed Air Storage	... 597
	16.3.3. Energy Storage via Flywheels	... 599
16.4.	Electrical Storage : The Lead Acid Battery	... 601
16.5	Chemical Storage	... 605
	16.5.1. Introduction	... 605
	16.5.2. Energy Storage via Hydrogen	... 605
	16.5.3. Ammonia	... 606
	16.5.4. Reversible Chemical Reactions	... 606
16.6.	Electromagnetic Energy Storage	... 609
16.7.	Thermal Energy Storage	... 612
	16.7.1. Sensible Heat Storage	... 612
	16.7.2. Latent Heat Storage	... 617
16.8.	Biological Storage	... 621
16.9.	Distribution of Energy	... 622
	16.9.1. Introduction	... 622
	16.9.2. Gas Pipelines	... 622
	16.9.3. Electricity Transmission	... 624
	16.9.4. Batch transport	... 625
	16.9.5. Heat	... 625
	16.9.6. Chemical Heat Pipe	... 627
17.	**Energy Conservation**	**628—678**
17.1.	An Economic Concept of Energy	... 628
17.2.	Principles of Energy Conservation and Energy Audit	... 630
17.3.	Energy Conservation Approach/Technologies	... 633
17.4.	Co-generation	... 635
17.5.	Waste Heat Utilization	... 638
17.6.	Combined Cycle Power Generation	... 642
17.7.	Heat Recuperators	... 647
17.8.	Heat Regenerators	... 652
17.9.	Heat Pipes	... 653
17.10.	Stirling Engine	... 661
17.11.	Heat Pumps	... 671
17.12.	Energy Storage	... 675
17.13.	Renewable Energy Sources/Devices	... 675
17.14.	Instrumentation and Control	... 677
Appendices		**679—690**
Index		**691—696**

CHAPTER 1

An Introduction to Energy Sources

1.1. General

Energy is the primary and most universal measure of all kinds of work by human beings and nature. Every thing what happens in the world is the expression of flow of energy in one of its forms. Most people use the word energy for input to their bodies or to the machines and thus think about crude fuels and electric power.

The energy sources available can be divided into three types:

1. *Primary Energy Sources.* Primary energy sources can be defined as sources which provide a net supply of energy. Coal, oil, uranium etc. are examples of this type. The energy required to obtain these fuels is much less than what they can produce by combustion or nuclear reaction. Their energy yield ratio is very high. The yield ratio is defined as the energy fed back by the material to the energy received from the environment. The primary fuels only can accelerate growth but their supply is limited. It becomes very essential to use these fuels sparingly. Primary fuels contributes considerably to the energy supply.

2. *Secondary fuels* produce no net energy. Though it may be necessary for the economy, these may not yield net energy. Intensive agricultural is an example wherein terms of energy the yield is less than the input.

3. *Supplementary sources* are defined as those whose net energy yield is zero and those requiring highest investment in terms of energy Insulation (thermal) is an example for this source.

Coal, natural gas, oil and nuclear energy using breeder reactor are net energy yielders and are *primary sources of energy. Secondary sources* are like solar energy, wind energy, water energy etc. Solar energy can be used through plants, solar cells and solar heaters. Solar tower is another emerging technology. Solar drying and solar heating are economical applications when passive methods are used. Because of the dilute nature of solar energy it is difficult to classify the source as a primary one. Better sources are wind, tide, wave and hydroelectric applications. Geothermal and ocean thermal are the other sources which may well prove worthwhile. It may be necessary in future to develop the secondary sources like solar, wind etc.

1.2. Energy Consumption as a Measure of Prosperity

Energy is an important input in all sectors of any country's economy. The standard of living of a given country can be directly related to per capita energy consumption. Energy crisis is due to the two reasons; firstly that the population of the world has increased rapidly and secondly the standard of living of human beings has increased. If we take the annual per capita income of various countries and plot them against per head energy consumption, it will appear that the per capita energy consumption is a measure of the per capita income or the per capita energy consumption is a measure of the prosperity of the nation. The per capita income of U.S.A. is about 50 times more than per capita income of India, and so also is the per capita energy consumption. The per capita energy consumption in U.S.A. is 8000 kWh per year, whereas the per capita energy consumption in. India is 150 kWh. U.S.A. with 7% of world's population consumes 32% of the total energy consumed in the world, whereas India, a developing country with 20% of the world's population consumes only 1% of the total energy consumed in the world. Therefore one might conclude that to be materially prosperous, a human being needs to consume more and more energy than his own.

Developing countries, at present export primary products such as food, coffee, tea, jute and ores etc. This does not give them the full value of their resources. To get better value, the primary products should be processed to products for export. This meeds energy. Assuming the present consumption of energy is estimated to be of 10 million megawatts, foy the year 2000 A.D. this figure would be about 4 times. This assumes that the present pattern of consumption, in which the relative energy consumption of countries remain the same, *i.e.*, the per capita energy in developed countries remain much more than in the developing countries. If the standard of living in the developing countries is improved and approaches that of the developed countries, the energy requirement in the world in the year 2000 A.D. will be much more than estimated above.

1.3. World Energy Futures

If present trend continues, the world in the year 2000 A.D. will be more crowded than that of today. The world population may reach 7 billions by 2000 A.D. The conventional sources of energy are depleting and may be exhausted by the end of the century or beginning of the next century. Nuclear energy requires skilled technicians and poses the safety as regards to radioactive waste disposal. Solar energy and other non-conventional energy sources are the sources, those are to be utilized, in future.

Conclusions of the study on alternate energy strategies are:

1. The supply of oil will fail to meet increasing demand before the year 2000, even if energy prices rise 50 per cent above current levels in real terms. Additional constrains on oil production will hasten this shortage, thereby reducing the time available for action on alternatives.

2. Demand for energy will continue to grow even if Government adopt vigorous policies to conserve energy. This growth must increasingly be satisfied by energy resources other than oil, which will be progressively reserved for uses that oil can satisfy.

3. The continued growth of energy demand requires that energy resources be developed with the utmost vigour. The change from a world economy dominated by oil must start now. The alternatives require 5 to 15 years to develop, and the need for replacement fuels will increase rapidly as the last decade of the century is approached.

4. Electricity from nuclear power is capable of making an important contribution to the global energy supply although worldwide acceptance of it, on a sufficiently large scale yet to be established. Fusion power will not be significant before the year 2000.

5. Coal has the potential to contribute substantially to future energy supplies. Coal reserves are abundant, but taking advantage of them requires an active programme of development by both producers and consumers.

6. Natural gas reserves are large enough to meet projected demand provided the incentives are sufficient to encourage the development of extensive and costly inter continental gas transportation systems.

7. Although the resource base of other fossil fuels such as oils sands, heavy oil and oil shale is very large, they are likely to supply only small amounts of energy before the year 2000.

8. Other than hydroelectric power, renewable resources of energy *e.g.*, solar, wind, wave are unlikely to contribute significant quantities of additional energy during the century at the global level, although they could be of importance in particular areas. They are likely to become increasingly important in the 21st century.

9. Energy efficiency improvements, beyond the substantial energy conservation assumptions already built into our analysis, can further reduce energy demand and narrow the prospective gaps between energy demand and supply. Policies for achieving energy conservation should continue to be key elements of all future energy strategies.

It was concluded that world oil production if likely to level off-very shortly and that alternative fuels will have to be meet growing energy demand. Large investments and long lead times are required to produce these fuels on a scale large enough to fill the prospective shortage of oil, the fuel that now furnishes most of the world's energy. The task for the world will be to manage a transition from dependence on oil to greater reliance on other fossil fuels, nuclear energy and later, renewable energy system.

1.4. Energy Sources and their Availability

1.4.1. Introduction. Today, every country draws its energy needs from a variety of sources. We can broadly categorize these sources as *commercial* and *noncommercial.* The commercial sources include the fossil fuels (coal, oil and natural gas), hydroelectric power and nuclear power, while the non-commercial sources include wood, animal waste and agricultural wastes. In an industrialized country like, U.S.A., most of the energy requirements are met from commercial sources, while in an industrially less developed country like India, the use of commercial and non-commercial sources are about equal.

1.4.2. Commercial or Conventional Energy Sources

Major Sources of energy include:

(1) *Fossil fuels i.e.* solid fuels (mainly coal including anthracite, bituminous, and brown coals lignits and peals), liquid and gaseous fuels including petroleum and its derivatives and natural gas.

(2) Water power or energy stored in water.

(3) Energy of nuclear fission.

Minor sources of energy include sun, wind, tides in the sea, geothermal, ocean thermal electric conversion, fuel cells, thermionic, thermoelectric generators etc.

Wood was dominant source of energy in the pre-industrialization era. It gave way to coal and coke. Use of coal reached a peak in the early part of the twentieth century. Oil get introduced at that time and has taken a substantial share from wood and coal. Wood is no more regarded as a conventional source. Hydroelectricity has already grown to a stable level in most of the developed countries. A brief account of the various important sources of energy and their future possibilities is given below.

The percentage use of various sources for the total energy consumption in the world is given in Table 1.1.

Table 1.1

Coal	32.5%	92%
Oil	38.3%	
Gas	19.0%	
Uranium	0.13%	
Hydro	2.0%	
Wood	6.6%	8%
Dung	1.2%	
Waste	0.3%	

Coal, oil, gas, uranium and hydro are commonly known as *commercial* or *conventional energy sources.* Looking at the percentage distribution one finds that world's energy supply comes mainly from fossil fuels. The heavy dependence on fossil fuels stands out clearly. One of the so most significant aspects of the current energy consumption pattern in many developing countries is that non-commercial sources such as firewood, animal dung and agricultural waste represent a significant 8% of the total energy used in the world. Then constitute about 4 times the energy produced by the hydro and 60 times the energy produced by nuclear sources. In some developing countries non-commercial energy sources are a significant fraction of the total resources. This dependence of the developing countries is likely to continue unless replaced by other alternative sources of energy.

(*i*) Coal. Since the advent of industrialization coal has been the most common source of energy. In the last three decades, the world switched over from coal to oil as a major source of energy because it is simpler and cleaner to obtain useful energy from oil.

Modern steam boilers burn coal in any of its forms as a primary fuel. Coal developed vegetable matter which grew in past geological ages. Trees and plants falling into water decayed and produced peat bogs. Gigantic geological upheavals burried these bogs under layers of silt. Soil pressure, heat and movement of the earth's crust distilled off some of the bog's gaseous matter to form brown coal, or lignite. Continuing subterrane an activity reduces the coal's gaseous content progressively to form different ranks; peat lignite, bituminous and anthracite.

With the commissioning of an additional 500 MW unit at the *Korba thermal power station,* on March 23,1989, the power station has become the largest power station of India. The plant is located on the west bank at the Hardeo river near Korba in Bilaspur district of M.P. The project is the second in the series *of super thermal power stations* being set up by the National Thermal Power Corporation (NTPC).

The gigantic complex has been set up in two stages. In the first stage three units of 200 MW were set up. In the second stage three units of 500 MW have been set up. With the commissioning of the last 500 MW unit, the plant has achieved its ultimate capacity of 2,100 MW. The 500 MW generators have been provided by BHEL. The whole project has costed ₹ 1875 crore. World Bank has assisted to the tune of US $600 million.

Madhya Pradesh, Maharashtra, Gujarat and Goa are benefitted from the project.

According to estimates coal is abundant. It is enough to last for 200 years. However, it is low in calorific value and its shipping is expensive. Coal is pollutant and when burnt it produces CO_2 and CO. Extensive use of coal as a source of energy is likely to disturb the ecological balance of CO_2 since vegetations in the world would not be capable of absorbing such large proportions of carbon dioxide produced by burning large quantities of coal.

***(ii)* Oil.** Almost 40% of the energy needs of the world are fed by oil. The rising prices of oil has brought a considerable strain to the economy of the world more, so in the case of the developing countries that do not possess oil reserves enough for their own consumption. With today's consumption and a resource amount of 250,000 million tonnes of oil, it would suffice for about 100 years unless more oil is discovered. The question is whether an alternative to oil would then be available, the world must start thinking of a change from a world economy dominated by oil.

Refining petroleum or crude oil produces our fuel oils. India is not particularly rich in petroleum reserves. The potential oil bearing areas are located in Assam, Tripura, Manipur, West Bengal, Ganga valley, Punjab, Himachal Pradesh, Kutch, eastern and western coastal area (in Tamil Nadu, Andhra Pradesh and Kerala). Andaman and Nicobar Islands, Lakshadweep, and in the continental shelves adjoining these areas.

Diesel power plants in India are installed in isolated places and the total installed capacity is estimated as 0.35 million kW, *i.e.* less than 2% of the total installed capacity in the country. No addition to this is expected in near future.

***(iii)* Gas.** Gas is incompletely utilized at present and huge quantities are burnt off in the oil production process because of the non-availability of ready market. The reason may be the high transportation cost of the gas. To transport gas is costlier than transporting oil. Large reserves are estimated to be located in inaccessible areas.

Gaseous fuels can be classified as:

(1) Gases of fixed composition such as acetylete, ethylene, methane etc.

(2) Composite industrial gases such as producer gas, coke oven gas, water gas, blast furnace gas etc.

***(iv)* Agriculture and organic wastes.** At present small quantities of agricultural and organic wastes consisting of draw saw dust, bagasse, garbage, animal dung, paddy husk and cornstem accounting a major energy consumption. Most of the remaining material was burnt or left, unused causing considerable environmental problems.

1. The waste should be utilised near the source, in order to reduce the transportation cost.

2. Appropriate equipment for burning, or extracting energy from the materials should be developed to suit the local conditions and meet the requirements of the rural areas.

3. Other non energy uses of the material should also be considered.

Considering the availability and the location of material produced, these resources are regarded as an important energy supply for the rural areas in the near future.

***(v)* Water Power.** Water power is developed by allowing water to fall under the force of gravity. It is used almost exclusively for electric power generation. In fact, the generation of water power on a large scale became possible around the beginning of the twentieth century only with the development of electrical power transmission. Prior to that, water power plants (Hydroelectric plants) were usually of small capacities usual less than 100 kW.

Potential energy of water is converted into mechanical energy by using prime moves known as hydraulic turbines. Water power is quite cheap where water is available in abundance. Although capital cost of hydroelectric power plants is higher as compared to other types of power plants but their operating costs are quite low, as no fuel is required in this case.

Dehar power house of Beas Sutlej link located on the right bank of Sutlej river is equipped with 6 units of 165 MW each, which is the largest size in country.

Hydroelectric power is one of the indirect ways in which solar energy is being used. Thus, the main factor in its favour is that it is the only renewable non-depleting source of the present commercial sources.

In addition it does not create any pollution problem. The development rate of hydropower is still low, due to the following problems.

1. In developing a project, it will take about 6-10 years time for planning, investigation and construction.

2. High capital investment is needed, and some parts of the investment have to be derived from foreign sources.

3. There are growing problems on relocation of villages involved, compensation for damage, selecting the suitable resettlement area and environmental impact.

Because of long transmission line to the villages with low load factor, the electric power will be available to the people in rural areas may not be economical and the setting up of isolated diesel generation plants will also experience high losses with the existing electric tariff rates. This leads to the development of *mini* or *micro* hydroelectric projects to supply the electric power to remote areas. These projects may operate as isolated systems or connected to the main grid where it is feasible. The importance or microhydroelectric projects have been observed in some parts of the country with availability of river flow throughout the year with a possibility of medium to higher head development. In order to reduce the cost of development to the acceptable figure, several measures have been considered as follows:

(*a*) Development of low cost turbines and generators.

(*b*) Participation of villages in the development and operation of the project.

(*c*) Using the appropriate technology and tolerable substandard requirement and project civil work component at the beginning stage.

(vi) **Nuclear Power.** According to modern theories of atomic structure, matter consists at minute particles known as atoms. These atoms represent enormous concentration of binding energy. Controlled fission of heavier unstable atoms such as U^{235} , Th^{232} and artificial element Pu^{239}, liberate large amount of heat energy. This enormous release of energy from a relatively small mass of nuclear fuels makes this source of energy of great importance. The energy released by the complete fission of one kg of U^{235}, is equal to the heat energy obtained by burning 4500 tonnes of high grade coal or 2200 tonnes of oil. The heat produced by nuclear fission of the atoms of fissionable material is utilized in special heat exchangers for the production of steam which is then used to drive turbogenerators as in the conventional power plants.

However there are some limitations in the use of nuclear energy namely high capital cost of nuclear power plants, limited availability of raw materials, difficulties associated with disposal of radioactive waste and shortage of well trained personnel to handle the nuclear power plants.

The uranium reserves in the world at present are small. These reserves are recoverable but are expensive. Further it is estimated that uranium reserves have only 3% of the energy contained in the oil reserves. A country like France produces about 30% of its total energy by nuclear methods, whereas a country like India has uranium sufficient enough only to produce 6×10^6 kW, a mere 1% of its current energy requirements.

Development of fast breeder reactor, which is not yet free from technical difficulties, will decide the future of nuclear power. Controlled fusion may also add brighter prospects to the use of nuclear energy. India has considerable resources of nuclear fuel which would help development of nuclear power in the country. The indicated and inferred reserves of uranium at two locations; Jaduguda and Narwapahar and Bhattin (Bihar) total about 33000 tonnes. There are other important deposits in Singhbhum (Bihar) and minor deposits in M.P., H.P., U.P., Rajasthan and are in the exploratory stage. Uranium may be in a position to sustain about 5000 to 10000 MW of nuclear power for its life time. In India, abundance of Thorium deposits are available from monazite sand in the west coast. Therefore, India's interests lie in Thorium Breeder Reactors.

Nuclear power is having considerable potential. In countries like France and Belgium, it is contributing to the extent of 70 percent and 58 percent to the overall power generated. In India it is still low and by the year 2000, a target of 10,000 MW for nuclear power is set. Three systems are considered for nuclear power generation. The first is based an natural uranium yielding power and plutonium. The second employs fast breeder reactor using plutonium and depleted uranium. The third is by fast breeder reactor using thorium and converting it to uranium. The first method alone has been commercialised in India and the other methods may take some more time for large scale use. Indian ore is poor in uranium content (0.025-0.067 percent) and a 235 MW plant requires about 23 tonnes of uranium besides initial charge of 60 tonnnes. Indian's reserves of uranium ore are estimated at about 50000 tonnes as oxide and of thorium 450,000 tonnes as thoria. Consequently the third method appears to have great potential in view of the large deposits of thoria as monozite sands. A number of scientific and technological hurdles have to be crossed till the stage of thorium based fast breeder reactor is reached. Design and indigenous fabrication capabilities have also to be achieved.

There are following nuclear power plants in India at the moment. The 400 MW (2 × 200) Tarapore (Maharastra) nuclear power station was commissioned in 1969, which uses enriched uranium as a fuel and boiling water reactors are employed. The plant at Rana Pratap Sagar Kota (Raj.) which also has capacity of 400 MW (2 × 200), uses natural uranium as fuel and pressurized heavy water reactors (CANDU). The station at Kalpakkam (Tamil Nadu) has the capacity at 440 MW (2 × 220), station at Narora has the capacity of 470 MW (2 × 235 MW), while the station at Kakrapar (Gujarat) has the Capacity of 940 MW (4 × 235 MW), the reactors at these, are similar to those at Rana Pratap Sagar.

***(vii)* Thermal (burner) and Breeder Reactors.** As stated above, nuclear fission involves splitting the nucleus of heavy atoms, like uranium or plutonium, in a controlled nuclear chain reaction. During fission, heat is released and this can be used to generate high pressure steam to drive turbo-generators and produce electricity. The current generation of 'thermal' or 'burner' nuclear reactors is only able to utilize a tiny fraction of the uranium fuel. The nuclear chain reaction is sustained by uranium-235 which constitutes little more than 0.7% of natural uranium. Thermal reactors also make limited use of the more abundant uranium-238 isotope. During the fission process a small proportion of that present in the fuel is converted (by neutron capture) into fissionable plutonium-239, and some of this fissions to produce heat.

Fast breeder reactors are of different design to thermal reactors. They burn enriched uranium (composed of say 20-30% uranium-235) or plutonium which has been separated from spent thermal reactor fuel. In theory they can actually produce more fissible material than they burn by 'breeding' fuel in a special zone around the core. As the nuclear chain reaction progresses, neutron bombard the blanket region, which is composed of 'Fertile'uranium-238 and convert some of this uranium into fissionable plutonium. This uranium can subsequently be extracted by reprocessing, and fabricated into new fuel assemblies.

In practice, breeder technology is not yet commercially developed. Existing prototypes are only able to achieve slow breeding rates, and would take in excess of 25 years to produce sufficient material to fuel a second, identical reactor. Another fertile, thorium-232, could also be used to provide fuel. It can be transformed by neutron capture into fissionable uranium-233. However,, there has been little work on this alternative fuel cycle, and it is not known whether it would be practicable.

There are major risks associated with the generation and storage of large quantities of highly noxious radioactive waste created by nuclear fission. The waste must be kept isolated from the biosphere virtually indefinitely regardless of mechanical failure, careless or malicious action, or natural disaster. Despite over a quarter of a century of research there is still no satisfactory solution to the problem of disposal.

The separation and use of plutonium presents a different set of hazards: plutonium is both highly toxic and the raw material for atomic bombs. A large thermal nuclear reactor produces several hundred kilograms of plutonium a year during normal operation. A commercial breeder would require about *three thousand* kilograms for its initial fuel inventory-sufficient plutonium for several *hundred,* nuclear weapons. Dependence on plutonium entails grave physical and security risks, and the prospects of theft by terrorist groups or foreign states intent on acquiring nuclear weapons. Plutonium is already in demand by the weapons states, who wish to increase their nuclear arsenals and by other countries eager to join the growing nuclear club.

(viii) **The Nuclear option.** One response to the problem of increasing fossil fuel dependency has been to advocate a rapid expansion of nuclear power. However, even if thousands of large nuclear reactors could be built over the comming decades, nuclear power would still only make a small contribution to meeting world energy demand. After more than a quarter of a century of development, nuclear power provides only a few percent of the world's electricity which itself only accounts for a small proportion of the total energy demand. Furthermore, the nuclear path is fraught with dangers. The intractable nature of many of the environmental as well as the social, political and technical problems, and the continued escalation in the costs, have led to widespread disenchantment with nuclear technology.

It is sometimes suggested that *nuclear fusion* has better prospects, but this may be as much wishful thinking as the early dreams of atomic electricity too cheap to be worth metering. Nuclear fusion is fundamentally different from nuclear fission; it involves fusing together of light atoms rather than the breaking apart of heavy ones. To ignite and sustain a fusion reaction between say, deuterium and tritium (two forms of hydrogen), it would be necessary to heat the fuel to a temperature in excess of 100 million degree centigrade-hotter than the sun—and to confine the resulting plasma for sufficient time to be able to extract useful amount of energy. Even if this daunting problem can be solved, the cost of the machinery that will be required is likely to make fusion wholly uneconomic.

1.4.3. New Energy Technologies. Numerous studies are going on around the world in this direction and it would be unwieldly to summarize all of them here. Only a few selected items will be reviewed briefly.

Coal. The first major break through is the application of fluidized bed technology for the coal gasification, carbonization and combustion. Vast improvement in performance and efficiency are achieved. The technology is already commercialized. Presently pressurized fluidised bed technology is being developed to further improve the performance. The fluidised bed technique has helped to utilise the low rank as well as high ash coals.

1.5. Principle of Fluidization and Fluidised Bed Combustion

When a bed of fine solid particles is subjected to an evenly distributed upward flow of gas, the bed remains static. At higher gas velocities the drag forces on the particles cause them to become suspended in the gas stream or fluidized and such a suspension resembles a boiling liquid and it is called a minimally fluidised bed. When the gas velocity is further increased the bed becomes highly turbulent and rapid mixing of the particles takes place and at this stage the bed of solid attains a pseudo-fluid state (Fig. 1.1). When the solid particles in the fluidised bed are preheated to the ignition temperature of the fuel it will burn imparting heat to the bed material through the volume of the bed. In order to maintain the bed temperature well below the ash fusion temperature, a large amount of heat is extracted by heat transfer coils immersed in the bed.

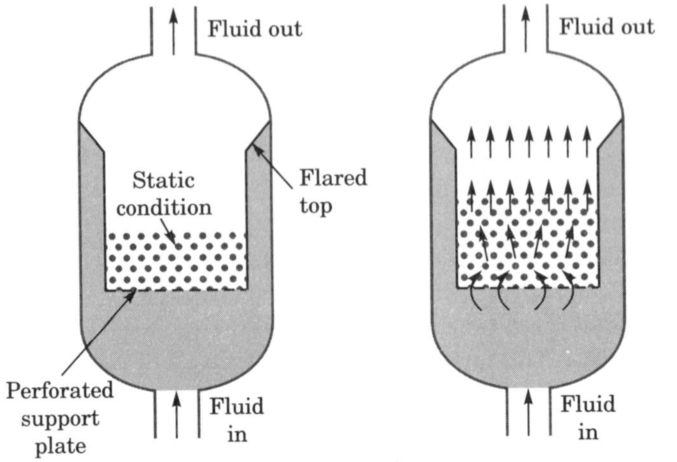

(*a*) Particles in static state (*b*) Particles in fluidised state

Fig. 1.1. Schematic of a fluidised bed

Thus fluid bed combustion (*FBC*) is an attractive process for power generation. It is a clean coal technology which can utilise *high ash coal, washery rejects, biomass* and *agricultural residues.* The advantages of fluid bed operation like different heat transfer and uniform bed temperature are advantageous for power generation. Fluidised bed combustors operating at atmospheric pressure have been in use for the past few years. Repowering and retrofitting existing power plants are being done in USA. The existing steam turbine and auxiliaries are retained and the existing boiler is replaced by a FBC boiler. A160 MW has been set up by TVA, Panducah which, it is stated, is the largest unit set up so far. These plants will be on conventional steam turbines with heat exchangers installed in the bed or outside. In India also such plants are used for steam generation. FBC power plants are being set up by coal India Limited and Tata Iron and Steel Co. A few companies like BHEL, IS GEC-Thompson, Thermax are offering FBC for burning different fuels including biomass.

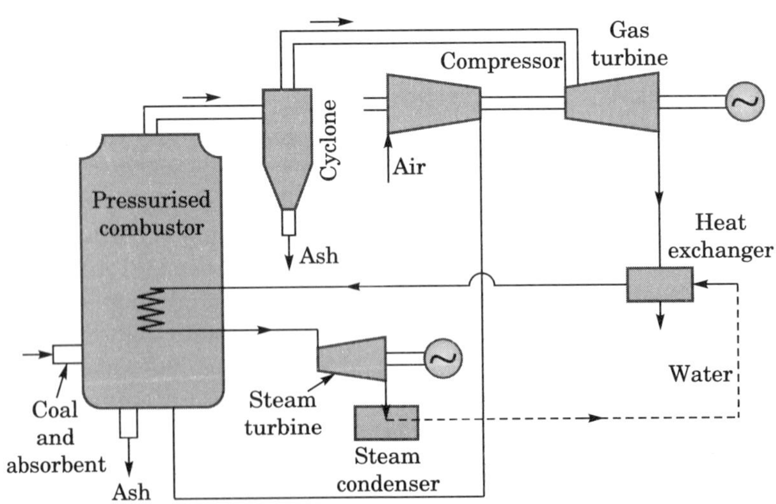

Fig. 1.2. Pressurised fluid bed combustion of coal.

If the FBC is operated under pressure, pressurised fluid bed combustion (PFBC) results and combined cycle system can be evolved. The flue gases are expanded to generate power in a gas turbine. Steam is generated using the hot exhaust gases from turbine. A flow sheet of such a plant is shown in Fig. (1.2). By using an adsorbent like calcium oxide along with coal the sulphur in the coal is fixed in the bed and there is no emission of SO_2. Due to moderate bed temperature of 900°C, NO_x formation is also minimised. The only problem may be the particulate emission but even this can be avoided by using cyclone and special filters. PFBC are more compact than FBC. However development of PFBC units is yet to be perfected. PFBC plants being more compact than a FBC result in reduced capital costs and cost of production of electricity.

FBC plants use a bubbling bed. A development in the current decade has been the circulating fluid bed combustors (CFBC) which on account of high velocity and circulation of solids results in better performance. A number of CFBC plants are being set up for steam and power generation. Two stage CFBC plant in which coal is gasified partially in the first bed and the resulting char is subsequent burnt in the second bed is being developed. The gas from the first stage can be burnt in a gas turbine, resulting in combined cycle power plant. FBC plants cause minimum pollution and even here cleaning the gases not without cooling them is being attempted. This hot gas clean up will improve the efficiency.

A novel process of coal gasification is the electrogasification. In this process the conventional water electrolysis carried out in presence of crushed coal particles, instead of gettihg H_2 and O_2 in the two electrode chambers H_2 and O_2 and a small amount of CO were obtained at a lower voltage. It has been observed that coal of any rank including lignite can be gasified at room temperature, and there is possibility of obtaining hydrocarbon in place of CO_2/CO with suitable changes in the experimental conditions. The power input to output bears a favourable ratio.

1.5.1. Advantages of Fluidised Bed Combustion Boilers

1. Ability to burn fuels containing very high inerts, washery rejects containing 73% ash and 1% moisture have been successfully burnt in the prototype FBC boiler at BHEL Tiruchy.

2. Fuel flexibility burning fuels ranging from fuel having HHV (higher heating value) of 10,500 kcal/kg down to rejects having heating value of 1,900 kcal/kg in the same combustor, (demonstrated by BHEL Triuchy).

3. Higher combustion efficiency of 90 to 92% and boiler efficiency of 80% based on higher heating value was achieved. Even with washery rejects a higher boiler efficiency of 75% was achieved.

4. As the bed is maintained between 850°C and 950°C ash does not get heated to the initial deformation temperature. Hence no clinkering or slagging or hard deposits on heat exchanger tubes.

5. Ash is less erosive hence life of boiler second pass and ID (induced draught) system is increased.

6. Because of low temperature combustion, corrosion caused by alkali compounds in ash significantly reduced.

7. Require much less boiler plan area than a stoker.

8. Uniform temperature throughout the furnace volume.

9. Reduced emission of harmful nitrous oxide.

10. Sulphur dioxide emission can be reduced to acceptable level with less expense.

11. Operation is as simple as that of an oil fired boiler.

Oil. On the atomisation and combustion of petroleum fractions extensive work have been reported to improve the energy efficiency. High temperature combustion and flame propagation is now better known from the mechanical point of view. Electric ignition systems have been devised for proper ignition/combustion. Improved efficiencies in the range of 5-15%, (additional) are reported. Oil burner design have been improved in terms of primary air, secondary air mixing and combustion. Another noteworthy development in this area is the finding that the liquid is better atomised at low/zero nozzle pressures if a high voltage of the order of 20 kV, dc is applied to the nozzle. It was observed that this technique generates very fine droplets of high surface area and the power consumption was much smaller than that of the conventional mechanical pressure atomisation. Attempts are underway to use this principle for mass transfer operations where high interfacial area is needed for improved rates with reduced energy requirements.

Gas. The earlier concept of simply burning away the natural gas at the oil well head itself has now changed. It has become economical to run long distance pipe lines for the gas to be transported several hundred kilometers to the place where it can be used. This is very significant improvement. Incidentally the natural gas can be more profitable employed as a raw material to produce several important chemicals which have been traditionally obtained from petroleum fractions.

Biogas from rural, urban and industrial waste is another urea presently under development. It is now known clearly that any organic waste can be employed for the production of biogas.

1.6. Non-conventional Sources

While fossil fuels will be the main fuels for thermal power, there is a fear that they will get exhausted eventually in the next century. Therefore other systems based on non-conventional and renewable sources are being tried by many countries. These are solar, wind, sea, geothermal and biomass. (There have been many attempts to find a suitable generic term to describe the whole range of technologies designed to tap the earth's natural energy flows. These technologies have variously been called 'alternative', 'appropriate', 'natural', 'new'. However the term *'renewable'* has gained the most widespread acceptance).

Solar energy can be a major source of power. Its potential is 178 billion MW which is about 20,000 times the world's demand. But so far it could not be developed on a large scale. Sun's energy can be utilised as thermal and photovoltaics. The former is currently being used for steam and hot water production.

Wind energy uses the high wind velocity available in certain parts. California State in USA is generating 500 MW from 900 wind turbines based on wind mills. Wind energy is used for pumping the water or power generation. About 0.7 million wind pumps are in operation in different countries. A minimum wind speed of 3 m/s is needed. This is considered to have a high efficiency. Coastal, hilly and valley areas are suitable for this process. Potential in India is estimated between 20,000 and 25,000 MW. Coastal areas of Gujarat, Maharashtra and Tamil Nadu are considered as favourable. A number of experimental stations have been set up. The maximum power generated from any single unit is about 1 MW.

Geothermal energy derives the heat in the centre of the earth and it is stated that potential to the extent of 3400 MW exists in New Zealand, USA, Japan and Iceland. 700 MW of power is generated in Phillippines and China. India does not appear to have any major exploitable source. Geothermal energy can also be used for cooling by using heat for vapour absorption system.

Energy from seas can be utilised as wave, tidal or ocean thermal energy. U.K. and Japan are pioneers in this area. About 13 kW per meter height of the wave can be generated. A plant to make 445000 kWh/yr of energy is being set up in Kerala State. This station will generate 75 kW in calm months and 120-150 kW in June-September. Ocean thermal energy conversion utilises the temperature difference between warm surface sea water at about 28°C and the cold deep sea water at 5-7°C at a depth of 800-1000 m in tropical areas. India's potential is large along the coastal length from Mumbai to Vishakapatnam. Tidal energy is energy that can be trapped from sea. France and USSR are developing this source at a level of 240 and 400 MW respectively, In India, the Gulf of Kutch, Gulf of Cambay and Sunderbans are potential sites. The total potential from sea is estimated as 50,000 MW from ocean thermal energy conversion, 40,000 MW from wave energy and 8000 MW from tidal.

Biomass is another renewable source of energy in the form of wood, agricultural residues, etc. The potential for agricultural residues alone is estimated as 480 mt with residues from food grains contributing about 100 mt. These can be burnt directly to generate steam for use in steam turbine for power generation or they can be gasified and the gas used in an internal combustion engine for agricultural pumping or power generation. Power generation is being tried on a small scale upto 1 MW or so, but large scale application is yet to be shown. The main problem with agricultural residues is their collection. In the case of forest wood, the requirement of fuel wood alone is between 250 and 300 mt and this quantity is not readily available. Efforts are being made for cultivating quick growing trees for use in power generation. Similarly biogas obtained by anaerobic digestion of animal dung and sewage is also suitable for use in Internal Combustion Engines. Biomass may prove a useful fuel for localised power generation in rural areas where electric transmission lines have not reached.

1.6.1. Solar Energy. Solar energy has the greatest potential of all the sources of renewable energy and if only a small amount of this form of energy could be used, it will be one of the most important supplies of energy specially when other sources in the country have depleted.

Energy comes to the earth from the sun. This energy keeps the temperature of the earth above that in colder space, causes current, in the atmosphere and in ocean, causes the water cycle and generate photosynthesis in plants.

The solar power where sun hits atmosphere is 10^{17} watts, whereas the solar power on earth's surface is 10^{16} watts. The total world-wide power demand of all needs of civilization is 10^{13} watts. Therefore, the sun gives us 1000 times more power than we need. If we can use 5% of this energy, it will be 50 times what the world will require. The energy radiated by the sun on a bright sunny day is approximately 1 kW/m^2, attempts have been made to make use of this energy in raising steam which may be used in driving the prime movers for the purpose of generation of electrical energy. However on account of large space required, uncertainty of availability of energy at constant rate, due to clouds, winds, haze etc., there is limited application of this source in the generation of electric power. Now-a-days the drawbacks as pointed out that energy cannot be stored and it is a dilute form of energy, are out dated arguments, since the energy can be stored by producing hydrogen, or by storing in other mechanical or electrical devices, or it can be stored in containers of chemicals called eutectric or phase changing salts. These salts which store large quantities of heat in a relatively small volume, melt when they are heated and release heat later as they cool and crystallize. The energy can be concentrated in solar furnaces, for example which can achieve temperatures in the region of 5000°C. The facts speak in favour of solar energy, as we have seen in analysis of commercial energy sources, that world's reserves of coal, oil and gas will be exhausted within a few decades. Nuclear energy involve considerable hazards and nuclear fusion has not yet overcome all the problems of even fundamental research, compared with these technologies, the feasibility of which is still uncertain and contested, the technical utilization of solar energy can prove very useful. Utilization of solar energy is of great importance to India since it lies in a temperature climate of the region of the world where sun light is abundant for a major part of the year.

The basic research in solar energy is being carried in universities and educational and research institutions, public sector institution, Bharat Heavy Electricals Limited and Central Electronic Limited are carrying out a co-ordinated programme of research in solar energy.

The applications of solar energy which are enjoying most success today are:

(1) Heating and cooling of residential building.

(2) Solar water heating.

(3) Solar drying of agricultural and animal products.

(4) Solar distillation on a small community scale.

(5) Salt production by evaporation of seawater or inland brines.

(6) Solar cookers.

(7) Solar engines for water pumping.

(8) Food refrigeration.

(9) Bio conversion and wind energy, which are indirect source of solar energy.

(10) Solar furnaces.

(11) Solar electric power generation by—

(*i*) Solar ponds.

(*ii*) Steam generators heated by rotating reflectors (heliostat mirrors), or by tower concept.

(*iii*) Reflectors with lenses and pipes for fluid circulation (cylindrical parabolic reflectors).

(12) Solar photovoltaic cells, which can be used for conversion of solar energy directly into electricity or for water pumping in rural agricultural purposes.

The heat from solar collectors is directly used for warming the living spaces of a building in conventional ways *e.g.,* through radiators and hot air registers. When the building does not require heat, the warmed air or liquid from the collector can be moved to a heat storage container. In the case of air, the storage is often a pile of rocks or some other heat-holding material, in the case of liquid, it is usually a large, well insulated tank of water, which has considerable heat capacity. Heat is also stored in containers of chemicals called *eutectic* or *phase changing salts.* These salts, which store large quantities of heat in a relatively small volume, melt when they are heated and release heat later as they cool and crystallize. When the building needs heat, the air or water from its heating system passes through the storage is warmed, and is then fed through the conventional heaters to warm the space. For sunless days or cloudy days, an auxiliary system as a back-up, is always required. The same is true for solar cooling systems.

The heat from solar energy can be used to cool buildings, using the absorption cooling principle operative in gas-fired refrigerators. Presently available equipment, however usually requires extremely high operating temperatures for above those for efficient solar collection. A great deal of current research is being devoted to developing systems requiring lower operating temperatures, but it will probably be several years before solar collectors will be commercially viable.

Solar energy units for heating domestic water are commercially available and are used by millions of people in various parts of the world, for example in Australia, Israel, Japan etc. A solar water heater commonly comprises a blackened flat plate metal collector with an associated metal tubing, facing the general direction of the sun. The collector is provided with a transparent glass cover and a layer of thermal insulation beneath the plate. The collector tubing is connected by a pipe to an insulated tank that stores hot water during non-sunny periods. The collector absorbs solar radiation and by transfer of resulting heat to the water circulating through the tubing by gravity or by a pump, hot water is supplied to the storage tank.

Solar water heating systems for domestic, industrial and commercial applications are at present available. In commercial establishments, there is great potential especially in hotels, hospitals, guest houses, tourist bungalows, canteen etc. For industrial applications solar water heating system can meet the low and medium temperature process heat requirements hot water upto 90°C, hot air upto 110°C and low pressure steam upto 140°C. These are specially useful in engineering, textile, chemicals, pharmaceutical, food processing, sugar, dairy and other industries. Hot water systems have relevance for many agricultural and village industries, such as for handloom fabrics, sericulture, leather tanning and hand made paper.

The basic method of *solar distillation* is to admit solar radiation through a transparent cover to a shallow, covered brine basin; water evaporates from the brine and the vapour condenses on the covers which are so arranged that the condensate flows there from into collection troughs and hence into a product-water storage tank. In arid, semi arid, or coastal areas, there is abundant sun light that can be used for converting brackish or saline water into potable distilled water.

A traditional and wide-spread use of solar energy is for drying particularly of agricultural products. This is a process of substantial economic significance in many areas. The process is of special interest in the case of soft fruits; these are particularly vulnerable to attack by insects, as the sugar concentration increases during drying. Fruit dryer in which fruit is placed, in carefully designed racks to provide controlled exposure to solar radiation often improves product quality and saves considerable time. A simple cabinet dryer consists of a box, insulated at the base, painted black on the inside and covered with an inclined transparent sheet of glass. Ventilation holes are provided at the base and at the top of the sides of the box to facilitate a flow of air over the drying material, which is placed on perforated trays in the interior of the cabinet base.

Large drying systems like grain, paddy, maize, cash crops like ginger, cashew, pepper, etc., spray drying of milk; timber and veneer drying; tobacco curing; fish and fruit drying, etc. have also been developed.

Solar refrigeration is intended for food preservation (or storage of biological and medical materials) and deserves top-priority in country like India. Solar air conditioning can be utilized for space cooling. Solar assisted heat pumps would provide both cooling and heating.

Cold storages are very important for preservation and conservation of food articles.

There are two methods of solar refrigeration.

(*a*) *Vapour absorption refrigeration* system that utilizes low grade thermal energy obtained from flat plate collectors with a little modification.

(*b*) Concentrating (focusing) collectors to supply heat at a higher temperature to a heat engine which then drives the compressor of a conventional refrigerator.

Solar refrigeration with an absorption system is a better way of direct utilization of energy. The vapour absorption system replacing the compressor by a generator absorber assembly can work with wide range of absorbents and refrigerants. In absorption system motive power required is very small, but. still C.O.P. of the system is low.

1.6.2. Electricity from Solar Energy. Electricity can be produced from the solar energy by photo voltaic solar cells, which convert the solar energy directly to electricity. The most, significant applications of photo voltaic cell in India, are the energisation of pump sets for irrigation, drinking water supply and rural electrification covering street lights, community TV sets, medical refrigerators and other small power loads.

Electricity is directly generated by utilising solar energy by the photo voltaic process. When photons from the sun are absorbed in a semiconductor, they create free electrons with higher energies than the electrons which provide the bonding in the base crystal. Once these free electrons are created, there must be an electric field to induce these higher energy electrons to flow out of the semiconductor to do useful work. The electric field in most solar cells is provided by a junction of materials which have different electrical properties. The photovoltaic effect can be described easily for *p-n* junction in semi-conductor materials of solar cells which are silicon, cadmium, sulphide/copper sulphide, Gallium Arsenite etc.

In a *solar thermal power production* system the energy is first collected by using a solar pond, a flat plate collector, focusing collector or heliostates (turnable mirrors). This energy is used to increase the internal energy or temperature of a fluid. This fluid may be directly used in any of the common or known cycles such as Rankine, or through a heat exchanger to heat a secondary fluid (working fluid) which is being used in the cycle to produce mechanical power from which electrical power can be produced easily.

Solar thermal power cycles can be broadly classified as low medium and high temperature cycles. Low temperature cycles generally use flat plate collectors or solar pond, maximum temperatures are limited to above 90 to 100°C. Medium temperature cycles work at maximum temperatures ranging from 150 to 300°C, using concentrating or focusing collectors. High temperature cycles work at maximum temperatures above 300°C.

In solar tower concentration system (Tower power concept), the incoming solar radiation is focussed to a central receiver or a boiler mounted on a tall tower using thousands of plane reflectors which are steerable about two axes and are called heliostates.

Results to-date show solar energy to be quite competitive with other sources of energy, if the solar tower plant size is about 100-200 MWe, with 3-6 hours thermal storage.

Over the last few years, few experiment power plants have been built, or under construction in U.S.A., France, Italy and Japan.

1.6.3. Wind Energy. Energy of wind can be economically used for the generation of electrical energy. Winds are caused from two main factors:

1. Heating and cooling of the atmosphere which generates convection currents. Heating is caused by the absorption of solar energy on the earth's surface and in the atmosphere.

2. The rotation of the earth with respect to atmosphere, and its motion around the sun.

The potential of wind energy as a source of power is large. The energy available in the winds over the earth's surface is estimated to be 1.6×10^7 MW, which is of the same order of magnitude as the present energy consumption on the earth.

Wind energy which is an indirect source of solar energy conver-sion can be utilized to run wind mill, which in turn drives a generator to produce electricity. Wind can also be used to provide mechanical power, such as for water pumping. In India generally wind speeds obtainable are in the lower ranges. Attempts are, therefore, on the development of low cost, low speed mills for irrigation of small and marginal farms for providing drinking water in rural areas. The developments are being mainly concentrated on water pumping wind mills suitable for operation in a wind speed range of 8 to 36 km per hour. In India, high wind speeds are obtainable in coastal areas of Saurashtra, Western Rajasthan and some parts of Central India. In these areas, there could be a possibility of using medium and large sized wind mills for generation of electricity and feeding the same into the grid.

Many types of wind mills have been designed and developed. However, only a few have been found to be practically suitable and useful. Some of these are:

1. Multiblade type wind mill.
2. Sail type wind mill.
3. Propeller type wind mill.
4. Savonins type wind mill.
5. Darrieus type wind mill.

The first three are the examples of horizontal axis wind mills, while the last two have a vertical axis.

Vertical axis machines are of simple design as compared to the horizontal axis.

Some *characteristics* of wind energy are stated below:

(*i*) It is a renewable source of energy.

(*ii*) Like all forms of solar energy, wind-power systems are non-polluting, so it has no adverse influence on the environment.

(*iii*) Wind energy systems avoid fuel provision and transport.

(*iv*) On a small scale, upto a few kilowatt system, is less costly. On a large scale, costs be competitive with conventional electricity and and lower costs could be achieved by mass production.

But with wind energy following problems are associated:

1. Wind energy available is dilute and fluctuating in nature. Because of the dilute form, conversion machines have to be necessarily large.

2. Unlike water energy, wind energy need storage means because of its irregularity.

3. Wind energy systems are noisy in operation ; a large unit can be heard many kilometers away.

4. Large areas are needed to install wind farms for electrical power generation.

In India the interest in the wind mills was shown in the last fifties and early sixties. Apart from importing a few from outside, new designs were also developed but it was not sustained. It is only in the last 12–15 years that development work is going in many institutions. An important reason for this lack of interest in wind energy must be that wind velocities in India are relatively low and vary appreciably with the seasons.

Data quoted by some scientists that for India wind speed value lies between 5 km/hr to 20 km/hr. These low and seasonal winds imply a high cost of exploitation of wind energy. Calculations based on the performance of a typical wind mill have indicated that a unit of energy derived from a wind mill will be at least several times more expensive than energy derivable from electrical distribution lines at the standard rates, provided such electrical energy is at all available at the wind mill site. The above argument is not fully applicable in rural areas for several reasons. First electric power is not and will not be available in many such areas due to the high cost of generation and distribution to small dispersed users. Secondly there is possibility of reducing the cost of the wind mills to suitable design. Lastly, on small scales, the total first cost for serving the felt need and low maintenance costs are more important than the unit cost of energy. In our country, as stated earlier, high wind speeds are obtainable in coastal areas of Saurashtra, Western Rajasthan and some part of the Central India, where there could be possibility of using medium and large size wind mills of generation of electricity.

Many projects on the wind mill systems for water pumping and for production of small amount of electrical power have been taken up by the various organisers in our country. Following are some of the developments.

1. CAZRI wind mill at Jodhpur (Rajasthan).

2. WP-2 water pumping wind mill by NAL Bangaluru.

3. MP-1 sail wind mill at NAL Bangaluru.

4. Wind mills at Central Salt and Marine Chemicals Research Institute Bhavnagar (Gujarat).

5. 12 PU 500 wind mill at NAL Bangaluru.

6. Madurai wind mill at Madurai (Tamil Nadu).

7. Tayabji wind mill at Tilonia near Ajmer (Rajasthan).

8. Sholapur wind mill at Sholapur (MS).

12 PU 500 wind mill, designed by NAL Bangaluru, can pump at the rate of about 5 to 6 thousand litres of water per hour over a total head of 5 metres, when the wind speed is in the range of 12 to 14 km per hour. It can develop more power at higher wind speeds upto about 32 km/hr.

The MP-1 sail type wind mill, which is also simple in construction, has same out-put as that of 12 PU 500. The rotor of it, is made of canvas sails and is of

7.5 metres in diameter. These two type of wind mills indicate promise for large scale exploitation and commercialization. The Department of Non-convention Energy Sources (DNES) Government of India has an important land mark in the country's programme towards the utilization of renewable energy, was the commissioning of four wind farms at Mandavi (Gujarat), 1.15 MW, Tuticorin (Tamil Nadu) 550 kW, Okha (Gujarat) 550 kW, Puri (Orissa) 550 kW and Deogarh (Maharashtra) with a capacity of 550 kW.

During the Seventh plan, nine wind-farm projects of aggregate capacity of 10.10 MW have so far been commissioned at Okha, Mandvi and Okha-Mandhi in Gujarat; Tuticorin and Kayattar in Tamil Nadu; Puri in Orissa; Deogad in Maharastra; Tala Cauvery in Karnataka; and Tirumala in Andhra Pradesh (DNES report).

Barring a few minor problems, the projects are successfully generating and supplying electric power to the respective state grids. Projects of aggregate capacity of 24.10 MW are under construction. During the seventh plan, about 2500 water pumping wind mills have been installed under the wind pump demonstration/field testing programme. Wind pumping technology has been upgraded to cover the deep-well wind pumping applications. Efforts have been made to bring about indigeneous production of various wind machines. A wind energy centre has been set up at National Aeronautical Laboratory Bangaluru, to provide technological inputs pertaining to design, development, testing certification, etc. A large data base for wind resource assessment has also been established.

India has a potential of 20,000 MW of wind power. DNES plans to harness 400 MW of wind energy during eighth plan period.

1.6.4. Energy from Bio-mass and Bio-gas. The potential for application of bio-mass as an alternate source of energy in India is very great. We have plenty of agricultural and forest resources for production of biomass. Biomass is produced in nature through photosynthesis achieved by solar energy conversion. As the word clearly signifies, Biomass means organic matter. In simplest form the reaction is the process of photosynthesis in the presence of solar radiation, can be represented as follows

$$H_2O + CO_2 \xrightarrow{\text{Solar energy}} CH_2O + O_2$$

In the reaction, water and carbon dioxide are converted into organic material *i.e.*, CH_2O, which is the basic molecule of forming carbohydrate stable at low temperature, it breaks at high temperature, releasing an amount of heat equal to 112,000 cal/mole (469 kJ mole).

$$CH_2O + O_2 \longrightarrow CO_2 + H_2O + 112 \text{ kcal/mole.}$$

The absorbed energy of photons should be at least equal to this amount. It is therefore, possible to produce large amount of carbohydrate by growing say, algae, under optimum conditions in plastic tubes or in ponds. The algae could be harvested, dried and burned for production of heat that could be converted into electricity by conventional methods. The bio-mass is used directly by burning or is further processed to produce more convenient liquid and gaseous fuels.

Bio-mass resources fall into three categories:

1. Bio-mass in its traditional solid mass (wood and agricultural residue), and

2. Bio-mass in non-traditional form (converted into liquid fuels).

The first category is to burn the bio-mass directly and get the energy. In the second category, the biomass is converted into ethanol and methanol to be used as liquid fuels in engines.

3. The third category is to ferment the bio-mass anaerobically to obtain a gaseous fuel called bio-gas (Bio-gas → 55 to 65% Methane, 30—40%. CO_2, and rest impurities *i.e.*, H_2, H_2S and some N_2).

Bio-mass resources include the following:

(*i*) Concentrated waste—municipal solids, sewage wood products, industrial waste, manure of large lots.

(*ii*) Dispersed waste residue—crop residue, legging residue, disposed manure.

(*iii*) Harvested bio mass, standby biomass, biomass energy plantation.

1.7. Energy Plantation

For large scale production of electrical power, the use of fire wood as a fuel for the boilers of a conventional power plant is suggested. This approach is called the "energy plantation" scheme, in which selected species of trees would be planted and harvested over regular time period, on land, near the power plant. A large area is required for it. Trees which are suggested for use in India are eucalyptus, casuarina and babool.

1.7.1. Bio-Gas. The main source for production of bio-gas is wet cow dung or wet livestock (and even human) waste, to produce bio-gas. The production of bio-gas is of particular significance for India because of its large cattle population. The total cattle population in the country is about 250 million. Some of the other sources of bio-gas are:

(*i*) Sewage, (*ii*) crop residue, (*iii*) vegetable wastes, (*iv*) water hyacinth, (*v*) poultry droppings, (*vi*) pig-manures, (*vii*) Algae, (*viii*) Ocean-kelp.

In big cities, sewage source is the main source for production of bio-gas. Bio-gas thus obtained can be used to run pumps to pumpout the sewage water itself. Pilot plants for such purpose, capable of handling sewage, have already been developed and installed in some areas. The sewage bio-gas is found to contain 84 per cent Methane, which is already pointed out, is a high quality fuel. Methane could be economically used to run engines to drive electric generators.

In the rural sector, bio-gas finds great applications in cooking, lighting, mechanical power and generation of small electricity. The gas can be used with advantage to improve sanitary conditions and also to check environmental pollution. Bio-gas can be used solely or with diesel in I.C. engines, for production of power. For converting I.C. engines of diesel or petrol/kerosene type, to gas engines a special attachment has to be fixed up. Ruston and Hornsby has developed 5 HP engine to work on bio-gas. Remarkable progress has been made in our country in respect of bio-gas plants. More than 7.3 lakh family based on bio-gas plants were installed during Seventh Plan Period, few community and institutional bio-gas plants have also been installed. Some examples of successful institutional Bio-gas plants commissioned by DNES during the year 1985-86 are at Muradnagar (U.P.), Rishikeah (U.P.), Sanganer (Raj.), Sihar (Raj.), Pondicheri, Bhopal (M.P.) etc. About 70 community/institutional type bio-gas plants are set during the year 1985-86.

1.7.2. Ocean Thermal Energy Conversion. This is also an indirect method of utilizing solar energy. A large amount of solar energy is collected and stored in tropical oceans. The surface of the water acts as the collector for solar heat, while the upper layer of the sea con stitutes infinite heat storage reservior. Thus the heat contained in the oceans, could be converted into electricity by utilizing the fact that the temperature difference between the warm surface waters of the tropical oceans and the colder waters in the depths is about 20—25°K. Utilization of this energy, with its associated temperature difference and its conversion into work, forms the basis of ocean thermal energy conversion (OTEC) systems. The surface water which is at higher temperature could be used to heat some low boiling organic fluid, the vapours of which would run a heat engine. The exit vapour would be condensed by pumping cold water from the deeper regions. The amount of energy available for ocean thermal power generation is enormous, and is replenished continuously. Several such plants are built in France after World War II (the largest of which has a capacity of 7.5 MW). With a 22°K temperature difference between surface and depths, such as exists in warmer ocean areas than the north sea, the Carnot efficiency is around 7%. This is obviously very low.

All the systems for OTEC method work on a closed Rankine cycle and use low boiling organic fluids like ammonia, propane, R-12, R-22 etc.

A schematic diagram of Rankine cycle OTEC plant is shown in Fig. 1.3. The warm surface water is used for supplying the heat input in boiler, while the cold water brought up from the ocean depths is used for extracting the heat in the condenser.

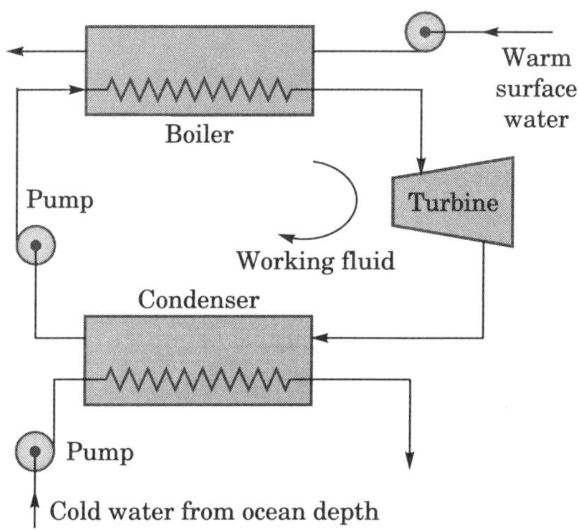

Fig. 1.3. Schematic diagram of Rankine cycle OTEC plant.

In India, Department of Non-conventional energy sources (DNES) has proposed to install a 1 MW OTEC plant in Lakshadweep Island at Kavaratti and Minicoy. Preliminary oceanographic studies on the eastern side of Lakshadweep Island suggest the possibility of the establishment of shore based OTEC plant at the Island with a cold water pipe line running down the slope to a depth of 800-1000 m. Both the Islands have large lagoons on the western side. The lagoons are very shallow with, hardly any nutrient in the sea water. The proposed OTEC plant will bring up the water from 1000 m depth which has high nutrient value. After providing the cooling effect in the condenser, a part of sea water is proposed to be diverted to the lagoons for the development of aqua culture.

1.7.3. Tidal Energy. The tides in the sea are the result of the universal gravitational effect of heavenly bodies like sun and moon on the earth. Due to fluidity of water mass, the effect of this force becomes apparent in the motion of water, which shows a periodic rise and fall in levels which is in rhythms with the daily cycle of rising and setting of sun and moon. This periodic rise and fall of the water level of sea is called *tide*. These tides can be used to produce electrical power which is known as tidal power. When the water is above the mean sea level, it is called *flood tide* and when the level is below the mean sea level, it is called *ebb tide*.

The use of tides for electrical power generation is practical in a few favourably situated sites where the geography of an inlet or bay favours the construction of a large scale hydroelectric plant. To harness the tides, a dam would be built, across the mouth of the bay. It, will have large gates in it and also low head hydraulic reversible turbines are installed in it. A tidal basin is formed, which gets separated from the sea, by dam. The difference in water level is obtained between the basin and sea. The constructed basin is filled during high tide and emptied during low tide passing through sluices turbine respectively. This principle is explained in Fig. 1.4. By using reversible water turbines, turbine can be run continuously, both during high tide and low tide. The turbine is coupled to generator, potential energy of the water stored in the basin as well as energy during high tide, is used to drive the turbine, which is coupled to generator, generating electricity.

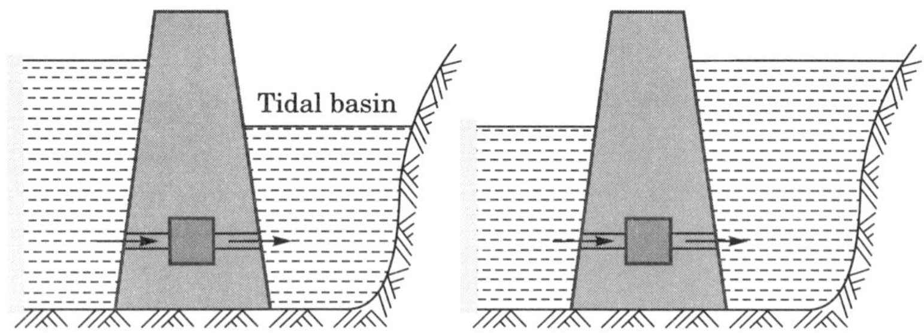

(*a*) High tide (*b*) Low tide

Fig. 1.4. Principle of tidal power generation.

Above arrangement of harnessing tidal energy, is known as single basin plant. The plant continuous to generate power till the tide reaches, its lowest level. Again a minimum head will be reached when it pays to shut down the turbine and open the bypass valves to drain the remaining basin water to sea. Single basin plant can not generate power continuously, though it might do so by using a pumped storage plant, if the load is supplied fluctuates considerably. To overcome this difficulty, two basin plants could be used to generate power continuously without interruption.

Such plants can be constructed only in selected places where the height of the tide is sufficient to justify economy.

A tidal power plant has been completed at Rance in France with rating of 240 MW in 1966. It works on single basin, two way system with pumping.

The possible sites for tidal power plants in India are suggested at Gulf of Cambay, Gulf of Kutchh in Gujarat and Sunderban area in West Bengal. However, no such plant has been constructed so far in our country.

1.8. Geothermal Energy

This is the energy which lies embedded within the earth. According to various theories the earth has a molten core. The fact that volcanic action takes place in many places on the surface of the earth, supports these theories. The steam and hot water comes naturally to the surface of the earth in some locations of the earth. For large scale use bore holes are normally sunk with depth upto 1000 m, releasing steam and water at temperatures upto 200 or 300°C and pressures upto 30 kgf/cm^2 (3000 kN/m^2). Two ways of electric power production from geothermal energy has been suggested. In one of these heat energy is transferred to a working fluid which operates the power cycle. This may be particularly useful at places of fresh volcanic activity. Where the molten interior mass of earth vents to the surface through fissures and substantially high temperatures, such as between 450 to 550°C can be found. By embedding coil of pipes and sending

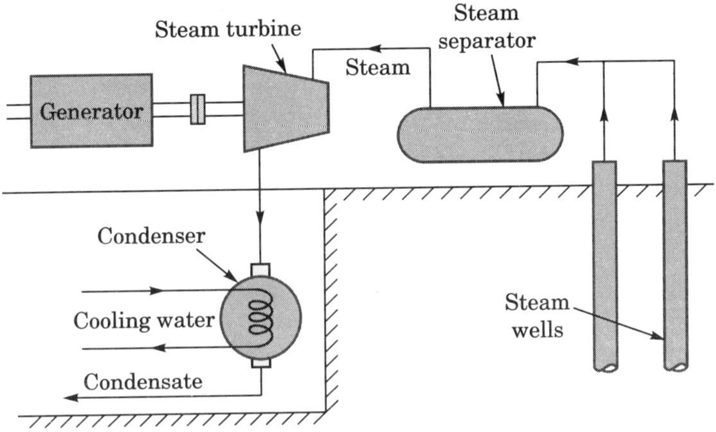

Fig. 1.5. A schematic of geothermal power plant (wet steam).

water through them can be raised. In the other, the hot geothermal water and/or steam is used to operate the turbines directly. From the well-head the steam is transmitted by pipe lines upto 1 m in diameter over distances upto about 3 km to the power station water separators are usually required to separate moisture and solid particles from steam.

At present only steam coming out of the ground is used to generate electricity, the hot water is discarded because it contains as much as 30% dissolved salts and minerals, and these cause serious rust damage to the turbine. The water, however contains more than 1/3 of the available thermal energy. Research is being carried out to build turbines which can withstand the corrosive effects of hot water coming out of wells.

Significant developments in the use of geothermal energy are expected in several countries of the world including India, in the coming years. This will be advantageous both from the point of view of conserving fossil and of pollution control in the atmosphere. Substantial sources of this form of energy are known to exist in U.S.A., Japan, USSR, New Zealand, Italy and Maxico. In India, the State of Himachal Pradesh is reported to possess geothermal energy in exploitable amount. World's first geothermal power station was established at Lardarello in Italy in 1905 and it now produces more than 460 MW of power. Other chief installation of geothermal energy is at Wairaki, New Zealand (250 MW).

A 7.5 tonnes capacity cold storage pilot plant based on geothermal energy is installed at Manikarah Himachal Pradesh under the sponsorship of DNES.

1.9. Hydrogen Energy

Hydrogen as an energy can play an important role as an alternative to conventional fuels provided its technical problems of production, storage and transportation can be resolved satisfactorily and the cost could be brought down to acceptable limits. One of the most attractive features of hydrogen as an energy carrier is that it can be produced from water which is abundantly available in nature. Hydrogen has the highest energy content per unit of mass any chemical fuel and can be substituted for hydrocarbons in a broad range of applications, often with increased combustion efficiency. Its burning process is non-polluting and it can be used in fuel cells to produce both electricity and useful heat.

1.10. Fuel Cells

It may be defined as *an electro chemical device* for the continuous conversion of the portion of the free energy change in a chemical reaction to electrical energy. It is distinguished from a battery in that it operates with continuous replenishment of the fuel and the oxidant at active electrode area and does not require recharging.

Main components of a cell are (*i*) a fuel electrode, (*ii*) an oxidant or air electrode and (*iii*) an electrolyde.

Hydrogen as a fuel has so far given the most promising results, though cells consuming coal, oil or natural gas would be economically much more useful for large scale application. Some of the fuel cells are hydrogen, oxygen, (H_2, O_2), Hydrazine (N_2H_4, O_2), carbon/coal (C, O_2), methane (CH_4, O_2) etc.

Hydrogen oxygen fuel cells (Hydrox), are efficient and the most highly developed cell. A low pressure hydrogen oxygen cell is illustrated in Fig. 1.6. Two porous carbon or nickel electrodes, are immersed in an electrolyte. Catalyst is embedded in Nickel electrodes. The electrolyte is typically 30% KOH because of its high electrical conductivity and it is less corrosive than acids.

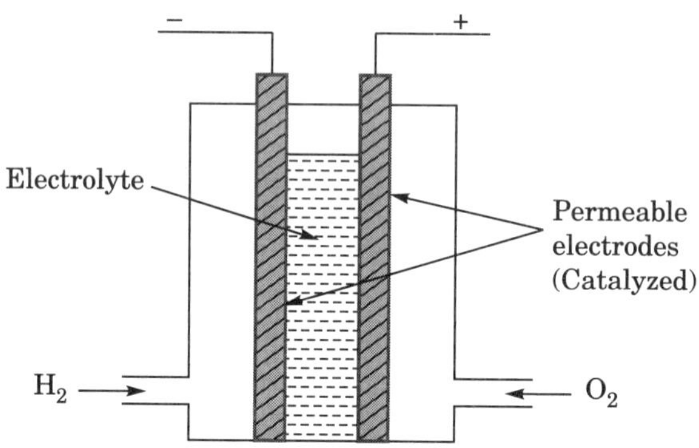

Fig. 1.6. A simple fuel cell.

Cells reactions are:

$$H_2 \xrightarrow{\text{Catalyst}} 2\,H^+ + 2e^-$$

$$2H^+ + OH^- \text{ (hydroxyl ions in electrolyte)} \longrightarrow 2H_2O$$

$$O_2 \rightleftharpoons 2O$$

$$O + H_2O + 2e^- \longrightarrow OH^-$$

H_2 is fed to one electrode and is absorved gives free electrons and also reacts with hydroxyl ions of the electrolyte to form water. The free electrons travel towards oxygen electrode through the external circuit. The two electrons arriving by the external circuit and one molecule of water to form 2 OH^- ions. These OH^- ions migrate towards to H_2 electrode and are consumed there. The electrolyte remains invarient. It is prime requirement that the composition of electrolyte should not change as the cell operates. The cell operates at or slightly above atmospheric pressure and at a temperature about 90°C. These type of cells are called low temperature cells. In high pressure cells pressure is upto about 45 atmospheric and temperature upto 300°C.

A single hydrogen-oxygen cell can produce an emf of 1.23 volt at atmospheric pressure and 25°C. By connecting a number of cells, it is possible to create useful potential of 100 to 1000 volts and power levels of 1 kW to nearly 100 MW.

Some of the advantages of fuel cells are:

1. It is a direct conversion process and does not involve a thermal process, so it has high operating efficiency. Present day fuel cell efficiency is 38% and it is expected to reach 60%.

2. The unit is lighter, smaller and needs less maintenance.

3. Fuel power plants may further cut generation costs by reducing transmission losses.

4. Little pollution, little noise, so that it can be readily acceptable in residential areas.

The primary drawbacks of fuel cells are, their low voltage, high initial costs and low service life.

1.11. Magneto Hydro-Dynamics Generator

The principle of Magneto Hydrodynamics (MHD) power generation enables direct con-version of thermal energy to electrical energy. MHD power generation works on the principle that described by Faraday: When an electric conductor moves across a magnetic field, a voltage is induced in it which produces an electric current. In MHD generators, the solid conductors are replaced by a fluid which is electrically conducting. The working fluid may be either an ionised gas or liquid metal. The hot, partially ionized and compressed gas is expanded in a duct, and forced through a strong magnetic field, electrical potential is generated in the gas. Fig. 1.7. Electrodes placed on the side of the duct pick up potential generated in the gas. In this manner, direct current is obtained which can be converted into AC with the aid of an inverter.

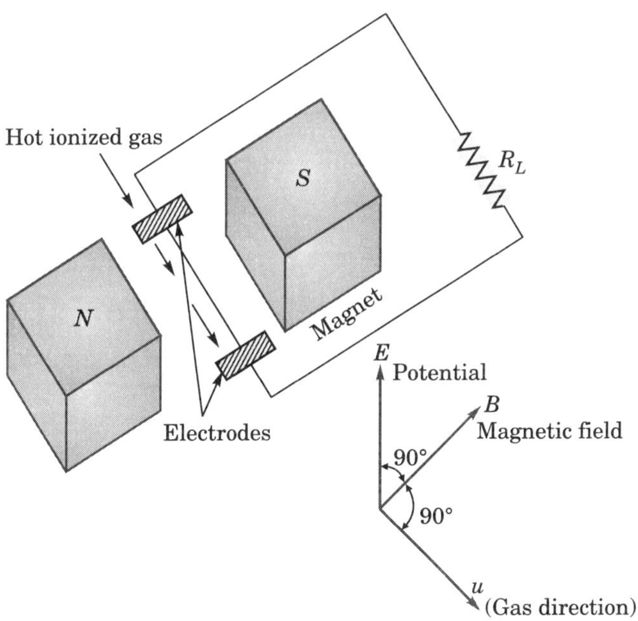

Fig. 1.7. Principle of MHD power generation.

Ionised gas can be produced by heating it to a high temperature. As the gas is heated, the outer electrons escape from its atoms or molecules. The gas particles acquire an electric charge and the gas passes into the state of plasma. High temperatures in excess of 2800°C are needed produce necessary ionization of the gas. However to achieve thermal ionization of products of combustion of fossil fuels or inert gases, extremely high temperatures are needed. Seeding the gas with potassium or cesium helps in ionization and reduces temperature requirement somewhat. The exhaust from MHD generator is at a temperature of about 2500 K and can be used as the heating medium for steam raising in a conventional boiler, thus suggesting the use of a combined cycle.

This system of power generation is simple, and has large power and temperature handling capacity. Having no moving parts, it has high reliability. It can be brought to full power from a standby conditions in 45 seconds. The output can be changed from no load to full load in fraction of a second.

An experimental power plant of 5 MW (Thermal input) has been commissioned at Tiruchirapalli.

1.12. Thermionic Converter

Another form of direct conversion of heat energy to electrical energy has been achieved in the thermionic converter. This consists of two electrodes help in a container filled with ionized cesium vapour. Heating one electrode 'boils out' electrons that travel to the opposite colder electrode. The positive ions in the gas neutralize the space-charge effect of the electrons that normally prevent the flow of electrons. Ionized gas offsets space-charge effect that tends to repel migration of electrons.

Electrons which are emitted by heating cathode, are migrated to cooler anode collector and flow through outer circuit to develop electric power. The anode materials for thermionic converters should have low work function. Barium and strontium oxides are preferred for this purpose. Cathode material should have higher work function, for it tungsten impregnated with a barium compound is a suitable material for this purpose.

Efficiencies of these tubes is of the order of 8 per cent but researchers expect to produce units working at thermal efficiencies of 30 percent. At this efficiency the emitting cathode will be sized at 100 sq cm for a 1 kW output. Potential will be d.c. about 2-3 volts. The present cathode is heated to about 1400°C.

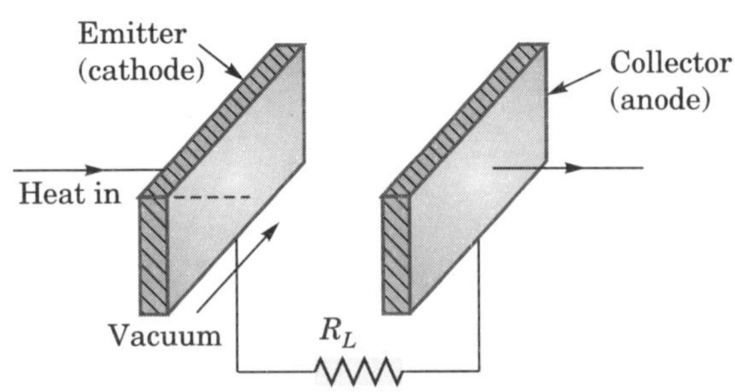

Fig. 1.8. Principle of thermionic converter.

This device is being designed for space power applications where high temperature operation is advantageous. It can be used for supplying power to boats and power tools, and irrigation pumping. These units can be connected in series or parallel for different plant voltages and capacities. These devices may be used as small power plants for peak load power in commercial power generation system.

1.13. Thermo Electric Power

This device converts heat directly into electric power. This eliminates the conversion of heat into kinetic energy of gas or steam flow. Its principle is based on the *Seebeck effect* which states that if two dissimilar materials are joined to form a loop and the two junctions are maintained at different temperatures, an emf will be developed around the loop. This principle is already in use in thermocouples to measure temperaturcs. The magnitude of the emf (E) developed by the above process is proportional to the temperature difference between the two junctions.

$$E \propto (T_2 - T_1) = \alpha (T_h - T_c)$$

where, T_h = Temperature of hot junction

T_c = Temperature of cold junction

α = Seebeck coefficient.

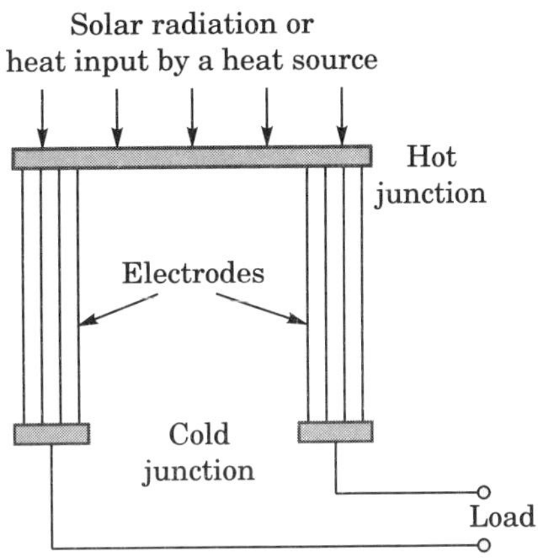

Fig. 1.9. Thermoelectric generator.

The hot junction is maintained at a temperature of T_h by the applied heat source which may be small oil or gas burner, a nuclear reactor, or direct solar radiation by paraboloidal concentrator and the cold junction is maintained at T_c by either water cooling or radiative heat transfer. The principle diagram is shown in Fig. 1.9.

This phenomenon offers one method of producing electrical energy directly from the heat of combustion, where fuel is cheap. This device can generutc power for standby or even base load plants. The thermal efficiency is of the order of 3%, it is not attractive for very special power generation. The new semiconductor materials are found those work more efficiently. These materials are also able to withstand at high temperature. Encouraging results have been obtained with ceramic type materials made of semiconductor 'doped' with materials to increase their conductivity to a level about half way between semiconductors and metal conductors.

These materials, known as mixed valence metals, include materials, such as manganese, iron, cobalt and nickel treated with oxygen, sulphur, selinium and tellurium. It is hoped that materials would work at 20 percent thermal efficiency. A peak load plant of 100 MW capacity can be developed where fuels are very cheap.

1.14 Renewable Energy Resources

Renewable energy sources include both 'direct' solar radiation intercepted by collectors (*e.g.*, solar and flat-plate thermal cells) and indirect solar energy such as wind, hydropower, ocean energy and biomass resources that can be managed in a sustainable manner. Geothermal fields tapped with present drilling technologies have a finite life but are sometimes considered renewable for planning purposes. Traditional methods of using biomass and derivatives such as wood and charcoal are highly inefficient, in contrast with modern techniques emphasing proper forest management sustained yield fuel wood plantations and efficient production.

If broadly interpreted, the definition of renewable resources also includes the chemical energy stored in food and nonfuel plant pruducts and even the energy in air used to dry materials or to cool, heat and ventilate the interiors of buildings. From an operational view point, the correct way to treat renewable energy is as a means to reduce the demand for conventional energy forms. Thus, in performing economic and financial analyses, there is no real distinction between renewable energy technologies and those designed to improve the efficiency of conventional energy use.

A further point is that cost-effective approaches to energy efficiency-ranging from no-or low-cost measures (*e.g.*, reducing excess air in boilers, shutting down equipment when not needed) to systems requiring moderate capital investment, such as heat recuperators, boiler replacements or cogeneration units, can improve the financial and economic feasibility of renewable as well as conventional energy systems. Improvements in the efficiency of energy use can be teamed with a variety of energy supply technologies, and this fact must be recognized when assessing the relative economics of renewable and conventional energy systems.

Three independent primary sources provide energy to the earth: the sun, geothermal forces and planetary motion in the solar system. In particular, direct solar radiation represents an enormous resource for a modern technological civilization. However, human capacity to harness these gigantic natural flows of energy to perform useful work depends largely on the economic feasibility of the required conversion in comparison with fossil fuel options and the extent to which large scale applications affect food production, climate and ecology.

1.14.1. Advantages of Renewable Energy. Even though renewable options are not likely to supply a substantial amount of energy to developing countries over the short term, they do have these advantages:

(1) Renewable energy is an indigeneous resource available in considerable quantities to all developing nations and capable, in principle, of having a significant local, regional or national economic impact. The use of renewable energy could help to conserve foreign exchange and generate local employment if conservation technologies are designed, manufactured, assembled and installed locally.

(2) Several renewable options are financially and economically competitive for certain applications, such as in remote locations, where the costs of transmitting electrical power or transporting conventional fuels are high, or in those well endowed with biomass, hydro or geothermal resources.

(3) Because conversion technology tends to be flexible and modular, it can usually be rapid deployed. Other advantages of modular over very large individual units include easy in adding new capacity, less risk in comparison with 'lumpy' investments, lower interest on borrowed capital because of shorter lead times and reduced transmission and distribution costs for dispersed rural locations.

(4) Rapid scientific and technological advantages are expected to expand the economic range of renewable energy applications over the next 8-10 years, making it imperative for international decision makers and planners to keep abreast of these developments.

Obstackles to the implementation of renewable energy systems

Experience with renewable energy projects in the developing countries indicates that there are a number of barriers to the effective development and widespread diffusion of these systems. Among these are:

(1) Inadequate documentation and evaluation of past experience, a paucity of validated field performance data and a lack of clear priorities for future work.

(2) Weak or non-existent institutions and policies to finance and commercialize renewable energy systems. With regard to energy planning, separate and completely uncoordinated organisations are often responsible for petroleum, electricity, coal, forestry, fuelwood, renewable resources and conservation.

(3) Technical and economic uncertainties in many renewable energy systems, high economic and financial costs for some systems in comparison with conventional supply options and energy efficiency measures.

(4) Skeptical attitudes towards renewable energy systems on the part of the energy planners and a lack of qualified personnel to design, manufacture, market, operate and maintain such systems.

(5) Inadequate donor coordination in renewable energy assistance activities, with little or no information exchange on successful and unsuccessful projects.

1.15. Prospects of Renewable Energy Sources

The one 'new' source of energy that promises to replace oil and gas, and ultimately coal is a different kind of fusion reactor—the sun. The total amount of incoming solar energy absorbed by the earth and its atmosphere in one year—3.8 $\times 10^{24}$ J—is equivalent to 15-20 times the amount of energy stored in all of the world's reserves of recoverable hydrocarbons. Indeed, if just 0.005% of this solar energy could be captured using fuel crops specially designed buildings, wind and water turbines, solar collectors have energy converters and the like, this would supply more useful energy over the year than is currently obtained by burning fossil fuels. Unlike capital energy resources, renewables can not be exhausted. The only limitation is the rate at which they are used—it is not possible to deplete any particular reservoir of energy (such as a column of moving air or falling water) faster than it is replenished.

Renewables already supply a major part of the world's energy needs. Biomass, for example, accounts for about *one seventh* of all fuel consumed, and supplies over 90% of that used in some third world countries:hydro generates *one quarter* of the World's electricity, and more than two thirds of that used in over 35 countries; and the sun contributes directly to space heating in virtually all buildings, through the walls and windows, although precise estimates of the size of this contribution are not available. However, over the last two decades there has been burgeoning interest in renewables from the more industrialised nations and this has led to growing capital investment.

Renewable energy technologies are in many ways more attractive than most conventional energy technologies.

(*i*) They can be matched in scale to the need, and can deliver energy of the quality that is required for a specific task, thus reducing the need to use premium fuels or electricity to provide low grade forms of energy such as hot water (which can be supplied in many other ways).

(*ii*) They can often be built on, or close to the site where the energy is required this minimises transmission costs.

(*iii*) They can be produced in large numbers and introduced quickly, unlike large power stations which have long lead times, often 10 years or more. Rapid planning and construction lowers unit cost and allows planners to respond quickly to changing patterns of demand.

(*iv*) The diversity of systems available also increases flexibility and security of supply. In contrast, over dependence on imported fuels makes a country more, vulnerable to political pressures from producer nations and multinationals. Generic faults in power plants, serious breakdowns, industrial action or simply bad weather can jeopardise the supply of electricity.

(*v*) While there are physical and environmental risks associated with the construction and operation of renewable energy technologies—as there are with all energy conversion systems—they tend to be relatively modest by comparison with those associated with fossil fuels or nuclear power. The failure of a solar panel or a remotely sited wind-turbine or wave energy converter might involve temporary inconvenience, but it will not, as a rule, endanger life or limb, nor cause lasting damage. The most serious consequences could be those associated with such events as the catastrophic failure of a large hydro-electric dam, fire in a biomass plantation, or the explosion of a methane digester.

The Table 1.2 contains estimates of the theoretical potential of the world's renewable energy resources and gives an indication of the size of the contribution that each makes to current energy demand. (Geothermal energy is included although it is not strictly a renewable source). The practical and economic potential of renewables is order of magnitude less than the theoretical potential. However, renewables could still meet total world energy demand many times over.

Table 1.2. World Renewable Energy Resources

Resource	*Form of delivered energy (Application)*	*Comment*
Solar Total solar radiation absorbed by the earth and its atmosphere is 3.8×10^{24} J/yr.	Low temperature heat (space heating water heating and electricity)	Millions of solar water heaters and solar cookers are in use. Solar cells and power towers are in operation.
Wind The kinetic energy available in the atmosphere circulation is 7.5×10^{20} J.	Electricty Mechanical energy (Pumping transport)	Several multi-megawatt wind turbines are in operation and many more in construction. There are numbers of small wind turbines and wind pumps in use.
Biomass Total solar radiation absorbed by plants is 1.3×10^{21} J/yr. The world's standing biomass has an energy content of about 1.5×10^{22} J.	High temperature heat (cooking, smelting etc.) Bio-gas (cooking, mechanical power etc.)	Biomass (principally wood accounts for about 15% of the world's (commercial fuel) consumption; it provides over 80% of the energy needs of many developing countries. There are millions of biogas plants in operation, most of them are in China.
	Alchol (transport)	Several thousand, million litres of alcohol are being produced notably in Brazil and the U.S. Production is increasing rapidly: many countries have launched liquid biofuel programes.
Geothermal The heat flux from the earth's through the surface is 9.5×10^{20} J/yr.	Low temperature heat (bathing, space and water heating)	Geothermal energy supplies about 5350 MW of heat interior for use in bathing principally in Japan, but also in Hungary, Ice land and Italy. Morethan a lakh houses are supplied with heat

Table cont...

Resource	Form of delivered energy (Application)	Comment
		from geothermal capacity is more wells. The installed than 2650 MW (thermal).
The total amount of heat stored in water or stream to a depth of 10 km is estimated to be 4×10^{21} J; that stored in the first 10 km of dry rock is around 10^{27} J.	Electricity	Installed capacity is more than 2500 MW but output is expected to increase more than seven fold by 2000.
Tidal Energy dissipated in connection with slowing down the rotation of the earth as a result of tidal action is around 10^{26}. J/yr.	Electricity	Only one large tidal barrage is in operation (at La Rance in France) and there are small schemes in Russia and China. Total installed capacity is about 240 MW and the output around 0.5 TWh/yr. In addition China has several small tidal pumping stations. Several large tidal schemes are being planned.
Wave The amount of energy stored as kinetic energy in waves may be of the order of 10^{18} J.	Electricity	The Japanese wave energy research vessel, the Kaimei, has an installed capacity of about 1 MW. There are, in addition several hundred wave powered navigational buoys. Designs after large prototype wave energy converters are being drawn up.
Hydro The annual precipitation land amounts to abouts 1.1×10^{17} kg of water. Taking the average elevation of land area as 840 m^2, the annually accumulated potential energy would be 9×10^{20} J.	Electricity	Large hydroschemes provide about one quarter of the world's total electricity supply and more than 40% of the electricity used in developing countries. The installed capacity is more than 363 GW. The technically usable potential is estimated to be 2215 GW or 19000 TWh/yr. There are no accurate estimates of the number of capacity of small hydroplants currently in operation.

The perspective studies carried out indicate substantial increase in the commercial energy requirements from now to the end of the century. The electricity generation requirements would increase from about 110 Twh at present to about 550 Twh by the year 2000. The requirements of oil are expected to increase from about 30 to 92 million tonnes and coal from 100 to 530 million tonnes.

Coal is the most important commercial source of energy in India. The resource assessment done in 1978 gives the total coal reserves of the country upto a depth of 600 m and in seams of thickness above 1.2 m as 85450 million tonnes. A significant amount of coals occur at depths greater than 600 m and in seams of thickness less than 1.2 m, thus the total reserve of coal are put at about 111,700 million tonnes.

The survey and explorations for oil and gas resources are presently being conducted mainly by the ONGC and to some extent by Oil India Limited. The net recoverable reserves of oil are estimated at about 300 million tonnes and natural gas of 73 million cubic metres.

Preliminary assessment of hydro-electric potential indicates that the exploitable potential is about 400 Twh of annual energy generation. Schemes with an energy potential of 40 Twh have already been constructed and are presently in operation. The bulk of untapped hydroelectric potential equipment to 360 Twh of annual energy generation is located in northern and north-eastern regions mainly along the rivers of Himalayan origin. The average load factor of hydroelectric plants presently in operation is about 42 per cent. India has modest reserves of uranium and substantial reserves of Thorium. The established uranium resources are considered capable of supporting a nuclear programme on natural uranium of about 8000 MWe. India's thorium reserves can support a very large programme of nuclear power development based on breeder reactor technology.

The intensity of use of electrical energy in the Indian economy has shown a steady increase. This trend necessitated substantial increase in the share of investment. The investment made in the power sector was of the order of ₹ 154,000 million in the sixth plan (1978–83). It would be difficult for the economy to sustain such increasing orders of investments in the future. It is therefore essential that the demand for electricity in terms of peak value and energy should be managed with a view to reduce the intensity of electricity used in various sectors. It is also essential to keep transmission and distribution losses and consumption of energy by auxiliaries in the power station to a minimum.

Under conditions obtaining in India hydroelectric power constitutes generally the most economical source for power generation. Accelerating its development will help in achieving the twin objectives of optimising investment and conserving energy resources. India has been following the policy of basing its thermal power development predominantly on coal. This policy would be continued in the future also. The location of thermal power plants is being gradually shifted closer to the coal mines to reduce the energy costs required for transportation of coals.

Due to the limited oil reserves India has to depend on substantial imports for meeting its future requirements. The bulk of the demand for oil is from transport sector, and in order to reduce the pressure from this sector it is necessary to explore possibilities of developing substitute fuels like biomass and producer gas. Accelerating the pace of electrification of rural areas and judicious pace of mechanisation of agriculture could help in reducing demand for oil.

Out of total non-commercial fuels the share of the fire wood is nearly 65 per cent. The availability of animal dung and agricultural waste is likely to increase in the future due to increased agricultural production and animal population. With the popularisation of bio-gas plants there might be an increase in the consumption of animal dung used as a fuel in the household sector.

Bio-gas and solar energy offer the greatest scope for the development of new energy sources. A number of prototype devices have been successfully developed for use of solar energy in areas like grain drying, water and space heating and power generation. These devices are presently under going field trials for their performance and reliability. Application of wind power for agricultural pumping with certain prototype models are under going field trials and evaluation. Explorations to establish the geothermal and tidal potential are in progress but they are not expected to contribute to any significant extent to the energy supplies by the turn of the century.

QUESTIONS

1. What are primary and secondary energy sources?
2. What are the conclusions on alternate energy strategies?
3. What are the conventional and non-conventional energy sources? Describe briefly.
4. Discuss briefly the possibilities of utilizing the following methods of power generation:
   (*i*) Solar energy
   (*ii*) Magneto hydrodynamics
   (*iii*) Fuel cells.
5. Describe the types of solar power plant. What are the limitations of a solar power plant?
6. What are the methods of direct energy conversion? Describe in brief.
7. Write short notes on:
   (*i*) Wind energy
   (*ii*) Tidal energy
   (*iii*) Bio-mass and bio-gas
   (*iv*) OTEC
   (*v*) Thermoelectric generator.
8. What are the prospects of non-conventional energy sources in India? Explain.
9. What is meant by renewable energy sources? Explain in brief these energy sources with special reference to Indian context.
10. What are the advantages and limitations of renewable energy sources?

UNIT—1

SOLAR-ENERGY

- Solar Radiation and its Measurement
- Solar Energy Collectors
- Solar Energy Storage
- Applications of Solar Energy

CHAPTER 2

Solar Radiation and Its Measurement

2.1. Introduction

In general, the energy produced and radiated by the sun, more specifically the term refers to the sun's energy that reaches the earth. Solar energy, received in the form of radiation, can be converted directly or indirectly into other forms of energy, such as heat and electricity, which can be utilized by man. Since the sun is expected to radiate at an essentially constant rate for a few billion years, it may be regarded as an inexhaustible source of useful energy. The major drawbacks to the extensive application of solar energy are:

1. The intermittent and variable manner in which it arrives at the earth's surface and
2. The large area required to collect the energy at a useful rate.

Experiments are underway to use this energy for power production, house heating, air-conditioning, cooking and high temperature melting of metals.

Energy is radiated by the sun as electromagnetic waves of which 99 per cent have wave lengths in the range of 0.2 to 4.0 micrometers (1 micrometer = 10^{-6} meter). Solar energy reaching the top of the earth's atmosphere consists of about 8% ultraviolet radiation (short wave length, less than 0.39 micrometer), 46% visible light (0.39 to 0.78 micrometer), and 46% infrared radiation (long wave length more than 0.78 micrometer).

2.2. Solar Constant

The sun is a large sphere of very hot gases, the heat being generated by various kinds of fusion reactions. Its diameter is 1.39×10^6 km while that of the earth is 1.27×10^4 km. The mean distance between the two is 1.50×10^8 km. Although the sun is large, it subtends an angle of only 32 minutes at the earth's surface. This is because it is also at a very large distance. Thus the beam radiation received from the sun on the earth is almost parallel. The, brightness of the sun varies from its centre to its edge. However for engineering calculations, it is customary to assume that the brightness all over the solar disc is uniform. As viewed from the earth, the radiation coming from the sun appears to be essentially equivalent to that coming from a black surface at 5762°K.

The rate at which solar energy arrives at the top of the atmosphere is called the *solar* constant I_{sc}. *This is the amount of energy received in unit time on a unit area perpendicular to the sun's direction at the mean distance of the earth from the sun.* Because of the sun's distance and activity vary throughout the year, the rate of arrival of solar radiation varies accordingly. The so called solar constant is thus an average from which the actual values vary upto about 3 percent in either direction. This variation is not important however, for most practical purposes.

The National Aeronautics and Space Administration's (NASA) standard value for the solar constant, expressed in three common units, is as follows:

1.353 kilowatts per square metre or 1353 watt per square metre.

116.5 langleys (calories per sq. cm) per hour, or 1165 kcal per sq. m per hour (1 langley being equal to 1 cal/cm^2 of solar radiation received in one day).

429.2 Btu per sq. ft. per hour.

The distance between the earth and the sun varies a little through the year. Because of this variation, the extraterrestrial (out side the earth's atmosphere) flux also varies. The earth is closest to the sun in the summer and farthest away in the winter. This variation in distance produces a nearly sinusoidal variation in the intensity of solar radiation I that reaches the earth. This can be approximated by the equation

$$\frac{I}{I_{sc}} = 1 + 0.033 \cos \frac{360(n-2)}{365} \quad \text{...(2.1)}$$

$$\simeq 1 + 0.033 \cos \frac{360 \times n}{365} \quad \text{...(2.2)}$$

where, n is the day of the year. As the distance between earth and sun varies a little through the year, due to it extraterrestrial radiation also varies.

It is also useful to know the spectral distribution of extraterrestrial solar radiation. Measurements of this distribution are made and recorded by M.P. Thekaekara. It will be noted from Fig. 2.1, which shows spectral distribution of solar radiation intensity at the outer limit of atmosphere, that the maximum value of 2074 W/m^2—μm occurs at a wavelength of 0.48 μm and that 99 percent, of the sun's radiation is obtained upto a wavelength of 4 μm. The percentage of radiation

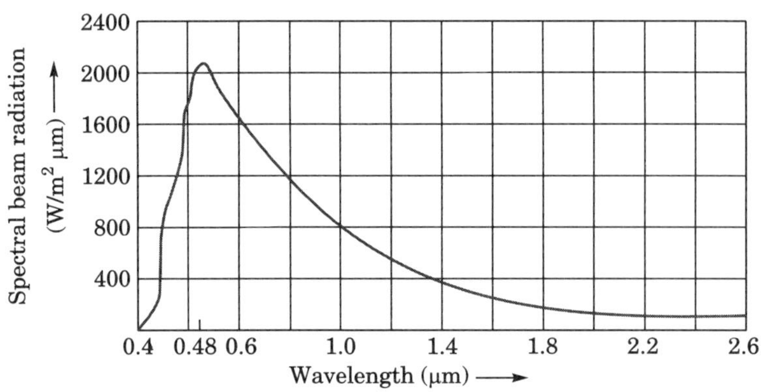

Fig. 2.1. Spectral distribution of solar radiation intensity.

obtained upto a certain wavelength is also given in Table 2.1.

Table 2.1

Wavelength (μ)	0—0.38	0.38—0.78	0.78—4.0
Approximate energy (W/m^2)	95	640	618
Approximate percentage of total energy	7%	47.3%	45.7%

2.3. Solar Radiation at the Earth's Surface

From the point of view of utilisation of solar energy we are more interested in the energy received at the earth's surface than in the extraterrestial energy. Solar radiation received at the surface of the earth is entirely different due to the various reasons. Before studying this it is important to know the following terms:

2.3.1. Beam and Diffuse Solar Radiation. The solar radiation that penetrates the earth's atmosphere and reaches the surface differs in both amount and character from the radiation at the top of the atmosphere. In the first place, part of the radiation is reflected back into the space, especially by clouds. Further more, the radiation entering the atmosphere is partly absorbed by molecules in the air. Oxygen and ozone (O_3) formed from oxygen, absorb nearly all the ultraviolet radiation, and water vapour and carbon dioxide absorb some of the energy in the infrared range. In addition, part of the solar radiation is scattered (*i.e.*, its direction has been changed) by droplets in clouds by atmospheric molecules, and by dust particles.

Solar radiation that has not been absorbed or scattered and reaches the ground directly from the sun is called *"direct radiation"* or *Beam radiation.* It is the radiation which produces a shadow when interrupted by an opaque object. *Diffuse radiation* is that solar radiation received from the sun after its direction has been changed by reflection and scattering by the atmosphere. Because of the solar radiation is scattered in all directions in the atmosphere, diffuse radiation comes to the earth from all parts of the sky. Fig. 2.2. The total solar radiation received at any point on the earth's surface is the sum of the direct and diffuse radiation. This is referred to in a general sense as the *insolation* at that point. More specifically, the insolation is defined as the total solar radiation energy received on a horizontal surface of unit area (*e.g.*, 1 sq. m) on the ground in unit time (*e.g.*, 1 day).

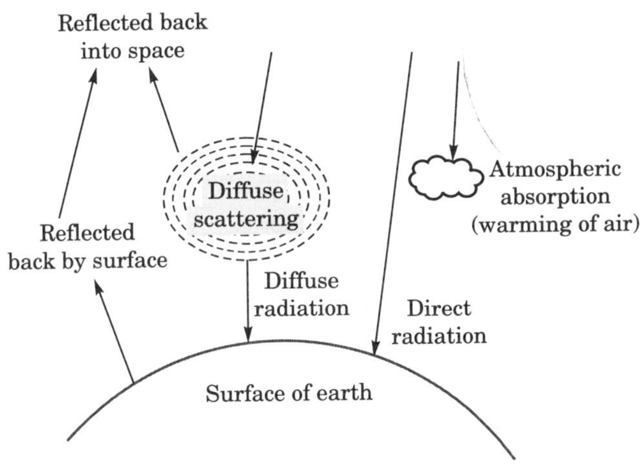

Fig. 2.2. Direct, diffuse and total radiation.

The insolation at a given location on the earth's surfaca depends, among other factors, on the altitude of the sun in the sky. (The altitude is the angle between the sun's direction and the horizontal). Since the sun's altitude changes with the date and time of the day and with the geographic latitude at which the observations are made, the rate of arrival of solar radiation on the ground is a variable quantity even in the time.

There are, nevertheless, some general points that can be made. The smaller the sun's altitude, the greater the thickness of atmosphere through which the solar radiation must pass the reach the ground Fig. 2.2. As a result of absorption and scattering, the insolation is less when the sun is low in the sky than when it is higher. However, when scattering occurs, diffuse radiation constitutes a larger fraction of the total received. On a clear, cloudless day, about 10 to 20 percent of the insolation is from diffuse radiation, the proportion increases upto 100 percent when the sun is completely obscured by clouds. When the humidity is high, the insolation on a cloudy day, consisting entirely of diffuse radiation, may be as high as 50 percent of the insolation on a clear day at the same time and place.

The intensity of the diffuse radiation seen by an observer on a clear day is not isotropic, but varies as a function of latitude time of the year, time of the day, atmospheric content, and other factors. This is discussed by a number of authors. The relationship is so complex that it is seldom considered theoretically. Most, often for solar system design purposes the diffuse radiation is assumed to be isotropic over the sky dome, as any other assumption is simply unwieldy for designers to use. Another justification often used for making this assumption is that the diffuse radiation is only a relatively small fraction (5 to 15 percent) of the total when the sun is not obscured clouds, and that most solar systems do not operate when the sun is obscured by clouds. Hence variations in the distribution of the diffuse radiation are not too important to most users.

2.3.2. Sun at Zenith. Position of the sun directly over head.

2.3.3. Air mass (m). It is the path length of radiation through the atmosphere, considering the vertical path at sea level as unity.

The air mass m is the ratio of the path of the sun's rays through the atmosphere to the length of path when the sun is at the zenith. Except for very low solar altitude angles, the air mass is equal to the cosecant of the altitude angle. Thus at sea level $m = 1$.

$m = 1$ when the sun is at zenith, *i.e.,* directly over head.

$m = 2$ when zenith angle is 60° (θ_z, the angle subtended by the zenith and the line of sight to the sun).

$m = \sec \theta_z$ when $m > 3$.

$m = 0$ just above the earth's atmosphere.

2.3.4. Attenuation of Beam Radiation. The variation in solar radia-tion reaching the earth than received at the outside of the atmosphere is due to absorption and scattering in atmosphere.

***(i)* Absorption.** As solar radiation passes through the earth's atmosphere the short-wave ultraviolet rays are absorbed by the ozone in the atmosphere and the long wave infrared waves are absorbed by the carbon dioxide and moisture in the atmosphere. This results in narrowing of the band width. In fact most of the terrestrial solar energy (*i.e.,* energy received by the earth) lies within the range of 0.29 μ to 2.5 μ.

***(ii)* Scattering.** As solar radiation passes through the earth's atmosphere the components of the atmosphere, such as water vapour and dust, scatter a portion of the radiation. A portion of this scattered radiation always reaches the earth's surface as diffuse radiation. Thus the radiation finally received at the earth's surface consists partly of beam radiation and partly of diffuse radiation.

It must be realized that scattering attenuates the radiation. The exact amount of scattering and consequential attenuation depends on the atmospheric conditions which vary from place to place and at a given place depend on the time of the day, the month of the year and the local weather.

Ozone absorbs mainly in the ultraviolet band. It absorbs almost completely the short wave radiation below 0.29 μm, and its transmittance is almost unity above wavelengths of 0.35 μm. Water vapour absorbs mainly in the infrared bands. At wavelength lower than 2.3 μm, the extra terrestrial solar radiation by H_2O and CO_2 in atmosphere, is strong. Hence for terrestrial application of solar energy, only wavelengths between 0.29 and 2.5 μm need be considered.

Fig. 2.3 shows spectral distribution curves. Under favourable atmospheric conditions, the maximum intensity observed at noon on an oriented surface at sea level is 1 kW/m^2. At an altitude of 1000 metres, the value rises to about 1.05 kW/m^2, and in higher mountains

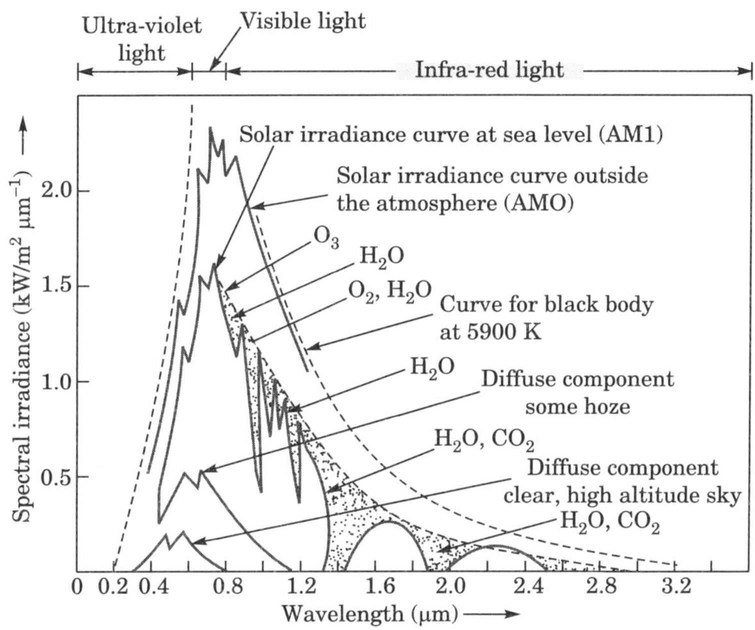

Fig. 2.3. The solar spectrum outside the atmosphere at a ground level.

values slightly above 1.1 kW/m^2 are obtained, compared with 1.353 kW/m^2 (the solar constant) in outer space. The latter value is sometimes called *air mass* 0 (AM0), whereas at sea level, the maximum possible intensity is called AM1. The upper curve applies at the outer limit of the atmosphere (AM0). The other lower curve applies to the earth's surface during clear days for a sea level location, for AM1. Dotted curve shows curve for a black body at 5900°K. The lower two curves are for diffuse components for some haze and clear sky conditions respectively. Considering the solar irradiation curve or lower and upper curve case, the *atmospheric transmission factor* is given as the area under lower curve divided by the area under the upper curve (solar constant). For m = 1.0, the, transmission factor is 0.633, and for m = 5.0, at the earth surface sea level, the transmission factor is 0.276. For different values of m, the transmission factor can be determined. Thus the length of the sun's rays through the atmosphere is of extreme importance in affecting reduction of solar intensity.

2.4. Solar Radiation Geometry

In solar radiation analysis, the following angles are useful:

ϕ_l = latitude of location — δ = declination

ω = hour angle — γ_s = solar Azimuth angle

s = slope — α = altitude angle

θ_z = zenith angle.

If θ is the angle between an incident beam radiation I and the normal to the plane surface, then the equivalent flux or radiation intensity falling normal to the surface is given by $I \cos \theta$. θ is called incident angle.

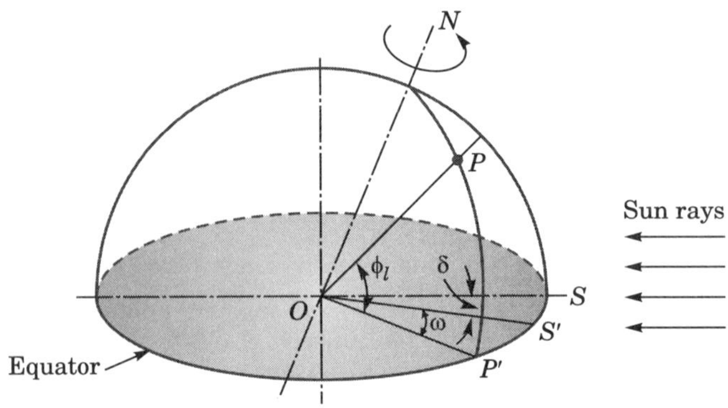

Fig. 2.4. Latitude ϕ_l, hour angle ω, and sun's declination δ.

Let us first define the above angles. The *altitude* ϕ_l of a point or location is the angle made by the radial line joining the location to the centre of the earth with the projection of the line on the equatorial plane. It is the angular distance north or south of the equator measured from centre of earth. As shown in Fig. 2.4, it is the angle between the line $\overline{OP}$ and the projection of $\overline{OP}$ on the equatorial plane. Point P represents the location on the earth surface and O represents the centre of the earth. By convention the latitude will be measured as positive for the northern hemisphere.

The *declination* δ is the angular distance of the sun's rays north (or south) of the equator. It is the angle between a line extending from the centre of the sun to the centre of the earth and the projection of this line upon the earth's equatorial plane.

This is the direct consequence of the tilt and it would vary between 23.5° on June 22 to 23.5° on December 22. At the time of winter solstice. (It is the latin world, meaning sun standing still. For minimum or maximum declination the sun appears to stand still) the sun rays would be 23.5° south of the earth's equator ($\delta = -23.5°$). At the time of summer solstice, the sun's rays would be 23.5° north of the earth's equator ($\delta = 23.5°$). At the equinoxes* the sun's declination would be zero. Fig. 2.5. shows approximately the variation of sun's declination throughout

---

*The Latin word "equinox" means *equal nights*. The nights are equal when the declination of the sun is zero.

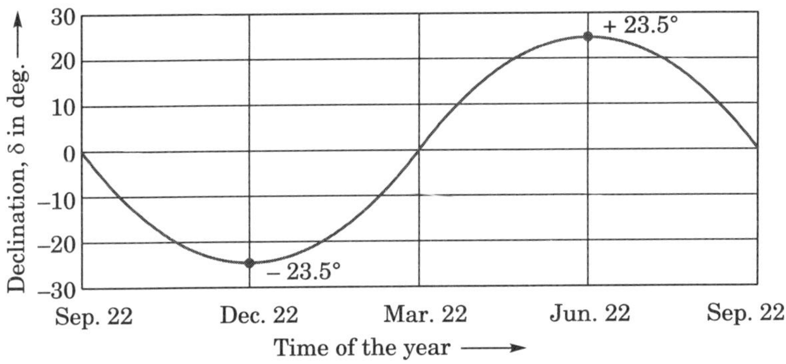

Fig. 2.5. Variation of sun's declination.

the year. Actually the declination varies from year to year, slightly. The value for a middle year between two consecutive leap year may be used. It is zero on the two equinox days of March 22 and September 22. The declination in degrees for any given day may be calculated from the approximate equation of Cooper (1989).

$$\delta \text{ (in degrees)} = 23.45 \sin \left[\frac{360}{365}(284 + n)\right] \qquad \text{...(2.3)}$$

where n is the day of the year, [*e.g.*, 21 June,1988 is the 173th (31 + 29 + 31 + 30 + 31 + 21) day of 1988 *i.e.*, n = 173].

The hour angle ω is the angle through which the earth must turn to bring the meridian of a point directly in line with the sun's rays. The hour angle ω is equivalent to 15° per hour. It is measured from noon based on the local solar time (LST) or local apparent time, being positive in the morning and negative in the afternoon. (The term LST will be defined a little later). It is the angle measured in the earth's equatorial plane, between the projection of $\overline{OP}$ and the projection of a line from the centre of the sun to the centre of the earth.

2.4.1. Altitude angle α (solar altitude). It is a vertical angle between the projection of the sun's rays on the horizontal plane and the direction of sun's rays (passing through the point). (Refer Fig. 2.6).

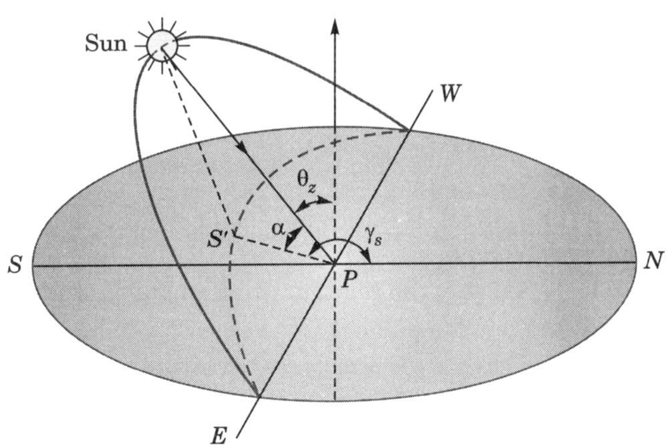

Fig. 2.6. Sun's zenith, altitude and azimuth angles.

2.4.2. Zenith angle θ_z. It is complimentary angle of sun's altitude angle. It is a vertical angle between the sun's rays and a line perpendicular to the horizontal plane through the point *i.e.*, the angle between the beam from the sun and the vertical

$$\theta_z = \frac{\pi}{2} - \alpha.$$

2.4.3. Solar Azimuth angle γ_s. It is the solar angle in degrees along the horizon east or west of north or it is a horizontal angle measured from north to the horizontal projection of the sun's rays. This angle is positive when measured west wise.

Altitude angle α, zenith angle θ_z, and solar azimuth angle γ_s, can be expressed in terms of ϕ_l (latitude angle), δ (declination), and ω (hour angle), which are also called basic angles. The expressions are:

$$\cos \theta_z = \cos \phi \cos \omega \cos \delta + \sin \phi \sin \delta \quad ...(2.4)$$

$$\cos \gamma_s = \sec \alpha (\cos \phi \sin \delta - \cos \delta \sin \phi \cos \omega) \quad ...(2.5)$$

and, $$\sin \gamma_s = \sec \alpha \cos \delta \sin \omega \quad ...(2.6)$$

These equations allow calculation of the sun's zenith altitude and azimuth angles, if the declination, hour angles and latitude are known. In applying these equations, attention must be given to correct signs for the latitude and declination and angle. If north latitudes are considered positive and south latitudes negative, the declination will be positive for the summer period between the vernal equinox and autumnal equinox (March 22 to September 22 approximately) and negative at other times. For non-horizontal surfaces, the other angles, such as incident angle, slope angle, surface azimuth angle, etc. are important angles.

2.4.4. The Slope (s). Is the angle made by the plane surface with the horizontal. It is taken to be positive for surfaces slopping towards the south and negative for surface slopping towards the north.

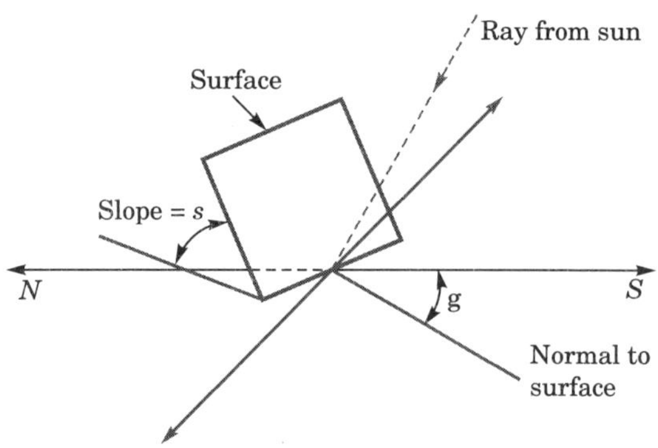

Fig. 2.7. Surface azimuth angle and slope defined.

2.4.5. Surface Azimuth angle (γ). It is the angle of deviation of the normal to the surface from the local meridian, the zero point being south, east positive and west negative.

2.4.6. Incident angle (θ). It is the angle being measured between the beam of rays and normal to the plane.

From Spherical geometry the relation between θ and other angles is given by the equation:

$$\cos \theta = \sin \phi_l (\sin \delta \cos s + \cos \delta \cos \gamma \cos \omega \sin s) + \cos \phi_l (\cos \delta \cos \omega \cos s - \sin \delta \cos \gamma \sin s) + \cos \delta \sin \gamma \sin \omega \sin s \quad ...(2.7)$$

where, ϕ_l = latitude (north positive)

δ = declination (north positive)

ω = hour angle, it is positive between solar mid-night and noon, otherwise negative.

At solar noon ω being zero and each hour angles equating 15° of longitude with morning positive and afternoon negative (*e.g.* $\omega = +15°$ for 11:00 and $\omega = -37.5°$ for 14:30) hour angle ω can be expressed mathematically as:

$$\omega = 15\,(12 - \text{LST}). \quad \text{...(2.8)}$$

Simpler versions of equation (2.7) are normally required. Some of these are as follows:

For *vertical surfaces, s* = 90°, equation (2.7) becomes

$$\cos\theta = \sin\phi\cos\delta\cos\gamma\cos\omega - \cos\phi\sin\delta\cos\gamma + \cos\delta\sin\gamma\sin\omega \quad \text{...(2.9)}$$

For *horizontal surfaces s* = 0°, $\theta = \theta_z$ zenith angle

$$\cos\theta_z = \sin\delta\sin\phi + \cos\delta\cos\phi\cos\omega \quad \text{...(2.10)}$$

$$= \sin\alpha$$

i.e.,

$$\cos\theta = \cos\theta_z = \sin\alpha \quad \text{...(2.11)}$$

For *surface facing due south* $\gamma = 0$,

$$\cos\theta_T = \sin\phi\,(\sin\delta\cos s + \cos\delta\cos\omega\sin s)$$

$$= \cos\phi\,(\cos\delta\cos\omega\cos s - \sin\delta\sin s)$$

(Incident angle θ is expressed as θ_T, denoting the surface as tilted one)

$$= \sin\delta\sin(\phi - s) + \cos\delta\cos\omega\cos(\phi - s) \quad \text{...(2.12)}$$

Vertical surfaces facing due south ($s = 90°$, $\gamma = 0$)

$$\cos\theta_z = \sin\phi\cos\delta\cos\omega - \cos\phi\sin\delta \quad \text{...(2.13)}$$

2.4.7. Day length. At the time of sun rise or sunset, the zenith angle, ϕ_z = 90°, substituting this in equation (2.10), we obtain sun rise hour angle (ω_s)

$$\cos\omega_s = -\frac{\sin\phi\sin\delta}{\cos\phi\cos\delta} = -\tan\phi\tan\delta$$

$$\omega_s = \cos^{-1}(-\tan\phi\tan\delta) \quad \text{...(2.14)}$$

Since 15° of the hour angle are equivalent to 1 hour, the day length (in hours)

$$t_d = \frac{2\,\omega_s}{15} = \frac{2}{15}\cos^{-1}(-\tan\phi\tan\delta) \quad \text{...(2.15)}$$

Therefore, the length of the day (t_d) is a function of latitude and solar declination.

The hour angle at sunrise or sunset on an inclined surface ω_{st} will be lesser than the value obtained by equation (2.15), if the corresponding incidence angle comes out to be more than 90°. Unders this condition, by putting $\theta = 90°$, in equation (2.7) or one of its simple versions. Thus for an inclined surface facing south substituting $\theta = 90°$, in equation (2.12), we obtain

$$\omega_{st} = \cos^{-1}[-\tan(\phi - s)\tan\delta] \quad \text{...(2.16)}$$

The corresponding day length (in hours) is then given by

$$t_d = \frac{2}{15}\cos^{-1}[-\tan(\phi - s)\tan\delta] \quad \text{...(2.17)}$$

2.5. Local Solar Time (or Local Apparent Time)

The time used for calculating the hour angle in all the equation above, is the local solar time or local apparent time, which does not coincide with the local clock time. The solar time can be obtained from the standard time observed on a clock by applying two corrections. The first correction arises because of the difference in longitude between a location and the meridian on which the standard time is based. The correction has a magnitude of 4 minutes for every degree difference in longitude. The second correction called the equation of time correction is due to the fact that earth's orbit and rate of rotation are subject to small perturbations. This correction is based on experimental observations and plotted in Fig. 2.7.

Thus

Local solar time (LST)

= Standard time ± 4 (Standard time longitude – longitude of location) + (Equation of time correction) ...(2.18)

The negative sign is applicable for the eastern hemisphere.

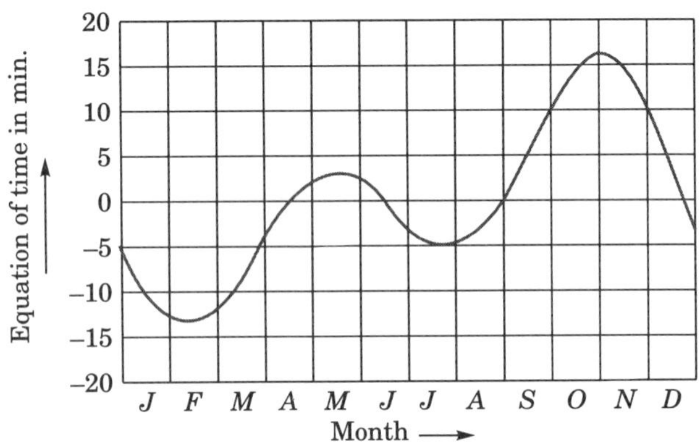

Fig. 2.8. Graph of equation of time (E) correction.

Example 2.1. *Determine the Local Solar time and declination at a location latitude 23° 15′ N, longitude 77° 30′ E at 12.30 IST on June 19. Equation of time correction is given from standard table or chart = (1′ 01″).*

Solution. ∴ The local solar time

= IST – (Standard time longitude – longitude of location)

+ Equation of time correction

$= 12^h\ 30^1 - 4\ (82°\ 30' - 77°\ 30') - 1'\ 01''$

Indian Standard Time (IST) is the local civil time corresponding to 82.5° E longitude

$= 12^h\ 30' - 4 \times 5 - 1'\ 01''$

$= 12^h\ 8'\ 59''$. **Ans.**

Declination δ can be obtained by Cooper's equation *i.e.,*

$$\delta = 23.45 \sin \left[\frac{360}{365}(284 + n)\right]$$

$$= 23.45 \sin \left[\frac{360}{365}(284 + 170)\right]$$

(n is the day of the here = 170 on June 19)

$= 23.45 \sin 86° = 23.43°$. **Ans.**

or $= 23°\ 25'\ 56''$.

Example 2.2. *Calculate the angle made by beam radiation with the normal to a flat collector on December 1, at 9.00 A.M., solar time for a location at 28° 35′ N. The collector is tilted at an angle of latitude plus 10°, with the horizontal and is pointing due south.*

Solution. $\gamma = 0$ since collector is pointing due south. For this case we have the equation

$$\cos \theta_T = \cos (\phi - s) \cos \delta \cos \omega + \sin (\phi - s) \sin \delta$$

Declination δ can be obtained with the help of Cooper equation on December1, $n = 335$.

$$\delta = 23.45 \sin \left[\frac{360}{365}(284 + n)\right]$$

$$= 23.45 \sin \left[\frac{360 \times (284 + 335)}{365}\right] = -22.11°.$$

Hour angle ω corresponding to 9.00 hour = 45°

Hence, $\cos \theta_T = \cos (28.58° - 38.58°) \cos (-22.11°)$
$\cos 45° + \sin (-22.11°) \sin (28.58° - 38.58°)$

$= \cos 10° \cos 22.11° \cos 45° + \sin 22.11° \sin 10°$

$= 0.6451 + 0.0653 = 0.7104$

or, $\theta_T = 44.72°$. **Ans.**

2.6. Solar Radiation Measurements

Measurements of solar radiation are important because of the increasing number of solar heating and cooling applications, and the need for accurate solar irradiation data to predict performance. Experimental determination of the energy transferred to a surface by solar radiation required instruments which will measure the heating effect of direct solar radiation and diffuse solar radiation. Measurements are also made of beam radiation, which respond to solar radiation received from a very small portion of the circum solar sky. A total radiation type of instrument may be used for measuring diffuse radiation alone by shading the sensing element from the sun's direct rays.

Two basic types of instruments are employed for solar radiation measurement:

(1) a *pyrheliometer,* which collimates the radiation to determine the beam intensity as a function of incident angle, and

(2) a *pyranometer,* which measures the total hemispherical solar radiation. The pyranometer measurements are the most common.

The total solar radiation arriving at the outer edge of the atmosphere is called the solar constant as already mentioned.

2.6.1. Pyrheliometers. A pyrheliometer is an instrument which measures beam radiation. In contrast to a pyranometer, the sensor disc is located at the base of a tube whose axis is aligned with the direction of the sun's rays. Thus diffuse radiation is essentially blocked from the sensor surface.

Most pyrhelimeters used for routine measurements operate on the thermopile effect and are similar to pyranometer in this respect. They differ in that mechanically they must follow the sun to measure only direct sunlight and avoid the diffuse component. In practice, direct solar radiation is measured by attaching the instrument to an electrically driven equatoral mount for tracking the sun. The diffuse component is avoided by installing a collimator tube over the sensor with a circular cone angle of about 5°.

Problems with pyrheliometer measurements are several fold; the aperture angle, the circum solar contributions and imprecision in the tracking mechanism. The first two problems are almost impossible to eliminate because of the inability to define the solar disk precisely and the finite dimensions of the instrument components. The practical matter of precise taking and sensor orientation are simply great. The use of correction factors in not only involved but somewhat unreliable.

The direct solar component on a horizontal surface may also be obtained using a shading ring, this is done by subtracting the shaded (diffuse) from the unshaded (global) reading.

Current practice in solar radiometry relies primarily on thermoelectric transducers. However, relatively low cost photovoltaic transducers are becoming more popular. To measure the direct solar radiation, the receiving surface must be normal to direct solar rays, *i.e.*, a line joining the sun and receiver. Three pyrheliometers have been in wide-spread use to measure normal incident beam radiation:

(*i*) the Angstrom pyrheliometer

(*ii*) the Abbot silver disc pyrheliometer

(*iii*) Eppley pyrheliometer.

The instruments provide primary and secondary standard of solar radiation measurements.

(*i*) **Angstrom compensation Pyrheliometer.** In this pyrheliometer, a thin blackened shaded manganin strip (Size 20 × 2 × 0.1 mm) is heated electrically until it is at the same temperature as a similar strip which is exposed to solar radiation. It is shown schematically in Fig. 2.9. Under steady state conditions

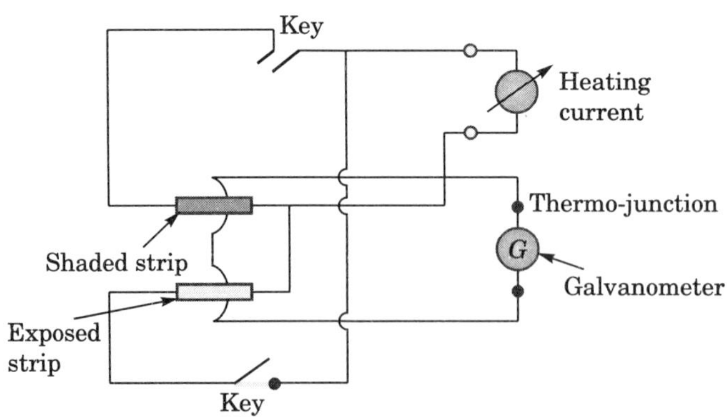

Fig. 2.9. Electric circuit for Angstrom Pyrheliometer.

(both strips at identical temperature) the energy used for heating is equal to the absorbed solar energy. The thermocouples on the back of each strip, connected in opposition through a sensitive galvanometer (or other null detector), are used to test for the equality of temperature. The energy H of direct radiation is calculated by means of the formula,

$$H_{DN} = Ki^2$$

where, H_{DN} = Direct radiation incident on an area normal to sun's rays

i = heating current in amperes K is a dimension and instrument constant

$= R/W\alpha$

where, R is the resistance per unit length of the absorbing strip (Ω cm). W is the mean width of the absorbing strip, and α is the absorbing coefficient of the absorbing strip.

(ii) **Abbot silver disk Pyrheliometer.** It consists essentially of a blackened silver disk positioned at the lower end of a tube with diaphragms to limit the whole aperture to 5.7°. A mercury in glass thermometer is used to measure the temperature at the disk. A shutter made of three polished metal leaves is provided at the upper end of the tube to allow solar radiation to fall on the disk at regular intervals and the corresponding changes in temperature of the disk are measured. The thermometer stem is bent through 90° so that it lies along the tube to minimize its exposure to the sun. The instrument must of course be celibrated against a primary standard, but their stability has been found to be very good and they are widely used for calibrating pyranometers.

(iii) **Eppley Pyrheliometer.** The sensitive element in an Eppley pyrheliometer is a temperature compensated 15 junction bismuth silver thermopile mounted at the base of a brass tube, the limiting diaphragms of which subtend an angle of 5.7°. A thermopile is basically a series arrangement of thermocouples used to develop a much greater voltage than is possible using only one. The tube is filled with dry air and is sealed with a crystal quartz window which is removable. A filter wheel is standard.

It is a stable instrument and can be used as a substandard. The instrument has found wide acceptance within the U.S.A. and many parts of the world.

2.6.2. Pyranometers. A pyranometer is an instrument which measures total or global radiation over a hemispherical field of view. If a shading ring is attached, the beam radiation is prevented from falling on the instrument sensor and in then measures only the diffuse component of the radiation. In most pyranometers, the sun's radiation is allowed to fall on a black surface to which the hot junctions of a thermopile are attached. The cold junctions of the thermopile are located iu such a way that they do not receive the radiation. As a result, an e.m.f. proportional to the solar radiation is generated. This e.m.f. which is usually in the range of 0 to 10 mV can be read, recorded or integrated over a period of time with regular calibration of about ± 2 percent can be obtained.

There are following types of pyranometers:

(i) Eppley pyranometer, *(ii)* Yellot solarimeter (photovoltaic solar cell, *(iii)* Moll-Gorczyheski solarimeter, *(iv)* Bimetallic Actiono-graphs of the Rabitzsch type, *(v)* Velochme pyranometer, *(vi)* Thermoelectric pyranometer etc.

First two types are described briefly in the following paragraphs.

(i) Eppley pyranometer. It is based on the principle as stated above that there is a difference between the temperature of black surfaces (which absorb most solar radiation) and white surfaces (which reflect most solar radiation). The detection of temperature difference is achieved by thermopile. It uses concentric silver rings 0.25 mm thick, appropriate coated black and white, with either 10 or 50 thermocouple junctions to detect temperature differences between coated rings. Later models use

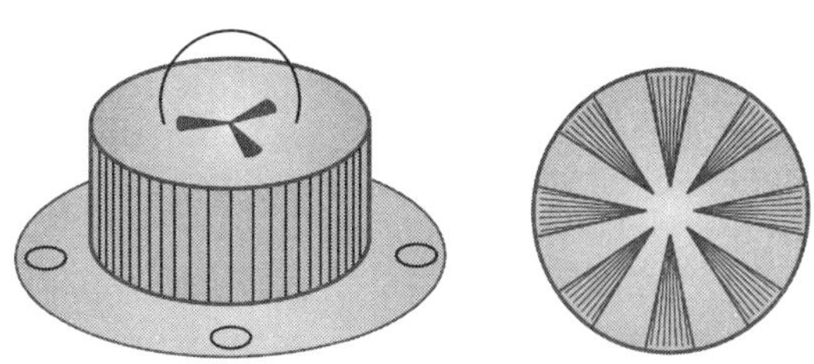

Fig. 2.10. Pyranometer with alternate black and white sensor segments.

wedges arranged in a circular pattern, with alternate black and white coatings. The disks or wedges are enclosed in a hemispherical glass cover. Similar instruments are manufactured in Europe under the name Kipp. The Eppley pyranometers, and similar instruments are calibrated in a horizontal position. Calibration of these instruments will vary to some degree if the instrument is inclined to measure radiation on other than a horizontal surface.

(*ii*) *Yellot Solarimeter* (Photovoltaic solar cell). Pyranometers have also been used on photovoltaic (solar cell) detectors. Silicon cells are the most common for solar energy. Silicon solar cells have the property that their light current (approximately equal to the short circuit current at normal radiation levels) in a linear function of the incident solar radiation. They have the disadvantages that the spectral response is not linear, so instrument calibration is a function of the spectral distribution of the incident radiation.

2.7. Sunshine Recorder

The duration of bright sunshine in a day is measured by means of a *sunshine recorder*. The sun's rays are focussed by a glass-sphere to point on a card strip held in a groove in a spherical bowl mounted concentrically with the sphere. Whenever there is a bright sunshine, the image formed is intense enough to burn a spot on the card strip. Through the days the sun moves across the sky, the image moves along the strip. Thus a burnt space whose length is proportional to the duration of sun shine is obtained on the strip.

2.8. Solar Radiation Data

Solar radiation data are available in several forms and should include the following informations.

1. Wheather they are instantaneous measurement or values integrated over some period of time (usually hour or day).
2. The time or time period of the measurements.
3. Whether the measurements are of beam, diffuse or total radiation, and the instrument used.
4. The receiving surface orientation (usually horizontal, it may be inclined at a fixed slope or normal).
5. If averaged, the period over which they are averaged (*e.g.*, monthly average of daily radiation).

Most of the data on solar radiation received on the surface of the earth are measured by solarimeter which give readings for instantaneous measurements at rate throughout the day for total radiation on a horizontal surface. Integrating the plot of rate of energy received per unit area per unit time over a whole day gives the langleys of radiation received on a horizontal surface.

It should be pointed out that solar radiation flux is generally reported in langleys per hour or per day (1 langley = 1 cal/cm^2). The unit 'langley' has been adopted in honour of Samuel Langley who made the first measurement of the spectral distribution of the sun.

For instance, the total daily solar radiation received in Kolkata (latitude 20° 32′ *N*) on the basis of yearly average is 680 langleys (*i.e.*, 680 cals/cm^2/day).

Average solar radiation data are also available from maps. Maps can be used as a source of average radiation if data are not available. Charts are also available for clear day horizontal radiation for any period for any latitude. Tables are also available for hours of sun shine for various locations.

Maps have been drawn by Mani and Chacko for India. They show the distribution of the average daily global radiation, and the average daily diffuse radiation. A typical daily record of the global and diffuse radiation measured on a clear day is shown in Fig. 2.11. In contrast to the smooth variation shown in the Fig. 2.11, a jagged variation with many peaks is obtained on a cloudy or a partly cloudy day. This is shown in Fig. 2.12. A solar designer is primarily interested in average values of radiation for a locations. The averaging is usually made over a month and tabulations showing the hourly variation of global and diffuse radiation, the amount received per day and the sunshine hours per day are prepared.

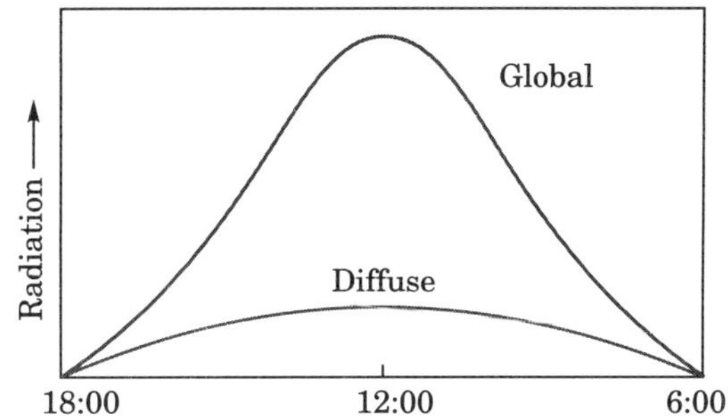

Fig. 2.11. Atypical daily record of global and diffuse radiation.

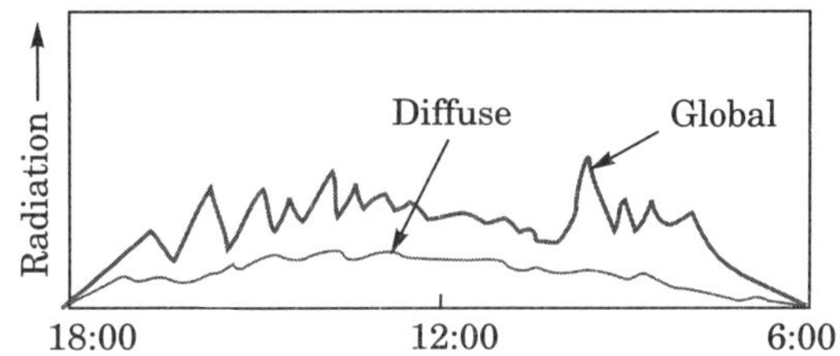

Fig. 2.12. A typical global and diffuse radiation on a cloudy day.

India lies between latitude 7° and 37° N, and receives an annual average intenaity of solar radiation between 16700–29260 kJ/m^2/day, (400–700 cal/cm^2/day). The daily solar insolation figures over the different places in India accurately available. Peak values are generally measured in April or May, with parts of Rajasthan and Gujarat receiving over 25100 kJ/m^2/day (600 cal/cm^2/day). During the monsoon and winter months the daily solar radiation decreases to about 16700 kJ/m^2/day (400 cal/cm^2/day).

The annual daily diffuse radiation received over the whole country is observed to be about 7300 kJ/m^2/day (175 cal/cm^2/day). The minimum values of diffuse radiation, measured over many parts of the country during November and December, are between 3135–4180 kJ/m^2 day, (75 and 100 cal/cm^2/day) while maximum values measured over the whole country are about 12550 kJ/m^2/day (300 cal/cm^2/day) specially in July in Gujarat.

2.9. Estimation of Average Solar Radiation

One of the earliest expressions, for monthly average horizontal solar radiation H_{av} was given by Angstrom (1924), which is

$$H_{av} = H_o' \left(a' + b' \frac{\bar{n}}{N} \right) \qquad ...(2.19)$$

where, a' and b' are arbitrary constants. (Freitz 1951, suggested that $a' = 0.35$ and $b' = 0.61$),

H_o = the monthly average horizontal solar radiation for a clear day.

$\bar{n}$ = average daily hours of bright sunshine for same period.

N = maximum daily hours of bright sunshine for the same period.

Values H_o' for use in equation 2.19 can be obtained from charts of Fig. 2.13. The day length can be obtained from a nomogram developed by Whillier (Fig. 2.14) or can be calculated from the equation

$$N = t_d = \frac{2}{15} \cos^{-1} (-\tan \phi \tan \delta).$$

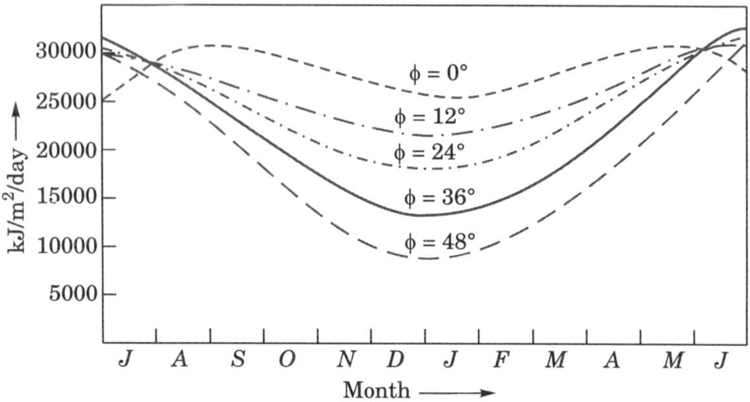

Fig. 2.13. Solar radiation on a horizontal plane for a clear day at various latitudes.

$$H_{av} = H_o \left(a + b \frac{\overline{n}}{N} \right) \quad ...(2.20)$$

where, H_o = the average monthly insolation at the top of the atmosphere.

a and b are the modified constants depending upon the location.

Constant a and b for various locations and climate conditions can be obtained from standard tables.

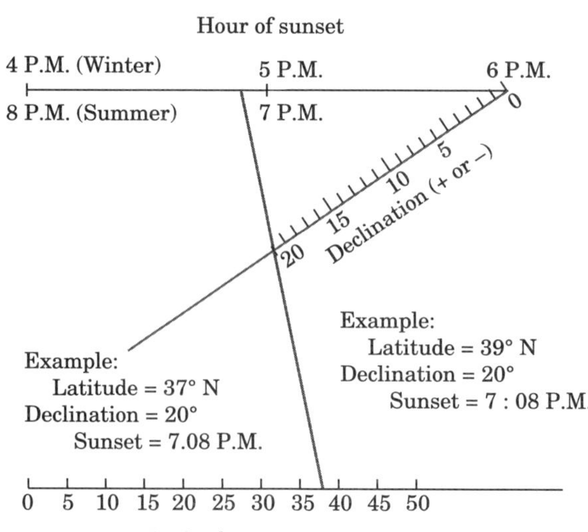

Fig. 2.14. Nomograph to determine δ and hour of sunrise/sunset in solar time (Whillier 1965).

Table 2.2 gives the values of constants a and b for some Indian towns.

H_o can be obtained from charts or it can be calculated by following empirical relation

$$H_o = \frac{24}{\pi} I_{sc} \left[\left\{ 1 + 0.033 \cos\left(\frac{360\,n}{365}\right) \right\} \left(\cos\phi \cos\delta \sin\omega_s + \frac{2\pi\omega_s}{360} \sin\phi \sin\delta \right) \right] \quad ....(2.21)$$

where, I_{sc} = solar constant per hour

n = day of the year

ω_s = sunrise hour angle.

The declination δ can be obtained from Cooper equation and the sunrise hour angle from the relation,

$$\omega_s = -\tan\phi \tan\delta.$$

Table 2.2. Constants a and b for Some Indian Towns

Location	*a*	*b*	*Location*	*a*	*b*
Ahmedabad	0.28	0.48	Manglore	0.27	0.43
Bangaluru	0.18	0.64	Shillong	0.22	0.57
Baroda	0.28	0.48	Srinagar	0.35	0.40
Bhavnagar	0.28	0.47	Trivandrum	0.37	0.39
Bhopal	0.27	0.50	Vishakhapatnam	0.28	0.47
Kolkata	0.28	0.42	Nagpur	0.27	0.50
Goa	0.30	0.48	New Delhi	0.25	0.57
Jodhpur	0.33	0.46	Poona	0.31	0.43
Chennai	0.30	0.44	Roorkee	0.25	0.56

Example 2.3. *Determine the average value of solar radiation on a horizontal surface for June 22, at the latitude of 10°N, if constants a and b are given as equal to 0.30 and 0.51 respectively, and the ratio*

$$\bar{n}/N = 0.55.$$

Solution. Declination δ for June 22 = + 23.5°, sunrise hour angle ω_s

$$= \cos^{-1}(-\tan\phi \tan\delta)$$

$$= \cos^{-1}(-\tan 10 \tan 23.5) = \cos^{-1}(-0.0766)$$

or $\omega_s = 94.39°$

n = day of the year = 172 (for 22 June).

We have the relation for average insolation at the top of the atmosphere.

i.e.,

$$H_0 = \frac{24}{\pi} I_{ac} \left[\left\{ 1 + 0.033 \cos\left(\frac{360\,n}{365}\right) \right\} \left(\cos\phi \cos\delta \sin\omega_s + \frac{2\pi\omega_s}{360} \sin\phi \sin\delta \right) \right]$$

Substituting the values, we get,

$$H_0 = \frac{24}{3.14} I_{ac} \left[\left\{1 + 0.033 \cos\left(\frac{360 \times 172}{365}\right)\right\}\right.$$

$$\left.\left(\cos 10 \cos 23.5 \sin 94.39 + \frac{2 \times 3.14 \times 94.39}{360} \sin 10 \sin 23.5\right)\right]$$

$$= \frac{24}{3.14} I_{sc} \text{ [} 1 - 0.033, 0.9816)(0.9848 - 0.9171$$

$$\times 0.9970 + 1.65 \times 0.1765 \times 0.3987)]$$

$$= \frac{24}{3.14} I_{sc} \text{ [(0.9675)(1.0143)]}$$

$$= \frac{24}{3.14} I_{sc} \text{ } 0.9816 \qquad ...(i)$$

The value of I_{sc} in S.I. units is 1353 W/m^2 or 4871 kJ/m^2 hr and MKS units is = 1165 kcal/hr m^2.

SI units

$$H_0 = \frac{24}{3.14} \times 1353 \times 0.9816 = 10143 \text{ W/m}^2 \text{ day}$$

$$\therefore \quad H_{ac} = 10143 \times (0.3 + 0.51 \times 0.55)$$

$$= 10143 \times 0.58 = 5884 \text{ W/m}^2 \text{ day.} \quad \textbf{Ans.}$$

MKS units

$$H_0 = \frac{24}{3.14} \times 1165 \times 0.9816 = \mathbf{8740 \text{ kcal/m}^2 \text{ day.}}$$

$$\therefore \quad H_{av} = 8740 \times 0.58 = 5070 \text{ kcal/m}^2 \text{ day.} \quad \textbf{Ans.}$$

2.10. Solar Radiation on Tilted Surfaces

The rate of receipt of solar energy on a given surface on the ground depends on the orientation of the surface with reference to the sun. A fully suntracking surface that always faces the sun receives the maximum possible solar energy at the particular location. A surface of the same area oriented in any other direction will receive a smaller amount of solar radiation. Because solar radiation is such a 'dilute' form of energy, it is desirable to capture as much as possible on a given area. We have seen in the preceding sections that the measuring instruments give the values of solar radiation falling on a horizontal surface. Because most of the solar collectors or solar radiation collecting devices are tilted at an angle to horizontal, it is therefore necessary to convert data for a hourly radiation (measured or estimated) on a horizontal surface of radiation on a tilted surface.

2.10.1. Beam Radiation. In most cases; the tilted surface faces due south *i.e.*, $\gamma = 0$, for this case,

$$\cos \theta = \sin \delta \sin (\phi - s) + \cos \delta \cos \omega \cos (\phi - s)$$

For horizontal surface ($\theta = \theta_z$)

$$\cos \theta_z = \sin \phi \sin \delta + \cos \phi \cos \delta \cos \omega.$$

It follows that the ratio of the beam radiation falling on the tilted surface to that falling on a horizontal (Fig. 2.18) surface is given by

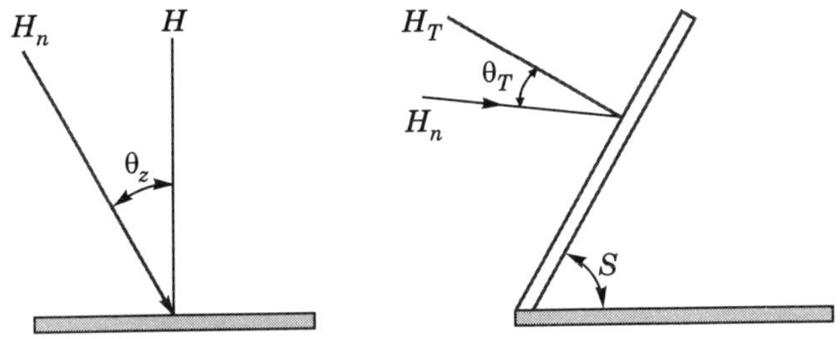

Fig. 2.15. Radiation on horizontal and tilted surface.

$$R_b = \frac{H_T}{H} = \frac{H_n \cos \theta_T}{H_n \cos \theta_z} = \frac{\cos \theta_T}{\cos \theta_z}$$

$$= \frac{\sin (\phi - s) \sin \delta + \cos (\phi - s) \cos \delta \cos \omega}{\sin \phi \sin \delta + \cos \phi \cos \delta \cos \omega} \quad \text{...(2.22)}$$

This ratio is called the *tilt factor* for beam radiation.

2.10.2. Total Radiation. Beam and diffuse component of solar radiation are absorbed in flat plate type collector. The angular correction factor has been determined for beam radiation as given by equation 2.22. Correction for diffuse radiation can be applied for clear days, for that it can be assumed that their origin is near the sun, that is, the scattering of solar radiation is mostly forward scattering. For such a case R may be assumed equal to R_b, where R is the correction factor for both direct and diffuse radiation.

$$R = \frac{H_T}{H}$$

or $$H_T = R(H_b + H_d) \quad \text{...(2.23)}$$

For cloudy or hazy days diffuse radiation can be assumed as uniformly distributed over the sky. The effective ratio of solar energy on the tilted surface to that on the horizontal surface is then

$$R = \frac{H_T}{H} = \frac{H_b}{H} R_b + \frac{H_d}{H} \quad \text{...(2.24)}$$

Conversion factor for diffuse radiation (R_d) is given by equation

$$R_d = \frac{(1 + \cos s)}{2}$$

A surface tilted at slope s from the horizontal sees $\frac{(1 + \cos s)}{2}$ of the sky dome.

The tilted surface also sees ground or other surroundings and if those surroundings have a diffuse reflectance of ρ for solar radiation, the reflected radiation from the surrounding on the surface from total solar radiation is

$$(H_b + H_d)(1 - \cos s)\frac{\rho}{2}.$$

Hence three components; the beam radiation, diffuse solar radiation and solar radiation reflected from the ground which the tilted surface sees, are considered above. This was first considered by Liu and Jordon. Hence combining the three terms.

$$H_T = H_b R_b + H_d \frac{(1 + \cos s)}{2} + (H_b + H_d)\frac{(1 - \cos s)}{2}\rho \qquad ...(2.25)$$

and
$$R = \frac{H_b}{H} R_b + \frac{H_b}{H}\frac{(1 + \cos s)}{2} + \frac{(1 - \cos s)}{2}\rho \qquad ...(2.26)$$

The values of diffuse reflectance as suggested by Liu and Jordon are as follows

ρ = 0.2 when there is no snow

= 0.7 when there is a snow cover.

Therefore for Indian conditions, a value around 0.2 is generally expected with surfaces of concrete or glass and can be used. Fortunately the reflected radiation term does not contribute much to the total because in India the value of the angle s would rarely exceed 30°.

QUESTIONS

1. (*i*) Define solar constant.
   (*ii*) What are the reasons for variation in solar radiation reaching the earth than received at the outside of the atmosphere?
2. Write notes on beam and diffuse radiation.
3. Define the terms:
   (*i*) Altitude angle (*ii*) Incident angle
   (*iii*) Zenith angle (*iv*) Solar azimuth angle
   (*v*) Latitude angle (*vi*) Declination angle
   (*vii*) Hour angle.
4. Calculate the angle made by the beam radiation with the normal to a flat-plate collector, pointing due south located in New Delhi (28° 38′ N, 77° 17′ E) at 9 : 00 hour, solar time on December 1. The collector is tilted at angle of 36° with the horizontal. [**Ans.** 45.8°]
5. Calculate the sunset hour angle and day length at a location latitude of 35° N, on February 14. [**Ans.** 80.5, 10 hr 44′]
6. What is the difference between a Pyrheliometer and a Pyranometer? Describe the principle of Angstrom type pyrheliometer.
7. Estimate the daily global radiation on a horizontal surface at Baroda (22° 13′ N, 73° 13′ E) during the month of March. If constants a and b are given equal to 0.28 and 0.48 respectively and average sunshine hours for day are 9.5. [**Ans.** 22, 485 kJ/m^2 day]

CHAPTER 3

Solar Energy Collectors

3.1. Introduction

A solar collector is a device for collecting solar radiation and transfer the energy to a fluid passing in contact with it. Utilization of solar energy requires solar collectors. These are general of two types:

(*i*) Non concentrating or flat plate type solar collector.

(*ii*) Concentrating (focusing) type solar collector.

The solar energy collector, with its associated absorber, is the essential component of any system for the conversion of solar radiation energy into more usable form (*e.g.*, heat or electricity). In the non-concentration type, the collector area (*i.e.*, the area that intercepts the solar radiation) is the same as the absorber area (*i.e.*, the area absorbing the radiation). On the other hand, in concentrating collectors, the area intercepting the solar radiation is greater, sometimes hundred of times greater than the absorber area. By means of concentrating collectors, much higher temperatures can be obtained than with the non-concentrating type. Concentrating collectors may be used to generate medium pressure steam. They use many different arrangements of mirrors and lenses to concentrate the sun's rays on the boiler. This type shows better efficiency than the flat plate type. For best efficiency, collectors should be mounted to face the sun as it moves through the sky.

3.2. Physical Principles of the Conversion of Solar Radiation into Heat

The fundamental process now in general use for heat conversion is the *green house effect*. The name come from its first use in green houses, in which it is possible to grow exotic plants in cold climates through better utilization of the available sunlight.

Most of the energy we receive from the sun comes in the form of light, a shortwave radiation, not all of which is visible to the human eye. When this radiation strikes a solid or liquid, it is absorbed and transformed into heat energy ; the material becomes warm and stores the heat, conducts it to surrounding materials (air water, other solids or liquids) or reradiates it to other materials of lower temperature. This reradiation is a long wave radiation.

Fig. 3.1 shows how temperature on earth is affected by the 'green house effect'. Visible sunlight is absorbed on the ground, at a temperature of 20°C, for example emits infra-red light at a wavelength of about 10 μm, but CO_2 in atmosphere absorbs light of that wavelength and back radiates part of it to earth. (CO_2 does not absorbs the incoming sunlight which has a shorter wavelength).

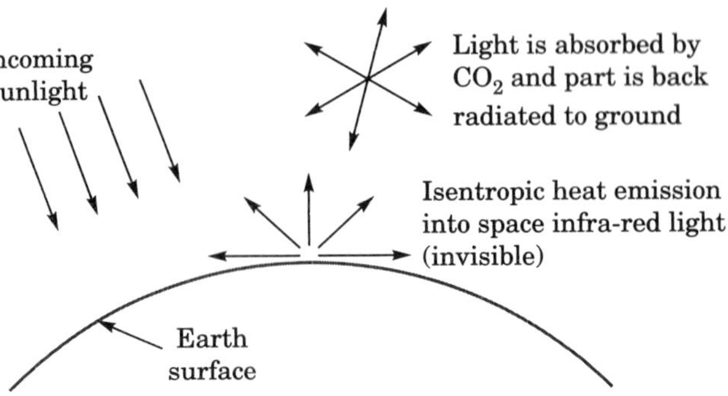

Fig. 3.1. The green house effect radiated to the CO_2 content of the atmosphere.

Hence the green house effect brings about an accumulation of energy of the ground.

Glass easily transmits short-wave radiation, which means that it poses little interference to incoming solar energy, but it is a very poor transmitter of long-wave radiation. Once the sun's energy has passed through the glass windows and has been absorbed by some material inside, the heat will not be reradiated back outside. Glass therefore, act as a heat trap, a phenomenon which has been recognized for sometime in the construction of green houses, which can get quite warm on sunny days, even in the middle of winter; this has come to be known in fact, as the 'green house effect'. Solar collectors for home heating usually called *flat plate collectors,* almost have one or more glass *covers,* although various plastic and other transparent materials are often used instead of glass.

In Fig. 3.2 a black-painted plate absorbs the incoming sunlight. About it, is fixed a plate of ordinary window glass. When the temperature of the black plate increases, its emits an increment of thermal heat in the form of infra-red light. The black absorber has the properties of a black body; ideal black bodies have not only the highest absorption rate but also the highest emission coefficient for all wavelengths of light. Emission increases with temperature, following T^4 law. The reemitted light if so progressively shorter wavelength and greater energy as the temperature of the blackbody increases. This is expressed by *Wien's law,* which may be written as:

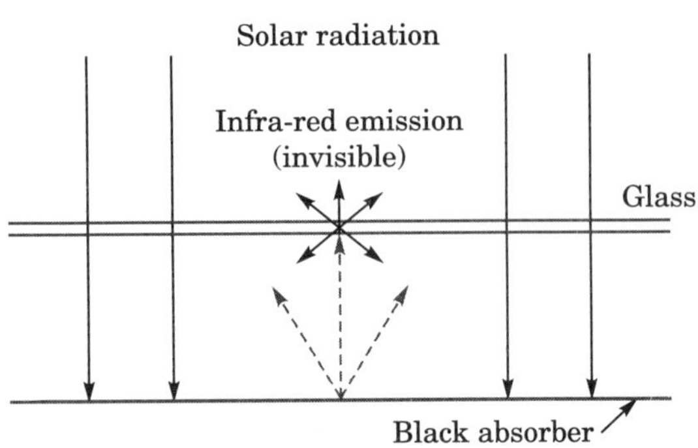

Fig. 3.2. Principle of green house effect.

$$\lambda_{max} \cdot T = \text{constant} = 2989\ \mu\text{m Kelvin}$$

T being the surface temperature of the blackbody and λ_{max} the wavelength at which light emission reaches a maximum. Some examples are given in Fig. 3.3.

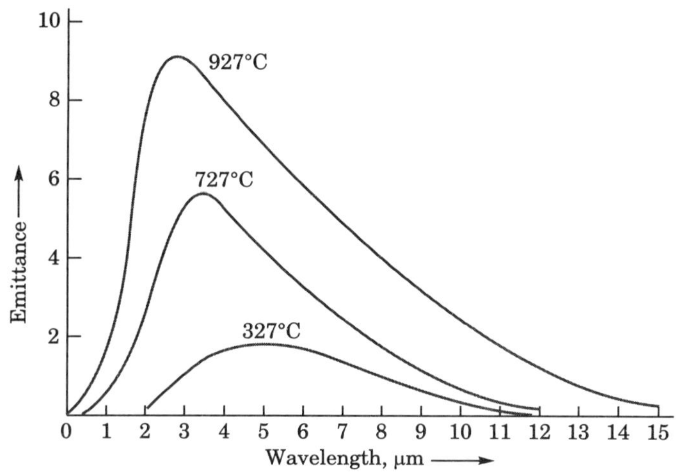

Fig. 3.3. Enuttance of black body at different temperatures.

The sun emits radiation like a "blackbody" whose surface temperature is about 5700°C, this corresponds to maximum emission of 0.5 µm. A blackbody at a room temperature emits radiation with a maximum at about 10 µm, which is within the spectrum of invisible of infra-red light. The ordinary glass plate fixed above the black plate in a green house has a spectral absorption which can be seen in Fig. 3.3. The profile of plastic is similar. Thus glass which is relatively transparent for visible light is absorbent for the infra-red light emitted by the black plate when it evacuates its thermal energy. The infra-red light absorbed by the glass is remitted in all directions, half of it is emitted to the outside and lost, the other half re-emitted towards the black plate which absorbs it again. More and more heat is accumulated in the way in the black plate, whose temperature thus increases. Equilibrium is reached when the energy gain by absorption of visible light is exactly balanced by the loss of energy through infra-red emission of the glass plate. With rising temperature, the wavelength of the infra-red emission becomes shorter. At 200°C (473°K) the maximum radiation is emitted at about 6 µm, compared with 10 µm at room temperature. Finally at about 500°C (773°K) the bulk of the radiation would be emitted at 4 µm, at which wavelength, glass is partially transparent for infra-red light.

It follows that an efficient green house effect is possible only below 500°C. However, unless concentration of sunlight is combined with the green house effect, the equilibrium temperature achieved are much lower because, practice, the equilibrium temperature is further reduced by heat losses from the black plate due to thermal conductivity and air convection.

3.3. Flat-Plate Collectors

Where temperatures below about 90°C are adequate, as they are for space and service water heating flat plate collectors, which are of the non-concentrating type, are particularly convenient. They are made in rectangular panels, from about 1.7 to 2.9 sq. m, in area, and are relatively simple to construct and erect, Flat plates can collect and absorb both direct and diffuse solar radiation, they are consequently partially effective even on cloudy days when there is no direct radiation.

Flat-plate solar collectors may be divided into two main classifications based on the type of heat transfer fluid used.

Liquid heating collectors are used for heating water and non-freezing aqueous solutions and occasionally for non-aqueous heat transfer fluids. Air or gas heating collectors are employed as *solar air heaters.*

The principal difference between the two types is the design of the passages for the heat for the transfer fluid.

The majority of the flat-plate collector have five main components as follows:

(*i*) A transparent cover which may be one or more sheets of glass or radiation transmitting plastic film or sheet.

(*ii*) Tubes, fins, passages or channels are integral with the collector absorber plate or connected to it, which carry the water, air or other fluid.

(*iii*) The absorber plate, normally metallic or with a black, surface, although a wide variety of other materials can be used with air heaters.

(*iv*) Insulation, which should be provided at the back and sides to minimise the heat losses. Standard insulating materials such as fibre glass or styro-foam are used for this purpose.

(*v*) The casing or container which enclose the other components and protects them from the weather.

(A) A Typical Liquid Collector

There are many flat-plate collector designs, but most are based on the principle shown in Fig. 3.4. It is the plate and tube type collector. It basically consists of a flat surface with high absorptivity for solar radiation, called the absorbing surface. Typically a metal plate, usually of copper, steel or aluminium material with tubing of copper in thermal contact with the plateas, are the most commonly used materials. The absorber plate is usually made from a metal sheet 1 to 2 mm in thickness, while the tubes, which are also of metal, range in diameter from 1 to 1.5 cm. They are soldered, brazed or clamped to the bottom (in some cases, to the top) of the absorber plate with the pitch ranging from 5 to 15 cm. In some designs, the tubes are also in line and integral with the absorber plate. For the absorber plate corrugated galvanized sheet is a material widely available throughout the world, Fig. 3.5 (*a*) and (*b*) show two ways in which it has been used.

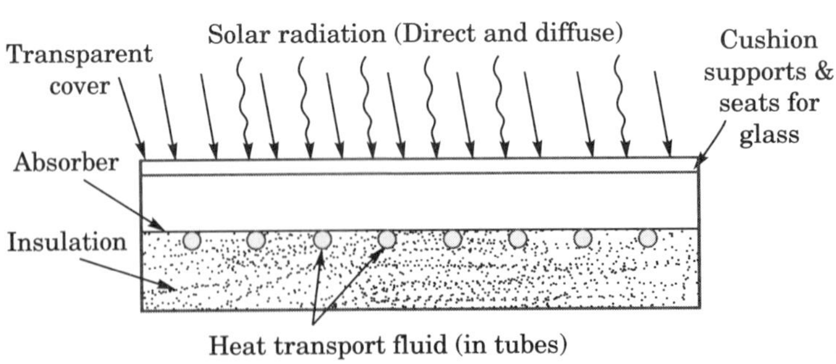

Fig. 3.4. Selection through typical flat-plate collector.

The use of conventional standard panel radiators shown in Fig. 3.5 (*c*) is one of the simplest practical applications. The methods of bonding and clamping tubes to flat or corrugated sheet are shown in Fig. 3.5 (*d*) and (*e*), while Fig. 3.5 (*f*) is the "tube in strip" or roll bond design, in which the tubes are formed in the sheet, ensuring a good thermal bond between the sheet and the tube.

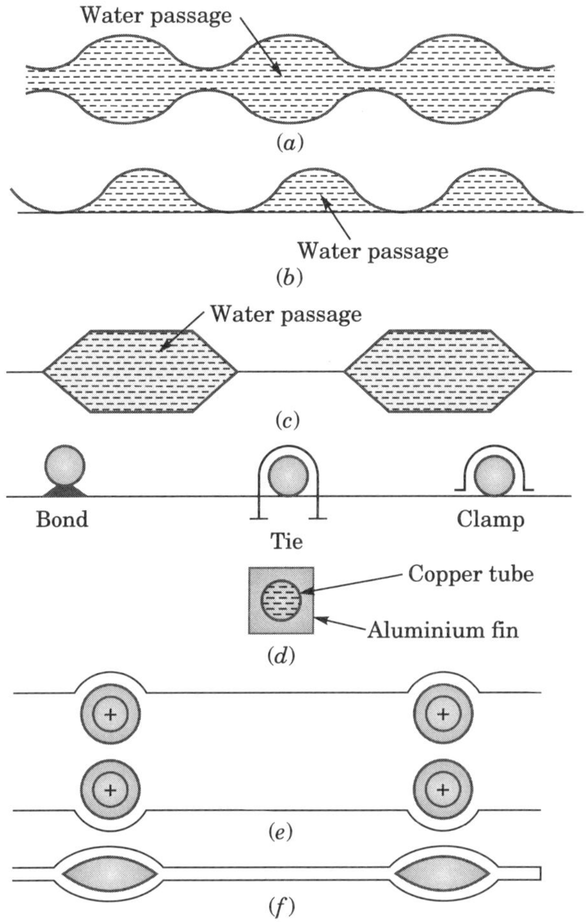

Fig. 3.5. Cross-section through collector plates.

Heat is transferred from the absorber plate to a point of use by circulation of fluid (usually water) across the solar heated surface. *Thermal insulation* of 5 to 10 cm thickness is usually placed behind the absorber plate to prevent the heat losses from the rear surface. Insulation materials is generally mineral wool or glass wool or fiberglass as stated above.

The front covers are generally glass (may be one or more) that is transparent to in coming solar radiation and opaque to the infra-red re-radiation from the absorber. The glass covers act as a convection shield to reduce the losses from the absorber plate beneath. Glass is generally used for the transparent covers but certain plastic films may be satisfactory. Glass is the most favourable material. Thickness of 3 and 4 mm are commonly used. The usual practice is to have 1 or 2 covers with a specific ranging from 1.5 to 3 cm.

Advantages of second glass which is added above the first one are:

(*i*) Losses due to air convection are further reduced. This is important in windy areas.

(*ii*) Radiation losses in the infra-red spectrum are reduced by a further 25%, because half of the 50% which is emitted outwards from the first glass plate is back radiated. It is not worth while to use more than two glass plates. This is due to the fact that each plate reflects about 15% of the incoming sunlight.

As we know that main purpose of the transparent cover of the flat-plate collector is to decrease heat loss without significantly reducing the incoming solar radiation. In the first place, the relatively still (or stagnant) air space between the cover and the absorber plate largely prevents loss of heat from the plate by convection.

Furthermore, if the cover is made of glass, it permits the passage of solar radiations with wavelengths less than 2 micrometers (μm) but it is largely opaque to the longer wavelength thermal infra-red. As a result, heat is trapped in the air space between the cover and the absorber plate in a manner similar to green house. The effect is to reduce the loss of heat from the absorber. However, since the enclosed air is inevitably warmer than the ambient air, there is some loss of heat to the surroundings from the top of the cover by convection, conduction and radiation. The rate of heat loss increases as the temperature of the air space rises; as will be seen shortly, this affects the overall efficiency of the solar collector.

A certain proportion of the incident solar radiation is lost by absorption in the glass cover plates, but the loss can be kept small by using a clear ("water white") glass with a low iron content. A much larger loss occurs as a result of partial reflection. Two glass plates may reflect some 15 per cent of the solar radiation coming from a perpendicular direction. The reflection loss increases as the direction of incidence departs from the perpendicular. The reflection of glass covers may be reduced by coating with thin films of certain substances (*e.g.*, magnesium fluoride) or by gentle etching with a solution of hydrofluoric acid. Such antireflective coatings add to the cost of the collectors but make them more efficient.

Transparent plastics have been used in place of glass, but they have some drawbacks. Most plastics are not as opaque as glass to the thermal infra-red radiation and so permit greater loss of heat from the absorber. They also suffer a decrease in transparency and sometimes breakup in the course of time due to heating and the action of solar ultraviolet radiation. Efforts are being made to develop better plastic materials that might be used in solar collectors.

For water streams the absorber plate can be any metal, plastic or rubber sheet that incorporates water channels, while for air systems the space above or below the collector plate serves as the conduit. The surface finish of the absorber plate may be a flat black paint with an appropriate primer. The primer coat should preferable be thin since a thick under coat of paint would increase the resistance to heat transfer. The primer should be of self etching type. If the primer is not a self etching type, the repeated thermal expansion and contraction of the plate may cause the paint to peel after a year or so. Several types of backed on or chemical finishes are also available. Black painted absorbers are preferred because they are considerably cheaper. The coatings applied on absorber plate are called "selective coatings" which reduces the amount of energy emitted by thermal infra-red radiation. These are good absorbers of radiation below about 2 μms wavelength, but they are poor emitter for longer wavelengths. A promising selecting coating is "black chrome" form of chromium sesquioxide (Cr_2O_3) which embedded chromium metal, in a layer 0.15 to 2 μm thick, electrode- posited on a nickel base.

The liquid heated is generally water. However sometimes mixtures of water and ethylene glycol are used, if ambient temperatures below 0°C are likely to be encountered.

Typical collector dimensions are 2 m × 1 m × 15 cm.

Heat Transport System. The heat generated in the absorber is removed by continuous flow of a heat-transport (or heat transfer) medium, either water or air. It is mainly in the design of the heat transfer system that flat plate collectors differ. When water is used, it is most commonly passed through metal tubes with either circular or rectangular cross-section; the tubes are welded to the absorber plate (or form integral part of it) so as to assure effective heat transfer of heat to the fluid. Some examples are already represented in Fig. 3.5. The tubes are connected to common headers at each end of the collector. In order to maximise the exposure to solar radiation, collectors are almost invariably sloped. Cooler water then enters at the bottom header, flows upward through the tubes where it is warmed by the absorber, and leaves by way of the top header. Fig. 3.6.

In one simple type of flat-plate collector, the absorber is a blackened sheet with close corrugations running from top to bottom. The water flows through the grooves formed by the corrugations. A problem with this design is that in cold weather, moisture may condense on the inside of the transparent cover plate and thus decrease the transmission of the solar radiation.

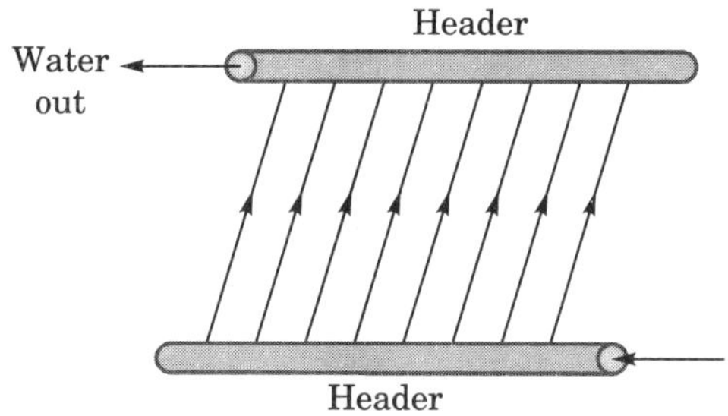

Fig. 3.6. Water flow in flat-plate collector.

Water is a very effective heat-transport medium, but it suffers from certain drawbacks, one is the possibility of freezing in the collector tubes in cold climates during cold nights. As stated earlier ethylene glycol is added to prevent freezing, but this generally adds to the complexity of the heating system. Furthermore, the antifreeze solution is less effective than water for heat removed from the absorber. In some cases, the water is drained from the collector tubes if freezing is expected, but difficulties have been experienced in refilling all the tubes in the morning.

Another problem arises from corrosion of the metal tubes by the water; this is aggravated if the water is drained at night thus allowing air to enter. The oxygen in air increases the rate of corrosion of most metals. Corrosion can be minimized by using copper tubing. Aluminium is a less expensive alternative, although periodic chemical treatment of water is desirable. Finally, leaks in a water (or anti freeze) circulation system require immediate attention.

(B) Typical Air Collectors or Solar Air Heaters

Fig. 3.7 shows a schematic flat-plate collector where an air stream is heated by the back side of the collector plate. Fins attached to the plate increase the

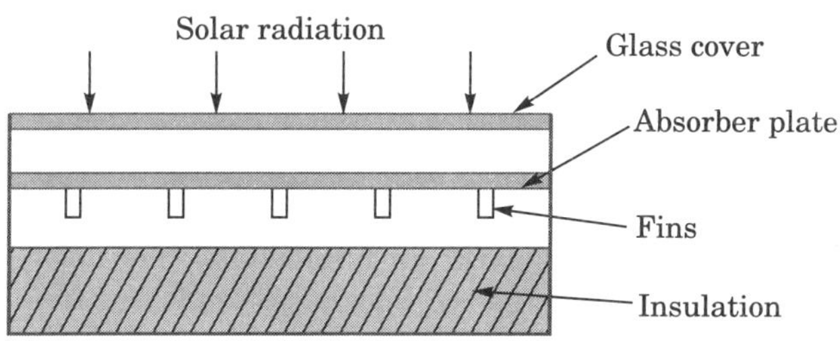

Fig. 3.7. Typical Solar Air Collector.

contact surface. The back side of the collector is heavily insulated with minerarl wool or some other material. The most favourable orientation, of a collector, for heating only is facing due south at an inclination angle to the horizontal equal to the latitude plus 15°.

Air has been used so far to a lesser extent as the heat-transport medium in solar collectors, but it may have some advantages over water. To decrease the power required to pump the necessary volume of air through tubes, wider flow channels are used. For example, the air may be passed through a space between the absorber plate and insulator with baffles arranged to provide a long (zig-zag) flow path Fig. 3.8.

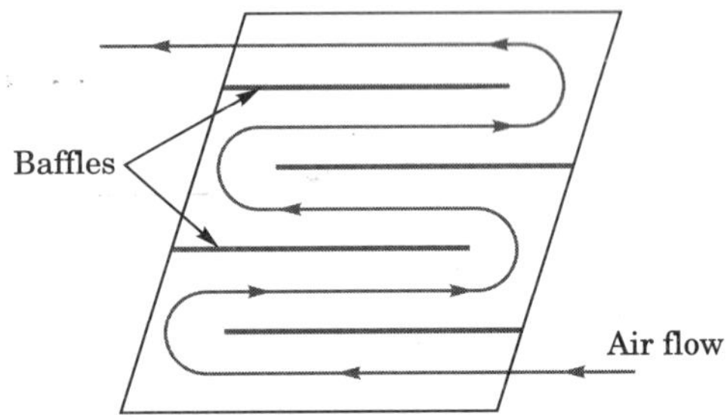

Fig. 3.8. Zig-zag air flow path in flat-plate collector.

The use of air as the heat-transport fluid eliminates both freezing and corrosion problems, and small air leaks are of less concern than water leaks. Moreover, the heated air can be used directly (or by way of heat storage) for apace heating. On the other hand, larger duct sizes and higher flow rates, with increased pumping power, are required for air than when water is the heat transport medium. Another drawback is that transfer of heat from air to water in a hot water supply system is inefficient.

But solar air heater has an important place among solar heat collectors. It can be used as subsystems in many systems meant for the utilization of solar energy. Possible applications of solar air heaters are drying or curing of agricultural products, space heating for comfort, regeneration of dehumidifying agents, seasoning of timber, curing of industrial products such as plastics.

Numerous variations in the design of collectors for heating air by solar energy are shown in Figs. 3.9 and 3.10. Air can be passed in contact with black solar absorbing surface such as finned plates or ducts as mentioned above, corrugated or roughened plates of various materials, several layer of metal screening and overlapped glass plates. Flow may be straight through, surpentine, above or below or on both sides of the absorber plate, or through a porous absorber material.

Basically air heaters are classified in the following two categories.

(1) The first type has a non-porous absorber in which the air stream does not flow through the absorber plate. Air may flow above and or behind the absorber plate, as shown in Fig. 3.9.

(2) The second type has a porous absorber that includes slit and expanded metal, transpired honey comb and overlapped glass plate absorber, as shown in Fig. 3.10.

(1) **Non-porous absorber plate type collectors.** A non-porous absorber may be cooled by the air stream flowing over both sides of the plate as shown in Fig. 3.9 (*a*). In most common design the air flows behind the absorbing surface. Air flow above the upper surface increases the convention losses from the cover plate and therefore is not recommended if the air inlet temperature rise at the collector are large, it is shown in Fig. 3.9 (*b*).

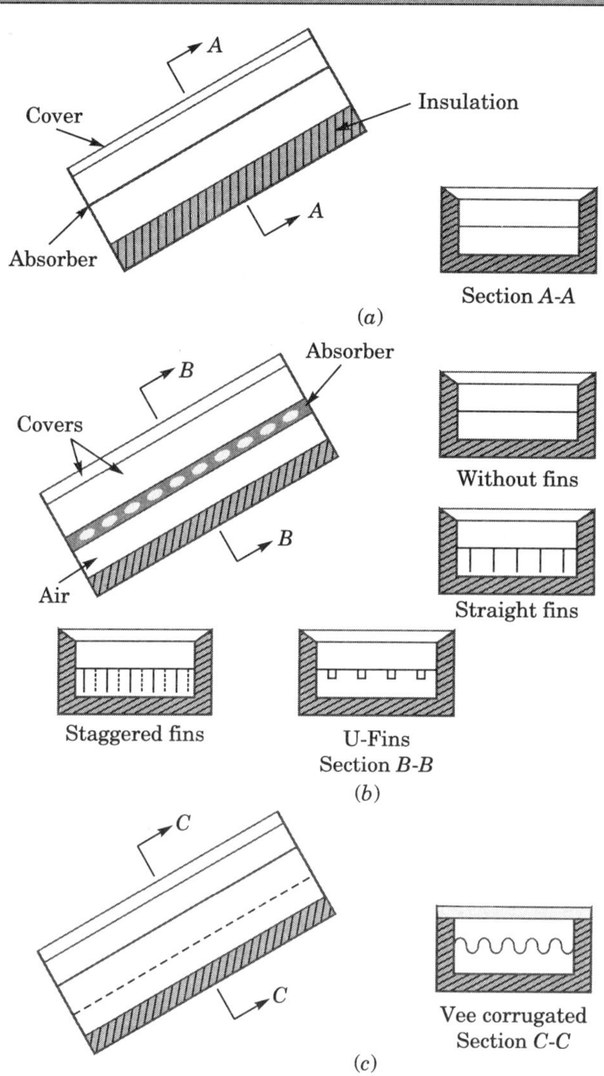

Fig. 3.9. Non-porous type air heaters.

Transmission of the solar radiation through the transparent cover system and its absorption is identical to that of a liquid type flat-plate collector. To improve collection efficiency selective coating may be applies provided there is no much cost. Due to low heat transfer rates, efficiencies are lower than liquid solar heaters under the same radiation intensity and temperature conditions.

Performance of air heaters is improved by:

(*a*) Roughning the rear of the plate to promote turbulence and improve the convective heat transfer coefficient, or

(*b*) Adding fins to increase the heat transfer surface. Usually turbulence is also increased which enhances the convective heat transfer.

A solar collector with V-corrugated copper foil is illustrated in Fig. 3.9 (*c*). Absorption of solar radiation is improved due to surface radioactive characteristics and the geometry of the corrugations, which help in trapping the reflected radiation.

(2) **Collectors with porous absorbers.** The main drawback of the non-porous absorber plate is the necessity of absorbing all incoming radiation over the projected area from a thin layer over the surface, which is in the order of a few microns. Unless selective coatings are used, radiative losses from the absorber plate are excessive, therefore the collection efficiency can not be improved. The pressure drop along the duct formed between the absorber plate and the rear insulation may also be prohibitive especially in the case of added fins to increase the heat transfer surface and turbulence rate. The difficulty with turbulence is the pressure drop across the collector. Too many surfaces and too much restriction to air flow will require a larger fan and a larger amount of energy to push the air through. The energy required for this cancels out saving from using solar energy, particularly if fan is electrical and it the amount of energy which is burned at the power plant to produce the electrical energy is included.

These defects are eliminated in a porous absorber type collectors in two ways.

(*a*) The solar radiation penetrates to greater depths and is absorbed gradually depending on the matrix density. The cool air stream introduced from the upper surface of the matrix is first heated by the upper layers which are cooler than the bottom layers. The air stream warm up, while traversing the matrix layers. The lower matrix layers are hotter than the upper ones, therefore, the air stream can effectively transfer hest from the matrix. Improper selection of the matrix porosity and the thickness may result in reduced efficiencies since the additional matrix layers beyond on optimum may no longer absorb the solar radiation and heat the air stream further.

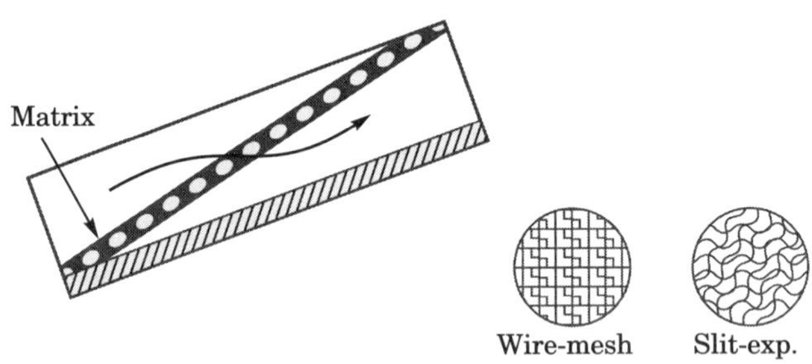

(*a*) Slit or expanded metal

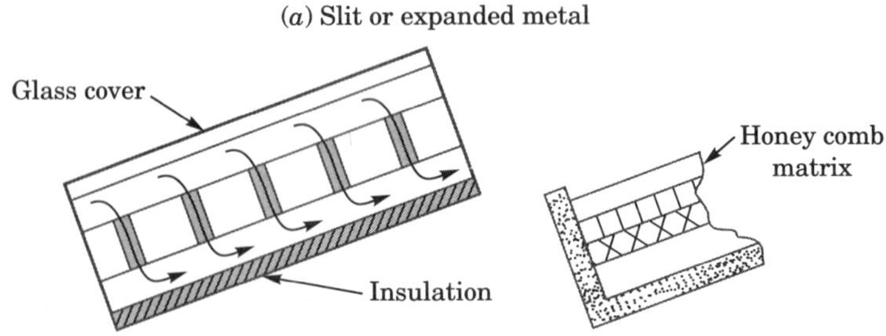

(*b*) Transpired honey comb

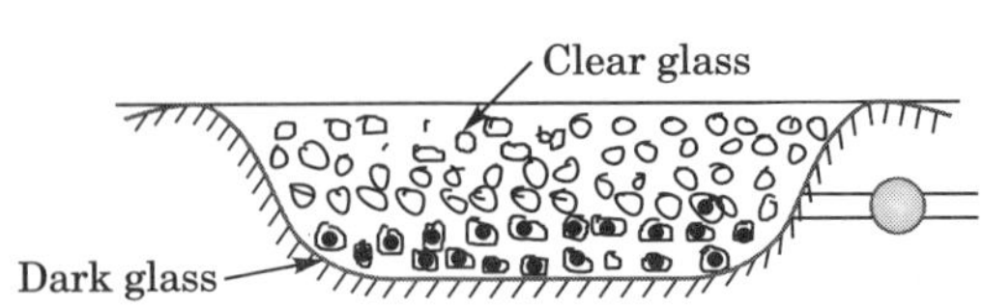

(*c*) Broken bottles absorber

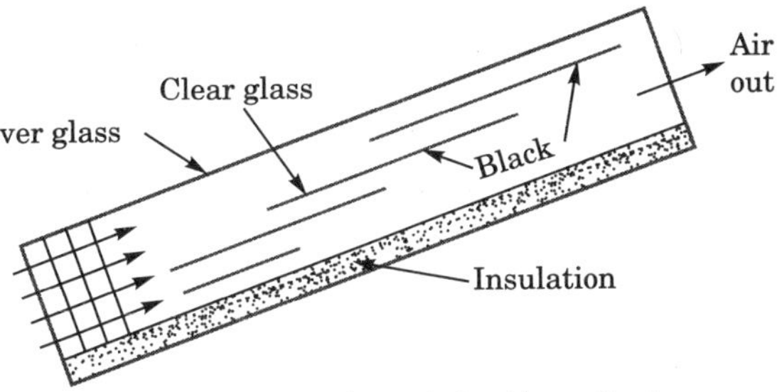

(*d*) Over-lapped-glass plate air-heating collector

Fig. 3.10. Sketches of porous absorber-type air heaters.

(*b*) The pressure drop for the matrix is usually lower than the non-porous absorber with flow behind the plate since flow per unit cross-section would be much lower. Although the matrix hinders the flow, the pressure drops reported for porous matrix absorbers are still lower than for the non-porous absorbers. The solar air heating utilizing a transpired honey comb [Fig. 3.10 (*b*)] is also very favourable from the pressure drop stand point since the flow cross-section is much larger.

Whillier has suggested a method of using crushed glass layers to absorb solar radiation and heat the air. A porous bed, as shown in Fig. 3.10 (*c*) made by forming layers of broken bottles (bottom dark top clear glass), may be readily used for agricultural drying purposes with minimal expenditure.

The overlapped glass plate air heater as shown in Fig. 3.10 (*d*) can be considered as a form of porous matrix, although overall flow direction is along the absorber glass plates instead of being across the matrix. Plate and air stream temperature increase gradually along the collector length and across from top to bottom. Thus thermal losses could be significantly reduced. The pressure drop is also significantly less than the non-porous flat-plate absorber design.

Applications of Solar Air Heaters. The solar air heaters, which supply hot air that could be mainly used for the following processes:

(*i*) Heating buildings.

(*ii*) Drying agricultural produce and lumber.

(*iii*) Heating green houses.

(*iv*) Air conditioning buildings utilizing desiccant beds or a absorption refrigeration process.

(*v*) Using air heaters as the heat sources for a heat engine such as a Brayton or Stirling cycle.

Advantages of Flat-plate Collectors

(*i*) They have the advantages of using both beam and diffuse solar radiation.

(*ii*) They do not require orientation towards the sun.

(*iii*) They require little maintenance.

(*iv*) They are mechanically simpler than the concentrating reflectors, absorbing surfaces and orientation devices of focusing collectors.

3.4. Transmissivity of Cover System

Transmissivity like reflectivity and absorptivity, is a function of the wavelength and the angle of incidence of the incoming solar radiation. Other variables which effect the transmissivity are the refractive index n and the extinction coefficient k of the medium but strictly speaking both n and k are also functions of the wavelength λ.

A surface with transmissivity $\tau = 0$ is known as an opaque surface. For such surfaces the sum of the absorptivity (α) and reflectivity (ρ) must be equal to unity ($\because \tau + \alpha + \rho = 0$). If however, $\tau = 1$, the medium is perfectly transparent. Most real materials are only partially transparent $0 > \tau \le 1$.

Transmissivity in partially transparent (*i.e.*, translucent) materials is dependent both upon the reflection and absorption of radiation. The problem is usually tackled in two stages:

(*i*) The transmissivity τ_ρ is first calculated considering reflection alone.

(*ii*) The transmissivity τ_α is then calculated considering absorption alone.

The transmissivity τ, allowing for both reflection and absorption is then given by

$$\tau = \tau_\rho, \tau_\alpha \quad \text{...(3.1)}$$

Since in the most applications there will be a slab of materials involving two faces, reflection at both faces shall have to be considered. Furthermore, the problem is complicated by the fact that the radiation undergoes multiple reflection at both these faces.

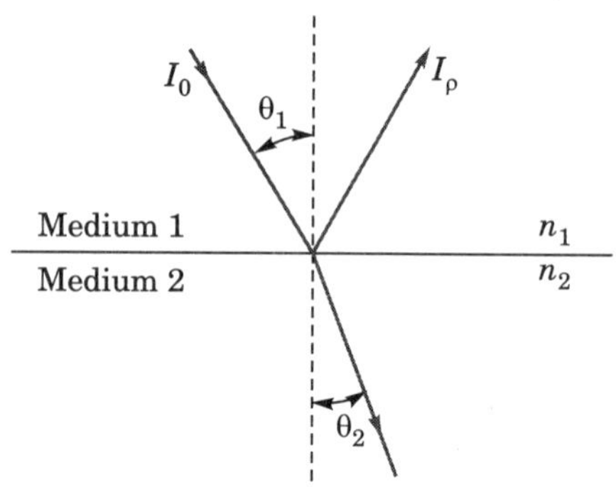

Fig. 3.11. Angle of incidence and reflection in two media.

Reflection at interfaces. Fresnel has given the relationship for the reflection of non-polarized radiation on passing from one medium to another (Fig. 3.11).

$$\frac{I_\rho}{I_0} = \rho = \frac{1}{2}\left[\frac{\sin^2(\theta_2 - \theta_1)}{\sin^2(\theta_2 + \theta_1)} + \frac{\tan^2(\theta_2 - \theta_1)}{\tan^2(\theta_2 + \theta_1)}\right] \quad \text{...(3.2)}$$

where, I_ρ = reflected beam intensity

I_0 = incident beam, radiation

θ_1 = angle of incidence

θ_2 = angle of refraction.

In the expression the two terms in the square brackets represent the reflection for each of the two components of polarization. The angles θ_1 and θ_2 are related to the indices of refraction.

$$\frac{n_1}{n_2} = \frac{\sin\theta_2}{\sin\theta_1} \quad \text{...(3.3)}$$

Thus if the angle of incidence and refraction indices are known from the above relations one can calculate the reflectance of the single interface.

For radiation at normal incidence,

$$\theta_1 = \theta_2 = 0$$

Equations (3.2) and (3.3) are combined to give

$$\rho = \frac{I_\rho}{I_0}\left[\frac{(n_1 - n_2)}{(n_1 + n_2)}\right]^2 \qquad ...(3.4)$$

Calculation of τ_ρ. Cover materials used in solar applications require the transmission of radiation through a slab or film of material and there are thus two interfaces per cover to cause reflection loss. In this situation, the depletion of the beam at the second surface is the same as that at the first, for each component of polarization, assuming cover interfaces are with air on both sides. Neglecting absorption in the slab, as shown in Fig. 3.12 and considering unit incident beam,

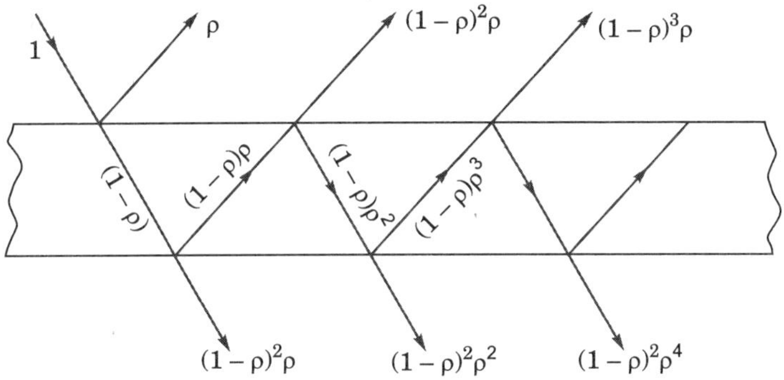

Fig. 3.12. Transmission through one cover.

$(1 - \rho)$ of the incident beam reaches the second interface. Of, this $(1 - \rho)^2$ passes through the interfaces and $\rho(1 - \rho)$ is reflected back to the first, and so on. Summing up the resulting terms, the transmittance for a single cover neglecting absorption is,

$$\tau_{\rho-1} = (1-\rho)^2 + (1-\rho)^2\rho^2 + (1-\rho)^2\rho^4 + ...$$

$$= (1-\rho)^2(1 + \rho^2 + \rho^4 + ...)$$

$$= (1-\rho^2)\sum_{n=\delta}^{n=\infty}\rho^{2n}$$

$$= (1-\rho)^2 . \frac{1}{1-\rho^2} = \frac{1-\rho}{1+\rho} \qquad ...(3.5)$$

In a similar manner it can be shown that for a system of n covers, all are of the same material,

$$\tau_{\rho,\,n} = \frac{1-\rho}{1+(2n-1)\rho} \qquad ...(3.6)$$

Calculation of τ_α. The absorption of radiation in partially transparent (*i.e.,* translucent) materials is described by the Bougers law

$$dI = kI\,d_x \qquad ...(3.7)$$

where, dI is the diminution in the radiation intensity, I is the local value of the intensity, k is the constant of proportionality and is called the extinction coefficient. It will be assumed to have a value independent of wavelength, x is the distance travalled by the radiation Equation (3.7) can be integrated to give

$$\int_0^L \frac{-dI}{I} = k\int_0^L d_x$$

$$[\log I]_0^L = [-k_x]_0^L$$

$$\log I_L - \log I_0 = -k_L$$

or $$\log \frac{I_L}{I_0} = -k_L$$

or $$\frac{I_L}{I_0} = e^{-k_L}$$

or $$\tau_\alpha = \frac{I_L}{I_0} = e^{-k_L} \quad ...(3.8)$$

Note that τ_α is the transmittance considering only absorption and L is the actual path of the radiation through the medium.

For n covers

$$\tau_\alpha = e^{-nkL} \quad ...(3.9)$$

when angle of refraction θ_2 is given, then the path traversed through the cover would be $L/\cos\theta_2$. Then equation (3.9) gets modified to form

$$\tau_\alpha = e^{-nkL/\cos\theta_2} \quad ...(3.10)$$

Now the transmittance allowing for both reflection and absorption can be obtained by the equation (3.1).

The value of *k for* glass varies from the about 0.01/cm (1 m^{-1}) for absolutely clear 'white glass' to 0.32/cm (32 m^{-1}) for poor quality glass with a greenish cast of the edges. Note that τ_α is the transmittance considering only absorption and L is the actual path of the radiation through the medium.

Transmittance-Absorptance Product. For solar collector analysis, it is necessary to evaluate the transmittance-absorptance product ($\tau . \alpha$). Of the radiation passing through the cover system and striking the plate, some is reflected back to the cover system. However, all this radiation is not lost since some is reflected back to the plate.

The transmissivity-absorptivity product is defined as the ratio of the radiation absorbed in the absorber plate to the radiation incident on the cover system and is denoted by the symbol ($\tau . \alpha$), an appropriate subscript b or d being added to indicate the type of incident radiation (*i.e.* beam or diffuse). As illustrated in Fig. 3.13, τ is the transmittance of the cover system as can be known from equation (3.1) and α is the angular absorptance of the absorber plate, $\tau . \alpha$ is absorbed

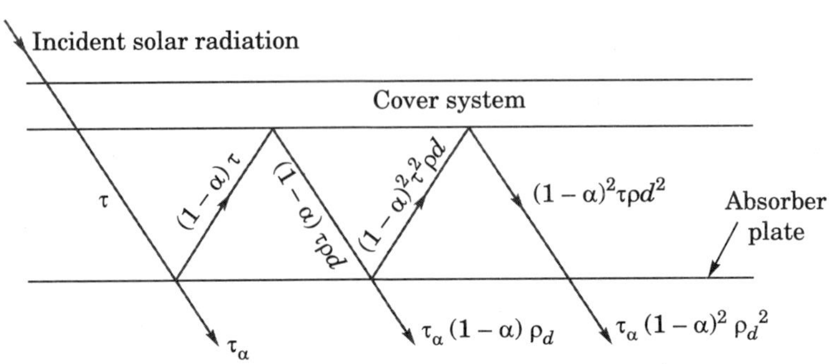

Fig. 3.13. Absorption of solar radiation by absorber plate.

by the absorber plate and $(1 - \alpha)\,\tau$, is reflected back to the cover systems. The reflection from the absorber plate is probably more diffuse than specular so that the fraction $(1-\alpha)\,\tau$ that strikes the cover plate is diffuse radiation and $(1-\alpha)\,\tau.\,\rho_d$ is reflected back to the absorber plate. The quantity ρ_d refers to reflection of the cover plate for incident diffuse radiation that may be partially polarized due to reflections at it is passed through the cover system.

Hence summation yields,

$$<\tau\alpha> = \tau\alpha + \tau\alpha\,(1-\alpha)\rho_d + \tau\alpha\,(1-\alpha)^2\rho_d^2 + \tau\alpha(1-\alpha)^2\,\rho_d^3 + \ldots$$

$$= \tau\alpha\,[1 + (1-\alpha)\rho_d + (1-\alpha)^2\rho_d^2 + (1-\alpha)^3\,\rho_d^3 + \ldots]$$

$$= \tau\alpha\,\frac{1}{1-(1-\alpha)\,\rho_d}$$

$$= \frac{\tau\,.\,\alpha}{1-(1-\alpha)\,\rho_d} \qquad \ldots(3.11)$$

Due to reflection $<\tau\alpha>$ will be greater $\tau\,.\,\alpha$.

For incident angle of 60°, ρ_d is approximately 0.16, 0.24 and 0.29 for one, two and three glasses respectively.

3.5. Energy Balance Equation and Collector Efficiency

The performance of solar collector is described by an energy balance that indicates the distribution of incident solar radiation into the useful energy gain and various losses. The thermal losses can be separated into three components:

(i) **Conductive losses.** An overall heat transfer coefficient value of less than 0.69 W/m^2 °K is suggested to minimize back losses.

(ii) **Convective losses.** Sizing the air gap between the collector covers at 1.25 to 2.5 cm reduces internal convective losses to the minimum possible level. Convection losses between glass plates can also be inhibited if a honeycomb type, cellular structure is placed between the absorber and the outer window plate. Evacuation of the space between the absorber and the outer cover has been proposed to reduce internal convection and conduction, but the cost of added supports and maintenance of a vacuum are excessive.

(iii) **Radiative losses.** Radiative losses from the absorber can be reduced by the use of spectrally selective absorber coatings. Such coatings have a high absorptance of about 0.9 in the solar spectrum and a low emittance, usually of the order of 0.1, in the infra-red spectrum in which the absorber radiates to the environment. Selective absorber coating, therefore decrease heat losses and increase collector efficiency.

Under steady conditions, the useful heat delivered by a solar collector is equal to the energy absorbed in the metal surface minus the heat losses from the surface directly and indirectly to the surroundings. This principle can be stated in the relationship:

$$Q_u = A_c\,[HR\,(\tau\,.\,\alpha)_e - U_L(t_p - t_a)] \qquad \ldots(3.12)$$

where, Q_u is the useful energy delivered by collector, Watts or kcal/hr.

A_c is the collector area, m^2.

HR is the solar energy received on the upper surface of the slopping collector structure;

W/m^2 or kcal/hr m^2.

H is the rate of incident beam or diffuse radiation on a unit area of surface of any orientation.

R is the factor to convert beam or diffuse radiation to that on the plane of collector.

Beam and diffuse radiations are considered separately. $(\tau . \alpha)$ for radiation is determined from the actual angle of incidence; $(\tau \alpha)$ for diffuse radiation may be taken as that for beam radiation at an incidence angle of 60°. The symbol *HR* is used to represent the sum of H_bR_b *and* H_dR_d.

τ is the fraction of incoming solar radiation that reaches the absorbing surface, *transmissivity* (dimensionless).

α is the fraction of solar energy reaching the surface that is absorbed, *absorptivity* (dimensionless).

$(\tau . \alpha)_e$ is effective transmittance absorptance product of cover system for beam and diffuse radiation.

U_L is *the overall heat loss coefficient.* It is the rate of heat transferred to the surroundings per square metre of exposed collector surface per degree celsius difference between average collector surface temperature and the surrounding air temperature, W/m^2 °C (kcal/hr m^2 °C).

t_p is the average temperature of the upper surface of the absorber plate,°C.

t_a is atmospheric temperature, °C.

A diagrammatic representation of terms in this relationship shown in Fig. 3.14. In order that the performance of a solar collector be as high as economically practical design and operating factors that increase the value *of HR* $(\tau . \alpha)$ in equation (3.12) and that reduce the value of U_L $(t_p - t_a)$ are selected. The greater the energy absorption in the metal surface and lower the heat loss from the surface, the higher is the useful recovery. If an unglazed absorber plate is used as the collector, the heat loss coefficient to the atmosphere U_L of 30 to 60 W/m^2 °C (35 to 50 kcal/hr. m^2 °C) is so large that an absorber temperature of 15 to 30°C above atmospheric temperature is the maximum achievable under full solar radiation of 1000 W/m^2 (860 kcal/hr m^2). Under these conditions, on useful heat is delivered

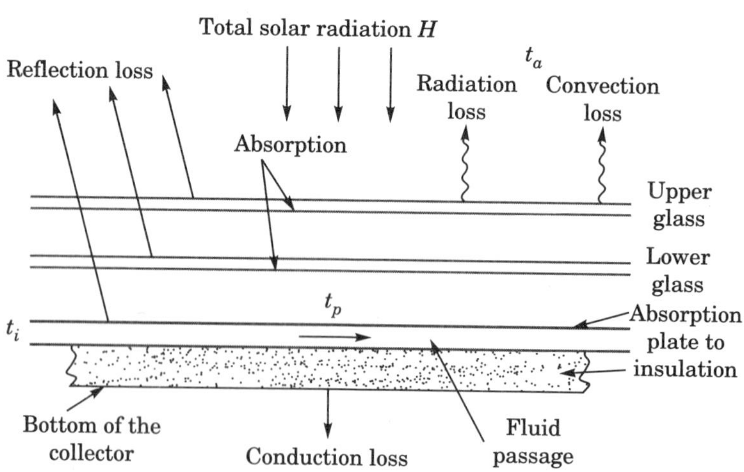

Absorbed energy = $A_c . HR . \tau . \alpha$

Effective heat loss = $A_c U_L (t_p - t_a)$

Fig. 3.14. Definition sketch for Equation (3.12).

from the collector because the heat loss is as large as the solar heat observed. When a fluid is circulated through the collector, useful heat output requires an even lower delivery temperature. Unless a low temperature application is involved, such as swimming pool heating, heat losses must, therefore be reduced.

To reduce the rate of radiation and convection loss as already stated, one or more transparent surfaces, such as glass, are placed above the absorber surface. One layer of glass can transmit as much as 92 per cent of the solar radiation striking it, while greatly reducing the heat loss coefficient U_L. This reduction is due to the suppression of convection loss by interposing a relatively air layer between absorber plate and glass, and by absorption of long wave radiation emitted by the hot metal absorber surface. The combined heat loss coefficient can be reduced to 5 to 10 W/m^2 °C (4.30 to 8.60 kcal/hr m^2 °C) by the use of one glass cover. Similar benefits can be achieved by use of certain transparent plastic materials.

The heat loss coefficient can be reduced further by using a second transparent cover with an air space between the two surfaces. Two convection barriers are then present, as well as two surfaces impeding radiation loss coefficient in the range of 4 W/m^2°C (3.85 kcal/hr m^2°C) are then typically obtained.

Radiation losses can be decreased by other techniques, such as reducing the radiation-emitting characteristics of the absorber. Thermal radiation emitted by the absorber plate may also be reduced reflecting it downward from the lower glass cover by employing an infra-red reflecting coating on the glass. A very thin, optically transparent layer of tin oxide or indium oxide deposited on the glass reflects thermal radiation back to the absorber plate. This coating absorbs some of the solar radiation, however, so the reduced thermal loss is largely offsat by reducing solar energy input to the absorber plate.

The foregoing discussion has been concerned with the methods for reducing U_L, the heat loss coefficient. By so doing, the total heat loss is minimised and collector efficiency is increased. It is evident from equation (3.12) that losses also decrease the difference between average plates temperature and air temperature decreases. The ambient (outside) air temperature is an uncontrollable factor, but the fact that it varies with time and with geographic location means that collector efficiency also depends on these factors. It is clear, that a collector is more efficient at lower plate temperatures than at high temperatures. But plate temperature depends on the temperature of the fluid being circulated in contact with the plate, the rate of fluid circulation and the type of fluid. Fluid temperature depends on conditions elsewhere in the heat utilization system, whereas the other factors in equation (3.12) depend on collector design, operating conditions, solar energy input, and atmospheric temperature.

The *energy balance equation* on the whole collector can be written as:

$$A_c\,[\{HR\,(\tau\,.\,\alpha)_b + HR(\tau.\,\alpha)_d\}] = Q_u + Q_l + Q_s \quad \text{...(3.13)}$$

where, Q_u = rate of useful heat transfer to a working fluid in the solar heat exchanger.

Q_l = rate of energy losses from the collector to the surroundings by re-radiation, convection and by conduction through supports for the absorber plate and so on. The losses due to reflection from the covers are included in the ($\tau.\,\alpha$) terms; and

Q_s = rate of energy storage in the collector.

Collector efficiency η_c, is the collector performance and is defined as the ratio of useful gain over any time period to the incident solar energy over the same time period.

$$\eta_c = \frac{\int Q_u \, d_T}{\int HR \, d_T} \qquad ...(3.14)$$

3.6. Thermal Analysis of Flat-plate collector and useful Heat Gained by the Fluid

Let us assume temporarily the negligible temperature gradient in the flow direction. Let W be the distance between the tubes (W = absorber plate length between tubes).

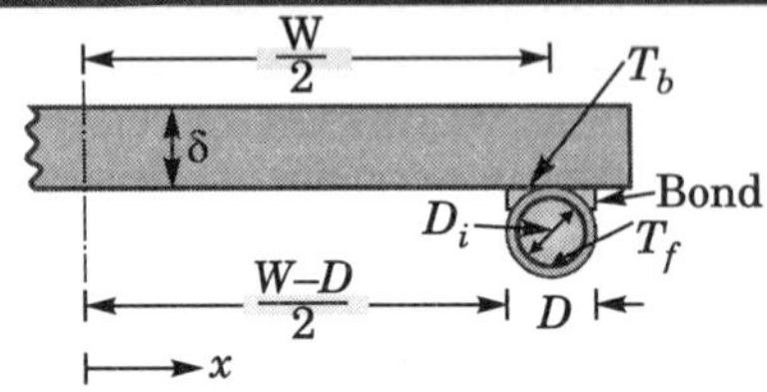

Fig. 3.15. Sheet and tube dimensions of a Liquid Flat Plate Collector.

Let D_i and D be the inside and outside diameters of the tubes, and δ is the sheet thickness as shown in Fig. 3.15. Because the sheet material is a good conductor, the temperature gradient through the sheet is negligible. Consider a section of the absorber plate with two adjacent fluid tubes. The temperature in the plate (T_p) will vary in x direction and it is assumed that the same distribution exists between any two tubes. Also assumed that sheet above the bond to be at same local base temperature T_b, *i.e.*, above the fluid tubes, the temperature will be constant, while in between the tubes, the temperature will pass through a maximum. The region between the centre line separating the tubes and the tube base can then be considered as a classical fin problem.

The fin length is equal to $\frac{W-D}{2}$ as shown in Fig. 3.16, and $W/2$ is distance from centre of collector to the centre of the tube. Considering heat, transfer occurring in the x direction. For an elemental region Δx and unit length in the flow direction, energy balance yields;

$$S\,\Delta x - U_L\,\Delta x(T - T_a) + \left(-K\,\delta\,\frac{dT}{dx}\right)_x - \left(-K\,\delta\,\frac{dT}{dx}\right)_{x+dx} = 0 \qquad ...(3.15)$$

or $$S\,\Delta x + U_L(T_a - T)\,\Delta x - K\,\delta\,\frac{dT}{dx} - \left[-K\,\delta\,\frac{d}{dx}(T + dt)\,\Delta x\right] = 0$$

or $$S\,\Delta x + U_L(T_a - T)\,\Delta x - K\,\delta\,\frac{dT}{dx} + K\,\delta\,\frac{dT}{dx} + K\,\delta\,\frac{d^2T}{dx}\,\Delta x = 0$$

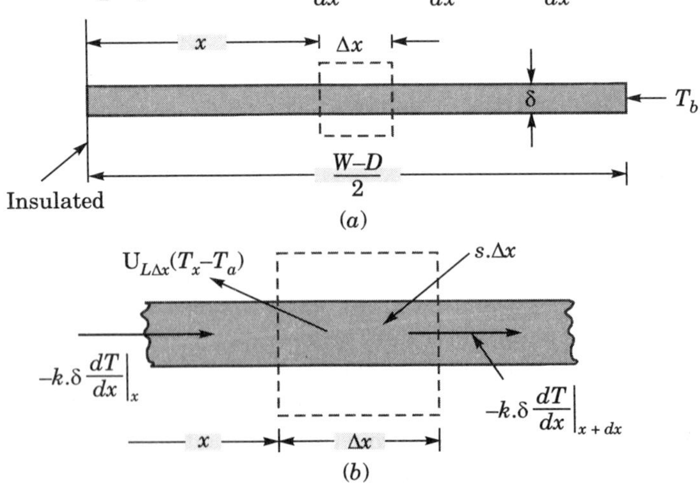

Fig. 3.16. Energy balance on fin element.

or $$S\,\Delta x + U_L(T_a - T)\,\Delta x = -K\,\delta\,\frac{d^2T}{dx_2}\,\Delta x$$

or $$S + U_L(T_a - T) = -K\,\delta\,\frac{d^2T}{dx^2}$$

or $$\frac{d^2T}{dx^2} = +\frac{U_L}{K\delta}\left(T - T_a - \frac{S}{U_L}\right) \quad \text{...(3.16)}$$

Expression 3.16 is of the second order differential equation, the two boundary conditions are,

$$(i)\ \left.\frac{dT}{dx}\right|_{x=0} = 0, \qquad (ii)\ T\Big|_{x=\frac{W-D}{2}} = T_b$$

Let $$m^2 = \frac{U_L}{K\delta},\ \text{and}\ \psi = T - T_a - \frac{S}{U_L}$$

The equation 3.16 becomes

$$\frac{d^2\psi}{dx^2} - m^2\,\psi = 0 \quad \text{...(3.17)}$$

which has the boundary conditions

$$\left.\frac{dT}{dx}\right|_{x=0} = 0,$$

$$\psi\Big|_{x=\frac{W-D}{2}} = T_b - T_a - \frac{S}{U_L}$$

The general solution is

$$x = C_1 \sinh mx + C_2 \cosh mx \quad \text{...(3.18)}$$

The constants C_1 and C_2 are found by substituting boundary conditions. Differentiating 3.18

$$\frac{d\psi}{dx} = C_1\,m \cosh mx + C_2\,m \sinh mx$$

Condition $\frac{d\psi}{dx} = 0$ at $x = 0$ gives $C_1 = 0$

From second condition

$$T_b - T_a - \frac{S}{U_L} = C_2 \cosh mx\,\frac{W-D}{2}$$

$$C_2 = \frac{T_b - T_a - \dfrac{S}{U_L}}{\cosh m\,\dfrac{W-D}{2}}$$

The equation 3.18 becomes

$$\frac{T - T_a - S/U_L}{T_b - T_a - S/U_L} = \frac{\cosh mx}{\cosh m\left(\dfrac{W-D}{2}\right)} \quad \text{...(3.19)}$$

The energy conducted to the region of the tube per unit of length in the flow direction is

$$q^1_{fin\ base} = -K\,\delta\,\left.\frac{dT}{dx}\right|_{x=\frac{W-D}{2}}$$

From equation (3.19),

$$\frac{dT}{dx} = \left(T_b + T_a - \frac{S}{U_L}\right) \frac{m \sinh mx}{\cosh m\left(\frac{W-D}{2}\right)}$$

$$\therefore \qquad q'_{fin\ base} = \frac{K\delta m}{U_L}[S - U_L(T_b - T_a)] \tanh\left(\frac{W-D}{2}\right) m \qquad \text{...(3.20)}$$

Equation 3.20 accounts for the energy collected to only one side of a tube, for both sides

$$\therefore \qquad q_{fin\ base} = \frac{2K\delta m}{U_L}[S - U_L(T_b - T_a)] \tanh m\left(\frac{W-D}{2}\right)$$

$$= 2\frac{1}{m}[S - U_L(T_b - T_a)] \tanh m\left(\frac{W-D}{2}\right)$$

$$\left[\because\ m^2 = \frac{U_L}{K\delta}, \therefore\ \frac{K\delta m}{U_L} = \frac{1}{m}\right]$$

$$= 2\left(\frac{W-D}{2}\right)[S - U_L(T_b - T_a)] \frac{\tanh m\left(\frac{W-D}{2}\right)}{m\left(\frac{W-D}{2}\right)}$$

$$= (W - D) F[S - U_L(T_b - T_a)]$$

where, $F = \left\{\dfrac{\tanh m\left(\frac{W-D}{2}\right)}{m\left(\frac{W-D}{2}\right)}\right\}$

and it is usually known as the *fin efficiency*. The useful gain of the collector also includes the energy collected above the tube region. The energy gain for the tube region

$$q_{tube} = D[S - U_L(T_b - T_a)] \qquad \text{...(3.21)}$$

Hence, the total energy gain of the collector tubes per unit length in the flow direction may be expressed as

$$q_u = q_{fin\ base} + q_{tube\ section}$$
$$= [(W-D)F + D][S - U_L(T_b - T_a] \qquad \text{...(3.22)}$$

Ultimately, the useful gain from the equation 3.21 must be transferred to the fluid. The resistance to heat flow to the tube from the plant may be consisted of three components:

(*i*) the resistance due to the bonding material between the plate and the tube;

(*ii*) the resistance due to the temperature gradient in the fluid at the tube wall;

(*iii*) the resistance due to the wall thickness of the tube. Hence

$$q_u = \frac{T_b - T_f}{\frac{1}{C_b} + \frac{1}{\pi D_i h_{fi}} + \frac{1}{C_w}} \qquad \text{...(3.23)}$$

where, C_b = the conductance of the bond
C_w = the conductance of the tubewall
and, h_{fi} = the local film heat transfer co-efficient.

The bond conductance is given as

$$C_b = \frac{K_b . b}{y} \quad ...(3.24)$$

where, K_b = bond conductivity
b = bond length
y = bond average thickness.

The bond conductance can be very important in accurately describing collector performance. Simply wiring or clamping of the tubes to the sheet results in a significant loss of performance. It is necessary to have good metal to metal contact so that the bond resistance is less than 0.33 m °C/W.

The useful energy gain of the fluid can then be expressed in terms of the known dimensions, the physical parameters and the local fluid temperature by solving the equation 3.23 for T_b and substituting to obtain q_u from Eq. 3.22.

$$q_u = WF^1 \,[S - U_L(T_f - T_a)] \quad ...(3.25)$$

where, $$F^1 = \frac{1/U_L}{W\left[\dfrac{1}{U_L\left[D+(W-D)F\right]}+\dfrac{1}{C_b}+\dfrac{1}{C_w}+\dfrac{1}{\pi D_i h_{fi}}\right]} \quad ...(3.26)$$

F^1 has been called the *collector efficiency factor*. Physical inter-pretation of F^1, for most of collector geometries will be clear from equation (3.26). The denominator is the heat transfer resistance from the fluid to the ambient air, this is equal to $1/U_0$ say. The numerator is the heat transfer resistance from the absorber plate to the ambient air, equal to $1/U_L$, F^1 thus is the ratio of these two heat transfer coefficients.

$$F^1 = \frac{U_0}{U_L} \quad ...(3.27)$$

F^1 is essentially a constant for any collector design and fluid flow rate. The heat transfer coefficient inside the tube may be assumed, 300 W/m^2 °C (for natural circulation) or 1500 W/m^2 °C (forced circulation).

For temperature distribution in the flow direction, consider the energy balance on a fluid element flowing through a pipe of length Δy which is receiving a uniform heat flux q_u, refer Fig. 3.17 so that

$$\dot{m}C_P\, T_f\,\Big|\, y - \dot{m}C_p\, T_f\,\Big|_{y+dy} + \Delta y,\, q_u = 0 \quad ...(3.28)$$

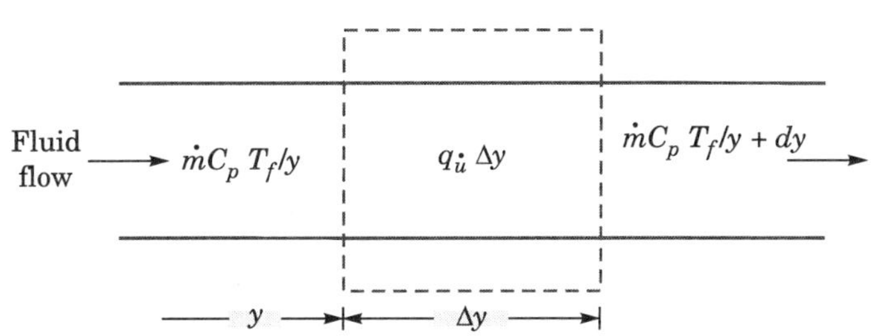

Fig. 3.17. Energy balance on fluid element.

Dividing throughout by Δy and finding the limit as $\Delta y \to 0$, and substituting in Eq. 3.28 for q_u

$$\dot{m}\, C_P \frac{dT_f}{dy} - WF^1 [S - U_L (T_f - T_a)] = 0 \qquad \text{...(3.29)}$$

If the assumption is made that F^1 and U_L are constant (and independent of y), then the solution of the differential equation for the temperature at any position (if subject to the condition that inlet fluid temperature is T_{fi}) is

$$\frac{T_f - T_a - S/U_L}{T_{fi} - T_a - S/U_L} = e^{-(U_L \cdot W F^1 y/\dot{m}\, C_p)} \qquad \text{...(3.30)}$$

If the collector has length L in the flow direction, then the outlet fluid temperature T_{fo} is found by substituting L for y in the equation (3.30).

$$T_{fo} = T_a + \left(\frac{S}{U_L}\right) - \left[\frac{S}{U_L} - (T_{fi} - T_a)\, e^{-U_L F^1 A_c/\dot{m} C_P}\right] \qquad \text{...(3.31)}$$

where, $A_c = WL$, the area of collector.

The total useful energy collection rate Q_u may be expressed as

$$Q_u = \dot{m}\, C_p (T_{fo} - T_{fi}) \qquad \text{...(3.32)}$$

Substituting for $T_{f,o}$ already derived, gives

$$Q_u = A_c F_R [S - U_L (T_{fi} - T_a)] \qquad \text{...(3.33)}$$

where, $$F_R = \frac{\dot{m}\, C_P}{A_c\, U_L}\left[1 - e^{-A_c U_L F^1/\dot{m} C_P}\right] \qquad \text{...(3.34)}$$

$$= \frac{G\, C_P}{U_L}\left[1 - e^{-U_L F^1/G C_P}\right]$$

$$\because \quad G = \dot{m}/A_c \qquad \text{...(3.35)}$$

F_R has been termed as the *heat removal factor* of the collector.

Example 3.1. *Data for a flat plate collector used for heating the building are given below:*

Factor	*Specification*
Location and latitude	*Baroda, 22°N*
Day and time	*1 January, 11: 30–12 : 30 (IST)*
Annual average intensity, of solar radiation	*0.5 langley / min*
Collector tilt	*latitude + 15°*
No. of glass covers	*2*
Heat removal factor for collector	*0.810*
Transmittance of the glass	*0.88*
Absorplance of the glass	*0.90*
Top loss coefficient for collector	*7.88 W / m² °C (6.80 kcal / hr m² °C)*
Collector fluid temperature	*60°C*
Ambient temperature	*15°C*

Calculate:
(i) Solar altitude angle, *(ii) Incident angle,*
(iii) Collector efficiency.

Solution. Solar declination δ

$$= 23.45 \sin\left[\frac{360}{365}(284 + n)\right]$$

$$= 23.45 \sin\left[\frac{360}{365}(284 + 1)\right]$$

$$= -23.00°$$

Solar hour angle $\omega_s = 0$, (at mean of 11 : 30 and 12 : 30)

Solar altitude angle α is given by

$$\sin \alpha = \cos \phi \cos \delta \cos \omega_s + \sin \phi_s \sin \delta$$
$$= \cos 22° \cos (-23°) \cos \theta + \sin 22° \sin (-23°)$$
$$= 0.85 - 0.146 = 0.7036$$
$$\alpha = 44.7°. \textbf{ Ans.}$$

Incident angle $\theta = \frac{\pi}{2} - \alpha = 90 - 44.7° = 45.3°$. **Ans.**

Now R_b is given equal to

$$= \frac{\cos(\phi - s)\cos\delta\cos\omega + \sin(\phi - s)\sin\delta}{\cos\phi\cos\delta\cos\omega + \sin\phi\sin\delta}$$

$$= \frac{\cos(22° - 37°)\cos(-23°)\cos 0 + \sin(22 - 37)\sin(-23)}{\cos 22\cos(-23)\cos 0 + \sin 22\sin(-23)}$$

$$= \frac{0.889 - 0.101}{0.85 - 0.146} = 1.40$$

Effective transmittance absorptance product is

$$<\tau . \alpha> = \frac{\tau . \alpha}{1 - (1 - \alpha)\rho_d}$$

where, ρ_d is diffuse reflectance for two glass covers, it is = 0.24

$$= \frac{0.88 \times 0.90}{1 - (1 - 0.90)\,0.24} = 0.811.$$

Solar radiation intensity (considering beam radiation only)

$$H_b = 0.5 \text{ ly/mm}$$
$$= \mathbf{0.5\ cal/cm^2.\ min}$$
$$= \frac{0.5 \times 10^4}{10^3} \times \mathbf{60\ kcal/m^2\ hr}$$
$$= 300 \text{ kcal/m}^2 \text{ hr}$$
$$= 350 \text{ W/m}^2 \text{ hr} \qquad [\because 1 \text{ kcal} = 1.163 \text{ watt}]$$

Now, $S = H_b R_b <\tau . \alpha>$

$$= 350 \times 1.40 \times 0.811 = 395 \text{ W/m}^2$$
$$= 340.62 \text{ kcal hr m}^2$$

Useful gain $q_u = F_R[S - U_L(T_{fi} - T_a)]$

$$= 0.810\,[340.62 - 6.80\,(60 - 15)]$$
$$= 0.810 \times 34.62$$
$$= 28.07 \text{ kcal/hr m}^2. \textbf{ Ans.}$$

or $$q_u = 0.810\ [395 - 7.88 \times 45]$$
$$= 33.21\ \text{W/m}^2\ \text{hr.}\ \textbf{Ans.}$$

Collection efficiency,

$$\eta_c = \frac{q_u}{H_b R_b} = \frac{28.07}{300 \times 1.40} = 0.06 = 6\%.\ \textbf{Ans.}$$

3.7. Concentrating Collector: Focusing Type

Introduction. Focusing collector is a device to collect solar energy with high intensity of solar radiation on the energy absorbing surface. Such collectors generally use optical system in the form of reflectors or refractors. A focusing collector is a special form of flat-plate collector modified by introducing a reflecting (or refracting) surface (concentrator) between the solar radiations and the absorber. These type of collectors can have radiation increase from low value of 1.5 to 2 to high values of the order of 10,000. In these collectors radiation falling on a relatively large area is focused on to a receiver (or absorber) of considerably smaller area. As a result of the energy concentration, fluids can be heated to temperatures 500°C or more.

An importance difference between collectors of the non-focusing and focusing types in that the latter concentrate only direct radiation coming from a specific direction, since diffuse radiation arrives from all directions, only a very small proportion is from the direction for which focusing occurs. The optical system directs the solar radiation on-to an absorber of smaller area which is usually surrounded by a transparent cover. Because of the optical system, certain losses (in addition to those which occur while the radiation is transmitted through the cover) are introduced. These include reflection or absorption losses in the mirrors or lenses and losses due to geometrical imperfections in the optical system. The combined effect of all losses is indicated through the introduction of a term called the *optical efficiency*. The introduction of more optical losses is compensated for by the fact that the flux incident on the absorber surface is concentrated on a smaller area. As a result, the thermal loss terms do not dominate to the same extent as in a flat-plate collector and the collection efficiency is usually higher.

Types of concentrating collectors. Concentrating or focusing collectors may be considered in two general categories: *line focusing* and *point focusing* types. In practice, the line is a collector pipe and the point is a small volume through which the heat transport fluid flows. Because the sun has a finite size, focusing does infact occur over a small area or volume rather than a line or point. As per the number of concentrating collector geometries, the main-types of concentrating collectors are:

(*a*) Parabolic trough collector

(*b*) Mirror strip reflector

(*c*) Fresnel lens collector

(*d*) Flat plate collector with adjustable mirrors

(*e*) Compound parabolic concentrator (C.P.C.).

Line Focusing Collectors: Parabolic Trough Reflector. The principle of the parabolic trough collector, which is often used in concentration collectors, is shown by the cross-section in Fig. 3.17, solar radiation coming from the particular direction is collected over the area of the reflecting surface and is concentrated at the focus of the parabola, if the reflector is in the form of a trough with parabolic cross-section, the solar radiation is focused alone a line. Mostly cylindrical parabolic concentratora are used, in which absorber is placed along focus axis. The collector pipe, preferably with a selective absorber coating, is used as an absorber. The dimension of parabolic trough or parabolic cylindrical collector can be vary over a wide range the length of a reflector unit may be roughly 3 to 5 m, and the width about 1.5 to 2.4 m, Ten or more such units are often connected end to end in a row, several rows may also be connected in parallel. Parabolic trough reflectors have been made of highly polished aluminium, of silvered glass or of a thin film of aluminized plastic, on a firm base. Instead of having a continuous form, the reflector may be constructed from a number of long flat strips on a parabolic base.

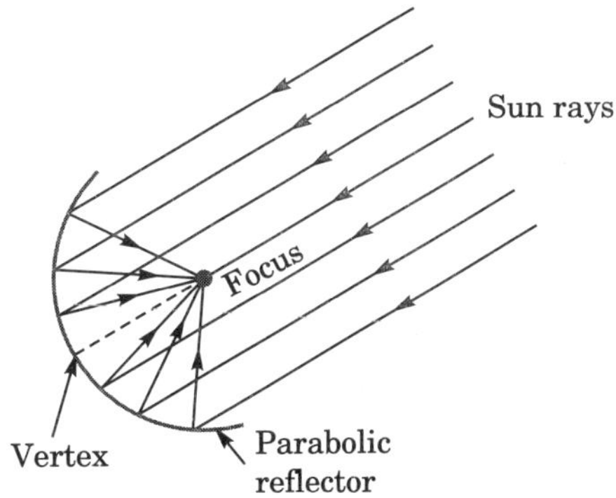

Fig. 3.17. Cross-section of parabolic-trough collector.

For the solar radiation to be brought to a focus by parabolic trough reflector, the sun must be in such a direction that it lies on the plane passing through the focal line and the vertex (*i.e.,* the base) of the parabola. Since the elevation of the sun is always changing, either the reflector trough or the collector pipe (absorber) must be turn continuously about its long axis to maintain the required orientation. Both schemes are used in different practical-designs. Either the trough/cylindrical reflector or the pipe is turned by partial rotation around a single axis parallel to the trough length. Trough type collectors are generally oriented in the east-west or north-south directions. For the east-west orientation, the collectors are laid flat on (or parallel, to) the ground. For the north-south orientation, however, the north end of the trough is raised so that the collectors are sloped facing south-just like flat-plate collectors. Ideally, the slope angle should be changed periodically; it is simpler, but less efficient, however to use a fixed angle design.

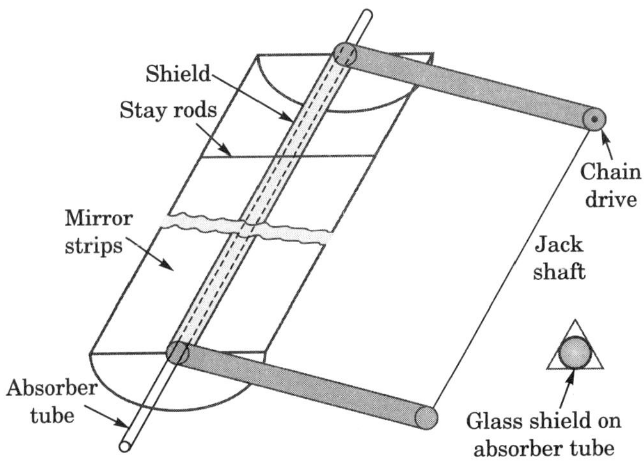

Fig. 3.18. A typical cylindrical parabolic system.

The north-south orientation permits more solar energy to be collected than the east-west arrangement, except around the winter equinox On the other hand, construction costs are higher for the north-south (sloping) type. Moreover, a system of such collectors requires a larger land area to allow for the shadowing effect of the sloping troughs. The increased separation distance between rows of collectors also results in increased pipe line costs and greater pumping and thermal losses. Finally the sun set position of an east-west reflector is essentially the same as the sunrise position, and little or no evernight adjustment is required. For the north-south orientation, however the trough (or receiver) must be turned through a large angle from sunset to sunrise. The choise of orientation in any particular instance depends on the foregoing and other considerations.

(*b*) **Mirror-Strip Reflector.** In another kind of focusing collector, a number of plane or slightly curved (concave) mirror strips are mounted on a flat base. The angles of the individual mirrors are such that they reflect solar radiation from a specific direction on to the same focal line (Fig. 3.19). The angles of the mirrors must be adjusted to allow for changes in the sun's elevation, while the focal line (for collector pipe) remains in a fixed position. Alternatively, as mentioned for parabolic trough collectors, the mirror strips may be fixed and the collector pipe moved continuously so as to remain on the focal line.

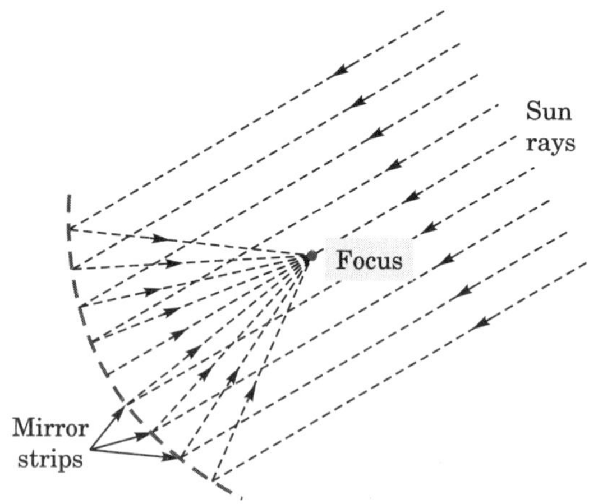

Fig. 3.19. Mirror-strip solar collector.

(*c*) **Fresnel Lens Collector.** In addition to the reflecting colle-ctors described above, a refraction type of focusing collectors has been developed. It utilizes the focusing effect of a Fresnel lens, as represented in cross-section in Fig. 3.20. For a trough-type collector, the lens is rectangle, about 4.7 m in overall length and 0.95 m in width. It is made in sections from cost acrylic plastic and can probably be produced in quantity at low cost. The rounded triangular trough serves only as a container and plays no role in concentrating the solar energy.

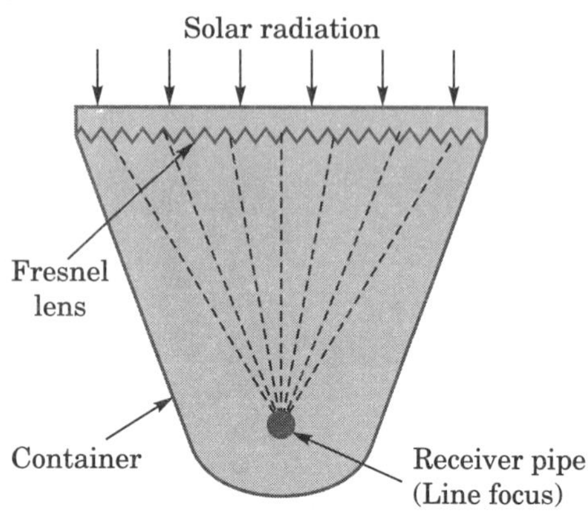

Fig. 3.20. Cross-section of Fresnel lens through collector.

To be fully effective, the Fresnel lens must be continuously aligned with the sun in two directions namely, both along and perpendicular to its length. This is achieved by orienting the troughs in the north-south direction with rotation about the length wise axis; in addition, the north ends of the troughs are raised to increase the slope as the sun's elevation decreases (and *vice versa*). The total solar radiation energy that can be collected annually is ahout 30 percent greater than for an east-west orientation.

Receiver pipe. The receiver pipe of a parabolic line focusing collector, shown in cross-section in Fig. 3.21 has the same general characteristics as a flat-plate collector. The solar radiation absorber is a central steel pipe with a treated surface. A selective absorber surface, such as the black chrome referred to earlier, may be advantageous. A hollow steel plug within the absorber pipe restricts the flow of the heat-transfer fluid to a narrow annular region. This results in a high flow velocity of the fluid and consequently a high rate of heat transfer from the absorber.

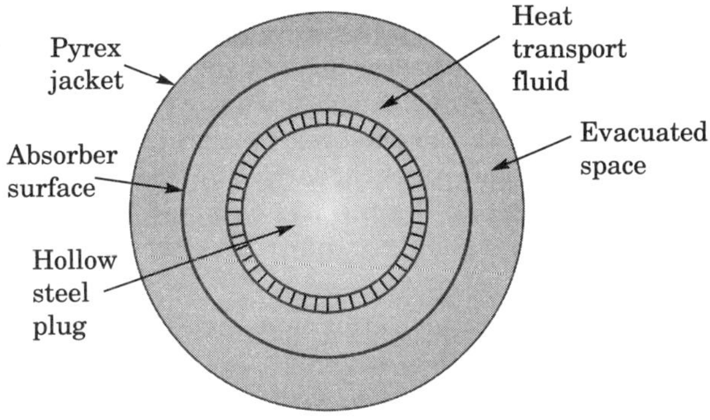

Fig. 3.21. Cross-section of solar energy pipe receiver.

The absorber pipe is usually enclosed in a glass (Pyrex) jacket in order to decrease thermal losses by convection and radiation. The space between the pipe and the jacket is sometimes evacuated to reduce convection losses. The diameter of the glass jacket may be about 5 cm and that of the absorber pipe about 3 cm. The annulus between this pipe and the plug may be as little as 2.5 mm wide.

In a Fresnel lens collector, the solar radiation is focused into the absorber from the top, rather than from the bottom as in the parabolic (reflection) type. A modified absorber design is then possible Fig. 3.22. Insulation at the bottom and sides of the absorber pipe and a flat-plate over the top reduce thermal losses. A stainless steel reflector adjacent to the pipe (absorber or receiver) reflects back emitted thermal radiation.

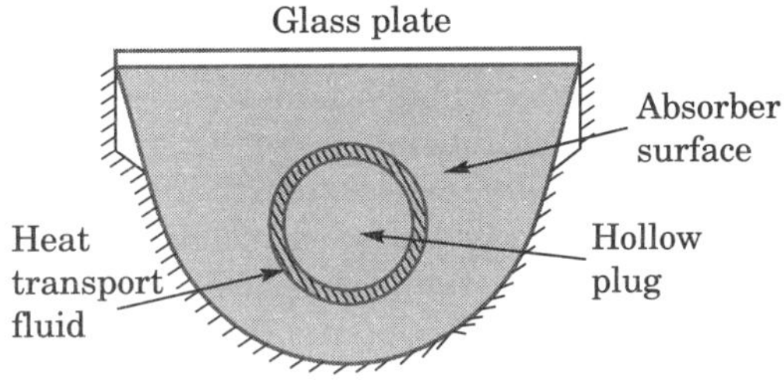

Fig. 3.22. Receiver for Fresnel lens collector.

Point Focusing Collector (Paraboloidal Type). A paraboloidal dish collector brings solar radiation to a focus at a point actually a small central volume. (Fig. 3.23). A dish 6.6 m in, diameter has been made from about 200 curved mirror segments forming a paraboloidal surface. The absorber, located at the focus, is a cavity made of a zirconium-copper alloy with a black chrome selective coating. The heat-transport fluid flows into and out of the absorber cavity through pipes bonded to the interior. The dish can be turned automatically about two axes (up-down and left-right) so that the sun is always kept in a line with the focus and the base (vertex) of the paraboloidal dish. Thus, the sun can be fully tracked at essentially all times.

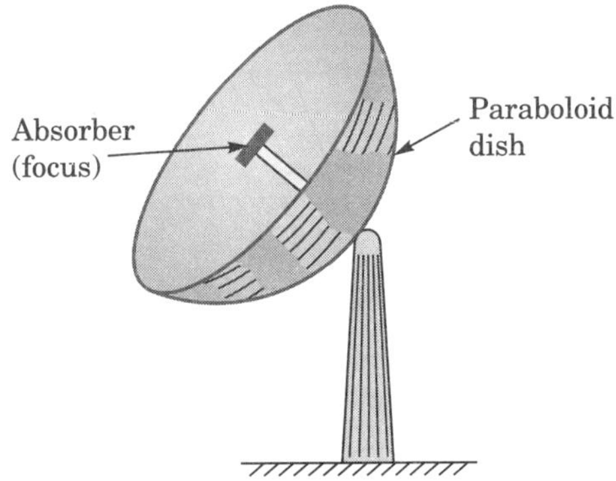

Fig. 3.23. Point focus solar collector (Paraboloid).

The concentration ratios (concentration ratio is the ratio of the area of the concentrator aperture to the energy absorbing area of the receiver, it determines the effectiveness of a concentrator), are very high in the case of parabolic system and therefore can be used where high temperatures are required. In a cylindrical parabolic system, the concentration ratio is lower than paraboloid counterparts. In both the cases, the receiver is placed at the focus *i.e.,* along the focal line in cylindrical parabolic or parabolic trough system and at the focus point in paraboloidal system.

Concentration ratios of about 30 to 100 or higher would be needed to achieve temperatures in the range 300 to 500°C or higher. Collectors designed for such high concentration ratios necessarily have small angles of field of view and hence need to track the sun continuously. A broad classification of such collector is:

(*i*) The linear focus collector in the form of a parabolic through or the ones employing faceted mirror strips.

(*ii*) Spherical and conical mirror (Axicon) with aberrated foci. The physical upper limit to the concentration ratios achievable with paraboloids and parabolic troughs is determined by their *f/d* ratios (focal length/diameter) and are about 10,000 and 100 respectively for the two cases. The concentration ratios achieved in practice are about 1/3 to 1/2 of the above values because of surface irregularities of the reflector, tracking errors etc.

(*iii*) Central receiver collector, such as the paraboloidal mirror and the tower power plant using heliostat mirrors.

A system equivalent to a very large paraboloidal reflector consists of a considerable number of mirror distributed over an area on the ground. Each mirror, called a *heliostat,* can be steered independently about two axes so that the reflected solar radiation is always directed towards an absorber mounted on a tower (Fig. 3.24). This type of collector is classified as *Central Receiver Collector.* This is mostly used in tower power plants for generation of electrical energy.

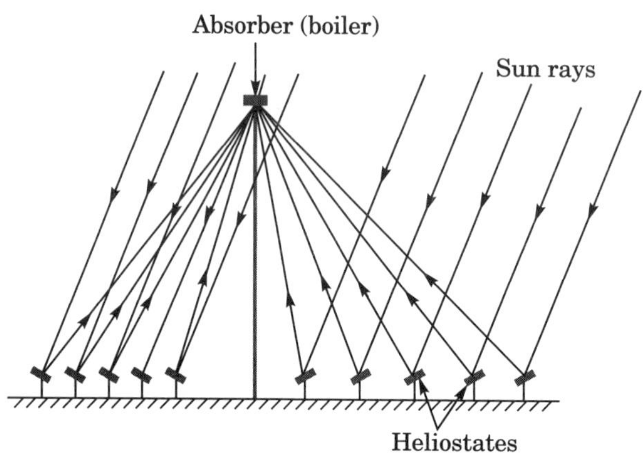

Fig. 3.24. Distributed heliostat point-focusing reflector (Central-Receiver).

In the typical central receiver, the mirror is composed of many small mirrors; each with its own heliostat to follow the sun. The heliostats are generally located in the horizontal plane, but when the situation is favourable, can simply follow the existing terrain. The basic difference between a single mirror concentrator and the heliostat system is that the heliostat system has a *dilute mirror.* This means that the entire surface within the system is not covered with mirror surface. This diluteness is generally termed as the *fill factor*. A central receiver with a fill factor of 40% means that 40% of the land area is covered by the mirrors.

In a central receiver optical system as shown in figure many small mirrors are separately mounted to act together like a dilute paraboloid. The basic problem associated with the central receiver is that the heliostat mirrors require non-linear drive rate in two coordinates to achieve the requirement of keeping the reflected image point on a fixed receiver. Along with the problem is the requirement that the heliostat be rugged enough to survive storms and operate successfully in a moderate wind.

Among all the steerable concentrators mentioned above paraboloids have the highest efficiency in terms of the utilization of the reflector area because in a fully steerable paraboloid there are no losses due to aperature effects. Also radiation losses are small because of the small area of the absorber at the focus. Both then they are the most difficult to fabricate and operate too. A practical size for aperature area would be about 50 m^2 from which 15 to 20 kW of useful energy could be extracted by thermal conversion processes.

Concentrating Collectors: *Non-Focusing Type*. The simplest type of concentrating collector is the mirror-boosted, flat plate collector. It consists of a flat plate facing south with mirorrs attached to its north and south edges (Fig. 3.25). If the mirrors are set at the proper angle, they reflect solar radiation on to the absorber plate. Thus, the latter receives reflected radiation in addition to that normally falling on it. The mirrors cut off part of the scattered radiation that would otherwise have reached the absorber plate, and only part of the scattered radiation falling on the mirrors will be reflected onto the absorber. Thus the concentration effect arises mainly from the increase in direct radiation reaching the absorber plate.

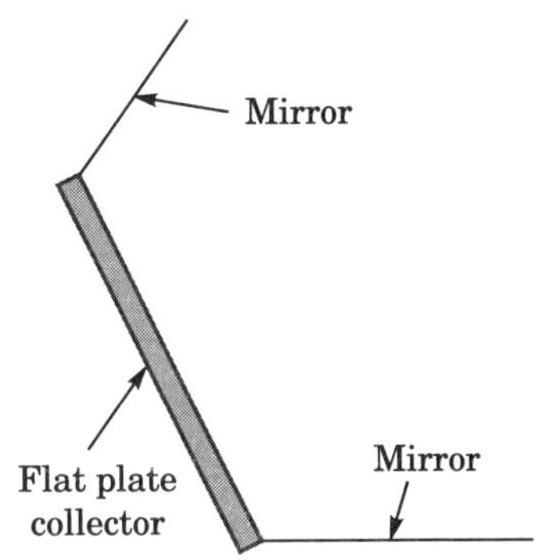

Fig. 3.25. Flat-Plate collector augmented with mirrors.

When a number of collectors are combined in two or more rows, as they often are, the rows must be set further apart in the north-south directions to allow for the additional sun shading caused by the mirror extensions. Furthermore, in order for the mirrors to be effective, the angles should be adjusted continuously as the sun's attitude changes. For these reasons, and they can provide only a relatively small increase in the solar radiation falling on the absorber, flat-plate collectors with mirrors are not widely used.

Compound Parabolic Concentrator (CPC). The CPC (or Winston Collector) is a trough-like arrangement of two facing parabolic mirrors (Fig. 3.26). Unlike the single parabolic trough reflector described earlier, the CPC is non-focusing, but solar radiation from many directions is reflected toward the bottom of the trough. Because of this characteristic; a large proportion of the solar radiation, including diffuse (scattered) radiation, entering the trough opening is collected (and

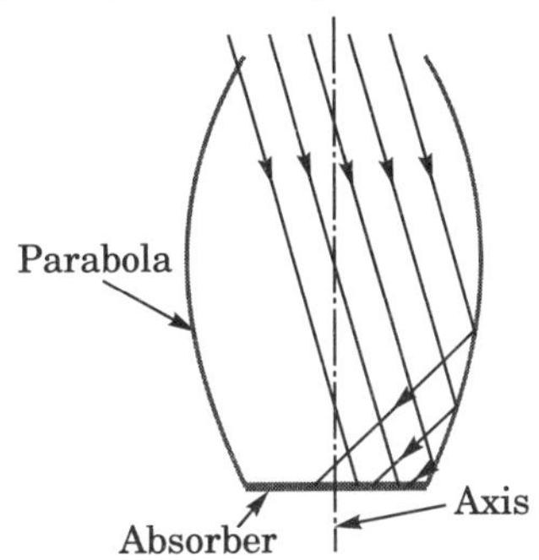

Fig. 3.26. Compound parabolic concentrator.

concentrated) on a small area. In addition to collecting both direct and diffuse radiations, an advantage of the CPC is that it provides moderately good concentration, although less than a focussing collector, in an east-west direction without (or only seasonal) adjustment for sun tracking.

It is possible to concentrate solar radiation by a factor of 10 without diurnal tracking, using this type of collector.

CPC reflectors can be designed for any absorber shapes. For example:

(*a*) Flat one sided absorber, (*b*) Flat two sided absorbers (fin),

(*c*) Wedge-like absorbers, or (*d*) Tubular absorbers.

For economic as well as for thermal reasons the fin and the tubular type of absorbers are preferable. With a concentric tubular absorber with an evacuated jacket, temperatures of about 200°C are achievable with Winston collectors. They are suitable for the temperature range of 100–150°C even if the absorber is not surrounded by a vacuum. It is claimed that Winston collectors are capable of competitive performance at high temperatures of about 300°C required for power generation, if they are used with selectively coated, vacuum enclosed receivers which decrease thermal losses from the collector. The *advantages* of this new type are:

(*i*) There is no need for tracking, as it has high acceptance angle, only seasonal adjustments are required. For concentration ratios of ~ 3, even seasonal adjustments may not be required.

(*ii*) The efficiency for accepting diffuse radiation is much larger than conventional concentrators, and

(*iii*) Its concentration ratio is equal to the maximum value possible for a given acceptance angle.

The parabolic trough or a linear parabolic collector is also more commonly known as the cylindrical parabolic collector. It has many commercial versions which are now available.

Out of all configurations described so far, the maximum concentration ratios are available with paraboloidal system, it is of the order of 10,000.

3.8. Advantages and Disadvantages of Concentrating Collectors over Flat Plate Type Collectors

Advantages. The main advantages of concentrator systems over flat-plate type collectors are:

1. Reflecting surfaces required less material and are structurally simpler than flat-plate collectors. For a concentrator system the cost per unit area of solar collecting surface is therefore potentially less than that for flat-plate collectors.

2. The absorber area of a concentrator system is smaller than that of a flat-plate system for same solar energy collection and therefore the insolation intensity is greater.

3. Because of the area from which heat is lost to the surroundings per unit of the solar energy collecting area is less than that for flat-plate collector and because the insulation on the absorber is more concentrated, the working fluid can attain higher temperatures in a concentrating system than in a flat-plate collector of the same solar energy collecting surface.

4. Owing to the small area of absorber per unit of solar energy collecting area, selective surface treatment and/or vacuum .insulation to reduce heat losses and improve collector efficiency are economically feasible.

5. Focusing or concentrating systems can be used for electric power generation when not used for heating or cooling. The total useful operating time per year can therefore be large for a concentrator system than for a flat-plate collector and the initial installation cost of the system can be regained by saving in energy in a shorter period of time.

6. Because the temperature attainable with concentrating collector system is higher, the amount of heat which can be stored per unit volume is larger and consequently the heat storage costs are less for concentrator systems than for flat-plate collectors.

7. In solar heating and cooling applications, the higher temperature of the working fluid attainable with a concentrating system makes it possible to attain higher efficiencies, in the cooling cycle and lower cost for air conditioning with concentrator systems than with flat-plate collectors.

8. Little or no anti-freeze is required to protect the absorber in a concentrator system whereas the entire solar energy collection surface requires anti-freeze protection in a flat-plate collector.

Disadvantages

1. Out of the beam and diffuse solar radiation components, only beam component is collected in case of focusing collectors because diffuse component can not be reflected and is thus lost.

2. In some stationary reflecting systems it is necessary to have a small absorber to track the sun image; in others the reflector may have to be adjustable more than one position if year round operation is desired; in other words costly orienting systems have to be used to track the sun.

3. Additional requirements of maintenance particular to retain the quality of reflecting surface against dirt, weather, oxidation etc.

4. Non-uniform flux on the absorber whereas flux in flat-plate collectors is uniform.

5. Additional optical losses such as reflectance loss and the intercept loss, so they introduce additional factors in energy balances.

6. High initial cost.

3.9. Performance Analysis of a Cylindrical Parabolic Concentrating Collector

Energy balances are used to describe the performance of focusing collector systems as done with flat-plate collectors. Consider a parabolic cylindrical collector as shown in Fig. 3.27. Useful energy gain by it, is obtained by energy balance at x per unit aperture* area.

$$q_u = \frac{H_b\, R_b\, \rho\gamma\, (\tau\,.\,\alpha)\, A_a}{A_a} - \frac{U_L A_r \left(T_{r,x} - T_a\right)}{A_a} \qquad ...(3.36)$$

where, q_u is heat gain per unit aperture area

$H_b R_b$ is the beam radiation intensity on a reference surface (generally horizontal) available from solar date of the place.

*The aperture (W) is the plane opening of the concentrator through which the solar radiation passes. For a cylindrical or linear concentrator, it is characterized by the width.

R_b is the ratio of the beam radiations on the reflector aperture to that on whatever reference surface H_b is measured, *i.e.*, $H_b R_b$ is the beam radiation on the plane of the aperture per unit time per unit area.

ρ is the specular reflectance of the reflector surface.

γ is the intercept factor *i.e.*, the fraction of the specularly reflected radiation that is intercepted by the absorber.

$(\tau . \alpha)$ is the transmittance absorptance product, the transmittance of the cover (when cover is present) and absorptance of the receiver.

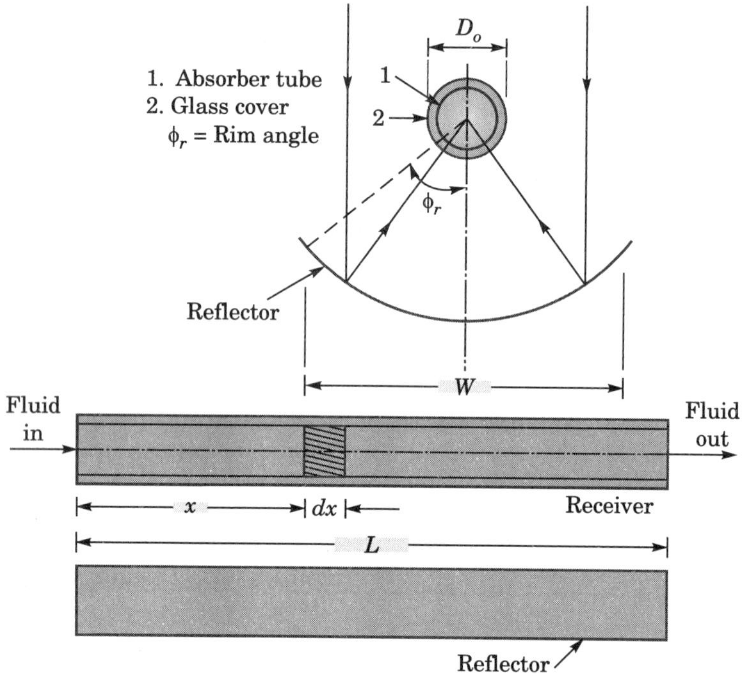

Fig. 3.27

A_a/A_r is the ratio of effective area of the aperture to area of the solar energy absorber, it is termed as the *concentration ratio.* A_a is the unshaded projected area of the reflector system which may be smaller than the total aperture area. The concentration ratio of the collector is given by

$$C_R = \frac{\text{Effective aperture area}}{\text{Absorber tube area}} = \frac{(W - D_o)L}{\pi D_o L}$$

$$= \frac{W - D_o}{\pi D_o}. \quad \text{...(3.37)}$$

For flat plate collectors $A_a = A_r$ and thus maximum energy flux per unit area of the receiver may be limited to about 1 kW/m^2 while in the case of focusing collectors A_a/A_r may vary from unity to several thousands and thus heat flux may be quite comparable to the conventional heat exchangers.

U_L is the surface (receiver) loss coefficient

$T_r x$ is the receiver surface temperature at x

T_a is the ambient temperature.

If the whole of the receiver is at a uniform temperature T_r, then total energy gain is,

$$Q_u = A_a H_b \rho\gamma R_b (\tau . \alpha) - U_L A_r (T_r - T_a) \quad ...(3.38)$$

Thermal losses are larger in flat-plate collectors, due to larger loss areas involved. But. in focusing collectors, by increasing A_a/A_r with fixed A_a and U_L, losses can be reduced (A_r is decreased). Thus focusing collectors permit higher temperature energy collection.

Focusing collectors differ in their thermal behaviour from the flat-plate type, because for receivers, the shapes are widely variable, the temperatures are higher, the edge effect are more significant, conduction terms are quite high, and the radiation flux on the receiver is not uniform. Therefore, it is not possible to present a single general method of estimating thermal losses. However the main features can be brought out by considering a representative parabolic cylindrical reflector and an uncovered receiver.

The generalized analysis of a focusing collector system is very similar to that of a flat-plate collector.

Heat gain per unit collector length L, can be expressed by energy balance which gives

$$q_u = \frac{A_a}{L} H_b R_b \rho\gamma(\tau\alpha) - \pi D_0 U_L (T_r - T_a) \quad ...(3.39)$$

(Since aperture area $A_a = W . L$, equation (3.38) can be written in above form)

The loss coefficient U_L is found from

$$U_L = \left(\frac{1}{h_{wind}} + \frac{1}{h_r}\right)^{-1} \quad ...(3.40)$$

where, h_{wind} = film coefficient due to wind

$= 5.7 + 3.8$ V, W/m^2 °C

V is the wind velocity in metre/sec.

h_r = radiation coefficient and can be determined from the equation;

$$h_r(T_r - T_a) = \sigma\varepsilon_r(T_r^4 - T_a^4)$$

where, σ = Boltzman Plank's constant

$= 5.67 \times 10^{-8}$ W/m^2 °C $= 4.876 \times 10^{-8}$ kcal/m^2 hr k^4

E_r = emissivity of the surface.

T_a = ambient temperature °C

$$h_r = \frac{\sigma\varepsilon_r\left(T_r^4 - T_a^4\right)}{T_r - T_a}$$

$$= \sigma\varepsilon_r(T_r + T_t)(T_r^2 - T_a^2)$$

$$T_r^2 - T_a^2 = (T_r + T_a)^2 - 2\,T_r\,T_a \simeq (2\overline{T})^2 - 2\overline{T}^2 = 2\overline{T}^2$$

where, $\overline{T} = \dfrac{T_r + T_a}{2}$, and it is assumed that T_r and T_a do not differ much.

$$\therefore \quad h_r = \sigma\varepsilon_r\, 2\,\overline{T}\, 2\,\overline{T}^2$$

$$= 4\sigma\varepsilon_r\overline{T}^3 \quad ...(3.41)$$

Since the heat flux in focusing system is high, the heat transfer resistance from the outer surface of the receiving tube to the fluid should include the tube wall. The overall heat transfer coefficient (based on the outside tube diameter) from the suroundings to the fluid is;

$$U_o = \left[\frac{1}{U_L} + \frac{1}{\frac{h_i D_i}{D_o}} + \frac{\frac{l_n \frac{D_o}{D_i}}{2k}}{D_o} \right]^{-1}$$

$$= \left[\frac{1}{U_L} + \frac{D_o}{h_i . D_i} + \frac{D_o\, l_n \frac{D_o}{D_i}}{2k} \right]^{-1} \quad \text{...(3.42)}$$

where, D_o and D_i are outer and inner diameters respectively,

h_i = the inside film heat transfer coefficient determined from forced convection (or natural convection) formulae.

k = the thermal conductivity of the tube material.

In terms of the energy transfer to the fluid at T_f,

$$q_u = U_L \pi D_0 (T_r - T_f)$$

$$= \frac{\pi D_o (T_r - T_f)}{D_o / h_i D_i + \frac{D_o l_n D_o / D_i}{2k}} \quad \text{...(3.43)}$$

where, U_L is the heat transfer coefficient from tube to fluid given by

$$U_L = \frac{1}{\frac{h_i D_i}{D_o}} + \frac{\frac{l_n D_o / D_i}{2k}}{D_o}$$

Eliminating T_r from equations (3.39) and (3.43)

$$q_u = \frac{U_o A_a}{U_L . L} \left[S - \frac{A_r}{A_a} U_L (T_f - T_a) \right]$$

where, $S = H_b R_b \, \rho \gamma \tau \alpha = H_b R_b \eta_{opt.}$

Optical efficiency is given by

$$\eta_{opt.} = \rho \gamma (\tau . \alpha)$$

$\because$ $\quad A_r = \pi D_o L$

Rewriting the above expression for q_u

$$q_u = F' \frac{A_a}{L} \left[S - \frac{A_r}{A_a} U_L (T_f - T_a) \right] \quad \text{...(3.44)}$$

where, F is the collector efficiency factor

$$F' = \frac{U_o}{U_L} = \frac{1/U_L}{\frac{1}{U_L} + \frac{D_o}{h_i . D_i} + \frac{D_o \, l_n \, D_o / D_i}{2k}} \quad \text{...(3.45)}$$

Energy balance for flow element gives (Fig. 3.28)

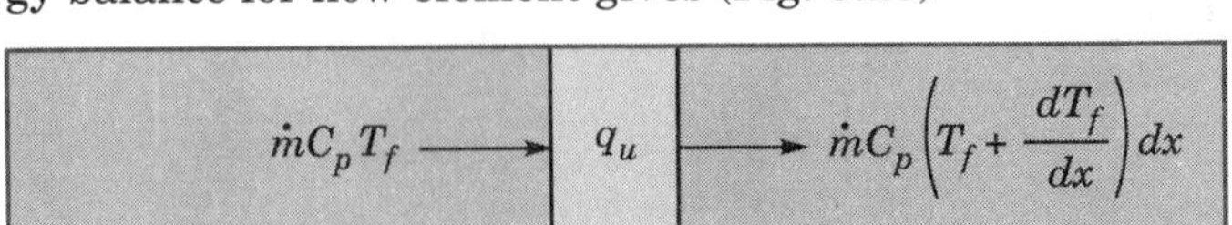

Fig. 3.28

$$\dot{m}C_P\left(T_i + \frac{\delta T_f}{dx}dx\right) - \dot{m}\,C_p\,T_f = q_u\,dx.$$

q_u is the energy per unit length.

or $$\dot{m}\,C_P\,\frac{\delta_{Tf}}{dx} = q_u = F'\,\frac{A_a}{L}\left[\left(S - \frac{A_r}{A_a}U_L\right)(T_f - T_a)\right]$$

or $$\dot{m}C_P\,\frac{d_{Tf}}{dx} - F'\,\frac{A_r}{L}\left[\frac{A_a}{A_r}S - U_L(T_f - T_a)\right] = 0$$

or $$\frac{d_{Tf}}{dx} + \frac{F'\,A_r\,U_L}{\dot{m}\,C_P L}\left[T_f - T_a - \frac{A_a}{A_r}\frac{S}{U_L}\right] = 0$$

or $$\frac{d}{dx}\left(T_f - T_a - \frac{A_a}{A_r}\frac{S}{U_L}\right) + \frac{F'\,A_rU_L}{\dot{m}C_PL}\left[(T_f - T_a) - \frac{A_a}{A_r}\frac{S}{U_L}\right] = 0$$

[The above equation is of the form $\frac{dy}{dx}\;K_1\,y = 0$, solution of which is of the form $y = K_2\,e^{-K_1x}$]

∴ Solution of above equation is

$$T_f - T_a - \frac{A_a}{A_r}\frac{S}{U_L} = C\,e^{\frac{-F'\,A_r\,U_L}{\dot{m}\,C_P\,L}} \qquad ...(3.46)$$

Boundary conditions:

(*i*) $T_f(at\ x = 0) = T_{fi}$ (*ii*) $T_f(at\ x = L) = T_{fo}$

T_{fo} = outer temperature of the fluid.

First boundary condition gives [from equation (3.46)].

$$T_{fi} - T_o - \frac{A_a}{A_r}\frac{S}{U_L} = C$$

i.e., $$\frac{T_f - T_a - \dfrac{A_a\,S}{A_r\,U_L}}{T_{fi} - T_o - \dfrac{A_a\,S}{A_r\,U_L}} = e^{\frac{-F'\,A_r\,U_L}{\dot{m}\,C_P\,L}.x} \qquad ...(3.47)$$

Second boundary condition yields

$$\frac{T_{fo} - T_a - \dfrac{A_a}{A_r}\dfrac{S}{U_L}}{T_{fi} - T_a - \dfrac{A_a}{A_r}\dfrac{S}{U_L}} = e^{\frac{-F'\,A_r\,U_L}{\dot{m}\,C_P}} \qquad ...(3.48)$$

Collector heat removal factor F_R is defined as,

$$F_R = \frac{\text{Actual gain}}{\text{Gain if collector surfaces were at inlet temp. } T_{fi}}$$

$$= \frac{\dfrac{\dot{m}}{A_r}C_P(T_{fo} - T_{fi})}{\dfrac{A_a}{A_r}.S - U_L(T_{fi} - T_a)} = \frac{\dot{m}\,C_P}{A_r\,U_L}\left[\frac{(T_{fo} - T_{fi})}{-(T_{fi} - T_a) - \dfrac{A_a}{A_r}\dfrac{S}{U_L}}\right]$$

$$= \frac{\dot{m}\, C_P}{A_r\, U_L}\left[\frac{T_{fo} - T_a - \dfrac{A_o}{A_r}\dfrac{S}{U_L} - \left(T_{fi} - T_a - \dfrac{A_a}{A_r}\dfrac{S}{U_L}\right)}{-\left(T_{fi} - T_o - \dfrac{A_a}{A_r}\dfrac{S}{U_L}\right)}\right]$$

$$= \frac{\dot{m}\, C_P}{A_r\, U_L}\left[1 - \frac{\left(T_{fo} - T_a - \dfrac{A_a}{A_r}\dfrac{S}{U_L}\right)}{\left(T_{fi} - T_a - \dfrac{A_a}{A_r}\dfrac{S}{U_L}\right)}\right] \qquad ...(3.49)$$

Making use of the equation (3.48),

$$= \frac{\dot{m}\, C_P}{A_r\, U_L}\left[1 - e^{\frac{-F'\, A_r\, U_L}{\dot{m}\, C_P}}\right] \qquad ...(3.50)$$

Total gain,

$$Q_u = F_R A_a\left[S - \frac{A_r U_L}{A_a}\left(T_{fi} - T_a\right)\right] \qquad ...(3.51)$$

Let F'' = collector flow factor

$$= \frac{F_R}{F'} = \frac{\dot{m}\, C_P}{A_r\, U_L\, F'}\left[1 - e^{\frac{-A_r\, U_L\, F'}{\dot{m}\, C_P}}\right] \qquad ...(3.52)$$

Collector efficiency is given by,

$$\eta_c = \frac{Q_u}{H_b\, R_b \times A_a} \qquad ...(3.53)$$

In equation (3.53) shading effect is neglected. The same analysis will be followed to a receiver that is covered, with effective value of transmittance-absorptanee product in the value of S.

Example 3.2. *Calculate the useful gain, exit fluid temperature and collection efficiency for a cylindrical parabolic concentrator system of 2 m width and 8 m length. The absorbing cylinder has a diameter of 6 cm and the transparent cover has a diameter of 9 cm optical properties are estimated as*

$$\rho = 0.85,\ (\tau\,.\,\alpha) = 0.77,\ \gamma = 0.94$$

Heat transfer coefficient from fluid inside to surroundings,

$$U_o = 5.2 \text{ kcal/hr m}^2\ °C\ (6.04 \text{ W/m}^2\ °C),$$

Heat transfer coefficient from absorber cover surface to surroundings

$$U_L = 6.0 \text{ kcal/hr-m}^2\ °C\ (6.98 \text{ W/m}^2\ °C).$$

The incident beam radiation on the aperture of the collector is 600 kcal/hr m² (698 W/m²) and the ambient temperature is 25°C. The collector is designed to heat a fluid, entering the absorber at 150°C, at a flow rate of 400 kg / hr. The fluid has C_p = 0.30 kcal/kg °C (1.256 kJ/kg °C).

Solution. A_r = area of the receiver pipe

$= \pi D_0 L$

$= \pi \times 0.06 \times 8 = 1.51$ m²

A_a = aperture area of the concentrator

$= (W - d_{co}) \times L$

where, W = width of concentrator

d_{co} = diameter of transparent cover

L = length of concentrator

$A_a = (2 - 0.09) \times 8 = 15.28 \text{ m}^2$

= collector efficiency factor,

$$F' = \frac{U_o}{U_L} = \frac{5.2}{6} = 0.86$$

Heat removed factor F_R is

$$= \frac{\dot{m}\, C_P}{A_r\, U_L}\left[1 - e^{-\frac{A_r\, U_L\, F'}{\dot{m}\, C_P}}\right]$$

$$\frac{\dot{m}\, C_p}{A_r\, U_L\, F'} = \frac{400 \times 0.30}{1.51 \times 6 \times 0.86} = 15.4012$$

$$F_R = 15.4012 \times 0.86 \times \left(1 - e^{-\frac{-1}{15.4012}}\right) = 0.8329.$$

The absorbed solar energy is

$$S = H_b R_b\, \rho\gamma\, (\tau.\alpha)$$
$$= 600 \times 0.85 \times 0.94 \times 0.77$$
$$= 370 \text{ kcal/hr m}^2 \text{ ...... (MKS)}$$
$$= 430 \text{ W/m}^2 \text{ ...... (SI).}$$

The values of F', F_R will be same in any unit, since they are factors (dimensionless).

Useful gain,

(In M.K.S. unit) $Q_u = A_a \times F_R \left[S - \frac{A_r\, U_L}{A_a}\left(T_{fi} - T_a\right)\right]$

$$= 15.28 \times 0.8326\left[370 - \frac{1.51 \times 6}{15.28}(150 - 25)\right]$$

= 3750 kcal/hr. **Ans.**

(In S.I. unit) $Q_u = 15.28 \times 0.8326\left[430 - \frac{1.51 \times 6.98}{15.28}(150 - 25)\right]$

= 4360 watt. **Ans.**

The exit fluid temperature can be obtained from

$$Q_u = \dot{m}\, C_p\, (t_{co} - t_{ci})$$

where, t_{co} = collector fluid temperature at outlet

d_{ci} = Fluid inlet temperature

$$t_{co} = t_{ci} + \frac{Q_u}{\dot{m}\, C_p}$$

$$= 150 + \frac{3750}{400 \times 0.30} \text{ or } \left[150 + \frac{4360}{400 \times \frac{1.256}{3.6}}\right]$$

= 181°C. **Ans.**

$$\eta_{collection} = \frac{Q_u}{A_a H_b R_b}$$

$$= \frac{3750}{16 \times 600} \times 100 \text{ or } \left(\frac{4360}{16 \times 698} \times 100\right)$$

$= 39\%$. **Ans.**

3.10. Selective Absorber Coatings

An efficient way to reduce thermal losses from the absorber plate of a solar heating panel is by using selective absorber coatings. An ideal selective coating is one that is a perfect absorber of solar radiation while being a perfect reflector of thermal radiation. Such a coating will make a surface, a poor emitter of thermal radiation. Hence a selective coating increases the temperature of an absorbing surface. If back losses of an absorbing surface are absent, the steady-state conditions give:

Solar flux absorbed = Thermal flux emitted.

The absorptance and emittance of radiation at a given wavelength are equal. However, at different wavelengths they can vary from near zero to near unity. Since 96% of the sun's radiation is concentrated in wavelength ranges of less than 2.5 (μm and 99% of the reradiation from a collector surface (operated at less than 400°K) is in wavelengths of more than 2.5 μm, it is possible to have a surface that will absorb all of the solar radiation while emitting very little.

A *selective surface* is a surface that has a high absorptance for short wave radiation (less than 2.5 μm) and a low emittance of long-wave radiation (more than 2.5 μm). Although a large number of experimental selective surface treatments and coatings have been tested, only a few have survived the laboratory.

A large number of non-selective coatings are available and are in widespread use as flat-plate collector coatings. These are primarily organic coatings such as *flat black paints*. Most of these coatings have absorptances exceeding 0.95 and emittances of 0.90–0.95. Although the emittances are high, use of these coatings may be justified economi-cally in some applications where high collector temperatures are not required, such as hot water systems or swimming pool heaters. Table 3.1 lists properties of some selective coatings.

Table 3.1. Selective coating properties

Coating	*Type*	*Absorptance* α	*Emittance* ϵ_p
Blank chrome	Electroplated	0.96	0.10
Black nickel	Electroplated	0.90	0.10
Black copper	Copper oxide	0.87–0.92	0.07–0.35
Black anodize	Aluminium oxide	0.94	0.07
Solar foil	Black chrome over copper	0.96	0.10
Enersorb (non-selectove)	Urethane paint	0.97	0.90
Nextel (non-selective)	Paint	0.98	0.89

Most surfaces that are good absorber for solar radiation are also good radiators for heat. If, for example, a non-selective surface has an absorptance of 0.95 for radiation effectively while at the same time radiating little heat. Most selective surfaces are composed of a very thin black metallic oxide on a bright metal base. The black oxide coating is thick enough to act as a good solar absorber, with an absorptivity as high as 0.96, but it is essentially transparent to long-wave thermal radiation emitted by an object at a temperature of several hundred degrees. Since bright metals have low emissivity for thermal radiation, that is, are poor heat radiators, and since the thin oxide coating is transparent to such radiation, the combination is a poor radiator of heat. As a result, the radiation loss from a selective surface is considerably lower than from a conventional, non-selective surface. The overall heat loss coefficient U_L is reduced when this type of surface is used.

A selective surface should also posses the following charac-teristics in addition to the above mentioned spectral characteristics:

(*i*) Its properties should not change with use. A significant degradation in the values of α and ϵ_p has been observed in the case of some selective coatings, largely due to the effect of exposure to atmospheric humidity.

(*ii*) It should be able to withstand the temperature levels associated with the absorber plate surface of a collector over extended periods of time. It should be able to withstand any short-term temperature rise which may occur when no useful heat is being removed.

(*iii*) It should be able to withstand atmospheric corrosion and oxidation.

(*iv*) It should be of reasonable cost.

QUESTIONS

1. Explain the principle of conversion of solar energy into heat.
2. What are the main components of a flat-plate solar collector, explain the function of each?
3. How solar air collectors are classified? What are the main applications of a drier?
4. Data for a flat-plate collector used for heating are given below:

Factor	*Specification*
Location and latitude	Coimbatore 11° 00′ N
Day and time	22 March, 14.30 – 15.30 (LST)
Average intensity of solar radiaton	...560 W/m^2
Collector tilt	...26°
No. of glass cover	...2
Heat removal factor for collector	...0.82
Transmittance of glass	...0.88
Absorptance of the plate	...0.93
Top loss coefficient (U_L) for collector	...7.95 W/m^2°C
Collector fluid temperature	...75°C
Ambient tempeature	...25°C

Calculate:

(*i*) Solar attitude angle, (*ii*) Incident angle, and

(*iii*) Collector efficiency. [**Ans.** 43.9°, 46.1°, 8.87%]

5. Enumerate the different types of concnetrating type collectors.
   Describe a collector used in power plant for generation of electrical energy.
6. Why orientation is needed in concentrating type collectors?
   Describe the different methods of sun tracking.
7. What are the advantages and disadvantages of concentrating collectors over a flat-plate collectors?

CHAPTER 4

Solar Energy Storage

4.1. Introduction

Solar energy is a time dependent and intermittent energy resource. In general energy needs or demands for a very wide variety of applications are also time dependent, but in an entirely different manner from the solar energy supply. There is thus a marked need for the storage of energy or another product of the solar process, if the solar energy is to meet the energy needs. This problem is specially severe for solar energy when it is used for beating in winter, because of its low availability during this period.

The need for energy storage of some kind is almost immediate evident for a solar electric system. An optimally designed solar-electric system will collect and convert when the insolation is available during the day. Unfortunately the time when solar energy is most available will rarely coincide exactly with the demand for electrical energy, though both tend to peak during the day light hours. There is also the problem of clouds with photovoltaic plants, and cloud cover for several days may result in substantially lowered electrical output compared to high insolation cloud-free days. Obviously during such days energy previously stored during high insolation times could be used to provide a continuous electrical output or thermal output. Thus the addition of storage can increase the reliability of being able to deliver power at an arbitrary needed time.

Storage of solar energy in a solar system may:

1. Permit solar energy to be captured when insolation is highest and then later used when the need is greatest. It can thus transform a diurnal solar energy input into a more uniform desired electrical or thermal output.

2. Make it possible to deliver electrical load power demand during times when insolation is below, normal or non-existent. Storage also makes it possible to deliver short peaks of power for exceeding the rated power capacity of the plant.

3. Be located close to the load, thereby minimizing the need for costly transmission and distribution facilities which would otherwise be required to meet peak load demands when storage not present there.

4. Improve the reliability of the solar thermal as well as solar electric system.

5. Permit a better match between the solar energy input and the load demand output than would be the case without storage.

The optimum capacity of an energy storage system depends in general, on the following factors:

(*i*) The expected time dependence of solar radiation availability.

(*ii*) The nature of loads to be expected on the process.

(*iii*) The degree of reliability needed for the process.

(*iv*) The manner in which auxiliary energy is supplied.

(*v*) The size of the solar thermal power system or solar-electric generator.

(*vi*) The cost per kWH of the stored energy.

(*vii*) The permissible capital cost allocated to storage.

(*viii*) Environmental and safety considerations.

(*ix*) An economic analysis that determines how much of the total usually annual loads should be carried by solar and how much by auxiliary energy sources.

Energy storage may be in the form of sensible heat of solids or liquid medium, as heat of fusion in chemical systems or as chemical energy of products in the reversible chemical reaction. Mechanical energy can be converted to potential energy and stored in elevated fluids. Product, of solar processes other than energy may be stored in tanks until needed.

The choice of media for energy storage depends on the nature of the process. For water heating, energy storage as sensible heat of stored water is logical. If air heating collectors are used, storage is sensible or latent heat effects in particular storage units are indicated, such as sensible heat in a pebble bed heat exchanger. If photovoltaic or photo chemical processes are used, storage mostly logical in the form of chemical energy or energy is stored in the batteries. Thus the energy may be stored in a variety of forms, *e.g.*, as heat, electrical, chemical, mechanical and magnetic.

4.2. Solar Energy Storage Systems

Solar energy storage systems may be broadly classified as under:

1. **Thermal Storage.** Energy can be stored by heating, melting or vaporization of material, and the energy becomes available as heat, when the process is reversed. Storage by causing a material to rise in temperature is called *sensible heat storage*. Storage by phase change, the transition from solid to liquid or from liquid to vapour is another mode of thermal storage, known as the *latent heat*

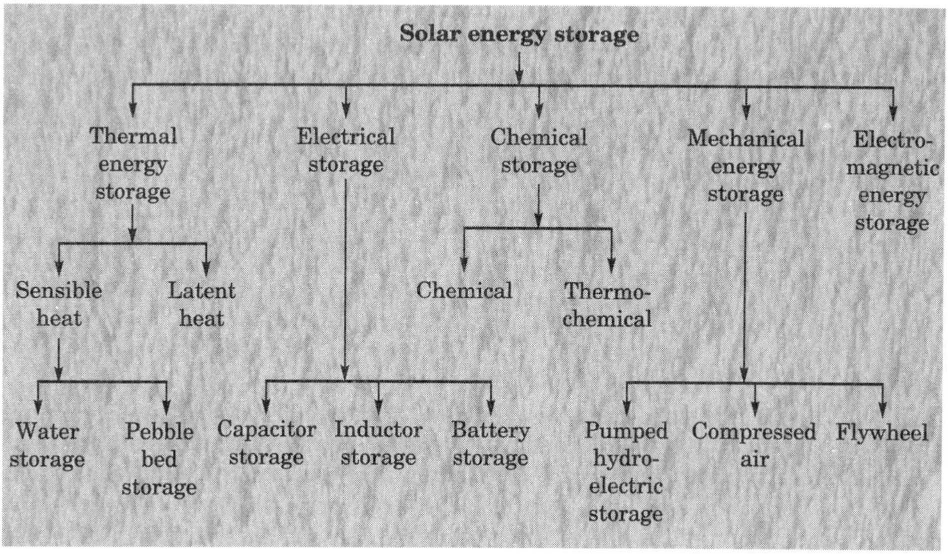

storage, in which no temperature change is involved. It is possible for both sensible and latent heat-storage to occur in the same material, as when solid is heated, then melted, then raised further in temperature.

Thermal energy storage is essential for both domestic water and space heating applications and for the high temperature storage systems needed for thermal power applications. Storage is also required in the process industries and horticultural. The choice of the storage material depends on the particular application and for many domestic applications, water and/or rock storage systems have been developed. Water and rock are typical examples of material which store energy as specific heat (sensible heat), but their use is limited by their comparatively low specific heats. The heat of fusion (latent heat) which is involved when a substance changes state from a solid to a liquid, provides an attractive method of storing a given amount of heat within a much smaller volume. Glauber's salt (Na_2, $SO_4.10H_2O$) is the least expensive and most readily available salt hydrate. For high temperature storage in the order of 200 to 300°C other salts have been used and the heat of hydration of inorganic oxides principally MgO and CaO.

(*a*) **Sensible heat storage.** Sensible heat storage involves a material that undergoes no change in phase over the temperature domain encountered in the storage process. The basic equation for an energy storage unit, operating over a finite temperature difference is:

$$Q_s = (m\ C_p)_s\ (T_1 - T_2) \quad ...(4.1)$$

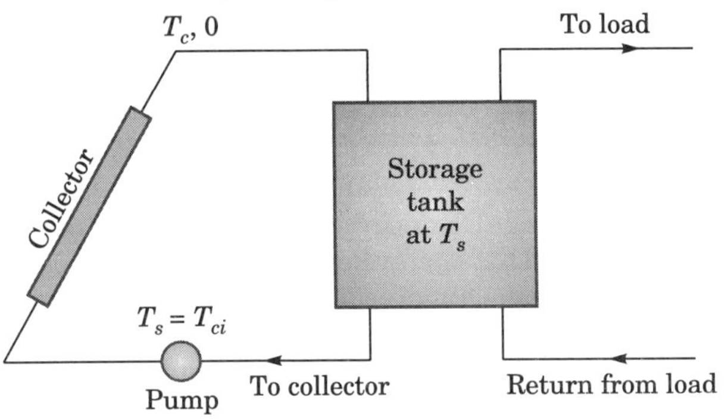

Fig. 4.1. Water tank storage unit. Energy is added by circulating water through collector and is removed by circulating water through load.

where Q_s is the total thermal energy capacity for a cycle operating between temperature limits T_1 and T_2 with m kg of storage medium of specific heat C_p. The temperature range over which such a system can operate is limited at the lower extreme by the requirements of the process, and the upper limit by the process of the vapour pressure of the liquid. The ability of store thermal energy in a given container of volume V is,

$$\frac{Q_s}{V} = \rho\ C_p\ \Delta T \quad ...(4.2)$$

where ρ is the density of the storage medium.

Thus the ability to store heat depends upon the product $\rho\ C_p$ water has the highest value, but other materials have low values. Materials which are generally used for this type of storage are (*i*) Water, (*ii*) Rock, gravel or crushed stone, (*iii*) Iron shot, (*iv*) Iron, red iron oxide or iron ore (magnetic), (*v*) Concrete, (*vi*) Refractory materials like magnesium oxide, aluminium oxide (alumina) and silicon oxide.

(A) Water storage. The most common heat transfer fluid for a solar system is water, and the easiest way to store thermal energy is by storing the water directly in a well insulated tank. The optimum tank size for flat-plate collector system is usually about 70 kg/m^2 (2 gal/ft^2). Water has the following characteristics for storage medium.

(*i*) It is an inexpensive, readily available and useful material to store sensible heat.

(*ii*) It has high thermal storage capacity.

(*iii*) Energy addition and removal from this type of storage is done by medium itself, thus eliminating any temperature drop between transport fluid and storage medium.

(*iv*) Pumping cost is small.

(B) Packed Bed Exchanger Storage. For sensible heat storage with air as the energy transport mechanism, rock, gravel, or crushed stone in a bin has the advantage of providing a large, cheap heat transfer surface. Its thermal capacity, however, is only about half that of water, and the bin volume will be about 3 times the volume of a water tank, that is heated over the same temperature interval. Water is superior because of its lower material cost and lower volume required per unit of energy stored. Rock does have the following advantages over water:

1. Rock is more easily contained than water.
2. Rock acts as its own heat exchanger, which reduces total system cost.
3. It can be easily used for thermal storage at high temperatures, much higher than 100°C; storage at high temperature where water cannot be used in liquid form without an experience, pressurized storage tank.
4. The heat transfer coefficient between the air and solid is high.
5. The cost of storage material is low.
6. The conductivity of the bed is low when air flow is not present.

A packed bed storage unit is shown schematically in Fig. 4.2, which uses the heat capacity of a bed of loosely packed particulate, material through which a fluid, usually air is circulated to add or remove heat from the bed. Essential features of a packed bed storage unit are a container, porous structure to support the bed and air distributors. In operation the flow is maintained through the bed in one direction during addition of heat and in the opposite direction during removal of heat. In this system, the heat addition and removal from the storage can not be carried out simultaneously. Insulation requirements at the outer surface of the packed bed are minimal, for short term storage, as the thermal conductivity in the radial direction is low.

Pebble bed exchanger have good heat transfer characteristics between air and the solids of the bed. This type of storage system has been used in the solar houses or with hot air collector system.

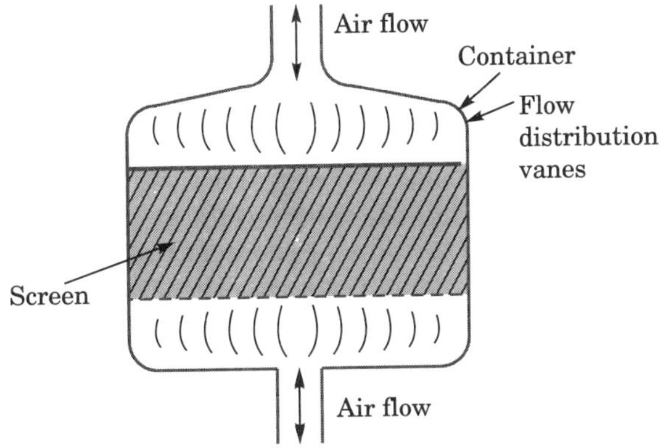

Fig. 4.2. Schematic of a packed bed storage unit.

(*b*) **Latent heat storage (phase change energy storage)**

In this system, heat is stored in a material when it melts and extracted from the material when it freezes. Materials that undergo a change of phase in a suitable temperature range may be useful for energy storage if the following criteria can be satisfied:

(*i*) The phase change must be accompanied by high latent heat effect.

(*ii*) The phase change must be reversible over a very large number of cycles without degradation.

(*iii*) The phase change must occur with limited super cooling.

(*iv*) Means must be available to contain the material and transfer heat into it and out of it.

(*v*) The cost of materials and its containers must be reasonable.

(*vi*) Its phase change must occur close to its actual melting temperature.

(*vii*) The phase change must have a high latent heat effect, that is, it must store large quantities of heat.

(*viii*) The material must be available in large quantities.

(*ix*) The preparation of the phase changing material for use must be relatively simple.

(*x*) The material must be harmless (non-toxic, non-inflammable, non-combustible, non-corrosive).

(*xi*) A small volume change during the phase change.

(*xii*) The material should have high thermal conductivity in both the phases.

If these criteria can be met, phase change energy storage systems can be of high capacities (relative to energy storage in sensible heat type systems) when operated over small temperature ranges, with substantial reduction in volume and weight.

Materials for phase change energy storage. There are several materials that undergo a change of phase. Glauber's salt (Na_2SO_4. $10H_2O$), water, $Fe(NO_3)_2$ $6H_2O$, and salt Eutectics are mostly used.

Glauber's ($Na_2SO_4.10H_2O$) (sodium sulphate decahydrate) phase changes from solid to liquid requires less energy than those from liquid to gas, but some solid to liquid changes, still provide useful amount of storage potential. Phase change materials involving water of dehydration have long been experimented with solar energy. One of these is Glauber's salt, which decomposes at about 32°C to given a solution plus Na_2SO_4 with a heat of fusion of 243 kJ per kg (56 kcal/kg) and has been proposed *mainly for storing domestic heat.* The reaction is

$$Na_2SO_4.\ 10H_2O + \text{energy} \rightleftharpoons Na_2SO_4 + 10H_2O$$

Energy storage is accomplished by the reaction proceeding from left to right on addition of heat. Energy extraction from storage is the reverse procedure, with the reaction proceeding from right to left and thermal effects reversed. It has been found that the performance decades on repeated cycling, with the thermal capacity of the system reduced. $Na_2SO_4.10H_2O$ has an incongruent melting point, and has its temperature increases beyond the melting point, it separates into liquid (solution) phase and solid Na_2SO_4. Since density of salt is higher than the density of the solution, a phase separation occurs. Attempts have been made to use gels or other agents to avoid phase separation. The above difficulties have been overcome by keeping the material in long thin containers and by mixing certain additives as mentioned.

A latent heat storage arrangement is shown in Fig. 4.3, in which, the storage material is placed in long thin containers, *e.g.*, cylinders, and the gas is passed through narrow spaces between the tubes.

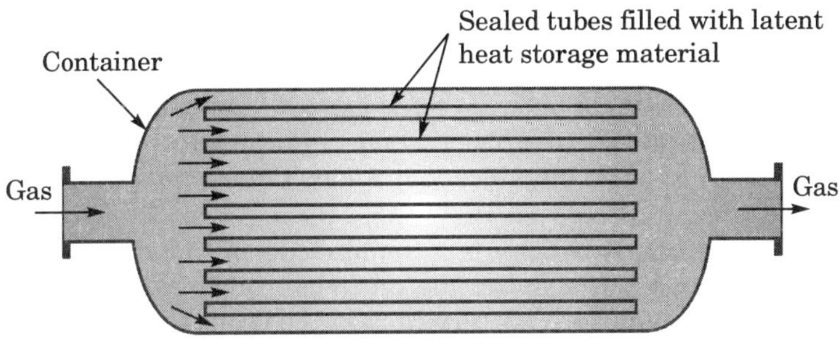

Fig. 4.3. A typical latent heat storage arrangements.

An advantage associated with this storage system is that it is more compact than a sensible heat system.

Some *organic compound* or *substances* serve as heat storage materials due to the desirable properties that possess in comparison with inorganic compounds. Some of these advantages include their ability to melt congruently, their self-nucleating properties, their capability to freeze without supercooling and their compatibility with conventional materials of construction. *Paraffin* and *fatty acids,* are the two candidates organic substances qualified as heat of fusion materials. Paraffins possess generally high heats of fusion and are available large temperature ranges. They are known to freeze without supercooling. Fatty acids are known to possess a reproducible melting and freezing behaviour and freeze with little or no supercooling. Hence these qualify as good phase change thermal energy storage materials with a major drawback of higher cost.

Refractory materials (MgO, Al_2O_3, SiO_2) are also suitable for high temperature sensible heat storage in addition to rock or pebble bed storage. Some thermal storage materials such as $ZnCl_2$, $Na(OH)_3$, NaOH, KOH—$ZnCl_2$ KCl—$MgCl_2$—NaCl, $MgCl_2$ NaCl, etc. are also used for the temperature range of 200–450°C.

2. **Electrical Storage.** Theoretically *Capacitors* could store large amounts of electrical energy for long periods. The total energy stored is

$$H_{cap} = \frac{1}{2} V\varepsilon E^2 \qquad ...(4.3)$$

where, V is the volume of the dielectric

ε is constant and

E is the electric field strength.

Electric field strength is limited by the breakdown strength E_{br} of the dielectric. Therefore, the electrical energy storable in a dielectric is limited. At present best dielectric material available is mica.

Practically, because the conductivity of a dielectric is never nil, there will always be leakage losses. At present, capacitive storage is economical for times no longer than 12 hours. To increase this time much research is needed. Therefore, unless better and cheaper dielectric materials for storing electric energy are found, capacitive storage on a large scale will remain uneconomical.

Whereas capacitors store electrical energy at high voltage and low current, inductor storage is at low voltage and high current. The energy stored in an

$$H_{ind} = \frac{1}{2} V\mu H_m^{\ 2} \quad ...(4.4)$$

where, μ is the permeability of the material and

H_m is the magnetic flux density.

For H_{ind} to be large, both μ and H_m should be large. Consequently, high magnetic fields are required. This will create large mechanical forces which should be supported by strong structures. The reverse operation of discharging the stored energy creates another problem, since it involves the opening of circuit carrying large currents.

Battery Storage. A battery is a combination of individual cells. A cell is the elemental combination of materials and electrolyte constituting the basic electrochemical energy storer. A battery can also be thought of as a black box into which electrical energy is put, stored *electrochemically,* and later regained as electrical energy. *Battery storage system may be included under chemical energy storage also.*

Secondary batteries are of chief interest for solar electrics, which are rechargeable, whereas primary batteries are non-rechargeable. Examples of secondary batteries are lead acid, nickel cadmium, iron air, Nickel-hydrogen, zinc-air, sodium sulpur, sodium-chlorine etc.

A generalized cell consisting of two electrodes called the anode and cathode immersed in a suitable electrolyte. When an electrical load is connected between the electrode charge separation occurs at the interface between on electrode and the electrolyte, freezing both an electron and an ion. The electron flows through the external load and ion through the electrolyte, recombining at the other electrode.

Energy efficiency of a battery is defined as (sometimes called watt hour efficiency)

$$\eta_{energy} = \frac{\int_0^{t_1} I_1 E_1 \, dt}{\int_0^{t_2} I_2 E_2 \, dt} \quad ...(4.5)$$

where, I_1 = Battery discharge current

E_1 = Battery discharge terminal voltage

I_2 = Battery charging current

E_2 = Battery charging terminal voltage

t_1 = Battery discharging time

t_2 = Battery charging time.

Two special cases of equation (4.5) are of interest to the system engineers. If the discharge and charge occurs at constant voltages then equations reduces to:

$$\text{Energy efficiency} = \frac{E_1 \int_0^{t_1} I_1 \, dt}{E_1 \int_0^{t_2} I_2 \, dt}$$

$$= \frac{E_1 \times \text{Amp. hour output}}{E_2 \times \text{Amp. hour input}} \quad ...(4.6)$$

The integrals might be evaluated graphically. Once found multiplication by the respective terminal voltage gives the watt hours of input or output. Equation (4.6) suggests why for some types batteries with nearly constant terminal voltage the chief interest might be in the I vs t curve for discharge or charge. The ampere hour output of a battery is a very common way of rating.

If the discharge and charge occurs at constant currents, then equation (4.5) becomes

$$\text{Energy Efficiency} = \frac{I_1 \int_0^{t_1} E_1 \, dt}{I_2 \int_0^{t_2} E_2 \, dt} \qquad ...(4.7)$$

If neither current nor voltage is held constant, then we must resort to equation (4.5) directly and either graphically or electronically evaluate the integrals to determine the energy efficiency for the battery. *Cycle life* is the number of times the battery can be charged and discharged under specified conditions. The cycle life of a battery may vary greatry with the depth of the discharge, deep discharge tending to result in short cycle life. However, deep discharge may be difficult to avoid in some battery applications as an electric vehicles.

3. Chemical Storage

(i) Storage in the form of fuel. It is possible to device a storage battery in which the reactant is generated by a photochemical reaction brought about by solar radiation. In this case the converter itself acts as a storage battery. The battery is charged photochemically and discharged electrically whenever needed. Some of the reactions that could potentially be useful for the storage of solar energy:

$$2NOCl + \text{photons} \longrightarrow 2\,NO + Cl_2$$

$$AgCl\,(s) + \text{photons} \longrightarrow Ag\,(s) + \tfrac{1}{2}Cl_2$$

$$NO_2 + \text{photons} \longrightarrow NO + \tfrac{1}{2}O_2$$

$$H_2C + \frac{1}{2}O_2 \xrightarrow{ZnO} H_2O_2$$

It is also possible to electrolyze water with solar generated electrical energy, store O_2 and H_2 and recombine in a fuel cell to regain electrical energy. Hydrogen is one of the most efficient and practical fuel.

Solar energy could be used by the anaerobic fermentation of algae for the production of methane CH_4. Methane is an excellent fuel which is stable at room temperature and which reacts with oxygen to provide high temperatures, releasing the solar energy in thermal form.

$$CH_4 + 2O_2 \longrightarrow 2H_2O + CO_2$$

Solar energy has been converted into chemical energy of methane with about 2% efficiency, therefore, one square kilometre of algae fields could produce an amount of methane storing 4 MW of converted solar energy. Algae could be grown in nutrient-loaded municipal sewage to produce methane. The grown algae would first have to be harvested, then treated to yield about 5% of solids, the rest being water. The solid material would then be subjected to anaerobic fermentation and after three weeks, half the organic material could be converted into methane.

Photosynthesis has been mentioned as a method of solar energy conversion. The products of reaction are oxygen and H_2CO which is a fraction of carbohydrates, in the presence of light and chlorophyll.

$$CO_2 + H_2O \longrightarrow H_2CO + O_2$$

The carbohydrates are stable at room temperature; but at high temperature the reaction is reversed, releasing the stored energy in thermal form.

(*ii*) *Thermo chemical energy storage (Reversible chemical reactions).* Thermochemical storage systems are suitable for medium or high temperature applications only. For storage of high temperature heat, some reversible chemical reactions appear to be very attractive. Advantages of thermochemical storage include high energy density storage at ambient temperatures for long periods without thermal losses and potential for heat pumping and energy transport over long distances. This type of storage is illustrated by a hypothetical reaction.

$$A + B \rightleftharpoons AB$$

The forward reaction takes place with absorption of heat from solar energy and the heat is stored in the form of products, when heat is desired, the products are to be remixed to allow the reversible reaction to take place with liberation of heat. The original reactant is regenerated completing the cycle. Both forward and reverse reactions takes place at constant but different temperatures, the forward reaction occurring at a higher temperature than the reverse reaction.

Following few chemical reactions are illustrated here from the stand point of their suitability based on the above criteria and are found suitable for use in solar applications.

Reaction	*Temp. of forward reaction °C*	*Temp, of reverse reaction °C*	*Energy stored per unit volume of storage material kg/m³*
(1)	(2)	(3)	(4)
$CH_4 + H_2O \rightleftharpoons CO + 3H_2$	780	610	209.4×10^3
$SO_3 \rightleftharpoons SO_2 + ½ O_2$	1025	590	460.6×10^3
$NH_4HSO_4 \rightleftharpoons NH_3 + H_2O + SO_2$	498	435	2143.7×10^3

In the above table first reaction is methane-syngas reaction, which is occurred in the presence of a nickel catalyst. Suitability of the reaction is that the technology required is available. Moreover in the reverse reaction (the methanation process) is being commercially used for the manufacture of methane on a large scale. The main problem is that the products CO and H_2 are gases and have to be stored at a pressure of about 10^7 N/m² (100 bars).

Second reaction is of decomposition of sulphur trioxide to sulphur dioxide and oxygen in the forward direction and recombination of sulphur dioxide and oxygen to form sulphur trioxide in the reverse direction, in the presence of a catalyst. This storage system has been suggested for use in 100 MW central tower solar power plant operating on Brayton cycle with helium as the working fluid.

(*iii*) *Hydrogen storage.* Energy can be both stored and transported as hydrogen, which serves as a secondary fuel. On wind-electric or photovoltaic systems, for example, with a d.c. output power can be fed directly into an electrolyzer tank which produces hydrogen and oxygen from ordinary water. The gases may be produced either under pressure or near atmospheric pressure and then *via* external pumps compressed to the desired pressure. The latter approach, however, requires auxiliary energy. The hydrogen and oxygen gas produced can be stored either in gas or liquid forms for a long time. It can, when needed, be quickly and easily converted again directly into electrical energy *via* the well known fuel cell. The system thus effectively stores the sun's energy as hydrogen and oxygen, and from this storage a smooth reliable power output may be taken for a limited time set by the hydrogen storage capacity.

Hydrogen can also be used as a fuel to drive automobiles and other useful engines, and there appears to be a growing number of people within the technical community advocating this application of hydrogen. They have extended this idea to encompass the entire energy system of the nation, and one hears and sees frequent references to an all hydrogen economy. Hence the chemical energy in hydrogen (and oxygen) can be converted into thermal, mechanical, or electrical energy.

4. Mechanical Energy Storage

(*i*) **Pumped hydroelectric storage.** Electric power in excess of the immediate demand is used to pump water from a supply (*e.g.,* like, river, or reservoir) at a lower level to a reservoir at a higher level. When the power demand exceeds the supply, the water in allowed to flow buck down through a *hydraulic turbine* which drives an electric generator. The overall efficiency of the *pumped storage,* that is, the percentage of the electrical energy used to pump the water is recovered as electrical energy is about 70%. The pumped hydroelectric storage is the most economical means presently available to electric utilities. Hence solar and wind energy in electrical form can be used by this system. There are relatively few suitable sites where there is a water supply at a lower level and a reservoir can be constructed at a higher level, but the use of natural or excavated underground caverns or excavated underground caverns as lower reservoirs, now being developed, should greatly increase the number of possible sites.

This method of solar energy storage has the advantage of avoiding the intermittent inherent to solar energy and of producing a.c. power directly. In this scheme, water acts only as the working fluid, once the initial volume of water is secured, the only flow needed would be that necessary to make up for the loss of water by evaporation (and seepage). Therefore, the scheme does not have to be near large bodies of water.

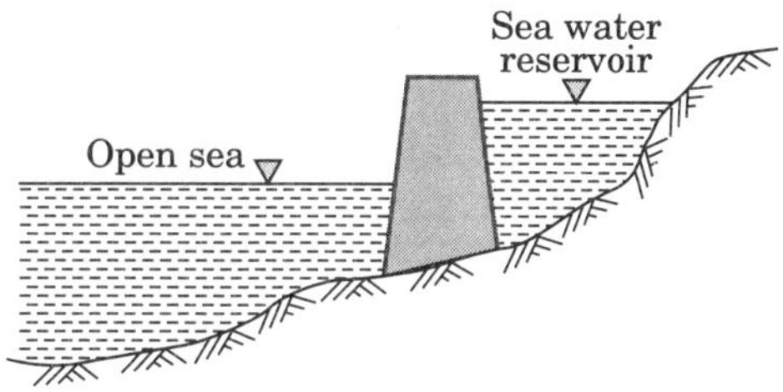

Fig. 4.4. Storage of solar energy by pumping water behind a dam.

(*ii*) **Compressed Air Storage.** Another alternative, for possibly storing energy in future solar-electric system is to store the energy in a volume of compressed air. A wind turbine, for example, could be created which would directly pump air into a suitable pressurized storage tank. Then later when the wind is not blowing the energy stored in the air could be utilized to drive an air turbine, whose shaft would then drive a generator, thus supplying the needed electric power when the wind is not blowing. Compressed air storage may also be applicable to other solar-electric conversion systems.

(*iii*) **Flywheel Storage.** A *flywheel* driven by an electric motor during off peak hours stores mechanical (rotational) energy as its speed is increased. The rotation of the flywheel can be used to operate a generator to produce electricity when required. The same machine serves as both a motor, when electricity is supplied, and as a generator, when the armature is rotated by the flywheel. Flywheel could conceivably be used for electric utility peaking units, for a storage of solar and wind energy, and for vehicle propulsion. The energy recovery efficiency is estimated to be upto 90 percent. Practical flywheel systems for such uses are being developed. The basic idea of flywheel energy storage, sometimes referred to as a 'super flywheel', is to accelerate a suitably designed physical rotor to a very high speed in a vacuum, as *via* an electric motor, at which state high energy storage densities are achieved. The energy is stored as kinetic energy, most of which can be electrically regained when the flywheel is run as a generator as stated above.

5. Electromagnetic energy storage (or energy storage via superconducting magnets). Electromagnetic energy storage requires the use of superconducting materials; these materials (metals and alloys) suddenly lose essentially all resistance to the flow of electricity when cooled below a certain very low temperature. If maintained below this temperature a superconducting metal (or alloy) can carry strong electric currents with little or no loss. Useful super conducting materials available commercially are a niobium titanium (N_b — T_i) alloy at temperatures below—263°C and a compound of niobium and tin (Nb_3Sn) below—255°C.

An electromagnetic field such as is produced by an electric current flowing through a wire (or wire coil), can store energy. Under ordinary conditions, losses result from the resistance of the wire, and energy must be supplied continuously to maintain the current. But if the wire coil were made of a superconducting material and kept at the required low temperature, resistance losses would be very small and, once initiated, an electric current would remain almost constant. Electrical energy supplied as direct current to the wire coil would then be stored in the electromagnetic field. By attaching the coil to a load, the stored energy could be recovered as electrical energy (direct current). This is the basis of the superconducting Magnetic Energy Storage (SMES) system under study of or possible eventual use in electric utility load levelling and/or peaking applications.

A number of problems are associated with superconducting storage. They include the operation and maintenance of a cryogenic (*i.e.*, low temperature refrigeration) plant for producing the liquid helium required for the very low superconductivity temperatues. Furthermore, special structures would be needed to withstand the strong magnetic field of an SMES unit. It appears at present that superconductivity storage would, at best, be economical only in large installations. On the whole SMES is rated as being a promising but long term prospect for energy storage.

4.3. Solar Pond

4.3.1. Introduction. A natural or artificial body of water for collecting and absorbing solar radiation energy and storing it as heat. Thus a solar pond combines solar energy collection and sensible heat storage.

The simplest type of solar pond is very shallow, about 5 to 10 cm deep, with a radiation absorbing (*e.g.*, black plastic) bottom. A bed of insulating material under the pond minimizes loss of heat to the ground. A curved cover, made of transparent fibre glass, over the pond permits entry of solar radiation but reduces losses by radiation and convection (*i.e.*, air movement). In a suitable climate, all the pond water can become hot enough for use in space heating and agricultural and other processes.

In shallow solar pond, as described above, the water soon acquires a fairly uniform, temperature. However, experience shows that the water in such a pond usually heats up only a few degrees, because of the natural convection currents which are set into motion as soon as heat is absorbed at the bottom. In a deeper pond also temperature variations generally exist. Loss of heat from the surface, especially at night, then results in circulation of water by convection. The situation is changed if the pond contains salt water at the bottom with a layer of fresh water above it. Because of its salt content, the solar pond bottom water is more dense than the cooler fresh water at the top, and hence it does not tend to rise. A relatively stable layer of heated salt water is thus produced at the bottom of the pond with a lighter layer of cooler fresh water, which acts as a heat insulator, above it. Thus a 'solar pond' is defined as an artificially constructed pond in which significant temperature rises are caused to occur in the lower regions by preventing convection. The more specific terms 'salt-gradient solar pond or 'non-convecting solar pond' are also often used, as to distinguish these ponds from 'shallow solar pond'.

Solar ponds promise an economical way over flat-plate collectors and energy storage by employing a mass of water for both collection and storage of solar energy. The energy is stored in low grade (60 to 100°C) thermal form which, in self, might be suitable for a variety of applications such as space heating, industrial process heat and to obtain mechanical and/or electrical energy.

Ponds have been studied experimentally and analytically at the National Physical Laboratory of Israel and by others. Tabour (1964) outlines the general concept and the major problems. Tabour and Merz (1965) carried out theoretical investigations of underlying physics of the solar pond and laboratory and field tests, to study the many factors affecting pond performance. At MIT, Stolzenbach (1968) developed numerical methods to predict temperature distributions within the solar pond. A group of Russian scientists at the Uzbek SSR Academy of Science (1973) has been actively engaged in very detailed and sophisticated research of the relevant physics. In India Dr. G.C. Jain (1973) has designed and is operating a solar pond for use in production of salt at the Central Salt and Marine Chemical Research Institute in Bhavnagar. Work has also been done at Pondicheri, where experimental studies have been conducted on a pond 100 m^2 in area and 2 m deep.

4.3.2. Principle of operation and description of non-convective solar pond. A solar pond is a mass of shallow water about 1 or 2 metres deep with a large collection area, which acts as a heat trap. It contains dissolved salts to generate a stable density gradient. Part of the incident solar radiation entering the pond surface is absorbed throughout the depth and the remainder which penetrates the pond is absorbed at the black bottom. If the pond were initially filled with fresh water, the lower layers would heat up, expand and rise to the surface. Because of the convective mixing and heat loss at the surface, only a small temperature rise in the pond could be realized. On the other hand, convection can be eliminated by initially creating a sufficiently strong salt concentration gradient. In this case, thermal expansion in the hotter lower layers is insufficient to destabilize the pond. With convection suppressed, the heat is lost from the lower layers only by conduction. Because of the relatively low conductivity, the water acts as an insulator and permits high temperature (over 90°C) to develop in the bottom layers. At the bottom of the pond, a thick durable plastic liner is laid. Materials used for the liner include butyl rubber, black polyethylene and hypalon reinforced with nylon mesh. Salts like magnesium chloride, sodium chloride or sodium nitrate are dissolved in the water, the concentration varying from 20 to 30 percent at the bottom to almost zero at the top.

In the salt-gradient solar ponds, dissolved salt is used to create layer of water with different densities—the more salt, the denser water. The concentration of the salt at the surface is low—usually less than 5 percent by weight and thus the water is relatively light. The salt concentration steadily increases with depth until at the bottom where it is very high, around 20 percent. Thus a solar pond has three zones with following salinity with depth:

(*i*) Surface convective zone or upper convective zone
(0.3—0.5 m), salinity < 5%.

(*ii*) Non-convective zone 1 to 1.5 m, salinity increases with depth.

(*iii*) Storage zone or lower convective zone
1.5 to 2 m, salinity ≃ 20%

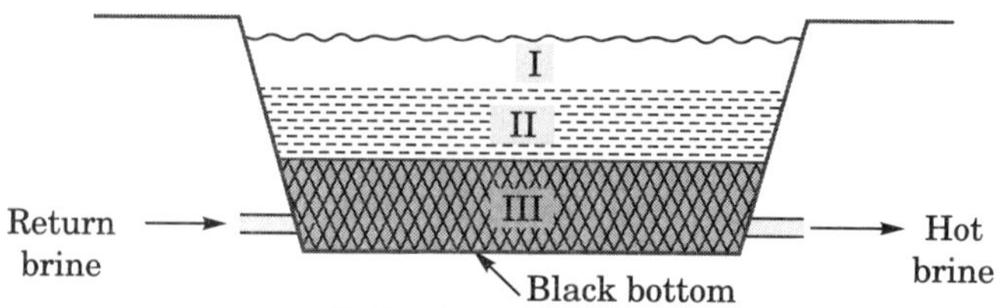

(*i*) Surface convective zone
(*ii*) Non-convective zone
(*iii*) Storage zone

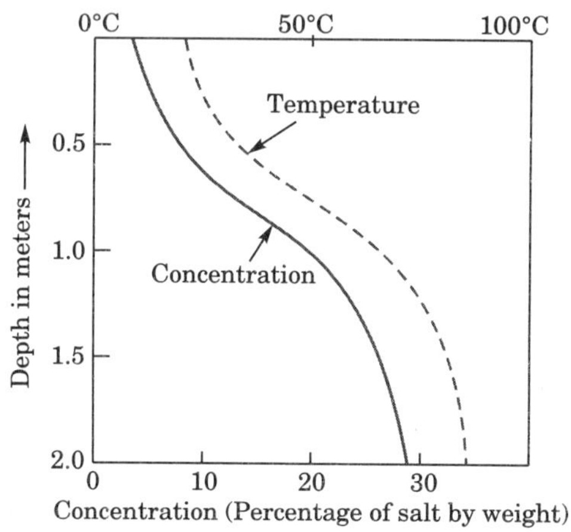

Fig. 4.5. Temperature and concentration profile for a typical pond.

In the above classification of zone, a solar pond of 2 metres depth is considered for, illustration, whereas depth may vary from 1 to 2 metres. At the bottom is the storage zone, which is typically one or two metres deep but can be as little as half a metre, or as much as several metres deep. The surface convective zone usually has a small thickness, around 10 to 20 cm. It has a low, uniform concentration which is close to zero, as well as a fairly uniform temperature, which is close to the ambient air temperature. The non-convective zone is much thicker and occupies more than half the depth of the pond. Both concentration and temperature increase with depth in it. It serves mainly as an insulating layer and reduces heat loss in the upward direction. This part acts as a thermal storage as some of the heat collection also takes place in this zone. The lower convective zone or storage zone is comparable in thickness to the non-convective zone. Both the concentration and temperature are nearly constant in this zone. It serves as the main heat collection as well as thermal storage medium. The deeper the zone, the more heat is stored.

The lowest zone traps heats for the long periods, damping the effects of daily and even seasonal change. This capacity for low cost storage is one of the chief advantages of salt-gradient solar ponds; they can be tapped for energy at night as well as during the day. Even during long periods of cloud cover or even ice cover the stored energy is still available. As salty water near the bottom heats up, it expands. However it cannot rise because it is denser than the less salty water above. Thus the solar pond is 'non-convecting', the warmed water stays trapped below.

Some heat is still lost by conduction to the surface, but this process is much weaker than convection. Lower waters may even warm upto and above the boiling point of pure water. The highest temperature ever recorded in a solar pond is 108°C, set in the summer of 1980 at the university of New Maxico, in Albuquerque.

Depending on location, water clarity and temperature, the solar pond can capture 10 to 20 per cent of the solar energy hitting its surface. Hence, each square metre of pond surface area can supply one half to two giga joules of thermal energy per year at temperatures from 40°C to 80°C. A flat plate collector of the same area would be twice as efficient but cost, ten times as much.

The salt used in a solar pond for creating density gradient should have the following characteristics:

1. It must have a high value of solubility to allow high solution densities;
2. The solubility should not vary appreciably with temperature.
3. Its solution must be adequately transparent to solar radiation.
4. It must be enviromentally benign, safe to handle the ground water.
5. It must be available in abundance near site so that its total delivered cost is low, and
6. It must be inexpensive.

Stability of non-convective layer in the presence of a temperature gradient can be maintained by creating, a sufficiently steep salt concentration gradient. In order to generate the initial salt concentration gradient required to prevent convection, the usual practice is to fill the pond with several layers of salt solution with the successive layers changing stepwise in concentration from near saturation at the bottom to fresh water at the top. For a typical pond of 1 metre deep, one might use six or eight layers. Less layers can be floated successively on top of the existing ones, or successively, more dense layers can be put in at the pond bottom flowing in as under currents and gradually lifting the whole pond. Because of the turbulent mixing and molecular diffusion, which occur during filling, the staircase type concentration profile evolves into a smooth nearly uniform concentration gradient. Typical value of salt concentration at the top surface is 20 kg/cubic metre, increasing to 300 and 260 kg/cubic metre, for magnesium chloride and sodium chloride respectively, at the bottom. It is necessary to add periodically concentrated solutions at the bottom, and wash the surface with fresh water to maintain the concentration gradient in the presence of diffusion effects.

4.3.3. Extraction of Thermal Energy. Energy is stored in low grade thermal form of the luwer convective zone. Convection in the zone is due to the process of heat extraction, accomplished by hot brine withdrawal and cool brine return. It is not practical to cover the bottom of the pond with an array of pipes acting as a heat exchanger due to two reasons.

(*i*) the cost will increase sharply in the case of large ponds of the size of about a square km; and (*ii*) unless there is convection around the pipes heat transfer from the stationary hot brine to the fluid in the pipes will be very poor.

Extraction of thermal energy stored in the lower layers of the pond can be easily accomplished without disturbing the non-convecting salt gradient zone above. Hot water can be extracted from a solar pond without disturbing the concentration gradient. This is achieved by installing the water outlet at the same height as the water inlet. Hot brine can be withdrawn and cool brine returned in a laminar flow pattern because of the presence of density gradient. For small or model ponds, heat exchangers consisting of pipes can be placed in the hot lower layers, but this entails not only the initial installation cost but the continued pumping losses associated with the heat transfer fluid.

Thermal energy from solar pond is used to drive a Rankine cycle heat engine. Hot water from the bottom level of the pond is pumped to the evaporator. Where the organic working fluid to vaporized (Refer Fig. 4.6). The vapour flows under high pressure to the turbine and thereby expanding through the turbine wheel and the electric generator linked to it. The vapour then travels to the condenser where cold water from the cooling tower condenses the vapour back it to a liquid. The liquid is pumped back to the evaporator where the cycle is repeated.

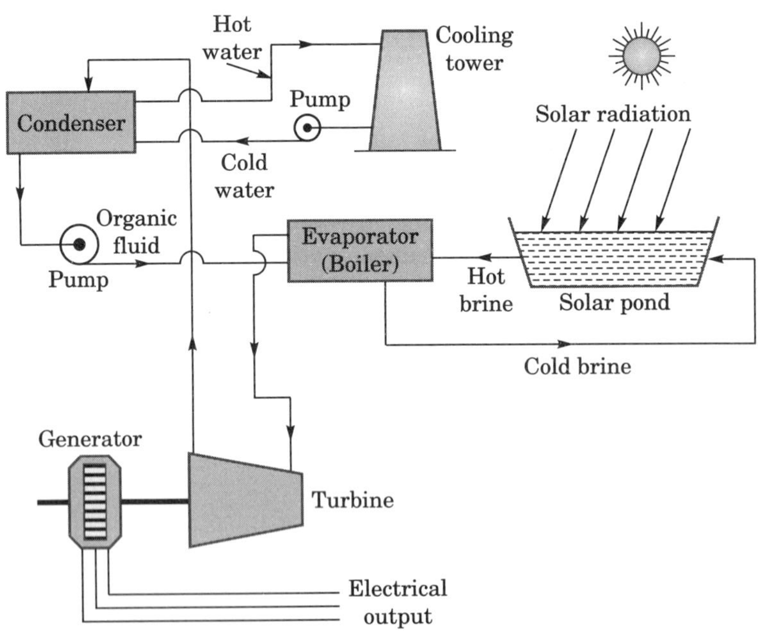

Fig. 4.6. Solar pond electric power plant with, cooling tower.

A 2000 sq. m solar point equipped with a 20 kW engine has been constructed in Australia.

4.3.4. Applications of Solar Ponds. Some of the main applications of a solar pond are discussed below:

(1) **Heating and Cooling of Buildings.** Because of the large heat storage capability in the lower convective zone of the solar pond, it has ideal use for heating even at high latitude stations and for several cloudy days. Many scientists have attempted and sized the solar pond for a particular required heating load for house heating. Calculations have shown that a solar pond with a 100 m diameter and 1 m deep lower convective zone is sufficient to drive either an absorption system or chillier capable of meeting 100 percent of typical cooling load of a 50 house community in Fortworth (U.S.A).

(2) **Production of Power.** A solar pond can be used to generate electricity by driving a thermo-electric device or an organic Rankine cycle engine—a turbine powered by evaporating an organic fluid with a low boiling point. The concept of solar pond for power production holds great promise in those areas where there is sufficient insolation and terrain, and soil conditions allow for construction and operation of large area solar ponds necessary to generate meaningful quantities of electrical energy. Even low temperatures heat that is obtained from solar pond can be converted into electric power. The conversion efficiency is limited due to its low operating temperatures (70-100°C). Because of low temperature, the *Solar pond power plant* (SPPP) requires organic fluid which have low boiling points such as halo-carbons (like Freons) or hydrocarbons (such as propane). A typical SPPP is shown in Fig. (4.6).

(3) **Industrial Process Heat.** Industrial process heat is the thermal energy used directly in the preparation and of treatment of materials and goods manufactured by industry. Several scientists have determined the economics of solar pond for supply of process heat in industries. According to them the solar pond can play a significant role supplying the process heat to industries thereby saving oil, natural gas, electricity, and coal. From the calculations it was concluded that, for crop drying and for a paper industry, for which economics have been determined, the heat from solar pond is highly competitive with oils and natural gas.

(4) **Desalination.** The low cost thermal energy can used to desalt or otherwise purify water for drinking or irrigation.

Multi-flash desalination units along with a solar pond is an attractive proposition for getting distilled water because the multi-flash desalination plant below 100°C which can well be achieved by a solar pond. This system will be suitable at places where portable water is in short supply and brackish water is available. It has been estimated that about 4700 m^3/day distilled water can be obtained from a pond of 0.31 km^2 area with a multi-effort distillation unit. The cost of distilled water appears to be high for industrialised countries but can be used in developing countries where there is a shortage of potable water. Moreover this type of desalination plant produces five times more distilled water than the conventional basin type solar still.

(5) **Heating animal housing and drying crops on farms.** Low grade heat can be used in many ways on farms, which have enough land for solar ponds.

Several small demonstration ponds in Ohio, Iowa and Illinois have been used to heat green houses and hogbarns.

(6) **Heat for biomass conversion.** Site built solar ponds could provide heat to convert biomass to alcohol or methane. While no solar ponds have been used for this purpose, it is an ideal coupling of two renewable-energy technologies.

QUESTIONS

1. Classify the methods of solar energy storage.
   Describe thermal energy storage system.
2. Describe in brief, the different energy storage methods used in the solar system.
3. What is principle collection of solar energy used in a non-convective solar pond?
   Describe a non-convective solar pond for solar energy collection and storage.
4. What are the main applications of a solar pond? Describe briefly.
5. Write short notes on:
   (*i*) Maintenance of stable density gradient in a solar pond.
   (*ii*) Heat extraction method from a solar pond.
   (*iii*) Chemical energy storage method.
   (*iv*) Hydrogen storage.
   (*v*) Electro-magnetic energy storage.

CHAPTER 5

Applications of Solar Energy

5.1. Introduction

The actual and proposed applications of solar energy may be considered in three general categories:

(*a*) *Direct Thermal Application* make direct use of heat, resulting from the absorption of solar radiation, for space heating (and cooling) of residences and other building, so provide hot water service for such burldings, and to supply heat for agricultural industrial, and other processes that require only moderate temperatures.

(*b*) *Solar Electric Applications** are those in which solar energy is converted directly or indirectly into electrical energy. Four general conversion methods are being investigated:

(*i*) *Solar thermal* methods involve production of high temperatures, such as are required to boil water or other working fluid for operating turbines which drive electric generators. These are considered under *solar thermal electric conversion.*

(*ii*) *Photovoltaic Methods* make use of devices (Solar Cells) to convert solar energy directly into electrical energy without machinery. This is discussed under solar *photovoltaic conversion.*

(*iii*) The conversion of solar energy into electrical energy without the use of machinery, by utilizing the *thermo electric, effect*, is also considered in chapter *Thermo electric conversion.*

(*iv*) *Wind Energy* is the form of solar energy that can be converted into mechanical (rotational) energy and hence into electrical energy by means of a generator. This *indirect* use of solar energy to generate electricity is described under the chapter of *Wind energy.*

Ocean thermal energy conversion depends on the difference in temperature between solar heated surface water and cold deep ocean water to operate a vapour expansion turbine and electric generator. This indirect use of solar energy is considered in chapter *Ocean Thermal Energy Conversion.*

(*c*) Energy from biomass and bio-gas, refers to the conversion into clean fuels or other energy related product of organic matter derived directly or indirectly from plants which use solar energy to grow. Biomass materials include agricultural, forest, and animal residues, as well as terrestrial and aquatic plants grown

*For detailed discussion on solar thermal and electric applications see author's book on "Solar Energy Utilization".

especially for the purpose. For a detail discussion see chapter on *Biomass and Biogas*.

Other Forms of Solar Energy. Water power, other than tidal power, also represents the utilization of solar energy. Heat from the sun causes the evaporation of surface waters, the vapour rises is condensed in the upper atmosphere, and falls as rain. The resulting water that collects at higher elevations has substantial (potential) energy than can be converted into electrical energy by means of a turbine generator at a lower level. The discussion on Hydro-electric power*, is beyond the scope of this book, only some brief discussion is there on *Micro hydel energy,* which is a non-conventional method of generating electric-energy.

It may be noted, too, that since fossil fuels-coal, oil, and natural gas-originate from living matter, their energy is really solar energy that has been converted and stored for millions of years. In fact, only nuclear energy, geothermal energy, and to a large extent, tidal energy, do not originate in the sun. Based on the above classification, direct solar energy applications are discussed below in the following order:

(1) Solar water heating. (2) Space heating.
(3) Space cooling.
(4) Solar energy: Thermal electric conversion.
(5) Solar energy: Photovoltaic electric conversion.
(6) Solar distillation. (7) Solar pumping.
(8) Agriculture and industrial process heat.
(9) Solar furnace. (10) Solar cooking.
(11) Solar production of hydrogen, and
(12) Solar green houses.

5.2. Solar Water Heating (or Hot Water Supply System)

The basic elements of a solar water heater are:

(*i*) Flat plate collector. (*ii*) Storage tank.
(*iii*) Circulation system and auxiliary heating system.
(*iv*) Control of the system.

The use of solar energy for heating water in many respects quite similar to its use for heating buildings. There are however, several aspects of solar water heating, that make it potentially better investment of energy, money and effort than solar building heating. For one thing, the demand for hot water is relatively constant throughout the year. Thus the collector and the other parts of the solar water heater vnll be working harder and longer to produce the saving in fuel that eventually must pay for the higher initial cost of the system. The solar building heating system, on the other hand, fully operational only during the coldest months of the heating season.

The simplest type of solar water heater is the *thermosiphon* system. Some typical and commercial designs of solar water heaters are:

(*i*) Natural circulation solar water heater (pressurized).
(*ii*) Natural circulation solar water heater (non-pressurized).
(*iii*) Forced circulation solar water heater.

*For detailed discussion see author's book on "Power Plant Technology".

(*i*) **Natural circulation solar water heater** (pressurized). A natural circulation system is shown in Fig. 5.1. It consists of a titled collector (south facing), with transparent cover glasses, a separate highly insulated water storage tank, and well insulated pipes connecting the two. The bottom of the tank is at least 1 ft. (0.3 m) the top of the collector, and no auxiliary energy is required to circulate water through it. Circulation occurs through natural convection, or thermosiphoning. As the water is heated in its passage through the collector, its density decreases and hence it rises and flows into the top of the storage tank, colder water from the bottom of the tank has a higher density and so tends to sink and enter the lower heater of the collector for further heating. The density difference between the hot and cold water thus provides the driving force (convection) for the circulation of water through the collector and the storage tank. Hot water is drawn off from the top of the tank as required and is replaced by cold water from the service system. As long as the sun shines the water will quietly circulate, getting warmer. After sunset, a thermosiphon system can reverse its flow direction and loss heat to the environment during the night. To avoid reverse flow, the top heater of the absorber is kept as stated above 0.3 m below the cold leg fitting on the storage tank. To provide heat during long, cloudy periods, an electrical immersion heater can be used as a backup for the solar system. A non-freezing fluid may be used in the collector circuit. The thermosiphon system is one of the least expensive solar hot-water systems and should be used whenever possible.

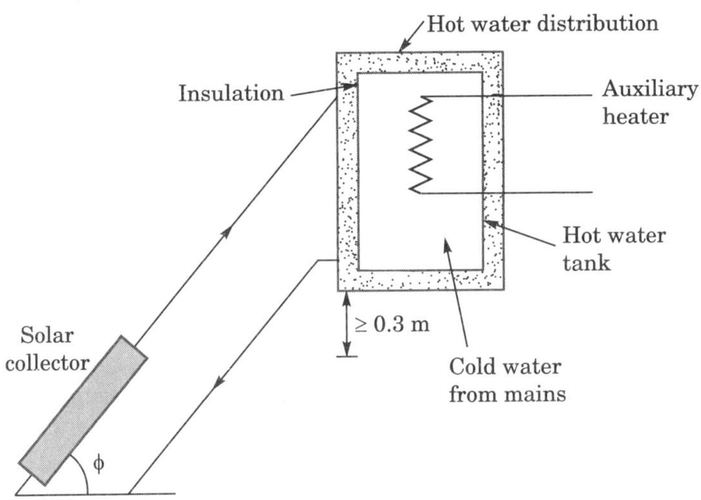

Fig. 5.1. Schematic of a neutral circulation solar water heater (pressurized).

Thermosiphon solar water heaters are passive systems and do not require a mechanical pump to circulate the water. Such heaters can be used extensively in rural areas, where electricity is expensive (or not available) and there is little danger of freezing.

(*ii*) **Natural circulation solar water heater** (non-pressurized). The pressurized system is able to supply hot water at locations of the storage tank. This creates considerable stress on the water channels in the collector which must be designed accordingly. The non-pressurized systems supply hot water by gravity flow only to users lower than tank. If pressurized hot water is required (for showers, or appliances) the difference in height will have to be large enough to meet the requirements. If the height of difference can not be accommodated, the only solution is to install a saparate pump and pressure tank. The stresses within non-pressurized system are lower which allows cheaper and easier construction. In this type also mechanical pump is not required as shown in Fig. 5.2, however, a oneway check

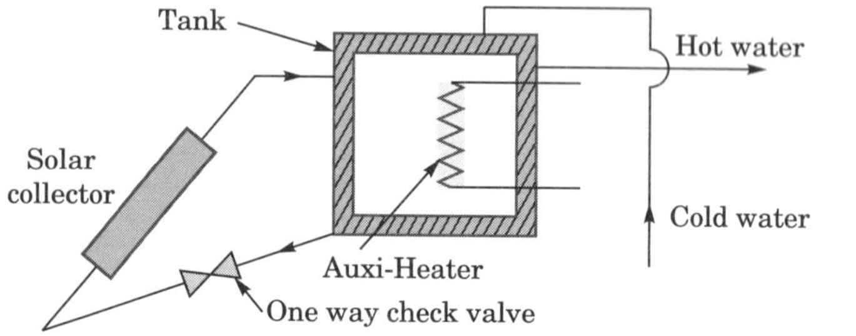

Fig. 5.2. Non-pressurized solar water heater.

valve may be desirable to prevent reverse circulation and thus loss of heat at night. A typical system for domestic water heating is shown in Fig. (5.3).

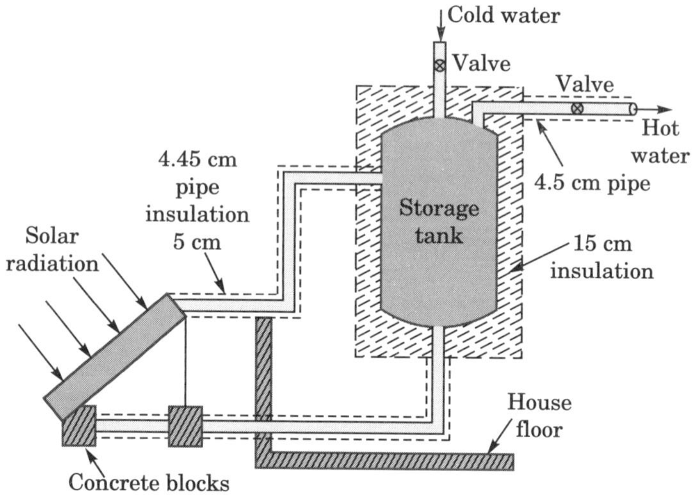

Fig. 5.3. A typical solar water heater.

(*iii*) **Forced circulation solar water heater.** Fig. 5.4 shows schematically an example of forced circulation system. By including an electric pump in the return

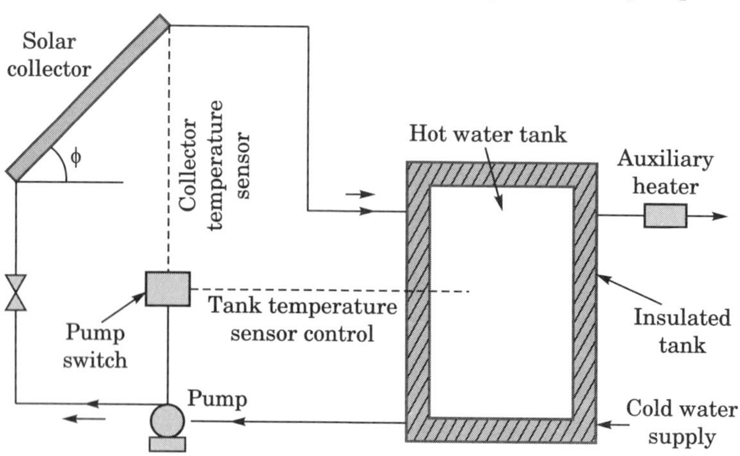

Fig. 5.4. Schematic of a forced circulation solar water heater.

circuit between the bottom of the storage tank and the lower header of the collector, the tank can be placed at a more convenient level (*e.g.,* in the house basement). This is now an *active system.* A control unit permits the pump to operate only when the temperature of the water at the bottom of the tank is below that of the water in the upper header.

A check valve is needed to prevent reverse circulation, and resultant night time thermal losses from the collector. In this example, auxiliary heater is shown as provided to the water leaving the tank and going to the load.

When there is a danger of freezing, the water may be drained from the collector; alternatively, a slow reverse flow of the warmer water may be permitted through the collector on cold nights. The freezing danger can be overcome, although at some increase in cost, by using an antifreeze solution as the heat-transport medium, as described earlier. The heat is then transferred to water in the storage tank by way of a heat exchanger coil. (See Fig. 5.5).

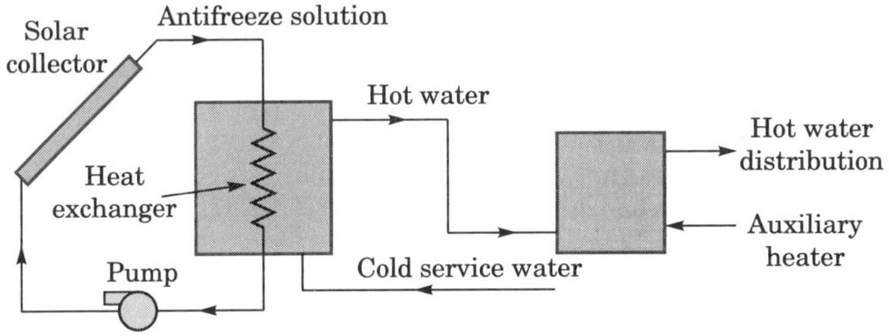

Fig. 5.5. Solar water heating system with antifreeze.

5.3. Space-Heating (or Solar Heating of Building)

Many different concepts have been proposed (and tested) for using solar energy in space heating of buildings. There are two primary categories into which virtually all solar heating systems may be divided. The first is *passive systems,* in which solar radiation is collected by some element, of the structure itself, or admitted directly into, building through large, south facing windows. The second is the *active systems* which generally consists of

(*a*) separate solar collectors, which may heat either water or air,

(*b*) storage devices which can accumulate the collected energy for use at nights and during inclement days, and,

(*c*) a back up system to provide heat for protected periods of bad weather. Heat is transferred from the collectors or from the storage means by conventional equipment, such as fan coil units, when hot or cold water is provided; fan, ducts, and air outlets, when the heat transfer medium is air; and radiant means when beating is the only task which must be accomplished.

Passive heating systems operate without pumps, blowers, or other mechanical devices; the air is circulated past a solar heated surface (or surface) and through the building by convection (*i.e.*, less dense, cooler air tends to rise while more dense, cooler air moves downward). In *active heating systems,* fans and pumps are used to circulate the air and often a separate heat absorbing fluid.

Passive solar thermal systems are more practical in locations where there is ample winter sunshine and an unabstructed southern exposure is possible. The building to be heated is an essential part of the system design. Active systems, on the other hand, can be adopted to almost any location and type of building; however they are more expensive than passive systems to construct and operate. An advantage of active solar systems is that the building air temperature can be controlled in the same way as with the conventional heating, but in most passive systems substantial temperature variations may occur in the course of the day.

In principle, it should be possible to provide all the heating (and cooling) needs of a building by solar energy. However, to do this, the heating system would have to be designed for minimum sunshine conditions and hence would be over-designed for the majority of the situations. In most cases, solar-energy systems provide roughly 50 to 75 percent of the annual heating requirements. The remainder is supplied by an auxiliary-heating systems using gas, oil, or electricity.

Solar Heating Systems

(A) Passive Heating Systems. An increasing number of residences using various passive systems have been built and are under construction mainly in U.S. Building with large windows facing the equator (South in the northern hemisphere or north in the southern hemisphere) and arranged to admit solar radiation into the building when the sun is low in the winter sky, have been termed 'Solar houses'. The gains to be realized from properly oriented windows are significant, but in cold climates losses during the periods of low radiation, nights and cloudy weather, must be controlled so net gains can be realized.

If a building is designed properly:

(*i*) It will function as a solar collector, collecting heat when the sun is shining and storing it for later use.

(*ii*) The building will function as a solar store house. It must store the heat for cool times when the sun is not shining, and store the cool for warm or hot periods when the sun is shining. Buildings which are made of heavy materials such as stone or concrete do this most effectively.

(*iii*) Building will function as a good heat trap. It must make good use of the heat (or cool) and let it escape only very slowly. This is done primarily by reducing the heat loss of the building through the use of insulation, reduction of infiltration and storm windows.

Wall and roof of the building must be oriented in such a way so as to receive solar radiation heat in the winter and shed it in the summer. A building can benefit from its orientation; for similar reasons, it also benefit from different ratios of its length to its width to its height. The optimum shape losses, the minimum amount of outward moving heat and gains maximum amount of solar heat in the winter, and retains the minimum amount of solar heat in the summer.

The basic design principles of passive solar space-heating systems, that is, without mechanical components, fall into the following five general categories:

(*i*) Direct gain
(*ii*) Thermal storage wall
(*iii*) Attached sun space
(*iv*) Roof storage
(*v*) Convective loop.

There are modifications within each of these categories and two or more may be combined in a single building.

Direct gain. In this system, the building has a south wall with a large number of windows; two layers of glass minimise heat losses. Solar radiation entering the windows falls on thick concrete, slate, or brick floors (and possibly walls) and is absorbed and stored as heat. The building air is then heated by radiation and convection from the floor and walls. Covering the windows with shutters or curtains at night reduces heat losses.

Thermal storage wall. A large wall like mass which absorbs solar radiation and stores heat, is placed directly behind large south-facing windows. The best known example of the storage wall design is the vertical *Troumbe wall,* made of concrete brick, or other masonry some 0.3 m or more thick. Dr. Felix Trombe, Director of solar energy laboratory at Odeillo (France) has used this principle. The system is passive, because the solar radiation is absorbed by a heavy concrete, south facing wall which is covered with double glazing. A Trombe wall arrangement is outlined in Fig. (5.6). The distance between glazing and concrete wall is about 10 to 20 cm. Concrete, wall is about 20 cm thick, painted black, which serves as both a radiation absorber and a heat storage medium. Heat radiated outward from the wall is trapped by the glass cover, so that the air in between is heated. The heated air rises and enters the adjacent room through vents at the top of the wall, the air circulates by natural convection and is returned by way of ducts at the bottom of the storage wall. Electric strip heaters are provided for standby purposes during excessively cold weather and during days with little sunshine.

Attached sunspace. A sunspace is any enclosed space, such as a green house or sun porch, with a glass wall on the south side. A sunspace may be attached (or built on) to a thick south wall of the building to be heated by the sun. Vents near the top and bottom of the wall, as in Fig. (5.6), permit circulation through the main building of the heated in the sunspace. Heat storage is provided by the thick wall, a concrete or masonry floor, water containers, and other materials in the sun space. Thus, an attached sunspace system combines features of direct gain and storage wall concepts.

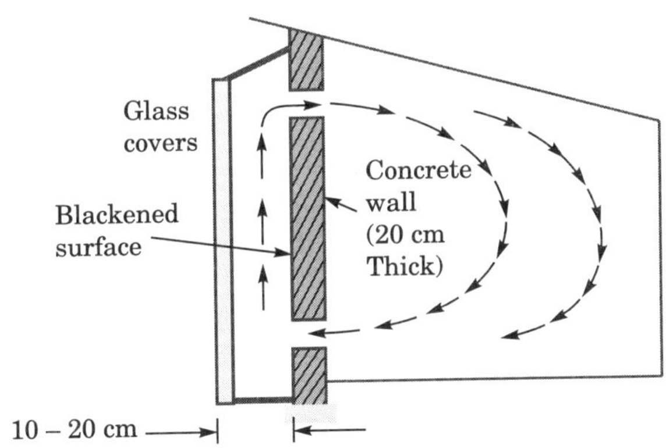

Fig. 5.6. A passive solar heating system.

Roof storage. A passive solar system, trade named Sky Therm, was designed for house having a flat roof located in a mild climate. The heat is absorbed and stored in water about 0.25 m deep contained in plastic bags held in blackened steel boxes on the house roof (Fig. 5.7). In a later design, a layer of clear plastic sealed to the top of the bag provides a stagnant airspace to reduce heat losses to the atmosphere. Heat is transferred from the heated water to the rooms below by conduction through a metal ceiling. Air circulation may be aided by means of electric fans, but this is not essential. To prevent loss of heat during the night, thermal insulator panels are moved, either manually or by a time controlled electric motor, to cover the water bags. In the day time, the panels, which are in sections, are removed and stacked one above the other.

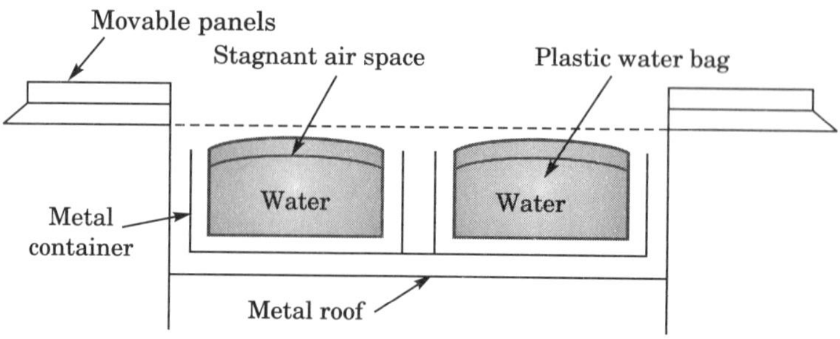

Fig. 5.7. Roof storage of solar heat.

Convective Loop. In most passive solar space heating systems, the heated air is circulated by convection, but the term convective loop is applied to systems that resemble the thermosiphon hot-water scheme described earlier. Such a convective loop heating system is outlined in Fig. (5.8). It includes a conventional flat-plate collector at a level below that of the main structure. A bed of rock, which may be located beneath a sun space, provides thermal storage. In normal operation, air passing upward through the collector is heated and enters the building through floor vents. The cool, denser air leaving the building returns to the bottom of the collector and is reheated. If more solar heat is available than is required for space heating, the floor vents may be partly (or wholly) closed. The heated air then flows through and deposits heat in the storage bed. Heat stored in this way may be used later, as needed, by transfer to the cooler air leaving the building.

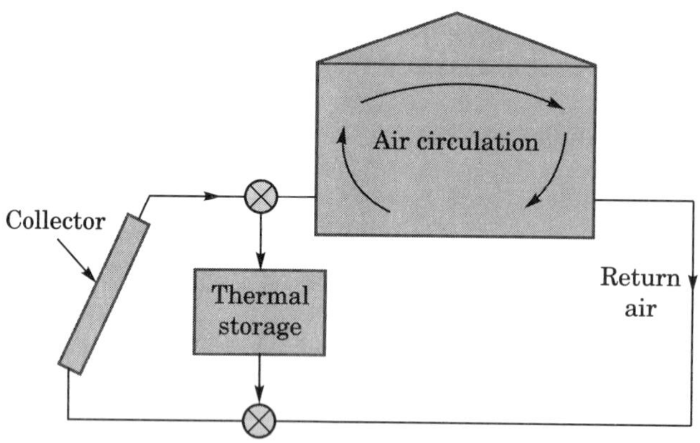

Fig. 5.8. Convective loop passive solar heating.

(B) Active Space-Heating Systems. Most of the hundreds of solar-heated residences built throughout the world use the active system, in which separate collectors are used together the solar radiation, transfer it to water or air, and store it in tanks of water or rock piles or both. The water and air are circulated by pumps or fans and conventional means are used to distribute the heat to the interior of the residences.

General Principles. Nearly all existing or proposed active solar space-heating and/or hot-water supply systems utilize three main components in addition to pumps and blowers: (1) a solar radiation collector with its associated heat transport (or heat-transfer) fluid, (2) a heat-storage medium, and (3) a distribution systems. The same arrangement of components can also be used to provide hot water for domestic and related use and, with the addition of other components, space cooling (air-conditioning). In the collector the solar radiation is collected and converted into heat. The heat-transport fluid removes the heat and carries it to the heat-storage system; the heat can then be withdrawn from the storage and distributed throughout the building.

Schematics of basic hot water and hot air heating systems are given in the Fig. (5.9) and (5.11) respectively and their individual components are considered in the following sections.

(1) Basic Hot Water System

An outline of an active heating system with a sloping flat plate collected located on the roof of the building is given in Fig. (5.9). This is a basic *hot water heating* system, with water tank storage and auxiliary energy source. Heat is transferred to the water in the storage tank, commonly located in the basement of the building. The solar heated water from the tank passes through an auxiliary heater, which comes on automatically when the water temperature falls below a prescribed level. For space heating, the water may be pumped through radiators or it may be used to heat air in a water to air heat exchanger.

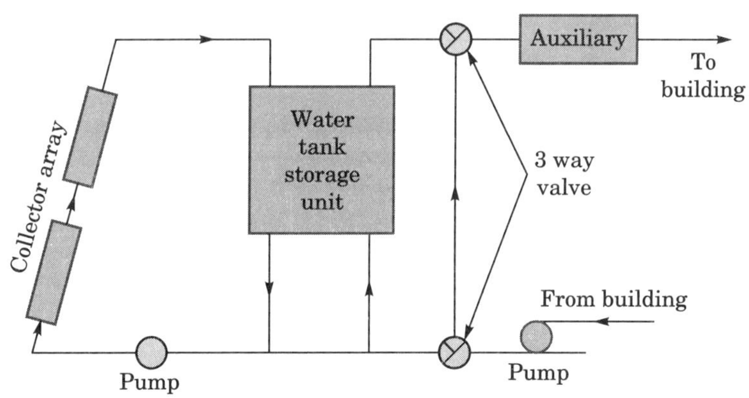

Fig. 5.9. Schematic of a basic hot water active system.

During normal operation, the three way valves are set to permit solar heated water to flow from the storage tank and auxiliary heater to the distribution system and back to the tank. If after several cloudy days, the heat in storage is depleted, the valves will adjust automatically to bypass the storage tank. In this way, auxiliary heating of the large volume of water in the tank is prevented. If the temperature in the heater at the top of the collector should fall below that at the bottom of the tank, the pump (at bottom left of figure) would be switched off automatically.

If in this system, the heat transport medium is an antifreeze solution, then there is a closed circuit of it, with the heat exchanger coil in the storage tank. This type of solar space heating system with hot water system is shown in Fig. 5.10.

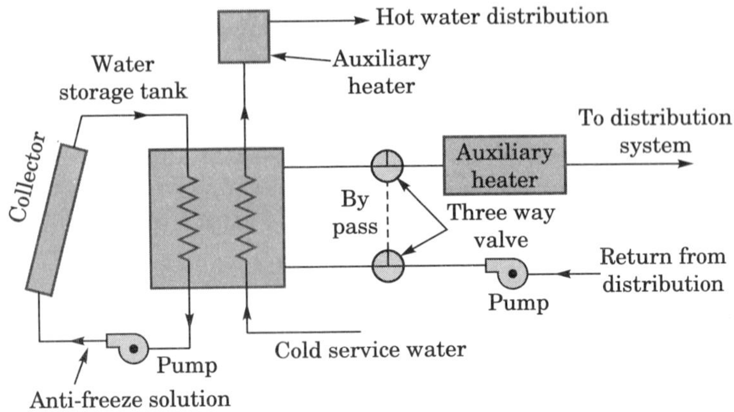

Fig. 5.10. Solar space heating and hot water system.

Advantages and disadvantages of basic hot water system are listed below:

Advantages

(*i*) In case of water heating, a common heat transfer and storage medium, water is used, this avoids temperature drop during transfer of energy into and out of the storage.

(*ii*) It requires relatively smaller storage volume.

(*iii*) It can be easily adopted to supply of energy to absorption air conditioners, and

(*iv*) Relatively low energy requirements for pumping of the heat transfer fluid.

Disadvantages

(*i*) Solar water heating system will probably operate at lower water temperature than conventional water systems and thus require additional heat transfer area or equivalent means to transfer heat into building.

(*ii*) Water heaters may also operate at excessively high temperatures (particularly in spring and fall) and means must be provided to remove energy and avoid boiling and pressure build up.

(*iii*) Collector storage has to be designed for overheating during the period of no energy level.

(*iv*) Care has to be taken to avoid corrosion problems.

(2) Basic Hot Air System

Schematic diagram of a basic hot air heating system is shown in Fig. (5.11). In this system the storage medium (pebbles or rock) is held in the storage unit, while air is the fluid used to transport energy from collector to the storage and to the building. By adjusting the dampers, the heated air from the collector can be divided between rock storage and the distribution system, as might be required by the conditions. For example, when the sun shines after several cloudy days it would be desirable to utilize the available heat directly in the distribution system rather than placing it in storage. Two three way valves can be used to bypass the storage tank, as explained above. An auxiliary source of heating is also provided. Auxiliary heating can be used to augment the energy supply to the building from the collector or storage if the supply of heat from it is inadequate.

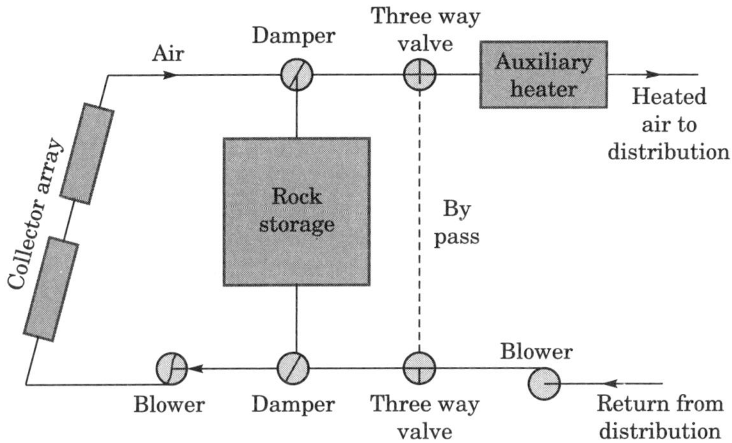

Fig. 5.11. Schematic diagram of a basic hot air heating system.

The position of the blower in figure is shown at the upstream of the collector and the storage, and it forces the air through these for heating. In this case slight leakage of heated air will take place. Blower can also be placed on the down stream side of the collector and storage, so that the pressure in the collector is not above ambient pressure, which might be advantageous in controlling leakage.

System based on this concept have number *of advantages,* compared with those based on use of water as a heat transfer medium.

(*i*) There is no problem with freezing in the collectors.

(*ii*) Corrosion problems are minimised.

(*iii*) Conventional control equipment for air heating is already available and can be readily used.

(*iv*) Problems of designing for over heating during periods of no energy removal are minimized, and,

(*v*) The working fluid is air and the warm air heating systems are in common use.

Disadvantages of air heating system are:

(*i*) Relatively higher power costs for pumping air through the storage medium.

(*ii*) Relatively large volumes of storage units.

(*iii*) Difficulty of adding absorption air conditioners to the system.

Collectors and Heat Transport. Because of their simplicity, fixed position, flat-plate collectors are almost invariable used in space heating applications of solar energy. Since they can collect diffuse as well as direct solar radiation, flat-plate collectors are partially effective even when the sun is not shining. Either air or water can be used as the heat-transport fluid. The plumbing system is less likely to have problems in a system using air, but larger ducts and a larger heat-storage volume are required.

Many types of flat-plate collectors are available. As a general rule, these collectors are fabricated in panels, commonly about 6 to 8 ft (1.8 to 2.4 m) long and 3 to 4 ft (0.9 to 1.2 m) wide. The number of panels (or the total collector area) depends on the space to be heated (or cooled) and the local climate, as well as on economic factors. In new home construction, the panels are mounted on a roof facing south (or south-southwest) with the optimum slope (*e.g,* latitude + 15° for heating). Otherwise, the panels with the required slope may be placed on the ground or on a flat roof, if one is available. The panels must then be spaced in such a manner that they do not shade one another significantly when the sun is low in the sky.

An essential piece of information for the design of a collector for space heating is the daily insolation, that is, the total amount of direct and diffuse solar radiation energy received per day on a horizontal surface of unit area on the ground. For a surface inclined at an angle to the horizontal equal to the latitude + 15°, the solar radiation received is higher in winter and lower in summer.

From a knowledge of the daily temperatures throughout the year, it is possible to determine the amount of thermal energy (heat) required for space heating in the winter and cooling in the summer. By using the daily insolation at the given location, an estimate can then be made of the area of a flat-plate collector, sloping at a prescribed angle, that will supply a certain proportion of the heat required.

The degree (or percentage) of dependence on solar energy for which a collector system is designed is determined by balancing the cost of the collectors against the estimated future costs of alternative energy sources. It is the future rather than the present costs that are important in this respect. The higher the expected cost of an alternative energy supply, the larger the economically acceptable collector area. Thus in deciding on this area, the local climatic and economic factors must be considered. Further more, the collector area should be proportional to the area of the building to be heated (or heated and cooled). A very rough rule for moderately cool climates is a collector area about 50 percent of the floor area.

Heat Storage. A means for storing heat is necessary in active solar energy systems to supply heating requirements during the evening and on cloudy days. The larger the heat-storage volume, the greater is the heat-storage capacity for a given material and the longer the sunless period for which heat could be available. But increasing the volume means increasing the cost and space requirements. As a compromise, heat storage tanks are generally designed to have a heat capacity equivalent to three days normal demand. The common heat storage materials are water and small pieces of rock. When water (or an antifreeze solution) is used as the heat transport fluid in the collector, a large, well-insulated tank of water provides the heat storage. The water may be pumped directly from the collector through the storage tank. It may pass through a heat exchanger coil in the tank when the heat-transport, liquid contains an antifreeze compound (*e.g.*, ethylene (glycol).

If air is the heat-transport medium, the heat is usually stored in pieces of rock or pebbles, roughly 2 in. (5 cm) across, contained in an insulated tank through which the air circulates. The efficiency of transfer of heat from the hot air to the rock pieces is greater for smaller sizes of the pieces. With decreasing size, however, the resistance to air flow increases and more pumping power is required to circulate the air. The selected size thus represents a compromise between two opposing factors.

A given volume of water can store substantially more heat than an equal volume of rock, assuming the same temperature increase in each case. Thus, water can store 4100 kilo joules per cubic metre per °C increase in the temperature on the other hand, the rock can store roughly 1460 kJ per cu m per°C, where the volume includes allowance for the spaces between the rock pieces. Hence, for the same heat storage, rock would occupy about three times the volume of water.

The following data provide an approximate guide to the relationship between the storage volume and the area of the flat plate solar collector for water and rock storage:

Water : 0.05 to 0.075 m^3 per m^2

Rock : 0.17 to 0.02 m^3 per m^2.

These numbers are average values for different climatic conditions; both smaller and larger volumes have been used in special circumstances.

If a home with a floor area of 150 sq. m (1600 sq. ft) has a solar collector area of 85 sq. m (915 sq. ft), the storage volume of water would be in the range of 4.2 to 6.3 cum (150 to 225 cu ft), whereas for rock, it would be 14 to 21 cum (500 to 750 cu ft). The larger volume in each case would be desirable in a colder climate.

A considerable increase in heat-storage capacity, with a corresponding decrease in volume, could be achieved by using a suitable salt-hydrate, instead of water or air, as the storage medium. A salt-hydrate is a salt in which the solid crystals include a number of water molecules. Special interest has been focused on the relatively inexpensive salt-hydrate sodium sulphate decahydrate ($Na_2SO_4 . 10H_2O$), com-monly known as Glauber's salt.

When heat is added to the salt, the temperature increases to 32°C, called the transition point. At this point, the temperature remains constant while the salt changes to an aqueous solution plus solid Na_2SC_4 (*i.e.*, without the water molecules); during this stage heat is stored, although the temperature does not rise. When all the Na_2SO_4 $10H_2O$ has been changed to Na_2SO_4, the, temperature can increase again as more heat is added. If heat is withdrawn from the system, the temperature falls to 32°C, where it remains while the stored heat is released and the Na_2SO_4 $10H_2O$ is completely regenerated; thus,

$$NO_2SO_410H_2O \underset{\text{Heat withdrawn at 32°C}}{\overset{\text{Heat supplied}}{\rightleftharpoons}} Na_2SO_4\,(\text{solid}) + H_2O(\text{solution})$$

In this way, the system offers the potential for storing heat at 32°C in about, one-third the volume of a water tank (or one-tenth of the volume of rock) for the same heat capacity.

Salt-hydrate heat storage has been tested in solar heating systems with air as the transport medium. The salt was contained in a number of stacked 5-gallon (19 litre) cans around which the air was circulated. The systems operated well for a time but gradually deteriorated.

The reason was that as the solid Na_2SO_4 formed (from the Na_2SO_4 $10H_2O$), it tends to deposit on the walls of the storage cans. As a result, the reverse process, in which the Na_2SO_4 $10H_2O$ is regenerated from the Na_2SO_4, did not occur completely. Hence, the storage capacity decreased as the system was operated. Further more, the solid deposited on the walls decreased the heat transfer efficiency. A possible means of overcoming both of these difficulties is by slow rotation of a drum containing the salt hydrate, the solid Na_2SO_4 formed would then remain in suspension in the solution.

Heat distribution. The distribution system for solar space heating is much the same as for any other heat source; in fact, existing systems for the gas or oil heat have been readily adopted, to solar heating.

When water is the heat-transport, fluid, it may be circulated through radiators in the building to be heated, and then back to the storage tank. Alternatively, a water to air heat exchanger can serve to heat air for distribution by way of conventional heating ducts. The air is blown (or drawn) across a pipe coil (heat exchanger) through which the hot water flows (Fig. 5.12). In a modification of this system, transfer of heat from hot water to air is achieved by passing the air through a container of rock pieces surrounding the storage tank.

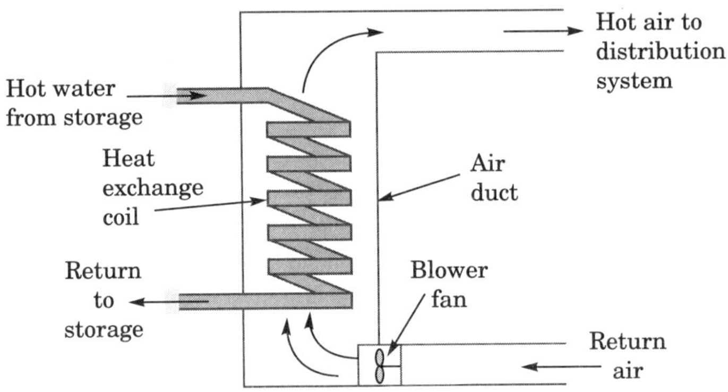

Fig. 5.12. Water-to-air heating system.

If air is the heat-transport fluid (with rock or salt-hydrate storage), the heated air can be circulated through the building and back to the storage system. The building space is then part of a closed circuit.

5.4. Space Cooling (or Solar Cooling of Buildings)

Introduction. The major current interest is in mechanical cooling (or air-conditioning) systems that depend on solar heat for their operation and are unaffected by atmospheric humidity. The two most common *refrigeration* techniques are vapor compression and absorption and, in principle, both could be adopted for use with solar energy, although the temperatures required are higher than those adequate for space heating. In the former procedure, solar heated water could vaporize propane or ammonia at a moderate pressure. The vapour could then drive a *turbine* which would in turn, operate a vapor compression cooling unit. Such a low pressure vapor turbine, however, would inevitably have a very low efficiency. Further-more, propane is highly flammable whereas ammonia forms a noxious gas.

Absorption cooling with solar energy, which is regarded as more practical, is possible with current technology although improvements in design would be desirable.

(1) Absorption air-conditioning

Two approaches have been to solar operation of absorption coolers. The first is a *continuous cooler* which is in construction and operation, is similar to conventional vapour absorption refrigeration system. The solar collector storage supplies the energy to the generator, where solar energy is available, otherwise it is supplied with the auxiliary energy source. The second approach is to use *intermittent coolers,* which are intermittent in operation and operate during the periods when solar energy is available. Intermittent coolers have not been used for air conditioning nor have been studied sufficiently for their application in solar air conditioning.

The two types of *absorption air-conditioners* available in the market, the lithium-bromide water ($LiBr—H_2O$) system and the ammonia-water ($NH_3—H_2O$) system. Of the two common absorption air conditioning systems, the $LiBr—H_2O$ is simpler since a rectifying column assures that no water vapour, mixed with NH_3

enters the evaporator where it could freeze. In the LiBr—H_2O system water vapour is a refrigerant. In addition, the aquaammonia system requires generator temperature of the order of 120°C to 150°C, such a temperature is higher than a flat-plate collector can provide, without special techniques. But this system could be possible with concentrating type collectors. The LiBr—H_2O system requires lower generator temperature of the order of 85° to 95°C which are achievable by a flat plate collector. Also the LiBr—H_2O system posses higher C.O.P. than the aqua-ammonia (NH_3—H_2O) system. The disadvantage of LiBr—H_2O systems is that the water can not exit in liquid form below triple point temperature, (*i.e.*, 0°C, practically, this system can not operate much below 4°C) since the refrigerant is water vapour. This is not a major disadvantage in air-conditioning applications.

The effective performance of an absorption cycle depends on the two materials that comprise the refrigerant-absorber pair. These two working fluids must have the following characteristics:

1. The absence of a splid phase absorbent.

2. A refrigerant more volatile than the absorbent, in order to be separated from the absorbent easily in the generator.

3. An absorbent that has small affinity for the refrigerant.

4. A high degree of stability for long-term operations.

5. It refrigerant that has a large latent heat so that the circulation rate can be kept at the minimum.

6. A low corrosion rate and nontoxicity for safety reasons.

The only disadvantage of the LiBr—H_2O pair is the possible problem with crystallization in the generator.

(A) LiBr—H_2O System. The absorption air-conditioning system is shown schematically in Fig. (5.13). The system consists of two parts *viz.*

(*i*) the solar collector and storage, and

(*ii*) the absorption air conditioner and the auxiliary heating.

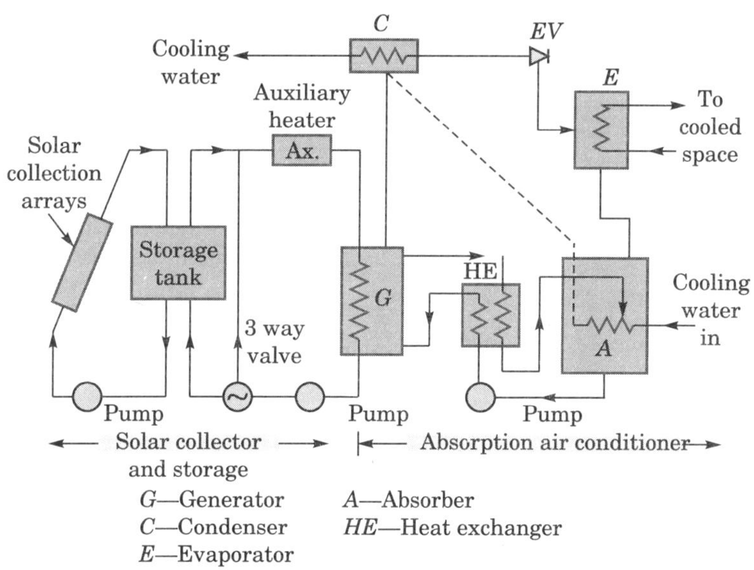

Fig. 5.13. Schamatic of solar operated absorption air conditioner.

The essential components of the cooler are, (*i*) generator (G), (*ii*) condenser (C), (*iii*) evaporator (E) (*iv*) absorber (A) (*v*) heat-exchanger (HE).

The operation of air-conditioners with energy from flat-plate collector and storage systems is the most common approach to the solar cooling today. In essence cooling is accomplished as the generator of the absorption cooler is supplied with heat by a fluid pumped from the collector storage system or from auxiliary. Heat is supplied to a solution of refrigerant in absorbent in the generator, where refrigerant is distilled out of the absorbent fluid. The refrigerant (now liquid) is condensed and goes through a pressure reducing valve to the evaporator where it operates and cools air or water for the cooling space. The refrigerant vapour goes to the absorber where it comes in contact with the solution which is weak in refrigerant and which flows from the generator. The vapour is absorbed in the solution, which is then returned to the generator. A heat exchanger is used for sensible heat recovery and greatly improves cooler C.O.P.

From the point of view of use of a conventional energy source, there is a single index of performance for rating cooling processes, that is the COP (Coefficient of performance), the ratio of the amount of cooling to the energy required. For solar operation there are two additional factors, the temperature required in the solar collector to drive the process and the ratio of cooling produced to solar energy incident on the collector. As solar processes are inevitably transient in their operation, the energy ratios and temperatures will vary with time and COP based on long term integrated performance provides an appropriate index of performance. Pumping to more absorbent solution may be by mechanical means or by vapour-lift pumping in the generator for low pressure systems like LiBr—H_2O system require water cooling of absorber and condenser. Systems of this type shown in the figure have been the basis of most of the experience to date with solar air conditioning.

The coolers used in most experiments todate are LiBr—H_2O machines water-cooled absorber and condenser. The pressure in the condenser and generator is fixed largely by temperature drops across heat transfer surfaces in the generator and condenser. The pressure in the evaporator and absorber is fixed by the temperature of the cooling fluid to the absorber and by the temperature drop across the heat transfer surfaces in the evaporator and the absorber. Thus, to keep the generator temperatures within the limits imposed by the characteristics of flat-plate collector, the critical design factors and operational parameters include effectiveness of the heat exchangers and coolant temperature. Common practice in solar experiments has been to use water cooled absorbers and condensers, which in turn requires a cooling tower.

Many of the machines used to date have nearly constant COP (ratio of the cooling produced to energy supplied to the generator) as the generator temperature varies over the operating range as long as it is above a minimum value. The COP of LiBr-H_2O coolers is often in the range of 0.6 to 0.8. The effect of variation in the solar energy supply. temperature to the generator is to vary the capacity of the cooler. Water is used as a coolant, and the generator temperatures may be in the range of 75 to 95°C.

Major problem areas of system using solar operated LiBr-H_2O coolers include the following:

Supply temperatures of the fluid to the generator must be higher than the generator temperatures. There is a 'squeeze' between the supply temperatures required and the upper temperature limits of 100°C for unpressurized water storage tanks. Operation at 100°C is difficult with many collector types, particularly flat plate collectors. Coolihg towers are needed in these systems.

Absorption coolers for present solar power projects are being designed to use temperatures at low as 80°C. In order to maintain reasonable efficiencies are high temperatures, collectors must (*i*) be built to withstand potentially high pressures within the system, (*ii*) have special transparent cover plates that admit the highest possible amount of incident radiation, yet be specially coated to reduce radiation of the energy as heat; and (*iii*) have absorber plates that are of high grade metal, such as copper, with special selective coatings to increase absorptance and decrease emittance (reradiation of heat). In many cases, too, the collector must be made larger for the summer cooling mode than for the winter heating mode.

(B) NH_3—H_2O coolers. The schematic diagram of an ammonia water cooler is similar to that of Fig. (5.13) above, except that a rectifying section must be added to the top of the generator to reduce the water vapour content of the vapour stream going to the condenser. The basic solution processes are similar to those of the LiBr—H_2O system, except that the pressures and pressure differences are much higher. Mechanical pumps are required to return solutions from the absorber to the generator. In many applications the condenser and the absorber are air cooled, with generator temperatures in the range of 125 to 175°C. In applications where water cooling is used, generator temperature may be in the range of 95 to 120°C.

There has been relatively little work done on experimental operation of NH_3—H_2O systems with solar energy. The generator temperatures required in today's commercial coolers using air cooled condensers and absorbers are too high for present flat-plate collectors. Two approaches can be taken to this problem: Increase the temperature range of operation of collectors or reduce the generator temperature of the cooler. Work on NH_3—H_2O system has been directed at development of cycles using higher concentration of NH_3 and high effectiveness components to reduce generator temperature requirements. Collector developments such as evacuated tubular collectors, CPC (compound parabolic concentrator) or other focusing collectors, or well designed flat plate collectors with selective absorbers and low-reflectance covers may provide means of supplying energy at higher temperatures. In this case, new means of energy storage in the range of 100 to 170°C would have to be available.

(2) Intermittent absorption cooling

A modified method for absorption cooling which operates inter-mittently rather than continuously, is based on the following principle. In it, the system consists of two vessels which function in two alternative modes. In one (regeneration) mode, one of the vessels is the generator and the other is the condenser of an absorption system [Fig. 5.14 (*a*)]. During this phase, heat is supplied to the generator by oil, gas, steam or solar energy. In the alternative (refrigeration) mode, the first vessel becomes the absorber and the other the evaporator [Fig. 5.14 (*b*)]. During this phase refrigeration occurs. The system operates in the regeneration mode for a few hours and is then changed (manually or automatically) to the refrigeration mode, and so on. This technique can also be used for food preservation in rural areas, where electric power is not readily available (Fig. 5.14).

In the refrigeration mode, heat is supplied to a dilute solution of lithium bromide in water contained in the generator unit. Water vapor at a moderately high pressure passes to the condenser unit and is condensed by the cooling water [Fig. 5.14 (*a*)]. When sufficient liquid water has collected in the condenser, the heat supply and cooling water are shut off and the refrigeration mode becomes operative. The lithium bromide solution in the absorber unit is cooled so that its vapour pressure is lowered. This causes the water in the evaporator to vaporize, and as a result cooling occurs. The relatively low pressure water vapor is then absorbed by the solution in the absorber unit [Fig. 5.14 (*b*)]. After some time, the initial conditions are restored, and the system reverts to the regeneration mode.

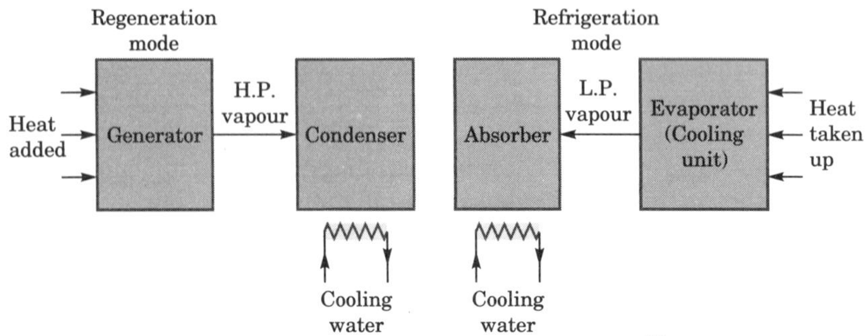

(*a*) (*b*)

Fig. 5.14. Intermittent absorption refrigeration

The other refrigerant absorbent combinations used in this system are ammonia water (NH_3—H_2O) and ammonia-sodium thiocynanate (NH_3—NasCN).

5.5. Solar Thermal Electric Conversion

Introduction. In general, the conversion of solar energy into electricity by way of thermal (or heat) energy. Heat can be converted directly into electrical energy by solar cell or thermionic or thermoelectric methods, but these techniques may not be suitable for use with the sun-generated heat. The most practical thermal electric procedure for solar energy is to utilize the energy to heat a working fluid (*e.g.,* a gas, water or other volatile liquid). The heat energy is then converted into mechanical energy in a *turbine* and finally into electrical energy by means of aconventional generator coupled to the turbine. This mechanical power production system is called *the solar thermal power production system.*

Solar thermal power genaration employs power cycles which are broadly classified as low, modium and high temperature cycles. Low temperature cycles generally use flat-plate collectors so that maximum temperatures are limited to about 100°C. Medium temperature cycles work at maximum temperatures ranging from 150 to 300°C, while high temperature cycles work at temperatures above 300°C.

For the low and medium temperature ranges, the thermodynamic cycles preferred is the Rankine cycle. For the high temperature range apart from the Rankine cycle, the Brayton and the Stirling cycles are also being considered. For the efficient conversion of heat energy into mechanical energy and hence into electricity, the working fluid should be supplied to the turbine at high temperature. To obtain such temperatures, above 175°C from solar energy requires the use of focusing/concentrating collectors. Since these collectors concentrate direct solar radiation, but essentially no diffuse radiation, they would be most effective in locations where there is ample sunshine. Two basic arrangements have been proposed for converting solar radiation into electrical energy: (1) the central receiver system and (2) the distributed collector system.

In the *central receiver system,* commonly known as the "power tower" design, an array of sun-tracking mirrors (heliostats) reflect solar radiation into a receiver mounted on top of a central tower. Solar energy absorbed in the central receiver is removed as heat by means of a heat transport fluid and converted into electrical energy in a turbine-generator.

The *distributed collector system* may consist of a number of parabolic trough-type (line focusing) collectors or of paraboloid dish type (point-focusing) collectors. The absorber pipes (or receivers) of the individual collectors are connected so that all the heated fluid is carried to a single location where the electricity is generated. The basic difference between the central receiver and the distributed collector systems is that in the former the solar energy falling on a large area is transmitted to a central point as radiation, but in the latter, the energy is carried as heat in a fluid.

Analysis of the two systems indicates that they may have different preferred applications. The energy losses in transmission by radiation are less than in transport as a hot fluid. Hence, higher temperatures should be attainable at the turbine inlet with the central receiver design than if the same amount of fluid were transported from distributed collectors. Furthermore, in a large distributed collector system, the cost of the long pipe lines and of the energy required to pump the heat-tranaport fluid through them would be considerable. However, these costs are offset somewhat by the high cost of heliostats and the central receiver tower.

It appears, therefore, that the central receiver system is preferable for the large scale generation of power for an electric utility. On the other hand, the distributed collector design may be more suitable for power plants of smaller electrical capacity, perhaps less than 2 MW. The so-called *total energy (or cogeneration) systems,* intended to supply both electric power and heat or process steam to an institution, small community, or industry, may fall into the later category.

Results todate show solar energy to be quite competitive with other sources of energy if the solar tower power plant is about 100-200 MWe, with 3-6 hours thermal storage.

Arizona public service company has proposed building of a 60 MW solar tower plant with a 175 m tall tower surrounded by 9000 heliostats each of 35 m^2 mirrors. At the central boiler a mixture of sodium chloride (60%) and potassium nitrate (40%) would be heated to temperatures above 550°C and the molten salt mixture would be transferred to a storage tank, followed by steam generation and turbine. The plant would have a potential to serve 3,70,000 barrels of oil each year.

Over the last few years, a few experimental tower power plants have been built or under construction in U.S.A., France, Italy and Japan. The largest of these is the solar thermal test facility at Albu-querque in New Mexico. It has a heliostats field of 222 mirrors each of 45 sq. m^2. area, formed by 25 mirror facets. The facets are mechanically shaped so that the reflected energy is focused. The heliostats are altazimuth mounted and their movement is controlled by a computer. The heliostats array is capable of concentrating in excess of 5 million thermal watts of power under optimum sun, heliostat and target conditions.

Thermal Electric Conversion Systems

In a solar thermal power production system the energy is first collected by using a solar pond, a flat plant collector or a focussing collector. This energy is used to increase the internal energy or temperature of a fluid. This fluid may be directly using in any of the common or known cycles such as Rankine, Brayton or

Stirling or passes through a heat exchanger to heat a secondary fluid (working fluid) which is being used in the cycle to produce mechanical power from which electrical power can be produced easily.

The following four systems are described in the following section:

(*i*) Low temperature cycles using flat plate collector or solar pond.

(*ii*) Concentrating collectors for medium and high temperature cycle.

(*iii*) Power tower concept or central receiver system.

(*iv*) Distributed collector system.

(1) *Low Temperature Systems*

The flat plate collector and solar ponds are classified as low temperature collectors because temperature achieved is of the order of 60°C to 100°C, with collection efficiency of 30 to 50%. If *Rankine cycle* solar thermal power production systems is employed, since the temperature of the fluid (water) is usually below 100°C (with solar pond the maximum temperature is limited approximately 80°C) and it is not possible to generate steam with flat plate collector or solar pond, so this can be used directly to run the prime mover. Therefore, some other organic fluid is used (Freon group etc.) which evaporates at low temperature and high pressure by absorbing the heat from the heated water. The vapour formed can be used to run a turbine or engine which may generate power, which will be sufficient to light the group of houses for rural areas and for irrigation purposes.

To convert solar energy into electricity through thermal conversion, researches in the world have done a considerable work in varying capacity power system. Organic fluid Rankine cycle has been extensively used in these studies, The development work in the generation of solar thermal power in higher range *i.e.,* 10 kW and above upto megawatts, is in line in various parts of the world. The systems have been developed by using a solar pond, flat plate collector, focusing collector (distribution type) or a heliostat. The heliostat systems are used normally in a very high range of solar thermal power production (around megawatts). (It will be described later).

A low temperature thermal electric power production scheme using solar pond is shown schematically in Fig. (5.15). Thermal energy from a solar pond is used to derive a Rankine cycle heat engine. Hot water from the bottom level of the

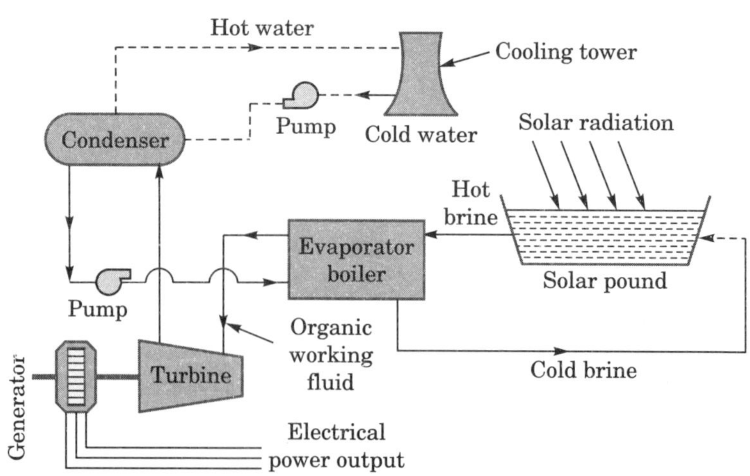

Fig. 5.15. Flow diagram of solar pond electric power plant.

pond is pumped to the evaporator where the organic working fluid is vaporized. The vapour flows under high pressure to the turbine and thereby expanding through the turbine wheel, and the electric generator linked to it. The vapour then travels to the condenser where cold water from the cooling tower condenses the vapour back it to a liquid. The liquid is pumped back to the evaporator where the cycle is repeated.

A 2000 sq. m. solar pond equipped with a 20 kW engine has been constructed at Australia. Another low temperature solar engine, using heated water from flat plate collector and butane as the working fluid is shown in Fig. (5.16), which is developed in France for lift irrigation.

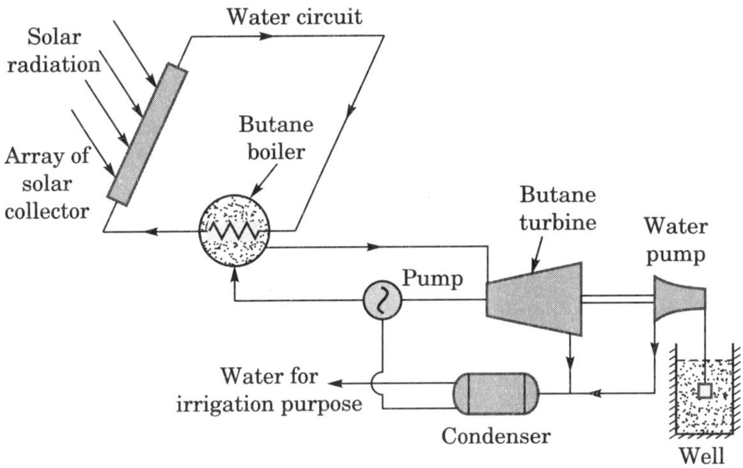

Fig. 5.16. Schematic of a low temperature solar power plant.

The system has array of flat-plate collectors to heat water upto nearly 70°C and in the heat exchanger, the heat of water is used for boiling butane. The high pressure butane vapour runs a butane turbine which operates a hydraulic pump which pumps the water from well and used for irrigation. The exhaust butane vapour from butane turbine is condensed with the help of water which is pumped by the pump. This condensate is fed to the heat exchanger or butane boiler.

Medium Temperature Systems with Concentrating Collectors. These systems generally employ an array of parabolic trough concentrating collectors, which give temperature above 100°C. General range of temperature is of the order of 250 to 500°C. As described earlier, a simple parabolic cylindrical concentrator for medium temperature system is shown in Fig. (5.17). It consists of a parabolic cylindrical reflector to concentrate sunlight on to a collecting pipe, within a pyrex or glass envelop. A selective coating of suitable material is applied to pipe to minimise infrared emission. Proper suntracking arrangement is made so that maximum sunlight is focused on the absorber.

High Temperature Systems. For the efficient conversion of heat energy into mechanical energy and hence into electricity, the working fluid should be supplied to the turbine at high temperature. To obtain such temperatures, above about 175°C, from solar energy requires the use of focusing or concentrating collectors. Since these collectors concentrate direct solar radiation, but essentially no diffuse radiation, they would be most effective in locations where there is ample sunshines (*e.g.*, in the western region of India). Two basic arrangements have been proposed for converting solar radiation into electrical energy:

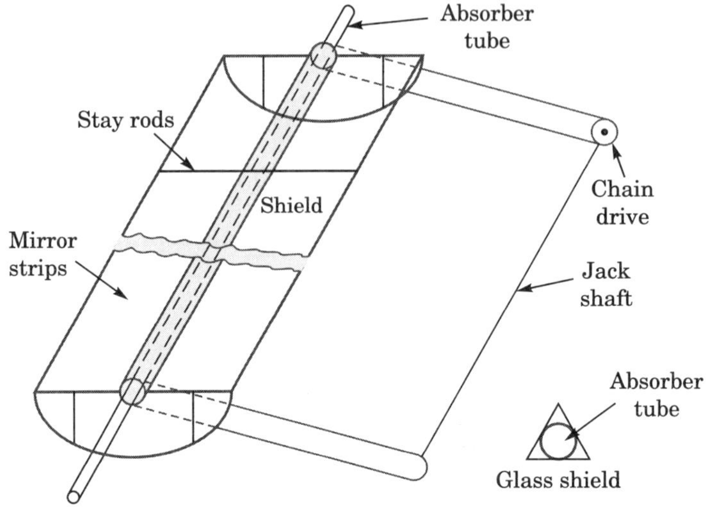

Fig. 5.17. A typical parabolic cylindrical concentrator.

(1) the central receiver system, and

(2) the distributed collector system or solar farms.

(1) Central Receiver Systems (Tower Power Plant)

Principle and Working. In this system as already stated, the incoming solar radiation is focused to a central receiver or a boiler mounted on a tall tower using Fig. (5.18) thousands of plane reflectors, which are steerable about two axes and are called heliostats. A schematic view of an electric power plant using gas turbine or gas turbine power plant working on Brayton gas power cycle is shown in Fig. (5.19). The mirrors are installed on the ground and are oriented so as to reflect the direct beam radiation into an absorber or recceiver (boiler) which is mounted on the top of a tower located near the centre of the field of mirrors to produce high temperature. This make it possible to position the boiler in the field of view of all mirrors, at all hours of the day. Beam radiation incident on boiler absorbed by black pipes in which working fluid circulates and is heated. The working fluid is allowed to drive a turbine and produce mechanical energy. The turbine which is coupled to an alternator produces electrical energy. As in many thermodynamics

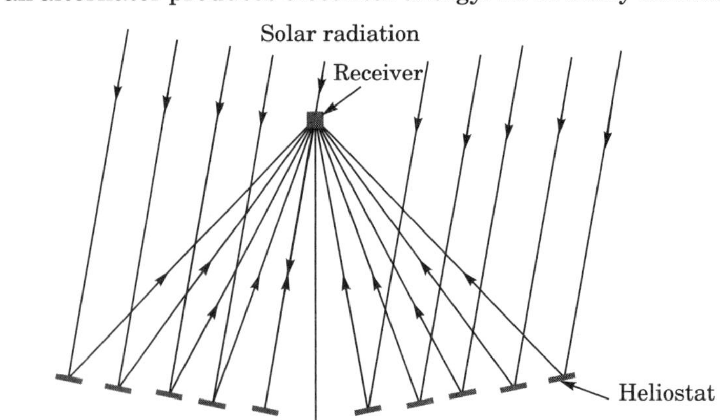

Fig. 5.18. Schematic arrangement of central receiver heliostat array.

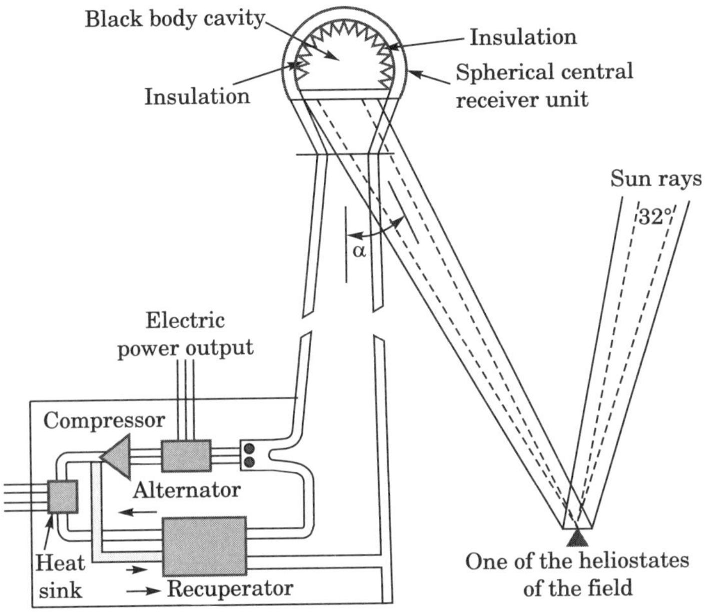

Fig. 5.19. Schematic of a Central Tower receiver associated with a field of flat mirrors and a gas turbine.

conversion, the heat sink is provided. A suitable heat storage is also provided to supply the heat energy during the periods of cloudiness.

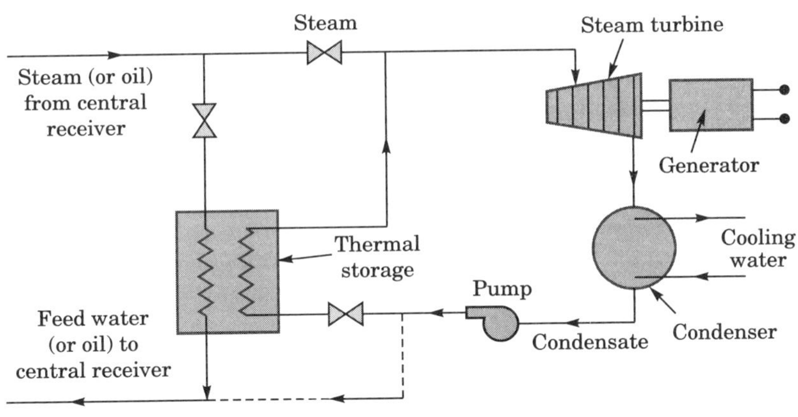

Fig. 5.20. Electric power generation using thermal storage.

Description of the system. This system can be subdivided into the following subsystems, namely.

(*a*) The tower with the central receiver on top of it,

(*b*) The heat conversion subsystem,

(*c*) The heat storage device, and

(*d*) The field of oriented mirrors.

Receiver Subsystem. The central receiver at the top of the tower has a heat absorbing surface (*e.g.*, panels coated with a heat absorbing material) by which the heat-transport fluid is heated. Two basic receiver configurations has been proposed; they are the *cavity* and *external receiver* pipe-types. In the cavity type, pipes line, the interior of a cavity; the solar radiation reflected by the heliostats enters through an aperture at the bottom of the cavity. On the other hand, in the external receiver type the absorber surfaces are on the exterior of a roughly cylindrical structure.

Heat losses by radiation and convection are generally less for the cavity type than for the external receiver configuration, but focusing may be more critical because, in order to be absorbed, the radiation must enter through the cavity aperture. Furthermore, the external receiver has the advantages of simplicity, modular panel construction, easier access for maintenance, a larger absorber area. The external receiver design has been recommended, for the 10 mW pilot plant at California.

Heat-Transport Subsystem. Water is a convenient heat-transport fluid, and in the form of steam it is the working fluid in the majority of existing turbines for generating electricity. Hence water will be used, at least in the earliest solar thermal electric plants. Liquid water *(i.e.,* turbine condensate) under pressure enters the receiver, absorbs heat energy, and leaves as superheated steam; typical steam conditions might be a temperature of 500°C and a pressure of about 100 atm (10 MPa). The steam is piped to ground level where it drives a conventional turbine generator system. As in all steam-electric plants, heat is rejected to the condenser cooling water.

In more advanced central receiver system, the heat transport fluid might be liquid sodium or a molten mixture of salts at about atmospheric pressure. At the bottom of the tower the high temperature liquid is circulated through a heat exchanger, where the heat would be transferred to water to generate stream at a high temperature and pressure. Thus a high efficiency of conversion into electricity could be achieved without the need for high pressure in the receiver. Another possibility is to use a gas as the heat transport medium and also as working fluid in a gas turbine. Thus two basic thermodynamic conversion methods *i.e.,* the Rankine and Brayton cycles are used for the subsystem that converts heat into electrical power.

In the Brayton cycle or gas turbine, a gas is compressed and heated and then expanded through a turbine, which is coupled to the generator. The exhaust gases may serve to preheat the compressed gas in a regenerator or recuperator. Air, helium, argon or other gases can be used. At present, an efficiency of 30% at 600°C is typical.

Energy Storage. As in all the power systems attention must be paid to the energy storage problem. It seems necessary to provide thermal storage, at least for short periods, so as to avoid the loss of time which would otherwise occur after an occasional shut-down, due to lack of sun shine while the plant reached its operating temperature again. Short term storage of heat can be provided by fire bricks, ceramic oxides (MgO...), fused salts ($NaNO_3$ melts at 260°C, Hitec a fused salt mixture is stable upto 540°C), sulphur (liquid between 113°C and 444°C) or lithium metal (liquid between 180°C and 1400°C), sodium metal (liquid between 98°C and 880°C). Some organic materials are also suitable upto 300°C. Current work is focused on storage by rocks, eutectic salts, and some synthetic organic materials of low vapour pressure, such as Gilotherm. The choice of a conventional storage material is determined by its energy density, thermal conductivity, corrosion characteristic, cost and convenience of use, as well by the operating temperature of the working fluid. The storage space must be well insulated against heat losses.

Mirrors. The flat mirror surface can be manufactured by metallization of float glass or flexible plastic sheets. In view of the life expectancy of plant (upto 30 years say), plastics seem less appropriate, because their optical and mechanical properties are liable to change with time. Float glass, metallized with silver or aluminium provides reflectivities of 93% and 82% to 86% respectively, subject to cleaning.

The mirror must be steerable. Without a supporting structure, for example a steel beam panel or a honeycomb core, the glass mirrors would not be capable of withstanding the wind loads that often occur in arid lands. The larger the area of the mirror, the greater must be the rigidity of the supporting structure. As a result the heliostats are heavy; a 50 m^2 mirrors weight about 2 tonnes.

The *characteristics* of the system are summarized below:

(*i*) All parts of it use known technologies.

(*ii*) The heat conversion subsystem comprising a turbine and an alternator may be of conventional type, thus avoiding the need of future development work.

(*iii*) Boiler or absorber is a light absorber, and low volume unit, resulting in low heat losses from it.

(*iv*) The heat need not be transported over long distances as compared to parabolic troughs, thus avoiding excessive plumbing and heat, losses, all the power is transferred to the central point by optical transmission.

(2) Distributed Collector System (or Solar Farms)

Solar farms have been proposed using parabolic trough concentrators or other type of concentrators to focus sunlight on to a central pipe surrounded by an evacuated quartz envelope. Heat collected by a fluid flowing through the pipes could be stored at a temperature above 500°C in a molten eutectic salt, and used as required to produce steam for electric power generation.

As noted earlier, distributed solar energy collectors may be preferred for thermal power plants of moderately small capacity (about 2 MW or less). Among the potential applications of such small plants, two under study are:

(1) total energy systems, and

(2) pumping water for irrigated farms.

Total Energy (or Opgeneration) Systems. The total energy concept is based on the sequential use (or cascading) or solar energy at two different temperature levels. First, a working fluid at a higher temperature can be used to drive a turbine and generate electricity in the usual manner. Then the heat discharged in the condenser cooling water at a lower temperature can be utilized for space heating and cooling. Alternatively, in a cogeneration system, the turbine discharge may be used to provide process steam for industry. For the discharged heat to be useful, the temperature should be higher than in a conventional steam electric power plant. As a result, the efficiency for electricity generation is decreased, but this can be more than offset by the economic value of the discharged heat.

Irrigation water pumping. Irrigation pumps of relatively low power can be driven directly by a vapour turbine using solar energy as described earlier. For deep wells, pumps of higher power are required, and these are preferably operated by electric motors. In a deep-well irrigation project, solar energy is used to heat a hydrocarbon oil to about 290°C by passage through an array of parabolic trough type concentrating collectors. The heat fluid passes by way of a thermal storage tank, to a heat exchanger where toluene is boiled to provide vapour for operating a turbine generator combination. Pumped water is used as a coolant in the turbine condenser, and the temperature is more like a conventional power plant, The electric power which can be generated is of the order of 150 kW drives electric motors to operate a deep-well pumps.

5.6. Solar Electric Power Generation: Solar Photo-Voltaics

Introduction. The direct conversion of solar energy into electrical energy by means of the photovoltaic effect, that is, the conversion of light (or other electromagnetic radiation) into electricity. The photovoltaic effect is defined as the generation of an electromotive force as a result of the absorption of ionizing radiation. Energy conversion devices which are used to convert sunlight to electricity by the use of the photovoltaic effect are called *solar cells.* A single converter cell is called a solar cell or, more generally, a *photovoltaic cell,* and combination of such cells; designed to increase the electric power output is called a *solar module* or *solar array.*

Photovoltaic cells are made of semiconductors that generate electricity when they absorb light. As photons are received, free electrical charges are generated that can be collected on contacts applied to the surfaces of the semiconductors. Because solar cells are not heat engines, and therefore do not need to operate at high temperatures, they are adopted to the weak energy flux of solar radiation, operating at room temperature. These devices have theoretical efficiencies of the order of 25 percent. Actual operating efficiencies are less than half this value, and decrease fairly rapidly with increasing temperature.

The best known application of photovoltaic cells for electrical power generation has been in space craft, for which the silicon solar cell is the most highly developed type. The silicon cell consists of a single crystal of silicon into which a doping material is diffused to form a semi-conductor. Since the early days of solar cell development, many improvements have been made in crystal growing and doping, electrical contact and cell assembly and production methods. Large number of cells have been manufactured with areas 2 × 2 cm, efficiencies approaching 10 percent, and operating at 28°C. The *efficiency* is the power developed per unit area of array divided by the solar energy flux in the free space (1.353 kW/m^2).

For terrestrial applications, silicon solar cells have shown operating efficiencies of about 12 to 15 percent. Though silicon is one of the earth's most abundant materials, it is expensive to extract (from sand, where it occurs mostly in the form SiO_2) and refine to the purity required for solar cells. The greater barrier to solar cell application lies in the costs of the cells themselves. Reducing the cost of silicon cells is difficult because of the cost of making single crystal. One very promising method is being developed to produce continuous thin ribbons of single-crystal silicon to reduce fabrication costs. Cells made from the ribbon have so far shown efficiencies of around 8 percent. Several other kinds of photocells are in the laboratory stage of development. Cadmium sulfide and CdS/Cu_2S cells are other possibilities. So far, efficiencies have been in the range of 3 to 8 per cent, and these cells have been less durable than silicon cells owing to degradation with exposure to oxygen, water vapour and sunlight, especially at elevated temperatures. The active part of the CdS cell is a thin polycrystalline layer of CdS, about 10 μm thick, on which a layer of Cu_2S compound perhaps 0.1 μm thick is grown. These cells can be made by deposition on long sheets of substrates, a process that might be adaptable to expensive mass production.

Photovoltaic cells could be applicable to either small or large power plants, since they function well on a small scale, and may be adaptable to local energy generation on building roof tops. The cost of energy storage and power conditioning equipment might, however, make generation in large stations the most economical method. Solar cells have also been used to operate irrigation pumps, navigational signals, highway emergency call systems, rail road crossing warnings, automatic meterological stations, etc., in location where access to utility power lines is difficult.

A PV (photo-voltaic) system consists of:

(*i*) Solar cell array (*ii*) Load leveller

(*iii*) Storage system (*iv*) Tracking system (where necessary).

In actual usage, the solar cells are interconnected in certain series/parallel combinations to form modules. These modules are hermetically sealed for protection against corrosion, moisture, pollution and weathering. A combination of suitable modules constitutes an array. One square metre of fixed array kept facing south yields nearly 0.5 kWH of electrical energy on a normal sunny day if the orientation of the array is adjusted to face the sun's rays at any time, the output can increase by 30 per cent. Solar PV system can produce an output only if sunlight is present. If it is required to be used during non-sunshine hours, a suitable system of storage batteries will be required.

Solar Cell Principles. The photovoltaic effect can be observed in nature in a variety of materials, but the materials that have shown the best performance in sunlight are the semiconductors as stated above. When photons from the sun are absorbed in a semiconductor, they create free electrons with higher energies than the electrons which provide the bonding in the base crystal. Once these electrons are created, there must be an electric field to induce these higher energy electrons to flow out of the semiconductor to do useful work. The electric field in most solar cells is provided by a junction of materials which have different electrical properties.

To obtain a useful power output from photon interaction in a semiconductor three processes ore required.

1. The photons have to be absorbed in the active part of the material and result in electrons being excited to a higher energy potential.

2. The electron hole charge carrier created by the absorption must be physically separated and moved to the edge of the cell.

3. The charge carriers must be removed from the cell and delivered to a useful load before they loose their extra potential.

For completing the above processes, a solar cell consists of:

(*a*) Semiconductor in which electron hole pairs are created by absorption of incident solar radiation.

(*b*) Region containing a drift field for charge separation, and

(*c*) Charge collecting front and back electrodes.

The photovoltaic effect can be described easily for *p-n* junction in a semiconductor. In an intrinsic semiconductor such as silicon, each one of the four valence electrons of the material atom is tied in a chemical bond, and there are no free electrons at absolute zero. If a piece of such a material is doped on one side by a five valence electron material, such as arsenic or phosphorus, there will be an excess electrons in that side, becomes an *n*-type semiconductor. The excess electrons will be practically free to move in the semiconductor lattice. When the other side of the same piece is dopped by a three valence electron material, such as boron, there will be deficiency of electrons leading to a *p*-type semiconductor. This deficiency is expressed in terms of excess of holes free to move in the lattice. Such a piece of semiconductor with one side of the *p*-type and the other of the *n*-type is called a *p-n* junction. In this junction after the photons are absorbed, the free electrons of the *n*-side will tend to flow to the *p*-side, and the holes of the *p*-side will tend to flow to the *n* region to compensate for their respective deficiencies. This diffusion will create an electric field E_F from the *n* region to the *p*-region. This field will increase until it reaches equilibrium for V_e, the sum of the diffusion potentials for holes and electrons. If electrical contacts are made with the two semiconductor materials and the contacts are connected through an external electrical conductor, the free electrons will flow from the *n*-type material through the conductor to the *p*-type material (Fig. 5.21). Here the free electrons will enter the holes and become bound electrons; thus, both free electrons and holes will be removed. The flow of electrons through the external conductor constitutes an electric current which will continue as long as more free electrons and holes are being formed by the solar radiation. This is the basis of photovoltaic conversion, that is, the conversion of solar energy into electrical energy. The combination of *n*-type and *p*-type semiconductors thus constitutes a photovoltaic (PV) cell or solar cell. All such cells generate direct current which can be converted into alternating current if desired.

The most normal configuration for a solar cell to make a *p-n* junction semiconductor is as shown schematically in Fig. (5.21). The junction of the '*p* type' and '*n* type' materials provides an inherent electric field which separates the charge created by the absorption of sunlight. This *p-n* junction is usually obtained by putting *p*-type base material into, a diffusion furnace containing a gaseous *n*-type dopant such as phosphorus and allowing the *n*-dopant to diffuse into the surface about 0.2 μm. The junction is thus formed slightly below the planar surface of the cell and the light impinges perpendicular to the junction. The positive and negative charges created by the absorption of photons are thus encouraged to drift to the front and back of the solar cell. The back is completely covered by a metallic contact

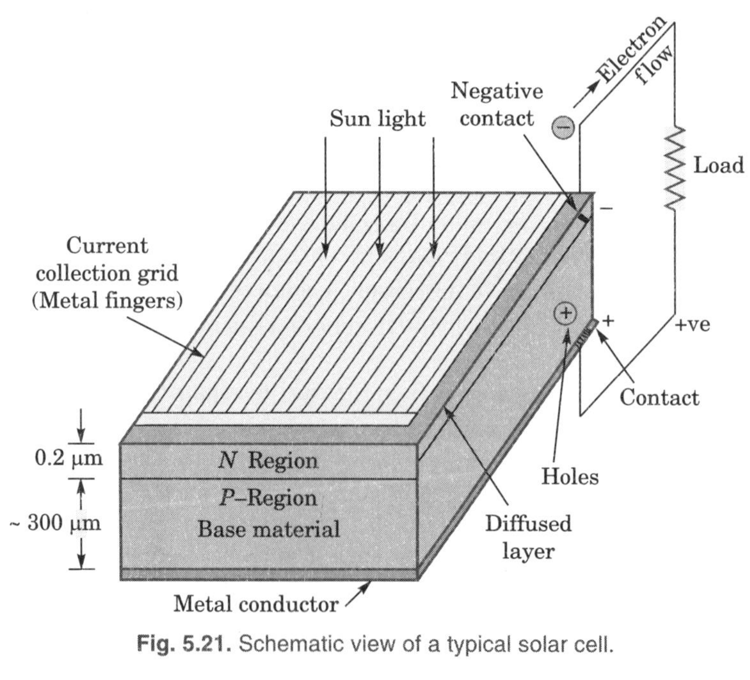

Fig. 5.21. Schematic view of a typical solar cell.

to remove the charges to the electric load. The collection of charges from the front of the cell is aided by a fine grid of narrow metallic fingers. The surface coverage of the conducting collectors is typically about 5 per cent in order to allow as much light as possible to reach active junction area. An anti reflective coating is applied on the top of the cell. Fig. 5.22 demonstrates how this *p-n* junction provides an electrical field that sweeps the electrons in one direction and the positive holes in the other. If the junction is in thermodynamic equilibrium, then the *Fermi energy,* must be uniform throughout, Since the Fermi level is near the top of the gap of an *n*-doped material and near the bottom of the *p*-doped side, an electric field must exist at the junction providing the charge separation function of the cell. Important, characteristic of the Fermi level is that, in thermodynamic equilibrium, it is always continuous across the contact between the two materials.

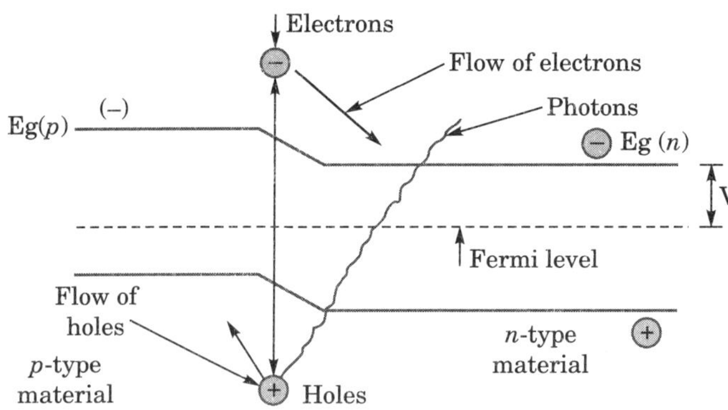

Fig. 5.22. *p-n* junction electric fields.

Each of the individual solar cells will produce power at about 0.5 V with the current directly proportional to the cell's area. The individual cells are connected in series-parallel combination to meet the voltage, power and reliability requirements of the particular application. Space cells are covered with transparent 'cover slips' to absorb the high energy particles in space that could cause damage in the cell and result in a degradation of output. For terrestrial applications, the solar cell panels have to be encapsulated to protect them from atmospheric degradation due to oxidation of the metal contacts, which would cause peeling and open circuits materials such as glass, acrylics or silicon epaxies are used to provide a clear, weather fight front covering for the panels.

Semiconductor Junctions. Modern solar cells make use of *semiconductor* materials, usually based on single-crystal silicon. When doped with phosphorus, arsenic, or antimony, the silicon becomes an *n*-type semiconductor as stated earlier; and when doped with boron, aluminium, indium, or gallium, it forms a *p*-type semiconductor. If a *p*-type semiconductor is brought into intimate contact with one of the *n*-type, they form a *p-n* (or *n-p*) junction. If the two semiconductor materials are derived from the same element (or compound), such as silicon, the system is referred to as a *homo-junction.* It is also possible for a *p-n* (or *n-p*) junction to be formed from two different semiconductor materials, such as cadmium sulfide (CdS) and cuprous sulfide (Cu_2S); this is known as a *hetero-junction.* The general behaviour at the junction, as outlined above, is the same regardless of the type. The *Schottky junction* consisting of a semiconductor and metal. This junction is formed by depositing a thin layer of a metallic conductor (*e.g.,* platinum) into a *p*- or *n*-type semiconductor. Schottky junction photovoltaic cells made with the so-called amorphous silicon are more efficient than homojunction *p-n* cells of the same material. Cost of it, is also less.

The MIS (metal insulator-semiconductor) solar cell is similar to the Schottky type except that a very thin layer (about 0.1 to 0.3 μm) of an insulator is deposited between the semiconductors or and the metallic conductor. A conversion efficiency of more than 17 percent has been reported for an MIS solar cell made with single crystal silicon.

Conversion Efficiency and Power Output

A solar cell usually uses a *p-n* junction its physical configuration is shown schematically in Fig. 5.21. Current and voltage relationship is given by

$$j_i = j_o \left[\exp\left(\frac{Ve}{KT}\right) - 1 \right] 1 \qquad ...(5.1)$$

where,

j_o = Saturation current also called the dark current and is obtained when a large negative voltage is applied across the diode.

V = The voltage across junction

e = The electronic charge

k = Boltzmann's constant

T = The absolute temperature

when light impinges on the junction, electron hole pairs are created at a constant rate providing an electrical current flow across the junction, the net current is thus the difference between the normal diode current and light generated current j_L. Simplified equivalent circuit for the cell is shown in Fig. 5.23. The internal series resistance R_s is mostly due to the high sheet resistance of the diffused layer which is in series with the junction. The light generated current acts as a constant current source supplying the current to either the junction or a useful load depending on the junction characteristic and the value of the external load resistance. The net current J is given by

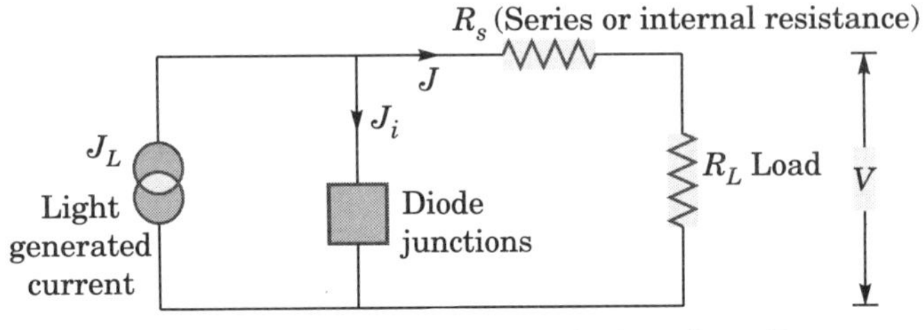

Fig. 5.23. The equivalent circuit of a solar cell.

$$J = J_L - J_i = J_L - J_0 \left[\exp\left(\frac{Ve}{kT}\right) - 1 \right] \qquad ...(5.2.)$$

The internal voltage drop in a cell can usually be minimized, and for ideal cell R_s may be assumed equal to zero *i.e.*, $R_s = 0$. With these the corresponding $J - V$ plot is given in Fig. 5.24. Open circuit voltage V_{0C} for the ideal cell is then given by

$$V_{0C} = \left(\frac{kT}{e}\right) \ln\left[\frac{J_L}{J_0} + 1\right] \qquad ...(5.3)$$

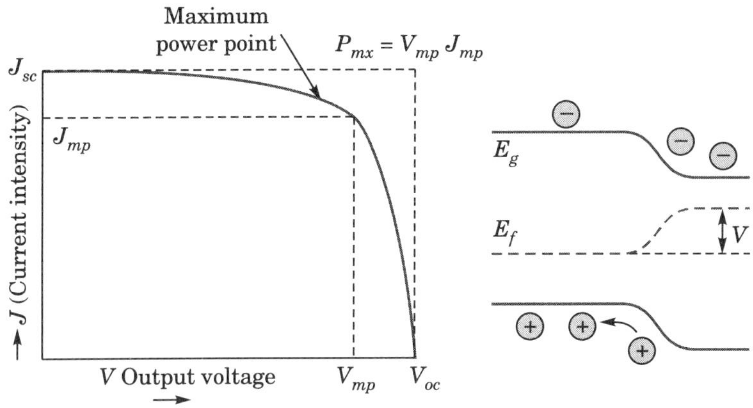

(a) Current voltage curve (b) Electron energy diargam

Fig. 5.24. A typical J — V plot for an ideal solar cell.

$\because$ $J_L >> J_0$, the 1 in the equation can be neglected. Then open circuit voltage

$$V_{0C} = \frac{kT}{e} \ln \frac{J_L}{J_0} \qquad ...(5.4)$$

In practice the open circuit voltage of the cell decreases with increasing temperature.

The maximum power that can be derived from the device is given by

$$P_{max} = V_{mp} . J_{mp} \qquad ...(5.5)$$

where V_{mp} and J_{mp} are the voltage and current at maximum power point as shown in Fig. 5.24 respectively. It can be seen that the maximum efficiency for the cell is obtained by dividing V_{mp} J_{mp} by the total power density of the sunlight P_{sun}.

Thus
$$\eta = \frac{V_{mp} J_{mp}}{P_{sun}}$$

$$= \left(\frac{J_L E_g}{e P_{sun}}\right) \underbrace{\left(\frac{J_{mp} V_{mp}}{J_L V_{0C}}\right)}_{\text{Fill factor}} \underbrace{\left(\frac{e V_{0C}}{E_g}\right)}_{\text{Voltage factor}} \qquad ...(5.6)$$

[where E_g = Forbidden energy gap]

The fill factor (FF) for a solar cell is defined as the ratio of two areas shown.

or
$$FF = \frac{J_{mp} V_{mp}}{J_L \times V_{0C}} \qquad ...(5.7)$$

Solar cell designers, strive to increase the fill factor values, to minimize internal losses. Maximum power can be defined in terms of V_{0C} and J_L and is given by

$$P_{max} = J_L \times V_{0C} \times FF \qquad ...(5.8)$$

A typical value of the fill factor for a good silicon cell is about 0.8.

The voltage factor $\frac{e V_{0C}}{Eg}$ is determined by the basic properties of the materials in the cell and is typically about 0.5 for a silicon cell.

For both theoretical and practical reasons, not all of the solar radiation energy falling on a solar cell can be converted into electrical energy. A specific amount of energy is required to produce a free electron and a hole in the sami-conductor material. In silicon for example, the energy minimum is 1.1 electron volts and this is available in radiation having a wavelength of 1.1 micrometers (μm or less). Consequently infrared radiation of longer wavelength has no photovoltaic effect in silicon but is largely absorbed as heat. Furthermore, even the energy of radiation with shorter wavelengths can not be used completely; energy in excess of that needed to free a bound electron is simply converted into heat.

For the foregoing reasons, only about 45 percent of the energy in the solar radiation at sea level is capable of producing electrons and holes in silicon. However, because of the electrical resistance of the semi-conductor material and other loss modes, the maximum practical efficiency for the conversion of solar energy into electrical energy in a silicon solar cell is estimated to be about 21 per cent. Photovoltaic semi-conductors with conversion efficiencies upto about 25 per cent or more are known (*e.g.* gallium arsenide), but it is uncertain if the extra conversion efficiency can compensate for the additional cost, except in special circumstance.

Because of internal losses arising from minute amounts of impurities, from defects in the silicon crystal, and from recombination of electrons and holes before they can be separated, and external losses from reflection, most commercial silicon solar cells have a conversion efficiency of roughly 13 to 14 percent (average about 12 per cent). However cells have been made with efficiencies of 18 percent, and this level will probably be attained in commercial cells.

The *power* output (in watts) of any generator of electricity, including a photovoltaic cell, is equal to the product of the voltage and current. Theoretically, a silicon solar cell should have a voltage of 1.1 volts, from 1.1 electron-volts energy of the free electrons produced. In practice, however, the maximum voltage is about 0.6 volt and this occurs on open circuit, when no power is produced. The maximum power of a silicon cell occurs at an output voltage of approximately 0.45 volt. In full sunlight, the current from a commercial cell is then roughly 270 amperes per sq. m of exposed surface. The power is thus about $0.45 \times 270 = 120$ watts (or 0.12 kW) per sq. m (*i.e.,* 13 watts/sq. ft.).

The rate at which solar energy reaches the top of the atmosphere (*i.e.,* the solar constant) is 1.353 kilowatts/sq. m (1.353 kW/sq. m). Part of this energy is reflected back to the space, and part is absorbed by the atmosphere. In full sunlight, the solar energy may reach the ground at a rate of roughly 1 kW/sq. m.

The electric power output of a photovoltaic cell is roughly proportional to the rate at which solar radiation falls on its surface. Hence the output of a cell of a given area can be increased by combining it with a concentrating collector. Tracking collectors of the line focus type can provide a concentration of a few hundred fold. With the compound parabolic (non focusing) concentrator (CPC), a concentration factor of about ten is possible without tracking. Since focusing collectors concentrate direct solar radiation only (but not diffuse radiation), they would be most useful in regions of high insolation. However, this limitation might not be applicable to the CPC, since it can collect diffuse (scattered) radiation.

Most of the solar energy that is not converted into electricity in a photovoltaic cell is absorbed as heat. In a commercial, single-crystal silicon cell, for example, with a conversion efficiency of about 12 per cent, more than 80 per cent of the incident solar energy appears as heat in the cell. As the temperature increases, the efficiency of an *n-p* homojunction silicon cell will decrease, for example, for a commercial cell it is about 8 per cent at 100°C. Cells made of gallium arsenide are superior to silicon cells for operation at elevated temperatures; their higher cost may then be justified when used with concentrating collectors. The conversion efficiency of a practical gallium arsenide solar cell is roughly 18 per cent at ordinary temperatures, 16 percent at 100°C and 12 percent at 200°C. Gallium arsenide is a good absorber of solar radiation and *n-p* homojunction can be made from thin films of polycrystalline material. High conversion efficiencies with concentrating collectors have been reported with cells made from a combination of gallium aluminium and gallium arsenide.

A Basic Photovoltaic System for Power Generation

A basic photovoltaic system integrated with the utility grid is shown in Fig. 5.25. It permits solarly generated electrical power to be delivered to a local load. It consists of:

(*i*) *Solar Array,* large or small, which converts the insolation to useful DC electrical power.

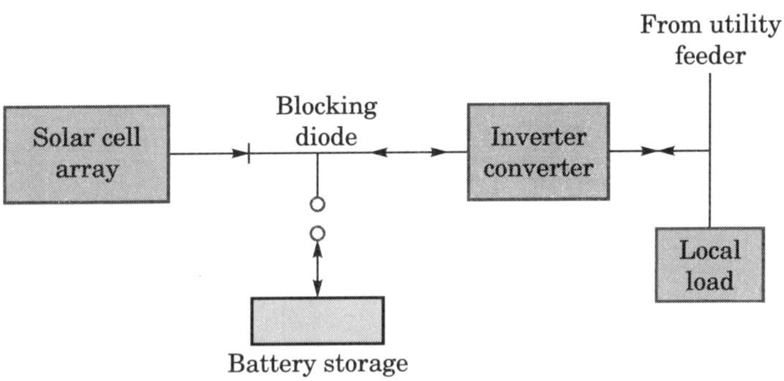

Fig. 5.25. Basic photovoltaic system integrated with power grid.

(*ii*) *A Blocking Diode* which lets the array-generated power flow only toward the battery or grid. Without a blocking diode the battery would discharge back through the solar array during times of no insolation (recall from Fig. 5.23 that the cell equivalent circuit has a forward biased diode in it).

(*iii*) *Battery Storage,* in which the solarly generated electric energy may be stored.

(*iv*) *Inverter/converter,* usually solid state which converts the battery bus voltage to AC of frequency and phase to match that needed to integrate with the utility grid. Thus it is typically a DC, AC inverter. It may also contain a suitable output step up transformer, perhaps some filtering and power factor correction circuits and perhaps some power conditioning, *i.e.,* circuitry to initiate battery charging and to prevent over charging. Power conditioning may be shown as a separate system functional block. This block may also be used in figure shown to function as a rectifier to charge the battery from the utility feeder when needed and when no insolation was present.

(*v*) *Appropriate Switches and Circuit Breakers,* to permit isolating parts of the system, as the battery. One would also want to include breakers and fusing protection (not shown) between the inverter output and the utility grid to protect both the photovoltaic system and the grid.

Solar Cell Modules (Solar Photovoltaic Arrays)

There may be tracking arrays or modules or fixed arrays. A tracking array is defined as one which is always kept mechanically perpendicular to the sun-array line so that all times it intercepts the maximum insolation. Such arrays must be physically movable by a suitable prime mover and are generally considerably more complex than fixed arrays. A fixed array is usually oriented east west and tilted up at an angle approximately equal to the latitude of the site. Fixed arrays are mechanically simpler than tracking arrays. Thus the array designs fall into two broad classes:

(*i*) *Flat-plate Arrays.* Wherein solar cells are attached with a suitable adhesive to some kind of substrate structure usually semi-rigid to prevent cells being cracked.

This technology springs from the space—related photovoltaic technology, and many such arrays have been built in various power sizes.

(*ii*) *Concentrating Arrays.* Wherein suitable optics, *e.g.,* Fresnel lenses, parabolic mirrors, compound parabolic concentrators (CPC), and others, are combined with photovoltaic cells in an array fashion. This technology is relatively new to photovoltaics in terms of hardware development, and comparatively fewer such arrays have actually been built.

Solar cell connecting arrangements. Cells may be connected in parallel to achieve the desired current and then stacked in series to achieve the desired voltage. The optimum operating voltage of a photovoltaic cell is generally about 0.45 volt at normal temperatures, and the current in full sunlight may be taken to be 270 amperes/sq. m. If the exposed area of a cell is 40 sq. cm or 40×10^{-4} sq. m the current, would be 1.08 amperes and the electric power output $0.45 \times 1.08 = 0.49$ watts, in full sunlight. A decrease (or increase) in the solar radiation has little effect on the voltage, but the current and power are decreased (or increased) proportionately.

By combining a number of solar cells in series (*i.e.,* in a string) the voltage is increased but the current is unchanged. For example, 110 volts, for operating commercial tools, motors, or domestic appliances, would require $\frac{110}{0.45} = 244$ cells in series. To increase the current output at the same time, several strings of 244 cells would be connected in parallel, as depicted in Fig. (5.26).

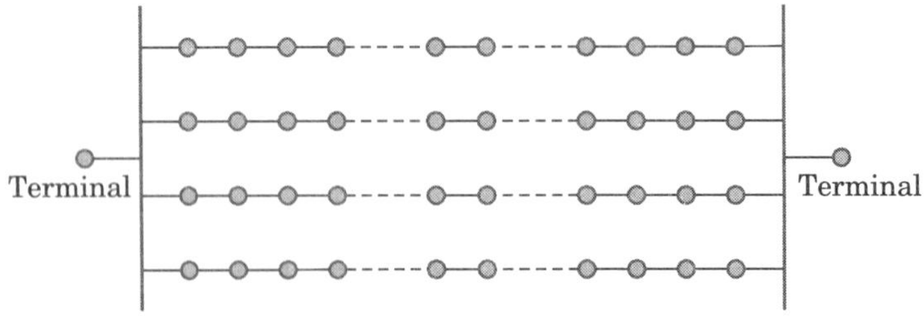

Fig. 5.26. Solar cell arrangement in series and parallel.

Suppose there were ten such strings in parallel, the current under optimum conditions would then be $10 \times 1.08 = 10.8$ amperes, and the power output would be $10.8 \times 110 = 1190$ watts or 1.19 kW. The so-called solar panels on space craft consist of modules (or arrays) of cells connected in series and parallel to produce the required voltage and power.

If even a single cell in a string should fail, the whole string would become inoperative. The cells in the remaining strings would maintain the voltage, but the current (and power) output of the system would be decreased by the loss of one string of cells. A short circuit in a cell would not disable the string, although there would be a slight drop in voltage. There is a possibility that the other cells might cause current to flow in the wrong direction through any string having a reduced voltage. This danger is eliminated by including a diode, which permits current flow in one direction only, at end of each string. Instead of a number of strings of cells in parallel, the current (and power) could be increased by locating a single string of cells at the focal line of a sun tracking, parabolic trough, concentrating collector. There would be some decrease in voltage because of the inevitable higher temperature of the solar cell material, but the current would in crease approximately in proportion to the concentration factor of the collector. Thus, with a concentration factors of 100, the current from a single string would be increased to about 100 amperes. If the number of cells in the string is sufficient to produce 110 volts, the total electric power output would be approximately 12 kW.

Battery Storage. The simplest means of storage on a smallar moderate scale is in electric *storage batteries,* especially as solar cells produce the direct electric current required for battery charging. The stored energy can then be delivered as electricity upon discharge. The common lead acid storage batteries, such as are used in automobiles, are not ideal for this purpose, but they are probably the best presently available. Extensive research in progress should lead to the develop-ment of more suitable batteries.

A possible alternative is to use the direct current from solar cells to decompose water (by electrolysis) into hydrogen and oxygen gases. These gases would be stored in a suitable form and utilized as needed to generate electricity in a *fuel cell.*

Inverters. These are the devices usually solid state, which change the array DC output, to AC of suitable voltage, frequency, and phase to feed photovoltaically generated power into the power grid or local load, as shown in figure. These functional blocks are sometimes referred to as power conditioning.

A general type of inverter circuit, which is found best suitable for the utility application is shown in Fig. (5.27). The current can be used in two modes: (1) as an inverter changing DC to AC or (2) as a rectifier changing AC to DC, thus charging the battery.

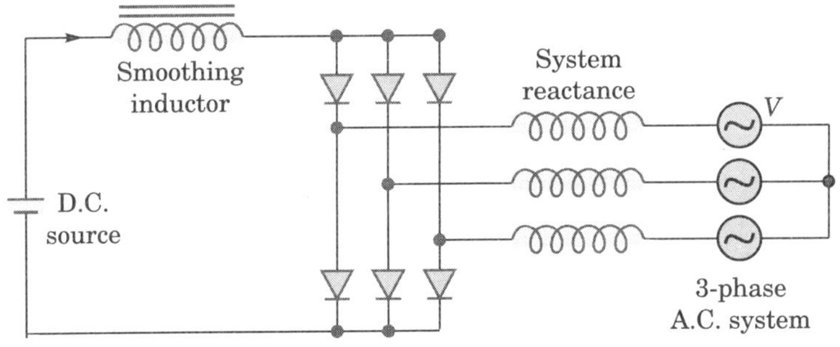

Fig. 5.27. Current-fed line commutated inverters.

It is clear that the system photovoltaics offers the options of DC power, AC power, hydrogen and oxygen fuels in either gas or liquid forms from which electricity can be generated. The system has many advantages and disadvantages.

Applications of Solar Photovoltaic System

Various solar photovoltaic systems have been developed and installed at different sites for demonstration, and field trial purposes. The *terrestrial* applications of these include provision of power supply to:

(*i*) water pumping sets for micro irrigation and drinking water supply,

(*ii*) radio beacons for ship navigation at ports,

(*iii*) community radio and television sets,

(*iv*) cathodic protection of oil pipe lines,

(*v*) weather monitoring, (*vi*) railway signalling equipment,

(*vii*) battery charging, (*viii*) street lighting.

The major application of photovoltaic systems lies in water pumping for drinking water supply and irrigation in rural areas. The photovoltaic water pumping system essentially consists of:

(*a*) a photovoltaic (PV) array, (*b*) storage battery,

(*c*) power control equipment, (*d*) motor pump sets, and

(*e*) water storage tank.

Advantages and Disadvantages of Photovoltaic Solar Energy Conversion

Advantages: (*i*) Direct room temperature conversion of light to electricity through a simple solid state device.

(*ii*) Absence of moving parts.

(*iii*) Ability to function unattended for long periods as evidence in space programme.

(*iv*) Modular nature in which desired currents, voltages and power levels can be achieved by mere integration.

(*v*) Maintenance cost is low as they are easy to operate.

(*vi*) They do not create pollution.

(*vii*) They have a long effective life.

(*viii*) They are highly reliable.

(*ix*) They consume no fuel to operate as the sun's energy is free.

(*x*) They have rapid response in output to input radiation changes; no long-time constant is involved, as on thermal systems, before steady state is reached.

(*xi*) They have wide power handling capabilities from microwatts to kilowatts or even megawatts when modules are combined into large area arrays. Solar cells can be used in combination with power conditioning circuitry to feed power into utility grid.

(*xii*) They are easy to fabricate, being one of the simplest of semi conductor devices.

(*xiii*) They have *high power to weight ratio,* this characteristic is more important for space applications than terrestrial, may be favourable for some terrestrial applications, The roof loading on a house top covered with solar cells, for example, would be significantly lower than the comparable loading for a conventional liquid solar water heaters.

(*xiv*) Amenable to on site installation *i.e.,* decentralized or dispersed power; thus the problem of power distribution by wires could be eliminated by the use of solar cells at the site where the power is required.

(*xv*) They can be used with or without sun tracking, making possible a wide range of application possibilities.

Their principal disadvantages are their *high cost,* and the fact that, in many applications, energy storage is required because of no insolation at night. Efforts are being made world-wide to reduce costs through various technological innovations.

5.7. Agriculture and Industrial Process Heat

Introduction. By supplying heat, as hot water, hot air or stream, for agricultural and industrial purposes, solar energy has the potential for replacing substantial amounts of fossil (non-renewable) fuels. Solar water heating and air heating are rather well established and well known by now. With all sophistications and modern techniques solar thermal energy systems are bulkier and still speak of only low efficiencies. In general in low grade thermal applications, where heat energy can as such be utilised. Solar energy devices could be used with advantages. Rising oil prices, combined with combustion generated pollution and fast depleting fossil fuels, have forced the developing countries like India to adopt atleast gradually, the low temperature applications of solar energy. A number of organisations are now offering a variety of solar systems for domestic, commercial and industrial use essentially in the low and medium temperature range. Apart from providing hot water or preheating boiler feed water as required, the solar systems are also being installed for more involved applications like maintenance of a certain temperature in process, tanks etc.

Solar energy for thermal applications in industries has proved to be economically viable at present for temperatures less than 100°C. With intensive development in the area of fixed and tracking concentrators, temperatures 0° upto 300°C will be feasible. The technology is expected to be matured in near future. In the present energy context, it is desirable to provide thermal energy below 300°C from solar. This is mandatory since high quality fuel, such as coal or oil, with high flame temperature when used for low temperature applications result in very low efficiency. Therefore the choice of solar energy for direct thermal applications upto 300°C in industries is quite obvious. The high quality fuel (coal, oil etc.) can be diverted for more important high temperature tasks. With this concept, the development of economically viable solar devices become more meaningful.

Agricultural and Industrial Applications and Potentials

These applications of solar energy may be considered in three general categories, according to the temperature range within which the heat is supplied.

- Low temperatures below 100°C.
- Intermediate temperatures 100 to 175°C.
- High temperatures above 175°C.

Low temperature applications. These are based on the use of flat-plate collectors, with either air or water as the heat transport medium. The general principles are the same as for space and water heating. The hot water may be utilized directly or the heat may be transferred to air. Provision must be made for storage of excess heat in water or in rocks or gravel. Among the potential applications of low temperature heat in the *agriculture* are the following:

Heating and cooling of commercial green houses.

Space heating of livestock shelters, dairy facilities and poultry houses.

Curing of bricks, plaster board etc. Drying grain, soybeans, peanut pods, fruits, tobacco, onions and kiln (Lumber) Solar energy can also be used to convert salty water (or other impure water) into potable water by distillation. Such solar stills have been operated for farm and community use, in several countries.

Intermediate Temperature Applications. Food processing, textile, laundry, and other industries often require both hot water and low pressure steam. For such applications, water can first be heated in flat-plate collectors followed by an array of parabolic through concentrating collectors. In this way, water under moderate pressure, can be heated to temperatures above 100°C. Some of the processes which fall in the category of intermediate temperature applications are the following:

Laundries, fubric drying, textile dyeing, food processing and can washing, kraft pulping (in paper industries), Laminating and drying glass fibre. Drying and baking in automobile industries, pickling (in steel industries) etc.

High Temperature Applications. Steam at temperatures above 175°C is used extensively in Industry particularly in the generation of electric power. The same general methods for producing high temperature steam with the aid of solar energy are applicable in all cases; they are described in the section of *Thermal Electric Conversion.* A high temperature application of solar energy is in pumping of irrigation water.

The several *advantages* of industrial applications over residential or commercial ones are:

(*i*) Industrial loads are mostly on continuous basis throughout the year.

(*ii*) Industrial plants have maintenance crew, or in small plants skilled people, who can attend to smooth operation of solar systems.

(*iii*) Total quantum of energy replaced by solar is significantly more causing higher reduction in oil imports and diversion of coal for high temperature tasks.

However there are some limitations:

(*i*) Intermittent availability of solar energy.

(*ii*) Instantaneous area. In all the cases roof area may not be adequate to accommodate required collector area. Additional costly land may have to be used. In some cases, roof have east west slopping, instead of north glazing type, rending placement of collectors to be costly and un aesthetic.

(*iii*) Industrial effluents can be harmful to the transparent covers and reflecting surfaces.

(*iv*) Through pay back period has come down to 3 to 5 years (hot water and air only), high initial capital investment is a major impediments.

5.8. Solar Distillation

Fresh water is a necessity for the sustenance of life and also the key to man's prosperity. It is generally observed that in some arid, semi arid and coastal areas which are thinly populated and scattered, one or two family members are always busy in bringing fresh water from a long distance. In these areas solar energy is plentiful and can be used for converting saline water into distilled water. The pure water can be obtained by distillation in the simplest *solar still,* generally known as the "basin type solar still". It is shown schematically in Fig. (5.28). Such solar stills have been operated for farm and community use in several countries. It consists of a blackened basin containing saline water at a shallow depth, over which is a transparent air tight cover that encloses completely the space above the basin. It has a roof-like shape. The cover, which is usually glass, may be of plastic, is sloped towards a collection trough. Solar radiation passes through the cover and is absorbed and converted into heat in the black surface. Impure water in the basin or tray is heated and the vapour produced is condensed to purified water on the cooler interior of the roof. The transparent roof material, (mainly glass) transmits nearly all radiation falling on it and absorbs very little; hence it remains cool enough to condense the water vapour. The condensed water flows down the sloping roof and is collected in troughs at the bottom. Saline water can be replaced in the operation by either continuous operation or by batches. Although there are numerous configurations of basin type units, their basic theory is identical. The basin type solar still has produced distilled water at a cost per unit of product lower than other types of solar equipment and is the only type in operation. Operating efficiencies of 35 to 50% for basin type still have been achieved in practical units, as compared with a theoretical maximum of slightly more than 60%.

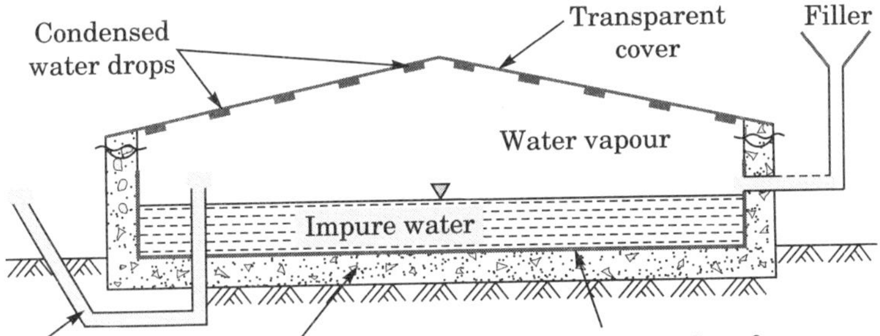

Fig. 5.28. Solar water still.

The performance rating and efficiency of the solar still is determined by plotting the graph of the amount of fresh water produced per unit of basin area in one day versus the solar radiation intensity over the same period. Such curves for several stills are drawn. Efficiency is defined as

$$\eta = \frac{w\Delta h}{H} \qquad \text{...(5.9)}$$

where, w = weight of distillate per square meter per day.

Δh = enthalpy change from cold water to vapour.

H = Solar radiation intensity per square meter per day.

Here area of the water surface is to be considered. Δh includes the latent heat of vaporization, which is being taken as average value 594.5 kcal/kg (2489 kJ/kg).

The performance of a solar still is generally expressed as the quantity of water produced by each unit of basin area in a day *i.e.*, cubic meters of litres of water per square meter of basin area per day. This quantity will vary with the design of the still, with the intensity of solar radiation and with the atmospheric conditions in the surroundings. The production rate depends primarily on the amount of solar radiation available but is affected by several other factors; like ambient air temperature, wind speed, atmospheric humidity, sky conditions etc., the effect of design parameters such as orientation of still, single sloped or double sloped, inclination of glass cover, insulation of the base etc., and the effect of operating parameters such as water depth in the tray, absorption emittance properties of the still, preheating of water etc.

Solar still installations may provide about 15 to 50 litres per day per 10 sq.m.

5.9. Solar Pumping

Solar pumping consists in utilizing the power generated by solar energy for water pumping, useful for irrigation.

Solar energy offers several features that make its utilization for irrigation pumping quite attractive. First, the greatest need for pumping occurs during the summer months when solar radiation is greatest. Second, pumping can be intermittent to an extent. During periods of low solar radiation when pumping decreases, evaporation losses from crops are also low. Finally, relatively in expensive pumped storage can be provided in the forms of ponds.

A number of recently constructed solar irrigation pump installations are now operational. The major obstacle to increase use of solar irrigation systems at this time is their relatively high capital cost. If the costs of solar pumps can be substantially reduced and assuming that conventional fuel costs continue to rise, solar irrigation could become economical, and increased use of such systems might be anticipated in future.

The basic system consists of the following components:

1. The solar collectors, may be
   - (*a*) Flat plate collectors or solar pond
   - (*b*) Stationary concentrator (CPC)
   - (*c*) Sun-tracking concentrators, (cylindrical parabolic trough concentrator or heliostats).
2. The heat transport system.
3. Boiler or heat exchanger.
4. Heat engine, it may be
   - (*a*) Rankine engine → Reciprocating engine / Vapour turbine
   - (*b*) Stirling hot gas engine
   - (*c*) Brayton cycle gas turbine
   - (*d*) Rotary piston engine.
5. Condenser.
6. Pump, it may be
   - (*a*) Reciprocating pump
   - (*b*) Centrifugal pump
   - (*c*) Diaphragm pump
   - (*d*) Rotary pump.

The solar pump is not much different from a solar heat engine working in a low temperature cycle. The sources of heat is the solar collector, and sink is the water to be pumped. A typical solar powered water pumping system is shown in Fig. (5.29). The primary components of the system are an array of flat-plate collectors and an Rankine engine with an organic fluid as the working substance. During operation a heat transfer fluid (pressurized water) flows through the collector arrays. Depending upon the collector configuration, solar flux and the operating conditions of the engine, the fluid will be heated in the collector to a higher temperature, the solar energy which is thus converted to the thermal energy. The fluid (water) flows into a heat exchanger (boiler), due to temperature gradient, and comes back to the collector. This water yields its heat to an intermediate fluid in the boiler. This fluid evaporates and expands in the engine before reaching the condenser, where it condenses at low pressure. The condenser is cooled by the water to be pumped. The fluid is then re-injected in the boiler to close the cycle. The expansion engine or Rankine engine is coupled to the pump and it could of course be coupled to an electric generation.

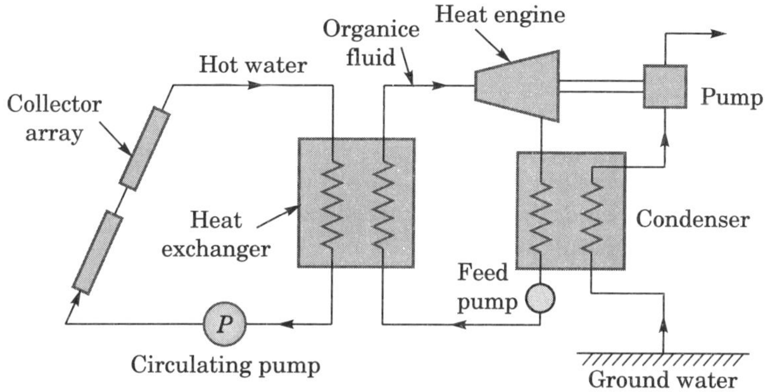

Fig. 5.29. Schematic of a solar pump.

The collector area to a large extend is determined by the overall efficiency of the system, η_o

$$\eta_o = \eta_e \times \eta_c$$

where, η_e is the efficiency of the engine

η_c is the efficiency of the collectors.

For a higher temperature in the boiler, results in a higher engine efficiency, but a higher temperature at the boiler requires a higher collector temperature, which decreases the collector efficiency. There is thus, an optimum range of operating temperatures for the solar pumping system which results in the maximum efficiency for the system.

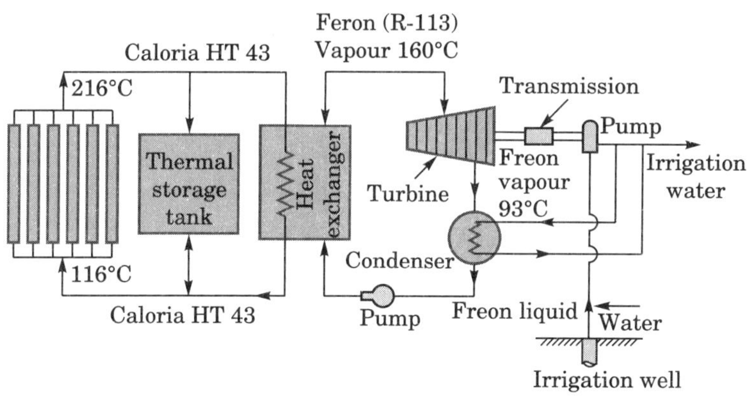

Fig. 5.30. A schematic of a turbine-driven pump using solar energy.

The working fluids used in the cycle are toluene (CP-25), mono chlorobenzene (MCB) trifluoro ethanol (TFE-100 and 85), hexafluro benzene (HFB), pyridine (CP-32), refrigerant-11 (R-11), refrigerant-113 (R-113), and thiopene (CP-34). R-113 is preferable to all, because of high cycle efficiency, non-toxic in nature and due to low cost.

A simplified outline of a turbine—driven pump system utilizing solar energy is shown in Fig. (5.30). In a particular system in New Mexico, the heat transport fluid (Exxon Caloria HT—43) is heated to 216°C in parabolic trough collectors with a total operature area of 624 sq. m. Part of the heated liquid is stored for use when the sun is not shining. The turbine working fluid (Freon type R-113) leaves the boiler and enters the turbine as vapor at a temperature of 160°C and 15 atm (1.5 MP_a) pressure. After expansion in the turbine, the vapor leaves at 93°C and 0.7 atm (0.07 MP_a); it is converted back to liquid in the condenser and returns to the boiler.

The irrigation pump operates at a rated power of 19 kW (25 horse power) and delivers water at 500 to 600 gal/min (32 to 38 litres/sec) from a well roughly 30 m deep. The energy efficiency (*i.e.*, percentage of solar energy collected that is converted into useful work) is 13 to 14 percent, this low value is largely a result of the relatively low temperature of the working fluid entering the turbine. Rankine efficiency will be within acceptable limits, if the temperature of the order of 200 to 400°C is obtained, using proper focusing collector system.

5.10. Solar Furnace

A solar furnance is an instrument to get high temperatures by concentrating solar radiations onto a specimen. Solar furnaces have long been used for scientific investigations. French scientist Lavoisier used it in 1774, with a lens as tall as man, for carrying chemical studies at high temperatures. In 1921, German scientist Strauble devised a solar furnace composed of a paraboloidal concentrator and a lens. Then he built another furnace. In which he used paraboloidal concentrator, 2 m in aperture and 86 cm in focal length, was fixed facing downward and the solar radiation was conveyed upward by heliostats (turnable mirrors). The first large solar furnace, with a thermal power of 45 kW was completed in France in 1952. A similar furnace with a power of about 35 kW, was constructed for the U.S. Army at Natick, Massachusetts, in 1958; thus was moved to the white Sands Missile Range, New Mexico, in 1976 where it is used to study thermal radiation effects. Solar furnaces have also been built in Japan and the U.S.S.R.

The world's largest solar furnace, with a design thermal power of 1000 kW, commenced operation at Odeillo in the French Pyrenees in 1973. It consists of 63 heliostats, each having an area of 45 sq. m. arranged in eight tiers. The heliostats reflect the solar radiation in a parallel beam toward a paraboloidal mirror 40 m high and 54 m maximum width. This mirror is made up of 9500 pieces which are curved to produce a minimum image of the sun at the focus. The maximum temperature, attained at the focal point of the mirror, is estimated to be over 3800°C; the maximum heat flux is 16,000 kW/sq. m.

Principle of Working

The principle of the solar furnace is outlined in Fig. (5.31). A number of heliostats (turnable mirrors) are arranged in terraces on a slopping surface (*e.g.*, on a hill side) so that, regardless of the sun's position, they always reflect solar radiation in the same direction onto a large paraboloid (or spherical) reflecting collector made up of many fixed mirrors) attached to the face of a structure. The collector then brings the radiation to a focus within a small volume (receiver). In figure a heliostat type furnace with horizontal optical axis is shown which is comparatively convenient and widely used in large furnaces. The most desirable mirror is that obtained by grinding and polishing a glass plate into an optical flat, aluminizing or silvering by vacuum evaporation, and cooling with a suitable film.

The change of elevation and that of azimuth can be obtained by the rotation of frame about a horizontal axis and about a vertical axis respectively. In order to rotate the frame, hydraulic or electric driving is used which is coupled with a servo system or a time system for sun following. The other method is to use many heliostats to convey the solar radiation into a concentrator.

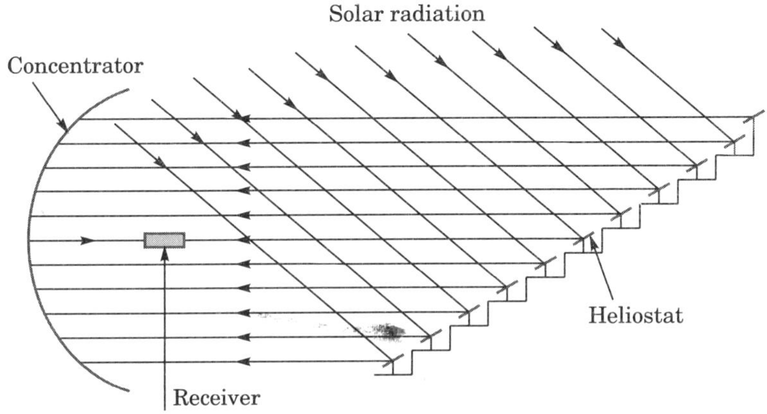

Fig. 5.31. Principle of solar furnace.

Uses. The solar furnace is an excellent means for studying properties of ceramics at high temperatures above the range ordinarily measured in the laboratory with flames and electric currents. Physical measurements include melting points, phase changes, specific heat, thermal expansion, thermal conductance, magnetic susceptibility and thermionic emission. Several useful metallurgical and chemical operations have been carried out at high temperatures in the solar furnaces. The melting and sintering of temperature ceramics such as zirconia is easily accomplished. Direct high temperature production or zirconia from zircon and alkali, beryllia from beryl, and tungstend from wolframite is carried out in solar furnaces. Purification of a refractory (Al_2O_3) by sublimation at high temperatures also has been carried out.

Advantages and Limitations of a Solar Furnace

Advantages. A solar furnace is a unique instrument which has the following characteristics or advantages:

(*i*) In a solar furnace heating is carried out without any contamination and temperature is easily controlled by changing the position of the material in focus.

(*ii*) It gives an extremely high temperature.

(*iii*) It provides very rapid heating and cooling.

(*iv*) Various property measurements are possible on an open specimen.

(*v*) Contamination by ions does not occur in fusion which might happen in the case of plasma or oxy hydrogen flame.

(*vi*) Proper desirable atmosphere can be provided to the specimen.

The main limitations with a solar furnace are:

(*i*) Its use is limited to sunny days, and to 4.5 hours only (maximum bright sun shine hours), and

(*ii*) highcost.

The high cost of solar furnace is still a barrier to their wide industrial use under present economic conditions. Numerous future applications, production of nitric acid and fertilizers from air, to give just one example are in sight. In the mean while, the enormous efforts going into space research has greatly accelerated the development of large, light weight, focusing collectors for producing high temperatures in outer space.

5.11. Solar Cooking

Introduction. In our country energy consumed for cooking shares a major portion of the total energy consumed in a year. In villages 95% of the consumption goes only to cooking. Variety of fuel like coal, kerosene, cooking gas, firewood, dung cakes and agricultural waste are used. The energy crisis is affecting everyone. It is affecting the fuel bills for those who use it for heating the houses and cooking their food. The poor of the developing countries who have been using drywood, picked up from the fields and forests as domestics fuel, have been affected in their own way, due to scarcity of domestic fuel in the rural areas. At present, firewood and cow dung cakes are the most important sources of fuel to cook food. Cowdung too precious to be allowed to be used for burning and cooking. It is very useful to improve the fertility of the soil, it should be used in proper way. The supply of wood is also fast depleting because of the indiscriminate felling of trees in the rural areas and the denudation of forests. There is a rapid deterioration in the supply of these fossil fuels like coal, kerosene or cooking gas. The solution for the above problem is the hardneesing of solar energy for cooking purposes.

Thus solar cookers have a very relevant place in the present fuel consumption pattern. Various designs of solar cookers have been developed in our country. The first solar cooker was developed in the year 1945 by Mr. M.K. Ghosh of Jamshedpur a freedom fighter. He developed a box type solar cooker with a reflecting mirror and a copper coil inside, on which the food materials used to be placed in pots. Mr. Ghosh also designed a parabolic reflector which was used for sometime as a boiler of Neera (palm juice). Later in 1953 NPL of India developed a parabolic solar cooker. The main reasons for non-acceptance of these devices was the cheap availability of cooking fuel during these days. The problem of harnessing and utilisation of solar energy arised after the fuel crisis of the 1970s, which also affected the rural areas.

Basically there are three designs of solar cooker:

(*i*) Flat plate box type solar cooker with or without reflector,

(*ii*) Multi reflector type solar oven and,

(*iii*) Parabolic disc concentrator type solar cooker.

Flat plate box type design is the simplest of all the designs. Maximum no load temperature with a single reflector reaches upto 160°C, In multi-reflector oven four square or triangular or rectangular reflectors are mounted on the oven body. They all reflect the solar radiations into the cooking zone in which cooking utensils are placed. Temperature obtained is of the order of 200°C. The maximum temperature can reach to 250°C, if the compound cone reflector system is used. With parabolic disc concentrator type solar cooker, temperatures of the order of 450°C can be obtained in which solar radiations are concentrated onto a focal point. Principle of operation of solar cookers is shown in Fig. 5.32.

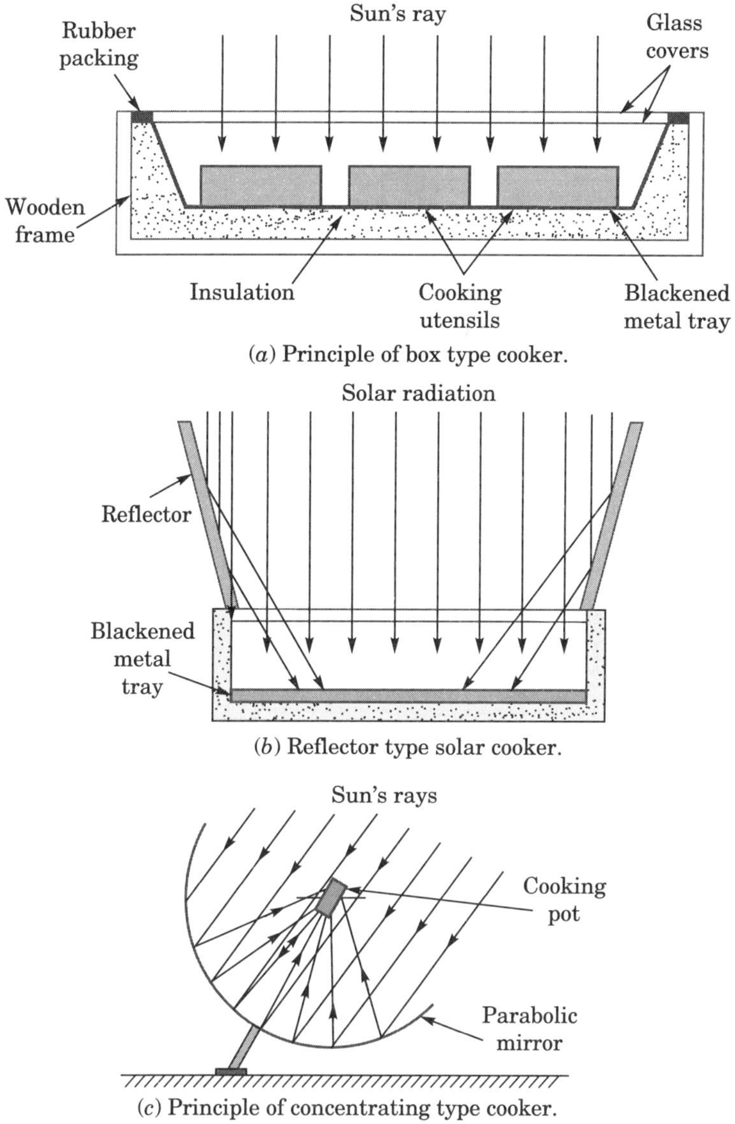

(*a*) Principle of box type cooker.

(*b*) Reflector type solar cooker.

(*c*) Principle of concentrating type cooker.

Fig. 5.32. Principle of operation of solar cookers.

Design Principle and Constructional Details of a Box Type Solar Cooker

The principle of operation of box type solar cooker is illustrated in Fig. (5.32) (*a*). The solar rays penetrate through the glass covers and absorbed by a blackened metal tray kept inside the solar box. The solar radiation entering the box are of short wavelength. The higher wavelength radiation is not able to pass through the glass cover *i.e.,* reradiation from absorber plate to outside the box is minimized by providing the glass cover. Two glass covers are provided to again minimize the heat loss. The loss due to convection is minimized by making the box air tight by providing a rubber strip all round between the upper lid and the box. Insulating material like glasswool, paddy husk, saw dust or any other material is filled in the space between blackened tray and outer cover of the box. This minimizes heat loss

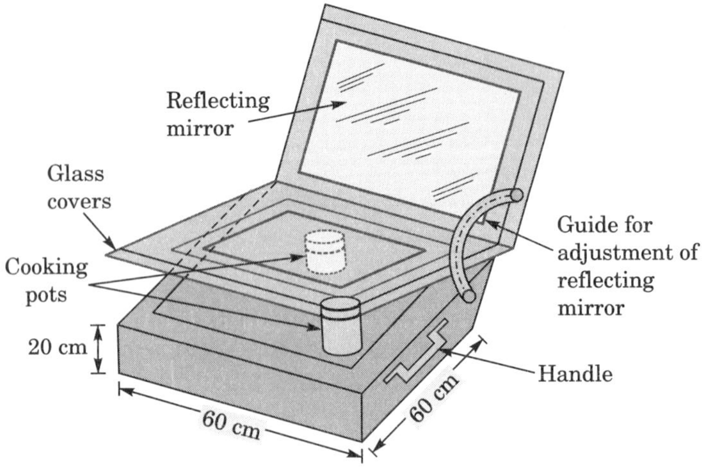

Fig. 5.33. Details of a box type cooker.

due to conduction. With this type of cooker is placed in the sun, the blackened surface starts absorbing sun rays and temperature inside the box starts rising. The cooking pots, which are also blackened are placed inside with food material, get heat energy and food will be cooked in a certain period of time depending upon the actual temperature attained inside. The temperature attained depends upon the intensity of solar radiation and material of insulation provided. The amount of solar radiation intensity can be increased by provided mirror or mirrors. The solar cooker is made up of inner and outer metal or wooden box with double glass sheet on it. Absorber tray (blackened tray) is painted black with suitable black paint like boiler interior paint. This paint should be dull in colour so that it can withstand the maximum temperature attained inside the cooker as well as water vapour coming out of the cooking utensils. The top cover contains two plain glasses each 3 mm thick fixed in the wooden frame with about 20 mm distance between them. The entire top cover can be made tight with padlock hasp. Neoprene rubber sealing is provided around the contact surfaces of the glass cover and the cooker box. A small vent for vapour escape, in provided in the sealing. Collector area of the solar cooker is increased by providing a plane reflecting mirror equal to the size of the box, and hinged on one side of the glass frame. A mechanism (guide for adjusting mirror) is provided to adjust the reflector at different angles with the cooker box. A15°C to 25°C rise in temperature is achieved inside the box when reflector is adjusted to reflect the sun rays into the box. In winter, when sun rays are much inclined to horizontal surface, reflector is a most useful addition.

Overall dimensions of a typical model are 60 × 60 × 20 cm height. This type of cooker is termed as family solar cooker as it cooks sufficient dry food materials for a family of 5 to 7 people.

The temperature inside the solar cooker with a single reflector is maintamed from 70°C to 110°C above the ambient temperature. This temperature is enough to cook food slowly, steadily and surely with delicious taste and preservation of nutrients. Maximum air temperature obtained inside the cooker box (without load) is 140°C, in winter and 160°C in summer. Depending upon the factors such as season and time of the day, type of the food and depth of the food layer, time of the cooking with this cooker ranges from 1 hr to 4 hrs. Meat should be allowed to stay for 3-4 hours. Vegetables take from ½ to 2½ hours. All types of Dals can be cooked between 1½ to 2 hours. Rice is cooked between 30 minutes and 2 hours. The best time of the day for cooking is between 11 am and 2 pm. Cooking is faster in summer than in winter due to high ambient temperature.

Following are the some *merits* of a solar cooker:

(*i*) No attention is needed during cooking as in other devices.

(*ii*) No fuel is required.

(*iii*) Negligible maintenance cost.

(*iv*) No pollution.

(*v*) Vitamins of the food are not destroyed and food cooked is nutritive and delicious with natural taste.

(*vi*) No problem of charring of food and no over flowing.

Limitations of a solar cooker are:

(*i*) One has to cook according to the sun shine, the menu has to be preplanned.

(*ii*) One can not cook at short notice and food can not be cooked in the night or during cloudy days.

(*iii*) It takes comparatively more time.

(*iv*) *Chapaties* are not cooked because high temperature for baking is required and also needs manupulation at the time of baking.

Box Type Solar Over (Multi reflector Type)

The hot-box type of cookers or solar ovens are ideal for baking purposes.

The solar ovens designed by Central Arid Zone Research Institute (CAZRI) Jodhpur, have been installed in villages for demonstration cum test purposes and performance has been found satisfactory. The solar oven which is shown in Fig. (5.34), is of multi-reflector type. This is rather a complicated type of cooker which

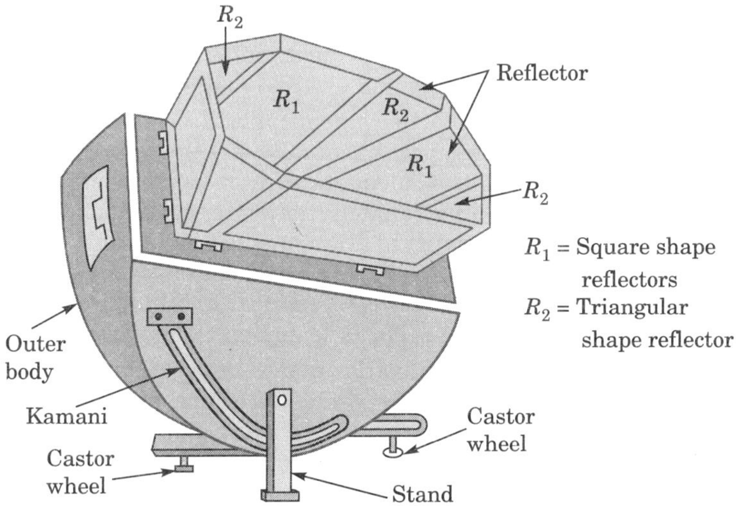

Fig. 5.34. Isometric view of solar oven.

has got many parts and also needs precision fitting. The solar oven consists of a well-insulated semi-cylindrical box made of sheet aluminium and wood. Two shells are made and the space between them, 7.5 cm, is filled with fibre glass insulation. The interior sheet is painted black. A door of the same insulating material is also made, for keeping and taking out food. The window (40 × 40 cm) of the oven consists of two transparent glass sheet (3 mm thick) with a spacing of 2.0 cm. Eight reflectors made of silvered glass mirrors, four of square shape and four of triangular shape have been used. The oven can be manually tilted and oriented towards the sun. A cradle like cooking platform is made in the oven which helps in keeping the vessel containing food horizontally irrespective of the sun.

On very clear days, maximum plate temperature in the oven reaches to 350°C in the summer season and 250°C in winter season. Practically all types of food preparations like cooking, roasting baking and boiling can be done within 25 to 75 minutes under clear sky conditions. It is reported by a villager using the solar oven that the *Bati,* a local preparation, prepared (baked) in this oven was more tasty than that made with conventional fuel (dung cake).

The main raw materials required are plane aluminium sheets, wooden plank, aluminium and M.S. angle iron, looking glass sheets, castor wheels etc. All are available indigenously. The cost of production of this solar oven is about ₹ 600. The main advantage of this solar oven is that its efficiency is high because its performance is not affected by wind and there are no chances of dust falling in the cooking pot which is a problem in the arid zone of India. Moreover, the food remains warm if kept inside the oven for hours together even after sunset.

From the research work and experiment and testing of the various types of the cookers, it has come to the conclusion that a box type cooker with a single reflector is the cheapest and most effective solar cooker for rural areas.

5.12. Solar Greenhouses

A greenhouse is a growth chamber which offers the possibilities of year round plant production. These are effective solar collectors. These can also be geared to the needs of the rural, urban and suburban populations. A greenhouse attached to a residence creates a pleasant improvement in the physical and mental environment of its occupants; designed in a truly passive solar collection manner with a well applied heat store, this type of solar collector (or power house) may also provide much of the required winter heat. Solar greenhouses are relatively easy to build with simple technology and low cost materials.

During the winter, in the northern hemisphere, the earth is slightly near to the sun and atmospheric interference is generally less than during the summer; it is therefore, often feasible to obtain adequate solar energy to sustain plant growth throughout the day and night. Unfortunately, the average green house is so designed that it loses most of the heat it gains to avoid overheating during the day, leaving no spare heat for night time use. A solar green house optimises the received sunlight and heat while reducing heat losses to a practical minimum with the object of providing stored heat for use over night time and on cloudy days.

Greenhouses provide crop cultivation under controlled environment. A green house is a structure covered with transparent material that utilizes solar radiant energy to grow plants and may have heating, cooling and ventilating equipments for temperature control.

The plant environment refers:

(*i*) Soil temperature (*ii*) air temperature (*iii*) air humidity (*iv*) Soil moisture (*v*) light (*vi*) air composition (*vii*) root medium composition (*viii*) protection from plant enemies (*ix*) exposure to rain (*x*) hail storm etc.

It has been shown that an ability to exercise control over the crop micro climate results in several fold increase in crop photosynthesis. It would then be possible to beneficially produce food even in those areas where and in those periods when field cultivation is not possible. With controlled plant micro-climate crops can be taken over a wider period of time as compared to normal season, thus making fresh food materials, like fruits and vegetables available for larger part of the year.

There is a phenomenal difference between the yields which can be achieved in green houses and those in open fields, especially with the aid of modern horticultural methods of plant growth. Nevertheless, energy consumption in food production is increasing faster than in most other industries; processed foods require more energy input to plant, cultivate, fertilise, harvest, package and transport than is available from the energy content of that food. Increasingly, fuel costs in this important industry are reaching levels at which unsubsidised growers are finding it hard to servive in business.

Advantages of Greenhouses. A greenhouse structure is less expensive to build than a fully insulated structure. It is indeed a live-in solar collector that provides.

(*i*) a source of inexpensive, good quality food that one grows one self,

(*ii*) a source of additional heat (temperature control) for the house attached to it,

(*iii*) a source of moderator for the humidity (humidity control) in the house.

A greenhouse can be separate structure sitting all by itself out in a field or can be an integral part of the houses. Greenhouses of one form or another have been built since long. They may be coverd with glass or glass substitutes when covered with plastic films, they are called plastic greenhouses (more popular in temperuture areas such as Mediterranean areas); when covered with glass fibre reinforced plastic panels, they are called fibre-glass greenhouses. But generally green house is synonymous with glass houses. Usually south-side is made of glass while north side is insulated. The sky radiation blocked by northside is only a small fraction of the total radiation in most regions. Green house architecture varies widely around the globe. In the countries of northern Europe where days are often cloudy, glass houses are employed. However, in other parts of the world where the sky is usually clear or only partly cloudy it is possible to use other designs which capture more heat than light.

Types of Greenhouses

Some broad classifications of greenhouses include the following:

(*i*) *Attached greenhouse,* which may be joined onto almost any suitable building structure.

(*ii*) *Porch* type greenhouses which may be designed as the entrance to a house, factory or office.

(*iii*) *Free standing* greenhouses, which may be situated on any convenient patch or piece of waste ground.

(*iv*) *Pit* type greenhouses, which are usually employed on differing level or sloping land scapes, and for the purpose of heat retention.

(*v*) *Cold frame* type of greenhouses which are simply hot-beds or plant facing frames equipped with a sloping roof.

Each of the foregoing types of greenhouse can be designed in a passive solar manner, making it an efficient solar collector and heat storage.

The *free standing* and *pit type* of greenhouses are not directly attached to any other building for greenhouse; on the other hand, they are separate structures sitting all by themselves. These two types have insulated walls on the north east and west sides. The walls are suitably designed to permit maximum reflection of the incoming radiation onto the plant canopy. Enough storage facilities are provided to meet the requirements. The pit type, however differs from the free standing type being partially sunk. So it has the advantage of being in contact with the temperature stability of the deep earth. A greenhouse which does not have a separate structure and is a part of a house. This type is known as *attached type.* As a matter of fact, it is quite cheaper to build a greenhouse as an addition to the existing house. It has the advantage that it can draw heat from an already heated structure when required and can return excess heat when conditions are sunny. Infact, if it is built in the right location with respect to existing windows and doors, it can be a very good heat source for the house. Further this type leads to the maximum return for only a moderate investment in time, energy and money.

There are various designs of attached type greenhouses, but one elongated along the east-west axis is the most efficient shape for solar energy collection. Usually the south wall of the building to which the greenhouse is attached as a thermal wall (masonry or water) and its large surface area is exposed to sunlight.

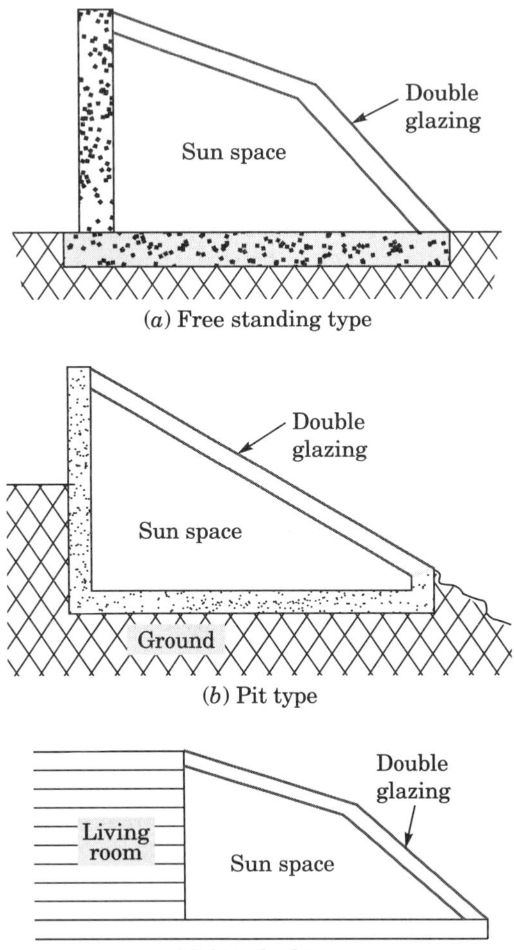

Fig. 5.35. Schematic of some major types of solar greenhouses.

Depending on climatological conditions, greenhouse structure can be divided into two categories:

(1) *Winter greenhouse* and (2) *Summer greenhouse.*

The *Winter greenhouse* is constructed in those countries where the outside ambient temperature is very low compared to the requirement of the plant growth obviously it is to be used for heating purposes. Hence attention to be given to catch enough of the incoming heat from the sun. So south side is usually made of glass, sometimes doubly glazed and interior surface of the northwall is painted white. Use of movable sulation to prevent unwanted heat loss, provision for storage facilities to account for the demand during sundown hours are some of the important features. Windows are provided in east and west walls for ventilation.

Summer greenhouses are constructed in those countries where the outside ambient temperature is very high compared to the plant growth temperature. Typically summer climate is characterised by high ambient temperature, clear sky and high humidity. So one has to design the green house such that inside temperature may not reach very high. Usually this type of green house has two sets of windows placed in the south as well as north wall. These windows can be used to cool the greenhouse by free convection. Another way is to duct the windows, in the south wall through which air is blown after it has evaporatively cooled by passing through evaporative pads. Also south wall is covered by removable insulation, whereas the required radiation can be admitted through east and west walls. The interior of north wall is usually coated with reflecting paint.

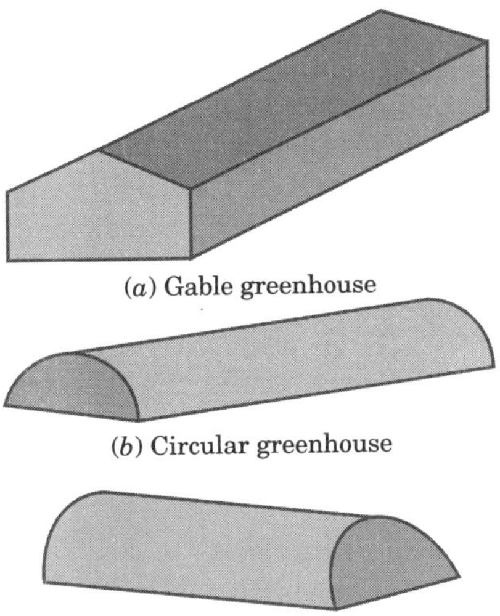

(*a*) Gable greenhouse

(*b*) Circular greenhouse

(*c*) Gothic arch greenhouse

Fig. 5.36. Some common shapes of green houses.

Some of the common shapes of greenhouses are shown in Fig. (5.36) and (5.37).

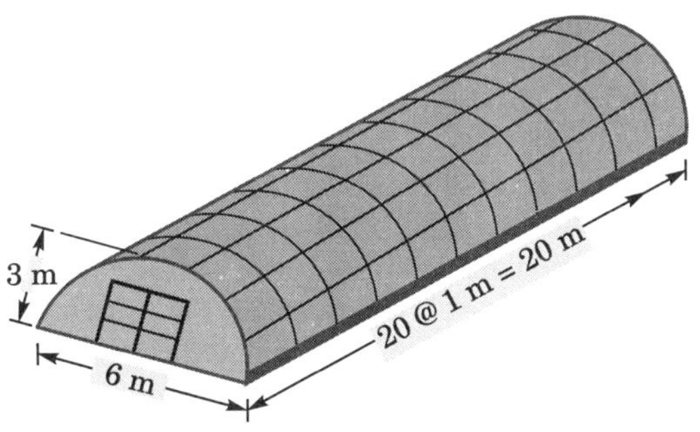

Fig. 5.37. Schematic diagram of pipe framed greenhouse.

Many innovative ideas have been incorpoted into the greenhouse designs. Both passive and active methods separately or in combination are made use of for meeting the energy needs of green houses. One such is soil bed heating from the solar energy. Network of pipes are laid in the soil beds and through these hot water can be circulated for soil heating.

Parameters for Plant Growth. In general, various parameters affect the plant growth, their discussions will help in the design analysis of the greenhouse construction.

(*i*) Light. It is essentially required in plant growth. At intensities of about 16500 lux, a good plant growth has been observed. However, plants grow quite well at intensities of 27500 lux (only a quarter of full sunlight). Plants are found to use only radiant energy in the visible and near visible portion of the spectrum.

(*ii*) Temperature. In plant grow, temperature is a dominant environmental factor. There is a different optimum temperature for each stage of plant development. This optimum value is related to other environmental factors. In controlled environments, however plants are grown at temperature between 10°C and 25°C.

(*iii*) Soil temperature. For most plants, the soil temperature of 20°C to 25°C has been reported to be optimum. This temperature determines the ability of a plant to absorb water from the soil since low soil temperatures are widely reported as deterimental in young plants, a high soil temperatures are recommended for rooting plants or germinating seeds.

(*iv*) Air movement. It influences transpiration, evaporation of water from soil, availability of carbon dioxide etc. This causes alteration of leaf size, internode length and other aspects of plant growth. In plant growth facilities, at speed of 0.8 to 2 cm/s, optimum growth has been reported.

(*v*) Humidity. It affects plant growth significantly while low as well as high value may cause the plant to be more susceptible to diseases due to pathagenic organisms, high humidity result in taller plants. The low humidities cause increase in evaporation rate and so more water is required. In many plant growth facilities humidities between 55% and 65% at 21°C to 25°C is maintained.

Design Principles

Orientation. If an attached green house is built on to a south facing wall in the northern hemisphere at latitudes greater than that of the tropic of cancer (23° 28′ N—the sun's most northerly overheat position), it will receive most of the available sunlight than heat which would fall on a free standing unobstructed green house. The most valuable direct beam radiation component would always emanate from the south and only half of the much less valuable sky dome's diffuse radiation component would be lost. The residence would gain a thermal barrier and protection from northerly winds. If the south facing wall were then painted white, it would reflect sunlight to a growing area at the rear of the greenhouse, enabling certain crops to be produced earlier than with a free standing greenhouses; if the south facing wall were painted black, it would instantaneously absorb more radiation and reflect less (acting as a heat store which could then slowly give up its stored heat to the green house). Growers involve in the cultivation of fruit or root crop, might well paint the south facing wall white to maximise the light availability; with greens or foliage plants, this may not be considered to be as important as the provision for heat storage.

The shape of a truly passive solar greenhouse is almost self determining, but will vary according to location and weather conditions. A useful theoretical rule of thumb which eliminates the calculation of the numerous variables is simply to add only 20° to the location latitude to determine the desired roof slope. For example, at 30°N, the roof angle to the ground should be 50°. If there is a predominate of clear sunny days during the winter, then no adjustment is theoretically necessary, but there is a predominance of cloudy winter days, the angle should be a little less steep to take advantage of the airy dome's diffuse radiation component, Beacause very steep roof angles lead to high and difficult structures, for practical purposes, it is better to add only 10° to the location or where there are severe structural constraints, merely to ensure that the roof slope is not, less than the given location latitude. To evaluate site obstructions, the sunlight availability protractors are used (these are the standard curves from which evaluation of the availability of sunshine is made).

Like any other solar collector, the solar greenhouse should be oriented in a southerly direction, but deviation from true south may be more generally tolerated and for plant growth may even be desirable. Since roof slopes for solar greenhouses are generally steeper than the roof slopes of dwelling houses, deviation from the true south can be as much as 20°; deviation upto 45° will result in not more than 20% decrease in the daily radiation total.

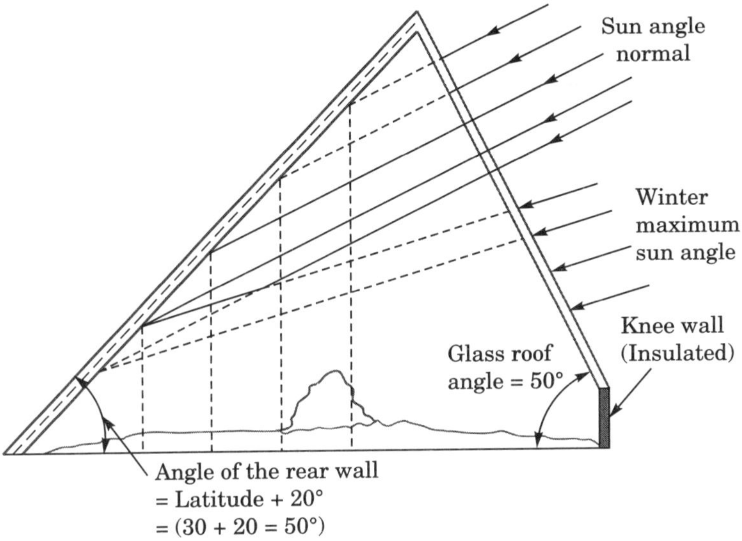

Fig. 5.38. Solar—Greenhouse—Optimised slope for orientation 30 N.

There is a good argument for orientation in a south-easterly direction which would allow the greenhouse to benefit more from the early morning sun, especially immediately after a long cold winter nignt and where the location experience clear morning. So many weather factors and the varying nature of crops can influence the siting of a greenhouse that it is essential for the user to study closely local weather patterns and meteorogical data to provide maximum protection from prevailing winds; he should ascertain weather scattered or diffuse radiation makes up a large part of the total available radiation, whether it is better to employ insulated opaque northerly corner walling to improve day lighting and whether the southerly corner can then he glazed to maximise the collection of winter solar radiation.

End walls. The south-east or south-west end walls should almost always be glazed. Glazed east-west end walls can allow light from the south glazing to pass straight through and thus there is an argument for insulating these walls.

Site. The designing begins with choosing of a suitable site for its location. It is necessary to understand how the sun travels through the sky at that place.

When combined with local whether data, the sun path study will give when and where the sun can be looked for heat gain. This will also help in deciding the requirment of heat storage, insulation and shading facilities.

Glazing. Two or more layer of glazing are required in cold climate. If however movable insulation is used at night, one layer of glazing can be used. Glass has been the most widely used material, though fibre glass, polyethylene, acrylics or polycarbonates are also used.

Ventilation. Adequate ventilation is a most for the successful operation of a greenhouse. Inside air changes per hour and more may be required under some conditions to keep plants dry enough to prevent from moulds and fungus infection. Outside air can be introduced into the interior through air tunnels at a depth of 2.5 to 3 m below the surface of the ground so that the growing area will be warmed in winter and cooled in summer. Exhaust vent to the outside can be used in warmer weather.

Increasing solar gain. The solar gain can be increased in many ways. One method is to provide exterior insulating shutter, which reduces heat loss from the greenhouse, thereby increasing the gain. The solar gain can also be incrased when the backside surface of greenhouse is silvered with reflective surface, which will act as a booster mirror and hence can increase the solar gain. The frame of south wall also can be coated with a highly reflecting white paint so as to reflect solar radiation into the greenhouse.

Heat storage. Thermal mass to store solar heat for later use is of prime importance for a solar green house. Where there is enough winter sun, the use of massive walls, floor and growing beds may be sufficient enough for the storage. For regions of low winter sunshine, provisions for remote heat storage are made. In this context use of water is widely practised. Water in drums, tins, cans, moist soil, etc are used for this purpose.

Greenhouse Environment and Control

A schematic of greenhouse environment interaction is given in Fig. (5.35).

A greenhouse glazing acts as a selective radiation filter in the sense that solar radiation could pass through it and enter into the greenhouse but the thermal radiation emitted by the objects within the greenhouse could not escape. This phenomenon results in the trapping of radiation within the green house and temperature rise is observed. This particular effect is often termed as 'Greenhouse effect' and it is the underlying principle of all solar thermal collectors. Multiple glazing reduces the greenhouse heat losses, but also reduces the amount of light entering the greenhouse. Each additional layer of glass will reduce the light transmission by about 13%. Clearly, there is an optimum thickness of multiple glazing which depends on the material used, the severity of the climate and the cost of the assembly.

A closed boundary of the green house structure also restricts mixing of greenhouse air with the ambient air, thereby reducing convective heat exchanges. This effect further enhances the retention of solar heat within the green house. It is the closed boundary of the greenhouse which permits CO_2 enrichment of the greenhouse air. The early developments in greenhouse technology took place in cold climate regions. Therefore, the attention was paid mostly to meet the greenhouse heating requirements. The present efforts in greenhouse cooling are directed in developing effective and economical evaporative cooling systems.

It has been observed that greenhouses make more effective use of water for crop production *i.e.,* the amount of water required to raise a unit quality of food is substantially less than that required for open field cultivation. The effect could be explained in the following way. The greenhouse crop transpires and the moisture diffused from the crop canopy to the greenhouse air continues to raise the green house air humidity because of its limited mixing with the outside air. The raised humidity of the greenhouse air, in turn, reduces the transpiration because the transpiration is a direct function of the humidity gradient between the plant cells and the contiguous air. The reduced moisture loss due to restricted transpiration has not been observed to adversely effect the greenhouse crop growth. Humidities in the greenhouses are, however, not allowed to raise above certain levels to control undesirable pathological effects. Obviously, the greenhouse space enjoy the amenability to environmental control practices because of its boundary, greenhouse effect, and reduced air circulation to the extent that commercial food production is made possible even under adverse climate conditions.

Greenhouse environmental control is need based, control of one condition may effect the other environmental-parameter, therefore, it is desirable than an integrated control package is developed. The recent developments, in microprocessor, based greenhouse environmental control practices are based on this reasoning. Various parameters of greenhouse aerial environment alongwith a partial list of control actions are listed in the follows table.

Table 5.1. Environmental Parameters and control action

Environmental parameter	*Control action*	*Parameter affecting crop growth*
Solar radiation	Ventilate	Light intercepted
Dry bulb temperature	Heating/cooling	Plant temperature
Wet bulb temperature	Spray water	Humidity gradient
Carbon dioxide-concentration	Air circulation enrichment	CO_2 gradient

A solar greenhouse is like any other building or structure losing most of its heat by conduction through the fabric and from air leakages through cracks by convection. Therefore, window and door seals should be as good, it not better, than those adopted in traditional buildings, construction methods. Unlike conventional buildings, greenhouses tend to be more exposed and air infiltration rates can be considerably higher at times of high wind pressures. High air infiltration rates cause unwanted air movement within the greenhouse, increasing the soil water evaporation which can then lead to the formation of condensation on the glazing, who consequent increases in heat losses and a reduction in the light transmission. Efficient and effective greenhouse seals and jointing are of the utmost importance to the economical control of environment.

Ground edge insulation must be provided in any greenhouse which seriously attempts to conserve energy. Greenhouses placed on concrete slabs without edge insulation lose more heat than those placed directly on soil.

Radiation losses alone are not usually an important factor, unless polyethylene is used instead of glazing. The most obvious way to reduce heat losses is to cutdown the amount of glazing, but a more practical approach is to devise methods of improving glass structures as heat retainers. In all but the mildest climate, double glazing should be used, as this halves the heat loss through glazing.

In the design of a solar green house, where heat losses have been reduced to a practical economic minimum, the regulation of moisture is an important function of the heating and ventilation systems. For example, if the internal temperature and relative humidity were to be maintained at 15°C and 75% with an outdoor air condition of 10°C and 90%, there would be a net removal of moisture with every cubic metre of air exhausted. With lower ambient levels of moisture content, the net removal of moisture will have an adverse effect on the soil condition unless moisture is added to provide a stable condition. At night time, the ambient temperature usually falls and the internal relative humidity may increase to 90%. Condensation of excess water vapour is then likely to occur on all cool surfaces, such as glass and uninsulated structural support members. Condensation drip on to plants and soil should be avoided because of the damage this may cause. Condensation drainage channels should be incorporated into the solar greenhouse structural design.

Two methods of greenhouse *heating* are in common use:

(*i*) Hot water or steam circulation.

(*ii*) Warm air space heaters.

The first method may be more expensive to install than the second, but the latter offers the additional advantage of better air distribution, *e.g.*, through perforated plastic tubes.

The following approximate formula may be used to assess the likely heating requirements.

The amount of heat required to maintain a given temperature within a glass house structure in kcal/hr.

= 6.8 × area of the glass wall and roof (m^2)

× temperature rise required above outside temperature (°C)

During summer, the heat build-up can be so great that the opening of windows and doors alone can not reduce the internal temperature sufficiently, unless, there is mechanically assisted means of ventilation, the grower may have to remove glazing panes to dispel the excess heat. Excess heat and consequent high air temperatures have an adverse effect on yield and the quality of the crops. The greenhouse temperature rise above the desired level should be limited to 4°C. This may require ventilation, which is achieved by fans.

For the efficient process of photosynthesis the adequate concentration of CO_2 within a greenhouse is essential. Enrichment of the CO_2 levels may considerably increase tomato yields, but whatever the crop, enrichment is necessary during daylight hours in the winter when ventilation rates are at a minimum. A steady state of CO_2 concentration should be about 300 ppm (parts per million), but growing plants will quickly reduce this level which can only be achieved by control at about 1000 ppm. A favoured method of CO_2 addition is by the combustion of low sulphur content paraffin.

Evaporative Cooling. Evaporative methods of greenhouse cooling have been employed for many years for use on hot days when there is a significant difference between the wet and dry bulb temperatures. Since the *wet bulb temperature* (and not the relative humidity) determines the temperature to which the air can be cooled solely by evaporation of water, a knowledge of the local meterological data, including average wet bulb readings, is essential.

The "wetted-pad" system of evaporative cooling is widely used. With this, a fibrous material in the form of self-supporting special fluted cellulose cells, is kept wet by water recirculating through it, thereby cooling the incoming air to within

1.5 to 2°C of wet bulb temperature of the outside air. The pad system also distributes the air uniformly and by virtue of its resistance, avoids turbulent air flow and delivers an even flow of cooled air into the greenhouse. The pad system units should be located on the side wall of the greenhouse which faces the prevailing winds during the summer, usually extending to about half of the length of the greenhouse. For very long structures, the pad should be centrally located and be sited at relatively low level, usually at about 0.5 m below knee wall height. Improving heat pump technology may progress a means of heat abstruction and cooling with desirable feature, of dehumidification.

Materials. Wood, aluminium and steel are the most popular materials for greenhouse frame. Wooden frames treated with preservatives can easily last for 5-10 years. Glass was the most popular glazing material until the development of clear plastic films and sheets some 30 years ago. Glass glazing requires higher initial investment and the chances of breakage are more in comparison to plastic glazing. Further, the leaks in glasshouses are more as compared to plastic greenhouses, Therefore, the environmental control in plastic greenhouse is more effective. The common plastic material for greenhouse glazing is clear UV resistant poly ethylene film (500-1000 gauge). These films are replaced within one or two years depending upon the local climate. Rigid clear panels of plastics such as corrugated fibre-glass can be employed for greenhouse glazing with as much as 20 years of life. The concern for greenhouse energy conservation is encouraging the developments of new plastic materials.

Concluding Remarks. Greenhouses are found in almost all type of climatic conditions. China now leads in having the largest area of more than 120,000 ha under covered cultivation. Japan is the second highest with over 80,000 ha area. Holland and Israel are exporters of green house products. In India the green house are available only in research and educational institutions.

From the economic analysis, it is concluded that the monetary return per unit land area is many times higher for greenhouse cultivation in comparison to the traditional agriculture. Greenhouse production is largely independent of fluctuating weather conditions and higher productivities, greenhouses can be viewed as food factories.

5.13. Solar Production of Hydrogen

There are three important problems of the world now a days, they are shortage of energy, depletion of fossil fuel sources and increasing pollution. People of the world with their increasing demand 2 of energy for their better standard of living, need energy source, which would overcome the above problems. Fossil fuels are finite in quantify and have adverse effect on environment, causing air pollution. Producing hydrogen for storing and transmitting solar energy seems the proper solution to these difficulties. Hydrogen can be used as a fuel, can be used to produce electricity, can be transmitted over long distances economically, and is a clean form of secondary energy.

Methods of producing hydrogen from solar energy

There are four basic methods:

(*i*) Direct thermal, (*ii*) Thermochemical,

(*iii*) Electrolytic, and (*iv*) Photolytic.

1. *Direct Thermal.* The dissociation of hydrogen and oxygen will start, if the water is heated upto about 3000°K or above. Water decomposition reaction is:

$$H_2O + \text{heat} \longrightarrow x_1H_2O + x_2H_2 + x_3O_2 \qquad ...(5.10)$$

where the substances are in gaseous form, and x_1, x_2 and x_3 are the mole fractions. Water should be decomposed at fairly high temperature (for equilibrium decomposition) combined with a reduced pressure. The energy for dissociation of hydrogen can be obtained from the solar energy. An optical system which collects solar radiation and concentrates it into a small area, can be utilized for this purpose. High temperatures can be obtained into a optical systems such as in large solar furnaces.

Advantages of this methods are:

(1) High thermal efficiency,

(2) Negligible environmental impact, and

(3) Intermediary chemicals are not required.

Because of high temperature requirements, it requires extensive research for commercial application.

2. *Thermochemical.* The difficulty of the high temperature requirements in direct thermal method, has led many scientists to search for thermo-chemical reactions to decompose water. In this methods, first water and one or more chemical elements or compounds react while heat is being added, resulting in a combination of hydrogen and/oxygen of water with compound and either hydrogen and oxygen being realised as a gas. Then, in one or more chemical reactions, the new chemical compound of the first reaction are reduced to their original composition with the help of other intermediary chemicals and/or heat, releasing oxygen and hydrogen. Hence inputs are heat, water and some work which is required to separate the resulting chemicals. Outputs are hydrogen, oxygen and low grade heat. The intermediary chemical elements and/or compounds are generated and recycled. For a complete thermochemical hydrogen production cycle, two or more chemical reaction steps are needed. For example.

Two step cycle:

$$\begin{aligned} H_2O + CH_4 &\longrightarrow CO + 3H_2 \\ CO + 2H_2 &\longrightarrow CH_4 + \tfrac{1}{2}O_2 \\ \hline \text{Sum}: H_2O &\longrightarrow CH_4 + \tfrac{1}{2}O_2 \end{aligned} \qquad ...(5.11)$$

Three step cycle:

$$\begin{aligned} H_2O + Cl_2 &\longrightarrow 2HCl + \tfrac{1}{2}O_2 \\ 2HCl + 2CrCl_2 &\longrightarrow 2CrCl_3 + H_2 \\ 2CrCl_3 &\longrightarrow 2CrCl_2 + Cl_2 \\ \hline \text{Sum}: H_2O &\longrightarrow H_2 + \tfrac{1}{2}O_2 \end{aligned} \qquad ...(5.12)$$

3. *Electrolyte.* In this method an electrolysis cell is used to produce hydrogen and oxygen from water. The cell consists of electrodes dipped in an electrolyte and connected to a d.c. supply. Water with some conducting chemicals is used as an electrolyte. When sufficient potential is applied between the electrodes to cause a current to flow, oxygen is liberated at the anode and hydrogen at the cathode.

Reactions are:

$$\begin{aligned} \text{Cathode}: 4H_2O + 4e^- &\longrightarrow 2H_2 + 4OH^- \\ \text{Anode}: 4OH^- &\longrightarrow 2H_2O + O_2 + 4e^- \\ \hline \text{Sum}: 2H_2O + \text{electrical energy} &\longrightarrow 2H_2 + O_2 \text{ (Heat)} \end{aligned} \qquad ...(5.13)$$

In this method, the solar energy is first converted to d.c. electric power, then hydrogen through electrolysis. Hence it is especially suited for coupling with ocean, thermal, wind, hydro and photovoltaic forms of solar energy since in these cases solar energy is converted to electricity.

4. *Photolytic.* In this phenomenon, sun's photons, under certain circumstances are absorbed by water molecules and when the energy absorbed a certain limit (68.3 kcal/mole of water), hydrogen can be released. Photons in the ultraviolet region of radiation spectrum passes the energies needed for the direct photolysis of water, in the presence of a catalyst. The function of such catalyst is to absorb the solar radiation and then transmit the energy to water in order to decompose it. The photolysis reaction with photocatalyst X can be expressed as

$$H_2O + 2X + \text{photon energy} \longrightarrow 2(\text{reduced X}) + 2H^+ + \tfrac{1}{2}O_2$$

$$2\,(\text{reduced X}) + 2H^+ \longrightarrow 2X + H_2$$

$$\text{Sum: } H_2O \longrightarrow H_2 + \tfrac{1}{2}O_2 \qquad \text{...(5.14)}$$

Note that photocatalyst X is not consumed, but is regenerated and available for reuse. Biological photocatalysts are also in existence.

Among the four basic methods for producing hydrogen from solar energy, the direct thermal method has the potential of highest thermal efficiency, followed by thermochemical, electrolyte and photolytic method.

QUESTIONS

1. Enumerate the different main applications of solar energy. Describe a hot water supply system.
2. How many solar house heating systems are there? What is the main difference between them?
3. Describe a passive solar space heating system.
4. With the help of a neat sketch, describe a solar heating system using water heating solar collectors. What are the advantages and disadvantages of this method?
5. With the help of a neat sketch describe a solar heating system using air heating solar collectors, with advantages and disadvantages of the system.
6. Describe the layout and working of a continuous solar cooling system.
7. What are the different approaches of thermal electric conversion system from solar energy?
8. Describe a central receiver system for a central solar thermal power station.
9. What is the principle of solar photovoltaic power generation? What are the main elements of a PV system?
10. What are the advantages and disadvantages of photovoltaic solar energy conversion?
11. Write notes on:
    (*i*) Agriculture and industrial process heat
    (*ii*) Solar distillation (*iii*) Solar pumping
    (*iv*) Solar cooking (*v*) Solar cell arrays.
12. Describe the principle of working of solar furnace. What are its main applications? What are the advantages and limitations of a solar furnace?
13. What do you mean by a green house? Enumerate the main types of green houses.
14. Describe briefly the different methods of producing hydrogen from solar energy.

UNIT—2

WIND-ENERGY

- Wind Energy
- Wind Energy Converters
- Generating Systems and Applications

CHAPTER 6

Wind Energy

6.1. Introduction

Chapter 5, dealt with the conversion of solar energy into thermal energy or electrical energy. We now deal with an indirect solar electric technology, *wind power.*

Wind results from air in motion. Air in motion arises from a pressure gradient. On a global basis one primary forcing function causing surface winds from the poles toward the equator is convective circulation. Solar radiation heats the air near the equator, and this low density heated air is buoyed up. At the surface it is displaced by cooler more dense higher pressure air flowing from the poles. In the upper atmosphere near the equator the air thus tend to flow back toward the poles and away from the equator. The net results is a global convective circulation with surface winds from north to south in the northern hemisphere.

It is clear from the above over simplified model that the wind is basically caused by the solar energy irradiating the earth. This is why wind utilization is considered a part of solar technology.

In actuality the wind is much more complex. The above model ignores the earth's rotation which causes a coriolis force resulting in an easterly wind velocity component in the northern hemisphere.

There is the further complication of boundary layer frictional effects between the moving air and the earth's rough surface. Mountains, trees, buildings, and similar obstructions impair stream line air flow. Turbulence results, and the wind velocity in a horizontal direction markedly increases with altitude near the surface.

Then there is the obvious fact of land and water with their unequal solar absorptivities and thermal time constants. During day light the land heats up rapidly compared to nearby sea or water bodies, and there tends to be a surface wind flow from the water to the land. At night the wind reverses, because the land surface cools faster than the water.

Local winds are caused by two mechanisms. The first is differential heating of land and water. Solar insolation during the day is readily converted to sensible energy of the land surface but is partly absorbed in layers below the water surface and partly consumed in evaporating some of that water. The land mass becomes hotter than the water, which causes the air above the land to heat up and become warmer than the air above water. The warmer lighter air above the land rises, and the cooler heavier air above the water moves into replace it. This is the mechanism of shore breezes. At night, the direction of the breezes is reversed because the land mass cools to the sky more rapidly than the water, assuming a clear sky. The second mechanism of local winds is caused by *hills* and *mountain sides*. The air above the slopes heats up during the day and cools down at night, more rapidly than the air above the low lands. This causes heated air the day to rise along the slopes and relatively cool heavy air to flow down at night.

It has been estimated that 2 percent of all solar radiation falling on the face of the earth is converted to kinetic energy in the atmosphere and that 30 percent of this kinetic energy occurs in the lowest 1000 m of elevation. It is thus said that the total kinetic energy of the wind in this lowest kilometer, if harnessed, can satisfy several times the energy demand of a country. It is also claimed that the wind power is pollution free and that its source of energy is free. Such are the seemingly compelling arguments for wind power, not unlike those for solar power. Although solar energy is cyclic and predictable, and even dependable in some parts of the globe, wind energy, however, is erratic, unsteady, and often not reliable, except in very few areas. It does, however, have a place in the total energy picture, particularly for those areas with more, or less steady winds, especially those that are far removed from central power grids, and for small, remote domestic and farm needs.

Conversion of the kinetic energy (*i.e.*, energy of motion) of the wind into mechanical energy that can be utilized to perform useful work, or to generate electricity. Most machines for converting wind energy into mechanical energy consist basically of a number of sails, vanes, or blades radiating from a hub or central axis. The axis may be horizontal, as in the more familiar windmills, or vertical, as it is in some cases. When the wind blows against the vanes or sails they rotate about the axis and the rotational motion can be made to perform useful work. Wind energy conversion devices are commonly known as *Wind turbines* because they convert the energy of the windstream into energy of rotation: the component which rotates is called the *rotor*. The terms turbine and rotor are, however, often regarded as being synonymous.

Because wind turbines produce rotational motion, wind energy is readily converted into electrical energy by connecting the turbine to an *electric generator*, The combination of wind turbine and generator is sometimes referred to as an *aero generator*. A step-up transmission is usually required to match the relatively slow speed of the wind rotor to the higher speed of an electric generator.

Although windmills have been used for more than a dozen centuries for grinding grain and pumping water, interest in large scale electric power generation has developed over the past 50 yrs. A largest wind generator built in recent times was the 800 kWe unit operated in France from 1958-1960. The flexible 3 blades propeller was about 35 m in diameter and produced the rated power in a 60 km/hour wind with a rotation speed of 47 rpm. The maximum power developed was 12 MWe.

Wind energy is one of America's greatest natural resources. The U.S. government could plan for installation of wind turbine generators with a total capacity of 1 GW by 1985. Windmills have been used for several centuries in countries like Netherland and Denmark where high velocity wind is available in abundent quantity. Nearly, 30,000 house mills capable of producing 100 MW and 3000 industrial windmills generating another 100 MW were operating in Denmark at the turn of this century.

In India the interest in the windmills was shown in the last fifties and early sixties. Apart from importing a few from outside, new designs were also developed, but it was not sustained. It is only in the last few years that development work is going on in many institutions. An important reason for this lack of interest in wind energy must be that wind, in India are relatively low and vary appreciably with the seasons. Data quoted by some scientists that for India wind speed value lies between 5 km/hr to 15-20 km/hr. These low and seasonal winds imply a high cost of exploitation of wind energy. Calculations based on the performance of a typical windmill have indicated that a unit of energy derived from a windmill will be at least several times more expensive than energy derivable from electric distribution lines at the standard rates, provided such electrical energy is at all available at the windmill site.

The above argument is not fully applicable in rural areas for several reasons. First electric power is not and will not be available in many such areas due to the high cost of generation and distribution to small dispersed users. Secondly there is possibility of reducing the cost of the windmills by suitable design. Lastly, on small scales, the total first cost for serving a felt need and low maintenance costs are more important than the unit cost of energy. The last point is illustrated easily: dry cells provide energy at the astronomical cost of about ₹ 300 per kWh and yet they are in common use in both rural and urban areas. This raises the question of that the felt needs are that a windmill might satisfy while large scale energy production at some favourable sites in India is a possibility that needs to be explored, there appears to be a definite need for small sources of mechanical energy in rural areas. For example, even casual polls among villagers quickly reveal that their first concern is invariably water for drinking, washing and irrigation; and lifting water is a task which a windmill can perform. For such a task, a windmill should produce about 100 W, considering that a pair of bullocks, often used for lifting water in villages, typically provides about 250 W power. Many projects on windmill systems for water pumping and for production of small amount of electrical power have been taken up by various organisers, such as National aeronautical laboratory Bangaluru, central salt and Marine Chemicals Research Institute Bhavnagar, Central Arid Zone Research Institute (CAZRI) Jodhpur etc.

Wind energy offers another source for pumping as well as electric power generation. India has potential of over 20,000 MW for power generation and ranks as one of the promising countries for tapping this source. The cost of power generation from *wind farms* has now become lower than diesel power and compable to thermal power in several areas of our country especially near the coasts. Wind power projects of aggregate capacity of 8 MW including 7 wind farms projects of capacity 6.85 MW have been established in diffrent parts of the country of which 3 MW capacity has been completed in 1989 by DNES. Wind farms are operating successfully and have already fed over 150 lakh units of electricity to the respective state grids. Over 25 MW of additional power capacity from wind is under implementation. Under demonstration programme 271 wind pumps have been installed upto February 1989. Sixty small wind battery chargers of capacities 300 watts to 4 kW are under installation. Likewise to stand alone wind electric generators of 10 to 25 kW are under installation.

6.2. Basic Principles of Wind Energy Conversion

6.2.1. The Nature of the Wind

The circulation of air in the atmosphere is caused by the non-uniform heating of the earth's surface by the sun. The air immediately above a warm area expands, it is forced upwards by cool, denser air which flows in from surrounding areas causing a wind. The nature of the terrain, the degree of cloud ever and the angle of the sun in the sky are all factors which influence this process. In general, during the day the air above the land mass tends to heat up more rapidly than the air over water. In coastal regions this manifests itself in a strong onshore wind. At night the process is reversed because the air cools down more rapidly over the land and the breeze therefore blows off shore.

The main planetary winds are caused in much the same way: Cool surface air sweeps down from the poles forcing the warm air over the tropics to rise. But the direction of these massive air movements is affected by the rotation of the earth and the net effect is a large countries-clockwise circulation of air around low pressure areas in the northern hemisphere, and clockwise circulation in the southern hemisphere. The strength and direction of these planetary winds change with the seasons as the solar input varies.

Despite the wind's intermittent nature, wind patterns at any particular site remain remarkably constant year by year. Average wind speeds are greater in hilly and coastal areas than they are well inland. The winds also tend to blow more consistently and with greater strength over the surface of the water where there is a less surface drag.

Wind speeds increase with height. They have traditionally been measured at a standard height of ten metres where they are found to be 20–25% greater than close to the surface. At a height of 60 m they may be 30–60% higher because of the reduction in the drag effect of the earth's surface.

6.2.2. The Power in the Wind

Wind possesses energy by virtue of its motion. Any device capable of slowing down the mass of moving air, like a sail or propeller, can extract part of the energy and convert is into useful work. Three factors determine the output from a wind energy converter:

(*i*) the wind speed;

(*ii*) the cross-section of wind swept by rotor; and

(*iii*) the overall conversion efficiency of the rotor, transmission system and generator or pump.

No device, however well-designed, can extract *all* of the wind's energy because the wind would have to be brought to a halt and this would prevent the passage of more air through the rotor. The most that is possible is for the rotor to decelerate the whole horizontal column of intercepted air to about one-third of its free velocity. A 100% efficient aerogenerator would therefore only be able to convert up to a maximum of around 60% of the available energy in wind into mechanical energy. Well-designed blades will typically extract 70% of the theretical maximum, but losses incurred in the gearbox, transmission system and generator or pump could decrease overall wind turbine efficiency to 35% or loss.

The power in the wind can be computed by using the concept of kinetics. The wind mill works on the principle of converting kinetic energy of the wind to mechanical energy. We know that power is equal to energy per unit time. The energy available is the kinetic energy of the wind. The kinetic energy of any particle is equal to one half its mass times the square of its velocity, or $\frac{1}{2} mV^2$. The amount of air passing in unit time, through an area A, with velocity V, is $A.V$, and its mass m is equal to its volume multiplied by its density ρ of air, or

$$m = \rho AV \quad \text{...(6.1)}$$

(m is the mass of air transversing the area A swept by the rotating blades of a wind mill type generator).

Substituting this value of the mass in the expression for the kinetic energy, we obtain, kinetic energy $= 1/2\ \rho AV.V^2$ watts

$$= \frac{1}{2} \rho A V^3 \text{ watts} \quad ...(6.2)$$

Equation (6.2) tells us that the maximum wind available the actual amount will be somewhat less because all the available energy is not extractable—is proportional to the cube of the wind speed. It is thus evident that small increase in wind speed can have a marked effect on the power in the wind.

Equation (6.2) also tell us that the power available is proportional to air density (1.225 kg/m^3 at sea level). It may vary 10-15 % during the year because of pressure and temperature change. It changes negligibly with water content. Equation also tells us that the wind power is proportional to the intercept area. Thus an aeroturbine C with a large swept area has higher power than a smaller area machine; but there are added implications. Since the area is normally circular of diameter D in *horizontal axis* aeroturbines, then $A = \frac{\pi}{4} D^2$, (sq. m.), which when put in equation (6.2) gives,

$$\text{Available wind power } P_a = \frac{1}{2} \rho \frac{\pi}{4} D^2 V^3 \text{ watts}$$

$$= \frac{1}{8} \rho \pi D^2 V^3 \quad ...(6.3)$$

The equation tells us that the maximum power available from the wind varies according to the square of the diameter of the intercept area (or square of the rotor diameter), normally taken to be swept area of the aeroturbine. Thus doubting the diameter of the rotor will result in a four-fold increase in the available wind power. Equation (6.3) gives us in sight into why the designer of an aeroturbine for wind electric use would place such great emphasis on the turbine diameter. The combined effects of wind speed and rotor diameter variations are shown in. Fig. 6.1. Wind machines intended for generating substantial amounts of power should have large rotors and be located in areas of high wind speeds. Where low or moderate powers are adequate, these requirements can be relaxed.

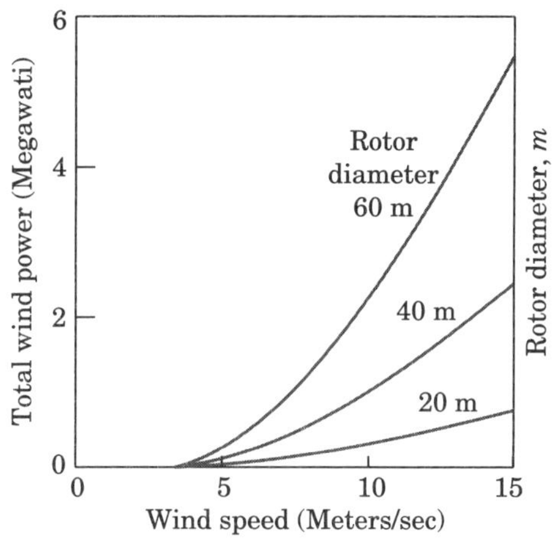

Fig. 6.1. Dependence of wind-rotor power on wind speed and rotor diameter.

The physical conditions in a wind turbine are such that only a fraction, of the available wind power can be converted into useful power. As the free wind stream encounters and passes through a rotor, the wind transfer some of its energy to the rotor and its speed decreases to a minimum in the rotor wake. Subsequently, the wind stream regains energy from the surrounding air and at a sufficient distance from the rotor the free wind speed is restored (Fig. 6.2 upper curve). While the wind speed is decreasing, as just described, the air pressure in the windstream changes in a different manner (Fig. 6.2, lower curve). It first increases as the wind approaches the rotor and then drops sharply by an amount Δp as it passes through and energy is transferred to the rotor. Finally the pressure increases to the ambient atmospheric pressure.

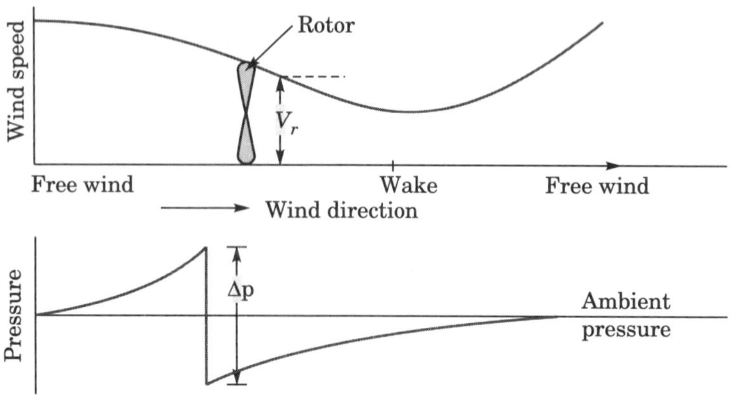

Fig. 6.2. Conditions in traversing a wind rotor.

The power extracted by the rotor is equal to the product of the wind speed as it passes through the rotor (*i.e.*, V_r in Fig. 6.2) and the pressure drop Δp. In order to maximize the rotor power it would therefore be desirable to have both wind speed and pressure drop as large as possible. However, as V is increased for a given value of the free wind speed (and air density), Δp increases at first, passes through a maximum, and then decreases. Hence for the specified free-wind speed, there is a maximum value of the rotor power.

The fraction of the free-flow wind power that can be extracted by a rotor is called the *power-coefficient*; thus

$$\text{Power coefficient} = \frac{\text{Power of wind rotor}}{\text{Power available in the wind}}$$

where power available is calculated from the air density, rotor diameter, and free wind speed as shown above. The maximum theoretical power coefficient is equal to 16/27 or 0.593. This value cannot be exceeded by a rotor in a free-flow wind-stream. (It can be exceeded under specific conditions, as will be seen later).

As an ideal rotor, with propeller-type blades of proper areodynamic design, would have a power coefficient approaching 0.59. But such a rotor would not be strong enough to withstand the stresses to which it is subjected when rotating at a high rate in a high-speed wind stream. For the best practical rotors, the power coefficient is about 0.4 to 0.45, so that the rotors cannot use more than 40 to 45% of the available wind power. In the conversion into electric power, some of the rotor energy is lost and the overall electric power, coefficient of an aerogenerator (*i.e.*, electric power generated/available wind power) in practice is about 0.35 (35 %).

Returning to equation (6.2), but now recognizing that V, in actuality, is not constant but is represented by a statistically 'noisy' wind speed time curve, $V_{(t)}$, then the instantaneous power, in the wind would be

$$P_{a(t)} = \frac{1}{2}\rho A[V(t)]^3 \text{ watts} \qquad ...(6.4)$$

Since we are normally more interested in average power, we must time average both sides of equation (6.4), signified by the bar below

$$\overline{P_{a(t)}} = \frac{1}{2}\rho A\,\overline{[V_{(t)}]^3} \text{ watts} \qquad ...(6.5)$$

Equation (6.5) tells us that for a non-steady state wind, it is necessary to cube the measured wind speeds and then take the average to find the average wind power available. It is immediately obvious that this non-steady state case is more complex than the simple steady state case, and it is why for the former case such great emphasis is placed on anemometry data at a proposed wind energy conversion system (WECS) site.

Transposing equation (6.5) results in

$$\frac{\overline{P_{a(t)}}}{A} = \frac{1}{2}\rho\,\overline{[V_{(t)}]^3}\ \text{watts/m}^2 \qquad ...(6.6)$$

which says that the average available wind power per unit area is directly related to the average of the wind speed cubed. This is one useful method of characterizing the potential specific power in the wind over geographic area.

There are clear advantages in selecting sites with annual mean wind speeds and building larger rather than smaller wind generators since:

(*a*) the power available in the wind increases as *cube* of the wind speed: doubling the wind speed increases the power available by eight-fold; and

(*b*) doubling the diameter of the turbine's rotor quadruples the swept area and hence the power output from the device (this law only applies to horizontal axis machines, for vertical axis machines the change in power output with diameter will be determined by the geometry of the rotor).

The way rotor diameter and wind speed affect power output can be seen in Fig. 6.1.

In practice a wind turbine's output will vary. There will be periods when there is insufficient wind for the machine to generate any power at all, and times when the wind speeds are so high that the machine has to be shut down to prevent damage.

Maximum Power. As stated above, that the total power can not be converted to mechanical power. Consider a horizontal-axis, propeller-type windmill, henceforth to be called a wind turbine, which is the most common type used today. Assume that the wheel of such a turbine has thickness $a\,b$, as shown in Fig. 6.3. Let P_i and V_i are the wind pressure and velocity at the upstream of the turbine, and P_e and V_e are pressure and velocity at downstream of the turbine. V_e is less than V_i because kinetic energy is extracted by the turbine.

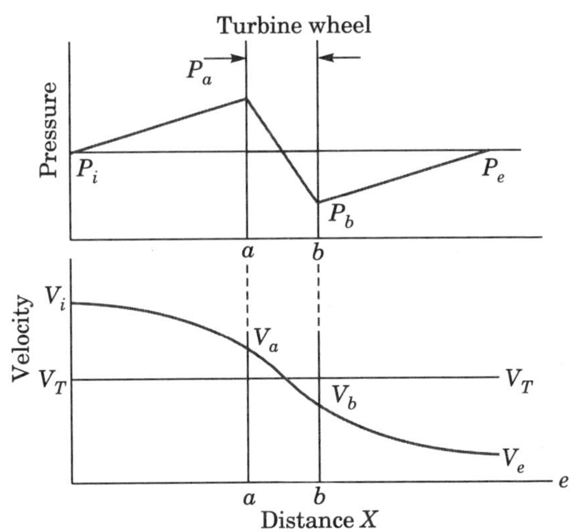

Fig. 6.3. Pressure and velocity profiles of wind moving through, a horizontal-axis propeller-type wind turbine.

Considering the incoming air between *i* and *a* as a thermodynamic system, and assuming that the air density remains constant (since changes in pressure and temperature are very small compared to ambient), that the potential energy is zero, and no heat or work are added or removed between i and a, the general energy equation reduces to the kinetic and flow energy-terms only:

Thus,
$$P_i v + \frac{V_i^2}{2g_c} = P_a v + \frac{V_a^2}{2g_c} \qquad \text{...(6.7 a)}$$

The general energy equation for steady state flow for unit mass (for a control volume) is:

$$Z_1 + \frac{V_1^2}{2g} + u + P_1 v_1 + \Delta q = Z_2 + \frac{V_2^2}{2g} + u_2 + P_2 v_2 + \Delta W_{sf}$$

where, Z_1 = potential energy, $\frac{V^2}{2g}$ = kinetic energy, u = internal energy

Pv = flow energy, Δq = net heat added

ΔW_{sf} = net steady flow mechanical workdone of the system

or
$$P_i + \rho\frac{V_i^2}{2g_c} = P_a + \rho\frac{V_a^2}{2g_c} \qquad \text{...(6.7 b)}$$

where v and ρ are the specific volume and its reciprocal, the density, respectively, both considered to be constant.

Similarly for the exit region *be*,

$$P_r + \rho\frac{V_e^2}{2g_c} = P_b + \rho\frac{V_b^2}{2g_c} \qquad \text{...(6.8)}$$

The wind velocity across the turbine decreases from a to b since kinetic energy is converted to mechanical work there. The incoming velocity V_i does not decrease abruptly but gradually as it approaches the turbine to V_a and as it leaves it to V_e. The $V_i > V_a$ and $V_b > V_e$, and therefore, from equations (6.7) and (6.8), $P_a > P_i$ and $P_b, > P_i$; that is, the wind pressure rises as it approaches, then as it leaves the wheel. Combining these equations,

$$P_a - P_b = \left(P_i + \rho\frac{V_i^2 - V_a^2}{2g_c}\right) - \left(P_e + \rho\frac{V_e^2 - V_b^2}{2g_c}\right) \qquad \text{...(6.9)}$$

It can be assumed that wind pressure at e can be assumed to ambient, *i.e.*,

$$P_e = P_i \qquad \text{...(6.10)}$$

As the blade width a. b is very thin as compared to total distance considered, it can be assumed that velocity within the turbine does not change much.

$$V_a \simeq V_t \simeq V_b \qquad \text{...(6.11)}$$

Combining equation (6.9) to (6.11) yields,

$$P_a - P_b = \rho\left(\frac{V_i^2 - V_e^2}{2g_c}\right) \qquad \text{...(6.12)}$$

The axial force F_x, in the direction of wind stream, on a turbine wheel with projected area, perpendicular to the stream A, is given by

$$F_x = (P - P_b)A = \rho A\left(\frac{V_t^2 - V_e^2}{2g_c}\right) \qquad \text{...(6.13)}$$

This force is also equal to change in momentum of the wind (from Newton's second law).

$$F_x = \Delta(\dot{m}\,V)/g_c$$

where, $\dot{m}$ = mass flow rate = $\rho A V_t$

Thus $$F_x = \frac{1}{g_c}\rho A V_t (V_i - V_e) \qquad ...(6.14)$$

Equating equations (6.13) and (6.14),

$$\rho A \frac{(V_i^2 - V_e^2)}{2g_c} = \frac{1}{g_c}\rho A V_t (V_i - V_e)$$

$$V_t = \frac{1}{2}(V_i + V_e) \qquad ...(6.15)$$

Considering the total thermodynamic system bounded by i and e, there are no changes in potential energy, the internal energy is there from T_i to T_e, and flow energy change is there from $P_i v$ to $P_e v$. In the system no heat is added or rejected *i.e.,* adiabatic flow. The general energy equation how reduces to the steady flow work W and kinetic energy terms,

$$W = kE_i - kE_e = \frac{V_i^2 - V_e^2}{2g_c} \qquad ...(6.16)$$

The power P is defined as the rate of work, from mass flow rate equation

$$P = m\frac{V_i^2 - V_e^2}{2g_c} = \frac{1}{2g_c}\rho A V_t (V_i^2 - V_e^2) \qquad ...(6.17)$$

Combining this with equation (6.15),

$$P = \frac{1}{4g_c}\rho A\,(V_i + V_e)(V_i^2 - V_e^2) \qquad ...(6.18)$$

Equation (6.17) reverts to equation (6.2) for P_{total} when $V_t = V_i$, and $V_e = 0$; that is, the wind velocity reduces to zero after leaving the turbine. This is not possible because the wind can not accumulate at turbine exit. It can be seen from equation (6.18), where V_e is positive in one term and negative in other, that too low or too high a value for V_e results in reduced power. There thus is an optimum exit velocity V_e opt, that results in maximum power P_{max}, which can be obtained by differenting P, and equating the derivative to zero, *i.e.,*

$$\frac{dp}{dV_e} = 0$$

or $$\frac{dp}{dV_e} = 3V_e^2 + 2V_iV_e - V_i^2 = 0$$

This is solved for a positive V_e to give V_e opt. (The quadratic has two solutions, *i.e.,* $V_e = V_i$ and $V_e = \frac{1}{3}V_i$, only second solution is physically acceptable).

Thus $$V_e \text{ opt} = \frac{1}{3}V_i \qquad ...(6.19)$$

Using the equation (6.18), for an ideal wind machine, with horizontal axis.

$$P_{max} = \frac{8}{27g_c}\rho A V_i^3 \qquad ...(6.20\ a)$$

$$= \frac{16}{27g_c}\frac{1}{2}\rho AV_i^3 = 0.593\left(\frac{1}{2}\cdot\frac{\rho VA_i^3}{g_c}\right)$$

$$= 0.595\, P_{total} \qquad ...(6.20\ b)$$

The *ideal, or maximum, theoretical efficiency* η_{max} (also called the *power coefficient)* of a wind turbine is the ratio of the maximum power obtained from the wind, to the total power available in the wind. The factor 0.593 is known as the Betz coefficient (from the name of the man who first derived it). It is the maximum fraction of the power in a wind stream that can be extract.

$$\text{Power coefficient} = C_p \equiv \frac{\text{Power output from wind machine}}{\text{Power available in wind}}$$

Thus C_p can not exceed 0.593 for a horizontal axis wind machine.

6.2.3. Forces on the Blades and Thrust on Turbines

As stated earlier, here blades of propeller-type wind turbine is considered. There are two types of forces which are acting on the blades. One is *circumferential force* acting in the direction of wheel rotation that provides the torque and other is the *axial force* acting in the direction of the wind stream that provides an *axial thrust* that must be counteracted by proper mechanical design.

The *Circumferential force, or torque T* can be obtained from

$$T = \frac{P}{\omega} = \frac{P}{\pi DN} \qquad ...(6.21)$$

where, T = torque *kgf* or *Newton (N)*

ω = angular velocity of turbine wheel, m/s

D = diameter of turbine wheel

$= \sqrt{\frac{4}{\pi}}\,.\,A, m$

N = wheel revolutions per unit time, s^{-1}

$\therefore$ The real efficiency $\eta = \dfrac{P}{P_{total}}$

or

$$P = \eta\,.\,P_{total} = \eta\,.\,\frac{1}{2g_c}\rho AV_i^3$$

For a turbine operating at power P, the expression for torque becomes

$$T = \eta\frac{1.\rho}{2g_c}\frac{AV_i^3}{\pi DN}$$

$$= \eta\frac{1}{2g_c}\frac{\rho.\pi}{4}\frac{D^2V_i^3}{\pi DN} = \eta\frac{1}{8g_c}\frac{\rho DV_i^3}{N}$$

$$= \eta\frac{1}{8g_c}\frac{\rho DV_i^3}{N} \qquad ...(6.22)$$

At maximum efficiency $\left(\eta_{max} = \dfrac{16}{27}\right)$, the torque has maximum value T_{max} which is equal to

$$T_{max} = \frac{2}{27g_c}\frac{\rho D V_i^3}{N} \qquad ...(6.23)$$

The axial force or thrust by equation (6.13)

$$F_x = \frac{1}{2g_c}\rho A(V_i^2 - V_e^2)$$

$$= \frac{\pi}{8g_c}\rho D^2(V_i^2 - V_e^2) \qquad ...(6.24)$$

The axial force on a turbine wheel operating at maximum efficiency where $V_c = \frac{1}{3}V_i$ is given by

$$F_x, max = \frac{4}{9g_c}\rho A V_i^2 = \frac{\pi}{9g_c}\rho A V_i^2$$

$$= \frac{\pi}{9g_c}\rho D^2 V_i^2 \qquad ...(6.25)$$

We see that axial forces are proportional to the square of the diameter of the turbine wheel, this limits turbine wheel diameter of large size.

6.2.4. Wind Energy Conversion

Traditional windmills were used extensively in the middle ages to mill grain and lift water for land drainage and watering cattle. Wind energy converters are still used for these purposes today in some parts of the world, but the main focus of attention now lies with their use to generate electricity. There is also growing interest in generating heat from the wind for space and water heating and for glass houses but the potential market is much smaller than for electricity generation.

The term "wind mill" is still widely used to describe wind energy conversion systems, however it is hardly an opt. description any more. Modern wind energy conversion systems are more correctly referred to as 'WECS', 'aerogenerators', 'wind turbine generators', or simply 'wind turbines'.

The fact that the wind is variable and intermittent source of energy is immaterial for some applications such as pumping water for land drainage—provided, of course, that there is a broad match between the energy supplied over any critical period and the energy required. If the wind blows, the job gets done; if it does not, the job waits.

However, for many of the uses to which electricity is put, the interruption of supply may be highly inconvenient. Operators or users of wind turbines must ensure that there is some form of back-up to cover periods when there is insufficient (or too much) wind available. For small producers, backup can take the form of:

(*i*) battery storage;

(*ii*) connection with the local electricity distribution system; or

(*iii*) a stand-by generator powered by liquid or gaseous fuels.

For utilites responsible for public supply, the integration of medium-sized and large wind turbines into their distribution net work could require some additional plant which is capable of responding quickly to meet fluctuating demand.

Small producers. Private citizens in several countries have won the right to operate wind generators and other renewable energy systems and to export power to the grid. For most small wind generators this requires that the output is 'conditioned' so that it conforms to the frequency and phase of the mains supply. Only a few small units are designed to maintain a constant rotational rate so that can be synchronized to the mains frequency and feed electricity directly into the grid. Most produce direct current (DC) or variable output alternating current (AC).

Power conditioning is readily achieved using an electronic blackbox called a 'synchronous' inverter, and although this is an expensive item of equipment, it does eliminate the need for batteries and for conversion of home appliances to run on DC.

Where there is no grid connection, electricity that is surplus to immediately requirements must be stored on site using heavy duty batteries. It can be recovered later when the demand exceeds the supply. An alternative is to dump it (by generating and dissipating heat) or better, to convert it into heat that can be stored, for example as hot water in a well insulated tank.

Large producers. Large and medium-sized wind generators are designed to give a stable and constant electrical output over a wide range of wind speeds and to feed current directy into the grid. They operate primarily as fuel savers, reducing the utility's total fuel burn.

The choice of generator type depends on the size of the local distribution grid and its associated generating capacity. An *induction generator* would normally be used where there is a significant amount. other generating capacity (which could provide the necessary reactive power for excitation). Induction generators are robust and reliable and squire minimal control equipment. For isolated networks where other local generating capacity is limited, and where a high degree of autonomous control is required, a *synchronous* generator is more appropriate. Synchronous generators are more complex and therefore more expensive than induction machines.

Lift and drag: The basis for wind energy conversion. The extraction of power, and hence energy, from the wind depends on creating certain forces and applying them to rotate (or to translate) a mechanism. There are two primary mechanisms for producing forces from the wind, *lift* and *drag*.

By definition lift forces act perpendicular to the air flow, while drag forces act in the direction of flow. Lift forces are produced by changing the velocity of the air stream flowing over either side of the lifting surface speeding up the air flow causes the pressure to drop, while slowing the air stream down leads to increase in pressure. In other words, any change in velocity generates a pressure difference across the lifting surface. This pressure difference produces a force that begins to act on the high pressure side and moves towards the low pressure side of the lifting surface which is called an airfoil. A good airfoil has a high lift/drag ratio, in some cases it can generate lift forces perpendicular to the air stream direction that are 30 times as great as the drag force parallel to the flow. The lift increases as the angle formed at the junction of the airfoil and the air-stream (the angle of attack) becomes less and less actute, upto the point where the angle of the air flow on the low pressure side becomes excessive. When this happens, the air flow breaks away from the low pressure side. A lot of turbulence ensues, the lift decreases and the drag increases quite substantially; this phenomenon is known as *stalling*. For efficient operation, a wind turbine blade needs to function with as much lift and as little drag as possible because drag dissipates energy. As lift does not involve anything more complex than deflecting the air flow, it is usually an efficient process.

The design of each wind turbine specifies the angle at which the airfoil should be set to achieve the maximum lift to drag ratio.

In addition to airfoils, there are two other mechanisms for creating lift. One is the so-called *Magnus effect,* caused by spinning a cylinder in an air stream at a high speed of rotation. The spinning slows down the air speed on the side where the cylinder is moving into wind and increases it on the other side; the result is similar to an airfoil. This principle has been put to practical use in one or two cases but is not generally employed. The second way is to blow air through narrow slots in a cylinder, so that it emerges tangentially; this is known as a *Thwaits* slot. This also creates a rotation (or circulation) of the air flow, which in turn generates lift. Because the lift drag ratio of airfoils is generally much better than those of rotating or slotted cylinders, the latter techniques probably have little practical potential.

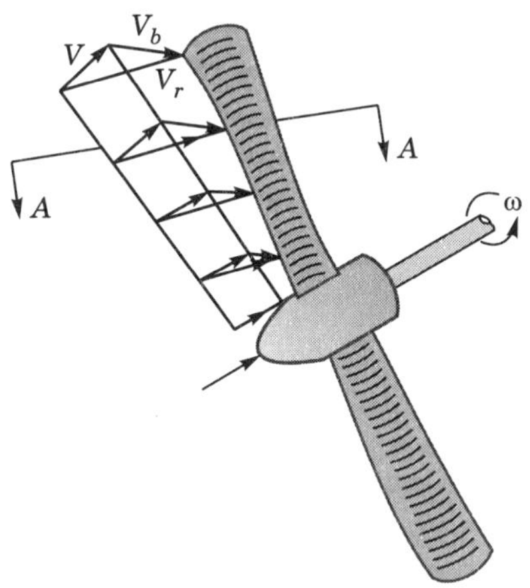

Symbols

V	free wind velocity
V_b	velocity of airfoil element ($\lvert V_b \rvert = \omega_r$)
V_r	resultant wind as 'seen' by airfoil element
F_L	lilt force (perpendicular to V_r)
F_D	drag force
θ	angle of twist
$\propto$	angle of incidence
ω	angular speed of rotor
r	distance of airfoil element from its axis of rotation.

Fig. 6.4(*a*)

The windmill blade 'sees' the resultant vector V_r. The blades need to be twisted because r varies in proportion to radius.

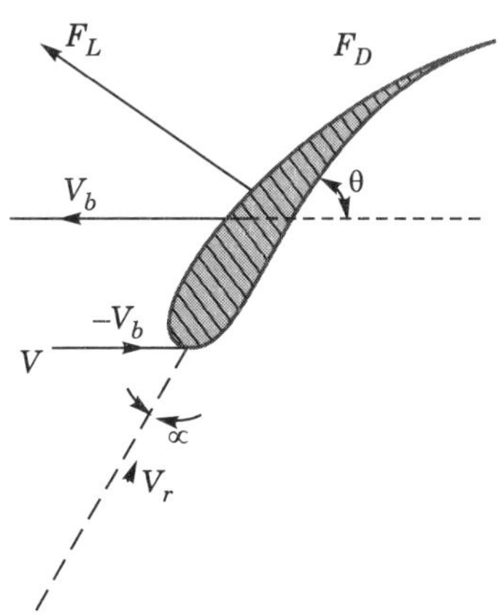

(*b*) Cross-section across *A—A* Fig. (*a*).

Fig. 6.4. Shape of a horizontal axis, wind turbine rotor blade.

Example 6.1. *Wind at 1 standard atmospheric pressure and 15°C has velocity of 15 m/s calculate:*

(i) the total power density in the wind stream,
(ii) the maximum obtainable power density,
(iii) a reasonally obtainable power density,
(iv) the total power, and
(v) the torque and axial thrust.

Given : turbine diameter = 120 m, and turbine operating speed = 40 r.p.m. at maximum efficiency. Propeller type wind turbine is con-sidered.

Solution.

For air, the value of gas constant

$$R = 0.287 \text{ kJ/kg K}$$

$$1 \text{ atm.} = 1.01325 \times 10^5 P_a.$$

Air density $\rho = \dfrac{P}{RT}$, $\rho = \dfrac{1.01325 \times 10^5}{287(288)} = 1.226 \text{ kg/m}^3$

(*i*) Total power $P_{total} = \dfrac{\rho A V_i^3}{2g_c}$

Power density $= \dfrac{P_{total}}{A} = \dfrac{1}{2g_c} . \rho\, V_i^3 = \dfrac{1}{2 \times 1} 1.226 \times 15^3$

$= 2068.87 \text{ W/m}^2$. **Ans.**

(*ii*) Maximum power density

$$= \frac{P_{max}}{A} = \frac{8}{27g_c} \rho A V_i^3$$

$$= \frac{8}{27 \times 1} 1.226 \times 15^3 = 1226 \text{ W/m}^2. \textbf{ Ans.}$$

(*iii*) Assuming $n = 35\%$

$$\because \quad \frac{P}{A} = n \frac{P_{total}}{A}$$

$= 0.35 \times 2068.87 = 724 \text{ W/m}^2$. **Ans.**

(*iv*) Total Power P = Power density × Area

$$= 724 \times \frac{\pi}{4} D^2 \text{ watt}$$

$$= 0.724 \times \frac{\pi}{4} \times 120^2 \text{ kW} = 8184 \text{ kW}. \textbf{ Ans.}$$

(*v*) Torque at maximum efficiency

$$T_{max} = \frac{2}{27g_c} \times \frac{\rho D V_i^3}{N}$$

$$= \frac{2}{27 \times 1} \times \frac{1.226 \times 120 \times 15^3}{40/60} = 55170 \text{ Nowton}. \textbf{ Ans.}$$

and maximum axial thrust

$$F_{x\,max} = \frac{\pi}{9g_c} \rho D^2 V_i^2 = \frac{\pi}{9 \times 1} 1.226 \times 120^2 \times 15^2$$

$= 1385{,}870$ Newton. **Ans.**

6.3. Wind Data and Energy Estimation

The seasonal as well as instantaneous changes in winds both with regard to magnitude and direction need to be well understood to make the best use of them in windmill designs. Winds are known to fluctuate by a factor of 2 or more within seconds (and thus causing the power to fluctuate by a factor of 8 or more). This calls for a proper recording and analysis of the wind characteristics.

There are various ways the data on wind behaviour is collected depending on the use it is intended to be put into. The hourly mean wind velocity as collected by the meteorological observations is the basic data used in a windmill designs. The hourly mean is the one averaged over a particular hour of the day, over the day, month, year and years. The factors which affect the *nature of the wind* close to the surface of the earth, they are:

(*i*) latitude of the place, (*ii*) altitude of the place,

(*iii*) topography of the place, (*iv*) scale of the hours, month or year.

Winds being an unsteady phenomenon, the scale of the periods considered is an important set of data required in the design. The *hourly mean velocity* (for many years) provides the data for establishing the potential of the place for tapping the wind energy. The *scale of the month* is useful to indicate whether it is going to be useful during particular periods of the year and what storage if necessary is to be provided for. The *data based on scale of the hour is* useful for mechanical aspects of design.

Since the winds near the surface of the earth are derived from large scale movement of atmospheric winds, the location height above ground level at which the wind is measured and the nature of the surface on earth have an influence on the velocity of wind at any given time. The winds near the surface of the earth are interpreted in terms of boundary layer concept, keeping in mind the factors that influence its development. The wind velocity at a given height can be represented in terms of gradient height and velocity :

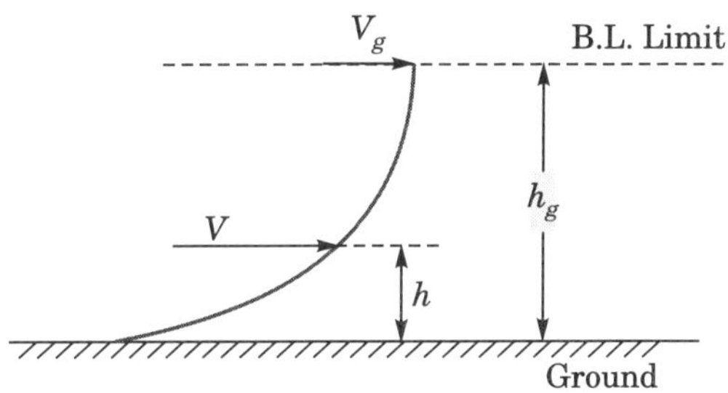

Fig. 6.5. Representation of gradient height and velocity.

$$\frac{V}{V_g} = \left(\frac{h}{h_g}\right)^n \quad \text{(refer Fig. 6.5)}$$

The values of V_g, h_g and n depends on the nature of the terrain, which are classified as

(*i*) Open terrain with few obstacles (open land, lake, shores, deserts, prairies, etc.)

(*ii*) Terrain with uniformly covered obstacles (wood lands, small towns, suburbs, etc.)

(*iii*) Terrain with large and irregular objects (large city centres, country with breaks of large trees etc.).

In as much as the height of the windmill rotor depends on the design wind velocity and cost of supporting structure. The above factors have a bearing on the design. Similarly, winds being an unsteady phenomenon, the scale of periods considered for this the temporal parameters (scale of hour, month and year) is an important set of data required in the design. While the hourly mean velocity (for many years) provides the data for establishing the potential of the place for tapping the wind energy. The scale of the month is useful to indicate whether it is going to be useful during particular periods of the year and what storage if necessary is to be provided for as already mentioned above. The data based on scale of the hour is useful for mechanical aspects of design. In addition to the data on the hourly mean velocity, two other informations required are:

- Spells of low wind speeds, and
- Gusts

The former again for providing storage or alternatives. The latter is required for structural design of the windmill as well as to provide safety measures against damage.

A number of criteria can be applied in estimating the importance of wind potential as a function of height and location. First of all careful siting is important because wind speed near the ground is greatly affected by houses, trees and similar features as stated above. Wind speed increases with height above ground, the if rate of increase being about the same at all locations. Therefore, if the wind speed at a given height is known, the speed at any other height may be calculated.

Surface wind data on a national or regional basis is usually presented in the forms of:

(*i*) *Isovents* or contours of constant average wind velocity (m/sec or km/hr.). The averaging period seen in the literature varies widely, but monthly, quarterly, and yearly averages are commonly seen. It is important to know what the data averaging period is when examining a given isovent contour map, for the winds change seasonally. Fig. (6.6) shows the wind map of India, in which the mean annual wind velocity zones are marked. It is seen that only some parts of the country

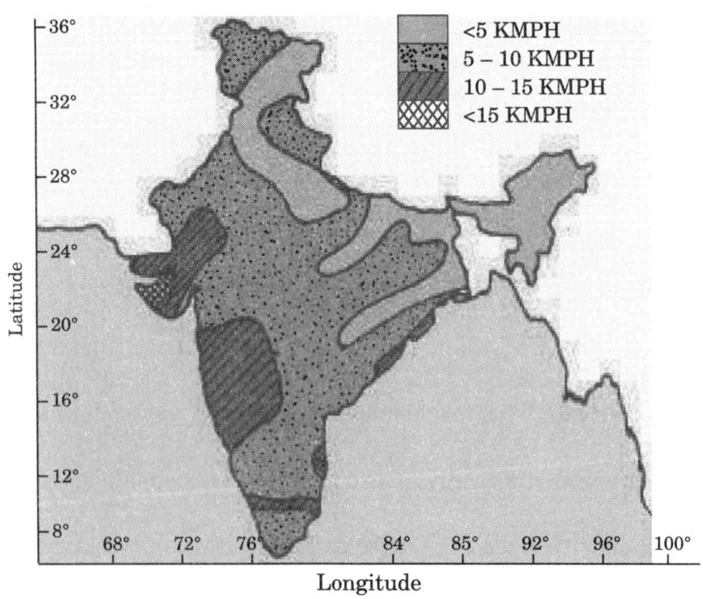

(*a*)

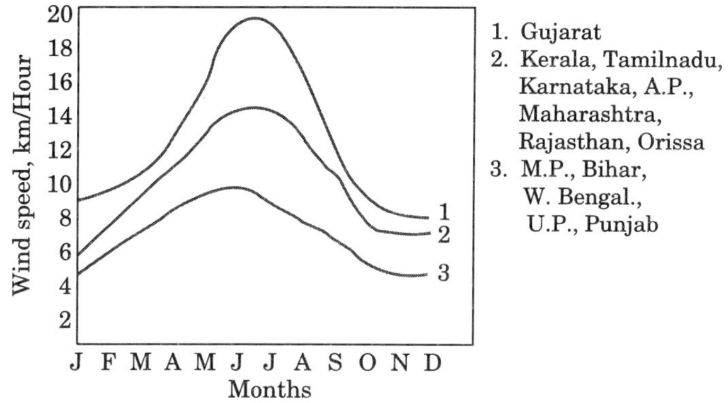

(*b*) Wind speeds in India (Average over 30 years)

Fig. 6.6. Wind map of India.

have reasonably good velocity (*i.e.,* from the point of view of wind mill operation). Even at other zones marked with low wind velocity one can always find 'spots high winds' depending on the topography.

(*ii*) *Isodynes* are contours of constant wind power (watts/m^2 of area perpendicular to the wind flow). Again it is important to know the averaging period. An example of this map is shown in Fig. (6.7).

Wind Surveys. Typical wind measurements at potential sites for wind machines usually require the following.

(1) *Instrumentation:*

3 cup anemometer and wind direction sensors.

Height of instruments; 10 m (30 feet) for preliminary data; 15 m to 45 m (50 to 150 feet) for long time data.

(2) *Data recording systems:*

Strip chart

Magnetic tap

(3) *Type of data:*

Wind speed and directional-hourly averages

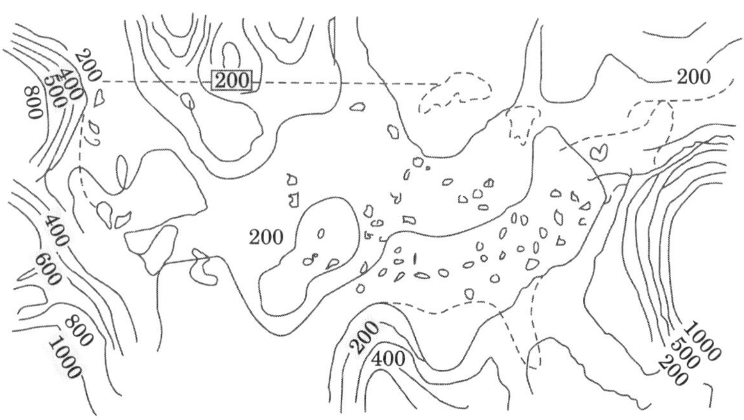

Fig. 6.7. A typical plot of available wind power (watts/m^2) yearly average.

(4) *Data reporting:*

Wind frequency curves

Daily, weekly, monthly.

Energy Estimation. The basic wind data of hourly mean wind velocity is recost into—number of hours in the year for which the speed equals or exceeds each particular value.

- annual hours of duration of various wind speed.

The first of these plotted amongst the hours in the year is called the velocity—duration curve. The replot of this on V^3 basis is called the power duration curve (V^3 being proportional to P).

These curves are useful for establishing the wind energy potential of a place, and the design wind speed.

The plot of the second form of data of number of hours for which mean wind velocity of a particular value is available is called the frequency duration curve. (Fig. 6.9). This curve is useful in deciding the design wind speed for a given site once the type of windmill is decided and its performance is known, the data is also useful in estimating the actual energy output of the plant.

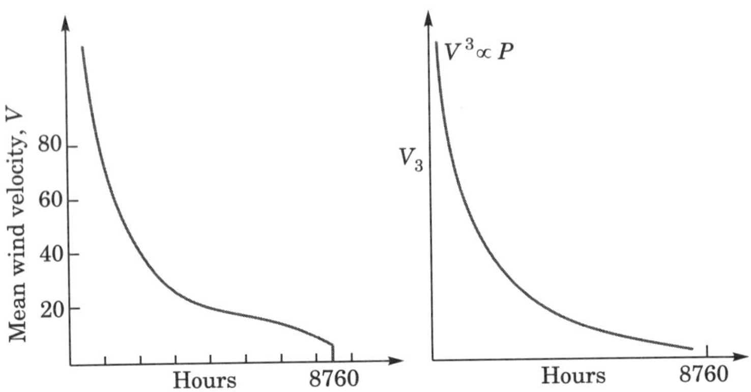

(*a*) Velocity duration curve (*b*) Power duration curve

Fig. 6.8

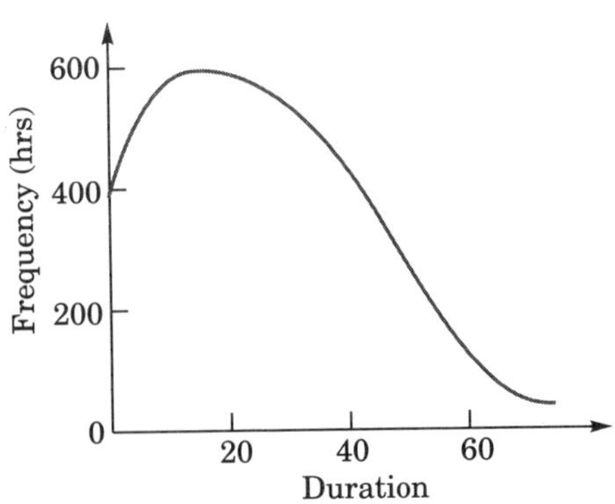

Fig. 6.9. Frequency duration curve.

The *three* speeds associated with the design of a windmill are:

V_c—cut in speed, the speed below which the wind mill does not operate.

V_d—design speed, the speed for which the rotor is designed.

V_f—furling speed, the speed at which the rotor is turned away from facing the wind or stopped otherwise with a view to protect the windmill.

All these speeds have a bearing on the mean wind velocity of the place. Another reference speed is the rated velocity V_{rat} at which the plant output is maximum.

The three speeds are marked on Fig. (6.10). The hatched area *a b c d e,* represents the annual energy output from an ideal plant. Area A B C D represents the output obtainable from an ideal plant, if the windmill were to operate at the design speed all the time. The ratio of areas *a b c d e*/A B C D is the annual load factor of the plant. In practice, the actual output will be smaller than that represented by area *a b c d e* due to the inability of the rotor to convert the entire kinetic energy available in the wind. This is represented by C_p, which varies as a function of velocity having the maximum value at the design wind speed. Fig. (6.11) shows a typical variation.

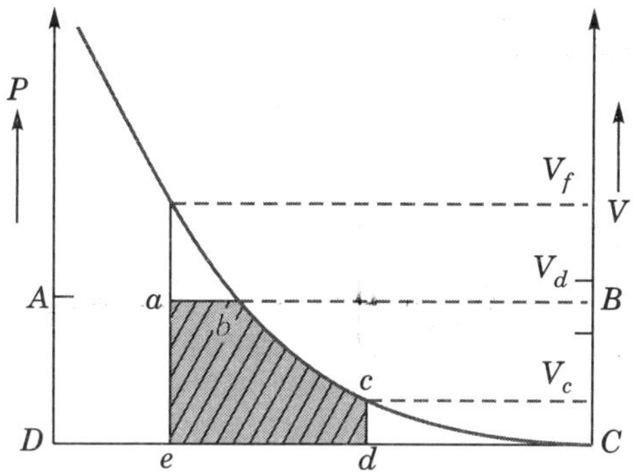

Fig. 6.10. Ideal plant output.

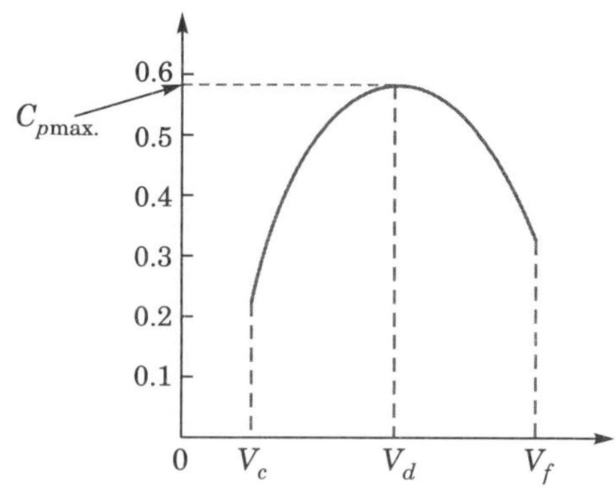

Fig. 6.11. Variation of C_p as a function of V.

Taking this into consideration, the annual output from a plant is given by

$$E = \int \eta_m \, C_p \, \rho A V^3 \delta_T \qquad ...(6.26)$$

where, δ_T is the time increment since the data is usually available in the form of number of hours for which each mean wind velocity is available, the expression takes the form

$$E = \sum_{c}^{f} \eta_m \, C_p \, \rho A V^3{}_{c-f} \delta_T \qquad ...(6.27)$$

As such the detailed knowledge of wind data, the type of windmill intended to be used and its performance characteristics are all required to establish the energy delivered by the plant at a given site.

6.4. Site Selection Considerations

The power available in the wind increases rapidly with the speed, hence wind energy conversion machines should be located preferable in areas where the winds are strong and persistent. Although daily winds at a given site may be highly variable, the monthly and especially annual average speeds are remarkably constant from year to year. The major contribution to the wind power available at a given site is actually made by winds with speeds above the average. Nevertheless, the most suitable sites for wind turbines would be found in areas where the annual average wind speeds are known to be moderately high or high.

The site choice for a single or a spatial array of WECS (wind energy conversion system) is an important matter when wind electrics is looked at from the systems point of view of aeroturbine generators feeding power into a conventional electric grid. If the WECS sites are wrongly or poorly chosen the net wind electric generated energy per year may be sub optimal with resulting high capital cost for the WECS apparatus, high costs for wind generated electric energy, and low returns on investment. Even if the WECS is to be a small generator not tied to the electric grid, the siting must be carefully chosen if inordinately long break even times are to be avoided. Technical, economic, environmental, social, and other factors are examined before a decision is made to erect a generating plant on a specific site. Some of the main considerations are discussed below.

(1) *High annual average wind speed.* A fundamental requirements to the successful use of WECS, obviously, is an adequate supply of wind as stated above. The wind velocity is the critical parameter. The power in the wind P_w, through a given cross-sectional area for a uniform wind velocity V, is

$$P_w = KV^3 \quad \text{...(6.28)}$$

where, K is a constant. It is evident, because of the cubic dependence on wind velocity that small increases in V markedly affect the power in the wind, *e.g.*, doubling V, increases P_w by a factor of 8. It is obviously desirable to select a site for WECS with high wind velocity. Thus a high average wind velocity is the principal fundamental parameter of concern in initially appraising a WECS site. For a more detailed estimate value, one would like to have the average of the velocity cubed.

Anemometer data is normally based on wind speed measurements from a height of 10 m. For the most accurate assessment of wind power potential it is absolutely essential that anemometer data be obtained at the precise site and hub height for any proposed WECS.

Strategy for siting is generally recognized to consists of:

(*i*) Survey of historical wind data,

(*ii*) Contour maps of terrain and wind are consulted.

(*iii*) Potential sites are visited.

(*iv*) Best sites are instrumented for approximately one year.

(*v*) Choose optimal site.

(2) *Availability of anemometry data.* It is another important siting factor. The principal object is to measure the wind speed which basically determines the WECS output power, but there are many practical difficulties with the instrumentation and measurement methods. The anemometer height above ground, accuracy, linearity, location on the support tower, shadowing and inaccurate readings therefrom, icing inertia of rotor whether it measures the horizontal velocity component or vertical, and temperature effects are a few of the many difficulties encountered. The anemometry data should be available over some time period at the precise spot where any proposed WECS is to be built and that this should be accomplished before a siting decision is made.

(3) *Availability of wind $V_{(t)}$ curve at the proposed site.* This important curve determines the maximum energy in the wind and hence is the principal initially controlling factor in predicting the electrical output and hence revenue return of the WECS machine. It is desirable to have average wind speed Vsuch that $V \geq 12 - 16$ km/hr (3.5 — 4.5 m/sec) which is about the lower limit at which present large

scale, WECS generators 'cut in' *i.e.,* start turning. The $V_{(t)}$ curve also determines the reliability of the delivered WECS generator power, for if the $V_{(t)}$ curve goes to zero there will be no generated power during that time. If there are long periods of calm the WECS reliability will be lower than if the calm periods are short. In making such reliability estimates it is desirable to have measured $V_{(t)}$ curve over about a 5 year period for the highest confidence level in the reliability estimate.

(4) *Wind structure at the proposed site.* The ideal case for the WECS would be a site such that the $V_{(t)}$ curve was flat, *i.e.,* a smooth steady wind that blows all the time; but a typical site is always less than ideal. Wind specially near the ground is turbulent and gusty, and changes rapidly in direction and in velocity. This departure from homogeneous flow is collectively referred to as "the structure of the wind."

(5) *Altitude of the proposed site.* It affects the air density and thus the power in the wind and hence the useful WECS electric power output. Also, as is well known, the winds tend to have higher velocities at higher altitudes.

One must be careful to distinguish altitude from height above ground. They are not the same except for a sea level WECS site.

(6) *Terrain and its aerodynamic.* One should know about terrain of the site to be chosen. If the WECS is to be placed near the top but not on the top of a not too blunt hill facing the prevailing wind, then it may be possible to obtain a 'speed up' of the wind velocity over what it would otherwise be. Also the wind here may not flow horizontal making it necessary to tip the axis of the rotor so that the aeroturbine is always perpendicular to the actual wind flow.

It may be possible to make use of hills or mountains which channel the prevailing winds into a pass region, thereby obtaining higher wind power.

(7) *Local Ecology.* If the surface is bare rock it may mean lower hub heights hence lower structure cost. If trees or grass or vegetation are present, all of which tend to destructure the wind, then higher hub heights will be needed resulting in larger system costs than the bare ground case.

(8) *Distance, to Roads or Railways.* This is another factor the system engineer must consider for heavy machinery, structures, materials, blades and other apparatus will have to be moved into any chosen WECS site.

(9) *Nearness of site to local centre/users.* This obvious criterion minimizes transmission line length and hence losses and costs. After applying all the previous siting criteria, hope fully as one narrows the proposed WECS sites to one or two they would be relatively near to the users of the generated electric energy.

(10) *Nature of ground.* Ground condition should be such that the foundations for a WECS are secured. Ground surface should be stable. Erosion problem should not be there, as it could possibly later wash out the foundations of a WECS, destroying the whole system.

(11) *Favourable land cost.* Laud cost should be favourable as this along with other siting costs, enters into the total WECS system cost.

(12) Other conditions such as *icing problem, salt spray* or *blowing dust* should not present at the site, as they may affect aeroturbine blades, or environmental is generally adverse to machinery and electrical apparatus.

The following guide lines may be useful (all figures are for a height of 20 m, which seems to be a reasonable minimum):

1. The best sites, for wind energy are found off shore and the sea coast. An average value on the coast is 2400 kWh/m^2 per year.

2. The second best sites are in mountains. A typical average value is 1600 kWh/m^2 per year.

3. The lowest level of the wind energy is found in plains where values are generally three or four times lower than that at the coast. A typical average is 750 kWh/m^2 per year.

As regards climates, other criteria interact with those discussed for:

1. In the humid equatorial region, there is virtually no wind energy, whether at sea, along the coast or in the mountains.

2. The amount of convertible wind energy is fair or good in day or hot climates, as well as in temperature and cold elimates.

3. In some warm, windy countries, wind energy may not be usable because of the frequency of cyclones (*e.g.,* Japan, Caribbean area).

Care should be exercised in choosing a suitable site for a wind machine since there is a significant amount of shear and compression in a normally horizontal wind stream as it passes over the topographical contours and the rough surface of the earth. This shear results in lower wind speeds near the surface than at the heights great enough for free wind to occur. In particular the free flow velocity, at heights large enough to be unaffected by the surface shear, is usually significantly larger than that of the winds either at the surface or at standard anemometer heights of 20 to 30 feet ($\simeq$ 7 to 10 m) where the wind is normally measured.

However, the wind shear, and consequently the available wind power at a given altitude, is also affected by the roughness of the earth's surface in a given location. If the area contains buildings, trees, wind machines, or other obstackles, the variation of the wind speed with altitude above ground level is usually greater for these obstructed areas than for the case of open water and flat plains. The *characteristics* of a good wind power site may be summarized as follows:

1. A site should have a high annual wind speed.
2. There should be no tall obstructions for a radius of 3 km.
3. An open plain or an open shore line may be a good location.
4. The top of a smooth, well rounded hill with gentle slopes lying on a flat plain or located on an island in a lake or sea is a good site.
5. A mountain gap which produces to wind funneling is good.

6.5. Basic Components of a WECS (Wind Energy Conversion System)

The main components of a WECS are shown in Fig. (6.12), in block diagram form. Summary of the system operation is as follows:

Aeroturbines convert energy in moving air to rotary mechanical energy. In general, they require pitch control and yaw control (*only in the case of horizontal or wind axis machines)* for proper operation. A mechanical interface consisting of a step up gear and a suitable coupling transmits the rotary mechanical energy to an electrical generator. The output of this generator is connected to the load or power grid as the application warrants.

Yaw control. For localities with the prevailing wind in one direction, the design of a turbine can be greatly simplified. The rotor can be in a fixed orientation with the swept area perpendicular to the predominant wind direction. Such a machine is said to be *yaw fixed.* Most wind turbines, however, are *yaw active,* that is to say, as the wind direction changes, a motor rotates the turbine slowly about the vertical (or yaw) axis so as to face the blades into the wind. The area of the wind stream swept by the wind rotor is then a maximum.

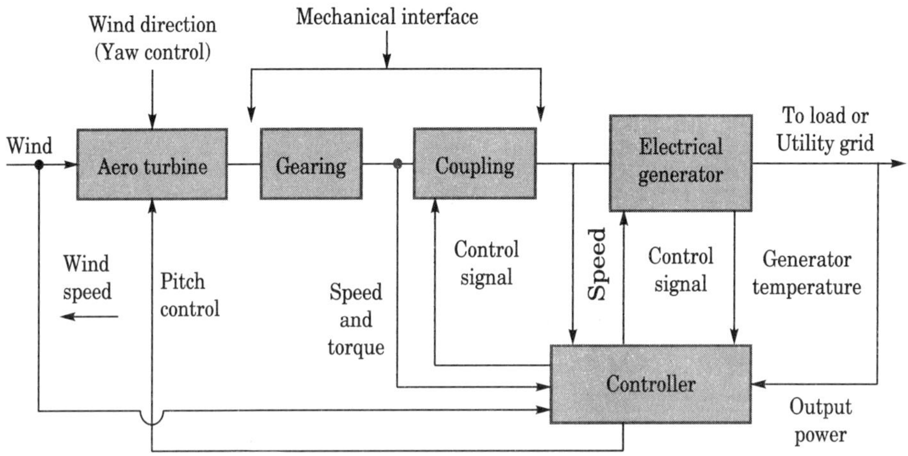

Fig. 6.12. Basic components of a wind electric system.

In the small turbines, yaw action is controlled by a tail vane, similar to that in a typical pumping windmill. In larger machines, a servomechanism operated by a wind-direction sensor controls the yaw motor that keeps the turbine properly oriented.

The purpose of the controller is to sense wind speed, wind direction, shafts speeds and torques at one or more points, output power and generator temperature as necessary and appropriate control signals for matching the electrical output to the wind energy input and protect the system from extreme conditions brought upon by strong winds electrical faults, and the like.

The physical embodiment for such an areo-generator is shown in a generalized form in Fig. (6.13). The sub components of the windmill are:

- wind turbine or rotor
- wind mill head

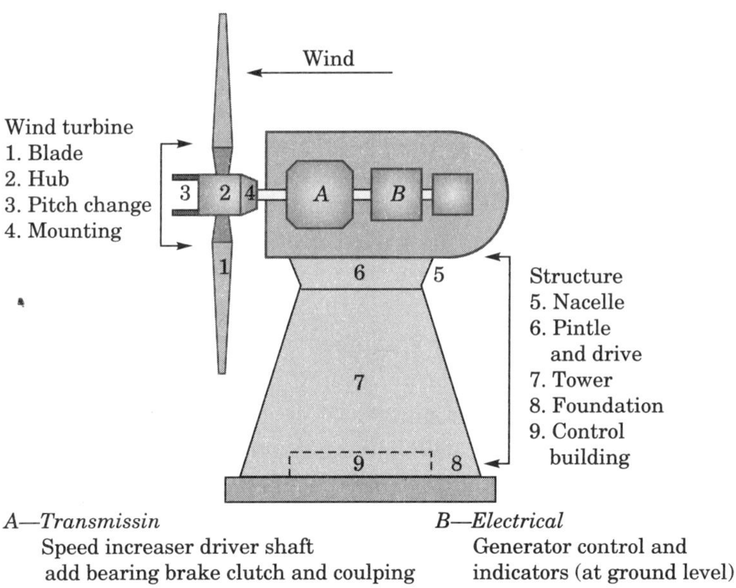

Fig. 6.13. Physical embodiment of wind-electric generating station.

- transmission and control

and • supporting structure

Such a machine typically is a large impressive structure.

Rotors

Rotors are mainly of two types:

(*i*) Horizontal axis rotor and (*ii*) Vertical axis rotor.

One advantage of vertical axis machines is that they operate in all wind directions and thus need no yaw adjustment.

The rotor is only one of the important components. For an effective utilization, all the components need to he properly designed and matched with the rest of the components.

The windmill head supports the rotor, housing the rotor bearings. It also houses any control mechanism incorporated like changing the *pitch* of the blades for safety devices and tail vane to orient the rotor to face the wind. The latter is facilitated by mounting it on the top of the supporting structure on suitable bearings.

Transmissions. The rate of rotation of large wind turbine generators operating at rated capacity or below, is conveniently controlled by varying the pitch of the rotor blades, but it is low, about 40 to 50 revolutions per minute (rpm). Because optimum generator output requires much greater rates of rotation, such as 1800 r.p.m, it is necessary to increase greatly the low rotor rate of turning. Among the transmission options are mechanical systems involving fixed ratio gears, belts, and chains, singly or in combination or hydraulic systems involving fluid pumps and motors. Fixed ratio gears are recommended for top mounted equipment because of their high efficiency, known cost, and minimum system risk. For bottom mounted equipment which requires a right angle drive, transmission costs might be reduced substantially by using large diameter bearings with ring gears mounted on the hub to serve as a transmission to increase rotor speed to generator speed. Such a combination offers a high degree of design flexibility as well as large potential savings.

Generator. Either constant or variable speed generators are a possibility, but variable speed units are expensive and/or unproved. Among the constant speed generator candidates for use are synchronous induction and permanent magnet types. The generator of choice is the synchronous unit for large aerogenerator systems because it is very versatile and has an extensive data base. Other electrical components and systems are, however, under development.

Controls. The modern large wind turbine generator requires a versatile and reliable control system to perform the following functions:

(1) the orientation of the rotor into the wind (azimuth of yaw);

(2) start up and cut-in of the equipment;

(3) power control of the rotor by varying the pitch of the blades;

(4) generator output monitoring—status, data computation, and storage;

(5) shutdown and cut out owing to malfunction or very high winds;

(6) protection for the generator, the utility accepting the power and the prime mover;

(7) auxiliary and/ or emergency power ; and

(8) maintenance mode.

Many combinations are possible in terms of the control system and may involve the following components:

(1) sensor—mechanical, electrical, or pneumatic:

(2) decision elements—relays, logic modules, analog circuits, a microprocessor, a fluidics, units, or a mechanical unit; and

(3) actuators—hydraulic, electric, or pneumatic. Arecommended combination of electronic transducers feeding into a micro-processor which, in turn, signals electrical actuators and provides protection through electronic circuits, although a pneumatic slip clutch may be required.

Towers. Four types of supporting towers deserve consideration, these are:

(1) the reinforced concrete tower,

(2) the pole tower,

(3) the built up shell-tube tower, and

(4) the truss tower.

Among these, the truss tower is favoured because it is proved and widely adaptable, cost is low, parts are readily available, it is readily transported, and it is potentially stiff. Shell-tube towers also have attractive features and may prove to be competitive with truss towers.

The type of the supporting structure and its height is related to cost and the transmission system incorporated. It is designed to withstand the wind load during gusts (even if they occur frequently and for very short periods). Horizontal axis wind turbines are mounted on towers so as to be above the level of turbulence and other ground—related effects. The minimum tower height for a small WECS is about 10 m, and the maximum practical height is estimated to be roughly 60 m.

The turbine may be located either unwind or downwind of the tower. In the unwind location (*i.e.,* the wind encounters the turbine before reaching the tower), the wake of the passing rotor blades causes repeated changes in the wind forces on the tower. As a result, the tower will tend to vibrate and may eventually be damaged. On the other hand, if the turbine is downwind from the tower as shown in figure, the tower vibrations are less but the blades are now subjected to severe alternating forces as they pass through the tower wake.

Both unwind and downwind locations have been used in WRC devices. Downwind rotors are generally preferred especially for the large aerogenerators. Although other forces acting on the blades of these large machines are significant, tower effects are still important and tower design is an essential aspect of the overall system design.

6.6. Classification of WEC Systems

1. First, there are two broad classifications:

(*i*) *Horizontal Axis Machines.* The axis of rotation is horizontal and the aeroturbine plane is vertical facing the wind.

(*ii*) *Vertical Axis Machines.* The axis of rotation is vertical. The sails or blades may also be vertical, as on the ancient Persian windmills, or nearly so, as on the modern Darrievs rotor machine.

2. Then, they be classified *according to size* as determined by their useful electrical power output.

(*i*) *Small Scale* (upto 2 kW). These might be used on farms, remote applications, and other places requiring relatively low power.

(*ii*) *Medium Size Machines* (2-100 kW). These wind turbines may be used to supply less than 100 kW rated capacity, to several residences or local use.

(*iii*) *Large Scale or Large Size Machines* (100 kW and up). Large wind turbines are those of 100 kW rated capacity or greater. They are used to generate power for distribution in central power grids. There are two sub classes:

(*a*) Single generator at a single site.

(*b*) Multiple generators sited at several places over an area.

3. As per the type of *output power,* wind aerogenerators are classified as:

(*i*) *DC output*

(*a*) DC generator

(*b*) Alternator rectifier

(*ii*) *AC output*

(*a*) Variable frequency, variable or constant voltage AC.

(*b*) Constant frequency, variable or constant voltage AC.

4. As per the *rotational speed* of the aeroturbines, these are classified as:

(*i*) *Constant Speed* with variable pitch blades. This made implies use of a synchronous generator with its constant frequency output.

(*ii*) *Nearly Constant Speed* with fixed pitch blades. This mode implies an induction generator.

(*iii*) *Variable Speed with fixed pitch blades.* This mode could imply, for constant frequency output:

(*a*) Field modulated system

(*b*) AC-DC-AC link

(*c*) Double output induction generator

(*d*) AC commutator generators

(*e*) Other variable speed constant frequency generating systems.

5. Wind turbines are also classified as per how the *utilization of output is made:*

(*i*) Battery storage.

(*ii*) Direct connection to an electromagnetic energy converter.

(*iii*) Other forms (thermal potential etc.) of storage.

(*iv*) Interconnection with conventional electric utility grids.

The system engineer seeking to integrate WECS will, naturally be most interested in the latter case but should be aware that WECS offer other options as well.

6.7. Advantages and Disadvantages of WECS

Advantages of wind energy are:

(*i*) It is a renewable source of energy.

(*ii*) Like all forms of solar energy, wind power systems are non-polluting, so it has no adverse influence on the environment.

(*iii*) Wind energy systems avoid fuel provision and transport.

(*iv*) On a small scale upto a few kilowatt system is less costly. On a large-scale costs can be competitive with conventional electricity and lower costs could be achieved by mass production.

Disadvantages of wind energy are:

(*i*) Wind energy available in dilute and fluctuating in nature.

(*ii*) Unlike water energy wind energy needs storage capacity because of its irregularity.

(*iii*) Wind energy systems are noisy in operation; a large unit can be heard many kilometres away.

(*iv*) Wind power systems have a relatively high overall weight, because they involve the construction of a high tower and include also a gearbox, a hub and pitch changer, a generator coupling shaft etc. For large systems a weight of 110 kg/kW (rated) has been estimated.

(*v*) Large areas are needed, typically, propellers 1 to 3 m in diameter, deliver power in the 30 to 300 W range.

(*vi*) Present systems are neither maintenance free not-practically reliable. However, the fact that highly réliable propeller engines are built for aircraft suggest that the present troubles could be overcome by industrial development work.

6.8. Wind Energy Collectors

6.8.1. Introduction. A windmill is a machine for wind energy conversion. A wind turbine converts the kinetic energy of the wind's motion to mechanical energy transmitted by the shaft. A generator further converts it to electrical energy, thereby generating electricity. The term 'windmill' which originally implied a mill for grinding grain becomes an obvious misnomer when applied to electric power generation. The term is still widely used however, *Aerogenerators* avoids the difficulty. Only in the last century have windmills been used to generate electric power.

Wind *aerogenerators* or wind turbine generators of WECS are generally classified as

- horizontal axis type, and
- vertical axis type

depending on their axis of rotation, relative to the wind stream.

Some authors refer to them also as wind axis rotors and cross wind axis rotors respectively. In the former types, the rotors are oriented normal to the direction of wind, while in the latter types, the effective surface of the rotor moves in the same direction as the wind.

Horizontal axis wind machines are further subclassified as single bladed, multibladed and by cycle multiblades type. Sail, wind, multibladed are example of horizontal axis wind machines, *Savonius* and *Darrius* rotor are example of vertical axis machines.

The vertical axis windmill or machine is again subdivided into two major types:

(*i*) Savonius or 'S' type rotor mill (low velocity wind).

(*ii*) Darrieus type rotor mill (high velocity wind) based on the working speed of the machine and the velocity ranges required by the machine for operation.

Vertical axis machines are of simple design as compared to the horizontal axis type.

6.8.2. Horizontal-Axial Machines

The common wind turbine with a horizontal (or almost horizontal) axis, is simple in principle, but the design of a complete system, especially a large one that will produce electric power economically, is complex. Not only must be individual components, such as the rotor, transmission, generator, and tower, be as efficient as possible, but these components must function effectively in combination. Some of the main design considerations will be considered later.

Some of the horizontal axis type wind machines are briefly described below:

1. *Horizontal axis using two aerodynamic blades.* In this type of design, rotor drives a generator through a step up gearbox. The blade rotor is usually designed to be oriented *downwind* of the tower. The components are mounted on a bed plate which is attached on a pintle at the top of the tower. This arrangement is shown schematically in Fig. (6.14). The rotor blades are continuously flexed by unsteady aerodynamic, gravitational and inertia loads, when the machine is in operation. If the blades are made of metal, flexing reduces their fatigue life with rotor the tower is also subjected to above loads, which may cause serious damage. If the vibrational modes of the rotor happen to coincide with one of the natural mode of the vibration of the tower, the system may shake itself to pieces. Because of the high cost of the blade rotors with more than two blades are not recommended. Rotors with more than two, say 3 or 4 blades would have slightly higher power coefficient.

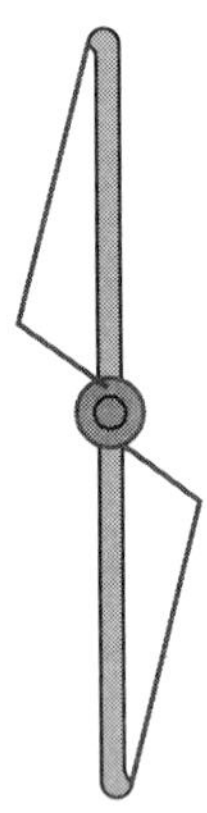

(*a*) Two blade prufile

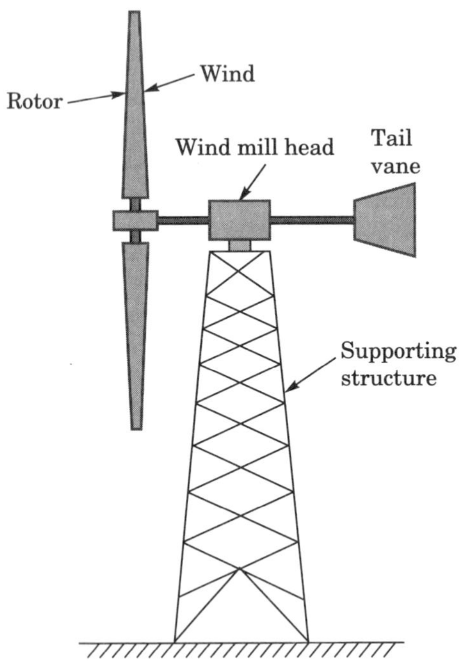

(*b*) Schematic of a horizontal axis type wind mills

Fig. 6.14

2. *Horizontal axis propeller type using single blade.* In this arrangement, a long blade is mounted on a rigid hub (Fig. 6.15). Induction generator and gearbox are also shown. If extremely long blades (above say 60 m) are mounted on rigid hub, large blade root bending moments may occur due to tower shadow, gravity and sudden shifts in wind directions. To reduce rotor cost, use of low cost counter weight is recommended which balance long blade centrifugally.

Advantages of one bladed rotor:

(*i*) Simple blade controls

- lower blade weight and costs
- lower gear box cost

(*ii*) Counter weight costs less than a second blade.

(*iii*) Counter weight call be inclined to reduce blade coning.

(*iv*) Pitch bearings do not carry centrifugal force.

(*v*) Blade root spar can be large diameter *i.e.,* more rugged.

Disadvantages:

(*i*) Vibration produced, due to aerodynamic torque.

(*ii*) Unconventional appearance.

(*iii*) Large blade root bending moment.

(*iv*) Starting torque reduced by ground boundary layer.

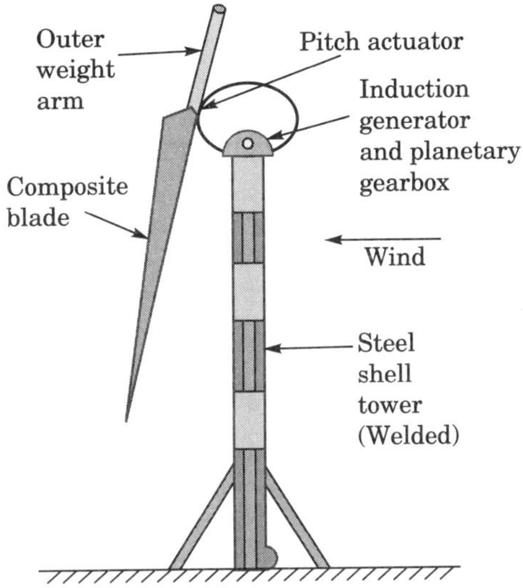

Fig. 6.15. Horizontal axis single blade wind mill.

(*v*) One-per-rev coriolis torque produced, due to flapping.

3. *Horizontal axis multibladed type.* This type of design for multiblades as shown in Fig. (6.16), made from sheet metal or aluminium. The rotors have high strength to weight ratios and have been known to service hours of freewheeling operation in 60 km/hr winds. They have good power coefficient, high starting torque and added advantage of simplicity and low cost.

Fig. 6.16. Multiblade propeller.

Fig. 6.17. Horizontal axis, Dutch type wind mill.

4. *Horizontal axis wind mill-Dutch type.* It is shown in Fig. (6.17), is one of the oldest designs. The blade surfaces are made from an array of wooden slats which 'feather' at high wind speeds.

5. *Sail type.* It's blade are shown in Fig. (6.18). It is of recent origin. The blade surfaces is made from cloth, nylon or plastics arranged as mast and pole or sailwings. There is also variation in the number of sails used.

Fig. 6.18. Blades of sail type wind mill.

The horizontal axis types generally have better performance. They have been used for various applications, but the two major areas of interest are electric power generation, and pumping water. The latter introduces some complexity into the design as the mechanical energy has to be transmitted over a distance. Also in some cases the rotor motion has to be converted to reciprocating motion.

6.8.3. Design Consideration of Horizontal-Axis Machines

The common wind turbine with a horizontal (or almost horizontal) axis is simple in principle, but the design of a complete system, especially a large one that will produce electric power economically, is complex. Not only must the individual components, such as the rotor, transmission, generator, and tower, be as efficient as possible, but these components must function effectively in combination. Some of the main design considerations are outlined below.

Rotor. A wind turbine's rotor may have any number of blades which may be made from wood, metal, or composites of several materials including glass reinforced plastics. Future blades may use carbon fibre which combines light weight and flexibility with immense strength.

In the most common type of machine the rotor is mounted on a horizontal (or near horizontal) shaft and connected either to a generator or some mechanical device such as a water pump or heat generator. The whole assembly is mounted on a tower so that it turns clear of the ground and away from surface drag and interference. From a performance point of view, the taller the tower the better because wind speeds increase with height.

Many turbine designs have been built over the years. These have included machines with a single, counter balanced blade or with multiple rotors, and turbines with two or more sets of counter rotating blades on the same shaft. Rotors have been constructed with cups, paddles, or sails rigged to rigid wooden spars as in the early type windmill. The rotor can be mounted upwind or down wind of the support tower.

Horizontal-axis rotors can be either lift or drag devices. Lift devices are generally preferred, since for a given swept area, high rotational speeds and more output power can be developed by lift than by drag forces. In general, a drag device can not move faster than the wind velocity, while a lift device can. Thus a litting surface can obtain higher top to wind speeds and, consequently, a higher power output to weight ratio and, for many applications, a lower cost to power output ratio.

Lift devices use slender blades with an aerofoil section that generate *aerodynamic lift* when placed in an air current; and drag devices are those, like the familiar multibladed farm wind pump, that rely essentially on *drag* forces to extract energy from the wind. Rotors that rely on lift are capable of turning at very high speeds as stated. Small rotors can spin at 400–500 revolutions per minute (rpm) with blade tip speeds of several hundred km per hour 8-10 times faster than the wind. Very large aerofoil rotors turn much more slowly, typically 20–40 r.p.m, but the blade tip speeds are comparable 65–110 m/s or 250–400 kmph.

Drag devices are less efficient wind energy converters and always turn more slowly than the wind. They work best in less windy inland locations. The faster they run the more they spill in the wind. Lift devices, on the other hand, are able to profit from the high power densities that are available at high wind speeds. The ratio of the power extracted by a lift device to that of a drag device is usually greater than 3 : 1 for the same swept area.

Drag devices are often capable of generating high torques (twisting forces) and are ideal for some uses such as water pumping. They are less suitable for electricity generation which requires very high rotational speeds (several thousand rpm). The reverse is generally true for lift devices. Small rotors can often rotate at such speeds that they can be connected directly to a generator. Larger rotors require step-up gearing.

One major disadvantage with rotors with slender aerofoil blades is that they will not start up without first being rotated to generate lift. They, therefore require some external source of power. Indeed, this is a general problem with rotors with a low 'solidity ratio' in which the blades occupy only a small part (10% or less) of the rotors swept area. The ratio of the projected area of the rotor (on a plane perpendicular to its axis of rotation) to the swept area of the rotor is known as the *'Solidity'* of the rotor. For multivane fan-type rotors such as the American farm wind mill, a typical solidity is 0.7. For high speed lift type propellers, on the other hand, the solidity is usually much lower, *i.e.*, 0.1 to 0.01.

Lift type rotors often use tapered and/or twisted blades to reduce the bending strains on *the roots of the blades.*

All horizontal-axis rotors depend mainly on lift to rotate them. Maximum efficiency is achieved when the angle of incidence approaches the maximum possible angle of attack prior to stalling (*i.e.,* when the lift drag ratio is as high as possible). Given that a rotor's tip travel faster than points nearer axis, the angle at which the wind meets the plane of rotation decreases in linear proportion to the radius. For this reason, an efficient wind rotor requires the blades to be twisted so that the optimum angle of incidence is obtained at all radii when the machine is running at its design speed. For the same reason, high-speed machines have blades set at a finer pitch than low-speed ones. The ratio of the speed of the rotor blade tips to the speed of the wind is called the tip-speed ratio. Every rotor has an optimum tip-speed ratio at which its maximum efficiency is achieved, and which also characterizes the rotor. The tip speed ratio computed numerically as:

$$\text{TSR (tip speed ratio)} = \frac{V_{tip}}{V} \qquad \text{...(6.29)}$$

where, V_{tip} = speed of the rotor tip

and, V = free wind speed.

The faster a rotor runs with respect to the wind speed, the less 'solidity' is required to intercept the entire stream-tube of wind passing through the rotor disc. In other words, a rotor with a high TSR needs less material and can have relatively slender blades. However, because it rotates at higher speeds and twist angles require only slight angles of attack at the tip, a high lift drag ratio is essential to produce a driving torque. Therefore high-speed low solidity wind rotors need good quality airfoils (similar to those used for aircraft). They must also be made with a good surface finish and the structural integrity necessary to withstand high rotational speeds. For this reason, the cost/m^2 of good, high speed rotors is often greater than that of low-speed rotors, despite the reduced quantity of material. Conversely, low TSRs require rotor discs with relatively solid blades or sails so that wind energy will not be lost through the gaps between them. These blades need coarser pitch setting, because lift is stronger in the plane of rotation and drag is weaker, the lift drag ratio is less critical. For this reason, wind mills with low TSRs tends to have many blades and the lifting surfaces can be much cruder without causing any serious loss of efficiency.

As the TSR increases, the number of blades decreases (see Table 6.1). Solidity is normally defined as the fraction of the total circumferential (either at the tip or sometimes at two thirds of the tip radius) that contains blades. Numerically it can be expressed as:

$$S = NC/\pi D \qquad ...(6.30)$$

where, N is the number of blades,

C is the average breadth of a blade

and, D is the diameter of the circle described by a blade.

Table 6.1 Blade number vs TSR

Tip speed ratio	*Number of blades*
1	6—20
2	4—12
3	3—8
4	3—5
5—8	2—4
8—15	1—2

It is usual to characterize a windmill rotor by comparing its curve of efficiency with its TSR. The efficiency is usually expressed as the C_p (power coefficient or performance coefficient). This is the fraction of wind energy passing through the rotor disc that is converted into shaft power *i.e.* (Eqn. 6.2)

$$C_p = 0.5\ \rho.AV^3$$

Every type of rotor has a unique C_p curve. The reason for the shape of the curve is that the optimum C_p occurs when the rotor blades are operating at their maximum LDR lift drag ratio (close to, but not quite, stalling). At lower TSR, the blades stall progressively and although they can generate a lot of lift, they also produce much more drag, which dissipates energy. At a higher than optimum TSR, they produce less than their maximum lift (and the lift forces act at a less favourable angle) because the angle of incidence decreases. At runaway speed, the angle of incidence approaches zero, and the lift and drag forces balance, so that no net shaft power is produced and the efficiency once again is zero. Windmills normally operate on the part of the curve to the right of the maximum C_p. They depend on an increasing wind speed to accelerate the rotor from rest through the peak C_p point.

Although the C_p—TSR curve is generally used to characterize wind rotors, in many ways the torque characteristic (the C_T—TSR curve) is of more interest

(Figs. 6.28 and 6.29) to those concerned with system design. A wind turbine will always operate at a speed that produces enough torque to balance the load's torque requirement exactly. Hence, if the characteristics of both the load and the rotor torque speed are known, the system performance can be defined.

Torque coefficient C_T, is defined as

$$C_T = \frac{T}{T_{max}} \quad ...(6.31)$$

where, T = shaft torque, and

T_{max} = torque at maximum efficiency.

The previous calculation (Art. 6.2.3) of thrust on a wind turbine provides a convenient opportunity to introduce definitions for the torque causing rotational shaft power. At this stage no attempt is made to analyze angular momentum exchange between the air and the turbine. However, it is obvious that if the turbine turns one ways the air must turn the other, and full analysis must eventually consider the vortices of air circulating down wind of the turbine.

The maximum conceivable torque T on a turbine rotor would occur if the maximum thrust could some how be applied at the blade tip furthest from the axis. For a propeller turbine of radius R

$$T_{max} = F_{max} R \quad ...(6.32)$$

and

$$F_{max} = \frac{1}{2} \rho A_1 V_i^2 \quad ...(6.33)$$

So

$$T_{max} = \frac{1}{2} \rho A_1 V_1^2 R \quad ...(6.34)$$

For a working machine producing a shaft torque T, the torque coefficient C_T is defined by Eqn. 6.31 *i.e.*

$$T = C_T T_{max}$$

We know the *tip speed, ratio* X is defined as the ratio of the outer blade tip speed V_t to the unperturbed wind speed V_i;

$$\lambda = \frac{V_t}{V_i} = \frac{R\omega}{V_i} \quad ...(6.35)$$

where R is the outer blade radius and ω is the rotational frequency.

From equation (6.34) substituting for R

$$T_{max} = \frac{\rho A V_i^2 (V_i \lambda)}{2\omega} \quad ...(6.36)$$

where P_a is the power in the wind from (Art. 6.2.2). The shaft power is the power derived from the turbine P_t, so

$$P_t = \tau \,.\, \omega \quad ...(6.37)$$

Now from equation $P_t = C_p P_a$, and thus using (6.31) and (6.36), equation (6.37) becomes

$$C_p P_a = C_T T_{max} \, \omega$$

$$C_p P_a = C_T P_a \, \lambda$$

$$C_p = \lambda C_T \quad ...(6.38)$$

Note that in practice power coefficient C_p and torque coefficient C_T will both be function of λ and are not constants.

By the Betz criterion the maximum value of C_p is 0.593, so in the 'ideal' case

$$C_{T\,max} = \frac{0.593}{\lambda} \quad ...(6.39)$$

The torque coefficient C_T is numerically equal to C_p/ω, where ω is the rotational speed radian/second. Machines with higher speeds have a slightly higher maximum C_p but a much lower C_T, particularly for starting. In fact, high solidity machines generally have higher starting than running torques, while the starting torques of low solidity machines are even lower than their quite low running torques.

The choice of rotor is based mainly on the pump's load characteristics. A positive displacement pump, such as the piston pumps used in the bore holes, demands a higher starting than running torque and, therefore, a high solidity rotor is essential unless some method of unloading the rotor to help it start is included. However, electricity generators need little torque to start them turning and they should be driven at high speeds, so that a high-speed, law soliding rotor is generally used for this type of load.

Number of blades. Wind turbines have been built with upto six propellers type blades but two-and three-bladed propellers are most common. A one bladed rotor with a balancing counterweight, has some advantages, including lower weight and cost and simpler controls, over the multi blade tupe. However, starting requires high wind speeds, and vibration and other forces can be large.

Turbines with three blades have been used in several WEC machines in order to avoid the vibrations experienced with two-bladed rotors. These vibrations are related to the turning (or yawing) of the rotor in order to face it into the wind. It appears, however, that the problem can be overcome by controlling the yaw rate. Because they cost less to fabricate in large sizes and are capable of operating with a high tip speed ratio, two-bladed systems are receiving major attention.

Blade Design. Wind turbine blades have an airfoil-type cross-section and a variable pitch (Fig. 6.19). They are slightly twisted from the outer tip to the root (*i.e.*, where the blade is attached to the hub) to reduce the tendency for the rotor to stall. In a few devices, the blades had a constant chords length *i.e.*, constant distance from edge to the other). As a general rule, however better performance is obtained with blades that are narraower at the tip than at the root (see Fig. 6.19).

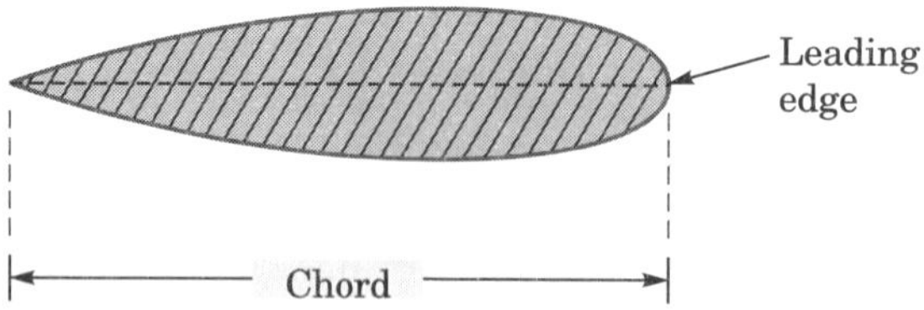

Fig. 6.19. Cross-section of airfoil-type wind-rotor blade.

As shown in Fig. (6.20), we see that the force that propels the blades of a conventional wind mill comes from the chord of the airfoil, being tilted away from the direction of motion. The motion causing the "wind due to motion" is the rotation of the blades. At the tip of the blades of a modern wind turbine the velocity is about six times the wind velocity. This means that the blades are set rather flat at a small angle with the plane of the rotation and almost at right angles to the direction of wind so that the effective wind properly approach from ahead of the leading edge. At other parts of the blade, between the tip and the axle, the velocity and the ideal set of the airfoil is at a greater angle to the plane of rotation. Ideally the blade should be twisted, but because of construction difficulties this is not always achieved.

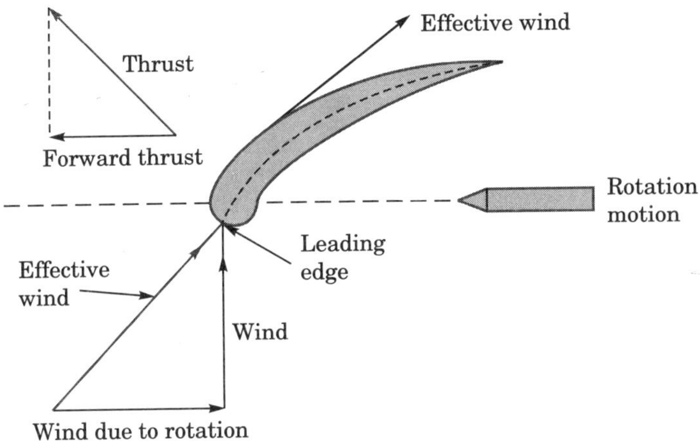

Fig. 6.20. Wind mill blade as an airfoil.

In the large two bladed wind turbines, the blades are inclined at a small angle called the *corning angles* to the vertical. This design decreases the bending load on the roots of the blades and lessens the danger of fouling the supporting tower under severe wind conditions. In the very large rotors consisting of a single piece mounted at the midpoint, there is no coning angle.

A fundamental problem in wind turbine design is to allow for the many forces to which the blades are subjected during normal operation, In addition to vibrational stresses resulting from rotation, there are several extraneous forces. There arise form wind turbulence, directional changes in the wind, variations of wind speed with height, wind gusts, gravitational forces, the pressure of the tower etc. The blades must be constructed and attached to the hub in such a manner as to with stand these forces. Consequently, aerodynamic performance is sacrificed to sum extent in the design of a rotor with adequate strength.

Although the wind power of a rotor increases with the square of the swept diameter, there is a practical limit to the size. In the first place, the mass of a blade increases rapidly with its dimensions. As a result, the wind power available per unit mass decreases as the dimensions of the rotor are increased; the tip speed ratio tends to decrease correspondingly. Further more, the wind drag forces generally increase with the blade size. The limiting dimensions depend on the design and constructional materials, but the maximum practical diameter of a two-blade rotor may perhaps be in the range of 90 to 110 m.

In order to provide both light weight and adequate strength aircraft industry techniques, as used, in connection with air place and helicopter propellers, have been drawn upon in the design and the construction of wind turbines. For small rotors in particular, the blades can be made of laminated wood, possibly covered with a thin skin of aluminium. Rotors upto 34 m in diameter have been fabricated from plastic reinforced with glass fibre. More conventional aircraft type blades are usually made from aluminium (or an aluminium alloy); they consist of a long central span with ribs at intervals, covered with an aluminium skin. The very largest rotor blades have been made of steel to provide adequate strength.

Yaw control. The area of the windstreams swept by the wind turbine is maximum, when blades face into the wind. This is achieved by control arrangement, is which when the wind direction changes, a motor rotates the turbine slowly about the vertical (or yaw) axis so as the face the blades into the wind.

For localities with the prevailing wind in one direction, the design of a turbine can be greatly simplified. The rotor can be in a fixed orientation with swept area perpendicular to the predominant wind director. Such a machine is said to be *yaw-fixed i.e.,* it can not be rotated around a vertical axis perpendicular to the wind stream. Most wind turbines, however, are *yaw-active*; that is to say, as the wind direction changes, they will "track" the changing direction of the wind. In this case, a motor rotates the turbine slowly about the vertical (or yaw) axis so as to face the blades into the wind. The area of the windstream swept by the wind in then a maximum.

In the smaller turbines, yaw action is controlled by a tail vane, similar to that in the American pumping windmill. In larger machines, a servomechanism operated by a wind direction sensor controls the yaw motor that keeps the *turbine* properly oriented.

Wind stream variation. Wind speed variability must be considered, especially in the design of the larger WEC machines. For a given turbine there is a minimum wind speed called the *cut-in speed* at which rotation is allowed to start the cut-in wind speed and the rated wind speed, at which the generator produces its rated power, the rotation rate is maintained constant by varying the output of the generator. At wind speeds exceeding the rated value, the rotor speed is held constant by automatic adjustment of the pitch of the blades. At very high wind speeds the blades are feathered, as in aircraft, and rotation ceases. The wind speed at which this occurs is called the *cut-out speed* (or furling speed), the electric power output of the generator is essentially constant.

The variation of the generated power with the wind speed is shown (approximately) by the full lines in Fig. 6.21, the dashed curve represents the theoretical variation without any control of the rotation rate of the turbine (*i.e.,* power proportional to the cube of the wind speeds. At wind speeds above the rated value, part of the available wind power is not used (or "spilled") as a consequence, of changing the pitch angle of the rotor. Consequently in this respect, the system is somewhat less efficient than one in which the rotor speed is allowed to increase with wind speed, within limits. However, machines of the letter type are less easily adapted to connection into an electric utility grid.

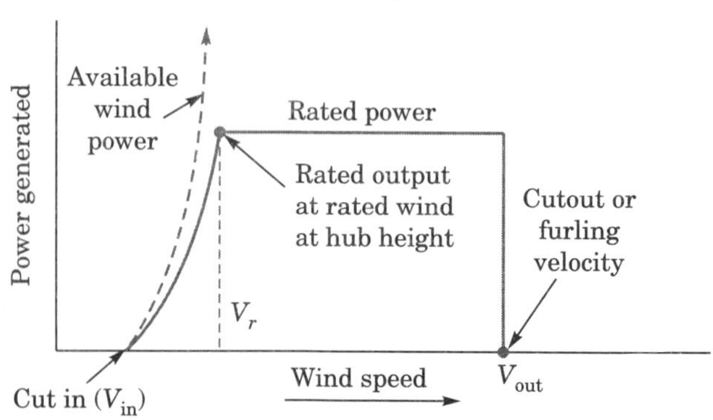

Fig. 6.21. Power generated and wind speed.

By designing the system to suit the wind conditions at a particular site, the wind power lost by spilling can be relatively small. At higher wind speeds, the speed of rotation of the blades (and of the hub to which they are attached) tends to increases. In fact, for a given wind speed, the rate of rotation must exceed a certain value if a stall, with loss of power, is to be avoided. However, a changing speed of rotation will generally not provide the optimum conditions for generating constant-frequency alternating current.

One solution to this problem is to allow the turbine speed to vary and to couple the hup with a mechanical or electrical system which can generate a current of constant frequency from a variable speed input. For example, variable frequency alternating current maybe generated and rectified to produce direct current. The direct current can then be used to generate constant frequency alternating current by means of an inverted. If the electricity produced by the WEC machine is to be used for battery charging, a direct-current generator (upto about 20 kW) may be connected directly to the rotor hub; the speed of rotation of the turbine is then immaterial provided it is not so low as to cause stalling.

Although it results in some loss of efficiency, an alternative solution, used in some aero generators, has been preferred. Between the cut-in wind speed and the rated wind speed, at which the generator produces its power, the rotation rate is maintained constant by varying the output of the generator. At wind speeds exceeding the rated value, the rotor speed is held constant by automatic adjustment of the pitch of the blades. At very high speed the blades are feathered, as in an aircraft, and rotation ceases. Over the range from the rated wind speed to the cut-out speed, the electric power out put of the generator as stated remains constant.

In order to take advantage of winds with speeds above average, the rated wind speed should be greater than average value but not so high that the generator is often operating below the rated power. The ratio of rated to average wind speeds should be about 1.8 for an average speed of 5 m/sec. (17.5 kmph) decreasing to 1.5 for average speed of 10 m/sec (35 kmph). Over the years many ingenious techniques have been developed for preventing over speeding and for dissipating the excess power that would otherwise be generated by the rotor in gale (strong winds). Over-speeding can literally tear the blades out by the roots. In comparatively small machines the tail vane can be hinged to turn the rotor out of the wind, thus effectively reducing the swept area. Alternatively, the whole assembly can be mounted slightly off-centre so that it swings aside when the thrust on the rotor exceeds a set rating. In very high winds the rotor can turn completely out of the wind. 'Spoilers' which unfold at high speed; can be used to exert a drag on the rotor; or the biases can be made from very flexible materials so that they bow back, twist and spill the wind. Most machines are fitted with a brake which can bring the rotor to full a stop.

In more sophisticated rotor designs the blades are 'Feathered' automatically—their pitch is changed in response to the wind speed. This approach is attractive since it permits the wind turbine to operate with virtually constant output over range of wind speeds and to continue to generate power even in a gale. In small machines 'feathering weights' can be placed on or near the hub, and these respond to centrifugal forces as the rotor's speed increases; as they lean outwards they automatically alter the pitch of the blades. The performance of large rotors is generally controlled by a computer which senses changes in the wind speed or electrical power and alters the pitch of the blades accordingly. Some-times just the tips of the blades move.

Wind turbines must be sufficiently robust to withstand cyclic gravity loads caused by the revolving blades and vibration caused by the buffeting of the wind. In poorly designed machines this can lead to increased wear, fretting and metal fatigue, and, in extreme cases, to blade or bearing failures or even complete structural collapse. One innovation that can reduce stress in large two-bladed rotors is the 'teetered hub' which allows the rotor to seesaw back and forth by a few degrees. This motion across the hub alleviates loads on both the blades and the tower and has enabled engineers to design much lighter, and hence cheaper structures.

Turbine Tower System. As stated earlier, the horizontal axis wind turbines are mounted on towers, and there are wind forces on the tower. Both upwind and downwind locations have been used so that tower design is an essential aspect of the overall system design.

6.8.4. Vertical-Axis Machines

Various types of vertical axis panemones have been developed in the past that use drag forces to turn rotors of different shapes. These include those panemones that use plates, cups, or turbines as the drag device, as well as the Savonius—S-shaped cross-section rotors which actually provide some lift force, but are still predominantly drag devices. Such devices have relatively high starting torques compared to lift devices because of their higher solidity, but have relatively low tip-to-wind speeds and lower power outputs per given rotor size, weight and cost.

As stated vertical axis rotors can be either drag-or lift-based. The cup anemometer is an example of a drag-based, vertical axis wind device. The drag on a cup is greater when its concave side faces the wind which causes the device to rotate. Lift also plays a small part: the cups crossing the wind experience a small lift because their convex surfaces deflect the wind and causes a pressure reduction. The main virtue of the cup anemometer is that it tends to rotate within a narrow range of TSRS under all conditions, so its rotational speed is closely proportional to wind speed. However, it can not carry a load with any efficiency, it has never been constructed on a large scale for use as a wind turbine. The Savonius rotor works on a principle similar to that of the cup anemometer but is adoated to produce shaft power. It also takes advantage of the lift generated as the curved outer surfaces of its scoops cross the air flow. There are also a variety of so-called Panamones; these are pure drag devices, in which one side of the rotor carries blades or sails square to the wind, while the other side produces reduced drag by shielding or furling the blades. Because drag devices tend to run at TSR below unity (since their tips can not readily travel faster than the wind), they are inevitably less efficient than lift-dependent devices. In addition, their high solidity makes them more material-intensive in relation to the wind area 'seen' by the rotor.

Although only a few are available commercially, cross flow or Darrieus wind turbines have lately been attracting some attention.

Vertical-axis rotors have the great advantage of not having to be turned into the windstream as the wind direction changes, because their operation is independent of wind direction, vertical axis machines are called *panemones* (from Greek words meaning "all winds"). Until recent times, these devices received little attention for the generation of electric power, partly because vertical-axis rotors have lower power coefficients at high tip speed ratios (see Fig 6.28). In addition, vertical-axis turbines are said to be more difficult to control in strong winds. It is now realized, however, that the lower efficiency of panemones may be more than offset by the simpler design and consequent lower construction (and maintenance) costs. Elimination of the need for yawing into the wind, for example, results in decreased stresses on blades, bearing, and other components. Moreover, the transmission and generator are on (or near) the ground rather than at the top of a tall tower.

Hence following are the three distinct advantages of vertical axis wind turbines over horizontal axis ones:

(1) They will react to wind from any direction and therefore do not need yawing equipment to turn the rotor into the wind.

(2) They can require less structural support because heavy components (like gearbox and generator) can be located at ground level. This configuration also eases installation and maintenance.

(3) Since the blades do not turn end over end, the rotor is not subjected to continuous cyclic gravity loads. (Fatigue induced by such action is a major consideration in the design of large horizontal axis machines).

Probably the single biggest disadvantage with vertical axis machines is that far less is known about them than horizontal axis ones. This handicap is rapidly being removed.

(a) The Savonius Rotor. Perhaps the simplest of the modern types of wind energy conversion systems is the Savonius rotor which works like a cup anemometer. This type was invented by S. J. Savonius in the year 1920. This machine has become popular since it requires relatively low velocity winds for operation.

It consists of two-half-cylinders facing opposite directions in such a way as to have almost an S-shaped cross section (refer Fig. 6.22).

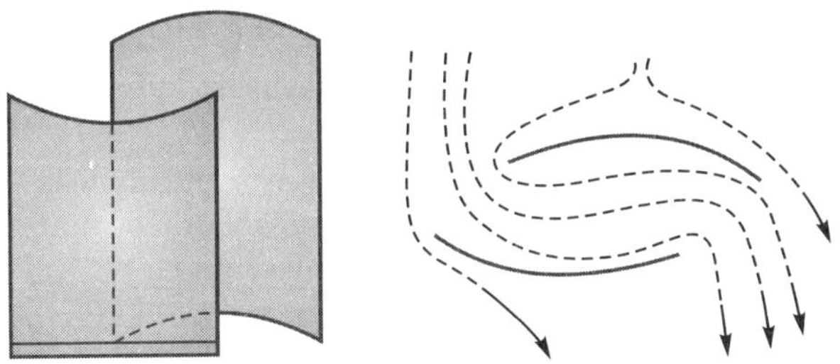

(a) (b)

Fig. 6.22. The Savonius rotor and its stream flow.

These two-semi circular drums are mounted on a vertical axis perpendicular to the wind direction with a gap at the axis between the two drums. Irrespective of the wind direction the rotor rotates such as to make the convex sides of the buckets head into the wind. From the rotor shaft we can tape power for use like water-pumping, battery charging, grain winnowing etc. However, instead of having two edges together to make an S-shaped they overlap to leave a wide space between the two inner edges, so that each of these edges is near the central axis of the opposite half cylinder, as shown in the figure. The main action of the wind is very simple; the force of the wind is greater on the cupped face than on the rounded face. In detail it is a bit more complicated. The wind curving around the back side of the cupped face exerts a reduced pressure much as the wind does over the top of an air-foil and this helps to drive the rotation. The wide slot between the two inner edges of the half cylinders, lets the air whip around inside the forward moving cupped face, thus pushing both in the direction of the rotation.

The ratio of the height to the overall diameter (aspect ratio) of the machine can be varied, but it is generally less than three to one. The power coefficient of the S rotor is low, but it might possibly be improved by changes in the design, number, and arrangements of the vanes. These matters need to be investigated.

Characteristics of Savonius Rotor

(*i*) Self starting (*ii*) Low speed

(*iii*) Low efficiency.

Advantages of Savonius Rotor

(1) A Savonius wind energy conversion system has a vertical axis which eliminates the expensive power transmission system from the rotor to the axis. Since it is a vertical axis machine it does not matters much about the wind direction. The machine performs even at low-wind velocity ranges.

(2) It has its low *cut in speed* (the wind speed required for switching electric power into the line); it produces power effectively in winds as slow as 8 km/hour, where as most-propeller type wind mills or machines require about 16 km/hour, for effective operation and large wind mill still more. This means that it is useful more of the time and is thus less dependent on storage of supplementary power.

(3) In common with all vertical-axis machines, the Savonius rotor has the advantage that the weight of the electric generator may be carried at ground level without the use of level gears. This is, however not a very important advantage.

(4) *Cost* of vertical axis wind turbine may be significantly lower than that of standard wind turbines.

(5) It has *simple structure,* hence easy to manufacture.

(6) *Yaw and pitch controls* are not needed to bring it into the wind or operate in high winds. A constant speed vertical axis wind turbine, automatically stalls at high wind speeds.

(7) *Ground level mounting,* for the generator and gearing permits easy access and maintenance, and reduces tower costs.

(8) *Overall weight* of the turbine may be substantially less than that of conventional systems. This is because of the small amount of material involved in relation to the swept area which translates to lower cost.

Disadvantages: (*i*) This type of machine is too solid, having so much metal or other material surface compared with the amount of wind intercepted. This not only leads to excessive weight for a large installation but also leaves the machine of the mercy of severe storm, since there is no way to reduce the effective area.

(*ii*) It is not useful for a very tall installation because a long drive shaft problems and also the bracing of the topmost bearing above the rotor of a very tall vertical-axis machine is awkward, requiring very long guy wires. In a conventional horizontal-axis wind electric machine with the generator a loft, the strength of the structure required to carry the added weight of the generator is small compared with that needed to servive a severe storm and the generator housing adds little to the area presented to the storm.

Areas of concern. The Savonius rotor has moderately good efficiency and a satisfactory starting characteristics, the latter being particularly important for use with a positive displacement pumps.

The rotor area requirement for getting the required amount of power is higher than any other systems. It is commonly, used for pumping, and to operate small agricultural machines like winnowers blowers, bird scarers, grinders etc.

The another use of this type of wind energy conversion system is to use this machine along with Darrieus rotors for starting purposes.

(*b*) *The Darrieus type machines* (High velocity wind). This machine was invented originally and patented in 1925 by G.J.M. Darrieus, a French engineer and his concept has recently been given serious consideration once again. In India at BHEL, Hyderabad and NAL Bangaluru research and development of the Darrieus wind energy conversion systems are in progress. This type of wind mills are already in use in Canada. As noted, a modern rapidly rotating propeller type windmill, by use of an efficient airfoil, effectively intercepts large area of wind with a small blade area. The Darrieus wind mill is a type of vertical axis machine that has the same advantage. An additional advangage is that it supports its blades in a way that minimizes bending stresses in normal operation.

It has two or three thin, curved (egg beater) blades with airfoil cross section and constant chord length (Fig. 6.23). Both ends of blades are attached to a vertical shaft. Thus the force in the blade due to rotation is pure tension. This provides a stiffness to help withstand the wind forces it experiences. The blades can thus be made lighter than in the propeller type. What happens in a serve storm is another question as yet unanswered. When rotating, these airfoil blades provide a torque about the central shaft in response to a wind stream. This shaft torque is being transmitted to a generator at the base of the central shaft for power generation.

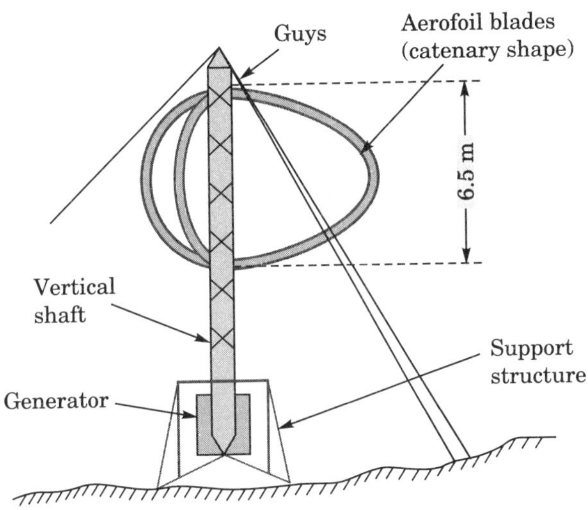

Fig. 6.23. Vertical axis wind mill.

Darrieus-type rotors are lift devices, characterized by curved blades with air foil cross-sections. They have relatively low solidity and low starting torques, but high tip to wind speeds and, therefore, relatively high power outputs per given rotor weight and cost. Various types of Darrieus rotor configurations have been conceived, including the F—Darrieus, D—Darrieus, the Y—Darrieus, and the □ —Darrieus. Such Darrieus rotors can be designed to operate with one, two, three, or more blades.

Darrieus rotors can also be combined with various types of auxiliary rotors to increase their starting torques. However such additions increase the weight and cost of the system, so trade offs in these characteristics must be considered in developing an optimum design for a given application.

Characteristics of Darrieus Rotor

(*i*) Note self starting (*ii*) High speed

(*iii*) High efficiency (*iv*) Potentially low capital cost.

As shown in Fig. 6.20 we see that the force that propels the blades of a conventional wind mill comes from the chord of the airfoil, being tilted away from the direction of motion, so that the thrust that is almost at right angles to the airfoil is tilted toward the forward direction and has a component in that direction, labelled "forward thrust". The remarkable thing about the Darrieus rotor is that it can not have the advantage of this tilt of the airfoil, and yet works without it, with the chord directly along the tangent to the circular path in the equatorial cross-section shown in Fig. 6.24. If the chord were tilted away from the tangent so as to tilt the thrust forward where the wind meets the airfoil on the wind ward side of the circle, as indicated by the broken lines in the figure then the other side of the circle, the wind would meet the other side of the aerofoil and the thrust would be filted backward to retard the motion.

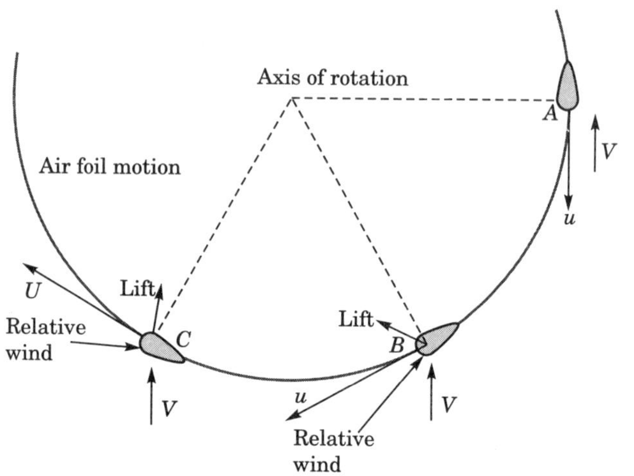

Fig. 6.24. Darrieus rotor principle.

The concept or Darrieus rotor principle (Fig. 6.24) is based on the principle by which a Bermuda-rigged sailing ship can sail across the wind at speeds greater than that of the wind, much like the blade of a horizontal-axis windmill. Lift causes a turning action stronger than the drag forces that impede it as long as the TSR is high enough to prevent the blades from stalling as they cross the wind. This concept patented by a Frenchman, Darrieus. Although modern Darrieus wind turbines usually cannot self-start under steady wind conditions, they do rotate slowly due to the drag difference between the rounded leading edges and the sharp trailing edges of the air-foils. If this is taking place when the wind speed drops below that of the rotor, the resulting increase in the TSR may be enough to unstall the blades. On one or two occasions when this possibility was not foreseen, the machines reached runaway speed with no load and accidentally destroyed themselves. However, there are ways to self-start a Darrieus wind turbine when desired, and methods of preventing over speed as well. There are also some types of high TSR, horizontal-axis turbines whose aerodynamics inhibit self-starting, these require either pitch changing or electrical motive power to run them up to a speed where the blades will unstall. The Darrieus turbine is usually shaped like an egg beater in profile, with blades curved in a form that minimizes the bending stresses caused by centrifugal forces—the so-called Troposkien' profile (Fig. 6.23).

Advantages. Advantages of such WEC system are:

(1) The major advantage of this design is that the rotor blades can accept the wind from any compass.

(2) Another added advantage is that the machine can be mounted on the ground eliminating tower structures and lifting of huge weight of machine assembly, *i.e.,* it can be operated close to the ground level.

(3) Since this machine has vertical axis symmetry, it eliminates yaw control requirement for its rotor to capture wind energy. A dual purpose and relatively simple shaft axis support is anticipated as well as ground level power output delivery due to presence of vertical shaft. This may in turn, allow easier access and serviceability.

(4) Airfoil rotor fabrication costs are expected to be reduced over conventional rotor blade costs.

(5) The absence of pitch control requirements for synchronous operation may yield additional cost savings.

(6) The tip speed ratio and power coefficient are considerably better than those of the *S*-rotor but are still below the values for a modern horizontal-axis, two-bladed propeller rotor.

Disadvantages. (1) Although a Darrieus machine has many directional symmetry for wind energy capture, it requires external mechanical aid for start up. Tests indicate that, with small machines, the problem can be solved by attaching *S*-rotors at the top and bottom of the vertical (rotational) axis. This approach does not appear to be feasible with larger machines, but if the wind power system connected to a utility grid, the generator can serve as a motor to start the turbine. The (alternating current) load can also provide a means for controlling the speed of the rotor regardless of the wind speed, so that variable pitch blades are not required. At very high speeds, stalling occurs and the rotation stops automatically.

(2) Rotor power output efficiency of a Darrieus wind energy conversion system is also somewhat lower than that of a conventional horizontal rotor.

(3) Because a Darrieus rotor is generally situated near ground proximity, it may also experience lower velocity wind compared to a tower mounted conventional wind energy conversion system of comparable projected rotor disc area. This may yield less energy output.

(4) Because a Darrieus rotor encounters greatly varied local flow conditions per revolution, greater vibratory stresses are encountered which will affect rotor system life. High tension cable it down of tower-shaft may require large extensive bearing for support.

(5) Finally since a Darrieus rotor cannot be yawed out of the wind or its blades feathered, special high torque braking system must be incorporated.

Other types of vertical-axis rotors include Magnus effect rotors, which consists of spinning cylinders. This concept was first demonstrated by experimentally by Magnus in 1912. When cylinders spun in a wind stream, translational forces are produced perpendicular to the wind stream by the Magnus effect (Fig. 6.25). Such a device can be used as a sail to propel ships or land vehicles (Fig. 6.26).

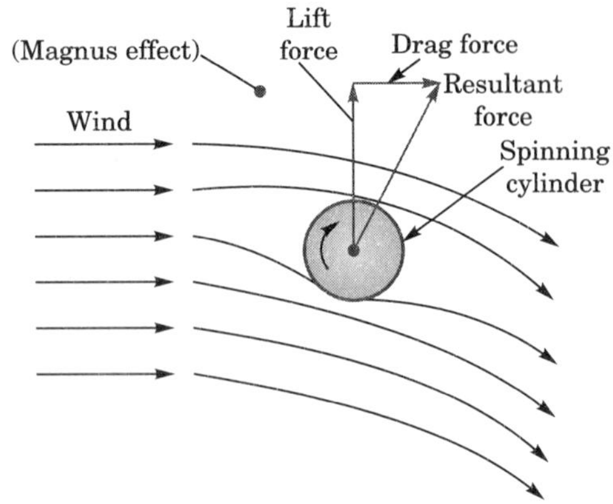

Fig. 6.25. The Magnus effect.

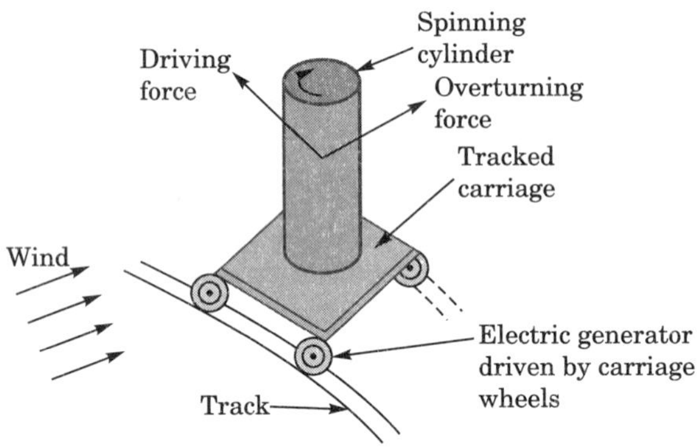

Fig. 6.26. The Magnus concept for generating electricity.

When a stationary horizontal cylinder is rotated about its axis in a crose wind, it will experience a *lift*-force. The effect is equally applicable to a vertical cylinder being rotated about its axis in a cross wind, It will experience a force perpendicular to its axis, which will cause it to move in a direction essentially perpendicular to that of the wind, The concept consists of several tall vertical cylinders that are rotated about their axes in the presence of wind. The resultant of the lift and drag forces, more in the direction of lift because of its larger value propels the cylinders horizontally along a track.

6.9. Analysis of Aerodynamic Forces acting on the Blade

Aerodynamic forces acting on a blade element tending to make it rotate are important parameters for a system engineer. This illustrates the basic principle of aeroturbine rotation.

First, we recognize that the whole general problem of forces on a wind generator blade are quite involved and beyond our scope here.

There are several basic types of blades an aeroturbine may have, *e.g.,* sails, planes and aerodynamic surfaces based on the air craft wing cross-section for which there are many kinds. The early history of wind-mills is based on the first two ; modern higher efficiency wind-electric generators are based on use of blades with aerodynamic surfaces.

Consider the aerodynamic blade shown in Fig. 6.27. The blade can be thought of as a typical cross-sectional element of a two-bladed aeroturbine. The element shown is at some radius *r* from the axis of rotation. It is moving to the left. Because the blade is moving in the plane of rotation it sees a tangential wind velocity, V_T, in the plane of rotation. This velocity component added vectorially to the impinging

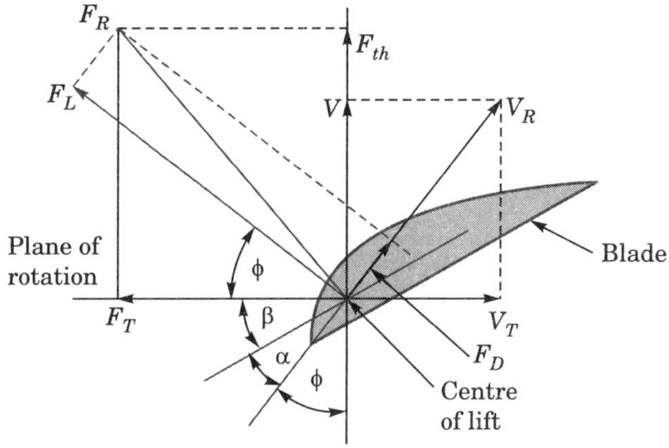

Fig. 6.27. Vector diagram of forces on a elemental blade section of an aeroturbine. (The blade is rotating to the left).

V = Impinging wind velocity.
V_T = Wind velocity in plane of rotation due to blade turning.
V_R = Resultant wind velocity seen by Aeroturbine blade.
F_L = Lift force (Normal to V_R).
F_D = Drag force (Parallel to V_R).
F_R = Resultant force on blade.
F_T = Torque producing component of F_R making aeroturbine rotate.
F_{th} = Thrust force component of F_R.
α = Angle of attack of blade.
β = Blade pitch angle.

wind velocity gives the resulting wind velocity, V_R, seen by the rotating blade element. At right angles to V_R, is the lift force F_L caused by the aerodynamic shape of the blade. The drag force, F_D is parallel to V_R. The vector sum of F_L and F_D is F_R which has a torque producing component, F_T and a thrust producing component. The former is what drives the aeroturbine rotationally and the latter tends to flex the blade and also overturn the aerogenerator. The vector diagram is centred on the centre of lift of the aerodynamic blade. As is well-known from aircraft wing theory, one of the critical parameters is α, the angle of attack of the aerodynamic element. It determines lift and drag forces and hence speed and torque output of the aeroturbine. These quantities can be varied by changing the blade pitch angle β, and this is the basic torque control method used on large variable pitch wind-electric generators. The torque would determine the AC output power if a synchronous generator was used.

Since V_T increases linearly as we go out radically, *r*, on an inclined aeroturbine blade, it is necessary to adjust β with *r* so as to always have a positive angle of attack and to maintain reasonable stress levels within the blade. This mean that at large *r*, β is made small while at small *r*, β is large. Thus the blade 'bites' the air more in close than near the tips. These considerations result in an aeroturbine blade with an apparent twist in it. The need for twisting windmills sails was recognised hundreds of years ago and widely used on Dutch windmills.

Having now achieved an elementary understanding of the basics of what turns the aeroturbine, we see that the airfoil orientation for an aeroturbine driven by the wind is exactly *opposite* to the orientation of a classical airplane propeller which is mechanically driven and whose function is to give a lift force in the axial direction. Thus, though aircraft wing theory is applicable to the operation of an aeroturbine, direct use of a classical aircraft propeller on an aeroturbine would not produce the most efficient aeroturbine because the aerodynamic surface is oriented backwards to what it should be.

6.10. Performance of Wind-machines

WECS efficiency is of interest to both aerogenerator designers and system engineers. As WECS is a capital intensive technology, it is desirable for the overall wind electric plant to have the highest efficiency possible, thus optimally utilizing capital resources and minimizing the busbar electric energy cost.

The overall conversion efficiency, η_0 of an aerogenerator of the general type is

$$\eta_0 = \frac{\text{Useful output power}}{\text{Wind power input}}$$

$$= \eta_A \cdot \eta_G \cdot \eta_C \cdot \eta_{Gen} \qquad \text{...(6.40)}$$

where, η_A = Efficiency of the aeroturbine

η_G = Efficiency of the gearing

η_C = Efficiency of the mechanical coupling

η_{Gen} = Efficiency of the generator

We immediately recognise equation (6.40) as an application of cascaded energy conversion, from which overall efficiency will be strongly determined by the lowest efficiency converter in the cascade. For the aerogenerator this is the aeroturbine; the efficiency of the remaining three elements can be made quite high but less than 100 per cent. It is now evident why so much emphasis is placed on the efficiency of the aeroturbine in wind literature.

Consider an arbitrary aeroturbine (Note that aeroturbine ≠ aerogenerator) of cross-sectional area A driven by the wind. Its efficiency would be:

$$\eta_A = \frac{\text{Useful shaft power output}}{\text{Wind power input}} = C_P$$

$$= \text{Coefficienct of performance.}$$

Thus the coefficient of performance of an aeroturbine is the fraction of power in the wind through the swept area which is converted into useful mechanical shaft power. The coefficient of performance is widely utilised throughout the recent wind literature. We have seen that C_p for horizontal axis wind machine has theoretical maximum value = 0.593 (Art. 6.2).

This theoretical efficiency limitation on a wind energy conversion system is loosely analogical similar to the thermodynamic Carnot efficiency limitation on a conventional thermal power plant.

We have also seen that the convertible power of energy is proportional to the cube of the wind speed. Thus if the wind speed decreases by 20%, the power output is reduced by almost 50%. The wind speed may very considerable from day to day and from season to season. The efficiency of a wind generator depends first and foremost on the design of an wind rotor and rotation speed, expressed as the ratio of blade tip speed to wind speed V_T/V if n is the rotation frequency, *i.e.*, rotations per second, if a rotor of diameter D meters, the tip speed is πnD m/sec.

The dependence of the power coefficient on the tip speed ratio for some common rotor types is indicated in Fig. 6.28. It is seen that the two-bladed propeller type of rotor can attain a much higher power coefficient (*i.e.*, it is more efficient) than the American multi-blade windmill and the classical Dutch four-bladed windmill. In practice two-bladed propeller (horizontal axis) rotors are found to attain a maximum power coefficient of 0.40 to 0.45 at a tip speed ratio in the range a roughly 6 to 10.

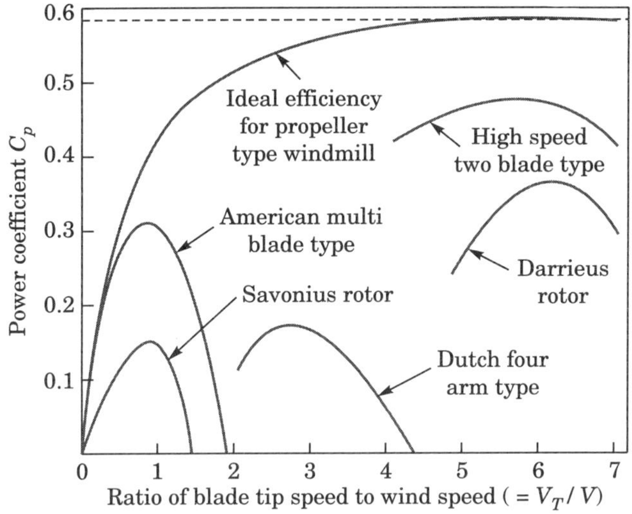

Fig. 6.28. Typical performance of wind machines.

Golding has derived the expression for aeroturbine efficiency as (refer Fig. 6.27).

$$\eta A = \frac{1 - k\dfrac{V_T}{V}}{1 + k\dfrac{V}{V_T}} = C_p \qquad ...(6.41)$$

where, $k = \dfrac{F_D}{F_L} = \dfrac{\text{drag}}{\text{lift}}$ ratio

V_T = wind velocity of blade element in plane of rotation due to blade turning.

V_T = impinging wind velocity.

Clearly if there were no drag, *i.e.*, $k = 0$, then the efficiency would be unity. In actuality k can be made very small, depending on the airfoil chosen and the angle of attack. Also equation (6.41) tells us the efficiency would be low if $\dfrac{V_T}{V}$ were very large or again if it were small. One suspicians that there exists an optimum ratio of $\dfrac{V_T}{V}$ (tip speed ratio).

If one assembles models of various types of aeroturbine blades and puts them in a wind tunnel and runs carefully controlled experiments of their efficiencies as functions of their tip speed ratios, then one obtains a family of curves similar to that shown in Fig. 6.28. We first note that there are rather large differences between the various kinds of aeroturbine blades. The Savonious rotor and the American multi blade (farm windmill) are obviously intended for low speed operation whereas the modern two blade type and Darrieus rotor type are intended for high speed operation more compatible with generating electric energy. The classical Dutch blade is intermediate. We also observe that all the blade performances fall below the ideal efficiency curve which has a maximum of 59.3%. It is interesting to observe that the modern two blade aeroturbine has a peak power coefficient which is a large fraction of the theoretical maximum, *i.e.* they are well designed.

The power delivered and speed at which it is delivered has a bearing on the windmill a application. When these two are matched at the delivery end, the windmill as an energy conversion system can have the maximum utility. Therefore these two performance variables have come into picture, *i.e.* one is C_p *vs* V_{tip}/V characteristics and other is torque *vs* V_{tip}/V characteristics. Figs. 6.28 and 6.29 shows trends in the variation of performance for some of the windmills as discussed above.

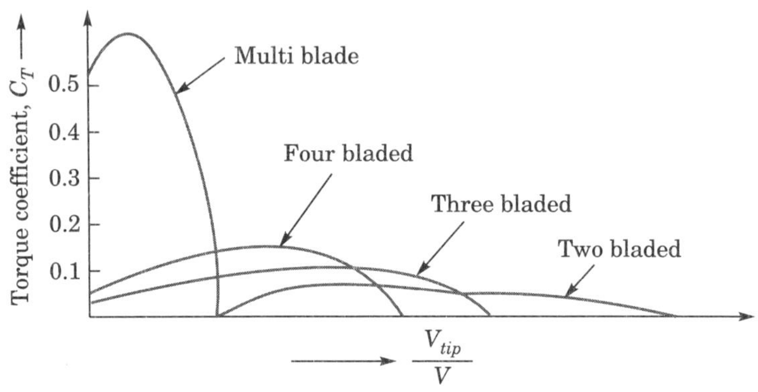

Fig. 6.29. Performance of wind machines.

It may be concluded that a two blade propeller has potentially the best performance of the system considered. This explains why virtually all large-scale systems built in the past employed two or three bladed machines. As the number of blades (and consequently the blade surface) increases, the C_p max decreases and occurs at lower V_{tip}/V ratio. This implies that for high performance, the blades rotate at high r.p.m. (and generally have air-foils for the surface). Consequently the torque is low. These types are preferred for electric power generation. The multi blade types with high starting torque on the other hand are more suitable for pumping water. In practice, it is impossible to build wind driven generators capable of operating at the same efficiency at all wind speeds. First, there is a minimum wind speed below which no power can be generated because of friction losses. Also, above a selected maximum speed, *i.e.*, the rated wind speed, the extracted power is held constant by stabilizing the rotor speed, for instance when the wind speed exceeds the maximum selected by the designer, the rotor blades are progressively turned on their axes to reduce the effective area facing the wind. As a result, the fraction of power extracted decreases as the wind speed increases, beyond the rated speed (refer Fig. 6.21). Finally at about 30 m/sec, the rotor is furled to avoid damage.

Only at intermediate wind speeds does the system efficiency reach its optimum and the power extracted then follows a V^3 law. The range of optimum operation depends on the engine which was selected so as to give optimum output over the year. In our example, this could be in the range of 10-14 m/sec, 14 m/s being the rated velocity maintained also at higher wind speeds; only there depending on the degree of sophistication, could between 70 and 85% of the convertible wind energy by transforming into kinetic energy by the rotor. Upto 20% of this energy would be lost in the gear type transmission, which connects the rotor shaft to the electric generator. The energy which is available for conversion over all wind speed is a function of the wind speed duration spectrum. There is a similarity with solar cells, the efficiency of which is strongly affected by the wavelength spectrum of the sun's radiation. For exploitable spectra, this spectral efficiency is comprised between 8% and 20%. Hence altogether, the total system efficiency of a wind generator amount to 3–7%.

Table 6.2

	Percentage of total available	*Percentage of maximum available*
Total wind energy	100%	
Maximum theoretical convertible (59.3% $\simeq$ 60%)	60%	100%
System efficiency (0.60) at rated wind speed. (Rotor effi. 0.75 transmission gear efficiency 0.08)	36%	60%
Energy integrated over the total speed spectrum (efficiency 0.08–0.2)	3–7%	5–12%

As a general rule, the conversion efficiency at a given location depends on what might be termed as wind quality; a steady wind is the ideal case, never encountered in practice, whereas a wind that varies greatly in speed is difficult to convert.

6.11. Generating Systems

6.11.1. Introduction

The basic components of a wind electric conversion system are shown in Fig. (6.30). Aeroturbines convert wind energy into rotary mechanical energy. A mechanical interface, consisting of a step-up gear and a suitable coupling transmits

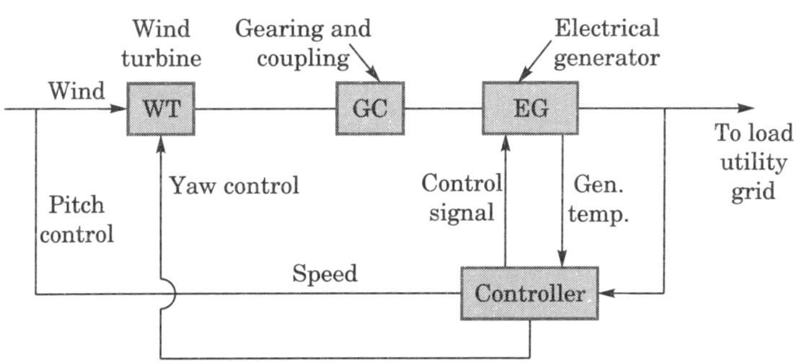

Fig. 6.30. Components of wind electric system.

the energy to an electrical generator. The output of this generator is connected to the load or system grid. The controller senses the wind direction, wind speed, power output of the generator and other necessary performance quantities of the system and initiates appropriate control signals to take suitable corrective actions. The system should be protected from excessive temperature rise of the generator, electrical faults and extra wind conditions.

The choice of an electrical generator and control method to be employed (if any) can be decided by consideration of the following three factors:

(*i*) the basis of operation *i.e.,* either constant tip speed or constant tip speed ratio.

(*ii*) the wind-power rating of the turbine and

(*iii*) the type of load demand *e.g.,* battery connection.

Wind power ratings can be divided into three convenient grouping, *small* to 1 kW, *medium* to 50 kW and *large* 200 kW to megawatt frame size.

Electrical generators types applicable to each of these ratings are:

Small—permanent, magnet, d.c. generators.

Medium—permanent magnet, d.c. generator, induction generator, synchronous generator.

Large—induction generator, synchronous generator.

The electrical control strategy employed for any particular scheme can be designed to effect control of the generator, the power transmission link or the load.

6.11.2. Schemes for electric generation

Several schemes for electric generation have been developed. These schemes can be broadly classified under three categories:

(*i*) Constant speed constant frequency systems (CSCF)

(*ii*) Variable speed constant frequency systems (VSCF)

and (*iii*) Variable speed variable frequency systems (VSVF).

(1) **Constant speed constant frequency system (CSCF).** Constant speed drive has been used for large generators connected directly to the grid where constant frequency operation is essential.

(*a*) **Synchronous Generator.** For such machines the requirement of constant speed is very rigid and only minor fluctuations about 1% for short durations (fraction of a second) could be allowed. Synchronisation of wind driven generator with power grid also will pose problems with gusty winds.

(*b*) **Induction Generator.** If the stator of an induction machine is connected to the power grid and if the rotor is driven above synchronous speed N_s ($N_s = 120\ f/p$), the machine becomes a generator and delivers constant line frequency power to the grid, (f = line frequency and p = number of poles for which the stator winding is made). The per unit slip is 0 and 0.05. The output power of wind driven induction generator is uniquely determined by the operating speed. The pull out torque (T_M) condition should not be exceeded. When this happens the speed continues to increase and the system may 'run away'. The torque speed characteristics of an induction machine in the motor and generating modes are shown in Fig. (6.31). Induction generators are basically simpler than synchronous generators. They are easier to operate, control and maintain, have no synchronisation problems and rae economical. However, they draw their excitation from the grid and hence impose reactive volt ampere burden. But static capacitors can be used to overcome this problem.

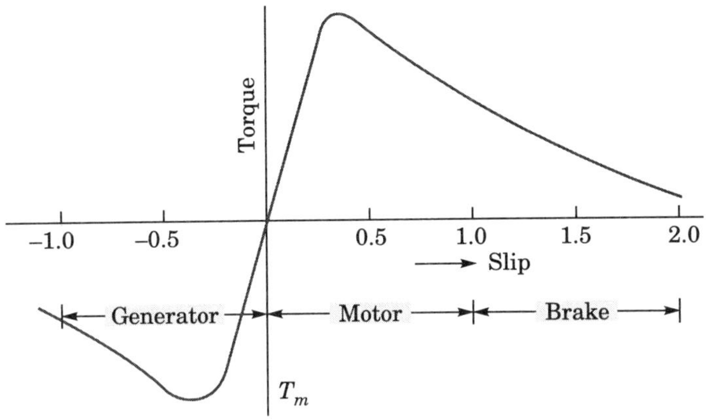

Fig. 6.31. Torque—slip characteristics of induction machine.

(*ii*) **Variable speed constant frequency scheme.** (VSCF Scheme). Variable speed drive is typical for most small wind generators used in autonomous applications, generally producing variable frequency and variable voltage output. The variable speed operation of wind electric system yield higher outputs for both low and high wind speeds. This results in higher annual energy yields per rated installed kW capacity. Both horizontal axis and vertical axis turbines will exhibit this gain under variable speed operation. The popular schemes to obtain constant frequency output are as follows:

(*a*) **AC—DC—AC link.** With the advent of high powered thyristors and high voltage d.c. transmission systems, a.c. output of the 3-phase alternator is rectified using a bridge rectifier and then converted back to a.c. using line commutated inverters. They utilise an a.c. source (power lines) which periodically reverses polarity and causes the commutation to occur naturally. Since frequency is automatically fixed by the power line, they are also known as synchronous inverters. The block diagram of the system is shown in Fig. (6.32).

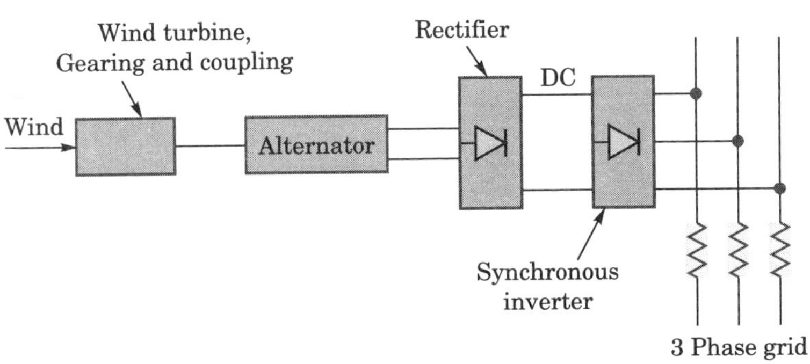

Fig. 6.32. Block diagram of wind electric scheme.

(*b*) **Double Output Induction Generator.** In this system a slipring induction motor is used as shown in Fig. (6.33). Rotor power output at slip frequency is converted to line frequency power by rectification and inversion output power is obtained both from stator and rotor and hence this device is called *double output* *induction generator.* Rotor output power has the electrical equivalence of an additional impedance in the rotor circuit. Therefore, increasing rotor outputs lead to increasing slips and higher speeds. Such an operation increases the operating speed range from N_s to $2\,N_s$, *i.e.,* slip varying from 0 to 1.0.

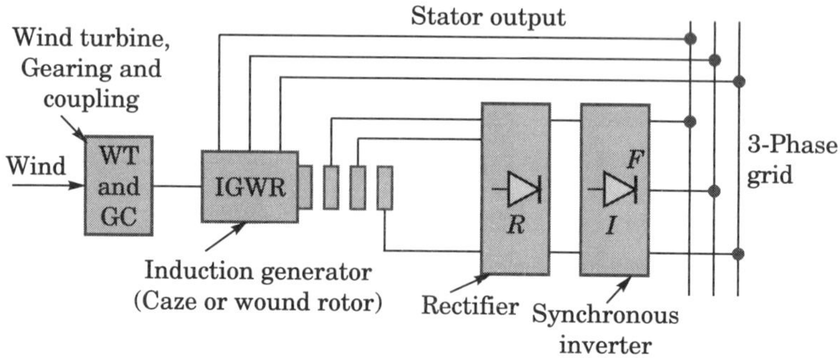

Fig. 6.33. Block diagram of double output wind driven Wound Rotor Induction Generator (IGWR).

(*c*) **A.C. commutation generator.** This system is also known as *Scherbius* system employs two polyphase windings in the stator and a commutator winding on the rotor. Basic problems in employing this device for wind energy conversion are the cost and the additional maintenance and the care required by the commutator and the brush gear.

(*iii*) **Variable Speed Variable Frequency (VSVF) Schemes.** Since resistive heating loads are essentially frequency insensitive, the a.c. generator can be effected at a variable frequency corresponding to the changing derive speed. For this purpose capacitor excited (self-excited) squirrel case induction machines can be conveniently used. Such a scheme is shown in Fig. (6.34). These systems are gaining importance for stand alone wind power applications. The magnitude and frequency of the emf depends on the value of the load impedance, prime-mover speed and excitation capacitance. Methods of analysing these variable—voltage variable frequency generators have been developed to predict the no-load and load performance characteristics. Computer programmes have also been formulated for this purpose.

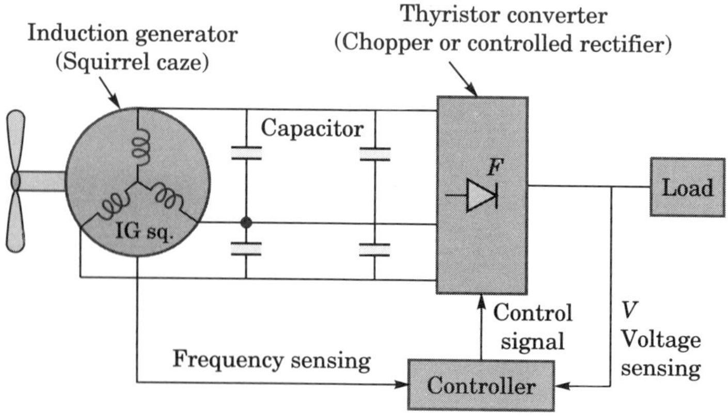

Fig. 6.34. Schematic diagram of capacitor self-excited (squirrel caze) induction generator.

The varying output voltage can be converted to constant d.c. using choppers or controlled rectifiers on constant a.c. using force—commutated inverters. AC converters and transducers can be introduced to monitor and control the desired performance quantities.

6.11.3. Generator Control

Control schemes which act on the generator alone are largely decided upon by the type of the generator employed.

Permanent magnet—there is no readily available means of controlling this type of machine directly. In order to vary the torque established by a permanent magnet alternator, it is necessary to change the armature current. Thus in normal operation the output voltage and frequency must be allowed to vary with wind speed.

D.C. generator—usually of the shunt or parallel field winding type in which a small variation of the field current achieved by means of a variable resistor connected in the field circuit will vary the terminal voltage and hence the power output. It should be noted that with the judicious choice of generator and load resistance a fairly good natural match can be obtained between the power and speed characteristics of the generator and the wind turbine thus obviating the need for any complicated control schemes and allowing variable speed operation of the turbine.

Induction generator. When used to supply power to an isolated load the machine may be made to self excite by means of a bank of capacitors connected to the terminals. Variation of this capacitance value varies with the terminal voltage and output frequency of this system.

When a *wound rotor induction generator* is used for a grid connection application, power control is achieved by varying the slip energy of the rotor circuit. This may be achieved through a variety of methods such as: rotor resistance control, cascading or variable speed slip-power recovery schemes.

Case rotor induction generators can be made to operate over a wider speed range by pole changing or pole amplitude modulating the main winding to achieve one, two or three separate speed ratings.

Synchronous machine power control is readily achieved through variation of the d.c. field excitation current. The frequency of the output voltage will be variable with wind speed for an isolated machine and fixed for a grid connected machine.

Transmission Control

The most effective method of varying the power flow in the transmission link between the generator and its connected load is by means of a silicon controlled rectifier device. This unit employs power electronic devices (thyristors) whose conduction periods (*i.e.*, on off time) can be controlled by apply ing delayed trigger pulses to the gate of each individual thyristor. Some common arrangements are shown in Fig. 6.35 (*a*, *b*, *c*).

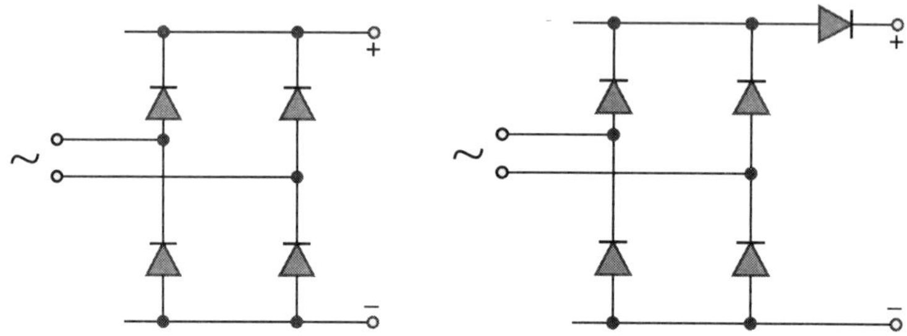

(a) Single phase controlled bridge. (b) Single phase uncontrolled bridge with d.c. chopper.

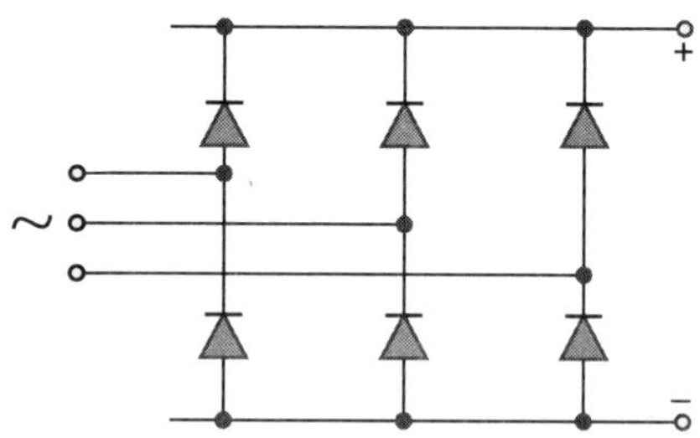

(c) 3 Phase fully controlled bridge.

Fig. 6.35. Transmission control.

Such schemes enables smooth control of the power in the system to be carried out but suffer from the disadvantages of being expensive since they must be rated to carry the normal full load power flow plus short time overload capability.

6.11.4. Load Control

Scheme employing switched load resistors enable stepwise load approximation to be made to the power (rotational speed) characteristic of the wind turbine. Operation within the maximum current loading of the generator can be set and the discrete resistance values selected so that for a particular wind speed, operation of the turbine is held fairly close to the C_p maximum value. This method of controls ideally suited to small stand alone wind turbines and has the merits of being cheap, simple and effective where the load demand is for the resistance heating.

6.12. Energy Storage

Operation of a wind turbine is not practical at very high or very low wind speeds. Consequently, if other sources, such as electric utility power, are not available, some form of energy storage is required. When the power generated exceeds the demand, the excess energy would be stored for use at other times. For WEC machines of low and intermediate electric power, battery storage is convenient.

Storage adds flexibility to use of WECS in that it permits peak shaving and capacity saving as well as fuel saving. However, it can be expensive, and each utility will have to evaluate whether or not to instal it. Storage makes it possible to deliver electric load power demand during times when wind is below normal or non-existent. Storage also makes it possible to deliver short peaks of power for exceeding the rated power capacity of the plant. It improves the reliability of the wind electric system over what it would be without storage. The energy may be used in a variety of forms, *e.g.*, as heat, mechanical, electrical chemical and magnetic.

For wind turbines with power outputs upto about 20 kW, direct-current generators can be used to charge batteries directly. For higher powers, alternating current, generators are required and the current must be rectified for battery charging. The chemical reaction taking place in the cell or battery when it is charged is reversed when the cell is discharged. Thus in the charged cell, electrical energy is *stored as chemical energy,* which can be recovered as electrical energy when the cell is discharged. Direct current from the batteries can be utilized to heat water for space heating and for domestic hot. water, and to operate lights and small tools and appliances. Conversion into alternating current by means of an inverter, may be necessary for large tools and appliances and for television sets.

Other kinds of storage may be more desirable in agricultural operations. For example, if the wind energy is to be used for heating green houses or drying crops, it can be stored as hot water (*Heat storage)*. Either direct or alternating current may then be used in resistance heaters without the need for batteries. Alternatively, the mechanical motion produced by the wind turbine can be converted directly into heat by frictional effects, such as by churning water.

In *water power storage* or mechanical storage, high-power aerogenerators are integrated with an electric utility, a favourable situation would be operation of several wind turbines in connection with a *hydroelectric power plant.* If the total power, wind and hydroelectric being generated should exceed the demand, the hydroelectric plant can be partly shutdown; alternatively, the excess power could be used to pump water from an auxiliary reservoir at the bottom of the dam back into the main reservoir. In this way, the overall capacity of the hydroelectric system would be increased.

Another alternative; for possibly storing energy, is to store the energy in a volume of *compressed air.* A wind turbine for example, could be created which would directly pump air into a suitable pressurized storage tank. Then later when the wind is not blowing the energy stored in the air could be utilized to drive an air turbine whose shaft would then drive a generator, thus supplying the needed electric power when the wind is not blowing. Wind energy, with a d.c. output the power can be fed directly into an electrolyzer tank which produces hydrogen and oxygen from ordinary water. The hydrogen and oxygen gases produced can be stored either in gas or liquid forms, and when needed, be quickly and easily converted again directly into electric energy via the well known fuel cell. Hydrogen can also be used *as a fuel* to drive automobiles and other useful engines.

6.13. Applications of Wind Energy

Energy extracted from the wind is initially energy in the form of rotary, translational, or oscillatory mechanical motion. This mechanical motion can be used to pump fluids or can be converted to electricity, heat, or fuel. Some of the most effective application are those that use energy derived directly from the wind, without further energy processing, conversion, or storage. However, if required, wind—derived energy can be converted to other forms of energy or can be stored through the use of compressed fluids, pumped hydrosystems, water-saver systems, batteries, hydrogen, flywheel, hot water, etc. Some energy is normally lost in each of these conversion or storage steps.

In any case, wind energy is one of the most flexible and tractable of all energy sources, since the mechanical energy derived directly from the wind can be readily and efficiently converted to other forms of energy. The efficiency of converting wind—derived mechanical energy to heat or electrical energy is usually much higher for instance, then the efficiency of converting solar or fuel-derived heat energy to mechanical or electrical energy, since the efficiencies that can be attained when converting heat to mechanical or electrical energy are limited by the relatively low Carnot cycle efficiencies, which, even under optimum conditions, usually do not exceed 30 to 35%.

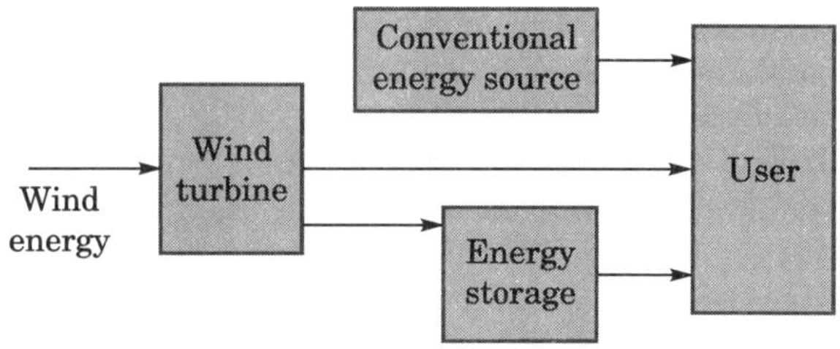

Fig. 6.36. Basic wind energy conversion system with energy storage.

Wind-turbine generators have been built in a wide range of power outputs from a kilowatt or so to a few thousand kilowatts. Machines of low power can generate sufficient electricity for space heating and cooling of homes and for operating domestic appliances. Low power WEC generators have been used for many years for the corrosion protection of buried metal pipelines.

Applications of somewhat more powerful turbines, upto about 50 kW, are for operating irrigation pumps, navigational signals (*e.g.,* light houses and buoys), and remote communication, relay, and weather stations, and for offshore oil drilling platforms.

Aerogenerators in the *intermediate power range,* roughly 100 to 250 kW, can supply electricity to isolated populations (*e.g.,* on islands), to farm co-operatives, and to small industries. Finally, the *largest WEC generators,* with rated powers of a few thousand kilowatts are usually planned for interconnection with an electric utility system. Present indications are that the optimum economic diameter of a wind turbine with a two-bladed propeller type rotor is about 110 M; the electric power output would range from 2000 to 5000 kW or 2 to 5 megawatts (MW), where 1 MW is 1000 kW or 1 million watts.

Pumping Applications. A typical wind powered pumping application is one that might use a horizontal—axis wind machine. An example is the ancient jib—sail design that is mostly wind used to pump irrigation water. Large number of water-pumping wind mills have been used on Indian farms. Other applications that are being developed include the pumping of water for aqueducts or for pumped-hydro storage of energy. In aqueduct systems, large scale wind driven units can provide power for the pumping of water from the main reservoir to auxiliary reservoirs in other parts of the aqueduct system. This can be done either by the direct mechanical pumping of the water or through the generation of electricity by the wind units, and the subsequent use of this energy to operate electrical water pumps incorporated in the aqueduct system.

There are two main families of wind turbines : wind pumps and wind generators. A pump generally involves a high-solidity rotor with a low TSR connected mechanically to a piston pump, which a generator is usually driven through a gear-box by a low-solidity rotor.

The two main end uses of wind pumps (irrigation and water supply) have very different technical operational and economic requirements. Irrigation designs are generally unsuitable for water supply duties, which may require heads 10-100 m high. Despite these limitations, however, water supply pumps are sometimes used for irrigation. Since many water supply wind pumps must run unattended for most of the time, their design should incorporate protection devices to prevent over speeding in storms and sturdy parts that required little attention. Consequently they are usually built of components manufactured from industrial steel and drive piston pumps *via.* reciprocating pump rods. This type of construction is expensive in relation to the output but the reliability and low maintenance are worth the price to someone like an Australian farmer, who may have a dozen wind pumps spread over a large area of semidesert to sustain several thousand head of cattle. In contrast, irrigation being seasonal, the mill may only be useful for a few months of the year. It also involves pumping much larger volumes of water through a low head. Because the intrinsic value of the water is low, it is essential to keep costs down. In addition, some one is usually present during irrigation to tend the machine. Therefore, windmills used for irrigation are generally indigeneous designs that are built by the farmer as a method of low cost mechanization.

If water-supply wind pumps are used on farms for irrigation, it may be difficult to provide piston pumps large enough to absorb the power from the windmill at low heads. Also, most wind pumps of this kind must be located over the pump in substantial, reinforced-concrete foundations. This usually makes them better suited for pumping from wells or bore holes than from open water, for which a surface-water suction pump can be used on the other hand, most indigeneous irrigation wind pumps, such as the Chinese models, use rotary pumps of one kind or another that are more suitable for low heads. These do not achieve mechanical forces as high as those of industrial wind pumps, which often pull their rods with a force equivalent to 1 *t* or more enough to uproot any carelessly installed pump. Furthermore, most indigeneous designs are much cheaper and easier to install because they are lower in height and do not require concrete foundations.

In pumped hydro applications, the wind units can be used to supply power to pump from an auxiliary reservoir below hydro electric dam back into the main reservoir above the dam. This enables the water stored in the main reservoir to be replenished when the wind is blowing, thereby adding to the capacity of the hydro-electric system to generate base load electric power.

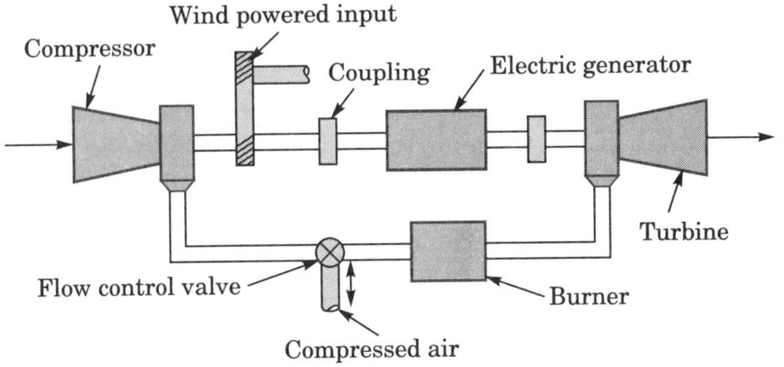

Fig. 6.37. Wind-assisted gas-turbine generating unit.

Wind power can also be used compress air for use in various applications, including the operation of gas turbines for generating electricity during the peak-demand periods of a public utility system. For this type of application, conventional gas turbines can be modified to separate the compressor, generator, and power stages by clutches. In one mode of operation, the motor generator operating as a motor and powered by a wind machine drives the air compressor. The compressed air is fed into a storage tank or into a large cavern, aquifer, or depleted natural gas well. Under this mode, the power turbine is inoperative, and no fuel is consumed.

In a second mode of operation, when the demand for power exceeds the supply of the base-load utility system, the compressor is disengaged, and the power turbine ia connected to the generator. The burner that drives the power turbine is fed fuel and compressed air from storage to generate power for the utility system.

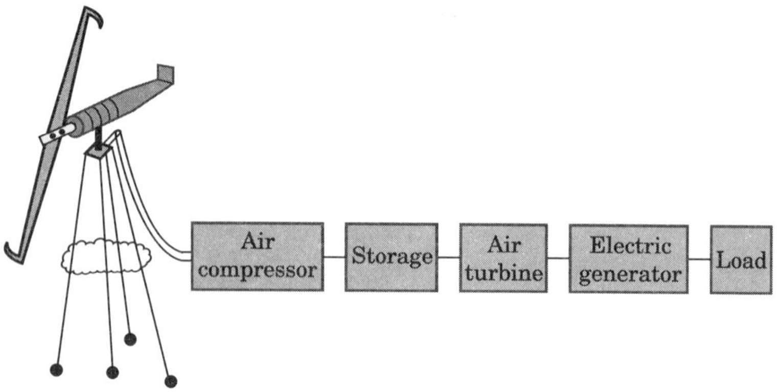

Fig. 6.38. System with compressed-air storage.

The temperature of air is raised when it is compressed without loss of heat (*i.e.*, adiabatic compression). In this case, less heat will need to be added to the air, when it is eventually used to drive a turbine at a given efficiency, than if its heat has been allowed to escape from the storage container and the temperature of the air had been allowed to drop to the ambient temperature (*i.e.*, isothermal storage). Adiabatic storage is obviously better, from the standpoint of energy conservation, than isothermal storage.

Wind-powered pumps can be used to desalinate water, using reverse osmosis units.

Wind-powered pumps can also be used to save fuel and electricity by compressing the working fluids used in heat pumps for space-heating applications, as discussed below.

Direct Heat Applications. Mechanical motion derived from wind power can be used to drive heat pumps or to produce heat from the friction of solid materials, or by the churning of water or other fluids, or in other cases, by the use of centrifugal or other types of pumps in combination with restrictive orifices that produce heat from friction and turbulence when the working fluid flows through them. This heat may then be stored in materials having a high heat capacity, such as water, stones, eutectic salts, etc., or the heat may be used directly for such applications as heating and cooling of water, and air-space for residential, commercial, industrial and agricultural buildings or for various types of industrial or agricultural process-heat applications. Heating system designed for heating of the building is done by using the electricity generated by the wind machine to provide resistive heating of water that is circulated through the buildings.

A home heating system that uses a wind-powered pump and a restrictive orifice to derive direct heat for a building, without first generating electricity also has been developed.

Examples of possible wind-powered agricultural process heat applications include green house applications, crop drying, milk rocessing, food processing, refrigeration, frost protection, ventilation, and waste processing.

Examples of typical industrial processes that might be able to use low temperature heat (*i.e.,* upto approximately 175°C) produced by wind energy include the following.

(1) Production of inorganic chemicals, including borax, bromine, chlorine, caustic soda, potassium chloride, and sodium metal.

(2) Production of plastic materials and synthetics such as polyethylene, polyvinyl chloride and polystyrene, for which approximately 45% of the process steam used is in the range of 100—175°C.

(3) Production of organic chemicals such as various types of alcohols and solvents, synthetic perfumes, flooring materials, rubber processing chemicals, etc.

(4) Food processing such as meat preparation and packing fruit and vegetable dehydration, wet corn milling, soybean oil milling etc., for which approximately 80% of the process heat is used at temperature less than 175°C. Of this about 25% is used for process heating, 15% for cooking, and dehydration.

(5) Textile processing, primarily steam or hot air, for drying, curing, and finishing of both yarns and textiles.

Electric Generation Applications. Wind power can be used in centralized utility applications to drive synchronous a.c. electrical generators. In such applications, the energy is fed directly into power networks through voltage step-up transformers.

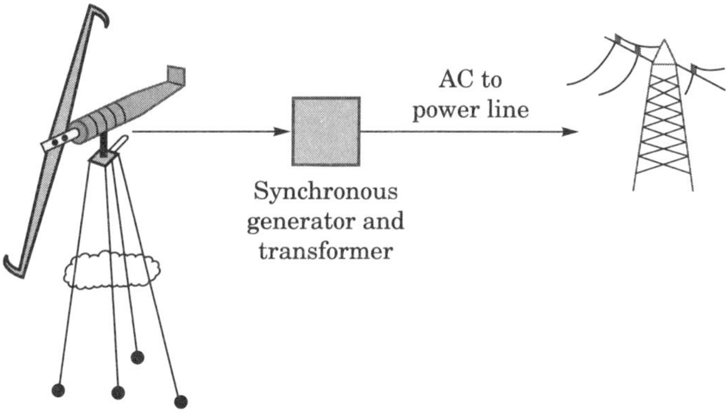

Fig. 6.39. System with direct feed to main powerline.

WECS units can be integrated with existing hydro electrical networks and used in a "water-saver" mode of operation. When the wind is blowing, electrical generation at the hydroelectric plants in the network can be reduced by an amount equal to that being produced by the "WECS units. Thus, part of the network load that is ordinarily produced by the hydroelectric generators is supplied by the wind turbines. Under these conditions some of the water that would have been used by the hydroelectric plant to supply the load is saved in the reservoir and made available for later use when the wind is not blowing. Additional hydroelectric generating facilities are provided at the hydro-plant to allow the water that was saved in the reservoir to be used at a greater rate when the wind was not biowing, thereby providing a *firm* generating capacity equal to the *firm* generating capacity of the hydro-plant plus the *average* generating capacity of the wind-powered plant.

In dispersed applications, wind power can be used to generate dc electrical power that, in turn, can be used for dc applications or space heaters, such as resistance heaters, or can be stored in batteries and then inverted for used by a.c. loads.

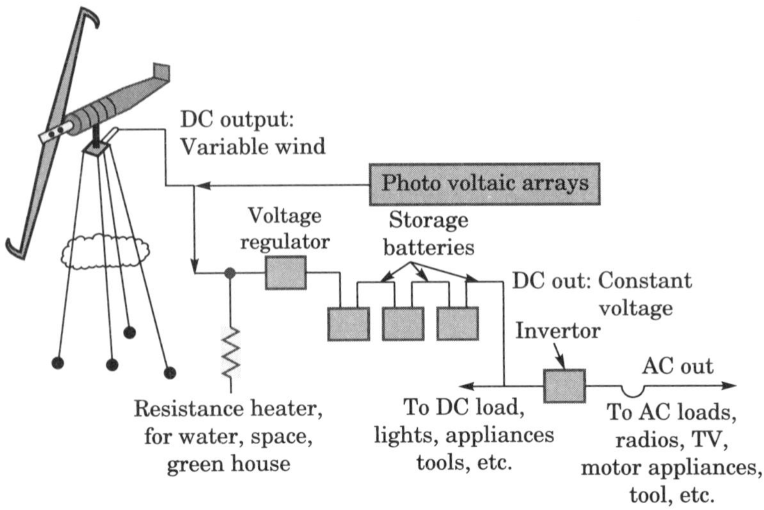

Fig. 6.40. System with battery storage.

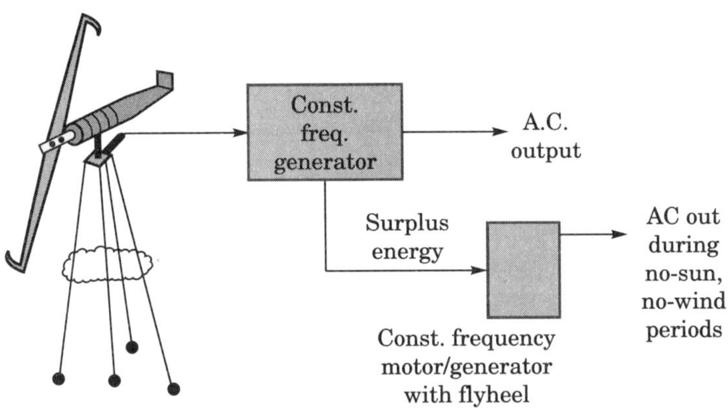

Fig. 6.41. Alternative for converting and storing wind energy: system with flywheeling storage.

There are a number of commercial units available for these types of dispersed operations and all can be used independently or tied into a utility network using an induction generator or a synchronous inverter.

In centralised or dispersed applications requiring constantly available sources of power, the energy can also be stored in the form of mechanical motion of a flywheel or as hydrogen and oxygen gases derived from the electrolytic dissociation of water. The hydrogen and oxygen can be stored in liquid form in tanks, or in gaseous form in tanks, caverns, aquifers, depleted natural gas wells, etc. The stored hydrogen can be used either as a fuel or direct space heating or industrial process heat, or it can be reconverted to electricity through the use of fuel cells, gas turbine generators that burn hydrogen, or by other means.

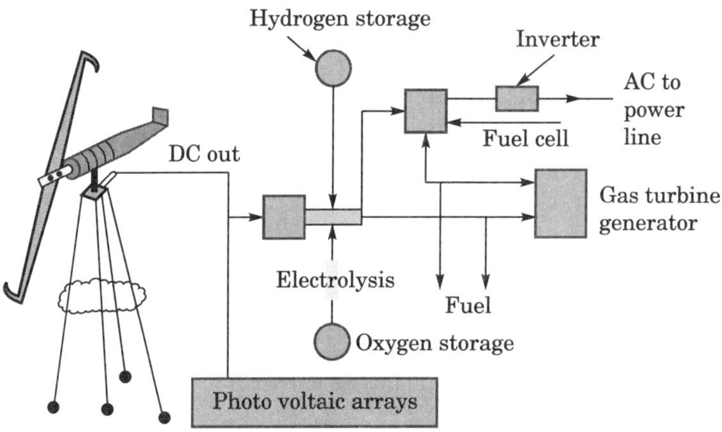

Fig. 6.42. Alternative for converting and storing wind energy system with hydrogen storage.

6.14. Interconnected System

There appear to be important advantages to use wind derived energy in combination with energy derived from other sources, such as conventional fuels, sunlight, ocean thermal differences, bioconversion fuels, etc. Since the wind blows intermittently in most locations there may be a need to store wind energy over long periods of time, perhaps upto ten days or more, if the energy is being used for isolated applications requiring continuous power. The cost of providing sufficient storage capacity for such applications can be reduced if the wind-derived power is interconnected with other sources of power. For instance, since in most locations the wind often blows when the sun is not shining, and *vice versa,* a system using wind energy collectors and sun energy collectors (solar thermal collectors or solar photovoltaic arrays) in combination can be expected to require less energy storage capacity than systems that use these type of collectors separately.

A combination of solar and wind energy systems are being experimented in a number of isolated locations not connected to the grid. Such a system is shown in Fig. 6.43. Depending on the frequency of the wind generator, consumer circuits could be switched on and off through a consumer control device to ensure that the frequency does not fluctuate by more than few percent of its rating. Similarly solar generator may consist of many sections linked individually to the battery and a number of sections kept in operation depends on the charge on the battery and wind speed. Many remote power systems throughout the world rely on diesel generators for the production of electrical power. Such applications include village electrification, irrigation and water pumping. Many of these applications incur high electric generating costs due to the price of diesel fuel. Coupling renewable energy systems, such as wind and photovoltaic (PV) and batteries with a diesel set can lower the cost of generation of electricity by reducing the amount of fossil fuel required. Designing such hybrid systems require two major considerations:

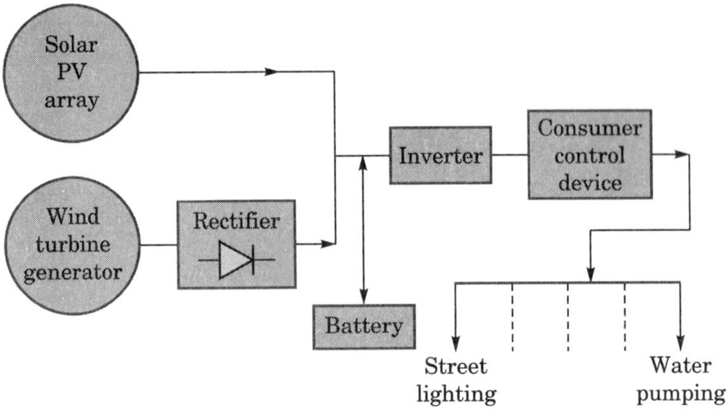

Fig. 6.43. Solar-wind system with storage battery supplying to various consumers.

(*i*) determining an optimal mix of powor generation sources to meet the required load while providing the lowest busbar cost of electricity.

(*ii*) arriving at a system operating strategy that maximises the use of renewable sources and allows the diesel set to run at full load when required.

6.15. Safety Systems

Safety systems of the wind turbines comprise the following features :

(*i*) *The computer.* The wind turbine is controlled by a computer which monitors the most important gauging instruments and compare the results. If errors are found the wind turbine is stopped.

(*ii*) *Emergency Stop.* If a situation arises which calls for the wind turbine to be stopped immediately, the emergency stop is used. The wind turbine will stop in few seconds by feathering the blades directly into the wind. It can not be started again before what caused the emergency stop has been rectified.

(*iii*) *Revolution Counters.* To prevent the rotor from racing, two revolution counters have been mounted on the shaft. These operates quiet independently and activate the emergency stop if the revolutions of the turbine exceed 24 r.p.m which is maximum.

In addition, the revolution counters transmit data to the com-puter where they are compared. If they differ, there is a fault in one of the revolution counters and the wind turbine is stopped. This will also occur if the blades are accelerated too fast.

(*iv*) *Wind Velocity.* This is measured and controlled by the com-puter in two ways. First gusts of wind are registered and if they are too strong the turbine is stopped. Then average wind speeds are measured over periods of 10 minutes, and the wind turbine is also stopped if these are too high.

(*v*) *The Parachutes.* Each blade tip has a parachute, which is activated if the rpm exceeds 28. An iron plumb bob, otherwise held in place by a magnet, is released from the blade trip, the centrifugal force exceeding the force of the magnet pulling out the parachute. This decrease the speed of the wind turbine considerable enough to stop it from racing. The parachute is an extra safety device should other fail. Till now they never been used.

(vi) Lightning Rods. The three blades and the mill or wind turbine cap are protected from lighting by these rods going from the tip of each blade to the ground.

6.16. Environmental Aspects

Wind turbines are not without environmental impact and their operation is not entirely risk-free. Following are the main effects due to a wind turbine.

(i) Electromagnetic interference. Interference with TV and other electromagnetic communication systems is a possibility with wind turbines as it is with other tall structures. TV interference is most likely in areas, where there is a weak signal because of the distance from the transmitter, where existing reception is none too good due to the surrounding hills and where the wind turbine is exposed in good position to receive and scatter the signals. Interference can be overcome by dispensing with aerials and sending TV signals by cable in areas that would otherwise be affected.

(ii) Noise. The noise produced by wind farms falls into two categories. The first type is a mechanical noise from the gearbox, generating equipment and linkages and the second type of aerodynamic in nature produced by the movement of the turbine blades. One component of the latter is the broad band noise which ranges upto several kilo hertz and the other is a low frequency noise of 15—20 Hz. Revolving blades generate noise which can be heard in the immediate vicinity of the installation, but noise does not travel too far.

(iii) Visual Effects. Megawatts power generating wind turbines are massive structures which would be quite visible over a wide area in some locations. A variety characteristics such as colour pattern, shape, rotational speed and reflectance of blade materials can be adjusted to modify the visual effects of wind turbines including the land scape in which they are installed.

Visual intrusion is probably the single most serious environmental problem associated with the on shore development of wind energy. Many of the best sites for wind turbines are in areas of outstanding natural beauty. Wind enthusiasts have argued that, like suspension bridges and some modern buildings, well-designed wind turbines can be aesthetically pleasing to the eye. They are more attrac-tive than microwave transmission towers and television aerials.

(iv) Bird life. Tall structures represents a potential collision hazard to bird life. However, studies indicate that majority of migrating birds either fly at much higher altitudes or take avoiding action. It is likely to affect night flying birds more and to be more of a problem with smaller, faster turning machines rather than very large rotors.

(v) Risk. The possibility of mishap is perhaps of greater concern to the public. The most serious failure from the safety point of view is the detachment of blade, or blade fragment which could be thrown a considerable distance and could damage people or property. A reliable control system to identify the fault situation rapidly and braking system to bring the rotor safely to rest could be the measures adopted to prevent such failures. Regular inspection and greater safeguards against over speeding should minimise the hazard. Most wind machines will be in rural areas which makes it less likely that anyone would be injured or killed by flying debris if there were an accident.

QUESTIONS

1. What is the basic principle of wind energy conversion?
2. Derive the expression for power developed due to wind.
3. Prove that in case of horizontal axis wind turbine maximum power can be obtained when:

$$\text{Exit velocity} = \frac{1}{2} \text{ wind velocity and}$$

$$P_{max} = \frac{8}{27} \rho AV^3 .$$

4. Wind at 1 standard atmospheric pressure and 15°C temperature has a velocity of 10 m/s. The turbine has diameter of 120 m and its operating speed in 40 r.p.m. at maximum efficiency. *Calculate:*
   (*a*) the total power density in the wind stream,
   (*b*) the maximum obtainable power density assuming $\eta = 40\%$,
   (*c*) the total power produced (in kW) and
   (*d*) the torque and axial thrust.
   [**Ans.** 613 W/m², 363 W/m², 245 W/m², 2770 kW, 16, 347 N, 616, 255 N]
5. Describe the main considerations in selecting a site for wind generators.
6. Describe with a neat sketch the working of a wind energy system (WECS) with main components.
7. How are WEC systems classified? Discuss in brief.
8. Discuss the advantages and disadvantages of wind energy conversion system?
9. Describe horizontal axis type aerogenerators.
10. Discuss the advantages and disadvantages of horizontal and vertical axis windmill. What methods are used to overcome the fluctuating power generation of windmill?
11. What are the advantages of vertical axis machines over horizontal type? Describe a rotor for relatively low velocity wind.
12. Write short notes on:
    (*i*) Applications of wind energy (*ii*) Savonius rotor
    (*iii*) Darrius rotor (*iv*) Wind energy storage.
13. Describe the different schemes for wind electric generation or describe the generating systems. Also describe the generator control schemes.
14. Describe the main applications of wind energy, giving neat sketches.

UNIT—3

ENERGY FROM BIOMASS

(BIO-ENERGY)

- — Biomass Conversion and Biogas Generation
- — Digester Design Considerations
- — Biomass as a Source of Energy and Thermal Gasification of Biomass

CHAPTER 7

Energy from Biomass

(Biomass and Bio-conversion)

7.1. Introduction

Biomass is organic matter produced by plants, both terrestrial (those grown on land) and aquatic (those grown in water) and their derivatives. It includes forest crops and residues, crops grown especially for their energy content on "energy farms" and animal manure. Unlike coal, oil and natural gas, which takes millions of years to form, biomass can be considered a *renewable energy source* because plant life renews and add's to itself every year. It can also be considered a form of solar energy as the latter is used indirectly to grow these plants by *photosynthesis*.

As the word-clearly signifies; biomass means organic matter and photo-chemical approach to harness solar energy means harnessing of solar energy by photosynthesis. Solar energy is stored in the form of chemical energy. Hence

Solar energy → Photosynthesis → Biomass → Energy generation.

Out of several sources of renewable energy like solar, wind, ocean thermal energy, tidal wave energy, Geothermal energy, nuclear energy, energy through biomass are important feature in our country. Biomass resources fall into three categories:

(*i*) Biomass in its traditional solid mass (wood and agriculture residue) and (*ii*) biomass in non-traditional form (converted into liquid-fuels). The first category is to burn the biomass directly and get the energy. In the second category, the biomass is converted into ethanol (ethyl alcohol) and methanol (methyl-alcohol) to be used as liquid fuels in engines. The *third* category is to ferment the biomass anaerobically to obtain a gaseous fuel called *bio-gas*. It is about this bio gas technology detailed discussion will be given in subsequent article. Biomass includes wood waste and bagasse, which have potential of generating substantial electric power. All these bio-mass are highly dispersed and bulky and contain large amount of water (50 to 90%). Thus, it is not economical to transport them over long distances, and conversion into usable energy must take place close to the source, which is limited to particular regions. However, biomass can be converted to liquid or gaseous fuels, thereby increasing its energy density and making feasible transportation over long distances.

Terrestrial crops include (1) sugar crops such as sugarcane and sweet sorghum; (2) herbaceous crops, which are non-woody plants that are easily converted into liquid or gaseous fuels; and (3) silviculture (forestry) plants such as cultured hybrid poplar, sycamore, sweet gum, alder, eucalyptus, and other hard woods. Current research focuses on the screening and identification of species that are available for short rotation growing and on the optimum techniques for planting, fertilization, harvesting, and conversion.

Animal and human waste are indirect crops from which methane for combustion and ethylene (used in the plastic industry) can be produced while retaining the fertilizer value of the manure.

Aquatic crops are grown in fresh, sea, and brackish waters. Both submerged and emergent plants are considered, including sea weeds, marine algae, and of particular interest, the giant California kelp.

Hence bio energy and energy plantations are the two main features which are discussed in the chapter. Bio-gas or methane is produced by the anaerobic decomposition of organic materials. This gas is produced from cowdung and other wastes such as cornhusks, leaves, straw, garbage, flesh of carcusses, poultry droppings, pig dung, human excreta and sewage. It is estimated that about 100 crore tonnes of fresh dung is available in the country, and 67.1 cubic metre of gas per tonne of wet dung can be produced. Aquatic biomass is another source to produce methane gas. It has been estimated that about two lakh hectares of land is covered by water hyacinth spreading West Bengal, Bihar, Assam, Andhra Pradesh, Tamil Nadu, Orissa and Kerala. It can absorb 80% of nitrogen and 60% of phosphorus from the secondary effluent in 5 days. The yield is 148 tonnes per hectare per year. A huge quantity of agriculture residues are available in the country which go waste every year. Against the total agricultural waste which could be around 200 million tonnes, the actual consumption as fuel might be around 40 million tonnes. It has been estimated that even for the smallest gas plant (2 cu. m.), one should have a regular supply of 45 kg of fresh dung everyday. The animal should preferably be stable bound otherwise the dung would be lost in the pasture. The family size gas plant will cook a day's meals for a small family. The rate of gas production will be highest if the mixture in the digester contains solids between 7-9%. Cow dung originally contain about 18% of solids. The bio-gas can also be used to run diesel engines which operate with the mixture of bio gas and diesel oil it can reduce the consumption of diesel oil by about 80% and engine run faster by 43% of extra power with this mixture.

Bio-fuels. The energy stored in dry biomass like wood and straw is most easily released by direct combustion—although dry materials can also be converted into liquid and gaseous fuels (for later combustion) by a variety of techniques (which are discussed in section 7.2). Biomass that is wet, or has a moisture content like sewage sludge and vegetable matter can be dried and burnt. However, it requires considerable energy to drive off the water, and this diminishes the value of the biomass as fuel. It is usually more appropriate to convert wet organic matter directly into premium fuels by a wet process like digestion or fermentation.

Fuels derived from biomass are easily handled and burnt, whereas raw biomass is often wet and of inconsistent quality or variable composition. Most importantly, refined biofuels can be used in existing fuel systems, often with little or no modification, and are compatible with current patterns of energy use. This improves their economic potential and the speed with which they can be introduced. Refined biofuels have high calorific values, and can be stored for long periods without deterioration. By contrast, untreated biomass readily absorbs water and rots.

Table 7.1. The Heat Content of Various Fuels

	Fuel	Heat Content (approximate)	Unit
Gas	Hydrogen	12	MJ/m^3
	Fuel gas	5—20	MJ/m^3
	Biogas	20—25	MJ/m^3
	Methane	38	MJ/m^3
Liquid	Methanol	21	GJ/t_e
	Ethanol	27	GJ/t_e
	Crude oil	44	GJ/t_e
Solid	Refuse	9	GJ/t_e
	Straw	16	GJ/t_e
	Wood (air dried)	12—15	GJ/t_e
	(kiln dried)	20	GJ/t_e
	Char	20	GJ/t_e
	Coal	22—32	GJ/t_e

Solids. Three solid biofuels—wood, straw and refuse are being burnt on an increasing scale in many countries to provide useful heat.

Wood. In the form of cut logs, wood chips and sawdust is currently used for domesting heating and to provide process heat in the timber and furniture industries. Wood though it is a relatively bulky fuel, it has a low ash and sulphur content and burns easily. It requires four volumes of air-dried wood to deliver an equivalent amount of heat to that from one volume of coal.

The market for domestic wood stoves is thriving, and several industrial countries are re-examining the potential for wood for industrial use and electricity generation. For industrial wood burners and boilers, the wood is usually shredded, and this facilitates drying, handling and transport—although green wood, with high moisture content, can be burnt in industrial fluidised bed boilers.

Straw. Straw-burning stoves and furnaces are becoming increasingly common in many countries. They burn various sizes of baled or chopped and compressed straw, and provide heat for crop drying and space and water heating. For grain drying on the farm there is a good match between supply and demand; this minimizes storage costs.

Municipal Refuse is far from an ideal fuel. It is messy to handle and has a low and variable energy content on average only about one third that of coal. It is dried, sorted and shredded, and can then be burnt to provide heat and power. However, precautions must be taken to ensure that corrosive combustion products do not damage the boiler; flue gases must also be conditioned before release to remove noxious products. The economics are significantly improved by recovering potentially valuable materials like metals and cullet (broken glass). The fuel is enhanced if the shredded material is compacted into pellets. Refuse derived fuels can be burnt directly or combined with other fuels such as residual oil or coal.

Liquids. Alcohols and vegetable oil are now replacing petrol and diesel as transport fuels in several countries and this process is likely to accelerate as oil prices rise. Other liquid biofuels are being investigated such as hydro-carbons from *Euphorbia* plants that can be grown in desert regions. Euphorbias are related to rubber trees; their latex is rich in petroleum like compounds.

Methanol and *Ethanol* can be blended with unleaded petrol and burnt in present day internal combustion engines. Minor engine modifications are necessary for blends containing more than about 20% alcohol, or for almost pure alcohol, these include on increased compression ratio and altered timing and carburation. Pipes, pumps and fuel tanks must also be protected because alcohols, particularly methanol, contain traces of water and corrosive, organic impurities. With 98%, alcohol, the fuel must be preheated by circulating hot air or water from the engine exhaust. A small electric heater is necessary for starting in cold climates. Diesel engines have been modified for dual fuel use.

Alcohols have a high octane rating, but a lower calorific value than petrol. Methanol contains 25% less energy per gallon than ethanol, and 50% less than petrol. However, alcohols can improve engine performance and are for cleaner than petrol or diesel; they burn with higher efficiency and at lower temperatures and are free from lead and sulphur. This reduces noxious emissions dramatically.

Vegetable oils. From crushed seeds and nuts (for example sunflower and rape seed, peanuts, palm, soya and corn) can be burnt in unmodified diesel engines. They can be blended with diesel fuel or used directly. Research on improving engine performance is underway in several countries. Vegetable oils offer an attractive fuel for tractors and other agricultural machinery and are already in use by farmers in, for example, South Africa and Brazil. Enough oil can be obtained from the sunflower crop of one field to fuel a tractor ploughing ten fields of smaller size.

Gases. The production of biogas from dung and agricultural wastes is now extensive in some developing countries. Surprisingly, its potential in the industrial world has only recently been recognised.

Biogas as already stated, is a mixture of methane and carbon dioxide with traces of hydrogen sulphide. The methane content varies between 50 and 70% depending on source. It has an energy content equivalent to about two thirds that of natural gas and can be burnt, as produced in stationary engines or turbines to generate heat and mechanical or electrical energy. It is impracticable to store large volumes of biogas at low pressure. It is therefore most economically used as it is produced for example, to meet space and water heating or cooking needs on farms. Biogas can be compressed and stored, or its fuel quality improved by removing carbon dioxide by chemical treatment.

Methane is a clean gas with a high calorific value. It can be used on site, fed into the gas mains—natural gas is 95% methane or converted into methanol by treatment with a catalyst at high temperature and pressure. It can also be compressed for use in limited range vehicles like tractors and fork lift trucks. However, to store or transport the energy equivalent of a single gallon of petrol as compressed gas requires a large tank. At 115 atmospheric pressure, the tank's dimensions would need to be 0.45 m in diameter and 1.5 m high.

Fuel gas is produced by the thermal degradation of biomass. Its main combustible components are carbon monoxide and hydrogen, but its composition and calorific value vary according to the way it is produced and the raw material used. Fuel gas from an air—blown gasifier, for example, may have a calorific value of around 5 MJ/m^3, whereas that from an oxygen fed system may be as high as 20 MJ/m^3.

It is generally uneconomic to transport low and medium calorific value gases over long distances. They are best burnt on site to provide heat and mechanical or electrical energy or converted into more useful substances like methane and methanol. Fuel gas could also be used to [illegible] petrol [illegible] world

War II over a million cars, trucks and buses run on gas that was manufactured from wood and other solid fuels in a small gasifier-towed in a trailer behind the vehicle.

Energy Plantation is the method of tapping maximum solar energy by growing plants. Energy farms are ideal solar collectors requiring virtually no maintenance, it is economical and non-polluting. It uses an established technology and it stores energy. Photosynthesis occurring in naturally, stores more than ten times much energy annually, in plant farm than is consumed by all mankind. But very little of this energy is tapped.

In India about 23% of total land area is under forests. Area under the social forestry of quick growing trees has increased. Fuel wood accounts for about 60% of all energy consumed in the country, Social forestry programme comprises the schemes (*a*) mixed plantation on waste lands, and (*b*) reforestration of degraded forests.

Jojaba an ever green shrub around 1.7 m in height grows wild in the semi-arid region of Mexico and U.S.A. Its seeds contain about 50 to 80% of oil and its plantation in U.S.A. and Israel has been successfully done providing 1,420 plants per hectare producing 1970 kg of dry seed per annum. The tree species namely Acacia, Tortilla, Albizzia, Lebbak, Prasois, Juliflora and likewise have been identified adaptable to the hot-arid regions in our country. The plants namely Erythrina and Leocaena which are known to be fast growing plants are proposed for the subtropical regions. Ethyl alcohol, the most promising compound, for mixing with gasoline, can be easily prepared from the starch and carbohydrates available from plants on other sources of bio-mass. Sycamore is a promising tree that yield upto 16 tonnes/acre per year. All of it is used except the foliage, which contains the nutrients and is returned to the soil. A harvested sycamore produces a number of sprouts that are themselves ready for harvesting in 2 to 3 years.

The utilization of bio gas as a source of energy goes hack to the beginning of this century when sewage sludge was anaerobically digested and the resulting gas collected and utilized in the sewage treatment plant itself fur heating the sludge dtiring digestion. The idea of production of bio gas from domestic and farm yard wastes and its utilization in rural areas as a source of energy originated in India in the late thirties with the khadi movement which was concerned with the scale of tree felling in rural areas for the fire wood. It was also argued that burning of the cow dung cakes is wasteful of a valuable resource which could be better utilized for fertilizing the fields. Anaerobic digestion of the wastes results not only in valuable bio gas production but also in a slurry whose fertilizer value is almost intact or even better since the fertilizing components in the digested slurry are directly utilizable by plants.

Upto 1990-91 over 14 lakh family size biogas plants have been set up in the country by DNES only. Assuming 85 % functionality of these plants for only 80% of the year, it is estimated that there is now an annual production of 1100 million cubic meter gas equivalent to 38.18 lakh tonnes a fuelwood per year, valued at ₹152.7 crores per annum. In addition, these plants are producing 183.6 lakh tonnes of enriched manure per year valued at ₹ 152 crores. Thus the benefits to society from the biogas plants are already in excess of ₹ 300 crores per year.

National programme on improved chulhas (NPIC) continues to ever-achieve the targets since inception. The cumulative numbers of improved chulhas installed all over the country are around 100 lakhs at the end of 1990-91, which are expected to result in saving of 42 lakh tonnes of wood equivalent per year valued at ₹ 168 crores per annum apart from numerous benefits for the environment of rural houses and villages and for women's welfare, due to the reduction in black smoke.

Under the Biomass programme energy plantation projects have been taken up with a view to fulfill the needs of fuel, fodder, and power generation together with good potential for rural employment. Very encouraging results have been obtained in production of fast growing species of biomass in arid areas. The department of non-conventional energy sources (DNES) has taken up projects worth 5 MW aggregate capacity split into mechanical and electrical application systems through gasifiers/stirling engines working on biomass at various locations in the country under a demonstration programme. Indigeneous capacity has been created for manufacturing gasifiers. and Stirling engines for generation of energy from various types of biomass. Upto the year 1988-89, 374 numbers of gasifiers stirling engines systems totalling about 4 MW capacity have already been installed in the country.

7.2. Biomass Conversion Technologies

A wide variety of conversion technologies is available for manufacturing premium fuels from biomass (see Table 7.2). Some are simple and well understood like digestion and fermentation; others like gasification have been tested in large pilot plants and are now being commercialised.

Table 7.2. Biomass Conversion Technologies

	Conversion Process	*Solids*	*Principles Products Liquid*	*Gases*	*Further Treatment*	*Premium Fuels*
Wet	Anaerobic Digestion			Methane^{b} and carbon dioxide	Carbon dioxide removal	Methane
	Fermentation		Ethanol		Distillation	Ethanol
	Chemical Reduction		Mixture of oils		Fractional distillation	Hydrocarbon liquids
Thermal Process	Liquefaction	Char	Pyroligneous	Fuel gas^{b}	Steam	Methane
	Gasification	Char	acids oils and	Fuel gas$^{a.c}$	reforming and/or	Methanol or
	Steam-gasification	Char	tars	Methane^{c}	shift reaction	Higher Alcohols
	Hydrogenation		Mixture of oils		Fractional distillation	Hydrocarbon liquids
	Oil Extraction		Vegetable oil		Esterification	Diesel Substitute

Key: a = low calorific value (5-10 MJ/m^3), b = medium calorific value (10-25 MJ/m^3),

c = high calorific value (30-45 MJ/m^3).

Each biomass resource—wood, dung, vegetable waste can be treated in many different ways to provide a wide spectrum of useful products. Domestic refuse, for example, can be dried and burnt to provide heat or converted into low calorific value gas by 'pyrolysis' (heating without air). Alternatively, it can be stirred into a slurry and digested to yield methane. Like-wise, liquid and gaseous fuels such as methanol and methane can be manufactured by several different routes and from a variety of feedstocks.

The choice of the process is determined by a number of factor— the location of the resource and its physical condition, the economics of competing processes, and the availability of a suitable market for the product.

Biomass conversion, or simply bioconversion can take many forms: (1) direct combustion, such as wood waste and bagasse (sugarcane refuge), (2) thermochemical conversion, and (3) biochemical conversion.

Thermochemical conversion takes two forms: gasification and liquefaction, *Gasification* takes place by heating the biomass with limited oxygen to produce low heating value gas or by reacting it with steam and oxygen at high pressure and temperature to produce medium heating value gas. The latter may be used as fuel directly or used in liquefaction by converting it to methanol (methyl alcohol CH_3OH), or ethanol (ethyl alcohol CH_3CH_2OH) or it may be converted to high heating value gas.

Biochemical conversion takes two forms. Anaerobic digestion and fermentation. *Anaerobic* digestion involves the microbial digestion of biomass. (An anaerobe is a micro-organism that can live and grow without air or oxygen, it gets its oxygen by the decomposition of matter containing it). It has already been used on animal manure but it also possible with other biomass. The process takes place at low temperature upto 65°C, and requires a moisture content of at least 80 per cent. It generates a gas consisting mostly of CO_2 and methane (CH_4) with minimum impurities such as hydrogen sulfide. The gas can be burned directly or upgraded to synthetic natural gas by removing the CO_2 and the impurities. The residue may consists of protein rich sludge that can be used as animal feed and liquid effluents that are biologically treated by standard techniques and returned to the soil.

Fermentation is the breakdown of complex molecules in organic compound under the influence of a ferment such as yeast, bacteria, enzymes, etc. Fermentation is a well-established and widely used technology for the conversion of grains and sugar crops into ethanol. About 500 million gal ethanol per year by 1985, were produced in the limited states by the use of surplus gain. It is intended for mixing with gasoline to produce *gasohol* (90% gasoline, 10% ethanol). This process requires high cost and high energy required. One scheme considered for reducing costs of ethanol production by fermentation is in finding less expensive grains or sugars and a process that requires less energy. Glucose produced by hydrolysis of an abundant carbohydrate polymer called *ligno cellulose* is being considered for the former.

Biomass energy concepts under study are resulting in the cultivation of large forests in areas not suitable for food production. The trees are to be harvested by automated means, then chipped and pulverized for burning in a power plant that would be located in the middle of the forest.

7.2.1. Wet Processes

Anaerobic digestion. Biogas is produced by the bacterial decomposition of wet sewage sludge, animal dung or green plants in the absence of oxygen. Feedstocks like-wood shavings, straw, and refuse may be used, but digestion takes much longer. The natural decay process, 'anaerobic decomposition' can be speeded up by using a thermally insulated, airtight tank with a stirrer unit and heating system. The gas collects in the digester tank above the slurry and can be piped off continuously. At optimum temperature (35°C) complete decomposition of animal or human faces takes around 10 days. Gas yields depend critically on the nature of the waste-pig manure, for example, is better than cowdung or household refuse. Each kilogram of organic material (dry weight) can be expected to yield 450-500 litres of biogas (9-12 MJ) at atmospheric pressure in a modern batch or continuous feed unit—one and a half to two digester volumes of gas per day. The residue left after digestion is valuable fertilizer. It is also rich in protein and could be dried and used as animal feed supplement.

Fermentation. As stated, ethanol (ethyl alcohol) is produced by the fermentation of sugar solution by natural yeasts. After about 30 hours of fermentation the brew (or 'beer') contains 6–10% alcohol and this can readily be removed by distillation. Traditionally, the fibrous residues from plant crops like sugar cane bagasse have been burnt to provide the heat. Suitable feed stocks include crushed sugar cane and beet, fruit etc. Sugar cane also be manufactured from vegetable starches and cellulose, maize, wheat grain, or potatoes, for example, must be ground or pulped and then cooked with enzymes to release the starch and convert it to fermentable sugars. Cellulose materials like wood, paper waste or atraw, require harsher pretreatment typically milling and hydrolysis with hot acid. One tonne of sugar will produce upto 520 litres of alcohol; a tonne of grain, 350 litres and a tonne of wood, an estimated 260 to 540 litres. After fermentation, the residue from grains and other feed stuffs contains high protein content and is a useful cattle feed supplement.

The hydrolysis and distillation steps require a high energy input; for woody feedstocks direct combustion or pyrolysis is probably more productive at present, although steam treatment and new low energy enzymatic hydrolysis techniques are under development. The energy requirement for distillation is also likely to be cut dramatically. Alcohol can be separated from the beer by many methods which are now under intensive development. These include solvent extraction, reverse osmosis, molecular sieves and use of new dessicants for alcohol drying. It may soon be possible to *halve* the energy required for alcohol production to produce a greater net energy gain.

Chemical reduction. Chemical reduction is the least developed of the wet biomass conversion processes. It involves pressure—cooking animal wastes or plant cellulosic slurry with an alkaline catalyst in the presence of carbon monoxide at temperatures between 250°C and 400°C. Under these conditions the organic material is converted into a mixture of oils with a yield approaching 50%. If the pressure is reduced and the temperature increased, the product is a high calorific value gas.

7.2.2. Dry Processes

Pyrolysis. A wide range of energy rich fuels can be produced by roasting dry woody matter like straw and wood chips. The material is fed into a reactor vessel or retort in a pulverised or shredded form and heated in the absence of air. (Air would cause the products of pyrolysis to ignite). As the temperature rises the cellulose and lignin break down to simpler substances which are driven off leaving a char residue behind. This process has been used for centuries to produce charcoal.

The end products of the reaction depend critically on the conditions employed; at lower temperatures—around 500°C—organic liquid predominate, whilst at temperatures nearer 1000°C a combustible mixture of gases results.

Liquefaction. Liquid yields are maximized by rapid heating of the feedstock to comparatively low temperatures. The vapours are condensed from the gas stream and these separate into a two phase liquor: the aqueous phase (pyroligneous acid) contains a soup of water-soluble organic materials like acetic acid, acetone and methanol ('wood alcohol'); the non-aqueous phase consists of oils and tars. These crude products can be burnt (with some difficulty), but it is usually more profitable to up-grade them to premium fuels by conventional refining techniques.

Other pyrolysis products include fuel gas—essentially carbon-monoxide and hydrogen and carbon char. The gas is generally burnt to maintain the temperature of the reactor; the char can be manufactured into briquetts for use as solid fuel.

Pyrolysis can also be carried out in the presence of small quantities of oxygen ('gasification'), water ('steam gasification') or hydrogen ('hydrogenation').

Gasification. Pyrolysis of wet biomass produces fuel gas and very little liquid. An alternative technique for maximising gas yields is to blow small quantities of air or oxygen into the reactor vessel and to increase the temperature to over 1000°C. This causes part of the feed to burn. Fuel gas from air-blown gasifiers has a low calorific value (around 5 MJ/m^3) and may contain upto 40% inert nitrogen gas overall yields of 80-85% can be expected. Fuel gas from oxygen-fed systems has a medium calorific value (10-20 MJ/m^3). This gas can either be burnt or converted into substitute natural gas (methane) or methanol by standard catalytic processes. Methanol yields of around 50% can be achieved from biomass.

Steam-gasification. Methane is produced directly from woody matter by treatment at high temperatures and pressures with hydrogen gas. The hydrogen can be added or, more commonly, generated in the reactor vessel from carbon monoxide and steam. Recent analyses suggest that steam gasification is the most efficient route to methanol. Net energy yields of 55% can be achieved although higher yields are likely in the future as the technology is developed.

Hydrogenation. Under less severe conditions of temperature and pressure (300-400°C and 100 atmospheres), carbon monoxide and steam react with cellulose to produce heavy oils which can be separated and refined to premium fuels.

Many countries are actively developing commercial processes for biomass liquefaction and gasification.

7.3. Photosynthesis

The most important chemical reaction on the earth is the reaction of sunlight mid green plants. Radiant energy of sun is absorbed by the green pigment chlorophyll in the plant and is stored within the plant in the form of chemical bund energy. Photosynthesis in the plants is an example of biological conversion of solar energy into sugars and starches which are energy rich compounds. So if plant fast growing trees having high photosynthesis efficiency we can harvest, and burn them to produce steam in a similar manner as in thermal power stations ultimates to produce the electric power. Such an "energy plantation" would be a renewable resource and an economical means of harnessing solar energy. However, photosynthesis concepts are less attractive as the average efficiency of solar energy conversion in plants is about 1% and the overall efficiency of the conversion sunlight to electricity would be about 0.3% compared to 10% for photovoltaic cells.

The process photosynthesis is extremely complex and not yet completely understood by scientists. (In Greek photo means light and synthesis means combination). In this reaction, water and CO_2 molecules broken down and a carbohydrate is formed with the release of pure oxygen. The process can be expressed as follows:

$$CO_2 + H_2O + \text{light} + \text{Chlorophyll} \longrightarrow \underset{\text{(Sugar)}}{(H_2CO)_6} + O_2 + \text{Chlorophyll}$$

or

$$6CO_2 + 12H_2O \longrightarrow C_6H_{12}O_6 + 6H_2O + 6O_2$$

The absorbed light is in the ultraviolet and infrared range. Visible light having a wavelength below 700 A° is absorbed by the green chlorophyll which becomes activated and passes its energy on to the water molecules. A hydrogen atom is then released and reacts with the carbon dioxide molecule, to produce H_2CO and oxygen. H_2CO is the basic molecule forming *carbohydrate,* stable at low temperature, it breaks at high temperature, releasing an amount of heat equal to 112,000 cal/mole.

$$H_2CO + O_2 \longrightarrow CO_2 + H_2O + 112 \text{ kcal/mole.}$$

The absorbed energy of photons should be atleast equal to this amount. It is, therefore, possible to produce largo amount of carbohydrate by growing say, algae under optimum conditions in plastic tubes or in ponds. The algae could be harvested, dried and burned for production of heat that could be converted into electricity by conventional methods.

Thus photosynthesis consists in building up of simple carbohydrates such as sugar etc. in the green leaf in presence of sunlight. The oxygen liberated is from H_2O molecule and not from CO_2. This process is called as carbon fixation or carbon assimilation. Photosynthesis is essentially a *reduction* and *oxidation* process.

The process of photosynthesis has two main steps:

(*i*) Splitting of H_2O molecule into H_2 and O_2 under the influence of chlorophyll and sunlight. This phase of reaction is called the *light-reaction,* In this phase of reaction, light absorbed by chlorophyll causes photolysis of water. O_2 escapes and H_2 is transformed into some unknown compounds. Thus solar energy is converted into potential chemical energy.

(*ii*) In the *second phase,* hydrogen is transferred from this unknown compound to CO_2 to form starch or sugar. Formation of starch or sugar are *dark reaction* not requiring sunlight.

The *conditions* necessary for photosynthesis are:

(1) *Light.* One of the important inputs for biomass production is the intensity of solar radiation only a part of this energy (40-45%), is of the appropriate wavelength (400-700 Å) to produce photosynthesis. The plants use radiations between 400 to 700 Å only a part of this energy is actually used in photosynthesis. This range of light is called photosynthetically active radiation (PAR). The upper limit of the photosynthesis efficiency is about 5%.

(2) CO_2 *Concentration.* Carbon dioxide is the primary raw material for photosynthesis. CO_2 constitutes about 0.03% of the atmosphere. However, if CO_2 availability is increased artificially, linear increase in the yield of several crops, upto a limit, have been observed. Hence one of the methods of increasing biomass is by supplying additional CO_2 to the plants. The main sources of CO_2 are:

(*i*) animal respiration, (*ii*) combustion of fuel,

(*iii*) the major source is the decay of organic matter by bacteria, and

(*iv*) ocean also is an important store of CO_2, much of which comes from photosynthesis by plants. Respiration of marine plants and animal releases CO_2 into the water.

(3) *Temperature.* Photosynthesis is restricted to the temperature range which can be tolerated by proteins, *i.e.,* 0°C to 60°C. Although photo chemical part is not affected by temperature, but biochemical part, controlled by enzymes, is highly sensitive to temperature.

Photosynthetic Efficiency

Efficiency of a solar energy utilization by biomass route obviously is very important parameter. The energy stored in the plants by way of carbon fixation in the form of chemical bond energy when expressed as a fraction of total insolation falling on the plant, it is called as photosynthetic efficiency. This can be explained by taking a specific case as follows:

Let us consider that the energy recevied by a surface at sea level is 100%.

Solar energy received .... = 100% ...(*i*)

Photosynthetically active radiation (PAR)... = 50% of (*i*)

Losses due to transmission and reflection from

leaf surfaces .... = 10 % of (*i*)

∴ Net energy available .... = 40% of (*i*)

23% of the net available energy is converted

into carbon fixation .... = 23% of net energy = 9.2

40% of thus fixed carbon energy is again lost due to simultaneous respiration process which is just the reverse process of photosynthesis.

∴ Final net trapped energy = 0.6 × 9.2 = 5.52% of insolation.

This table gives only a specific case. However in literature we find that photosynthetic efficiencies have been quoted from 0.1 to 5%. This variation is mainly because of two reasons:

(*i*) There is a wide variation in efficiency from plant to plant.

(*ii*) There are two basis which are being used for calculation of efficiencies. Some people calculate it on the basis of yearly insolation independent of the duration of maturity of a given crop. For example, even if a particular crop matures in 3 months, the chemical energy stored is presented as a fraction of the total insolation throughout 12 months. On the other hand some efficiencies stated are based upon the insolation received only in the duration of maturation of a particular crop.

7.4. Biogas Generation

Introduction. Biogas, a mixture containing 55-65% methane, 30-40% carbon dioxide and the rest being the impurities (H_2, H_2S, and some N_2), can be produced from the decomposition of animal, plant and human waste. It is a clean but slow burning gas and usually has a calorific value between 5000 to 5500 kcal/kg (20935 to 23028 kJ/kg) or 38131 kJ/m^3. It can be used directly in cooking, reducing the demand for firewook. Moreover, the material from which the biogas is produced retains its value as a fertilizer and can be returned to the soil. Biogas has been the material for its production, hitherto. It is not only the excreta of the cattle, but also the piggery waste as well as poultry dropping are very effectively used for biogas generation. A few other materials through which biogas can be generated are algae, crop residues (agro wastes), garbage kitchen wastes, paper wastes, sea wood, human waste, waste from sugarcane refinery, water hyacinth etc., apart from the above mentioned animal wastes. Any cellulosic organic material of animal or plant origin which is easily biodegradable is a potential raw material for biogas production.

Biogas is produced by digestion, pyrolysis, or hydrogasification. *Digestion* is a biological process that occurs in the absence of oxygen and in the presence of anaerobic organisms at ambient pressures and temperatures of 35-70°C. The container in which this digestion takes place is known as the *digester.*

Anaerobic digestion. Biogas technology is concerned to micro organisms. These are living creatures which are microscopic in size and are invisible to unaided eyes. These are different types of micro organisms. They are called bacteria, fungi, virus etc. Becteria again can be classified into two types–beneficial bacteria and harmful bacteria. Compost making production of biogas, vinegur, etc., are examples of beneficial bacteria. Bacteria causing cholera, typhoid, diphtheria are examples of harmful bacteria. This type of bacteria which causes disease both in animals and human beings is called *pathogen.*

Bacteria can be divided into two major groups based on their oxygen requirement. Those which grow in presence of oxygen are called *aerobic* while the others grow in absence of gaseous oxygen are called *anaerobic.* When organic matter undergoes fermentation (process of chemical change in organic matter brought about by living organisms) through anaerobic digestion, gas is generated. This gas is known as *bio-gas.* Biogas is generated through fermentation or bio-digestion of various wastes by a variety of anaerobic and facultative-organisms. Facultative bacteria are capable of growing both in presence and absence of air or oxygen.

Aerobic and anaerobic fermentation can be used to decompose organic matter. Normally aerobic fermentation produces CO_2, NH_3, and small amounts of other gases along with a decomposed mass and evolution of heat. Anaerobic fermentation produces CO_2, CH_4, H_2 and traces of other gases along with a decomposed mass. Aerobic fermentation is used when the main aim is to render the material hygenic and to recover the plant nutrients for reuse in the fields. The residue is rich in C, N_2, P, K and other nutrients. In a biogas plant the main aim is to generate methane and hence anaerobic digestion is used. Here the complex organic molecule is broken down to sugar, alcohols, pesticides and amino acids by acid producing bacteria. These products are then used to produce methane by another category of bacteria.

As already mentioned the treatment of any slurry or sludge containing a large amount of organic matter, utilizing bacteria and other micro-organisms under anaerobic conditions is commonly referred to as anaerobic digestion or simply digestion. This anaerobic digestion consists broadly of three phases:

(*i*) *Enzymatic hydrolysis.* Where the fats, starches and proteins contained in cellulosic biomass are broken down into simple compounds.

(*ii*) *Acid formation.* Where the micro organisms of facultative and anaerobic group collectively called as acid farmers, hydrolyse and ferment, are broken to simple compounds into acids and volatile solids. As a result complex organic compounds are broken down to short chemical simple organic acids. In some cases, these acids may be produced in such large quantities that the pH may be lowered to a level where all biological activity is arrested. This initial acid phase of digestion may last about two weeks and during this period a large amount of carbon dioxide is given off.

(*iii*) *Methane formation.* Where organic acids as formed above are then converted into methane (CH_4) and CO_2 by the bacteria which are strictly anaerobs. These bacteria are called methane fermentors. For efficient digestion these acid formers and methane fermentors must remains in a state of dynamic equilibrium. This equilibrium is a very critical factor which decides the efficiency of generation. It has been demonstrated that the methane formers are sensitive to pH changes. A pH value between 6.5 to 8 is the best for fermentation and normal gas production. If organic acids are formed at a faster rate than the limited population of methane formers can assimilate, then the accumulated acids will reduce the pH to levels unfavourable to methane formers.

In controlled waste digestion the environment must be maintained suitable for the continued growth of both acid forming and methane-forming bacteria. The proper environment requires the balance between the population of organisms, food supply, temperature, pH, and food accessibility. Digestion processes are being improved as the conditions which influence organic metabolism are better understood and better equipment and methods are available for controlling these conditions.

Basic processes and energetics. The general equation for anaerobic digestion is

$$C_xH_yO_z + \left(x - \frac{y}{4} - \frac{z}{2}\right) H_2O$$

$$\longrightarrow \left(\frac{x}{2} - \frac{y}{8} + \frac{z}{4}\right) CO_2 + \left(\frac{x}{2} + \frac{y}{8} - \frac{z}{4}\right) CH_4 \quad ....(7.1)$$

For cellulose this becomes

$$(C_6H_{10}O_5)_n + nH_2O \longrightarrow 3n\ CO_2 + 3n\ CH_4 \quad ...(7.2)$$

Some organic material (*e.g.*, lignin) and all inorganic inclusions do not digest. These add to the bulk of the material, form a scum and can easily clog the system. In general 95% of the mass of the material is water. The reactions are slightly exothermic, with typical heats of reaction being about 1.5 MJ per kg dry digestible material, equal to about 250 kJ per mole of $C_6H_{10}O_5$. This is not sufficient to significantly affect the temperature of the bulk material.

If the input material had been dried and burnt, the heat of combustion would have been about 16 MJ/kg only 10% of the potential heat of combustion need be lost in the digestion process. This is 90% conversion efficiency. Moreover very wet input has been processed to give a highly convenient and controllable gaseous fuel, whereas drying of 95% aqueous input would have taken a further 40 MJ per kg of solid input. In practice digestion is seldom left to go to completion because of the long time involved, and 60% conversion is common. Gas yield is about 0.2 to 0.4 m^3 per kg dry digestible input at STP, with throughput of about 5 kg dry digestible solid per m^3 of liquid.

It is generally considered that three ranges of temperature favour particular types of bacteria. Digestion at higher temperature proceeds more rapidly than at lower temperature, with gas yield rates doubling at about every 5°C increase. The temperature ranges are (1) psicrophilic, about 20°C, (2) mesophilic, about 35°C and (3) thermophilic, about 55°C. In tropical countries unheated digesters are likely to be at average ground temperature between 20 and 30°C. Consequently the digestion is psicrophilic, with retention times being at least 14 days. In colder climates the digesters have to be heated, probably by using part of the biogas output, and a temperature of about 35°C is likely to be chosen. Few digesters operate at 55°C unless the purpose is to digest material rather than produce excess biogas.

The *biochemical processes* occur in three stages, each facilitated by distinct sets of anaerobic bacteria:

1. Insoluble biodegradable materials, *e.g.*, cellulose, polysaccharides and fats, are broken down to soluble carbohydrates and fatty acids. This occurs in about a day at 25°C in an active digester.

2. Acid forming bacteria produce mainly acetic and propionic acid. This stage likewise takes about one day at 25°C.

3. Methane forming bacteria slowly, in about 14 days at 25°C, complete the digestion to ~ 70% CH_4, ~ 30% CO_2 with trace amounts of H_2 and perhaps H_2S. H_2 may play an essential role, and indeed some bacteria *(e.g., Clostridium)* are distinctive in producing H_2 as the final product.

The methane forming bacteria are sensitive to pH, and conditions should be mildly acidic (pH 6.6 to 7.0) and certainly not below pH 6.2. Nitrogen should be present at 10% by mass of dry input, and phosphorus at 2%. A golden rule for successful digester operation is to *maintain constant conditions* of temperature and suitable input material. As a result a suitable population of bacteria is able to become established to suit these conditions.

When comparison of methane percentage from different organic matter was done for example cowdung. Poultry dropping and dairy waste scum, then best result was observed in dairy waste. 75 to 79 methane percentage found in dairy waste biogas while in cowdung, biogas was only 65%.

Advantages of anaerobic digestion. There are number of advantages of anaerobic digestion.

1. *Calorific valve of gas.* One of the main benefits is the produc-tion of a biproduct the biogas which has a calorific value and can therefore, be used as an energy source to produce steam or hot water. Because in dairy industries energy source is very important for dairy use, so there is no problem of gas storage or supply, but gas can be directly useful in heat energy.

2. *New sludge production.* The conversion of organic matter to methane and carbon dioxide results in a smaller quantity of excess sludge.

3. *Stable sludge.* In the case of municipal digestion the main reason for their installation was to produce a non-putrescable and inoffensive sludge and in many cases only a proportion of the gas produced was utilised.

4. *Low running cost.*There is no airation in the anaerobic treatment naturally in this digestion, running costs are a quarter of the equivalent aerobic system.

5. *Low odour.* Since the system is enclosed the odours are contained. Compounds which are responsible for odour are broken down during digestion. The only slight odour of hydrogen sulphide normally presents in gas. However if the gas is burnt the problem will not arise.

6. *Stability.* A well adapted anaerobic sludge can be presented unfed for a considerable period of time without appreciable deterioration.

7. *Pathogen reduction.* Work has shown that passage of the effluent through the digester reduces the number of pathogens present, so reducing subsequent disposal problems.

8. *Value of sludge.* The cases where aerobic sludge is treated anaerobically the resultant sludge has a higher nitrogen content giving it increasing value as a fertilizer. It has also been reported that the sludge acts as a soil conditioner.

9. *Low nutrient requirement.* As a consequence of the low production of the bacterial solids the nutrient requirement is also low.

In addition using of biogas in industries will curtail the consumption of coal. If biogas is used instead of coal in boilers, it will lessen the air pollution.

7.5. Factors Affecting Biodigestion or Generation of Gas

The following are the factors that affect generation of biogas:

(1) pH or the hydrogen-ion concentration

(2) Temperature

(3) Total solid content of the feed material

(4) Loading rate (5) Seeding

(6) Uniform feeding (7) Diameter to depth ratio

(8) Carbon to Nitrogen ratio (9) Nutrients

(10) Mixing or stirring or agitation of the content of the digester

(11) Retentation time or rate of feeding

(12) Type of feed stocks

(13) Toxicity due end product

(14) Pressure

(15) Acid accumulation inside the digester.

1. *pH or hydrogen ion concentration.* pH of the slurry changes at various stages of the digestion. In the initial acid formation stage in the fermentation process, the pH is around 6 or less and much of CO_2 is given off. In the latter 2-3 weeks time, the pH increases as the volatile acid and N_2 compounds are digested and CH_4 is produced. To maintain a constant supply of gas, it is necessary to maintain a suitable pH range in the digester.

The digester is usually buffered if the pH is maintained between 6.5 to 7.5. In this pH range, the micro-organisms will be very active and biodigestion will be very efficient. If the pH range is between 4 and 6 it is called acidic. If it is between 9 and 10 it is called alkaline. Both these are deterimental to the methanogenic (Methane production) organisms. It should always be remembered that there should not be any sudden upset in the pH by the addition of any material which is likely to cause an imbalance in the bacterial population.

The ideal pH valves for digestion of sewage solids are reported to be in the range 7 to 7.5. But a slightly higher value of 8.2 has been reported to be optimum for digestion of raw animal or plant wastes.

2. *Temperature.* Methane bacteria work best at a temperature of between 35° to 38°C. The fall in gas production starts at 20°C and stops at a temperature of 10°C. At one experiment 2.25 cu m of gas was produced from 4.25 m^3, of cattle dung everyday when the digester temperature was 25°C. When the temperature was raised to 28.3°C, the gas production increased by 50% to 3.75 cu m/day.

There are two significant temperature zones in anaerobic diges-tion. These have been studied in some detail for digestion of sewage sludges for 90% digestion. Fig. (7.1) shows the time required for 90% digestion at various temperatures, and the two temperature zones. It has been established that two types of micro organisms, mesophilic and thermophilic are responsible for digestion at the two temperature ranges. The optimum mesophilic temperature lies at about 35°C, while the optimum thermophilic temperature is around 55°C. In temperature climates most of the sewage digestion tanks are heated to 35°C so as to reduce the time required for digestion and therefore the capacity of the tanks. The thermophilic range has not been put to use because of the problems associated with heating the tanks to such high temperatures. Heating of tanks designed mainly for collection of biogas may not be practicable, but it must be understood that temperature is a very important factor since it affects the bacterial activity directly. Any gross deviation from a normal operating temperature may result in the unsatisfactory performance of the digester.

The gas production starts falling very steeply when the temperature goes below 20°C and almost stops at 10°C. Generally it is easier to maintain the temperature of the digester at the mesophilic range rather than at the thermophilic range.

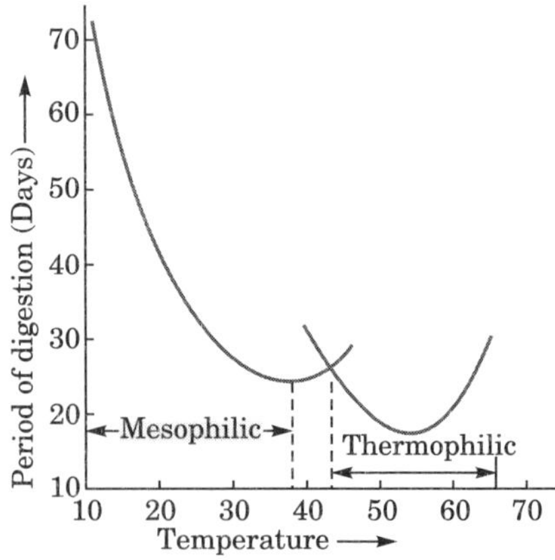

Fig. 7.1. Effect of temperature on digestion.

In addition to ambient temperature, other weather conditions also influence the gas generation *viz.*

(*a*) Wind velocity (chill factor)

(*b*) Sun shine directly available to keep the dome at the optimum temperature.

(*c*) Type of food given to cattle (in case of Gobar gas generation).

3. *Total Solid Content.* The cow dung is mixed usually in the proportion of 1 : 1 (by weight) in order to bring the total solid content to 8-10%. The raw cowdung contains 80-82% of moisture. The balance 18-20% is termed as total solids. The adjustment of total solid content helps in bidegesting the material at the faster rate, and also in deciding the mixing of the various crop residues weeds plants etc. as feed stocks in biogas digester.

4. *Loading Rate.* Loading rate is defined as the amount of raw material (usually kg of volatile solids) fed to the digester per day per unit volume. Most municipal sewage treatment plants operate at a loading rate of 0.5 to 1.6 kg of volatile soilds per m^3 per day. Much higher rates are possible, but as with most aspects of digesters, the optimum situation is a compromise. If a digester is loaded with too much raw material at a time, acids will accumulate and fermentation will stop. The main advantage of higher loading rate is that by stuffing a lot into a little space, the size and therefore the cost of the digester can be reduced. The results of a particular investigation for the loading range 1.2 to 5.3 kg volatile solids/m^3/day are given in the Table 7.3.

Table 7.3. Effect of Dung Loading on Gas Production

Expt. No.	*Dung loading kg/m^3/day*	*Vol. solids loading kg/m^3/day*	*Detention time, day*	*Gas production/unit vol. of digester m^3/day*	*Gas production/unit/vol. of matter added m^3/m^3*	*% reduction in volatile matter*
1.	8.0	1.17	50	0.022	0.20	22.8
2.	16.0	2.40	25	0.42	0.16	19.5
3.	24.0	3.76	17	0.52	0.14	14.3
4.	32.0	5.29	12	0.47	0.09	14.2

5. *Seeding.* Although the bacteria required for acid fermentation and methane fermentation are present in the cow dung, their numbers are not large. While the acid formers proliferate fast and increase in numbers, the methane formers reproduce and multiply slowly. It would be advantageous to increase the number of methane formers by artificial seeding with a digested sludge that is rich in methane formers. But beyond a certain seed concentration, the gas production will decrease, due to reduction of raw cow dung solids fed to the digester.

6. *Uniform Feeding.* One of the prerequisites of good digestion is the uniform feeding of the digester so that the micro-organisms are kept in a relatively constant organic solids concentration at all times. Therefore the digester must be fed at the same time everyday with a balanced feed of the same quality and quantity.

7. *Carbon Nitrogen ratio of the input material.* Besides carbon the quantity of N_2 present in the wastes is a crucial factor in the production of biogas. All living organisms require nitrogen to form their cell proteins from a biological view point, a digester is a culture of bacteria feeding upon and converting organic wastes. The elements of carbon (in the form of carbon hydrates) and nitrogen (as protein, ammonia nitrates, etc.) are the main food of anaerobic bacteria. Carbon is used for energy and nitrogen for building the cell structure. The bacteria use up carbon about 30 times faster than they use up nitrogen. Carbon and nitrogen should be present in the proper proportion. Other conditions (temperature, pH etc.) being favourable, a carbon, nitrogen (C/N) ratio of 30 (*i.e.,* 30 times more carbon than nitrogen) will permit digestion to proceed at an optimum rate. When there is too much carbon in the raw wastes, nitrogen will be used up first and carbon left over. This will make the digester slow down and come to a stop. In this case the bacteria will not be able to use all the carbon present and the breaking down of the organic matter will be inefficient. On the other hand if there is too much nitrogen, the carbon soon becomes exhausted and fermentation stops. The nitrogen left over will combine with hydrogen to form ammonia. This can kill or inhibit the growth of bacteria specially the methane producers. The optimum Carbon-Nitrogen ratio that best suits for maximum microbiological activity is 30 : 1.

8. *Diameter to Depth Ratio.* Research investigations reveal that gas production per unit volume of digester capacity was maximum when the diameter to depth ratio was in the range of 0.66 to 1.00. But reports from the field do not confirm this. Digesters of 16 ft depth and 4 to 5 ft diameter were reported to be working satisfactory. One reason may be that because in a simple unstirred single stage digester the temperature varies at different depths. The most actively digesting sludge is in the lower half of the digester and this is less affected by changes in night and day temperature.

9. *Nutrients.* The major nutrients required by the bacteria in the digester are, C, H_2, O_2, N_2, P and S, of these nutrients N_2 and P are always in short supply and therefore to maintain proper balance of nutrients an extra raw material rich in phosphorus (night soil) and N_2 (chopped leguminous plants) should be added along with the cow dung to obtain maximum production of gas.

10. *Mixing or stirring or agitation of the content of the digester.* Since bacteria in the digester have very limited reach to their food, it is necessary that the slurry is properly mixed and bacteria get their food supply. It is found that slight mixing improves the fermentation, however a violent slurry agitation retards the digestion.

In sewage treatment plants, the digesters, especially the first tank in two stage digestion, are equipped for mixing the tank contents either by sludge recirculation pumps, or gas recirculation or by using one or more deep draft tubes with propeller mixers at the bottom. But all these methods are costly to adopt, especially in the small plants that are constructed for digesting the individual house hold and farmyard wastes. However some amount of mixing is required if the digester is to operate efficiently. This can be achieved to some extent by designing the inlet and outlet arrangements in a proper manner so that incoming slurry tends to stir up the contents of the digester. If there is little or no mixing in the tank, the slurry in the tank will stratify into three distinct zones, scum at the top, digested sludge at the bottom (the actively digesting sludge lying on top of well digested sludge) and a thin liquid called supernatant between. Methane organisms in cultures are reported to develop in the sediment at the bottom and scarcely in the upper layers of the medium. But the scum at the top is rich in the volatile acids. Mixing brings the methane bacteria from the bottom of the digester into contact with the volatile acids so that further stabilization can proceed efficiently and methane production speeded.

11. *Retention time or rate of feeding.* The period of retention of the material for biogas generation, inside the digester is known as the retention period. This period will depend on the type of feed stocks and the temperature. Normal value of the retention period is between 30 and 45 days and in some cases 60 days.

By regulating the daily feed volume, the retention time can be controlled.

Periods for different materials to get well fermentation are:

(*i*) Cow and buffalo dung	50 days
(*ii*) Pig-dung	20 days
(*iii*) Poultry droppings	20 days
(*iv*) Night soil	30 days

As mentioned this period also depends on the ambient temperature, in tropical regions it can be even 30 days. Actually in India, 30, 40 and 50 days, retention is adopted depending on the region.

The retention period of a digester is calculated by dividing total capacity of the digester by the rate at which organic matter is fed into it methanogenic bacteria multiply at the slower rate. Methane organisms become double in 2-4 days or even in more days. Retention time should not be less than 2-4 days, otherwise bacteria may come out with the slurry affecting the whole process of biogas production. Hence in continuous type of plant the rate of feeding is an important factor. Maximum gas production is achieved by daily uniform feeding. The rate of feeding will depend upon the retention period and type of feed stocks. More gas will not be produced by excess feeding. Due to variation of retention period undigested slurry may come out. When pure cow dung is the feed material., 1/50 (one fiftieth) of the total digester mass serves as the daily feed.

12. *Type of feed stocks.* As already stated, all plant and animal wastes may be used as the feed materials for a digester. When feedstock is woody or contains more of lignin, then biodigestion becomes difficult. This cow and buffalo dung, human excreta, poultry droppings, pig dung, waste materials of plants, cobs, etc. can all be used as feed stocks. To obtain an efficient biodigestion, these feed stocks are combined in proportions. Predigestion and finely chopping will be helpful in the case of some materials. Animal wastes are predigested. Plant wastes do not need predigestion. Excessive plant material may choke the digester.

13. *Toxicity.* The digested slurry, if allowed to remain in the digester beyond a certain time, becomes toxic to the micro organisms and might cause fall in the fermentation rate.

Biological systems needs some trace elements like calcium, magnesium, potassium etc. Production of biogas is reduced when these elements or present in higher concentrations. Synthetic materials are toxic to methanogenic bacteria. Pesticides and disinfectants from farms can kill bacteria.

14. *Pressure.* Some work conducted at National Environmental, Engineering Research Institute (NEERI) Nagpur and other places indicated that the pressure on the surface of slurry also affects the fermentation. It has been reported to be better at lower pressures.

15. *Acid accumulation inside the digester.* Intermediate products such as acetic propionic butyric acids are produced, during the process of biodigestion. This causes in a decrease of the pH, especially when fresh feed material is added in large amount. These acids may be converted into methane by addition of neem cake. However the buffering nature of the digester should not be upset. Cowdung operated plants remain well buffered and the problem of acid accumulation does not-arise in the continuous fermenting systems. Acid accumulation is usually occured in batch digestion systems.

7.6. Classification of Biogas Plants

Biogas plants are mainly classified as:

(1) Continuous and batch types (as per the process).

(2) The dome and the drum types.

(3) Different variations in the drum type.

1. Continuous and batch types

(*a*) *Continuous plant.* There is a single digester in which raw material are charged regularly and the process goes on without interruption except for repair and cleaning etc. In this case the raw material is self buffered (like cow dung) or otherwise thoroughly mixed with the digesting mass where dilution prevents souring and the biogas production is maintained. The continuous process may be completed in a single stage or separated into two stages.

(*i*) *Single stage process.* The entire process of conversion of complex organic compounds into biogas in completed in a single chamber. This chamber is regularly fed with the raw materials while the spent residue keeps moving out. Serious problems are encountered with agricultural residues when fermented in a single stage continuous process. Refer Fig. 7.2.

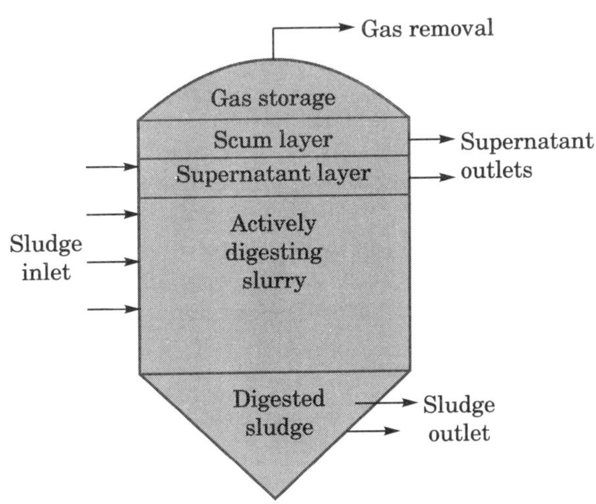

Fig. 7.2. Schematic of single process conventional digester.

(*ii*) *Double stage process.* The acidogenic stage and methanogenic stage are physically separated into two chambers. Thus the first stage of acid production is carried out in a separate chamber and only the diluted acids are fed into the second chamber where bio-methanation takes place and the biogas can be collected from the second chamber. Refer Fig. 7.3. Considering the problems encountered in fermenting fibrous plant waste materials the two stage process may offer higher potential of success. However, appropriate technology suiting to rural India is needed to be developed based on the double stage process.

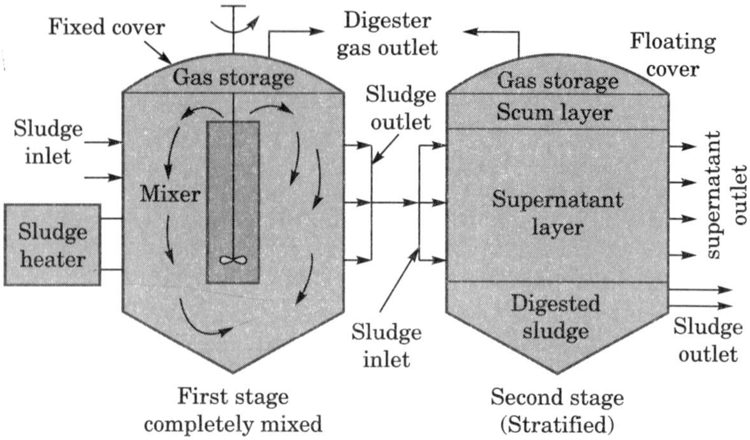

Fig. 7.3. Schematic of two-stage digestion process.

The main features of continuous plant are that:

(1) It will produce gas continuously;

(2) It requires small digestion chambers;

(3) It needs lesser period for digestion;

(4) It has less problems compared to batch type and it is easier in operation.

(*b*) *The batch Plant.* The feeding is between intervals, the plant is emptied once the process of digestion is complete. In this type, a battery of digesters are charged along with lime, urea etc. and allowed to produce gas for 40-50 days. These are charged and emptied one by one in a synchronous manner which maintains a regular supply of the gas through a common gas holder. Sometimes the freshly charged digester is aerated for a few days after which it is closed to atmosphere. The biogas supply may be utilised after 8-10 days. Obviously such a plant would be expensive to install and unless operated on large scale it would not be economical. Such systems have been generally installed in European countries. Their installation and operation being capital and labour intensive. They are totally unsuitable for Indian conditions, except when it is taken as a commercial venture.

The main features of the *batch plant* are:

(*i*) The gas production in it, is intermittent, depending upon the clearing of the digester.

(*ii*) It needs several digesters or chambers for continuous gas production, these are fed alternately.

(*iii*) Batch plants are good for long fibrous materials.

(*iv*) This plant needs addition of fermented slurry to start the digestion process. There may be a direct change to the acid phase in absence of the fermented slurry, which affects formation of methane.

(*v*) This plant is expensive and has problems comparatively, the continuous plant will have less problems and will be easy for operation.

2. The dome and the drum types. There are numerous models of a biogas plant mainly two main types are usually used:

(*i*) The floating gas holder plant and other is

(*ii*) Fixed dome digester.

The floating gas holder digester which is used in India is known as KVIC plant. The fixed dome digester is called the Chinese plant. There are different shapes in both the designs, cylindrical rectangular, spherical etc. Again, the digester may be vertical or horizontal. They can be constructed above or underneath the ground. The floating gas holder digester developed in India is of masonry construction with the gas holder made of M.S. plates. The gas holder is separated from the digester. Rusting of the gas holder as well as the cost of the gas holder are the main drawbacks of this system.

In the fixed dome digester, the gas holder and the digester are combined. The fixed dome is best suited for batch process especially when daily feeding is adopted in small quantities. The fixed dome digester is usually built below ground level and is suitable for cooler regions. Local materials can be used in this construction. The pressure inside the digester varies as the gas is collected. This is not found to cause any serious problems in small plants.

3. Different variations in the drum type. There are two main variations in the floating drum design. One with water seal and the other without water seal. Water sealing makes the plant completely anaerobic and corrosion of the gas holder drum is also reduced. The other variations are of materials used both in construction of the digester and the gas holder. Bricks and stones are the commonly used materials. Ferro cement rings are also used in the construction of digester, which are best suited for clayey soils and sandy tracks. Gas holders are also manufactured out of ferro cement, as M.S. sheets get corroded. Polyethylene is also used in the construction of gas holder. The latest design uses fibre glass reinforecd plastic.

The horizontal plants are suited for high ground water level or rocky areas. These are not recommended when retention period is 30 days. Cylindrical shape of the digester is preferred because cylinder has no corners and so that there will be no chances of cracks due to faulty construction. This shape also needs smaller surface area per unit volume, which reduces heat losses also. Moreover the scum formation may be reduced by rotating gas holder in the cylindrical digester.

7.7. Advantages and Disadvantages of Floating Drum Plant

Advantages:

(1) It has less scum troubles because solids are constantly submerged.

(2) No separate pressure equalizing device needed when fresh waste is added to the tank or digested slurry is withdrawn.

(3) In it, the danger of mixing oxygen with the gas to form an explosive mixture is minimized.

(4) Higher gas production per cu m of the digester volume is achieved.

(5) Floating drum has welded braces, which help in breaking the scum (floating matter) by rotation.

(6) No problem of gas leakage.

(7) Constant gas pressure.

Disadvantages:

(1) It has higher cost, as cost is dependent on steel and cement.

(2) Heat is lost through the metal gas holder, hence it troubles in colder regions and periods.

(3) Gas holder requires painting once or twice a year, depending on the humidity of the location.

(4) Flexible pipe joining the gas holder to the main gas pipe requires maintenance, as it is damaged by ultraviolet rays in the sun. It may be twisted also, with the rotation of the drum for mixing or scum removal.

7.8. Advantages and Disadvantages of Fixed Dome Type Plant

Advantages:

(1) It has low cost compare to floating drum type, as it uses only cement and no steel.

(2) It has no corrosion trouble.

(3) In this type heat insulation is better as construction is beneath the ground. Temperature will be constant.

(4) Cattle and human excreta and long fibrous stalks can be fed.

(5) No maintenance.

Disadvantages:

(1) This type of plant needs the services of skilled masons, who are rather scarce in rural areas.

(2) Gas production per cum of the digester volume is also less.

(3) Scum formation is a problem as no stirring arrangement.

(4) It has variable gas pressure.

7.9. Types of Biogas Plants

As stated earlier there are numerous models of a biogas plant. But they can be grouped under two broad heads—one with the floating gas holder and the other with a fixed dome digester. In floating gas holder plant, the gas holder is separate from the digester. But in the fixed dome digester, the gas holder and the digester are combined. The family size biogas plants available today in India are broadly of two types. The Khadi Village Industries Commission (KVIC) model and Janta model which are shown in Figs. 7.4 and 7.5. The KVIC plant is of steel drum type or floating gas holder design, in which the digestion takes place in a masonry well and the drum floats as the gas collects and is taken out from the top.

The Janta model or fixed dome digester (also called Chinese plant) is a drumless type similar in construction to the KVIC model except that the steel drum is replaced by a fixed dome roof of masonry construction. The floating gas holder digester developed in India is of masonry construction with gas holder made of M.S. plates. The drum in the KVIC model is the costliest component and its life is comparatively less (about 10 years). The dome roof in the Janta model requires specialised design and skilled masonry construction. A poorly constructed roof generally leads to leakage from top and junction of the roof with the digester wall, thereby causing drop in gas yield. The overall cost of both types varies from ₹ 5000 to ₹ 15,000 depending upon the capacity of the biogas plant and subsoil conditions.

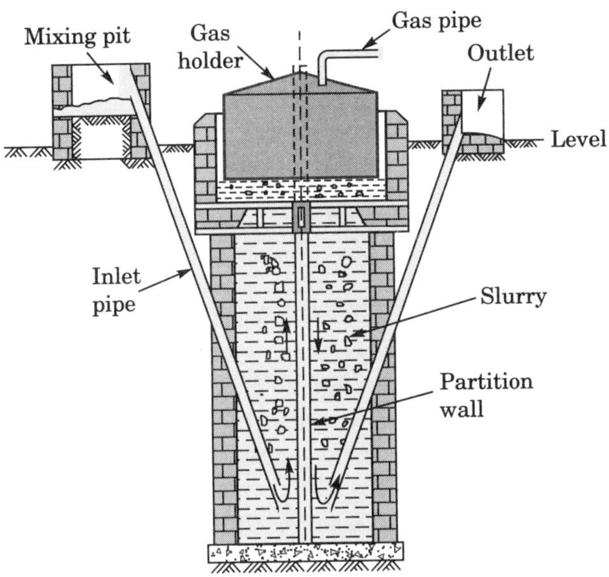

Fig. 7.4. Common circular digester with floating gas holder and no water seal (India). (KVIC digester).

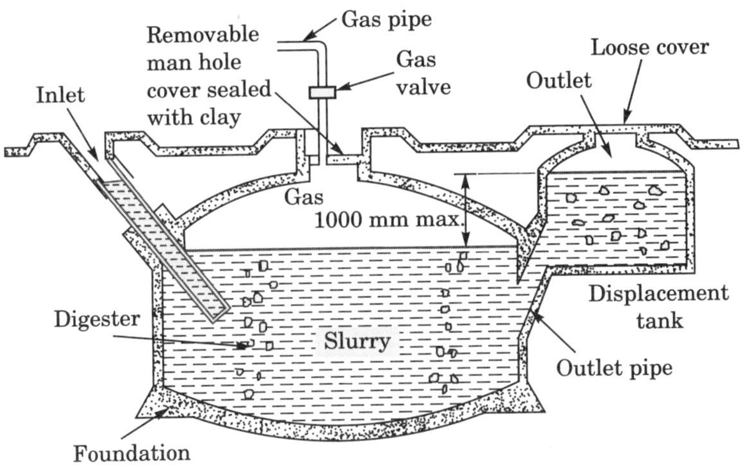

Fig. 7.5. Common circular fixed dome digester (China).

In addition to the aforesaid cost and construction material problems, there are constructional problems which the farmers or beneficiaries face. The construction of biogas plants especially in Janta type needs the services of skilled masons who are becoming rather scarce in rural areas. It is observed that plants constructed by unskilled masons or untrained workers have structurally failed or unable to retain dung slurry, gas or even both while the failure of such plants adversely affects plants owners. The prospective plant owners are seldom sure about the correct choice of the plant. Besides the construction of the plant, there are some operational and maintenance problems which almost hinder progress of biomass development.

Fig. 7.6 shows a flexible bag digester. The digester is made of plastic material and can be easily installed. The short life of the material due to the effect of ultraviolet rays is a main drawback.

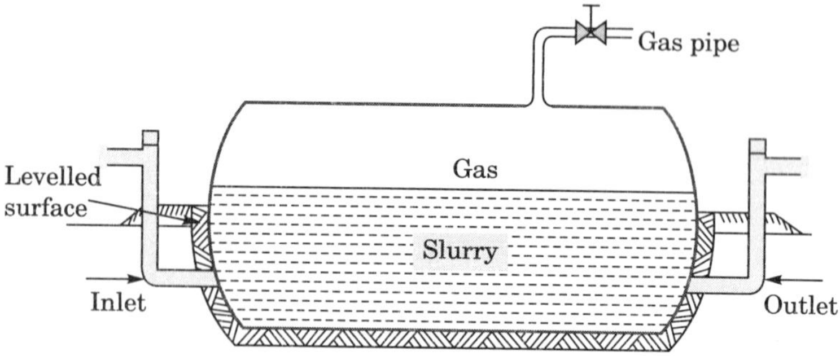

Fig. 7.6. Flexible bag type combined digester/gas holder.

Fig. 7.7 shows a digester suitable for high water-table. Here the digester diameter below the gas holder is increased so that the total depth can be reduced.

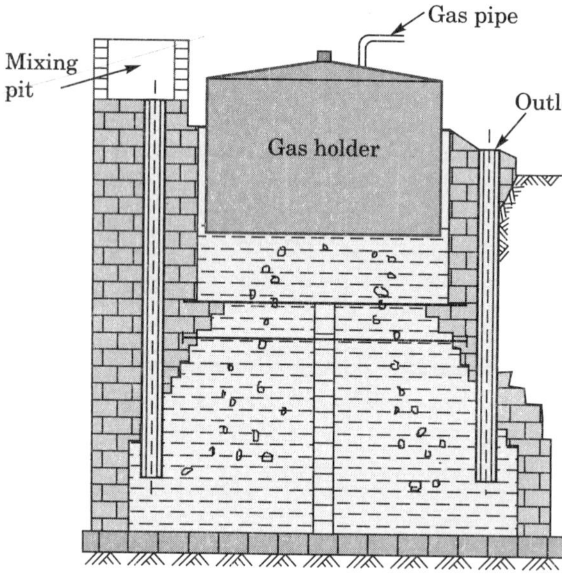

Fig. 7.7. Taper digester with floating gas holder (Nepal).

When absolute segregation of the slurry is required a floating gas holder with water seal is used. This construction is shown in Fig. 7.8.

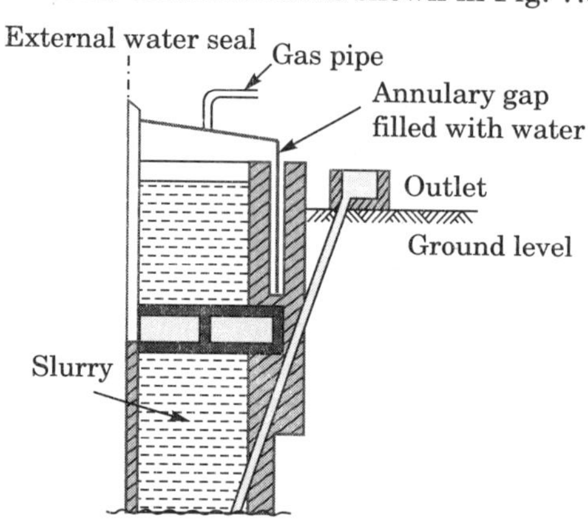

Fig. 7.8. Digester with floating gas holder and water seal (Pakistan).

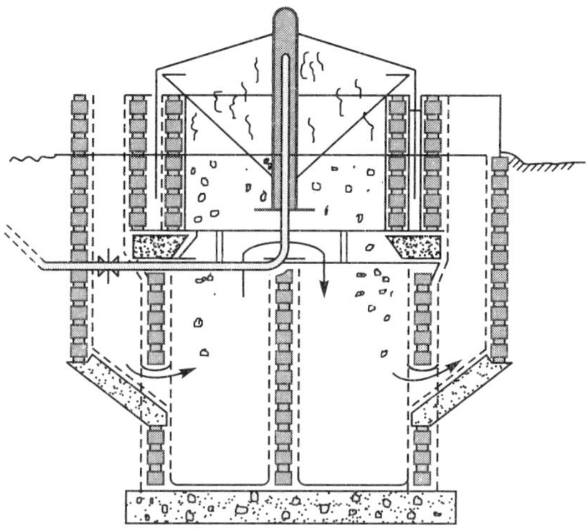

Fig. 7.9. Cross-section of biogas plant with water-seal arrangement and gas exit through centre guide frame.

In order to avoid the depth of the digester a square type of construction had also been tried. This is shown in Fig. 7.10. This type of plant did not prove successful due to clogging.

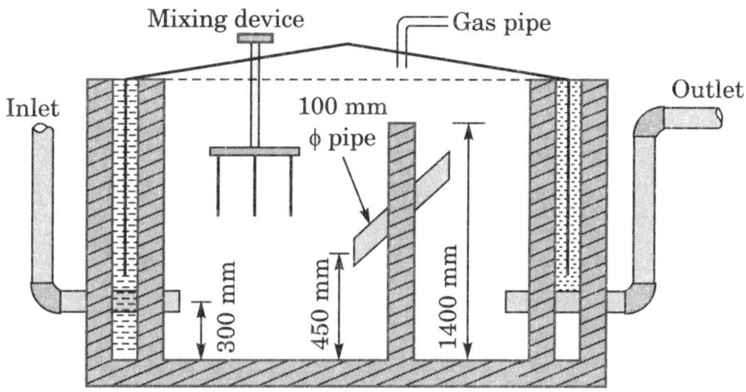

Fig. 7.10. Two-chamber rectangular digester with floating gas holder and water seal (Philippines).

A very simple type of digester using oil drums is shown in Fig. 7.11. These are best suited for research purposes.

In many situation separation of the gas from the digester has been found to be advantageous. Such arrangements are shown in Figs. 7.12 and 7.13. In this digesters gas is let to a separate chamber with the floating gas holder and water seal. Though there are number of conveniences in this arrangement it is comparatively costly.

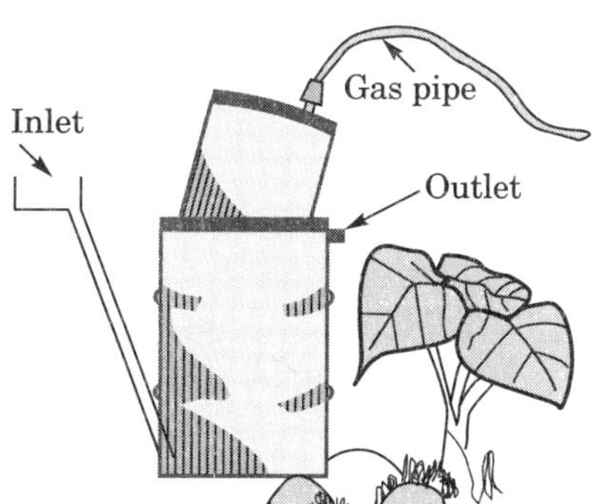

Fig. 7.11. Oil drum digester (Indonesia).

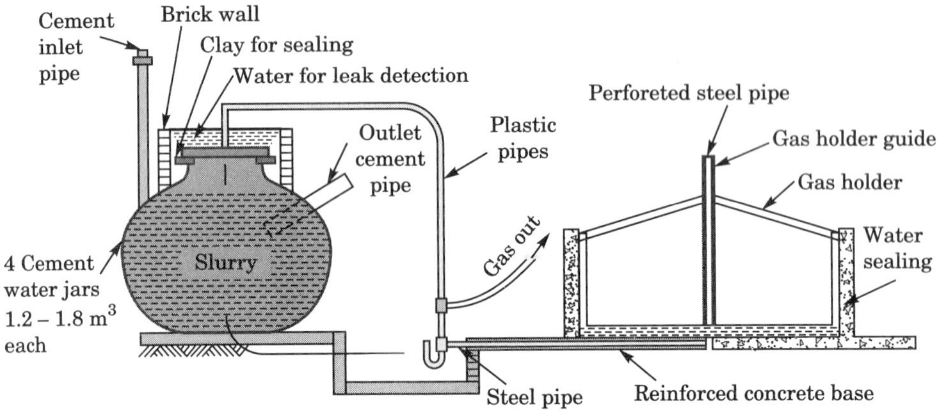

Fig. 7.12. Jar digester with separate gas holder (Thailand).

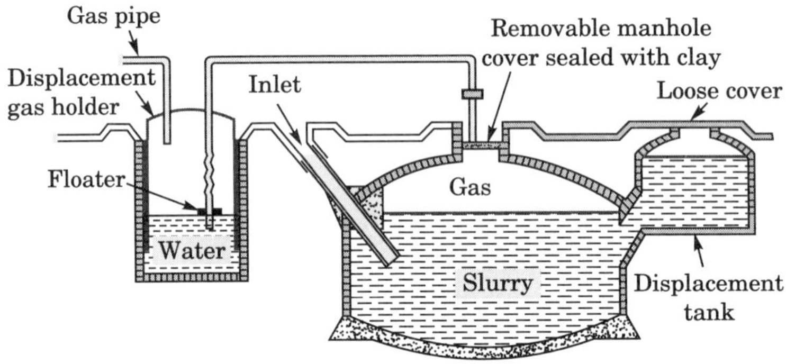

Fig. 7.13. Fixed dome digester with separate gas holder (Sichuan, China).

Gas removing system, the pipe arrangement and gas holder support system are shown in Figs. 7.14 and 7.15.

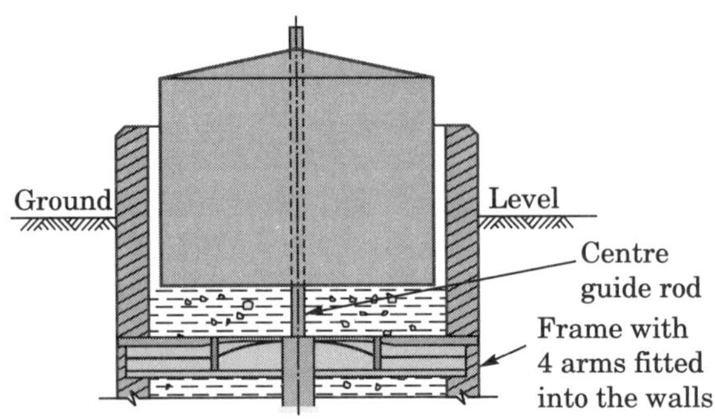

Fig. 7.14. Gas holder internal guide system.

The different design variations of family type biogas plants available at present in our country include:

1. KVIC (Khadi and Village Industries Commission) design.
2. PRAD (Planning, Research and Action Division) design (modification of Chinese design).
3. ASTRA (Application of Science and Technology to Rural areas) design.
4. Murugappa Chettiar Research Centre design.
5. Tamil Nadu Agricultural University dome type design.
6. Himachal Pradesh capsule design.
7. Kucha-Pucca Model of Punjab Agricultural University Ludhiana.
8. AFPRO (Action of Food Production) design.

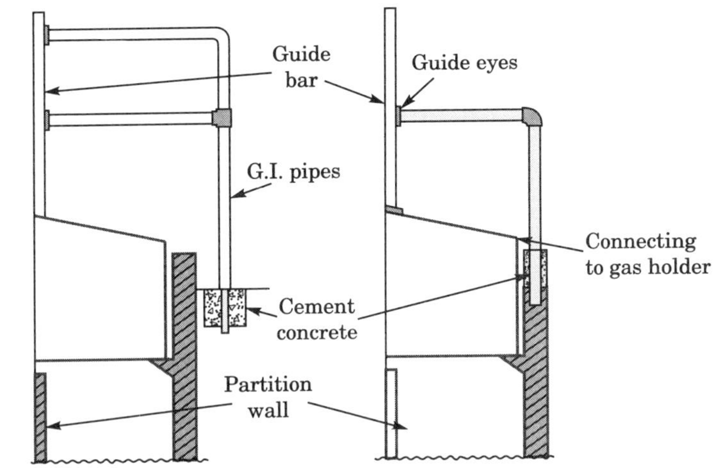

Fig. 7.15

9. Plug flow design.
10. Roorkee design.
11. Deen Bhandu design.
12. Fibre glass fixed dome design.
13. Mobile biogas plants.
14. Plastic emulsion coated, heavily insulated, temperature controlled Switzerland biogas plants.
15. Ganesh model.

The various designs listed above have their own merits and demerits. They are popular in different regions depending upon the choice of the beneficiaries.

The geometrical configurations of the various biogas plants listed above vary. As stated earlier, the digester of KVIC is a kind of a circular mini well, where as in Deen Bhandu, the concreted bottom is somewhat hemispherical. The fixed dome Chinese type is like that of a kettle design, while that developed by the Himachal Pradesh, is like a capsule module. There configuration modify the surface area available for microbial activity besides aiming at cost reduction.

Some more details of commonly used Biogas plants in India. As we have seen above that there are two types of biogas plant.

(1) Fixed Dome Biogas plant, examples are Janta Biogas plant and Deen baudhu Biogas plant.

(2) Floating gas holder plant, examples are Khadi and Village Industries type Biogas plant, Pragati Design biogas plant, Ganesh Biogas plant, Ferro cement digester biogas plant.

(1) **Janta Bioges Plant.** This was first developed by the planning, Research and Action Division Lucknow in 1978, It is an improved version of the chinese fixed dome biogas plant. The plant is shown in Fig. 7.5 , with its different parts.

The foundation of Janta biogas plant is laid at the base of the underground pit on a levelled ground which bear the load of the slurry as well as digester walls. Digester is cylindrical in shape constructed with bricks and cement. It holds the dung slurry for a retention time so that the biogas is produced from the slurry in the digester. It should be noted that the diameter and height ratio of the digester is kept 1.75 : 1. The gas is stored in gas portion, which is an integral part of plant, between dome and digester, where the usable gas is stored. The height of the gas portion is above the inlet and outlet openings to beginning of dome, and is equal to maximum volume of the gas to be stored (30-40% of plant capacity) and equal to volume of slurry to be displaced at inlet and outlet.

Dome is constructed over the gas portion, with volume of 60% of the plant capacity. It must be constructed very carefully integrating it with digester and gas portion so that no leakage of gas can take place. The gas outlet pipe is fixed at the top of dome for laying the line.

Inlet and outlet portions are constructed for putting the fresh slurry inside the plant and to take the digested slurry out. The inlet and outlet are of larger sizes, provided on each side of the digester, facing each other. The opening to the digester for feeding the waste material and effluent outlet from it are also of large sizes. The discharge of slurry out of the plant is due to pressure of the gas in the plant. Over the inlet portion, an inlet mixing tank is also constructed to mix the dung and water.

The plant have now been replaced by Deen-bandhu plant in the field.

(2) **Deen bandhu Biogas Plant.** This is also a fixed dome plant developed by Action for Food production, New Delhi, which is a low cost biogas plant. The principle of working of this plant is same as that of Janta model, except configuration of inlet entrance and digester.

The foundation of the plant is constructed in the segment of spherical shape as shown in Fig. 7.16. On the outer periphery of this foundation, the dome shaped digester is constructed with same base diameter. In this way the digester, gas portion and dome look as a single unit. The surface area of the biogas plant is reduced with same digester volume, reducing the earth work and cost of construction without sacrificing the efficiency. The higher compressive strength of the brick masonary and concrete makes it a safe structure as the plant is always under compression. A spherical structure loaded from the convex side will be under compression and therefore the internal load will not have any residual effect on the structure.

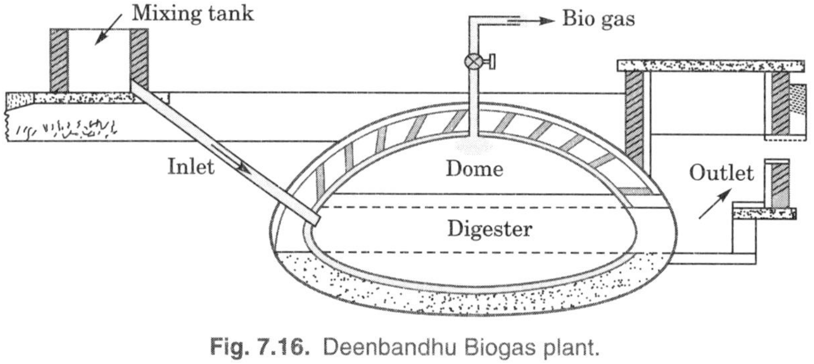

Fig. 7.16. Deenbandhu Biogas plant.

At the top of the foundation, a window opening is kept (outlet portion) for the outward movement of the digested slurry. The asbestos cement pipe of 15 cm diameter is used for inlet instead of separate opening. The pipe is embedded in the digester wall at a fixed position, just opposite to outlet opening, to avoid short circuiting of fresh material and digested slurry.

The volume of the outlet is increased to produce requisite gas pressure through the weight of the displaced slurry. At the top of the dome a gas outlet pipe is fixed as in the case of Janta plant.

Floating Gas Holder Plant. In floating gas holder plant the biogas is stored in a drum; which moves up and down on a guide, according to quantity of gas stored in it. The gas holder rests on a guide frame which is fixed in a digester walls.

The drum may be made of steel, ferro cement HDPC or fibre glass. The pressure of gas in the floating drum type biogas plant is always constant, which is equivalent to weight of gas holder over unit area. It can be increased also by putting additional weight on the gas holder.

The sides of the gas holder remains in the slurry to seal the leakage of gas to the atmosphere.

The cost of this plant is more, as the cost of the gas holder itself is high. Besides, it requires regular check and maintenance.

(1) **Khadi and Village Industries type Biogas Plant.** As stated earlier the KVIC type plant is a floating gas holder or drum type biogas plant which is in use in India earlier than all other types of biogas plants. The design was developed and perfected in India in the year 1954. This was taken up for propagation in the villages in the year 1962, by Khadi and Village Industries Commission, Mumbai, therefore, it is known as KVIC design. The design is available in sizes of 1 cum to 140 cum gas per day. In KVIC plant the gas is stored in mild steel drum of storage capacity of 30-40% of plant size at a pressure of about 10 cm of water column, which is sufficient to carry it upto a length of 20 meters to 100 meters, depending on the size of the plant.

(2) **Pragati Design Biogas Plant.** The design has been developed by United Socio-Economic Development and Research Programme (UNDARP) Pune, in order to have a cheaper floating drum biogas plant. In this design the depth of pit is less than KVIC plant so that it can be constructed in hilly and high water table areas. The cost of Pragati plant is 20%, less than KVIC plant.

The design shown in Fig. 7.17, indicates its different parts. The foundation of this plant is of conical shape, with difference of one feet between outer periphery and it's centre so as to reduce the earth and digester wall work. It is constructed at the base of the pit with cement, sand and concrete, keeping the site conditions is view so it can bear the load due to weight of slurry in the digester.

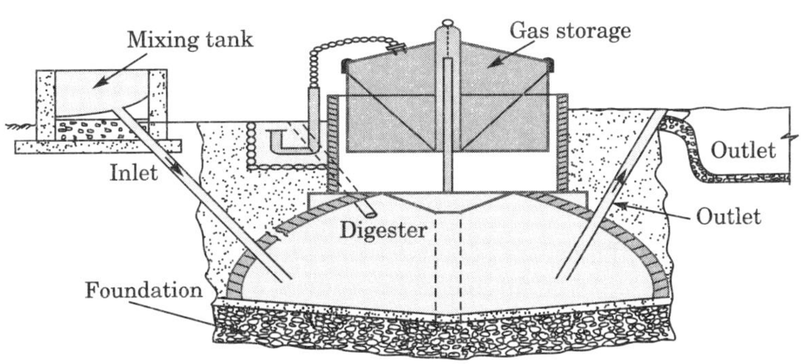

Fig. 7.17. Pragati Biogas Plant.

The digester of Pragati design plant start from the foundation in dome shape thereby reducing the constructional area, for same digester volume, thus reducing the cost of construction of the plant. The wall thickness of digester is kept 75 mm only. Dome shape construction takes place upto a collar base, where a central guide frame is provided. The digester wall above guide-frame is constructed in cylindrical shape.

Partition wall is constructed in the digester for 4 cum and bigger sizes so as to control the flow of slurry inside the digester. It divides digester into two parts separating inlet and outlet.

The inlet is through pipe, placed while constructing digester wall. It is used for feeding daily slurry into the digester and is generally of 100 mm diameter. The outlet pipe is also 100 mm in diameter, and fixed while constructing digester wall. The asbestos cement pipe can be used for inlet and outlet.

The guide frame is made of angle iron and steel pipe, is embedded in the digester wall at top of spherical portion of digester. The central guide pipe holds gas holder which is also made of M.S. sheet and angle iron. It floats up and down along pipe depending on the quantity of gas in the drum.

(3) **Ganesh Biogas Plant.** It is basically a KVIC plant constructed with bamboo and polythene sheet. It was developed in district Rampur of Uttar Pradesh. Only the digester is made of an angle iron frame, bamboo and polythene sheet.

The KVIC gas holder and guide frame are used in this design also. The cost of this plant is 70% of KVIC plant.

(4) **Ferro Cement Digester Biogas Plant.** Ferro-cement is a form of reinforced concrete made of a wire mesh, sand, water and cement. In this composite material layers of thinwise mesh distributed throughout the thickness of element are integrated with thick cement mortar over an inverted mould. In fact, it is reinforced mortar having high resistance to cracking and permeability.

In the ferro cement digesters the periphery is divided into 4 or 6 segments, which are casted at a central place over a prepared masonary or wooden mould. These precasted segments are then transported and erected at the plant site. This type of digester is an ideal substitute for brick digester in areas where good quality bricks are not available and also solves the problems of leakage through masonary joints. All the systems of the plant are of KVIC design and specifications.

7.10. Constructional Details of Some Main Digesters

1. **KVIC Digester of Gobar Gas Generator.** Construction of the gas plant can be understood broadly from Fig. 7.4. It mainly consists of two main parts : (1) Digester or pit, (2) The gas holder or the gas collector.

Digester. Also called as the fermentation plant, it is a sort of well of masonry work, dug and built below the ground level. The depth of this well varies from 3.5 metres to 6 metres, and diameter from 1.35 metres to 6 metres, depending upon the gas generating capacity and the quantity of raw material fed each day. The digested well is divided vertically into two semi-cylindrical compartments by means of a partition wall in the centre. The partition wall is lower than the level of the digester rim and hence it is submerged in slurry when the digester is full. Two slanting cement pipes reach the bottom of the well on either side of the partition wall. One pipe serves as the inlet and the other as outlet. An inlet chamber near the digester at surface level serves for mixing dung and water which is done mechanically or manually. The mixture of dung and water in proportion of 4 : 5 by volume, called *slurry,* flows down the inlet pipe to the bottom of the primary compartment of the digester. The digester is designed to hold 60 days raw material. This ensures enough stay time of the input material for complete digestion. The outlet chamber is again at surface level, just a few cms below the level of the inlet chamber. If both compartments of the digester are full and if more slurry is added from the inlet, then an equivalent amount of fermented slurry flows out of the outlet and discharged into the composite pit.

Gas holder. It is a drum constructed of mild steel sheets, cylindrical in shape with a conical top and radial supports at the bottom. It fits into the digester like a stopper. In sinks into the slurry due to its own weight and rests upon the ring constructed for this purpose. As the gas is generated the holder rises and floats freely on the surface of the slurry. As pipe is provided at the top of the holder for flow of gas for usage. To prevent the holder from tilting a central guide pipe is fitted to the frame and is fixed at the bottom in the masonry work. The pressure under which the gas is generated in this arrangement, varies between 7-9 cms of water column. The holder also acts as a seal for the gas. The cost of the holder constitutes almost 40% of the total cost of the digester.

Since the gas must be dried before it is sent to the oven for use, it is better to send the gas through a vessel filled with soda lime. This will help in better utilization of available calorific value. Without much pressure drop and difficulty, the gas can be sent to 15 to 30 m distances from the source point. To ensure leak proof joints the ends of the metal part are lined in such a way as to prevent shocks. Generally the pit is deep and narrow, but at places where the water level is low, the design has been modified and the volume has been taken horizontally. The floating drum is metallic and it is given an anticorrosive paint coating. If not properly maintained, the drum corrodes soon and the life of the plant very much gets reduced. Perhaps the cost and the maintenance factors of this type of digesters are prohibitive factors for being not very much liked by users. However, the construction is quite simple and the gas comes out at the constant pressure.

Maintenance of the plant. The only maintenance required is the painting of the gas collector at regular intervals. Many other materials like Ferrocement, fibre glass, reinforced polymer gas holders are being tried instead of M.S., but all of them are quite expensive. To avoid corrosion of the gas holder, a jacketted top of the pit is also suggested wherein the jacket some type of non-corrosive oil is filled once for all.

2. **Chinese Digester (Janta Biogas Plants).** Chinese have mainly gone into biogas technology for the sake of fertilizer with biogas as by product. When about 500,000 biogas plants exist in India, China has gone very much ahead with more than about 20 million plants (1988 data). The Chinese design is quite different from that of KVIC. The Chinese design contains a fixed dome for the collection of the gas and hence the gas availability is at variable pressures. This type of design is shown schematically in Fig. 7.5. The cost of the design is very much low and the construction is easier. As an illustration, It can be said that 2 cum plant of Chinese type can be constructed at a cost of about ₹ 3,000, where the KVIC may cost around ₹ 6,000 for the same capacity. The difference in cost may vary for different capacities but yet the chinese type is cheaper. The fixed dome is made of masonry and this replaces the floating drum of KVIC digester. Generally, the pressure of the gas that exists in KVIC plants is about 15 cm of water while a pressure of even 70 cm. of water in Chinese type of design is not uncommon.

Planning, Research and Action Division (PRAD) Lucknow has designed Janta plants modifying the design of chinese plants. It is expected that this type may be more popular due to reduction in cost.

Agricultural Tools Research Centre (ATRC) Bardoli (Gujarat) have been working on designs and developments of improved biogas manure plants based on fixed dome Chinese model (Janta design).

As we know KVIC design of biogas manure plant with the floating iron gas holder was the only popular design for more than 3 decades since the beginning of the biogas programme in India. Its limitations of high initial cost and maintenance of the expensive gas holder are well known. In the last decade, local modifications of the fixed dome chains model called Janta design is introduced in India. In few states the fixed dome biogas plant is becoming increasingly popular and fast substituting the KVIC Plants. Of course this design also has some limitations. Due to various reasons the number of plants installed in the country are quite less compared to its potential.

As we know the biogas is stored under a fixed dome of brick masonry or RCC, requirement of skilled and trained masons is a bottleneck in the whole implementation programme. For the lack of necessary skill and precision, leakage of biogas is observed from the gas portion of many fixed dome plants installed so far. RCC or brick masonry construction is not completely gas leak proof. Therefore, at least three layers of extra careful plasters are a must to prevent any leakage. The cement plaster work is very laborious and also adds to the cost of construction. Moreover quality and correct proportion of the raw materials during construction have to be maintained properly to achieve biogas impermeability. In areas where soil swell and shrink considerably there are chances of developing cracks in the brick masonry construction.

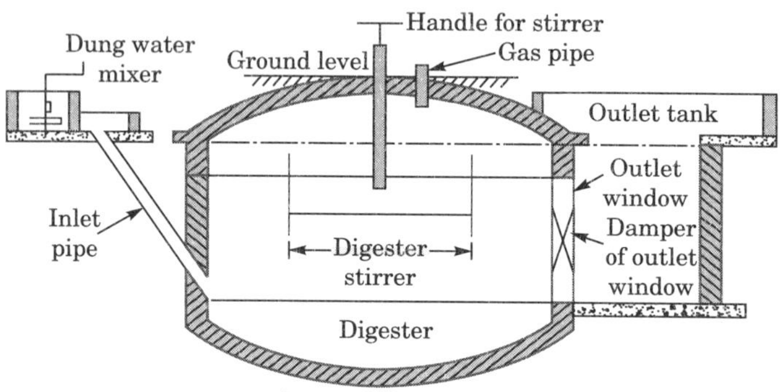

Fig. 7.18. Precast R.C.C. Biogas Manure Plant.

The above limitations are considered in the development of the fixed dome biogas plant at ATRC Bardoli (Gujarat) and suggested to cast precast RCC digester into three main elements as shown in Fig. 7.18.

The purposes of precasting the digester are:

(*i*) to achieve better technical results in controlled condition;

(*ii*) to relieve the farmers facing lot of troubles in procuring raw materials of the plant and getting skilled mason and laboures; and

(*iii*) to facilitate use of vibrator for compact and impermeable construction.

7.11. Bio gas from Plant Wastes

The subject of biogas production from fresh-plant wastes is not at all new. Biogas production was a common feature even 40 years ago on many European Farms. The biogas was used in kitchen oven, chicken hatching, washing machine, automobiles and other IC engines, room heating devices, refrigerators etc. These plants were essentially batch fermentation plants which had heating systems.

The process of biodigestion as already described is carried out generally in following two recognized systems:

(1) Batch fermentation and (2) Continuous fermentation.

In batch fermentation, the feeding is between intervals. The plant is emptied once the process of digestion is complete.

In continuous fermentation the feeding is done everyday and digested slurry equivalent to the amount of feed overflows from the plant.

The continuous process may be completed in a single stage or separated into two stages.

(*a*) *Single stage process.* The entire process of conversion of complex organic compounds into biogas is completed in a single chamber. This chamber is regularly fed with the raw materials while the spent residue keeps moving out. Serious problems are encountered in a single stage continuous process. This subject is discussed separately.

(*b*) *Double stage process.* The acidogenic stage and methanogenic stage are physically separated into two chambers. This the first stage of acid-production is carried out in a separate chamber and only the diluted acids are fed into the second chamber where bio-methanation takes place and the biogas can be collected from the second chamber. Considering the problems encountered in fermenting fibrous plant waste materials the two stage process may offer higher potential of success. However appropriate technology suiting to rural India is needed to be developed based on the double stage process.

Wet and Dry Fermentation

1. *Wet fermentation.* In this case the digester is largely filled with water so that the dry matter generally remains less than 10%. The material similar to cow dung ferment very well in this process. However, fresh plant materials being light, float on water forming a scum. This scum must be broken and the materials are submerged every few hours to maintain continuity of the process. This is the major problem encountered while fermenting agricultural waste by this process.

2. *Dry fermentation.* In order to prevent floatation of the plant materials on water, the amount of water in the digester is kept to its minimum which is just sufficient to keep the raw materials wet for its active fermentation. The total solids may be 25-30% with no free water. This process is called dry fermentation. The problem of floatation and scum formation may not arise, but the accumulation of acids and entrapment of the gas in plant materials is likely to occur. The plug flow or movement of the plant material in the digester may also not take place. Thus the problem of pH regulation, proper uniform culture, development and movement of the material pose serious problems in this process. Some of these problems may be less severe when dry fermentation is carried out in the batch fermentation process.

Problems in Straw Fermentation

1. *Scum Formation.* When cow dung is mixed with equal amount of water it forms a smooth slurry which is self-buffered flows smoothly from inlet to outlet and ferments well in any simple digester. The straw material floats on water. Water is essential for fermentation but it also helps in scum formation. Even after the submergence of the straw material in the biogas digester the rising gas bubbles increase the buoyancy of the straw particles, thereby it further helps in floatation. This scum become more compacted as the time progresses and may become

sufficiently strong. When a family size plant is fully loaded with straw material, the freshly submerged scum may reappear within a few hours during summer months. This phenomenon poses the greatest problem in successful straw fermentation in continuous fermentation system. Thus suitable manual stirring device which one man (or a house wife) can operate is needed to be developed.

2. *Movement in digester.* Automatic movement of the charged material inside the digester from its inlet to outlet due to the density gradient would be essential. While cow dung slurry moves smoothly allowing the gas bubbles to escape, the straw materials remain floating and may trap the gas. However it may move away from the feeding point as the slurry fed is daily pushed in. Since stirring (to break the scum) is essential, there is a chance that unfermented material may pass through the outlet, if the outlet is near the digester top. However, the density of the straw material approaches unity as it undergoes fermentation. It becomes less and less prone to floatation and tends to remain suspended in water. However, it does not settle down at the bottom.

Pilot Plants Using Plant Wastes. There are many wastes such as paddy straw, or wheat straw, or water hyacinth etc. which can be utilized to generate biogas. A domestic biogas plant of 0.4 m^3 capacity was developed and fabricated at Jyoti Solar Energy Institute Vallabh Vidyanagar (Gujarat), which could be placed inside the kitchen and save 50% LPG requirements of a family. This plant uses *water hyacinth* (Echhonia Gassipes), a water weed, available in many parts of India. It is a very fast growing aquatic weed. Its annual productivity is about 1050 tonnes per hectare of water surface. This waste can be fed into biogas generator to generate biogas.

Physico-chemical characteristics of water hyacinth

Physical characteristics	*Percentage*
(*i*) Moisture	92.87
(*ii*) Total solids	7.13
(*a*) Volatile solids	5.82
(*b*) Residue	1.31
Chemical characteristics	
(*i*) Carbon	32.51
(*ii*) Hydrogen	4.22
(*iii*) Nitrogen	1.78
(*iv*) Cellulose	25.00
(*v*) Lignin	10.99
(*vi*) Carbon to nitrogen ratio	18.26
(*uii*) Specific gravity	0.25

Moisture and total solids are on wet weight basis and all other analysis are on dry wt. basis. From the above, we see that water hyacinth has a very high content of moisture and 83% of its tolal solids are volatile. Its carbon to nitrogen ratio is 18.26 and cellulose content is 25%. Which shows that it has a good potentiality for biogas production.

The results of experiments on water hyacinth for bio gas production are:

(*i*) Biogas production per kg of
Wet water hyacinth 53.50 litres

(*ii*) Biogas production per kg of dried water hyacinth — 750.61 litres

(*iii*) Biogas production per kg of volatile solids of water hyacinth — 919.24 litres

(*iv*) Methane contents of the biogas generated from water hyacinth — 78%

From the above result it is seen that 1 kilogramme of wet water hyacinth produced 53.5 litres of biogas with 78% of methane content. Retention period 30 days observed.

Behaviour of water hyacinth under biodegradation is different, from that of cattle-dung. Cattle-dung has a specific gravity almost equal to water and remains wherever it has been fed into the digester white water hyacinth floats over water surface when fresh and as digestion proceeds partially and fully decomposed material settles down at the bottom. So, traditional biogas plants based on cattle-dung as feed material could not be used for water hyacinth (also for other plant wastes). It is also observed that deliberate efforts are required to bring an intimate contact of microbes with fresh and floating material for decomposition. The final decomposed material obtained is in powder form.

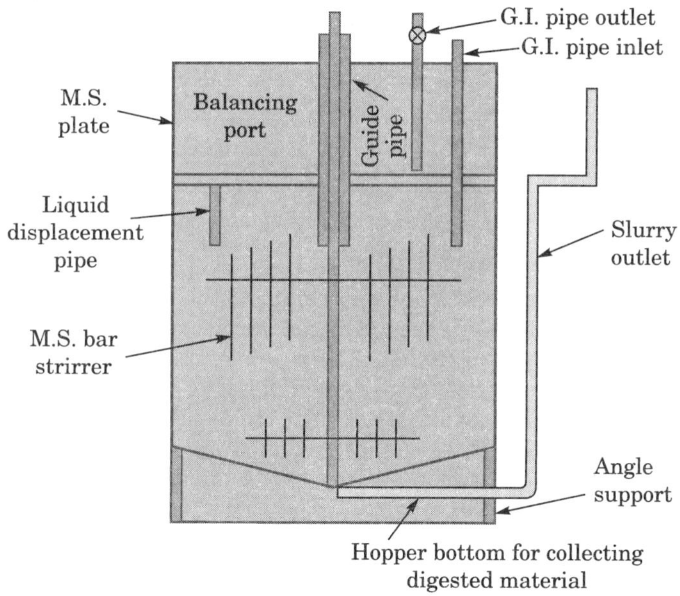

Fig. 7.19. A domestic biogas plant for water hyacinth.

The main design modifications done in traditional biogas plant working on cattle dung are:

(*i*) The inlet is provided near the top of the digester with proper sealing arrangement.

(*ii*) The slurry outlet is provided from the bottom of the digester. The bottom should be hopper to facilitate the discharge of digested slurry.

(*iii*) There is a stirring (mixing) arrangement to bring the intimate contact of microbes with substrate.

In this plant, the 550 gms chopped dried water hyacinth is fed daily with 20 litres of water. 400 litres of biogas (with 78% methane content) is generated. Chopped wet water hyacinth initially mixed with digested slurry from a continuously operated gobar gas plant.

It is concluded from this pilot plant that domestic biogas plant based on dried water hyacinth would be very useful for substituting the conventional fuels for cooking.

Kachra Gas Plant. A family size biogas plant (2-3 m^3 biogas per day capacity) based on continuous fermentation process was designed and tested at Gujarat Agricultural University Anand. This plant was named as "Kachra gas plant".

This is briefly described as follows.

This plant as shown is Fig. 7.20 all the modified arrangements as described earlier.

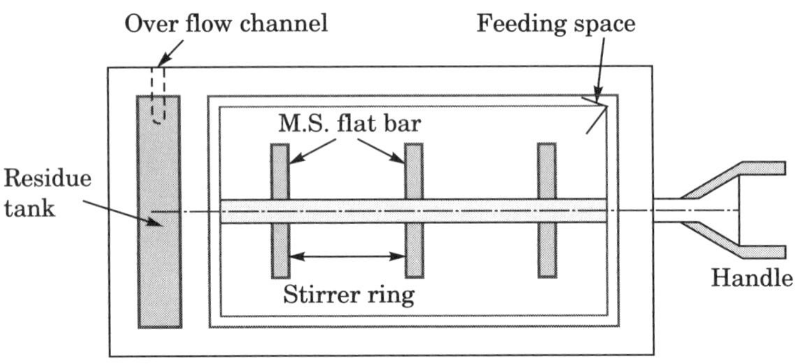

Fig. 7.20. Plan of a kachra gas plant.

Feeding. The feeding material must be chopped to few cm sized pieces when it consists of fresh water plant materials. Cow dung can be charged as such after breaking its clods by hand. The feeding material can be charged in any position. The plant was generally operated on singly type of materials for example, paddy straw or wheat straw or water hyacinth alone. The wheat or paddy straw was pushed air dry with a stick at the rate of 10 kg chopped material every day through the corner of the digester. Thus the problem of slurry making in the beginning is not involved.

Stirring. Stirring is the most important operation, since the material floats in a thick layer (30-40 cm). The stirring should be so designed that it should be able to submerge the floating material. In the present plant, the horizontal stirrer is provided which is mounted on a 4 cm dia. water pipe shaft.

Operation of Plant. The plant is initially filled with water in which few buckets of cow dung or dirty drain water or well rotten compost are added. The fresh fibrous plant material is processed through the chaff cutter and about 10 kg chopped material is spread on the ground. This is frequently sprayed with water to keep it moist. Thus the material is charged into the digester after 10 days of decomposition. About 1 kg urea may also be added to the digester. The regularly feeding and stirring schedule may then be followed regularly. The evolved gas should be let out into air for about a week. During this period the gas should never be tested for burning due to possible danger of its explosion and accident.

7.12. Community Biogas Plants

Besides the individual family size biogas plants, promotion of community and institutional biogas plants might also help in solving the energy problems.

The investment which is required for individual biogas plant, is beyond the reach of a very large section of farmers in poor countries. Though a capital cost upto 75% is provided by Deptt. of Non-conventional Energy Sources (DNES) in India, still the larger number of poor farmers are not in position to have individual biogas unit. Moreover this technology, specially in India, can benefit only those farmers who have more than 3-4 heads of cattle. Such farmers form only a small percentage in most of the developing countries. Another point in favour of community plants is that small individuals plants can not meet the cost of maintenance staff without which the success of the programme can not be achieved. The installation of bigger size plants for a group of families or a community is recommended to avoid all the difficulties mentioned above. These plants are more economical than individual plants.

A community plant system will nearly always have more gas than required for domestic use. The extra gas can be used to generate electricity, which can be used to light the village homes and streets in the first place and for pumping water, running chaff cutters, flour mills and other industries. This will reduce the fossil fuel requirement. It has been demonstrated that five lamps can be lighted with electricity generated from gas, used by one gas lamp. Anaerobic digesters maintained at 35°C generate 33.3% more gas in one third the digestion period, compared to those operated at ambient temperatures. Engines are being run on biogas to generate power and also to heat the digesters. A total energy module (by Fiat) consumes 10 cu m biogas (50,000 k cal) delivers 15 kWh power (13,000 k cal) and hot water at 60° to 80°C, (33,000 k cal). The total efficiency is 92%. These units have made it possible to operate biogas plants in cold countries successfully, where it was not economic to use them so far.

Several problems will have to be solved before the community plants can successfully operate. Quick method to detect quantity of earth mixed with cattle dung as well as total solids has to be worked out. Collection and transport of all wastes including cattle dung has to be managed. Pre-processing of agricultural waste and cattle dung, to remove earth silt gravel etc., has to be worked out. Cheap heating of the digester with solar heat and waste engine heat have to be standardized. Waste wood is available in all villages, its gasification will have to be done, so that a common mode of using all energy sources is established. Gas piping is costly and not easy to maintain. Correct gas pressure during cooking hours, mean that the gas pipes will have to be big enough to maintain peak hour flow without pressure drop. The measurement of gas is essential and gas metres are costly. Investment of gas pipes and metre may offset the economic advantage of bigger community plant.

Cost of materials of construction and of the land is reduced inversely by square of the capacity of the plant. Bigger sized plants are cheaper. These plants can be developed and can be made much more efficient and managed professionally. Organic wastes, like agricultural byproduce, sweepings, waste paper, leaves and night-soil from community latrins attached to the community plant can all be digested to obtain much more fuel gas and organic fertilizer. A system of community plants in a village can make the village absolutely, clean, get rid from smoke, flies and parastic worms. The families will now collect field refuse, waste wood etc, for use as domestic fuel, can sell it to the community plant or exchange it for gas.

The community biogas plant, requires through planning and perfect management since the failures, if any, will ultimately cause a setback to implementation of the entire programme. It would be advisable to have atleast two small size plants is place of one big size plant. Similarly, the plant using human excrete may be separate to enable the use of the gas for lighting rather cooking. This will make for easier acceptance of the community plants by the beneficiaries. Similarly, for organisational efficiency it would be better to limit community plants to some 50-60 households rather than an entire village community. This, of course, will not apply in case of big institutional plants meant for establishments like hostels, educational institutions, hospitals, big dairies, stables etc.

7.13. Materials Used for Bio gas Generation

Feed stock materials. The following organic matter rich feed stocks are found feasible for their use as input materials for biogas production:

Animal wastes:	Cattle dung, urine, goat and poultry droppings, slaughter house wastes, fishwastes, foetus wastes, leather and wood wastes, sericulture wastes elephant dung, piggery wastes etc.
Human wastes:	Faeces, urine and other wastes emanating from human occupations.
Agricultural wastes:	Aquatic and terrestrial weeds crop residue, stubbles of crops, sugarcane trash, spoiled fodder, bagasse, tobacco wastes oilcakes fruit and vegetable processing wastes, press mud, cotton and textile wastes, spent coffee and tea wastes.
Waste of aquatic origin:	Marine plants, twigs, algae, water hyacinth and water weeds.
Industrial wastes:	Sugar factory, tannery, paper etc.

The following three marine plants (waste of aquatic origin) are considered promising for biomass production.

(*i*) Water hyacinth (*ii*) Algae and (*iii*) Ocean kelp.

(*i*) *Water hyacinth.* It is a floating water plant which grows mainly in the rivers and canals of the world. Water hyacinth otherwise botanically called a "Eichhonia Gassipes" is one of the important sources for biogas production next to *animal waste.* Water hyacinth removes both nutrients and organic directly from water *via* their expensive root system, which can increase at their phenomental rate of 15% of their surface area per day, producing at least 20 tonnes net weight per hectare per day. Based on measured growth rates it is estimated that one hectare of water hyacinth could remove the nitrogen and phosphorus wastes of over 800 people per day. These concentrated nutrients can then be removed from the system by harvesting water hyacinth thereby achieving a dual goal of purifying the water and giving nutrients to the microbes for digestion for producing biogas.

Water hyacinth contains 95 per cent of water and only about 5 per cent of cellulose, lignin etc. Its growth is enhanced more in tropical weather and is ideal to grow in India. It grows extremely well in sewage ponds thereby cleaning the ponds. It sucks water so much, water would become scarce soon if not supplemented. Water hyacinth gives out 350 to 420 litres of biogas per kg of dry weight. It absorbs salts of copper, lead, silver cadmium and chromium. The biogas produced out of water hyacinth is slightly higher in methane content. Once it is planted in any water, it is extremely difficult to wipe out it traces. Besides being good food, fertilizer and biogas producer, its excellent for paper production.

(ii) Algae. Algae are single cell plants grow in profusion in lakes, tanks etc. There are several varieties of algae. Some are considered excellent food material since they have very high protein contein (upto 50%). Algae occur in the seas also but these species are not considered for biomass production since their yield rates are poor. Among fresh water algae, four varieties have high yields; chlorella, scenedesmus, ulothoria and spirutina. The best way of growing them would be in shallow inland ponds. Algae can be either burnt, directly to produce heat or anaerobically fermented to produce methane. Lower calorific value derived from methane is of the order of 3300 kcal per kg of dry algae.

(iii) Ocean kelp. Ocean kelp is a kind of sea weed which grows in the coastal areas and also in the high seas. The solar conversion efficiency is higher than that of current land crops. The attraction in this proposal is the utilization of sea areas which have no other use. Yield rates ranges from 300-500 wet tonnes/acre/year. Kelp can be used to produce biogas, burnt as fuel or fed to cattle.

In the case of all plants that flourish in water, they will have to be dried and chopped before being fed to the digester tank.

7.14. Selection of Site for a Biogas Plant

Following factors must be considered while selecting the site for a biogas plant.

(i) Distance. The distance between the plant and the site of gas consumption, should be less in order to achieve economy in pumping of gas and minimizing gas leakage. For a plant of capacity 2 m^3, the optimum distance is 10 m.

(ii) Minimum gradient. For conveying the gas a minimum gradient of 1% must be made available for the line.

(iii) Open space. The sunlight should fall on the plant as temperature between 15°C to 30°C is essential for gas generation at good rate.

(iv) Water table. The plant is normally constructed underground for ease of charging the feed and unloading slurry requires less labour. In such cases care should be taken to prevent the seepage of water and plant should not be constructed if the water table is more than 10 ft. (3 m).

(v) Seasonal run off. Proper care has to be taken to prevent the interference of run off water during the monsoon. Intercepting ditches or bunds may be constructed.

(vi) Distance from wells. The seepage of fermented slurry may pollute the well water. Hence a minimum of 15 m should be maintained from the wells.

(vii) Space requirements. Sufficient space must be available for day to day operation and maintenance. As a guideline 10 to 12 m^2 area is needed per m^3 of the gas.

(viii) Availability of water. Plenty of water must be available as the cowdung slurry with a solid concentration of 7% to 9% is used.

(ix) Source of cowdung / materials for biogas generation. The distance between the material for biogas generation and the gas plant site should be minimum to economise the transportation cost.

7.15. Digester Design Considerations

Digestion tanks may be of any convenient shape and provided with a cover to retain the gas. The cover may be a fixed one or floating. A number of factors are to be taken into account to arrive an optimum size of a biogas plant. These are:

1. the volume of waste to be digested daily,
2. the type and amount of waste available for digestion consistently,
3. period of digestion,
4. methods of stirring, the contents if any,
5. method of adding the raw waste and removing digested slurry,
6. efficiency of the collection of the raw waste,
7. the climate condition of the region,
8. the availability of other cellulosic fermentable waste in that area,
9. information about sub-soil condition and water table, and
10. type of the cover.

Generally no separate heating and stirring of the contents are provided for digesting cattle waste (Gobar Gas digester), Stirring arrangement is provided for farmyard waste and plant wastes as already discussed.

The *capacity* of the digestion tank may be formulated approximately as

$$\text{Capacity} = \frac{V_1 + V_2}{2} \cdot t$$

where, V_1 = the volume of raw waste added daily.

V_2 = volume of the waste after digestion.

and t = period of digestion, in days.

The required capacities of digestion tanks may be calculated on the basis of destruction 2/3 of the organic matter in the slurry and the conversion of 1/4 of what is destroyed into mineral matter and the remaining to gases. The period of digestion 't' is fixed by the time necessary to produce a satisfactory digested sludge. This depends mainly upon the temperature of digestion and it may be fixed from Fig. 7.1.

It is well known fact the production of biogas through anaerobic digestion of the biomass is dependent on the *temperature*. The production slows down considerably below 20°C and ceases altogether below this temperature. In the *mesophilic* range the usual range of gas production in the Indian biogas plants, methanogenic bacteria grow best at temperatures of 33-40°C. The rate of gas production approximately doubles up for every 10°C rise in temperature between 15°C and 35°C and doubles between 35°C and 55°C (thermophilic range). Hence a digester operating at 15°C has to have a volume four times greater than the one at 35°C and eight times the one at 55°C to yield the same daily production of the digester. However, calculations are simple.

For *sub-soil condition and high water table areas,* modifications in design are needed. If the clay type soils maximum shrinkage of soil is possible and hence at an interval of 61 cm height one concrete ring structure around the digester might help in strengthening the digester. The Ganesh model of biogas plant in which 4 to 5 well rings of sufficient diameter are placed one over the other and cemented together at the joints seems to work well in areas comprising of sandy soil.

In high water table areas, the Deenbandhu model whose entire structure goes upto 1.2 to 1.5 m below ground level is quite suitable. Besides this, construction of plants above ground with usual backing and soil grouting might also be considered in such areas. A leak proof two wall system with aqua proof cementing might also help in enhancing the gas output.

Shape of digesting tanks used in practice are diverse and no particular shape seems to posses advantage over all other. The digester of KVIC is a kind of a circular mini well, whereas in Deenbandhu, the concreted bottom is somewhat hemispherical. The mobile plants have a cubical pattern while some are cylindrical types. The fixed dome chinese type (Janta model) is like that of a kettle design while that developed by the Himachal Pradesh is like a KVIC biogas. In northern regions of our country, especially at high altitudes or hilly areas, the average ambient temperatures during winter months are rarely as high as 20°C and hence there are considerable heat losses from the slurry to the ambient resulting in such temperatures inside the digester where the gas production may altogether stop. The incorporation of the heating system is therefore desirable. For operation of the digester in the thermophilic range, the temperatures are required to be maintained at nearly constant value because the thermophilic bacteria are highly sensitive to fluctuation in the temperature.

In our country, however most digesters operate in the mesophilic range one should therefore try to adopt such methods which are easy and do not complicate the existing technology.

Retention time depends on the climatic conditions, and so size of the digester. For hot, tropical prone climate, 30 to 40 days retention time would be sufficient. In hot regions with a period of winter, the retention time may have to be 40 to 60 days. In temperature climate with a definite drop in temperature during the winter, the retention time may be 60 to 90 days. According to these factors the requirement of gas will have to be worked out and this has to be tallied with the availability of feed materials to decide the size of capsule module. These configurations modify the surface area available for microbial activity besides aiming at cost reduction. Two types of covers may be used for digestion tanks, which are already described earlier *i.e.*, one is fixed type and other is floating type.

Production Requirement. The gas requirement for various purposes is given in Table 7.4.

Table 7.4

Application	*Specification*	*Gas Consumption ft^3/hr*
Gas working	2″ dia burner	11.5
	4″ dia burner	16.5
	6″ dia burner	22.5
Gas lighting	1 mantle lamp	2.5 to 3
	2 mantle lamp	5
	3 mantle lamp	6
Refrigerator	18″ × 18″ × 12″	2.5
Incubator	18″ × 18″ × 18″	1.5 to 2
Boiling water	—	10/Gallon
Running engines	converted gasoline or diesel	16-18 p/hr

The figures given in above table depend on methane content and delivery pressure.

Suitability of Raw Material. If the availability of dung is less and will meet the user's fertilizers requirements but leave him short of gas, then he can substitute a portion of vegetable waste for an equal amount of dung (dry).

Table 7.5

Substance	*Nitrogen percentage*	*C/N ratio*
Urine	15 to 18	0.8
Cow manure	1.7	0.25
Poultry manure	6.3	—
Night soil	5.5 to 6.5	6 to 10
Meat scraps	5.1	—
Grass clipping	4.0	12
Sheep manure	3.75	—
Cabbage	3.6	12
Tomato	3.3	128
Mixed grasses	2.4	19
Wheat floor	1.5	—
Mustard	1.5	26
Potato taps	1.5	25
Wheat straw	1.3	128

This will enable him to get more gas for the same amount of fertilizer. In such cases it is necessary to maintain C/N ratio. This ratio should be maintained between 30 and 35 by properly varying the quantities of other biodegrable materials.

Normally the weight of dung in a dung vegetable mixture should be maintained above 50%. The amount of dry solids in cowdung is 18%. The nitrogen content in dry dung is 1.7%, and carbon content is 25 times that of nitrogen. By using the data given in the Table (7.5) one can get an optimum mixture of dung and other vegetable waste. The ultimate concentration of solids in the slurry should be between 7 and 9%.

Volume of Digester for Biogas Production using Cow dung only. At 15°C, 1 kg of dry dung gives 0.186 m^3 of gas. The ratio of dry dung to water in wet dung is 1 : 4. Normally 1 kg of wet dung is mixed with 1 kg of water to get the slurry. The duration of each cycle, depends on the temperature. Hence if the gas requirement and duration of cycle are known, then the amount of water and dung needed can be estimated. The volume of the digester is equivalent to the volume of slurry. A gas space of 10% of this volume could be provided in the digester. Hence the total volume of the digester can be taken as 1.1 times the volume of the slurry.

Digester sizing

The energy available from a biogas digester is given by:

$$E = \eta H_b V_b \qquad \text{...(7.3)}$$

where η is the combustion efficiency of burners, biolers etc. (~ 60%). (Some of the heat of combustion of the methane goes to heating the CO_2 of the biogas and is therefore unavailable for other purposes. The net effect is to decrease the efficiency).

H_b is the heat of combustion per unit volume (calorific value) of biogas (20 MJ/m^3 at 10 cm water gauge pressure, 0.01 atmosphere), and V_b is the volume of biogas.

Alternatively

$$E = \eta H_m F_m V_b \qquad ...(7.4)$$

H_m is the heat of combustion of methane (56 MJ/kg, 28 MJ/m^3 at STP) and F_m is the fraction of methane in the biogas (~ 0.7). The volume of biogas is given by

$$V_b = Cm_0 \qquad ...(7.5)$$

C is the biogas per unit dry mass of whole input (0.2 to 0.4 m^3 per kg) and m_0 is the mass of the dry input (*e.g.*, 2 kg per day per cow). The volume of fluid in the digester is given by

$$V_f = m_0/\rho_m \qquad ...(7.6)$$

where ρ_m is the density of dry material in the fluid (~ 50 kg/m^3).

The volume of the digester is given by

$$V_d = V_f t_r \qquad ...(7.7)$$

where V_f is the flow rate of the digester fluid and t_r is the retention time in the digester (~ 8 to 20 days).

Example 7.1. *The following data are given for a family biogas digester suitable for the output of five cows: the retention time is 20 days, temperature 30°C, dry matter consumed per day = 2 kg, biogas yield is 0.24 m^3 per kg. The efficiency of burner is 60%, methane proportion is 0.8. Heat of combustion of methane = 28 MJ / m^3.*

Calculate: (i) the volume of biogas digester,

and (ii) the power available from the digester.

Sol. Mass of the dry input m_0 = 2 kg/day × 5

= 10 kg/day ...(*i*)

We have the relation Eqn. (7.6), fluid volume V_f is $= \dfrac{m_0}{\rho_m}$

$$= \frac{10 \text{ kg/day}}{50 \text{ (kg/m}^3)}$$

= 0.2 m^3/day ...(*ii*)

From expression $V_d = V_f t_r$, *the digester volume is*

$V_d = 0.2 \times 20 = 4.0$ m^3 **Ans.**

Volume of biogas is $V_b = Cm_0$

= biogas yield input × mass of dry input

= 0.24 × 10 = 2.4 m^3/day

The power available from the digester is

$E = \eta H_m F_m V_b$

= 0.6 × 28 × 0.8 × 2.4

= 32.25 MJ/day = 8.80 kWh/day

= 370 W (continuous thermal) **Ans.**

Design of a community biogas plant for a village. Biogas is largely used for cooking, biogas for lighting, driving the engine, etc. has still to become popular. Hence the requirement of a family will have to be determined on the basis of cooking needs. The energy required for cooking will vary according to the eating habits, type of cooking (boiling, frying or baking), type of food, number of meals, number of persons, the type of vessels etc. However the average gas consumption for cooking purposes per day can be between 0.28 to 0.42 cum. For lighting the mantle lamp of 100 candle power, it will need 0.11 to 0.15 cum of gas per hour. Similarly for engine it will be 0.45 to 0.54 cum per horsepower per hour.

The following table gives the amount of dung required to produce 1 cu m of gas per day under average conditions calculated on the basis of 40 to 60 days retention period:

Cattle dung .... 32 kg/day
Pig manure .... 20 kg/day
Poultry manure .... 12 kg/day

In the case of human excreta, the thumb rule may be that one adult will provide enough material for the production of 0.03 cu m of gas per day *i.e.,* about 35 adult will provide about 1 cu m of gas.

Exact quantity of excreta from different animals and human beings, is difficult to calculate. The quantity will depend on the age of the animal, its feed etc., but the following average value may be taken for estimating the total feed available.

Table 7.6

Cows	10 kg per day
Oxes	12 kg per day
Buffalo	15 kg per day
Calves	5 kg per day
Horses	10 kg per day
Goat/sheep	5 kg per day
Pigs	2 kg per day
Human excreta per person	0.4 kg per day
Chicken	0.18 kg per day

A schematic diagram of the community plant is shown in Fig. 7.21.

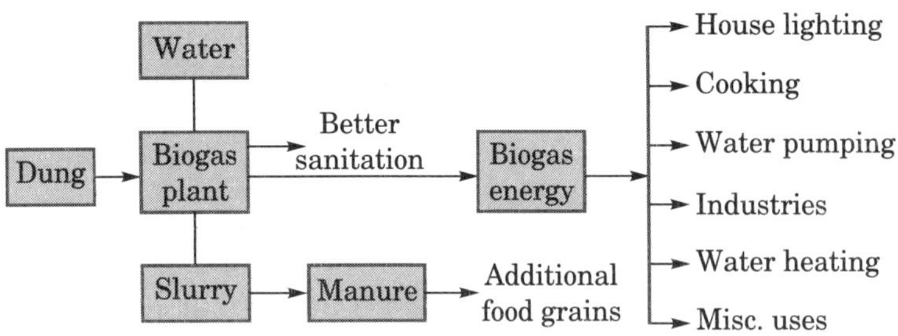

Fig. 7.21. Flow sheet of a Community Biogas plant for a village.

Design calculations. A village consisting of 98 families, each family consisting of 5 persons (adults). Two children are equivalent to one person.

Village survey report gives the following informations about cattles.

Table 7.7

Animals	*No.*
Cows	102
Oxes	124
Buffalo	52
Pig	3
Total	281

A community biogas plant is to be designed only for cooking and house lighting.

Gas required for cooking/person/day = 0.227 m^3

Gas required for lighting 100 C.P. lamp per hour = 0.126 m^3

Each family is allotted 2 lamps, which would burn 2 hours daily.

Calculation of total quantity of dung available

Cow 10 kg/day/cow

Total cow dung = 10 × 102 = 1020 kg/day

Ox 12 kg/day/ox

Total dung from oxes = 12 × 124 = 1488 kg/day

Buffalo 15 kg/day/buffalo

Total dung = 15 × 52 = 780 kg/day

Pig 2 kg/day/pig

Total 2 × 3 = 6 kg/day

Total quantity of dung available = 1020 + 1488 + 281 + 6

= 2795 kg/day = 2800 kg/day say

We can approximately assume that:

Gas produced from 1 kg of gobar in winter

(Nov. to Feb.) = 42 litres

and gas produced from 1 kg of gobar in summer

(March to October) = 55 litres.

Calorific value of the biogas/cu m = 4713 kcal

(With burning efficiency of 60%)

density of slurry = 1090 kg/m^3

Total gas produced per day in winter = 42 × 2800

= 117600 lit/day = 117.600 m^3/day

Total gas produced per day in summer

= 55 × 2800 = 154000 lit/day

= 154 m^3/day.

It is advisable to design on minimum gas generating capacity *i.e.*, 117.6 m^3/day.

Total gas required in village (there are 490 people) for cooking

= 0.227 × 490 = 111.23 m^3

Gas required for lighting/day (100 CP. lamp two hours daily burning per connection)

= 2 × 0.126 m^3

Total gas requirement for 98 families/day for lighting

= 2 × 0.126 × 98 = 24.69 m^3

Total gas required for cooking and lighting

= 112.23 + 24.69 = 136.92 m^3

The gas produced is 138.60 m^3/day (Assuming all the cattle dung is available)

Surplus gas = 117.60 – 136.92 = – 19.32 (Neglecting –Ve) = 19.32 m^3

Out of which we can use for 5 street lamps.

Gas required = 5 × 0.252 = 1.16 m^3.

Rest (1.68 – 1.16) = 0.52 m^3, can be accounted for leakage etc.

For the sake of convenience in distribution, we can have 4 digesters.

The quantity of dung for each digester = $\frac{2800}{4}$ = 700 kg.

When unit weight of dung is mixed with unit weight of water (to get optimum solid concentration for digestion) to make the slurry (the density of the slurry = 1090 kg/m^3).

$$\text{Volume of the daily charge} = \frac{\text{Weight of (dung + water)}}{\text{Density of slurry}}$$

$$= \frac{700 + 700}{1090} = \frac{1400}{1090} = 1.28 \text{ m}^3.$$

Digester volume = volume of the daily charge × Retention time

$= 1.28 \times 30 = 38.4 \text{ m}^3.$

This is the minimum working volume. The actual volume should be about 10% more to provide some empty space at the top for proper disengagement of the gas.

Actual digester volume $= 1.1 \times 38.4 = 42.24 \text{ m}^3$

A cylindrical shape for the digester may be chosen.

The hydrostatic pressure due to sturry (hydrostatic pressure = depth of slurry × density of slurry) acting on the inner surface exerts tensile load on the digester wall. The part of the digester below the ground level is subjected to heavy compressive load due to the earth pressure, which increases with depth. A circular shape has a better load bearing capacity.

The height (H) to diameter (D) ratio is taken as unity for any large capacity plant.

$$\text{Volume} = \frac{\pi}{4} D^2 . H$$

$$H^3 = \frac{\text{volume}}{0.785} = \frac{42.24}{0.785} = 53.88$$

$$H = 3.5 \text{ m.}$$

Height of the digester = 3.5 m $\simeq$ 4 m

Diameter of the digester = 3.5 m $\simeq$ 4 m.

Gas holder calculation. Since the gas will be used only for 4 hours in a day the volume of the gas holder must be at least 70% of the daily gas production. Production of gas/kg of fresh dung = 0.06 m^3 for a retention period of 30 days (Standard data).

Vol. of the gas to be retained by the gas holder:

$$= \frac{700 \times 0.06}{2} = 21 \text{ m}^3$$

Assume a diameter of 4 m

volume = 0.785 d^2h, h = 1.970 m.

Total slurry = volume fed in = Total dung + water

= 2800 + 2800 = 5600 kg.

Conclusions: The plant will meet the cooking and lighting requirement of 98 families. It is simple in design. If other requirements like street lighting, power for water pumping are to be met, then total weight of the dung is to be increased by other means *i.e.,* by increasing number of cattles or it can be met by plant wastes, and accordingly design of biogas plant is done.

Size of the Bio-gas Plants

The size of the bio-gas plants depends upon the size of the family, the purpose for which biogas is utilized, and the quantum required for various operations connected with it. The national project on biogas development pattern for subsidy promotion includes 2, 3, 4, 6, 8,10,15, 20 and 25 m^3 capacity biogas plants. The commission on Additional Sources of Energy has also approved one hundred % subsidy for promotion of community biogas plants of 45 m^3/day, and above capacity, meeting the fuel needs of more than 25 families in rural areas. With regard to institutional biogas plants, the size limitations should be fixed depending upon the availability of wastes in the institution. A capital cost upto 75% is provided by Deptt. of Non-conventional Energy Sources (DNES) New Delhi and the balance 25% is to be borne by the recipient organisation.

7.16. Methods for Maintaining Biogas Production

The techniques usually suggested for maintaining the biogas production are briefly summarized as follows:

(1) *Insulating the gas plant.* For reducing heat losses from the digester, the external surface of the digester is adequately insulated using different materials like mineral wool, aluminium cladding, fibre glass, straw etc. For this reason, the Janta type of plants are usually constructed below the ground level. Conversely, in the floating dome design, almost 54% of the total heat losses occurs from the top of the gas holder, therefore simple insulation like readily available dry agriculture residues (paddy straw mat) on the top of the gas holder could reduce this loss considerably.

(2) *Composting.* The heat released in aerobic composting of agriculture residues around the annular ring in the upper part of the digester could be utilized to raise the digester operating temperatures. From experiments performed on biogas plants, it is reported that a temperature rise of 8-10°C above ambient temperature even during the coldest season, as a result of heat released from composting manure and straw. Under optimum conditions of moisture, composting is complete in 3-4 weeks and the released heat varies with time remarkably. Small quantity of water in the straw around the digester is added later when the drop in temperature occurs and therefore the temperature of the composting material and digester contents can be kept relatively at constant temperature.

In this method the whole digester slurry is heated from the exothermic reaction of the compost and the escape of heat of radiation from the top of the digester is prevented by providing thick paddy straw mat on the top of the gas holder. (Refer Fig. 7.22). In the plant as shown in figure a composed pit 1 m × 1 m all round the digester is excavated. It is lined with 13 cm thick wall. Then it is filled with alternate layers of paddy straw and dung. The thickness of each layer is about 15 cm and 5 cm respectively. Each layer must be thoroughly moisturised by sprinkling water. The decomposition of the organic matter and paddy straw will start within 2-3 days and sufficient heat would be produced which will raise the temperature of the digested slurry. In one experiment temperature of the compost was observed to be about 70°C in the beginning, which after 15-20 day came down to 63°C, and further after two it was about 51°C, It was further observed that the temperature of the digester slurry remained always 40°C.

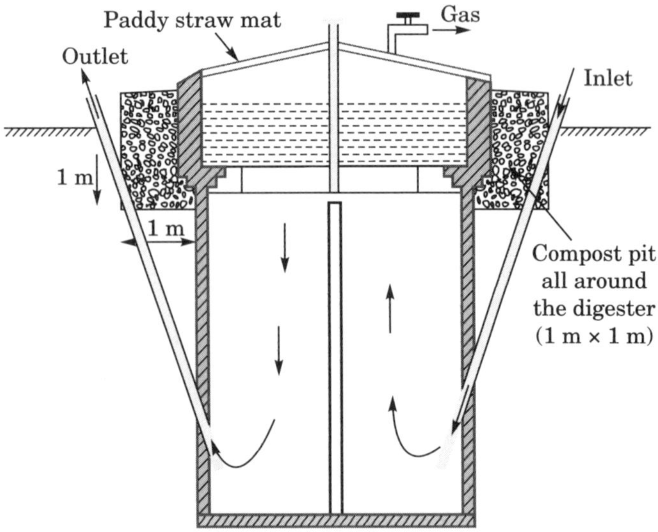

Fig. 7.22. Compost heating of biogas plant.

This the least expensive method for getting more biogas during winter.

(3) *Hot water circulation.* This system has been reported to be efficient for maintaining temperatures of the fermented slurry at the desired level. The cost of system, however comes higher.

(4) *Use of chemicals.* Use of chemicals such as urea and urine have been suggested by some workers to increase the digester temperature. The actual heat gains and the corresponding economics in raising the temperature have not been worked out yet. Addition of urine of animals was also observed to enhance the gas production.

(5) *Solar energy systems.* The solar heat to the biogas digester could be provided in two distinct ways namely (*i*) the active and (*ii*) the passive means. The active systems involve heating of the digester feed or direct of the digester contents. In the first method, the incoming feed (or water which is used to dilute the slurry) is preheated using solar energy during the day and is fed to the digester when it attains the desired temperature.

Attemps have also been made to construct a shallow solar pond water heater over the gas holder with an idea to have a solar water heater so that hot water could be used for hot charging. The production rate is increased by 10-11% from this method. In another method the hot water from the solar pond is circulated through the coils in the digester by a biogas powered pump. In general water is heated in flatplate collectors and is thermosiphoned through the digester.

The passive method involves the construction of a "green house" around the digester to capture the radiant heat energy. In an experiment, a passive green house built over an under ground digester raised the temperature by 10°C.

In another method a solar canopy is used to enhance the biogas production for the low temperature in winter, polythene or plastic sheet is used for canopy. This method is not effective at higher hills, because of snow and low temperature, where the polythene or plastic sheet used for canopy could not last long. The only way to effectively insulate the biogas plant in the hilly areas to enhance the biogas production is to construct a house or cattle shed on the biogas plant. Very favourable results have been obtained in China, Turky and in the Kulu hills of Himachal Pradesh.

7.17. Problems Related to Bio-gas Plants

Some problems are natural and some are created by the persons biogas plant owners but all are controllable.

1. Handling of effluent slurry is major problem if the person is not having sufficient open space or compost pits to get the slurry dry. Use of press filters and transportation is expensive and out of reach of poor farmers. For a domestic plant, 200 litres capacity oil drums can be used to carry this effluent to the fields but this will require some human-animal labour or consumption of diesel if a auto vehicle is used.

2. The gas forming-methanogenic bacteria are very sensitive towards the temperature compared to those of non-methanogenic. During winter as the temperature fails, there is decrease in the activity of methanogenic bacteria and subsequently fall in gas production rate. Many methods have been suggested to overcome this temperature problem 33 described earlier, *e.g.*,

(*a*) Use of solar heated hot water to make a slurry of influent but the temperature of water should not exceed 60°C otherwise the mesophilic bacteria will die.

(*b*) Circulation of hot water obtained either from solar heater or I.C. engine heat exchanger, through pipes inside the digester.

(*c*) Green house effect also give good results but it is costlier and after few years the polythene sheet used in it becomes opaque.

(*d*) Manual or auto stirring of digester slurry and bubbling of generated gas from the bottom of the digester also helps in gas production rate in winter as well as in summer.

(*e*) Addition of various nutrients for bacteria.

(*f*) Covering the biogas plant by straw bags during night hours.

3. Due to lack of proper training to the bio-gas plant owners for the operation of plant, a lot of problems arises. It has been noticed that many persons increase the loading rate and some also do not try to mix the cattle dung with water, keeping in mind more gas production. Due to this, the flow of slurry from inlet towards outlet is very slow or even stops. This may cause accumulation of volatile fatty acids and drop in pH and then failure of digester. Also it is not possible to stir the digester content of high solid concentration.

4. Some persons add urea-fertilizer in large quantities due to which toxicity of ammonia nitrogen may cause a decrease in gas production.

5. pH and volatile fatty acids play an important role in anaerobic digestion and should remain under optimum range otherwise this may cause upsetting of digester and even its failure. pH can be checked from time to time by the use of cheep and easily available pH paper but volatile fatty acids can only be determined in a laboratory having its testing facilities. For controlling pH in optimum range, it tends to fall below 7.0, lime has been suggested, as it is easily available cheap material and does not harm the activity of bacteria.

6. Leakage of gas from gas holder especially in case of Janta type biogas plants is a major and very common problem. When there is quite enough gas in a gas holder, the leakage should be checked by using water and the points marked and then get repaired. During repairing there should be no gas inside the gas

holder and the stop cock remains open till repaired points get dry. Quality of constructing material such as cement is important.

7.18. Starting a Biogas Plant

Starting any type of newly constructed biogas plant or starting a digestion in biogas plant is common to all types of biogas plants. When the construction is over then it is necessary to decide whether the digester is to run on agricultural / forest residue or on animal/ human waste. If the digester is to be fed with animal/ human waste then it is not necessary to be inoculated because it contains necessary bacteria although not in proper number as required. These bacteria get activated under favourable environment and start working. It has been found experimentally that if wet cattle dung is kept for anaerobic digestion mixing with water in 1 : 1 ratio (W/V) at 37°C, the usual gas production starts on 20th day, but if the influent is mixed with effluent slurry obtained from another biogas plant operating on some feedstock then the gas production starts earlier. The gas production rate increases with increase in effluent slurry mixed in with influent slurry.

If the digester is to be operated on agricultural/forest residues then inoculation in necessary otherwise there will be no is gas formation. Some experiments has been done using agro-residues and it was found that the burnable gas having more than 50% of methane could be obtained only within 2 days of starting the digester when the influent was mixed effluent slurry in 1 : 1 ratio (V/V), obtained from a cattle dung fed domestic biogas plant.

7.19. Filling a Digester for Starting

For starting a digester its air should be taken out by filling it with influent upto its over flow in every type of biogas plant. However in case of fixed dome (Janta model) there are some chances that the slurry may enter the gas outlet pipe whereas in floating dome type biogas plant there will be some oxygen (air), at zero level of gas holder and there may be explosion due to the mixture of evolved gas and air. To overcome this problem, the space above digester slurry should be filled with some inert gas such as nitrogen or with biogas. The later can be obtained from another biogas plant by using taken. In animal/human waste bacterial metabolism will tend to use the oxygen in the digester head-space and so leave the atmosphere of nitrogen. In the digester, in which influent has been inoculated with effluent sluny, the gas production will be soon and the percentage of oxygen in digester over head-space will become very less, thus minimising the chances of explosion. Water can also be used to take out the oxygen of digester and then new material can be fed slowly.

7.20. Fuel Properties of Bio Gas

Biogas generated by anaerobic fermentation of organic wastes, essentially contains methane and carbon dioxide in large proportion and has traces of other gases. The importance properties are as follows:

Table 7.8

Composition	*(% volume)*
Methane	50–60
Carbon dioxide	30–45
Hydrogen	5–10
Nitrogen	0.5–0.7
Hydrogen sulphide and oxygen	Traces.

Calorific value	
60% methane	: 22.350 to 24.22 MJ/m^3
without CO_2	: 33.525 to 35.390 MJ/m^3
Octane rating without CO_2	: 130
Octane rating with CO_2	: 110
Ignition temperature	: 650°C
Air to methane ratio for complete combustion (by volume)	: 10 to 1
Explosive limits to air (by volume)	: 5 to 15

7.21. Utilization of Bio-gas

The main products of the bio-gas plant are fuel gas and organic manure. Biogas is a flammable gas. Methane is the only combustible portion in the gas and hence around 60% by volume is only usable for combustion.

The biogas can be utilized effectively for household *cooking, lighting, operating small engines, utilizing power for pumping water, chaffing fodder* and *grinding flour* by using the already known technology.

The utility of the gas in burners is well established. There are millions of homes today using this gas as fuel solely for cooking purposes. The burning of cattle dung has not only led to national wastage of organic manure but has also caused health hazard problems. It has been estimated that 30 million *chullas,* besides consuming 133 million tonnes of wood, are amongst the major causes of high incidence of respiratory diseases and trachoma of eyes of the women folk. The only way of saving the house wife from the irritating smoke of the dung cakes and wood is to popularise the use of biogas as far as possible.

Low cost burners have been designed by engineers. A nozzle is needed for lamp and stove, which consists of hole in the size of needle point (0.5 mm dia.) and the other end of the nozzle is connected to the gas supply hose from the digester. After the biogas enters the stove or lamp, it will spray out from the nozzle at a very high velocity and the air surrounding this gas streams becomes a low pressure area. Therefore air is drawn into the mixing chamber through the air inlets to mix with the biogas. From the mixing chamber the bio-gas air mixture rushes to the openings at the fire sieve plate for combustion. The brightness and force of combustion of the stove and the lamp depend on the biogas pressure, the mixing ratio of biogas with air (approx. 1 : 10) and the conditions of mixing between biogas and air. A biogas lamp needs a mantle, which is made of a Ramic fibre (the same fibre from which grass cloth and linen is made). The Ramic fibre turns to ashes when burned and forms a layer of thorium oxide, which emits a dazzling white light at high temperatures. As biogas is almost half as light as air and as hot air flows upward, the brightness of the standing biogas lamp is greater than that at the hanging lamp. A biogas lamp of a luminosity of about 60 watts equivalent electrical light can function for six to seven hours if one cu. metre gas is available.

One horse power engine can work for two hours roughly with a cu. metre of gas. This quantity of gas can cook three meals for a family of about five. It may be noted that gobar gas can not be bottled into cylinders like L.P.G.

It is highly economical to build power houses, at the places of generation of biogas so that electricity can be produced and the same connected to the grid. This certainly will avoid the power cut situations heavily. Electricity could be produced directly by using bioorganic matter as the source of fuel in a biochemical fuel cell.

Air is injected at the cathode as an *oxidiser* and bioorganic matter at the anode as the fuel. The electrolyte is usually an organic solution or an acqueous medium such as potassium hydroxide. The bacteria create new losses by consuming about half of the fuel. The consumption is necessary for their nutritional requirements. The electrical power output of a biochemical fuel cell is proportional to the bacterial metabolism rate. It is possible to scrub the biogas by treating with sodium carbonate and methane. An experiment at IIT, Chennai has shown that the flame temperature of the gas can be increased upto 1000°C. This may be ideal for using the gas for *brazing* in workshop.

Biogas can be used to operate both Cl (diesel) and SI (petrol) engines. C.I. engines can run on dual fuel (biogas + diesel) and pilot injection of diesel is necessary for igniting the mixture of air and biogas inside the cylinder. However the initial starting of diesel engine is done on pure diesel. Spark ignition (SI) engine can be operated on biogas after initially starting on petrol. The quantity of gas required for running oil engines must be sufficient on an average 425 litres of gas is required per horse power per hour. Water pump or generator can be connected to the engine. If a 5 HP engine is to be used for say 8 hours atleast 18 cubic metres of gas would be required per day. That means at least 30 to 35 animals are required for this purpose only. Over and above this gas required for cooking or lighting etc. shall also have to be provided for.

The existing diesel engines can be directly converted to use biogas, with a slight modification, saving thereby about 80% of diesel oil. It is possible to reduce the diesel oil consumption by further research. Biogas can be solely used in SI engine by suitably modifying the carburettor.

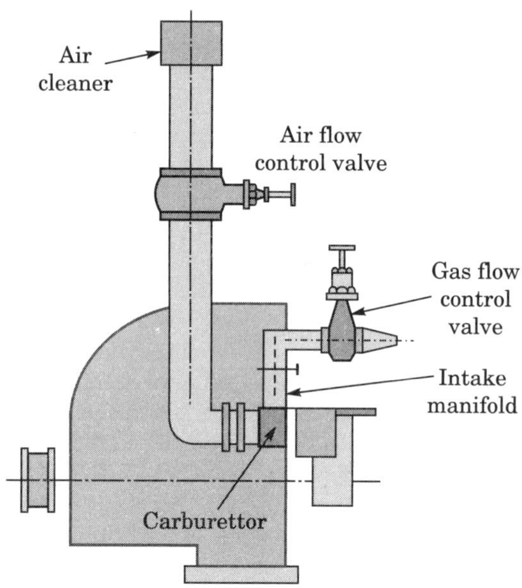

Fig. 7.23. Modification required for operating a stationary SI engine on biogas.

Modification of SI engine: SI engines can run completely on biogas, however, the engines are required to be started on petrol at the beginning, conversion of SI engine for operation on biogas includes provisions for the entry of biogas, throttling of intake air and advancing the ignition timing.

Biogas can be admitted to a stationary SI engine through the intake manifold and air flow control valve can be provided on the air cleaner pipe connecting the air cleaner and carburettor for throttling the intake air, as shown in Fig. 7.23. In this case the intake air is required to be manually throttling in the initial stage.

Modification of CI engine: CI engine can operate on dual fuel and the necessary engine modifications include provision for the entry of biogas with intake air, advancing the injection timing and provision of a system to reduce diesel supply.

The entry of biogas and mixing of gas with intake air can be achieved by providing a mixing chamber below the air cleaner which facilitates through mixing of biogas with air before entering into the cylinder. The arrangement as shown in Fig. 7.24 is largely used in stationary engines commercially available in India. The capacity of mixing chamber may be kept equal to the engine displacement volume. The pilot injection of diesel in the cycle is required to be advanced for smooth and efficient running of engine on dual fuel. The admittance of biogas into the engine at the initial stage increases engine speed and therefore, a suitable system to reduce the diesel supply by actuating the control rack needs to be incorporated.

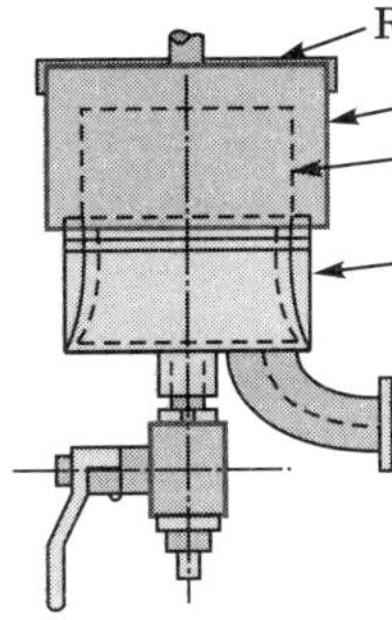

Fig. 7.24. Details of mixing chamber and air filter box.

It is concluded that:

(*i*) Biogas is a substitute for conventional engine fuels with little modification in both SI and CI engines. Petrol replacement of the order of 100% and diesel replacement of about 80% is possible using biogas.

(*ii*) SI engines develop 85% of rated power whereas CI engines develop full power on biogas. Thus application of biogas in CI engine is a better alternative.

(*iii*) By reducing the CO_2 content in biogas, the engine performance is improved.

(*iv*) The ignition timing of SI engine using biogas a fuel shall be advanced by 4-5° for better engine performance.

(*v*) The injection timing CI engines operating on dual fuel (biogas + diesel) shall be kept between 31-33° before TDC for better performance.

(*vi*) It is economical to use biogas in engine keeping in view the present trend of increase in prices of conventional fuel and their shortage.

In sewage treatment plants the gas is utilized as fuel for the boilers that supply hot water for heating the digesters, for running gas engines which may be coupled to pumps, blowers or generators. The hot water boilers are fired with biogas through natural-gas burners. In view of the characteristics of biogas, the gas engine in which it is used, must be specially adapted to it. If more than 0.25 to 1.0% of hydrogen sulphide is present in the biogas, the gas should be scrubbed before being burnt in a gas engine, since hydrogen sulphide make the biogas corrosive. For scrubbing the gas should be passed through a chamber filled with ferric oxide sponge.

The other main product of the biogas plant in the *organic manure*. This comes out at outlet as slurry which is quite rich in nitrogen and humus. It is in fully digested condition and is in a finely divided condition. It can be applied directly to the farm by mixing with irrigation water. This way maximum benefit is derived from the manure because nitrogen content of fresh slurry is over 2% and it is in a condition which mixes with the soil very well.

When the slurry cannot be used with irrigation water it can be used for rapid fermentation of compost. The outlet slurry is led into channels which connects a number of pits. The vegetable refuse like grass, leaves, corncobs and all kind of waste material is dumped in the pit in a layer on top of which the outlet slurry is allowed to spread on top of it, again further waste material could be piled and the process repeated until the pit is full. Another pits are also filled in the same manner. It is found that the large content of bacteria and the nutrient material in the gas plant slurry accelerate the process of composting. By providing number of pits, the disposal of the manure becomes easy.

The gobar gas manure can also form a good organic base for enriched manure, *i.e.*, by enriching the manure with chemical fertilizer like ammonium sulphate, superphosphate etc. a very fine organic base manure mixture could be produced.

7.22. Biomass as a Source of Energy

7.22.1. Introduction

Through the process of photosynthesis, plants are able to convert solar energy into biomass which can be used as a source of fuel. Recently, in more industrialized nations, human activity has taken advantage of photosynthesis only for the production of food and fibre. However, the majority of the world people, the poor in the under developed countries, still today provide most of their space heating and cooking energy from biomass materials, mainly wood and animal dung. As a result of projection of energy shortages in the years to come, interest in alternative fuel sources has increased considerably. Biomass is one of those energy sources being considered. The specific production of biomass for fuel could occur on energy farms. Here, we do not consider the presently available waste and residue biomass resources a very important and sizeable source.

There are basically three types of energy crop systems, annual energy crops (grains and grasses), silviculture energy farms (trees), and aquatic energy farms (both land based and open ocean). Since we have had extensive agricultural experience with annual land crops, the present order of feasibility of these types of energy farms is as just mentioned.

The land-based farm concepts range form the use of conventional crops on prime agriculture land and short rotation tree farms on marginal lands to the selection of novel and native species of plants. Prime land for energy farming could become available by reducing grain exports or extending the open-range grazing of cattle before shipping to feedlots. The land-based aquatic systems include the possibility of production of biomass as an integrated part of a waste-water treatment system. Sewage or other wastes would provide the required nutrients and water for submerged plants (*e.g.*, algae); addition of carbon dioxide to the system is required for higher biomass yields. Power plant flue gases may be considered as a potential source of carbon dioxide. Highly saline deserts waters could be used to produce nitrogen-fixing aquatic plants that would be an aerobically digested to form synthetic gas. Again this could be integrated with a nearby power plant for returning carbondioxide to the system.

The open-ocean concept is the least well developed. At this point it consists of cultivating certain species of giant kelp in the open ocean on artificial supports. The nutrients would be supplied by unwelling nutrient rich deep bottom waters.

The kelp (water hyacinth also) would be harvested and anaerobically digested to produce natural gas. This system as envisioned by its developers, would be integrated with mariculture of fish (for human consumption); production of animal food and chemicals would be important coproducts.

There are basically three major factors that influence the amount of energy the nation can derive from energy farms: land/water availability, and productivity, and economics. These factors differ for each type of biomass system. They are dependent on location, the level of technical development, and/or the consumptions and extrapolations used. Biomass as a source of energy has several advantages over dwindling fossil resources it is renewable environmentally clean and easily adaptable. The distribution of major energy farms in rural India are non-conventional energy 65%, human and animal 15%, and conventional 20%. Cooking consumes 64% of the total energy, agriculture 22%, village industries 7%, lighting 4%, and transportation 3%. The major contribution to fuel comes from firewood constituting 68.5% followed by oil products 16.9%, animal wastes 8.3%, coal 2.3% and others 3.4%.

According to the Report of Firewood study committee, fuel wood requirement is 133 million tonnes and annual availability is meagre 49 million tonnes. Our total forest area is 75 million hectares which is only 22.8%, of that geographical area. The per capita availability is likely to decrease from 0.29 ha to 0.17 ha up to beginning of the twenty first century. Hence under these conditions, it is imperative to enhance the productivity per unit area or to utilise the waste lands, estimated to be 936.91 ha in lakhs (non-forest areas only) optimally. In this context a lot has been said about fast growing species, energy plantations and less about fast growing shrubs/plants with high biomass yields, which are compatible to rural areas and may be easily grown on waste lands. Research centre at IIT Delhi has been actively engaged in *promoting* the plantation of high yielding varieties like *Bougainvillea, S. Munjha* L. camara and integrated approach for total utilisation of biomass produced by them. In the different biomass research centres at different regions of the India, a large number of species have been studied. Mainly are; Albizzia lebbek, Terminalia arjuna. Sesbania aegyptiaca, Leucaena (pheno type), Prosopis juliflora, Sesbania grandiflora, Acacia auriculiformis, non-legumes (ceiba pantandra, casuarina equisetifolia and Eucelyptus hybrid), Jojoba (Simmondsia chinensia Link Schnd, Sargassum Tenerimum, Sweet Sorghum) etc.

7.22.2. Energy Plantation

Growth of plant material for their fuel value offers a renewable source of liquid fuel and organic chemicals. Energy plantations can be considered as long term alternatives to fossil and nuclear energy sources. When land plants are grown purposely for their fuel value, by capturing solar radiation in, is called *"Energy plantation"*. It is worth considering this possibility seriously because, plants as they grow, serve as a convenient single service disposal solar energy storage devices, from which the accumulated energy can be released, at will, at temperatures comparable with those achievable from fossil fuels. More energy is derived from materials of forest origin than from all the thermal power plants in the country. Yet little consideration is given to harvesting plants as fuels. It is necessary to think in terms of "calorie plants" not of trees grasses or farm crops as such ; this way of thinking will permit us to define the character and culture of plants for optimising the fuel harvest in, the so called "Energy plantation".

"Energy plantations" by design are managed and operated to provide substantial amounts of usuable fuel continuously throughout the year at the costs competitive with other fuels. Annual plants, typical of the important farm crops, are unsuitable for providing a year-round supply of fuel. Mature trees, as used by the forest products industry, grow too slowly to provide the high yields desirable for 'Energy plantations'. Wild or unmanaged vegetations vary greatly in form and quantity between widely separated sites to be a reliable and manageable source of fuel. The form of the "Energy plantation" fuel must be acceptable to its intended users.

Soil Conditions and Species of Vegetations

Choosing 'calorie plants' depend on the climate and soils, where the energy plantations are to be located. Lands that are primarily suitable for agricultural crops will not be devoted for growing fuels on large scale. However the 'calorie crops' will still required land and climate suitable for their growth like adequate moisture, a moderate growing season, a few cms of top soil capable of retaining moisture and being fertilised. Large hectares of land do exist in our country with climate and soil conditions (favourable) for such plantations. To identify such areas for a particular species of 'calorie crops' it may be necessary to do a survey. There are areas in our country where soils and climates are suitable for intensive culture of variety of soft and hard woods.

Whatever be the candidate for the calorie crop chosen for a particular zone, they must effectively capture solar energy on a small scale and also be manageable on a large scale. Depending upon the type of candidate, tree and trees life cycles, intensive forest practices allow a greater number of trees to be planted per unit area, thus reducing the space between them. The geometry of leaf-bearing branch should be such that maximum surface area faces the sun.

Design and Management of Energy Plantation

Management of large scale energy plantation may require complete mechanisation or semi-mechanisation. In India, where large labour force exists, and also to give meaningful livelihood to the large number of labourers, it is essential that mechanisation is kept a minimum. High yields of energy plantation fuels from trees are possible, where candidate trees are planted with denser that conventional spaces and harvested on shorter than conventional cycles.

A computer model has to developed with the existing data of the candidate tree and particular site to predict combinations of harvest cycles and planting densities, which will optimize the harvest of fuel and the operating cost. Typical calorie crops includes 12,000 to 24,000 trees/hectare. The plantation is harvested initially after one to three years and subsequently on two or four years cycle for five or six harvested. In some experimental area bio-mass of 17 to 25 dry tonnes/ha/yr has been achieved. This means plantation of 50 million hectares can yield fuel equivalent to 3,000 trillion kilo calories, scheduling all operations, from site preparation to plantation regeneration, ultimately depends on the harvested cycle selected to optimise plantation yields.

In India the annual average solar insolation is over 400 cal/cm^2/day ordinary vegetation grow in the country is collecting and storing, fuel equivalent to 0.5 to 0.77% of the annual solar radiation on the land. This sunshine conversion rate is equivalent to collecting and storing between 100 to 175 trillion kcal/ha/yr. It seems therefore that, a solar energy conversion rate of about 0.9% can be achieved widely by optimizing plant material production, for fuel value by methods like, double cropping, mix species planting or crop rotation. The technology for all these operations in general are available in the various sectors of our agricultural economy. A rough estimate of the cost of the heat equivalent of the crop from energy plantation indicates, that it is lower than the current cost of hydrocarbon fuels and slightly more than the cost of coal in areas close to the coal mines.

7.22.3. Advantages of Energy Plantation

Storage. Fuels produced by plant growth will have two main advantages over directly harnessing the solar energy at the surface of the earth either by thermal or photovoltaic means, *i.e.,* neither of these two means can be counted upon for a continuous supply of energy unless it is equipped with elaborate storage units. Whereas plants, as they grow they serve as their own energy accumulator. There are practically no storage losses. This feature allows a considerable amount of flexibility in the reaping schedule for fuel crop. The flexibility is particularly wide if the crop maturity cycle extends over years. On the contrary similar flexibility is not available if crop matures in a few months. Perennial crops are always preferred for this purpose due to their continuous growth with minimum of efforts.

Economics. Economics of energy obtained through plantation will essentially depend upon:

(*i*) Cost of fuel mass produced in the comparison to conventional fuel.

(*ii*) Cost of processing required, if any.

Cost of fuel will depend on:

(*i*) Cost of land used, *i.e.,* in turn the yield of fuel per unit area/yr.

(*ii*) Cost of plantation per unit area/yr.

Cost due to land depends upon the duration of maturity cycle, *i.e.,* smaller maturity cycle means less cost. Cost due to land can be considerably reduced in having multiple crops and denser plantation. Under optimised conditions the cost of energy obtained through plantation works out to be lesser than present fossil fuel cost on heat value basis as obtained by direct burning of wood.

Air Pollution. Vegetable matter contains less than 0.1% sulphur and the problem of SO_2 pollution by the combustion of biomass is practically non-existent.

Ash. Burnt vegetable matter leaves ash which is rich in plant nutrient-minerals unlike ash of fossil fuels. It can directly be used as a manure.

Ecological condition. Energy plantation will convert large tracks of semi-barren land into green belt and thus ecological conditions can be restored.

CO_2 balance of the Earth. Utilization of biomass for energy either by direct combustion or through production of liquid and gaseous fuels produces CO_2. However CO_2 thus produced does not disturb the CO_2 balance in the atmosphere. Growth of biomass consumes CO_2 as much as produced by consumption of biomass. Combustion of fossil fuel releases the CO_2 in very large amounts and also they consume huge amounts of O_2 from the atmosphere.

7.22.4. Plants Proposed for Energy Plantation

The key consideration in biomass energy farming is the choice of plant type and species to be cultivated. The choice of the species will depend on location, soils water quality and a host of other factors. Some of the plants are described below.

Casuarina. Casuarina is a large evergreen tree that looks like a feathery conifer, and is hardly plant well adapted to coastal areas. Its botanic name is casuarina equisetifolia L.

These species reach a height of 15-25 m with trunk diameter of about 18 cm but when allowed to grow can attain heights of upto 50 m with trunk diameter of 30 cm.

Casuarina equisetifolia is indigeneous to north and north West Australia, some pecific islands and from Indonesia and Malaysia to India and Sri Lanka. It has been introduced for fire wood, beautification and other purposes in India, Pakistan, Africa and West Indies.

It grows well in warm, subtropical and tropical climates. Although the monthly mean temperature in its native areas is 10°–30°C, they adapt to a wide range of temperatures. Being low land species they can be planted from sea-level upto 1500 m. It has been planted successfully in areas with annual rainfall as little as 200–300 mm or as much as 5,000 mm. It grows poorly on heavy soils such as clays.

Casuarina equisetifolia grows all over Decean in India and the whole tree is harvested after 5-7 years, the yield of loppings from the second year being 3-5 tonnes/ha/yr. In Tamilnadu, casuarina is grown under varied conditions: (*i*) In garded lands with Groundnut, maize, sorghum, etc. as inter crops, (*ii*) In dry lands, (*iii*) In sandy coastal belts with little after-care and (*iv*) In sandy coastal belts with minimum cultural operations. The rotation period and the yield varies with the mode of cultivation. The usual rotation cycle ranges from 4-8 years. Casuarina has yielded upto 250 tonnes/ha in four years in Pondicherry. In Thana (Maharashtra), it has recorded a yield of upto 285 tonnes/ha in four years under irrigated conditions. In Karnataka state, casuarina equisetifolia is the major species planted for firewood. This specie is considered to yield the best firewood in the world. Experimental study is being carried out at chettiar research centre at Chennai, Sholapuram research centre, at INCOR (Institute for Offshore and Coastal Research) Vishakhapatnam etc.

The main uses of casuarina are:

(1) It is used as wood. (2) It is used for a pulp.

(3) Used for wind break. (4) In erosion control.

Eucalyptus Plantation. Eucalyptus plantation as a source for fuel wood has been practised for quite sometimes in India. It is normally included in the category of fast growing species for fuel wood. Species are about 140, mostly Australian excepting a few in New Guinea and the Indian Archipelago.

Eucalyptus globulus shows remarkable early growth on favourable sites (about 15 m in 3 yrs and 30 m in 10 years in India), and attains nearly 60-70 m in height and 2.3 m in diameters, with a straight trunk as long as two-thirds of its/ total height.

Eucalyptus wood has found many uses. For example, the wood has been used in place of steel for railroad tracks. Major uses are for more typical timber products: fence posts, lumber, paper pulp, fuel, and certain chemicals and oil extracted from leaves. An additional by product from Eucalyptus farms is honey.

Its production or yield rate (dried) is about 24.1 tonne/acre/yr.

Sorghums. It is a good candidate for energy crops because it has very high yield. It requires less water for normal growth and resulting in lower irrigation costs. Its yield is 10 to 30 tonnes/acre. Under the project "Production of ethanol from sweet sorghum" sponsored at NARI (Maharashtra) 12 strains were tested, out of which 4 strains of Sacchar omyces cerevisiae were good producers under the conditions established on an average they produced 10% alcohol of all the cultivers of sweet sorghum used, none showed any inhibition in the production of alcohol.

The main species which are being experimented in different zone of India are:

Terminalia arjuna,
Sesbania aegyptiaca
Leucaena
Sesbania grandiflora
Prosopis juliflora
Jojoba
Sargassum Tenerimum (a species of brown seaweed)
Grrevillea robusta
Guazuima ulmifolia

Table 7.9. Common Species Recommended

Name the species	*Yield* *m^3/ha/yr*	*Calorific value* *Kcal/kg*
Casuarina equisilifolia	7–10	4950
Eucalyptos camaldulensis	7–10	4800
Leucaena leucocephala	30–40	4200–4800
Prosopis Juliflora	5–6	High
Sesbania Sp.	20	–
Acavia auriculiformis	10–12	4800–1900
Albizia lebbek	5	5200
Cassia siamea	5–10	–

The species are found of high yield rate.

7.23. Methods for Obtaining Energy from Biomass

We have seen that the photo synthetic processes of plants and certain bacteria use solar energy to convert CO_2 and water into starch, cellulose or sugars with the evaluation of oxygen. Solar energy thus stored is *biomass.* The biomass energy from energy farms can be converted into various energy forms. There are significant costs for biomass transportation. Biomass, because of its low density and low energy content per unit weight, does bear a significantly higher cost in transportation than natural gas or coal. Consequently, it would be unreasonable to expect to grow on energy crop and transport it a great distance to be converted into an end use energy form. Instead, the biomass produced would be converted *on site* into electricity, synthetic natural gas, a pyrolytic char, or oil, which is more, energy dense and more economical to transport. Also, on site utilization of biomass energy would allow co-generation from waste heat, maximizing overall efficiency. A most attractive concept is the location of biomass-fueled industries next to the energy farm, in this manner there are minimal cost involved in the transportation for the end user, and the fuel can be used in a more efficient end use form. For example, the biomass can be converted through direct combustion to steam or electricity, or gasification to substitute for natural gas or as a chemical feed stock.

The processes by which the products of energy farms can be converted into various energy forms will be discussed only briefly here. (Fig. 7.25).

Combustion. One process now in commercial operation that uses biomass to produce energy is combustion. Direct combustion requires biomass with a moisture content around 15% or less, so it may require drying prior to combustion for most of the crops. The combustion produces steam both for process use and for electricity. The exact economics of the combustion of wood for steam production varies at the many papermills, sawmills and various other industries that utilize this technology. However, the fact that these industries have the availability of conventional fuels such as fuel oil and coal and yet choose to use wastes, demonstrates the fact that the combination of the disposal credit for the waste material and its immediate availability makes it very competitive with existing fuels.

Electricity can also be produced by installing require capacity power plant near the industry.

Anaerobic digestion. Anaerobic digestion and combustion are probably the two most promising existing technologies because they have the longest history of use with established economics and are in current use with biomass wastes commercially. Anaerobic digestion process is described already in detail in previous sections. Anaerobic digestion plants are simple in construction with low capital outlay. Anaerobic digestion of organic wastes may constitute an effective device for pollution control in addition to energy generation. The main advantage of this biochemical process is that it utilizes biomass with water content as high as 99%. Another advantage is that small units are available, which can be operated at individual farms. Furthermore, the residue has fertilizer value. The primary limitation with this process is that the large quantity of waste water is to be disposed of after digestion.

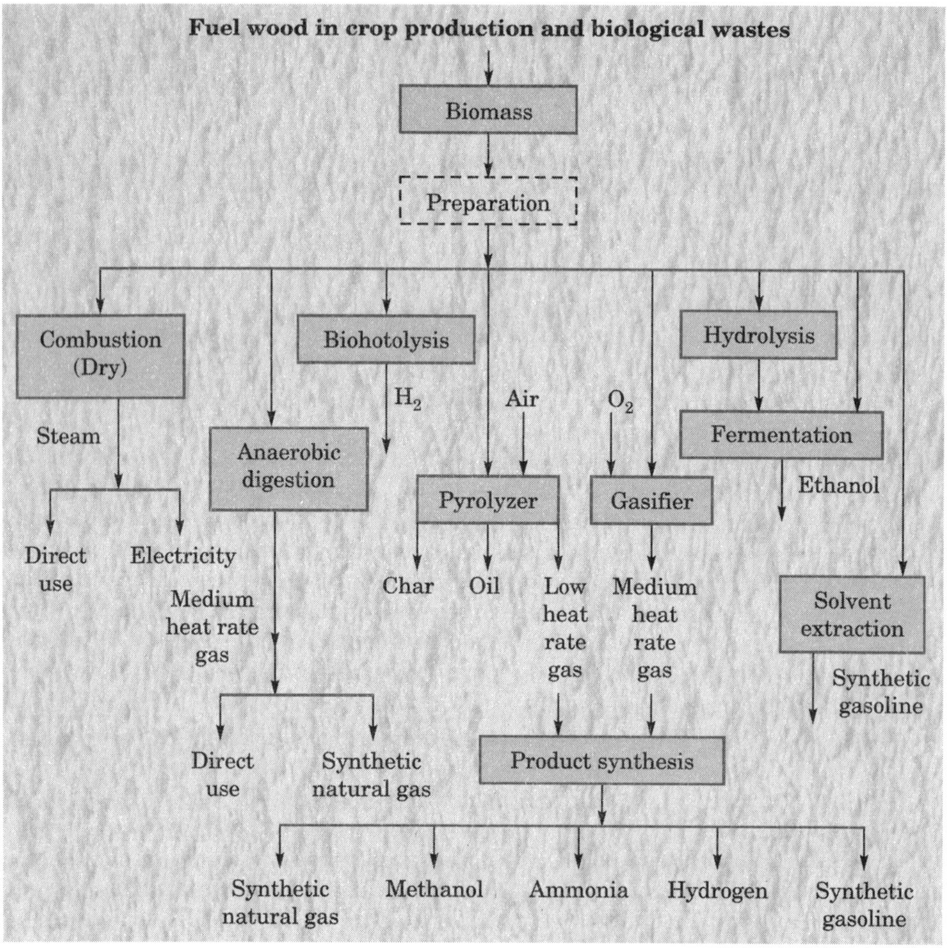

Fig. 7.25. Possible energy conversion routes and products from biomass.

Pyrolysis. It is an irreversible chemical change caused by the action of heat in absence of oxygen. This process may yield either solid, liquid or gaseous fuel. Without oxygen the energy splits the chemical bonds and leaves the energy stored in biomass. The reactions are complex in the process. Pyrolysis of cow manure, wood-saw dust, liberates H_2, N_2, CO, CO_2, C_2H_4, C_2H_6 and C_7H_8. Hydrogen and carbon monoxide can be converted into methanol, gasoline, diesel ammonia for fertilisers, drugs, building or bonding materials and synthetic textiles.

The main advantages of pyrolysis include, compactness, simple equipment, low pressure operation, negligible waste product and high conversion efficiency of the order of 83%.

Hydrolysis and Ethanol Fermentation

Hydrolysis is the technology which converts cellulose to alcohols through fermentation. Ethyl alcohol can be produced from a variety of sugar by fermentation with yeasts. Molasses is diluted with water to a sugar content of about 20% by weight, acidified to pH 4.5 and then mixed with yeast culture (5% by volume) in a fermentor. Ammonia is used to reduce acidity. When 8-10% alcohol is accumulated, then liquid is distilled, fractionated and rectified 2.5 litres of can molasses produces about one litre of alcohol. Alcohol fermentation of a large variety of substrates has been obtained. They include: corn barley mashes, sugarcane, molasses and chees way, sugar beet molasses and potatoes.

Gasifier. A most promising conversion technology is pyrolysis-gasification. Considerable experience in commercial operations with this technology took place in Europe and continuous in numerous Third World countries for generation of low heating value gas and electric production from biomass waste materials. There is a considerable new research and development under way, and several commercial manufacturing operations are beginning to produce gasifiers. This technology'is expected to be available within the next few years on a commercial basis. It appears to be economically competitive with natural gas at todays prices using biomass *wastes* and would be economically competitive in near future if its feed stocks were dry annual energy crops.

7.24. Thermal Gasification of Biomass

7.24.1. Introduction

The word *gasification* implies con verting a solid or liquid into a gaseous fuel without leaving any solid carbonaceous residue.

Gasifier. It is an equipment which can gasify a variety of biomass such as wood waste, agricultural waste like stalks, and roots of various crops, maize cobs etc. The gasifier is essentially a chemical reactor where various complex physical and chemical processes take place. Biomass gets dried, heated, pyrolysed, partially oxidised and reduced, as it flows through it.

The gas produced in the gasifier is a clean burning fuel having heating value of about 950-1200 kcal/m^3. Hydrogen (18-20%), and carbon-monoxide (18-24%) are the main constituents of the gas.

Methods of converting biomass into useful forms of energy can be broadly classified as:

(*i*) biochemical and (*ii*) thermochemical.

The biochemical route of conversion of biomass to other useful forms is a low energy process and relies upon the action of bacteria which degrade complex molecules of biomass into simpler ones. The production of biogas, a mixture of methane and carbon dioxide, from animal dung, by means of anaerobic digestion, is an important example of this process. The biochemical methods of conversion are discussed, in previous sections. In thermochemical methods, the biomass is raised to high temperatures and, depending on the quantity of oxygen supplied, processes such as pyrolysis, combustion, and gasification occur. Direct burning of biomass in stoves and open fires is an example of combustion in which the supply of oxygen is generally higher than that required stoichiometrically. Pyrolysis and gasification occur under the conditions of lower rates of oxygen supply. Preparation of charcoal from wood and incineration of municipal solid waste are examples of pyrolysis. Under some conditions of temperature and oxygen supply, a gaseous mixture, rich in carbonmonoxide and hydrogen, referred to as producer gas, is formed. This process is called thermal gasification and is discussed in detail in the following sections.

Thermal gasification of biomass

In what is known as a gasifier, a solid fuel is converted by a series of thermochemical processes like drying, pyrolysis, oxidation, and reduction to a gaseous fuel—producer gas. If atmospheric air is normally used as the gasification agent, the producer gas consists mainly of carbonmonoxide, hydrogen and oxygen. A typical composition of the gas obtained from wood gasification on volumetric basis is as follows:

Carbon monoxide	18–22%
Hydrogen	13–19%
Methane	1–5%
Heavier hydrocarbons	0.2–0.4%
Carbon dioxide	9–12%
Nitrogen	45–55%
Water vapour	4%

The gas can be used for generation of motive power either in dual fuel engines or in diesel engines with some modifications. Engines operating on a spark ignition system (*e.g.*, petrol engines) can be made to run entirely on producer gas, whereas those using compression ignition systems (*e.g.*, diesel engines) can be made to operate with about 60–80% fuel oil replacement by the gas. In larger systems, the gas can be burnt directly (*e.g.*, in an industrial oil fired boiler).

As mentioned earlier, complete combustion takes place with excess air or at least 100% theoretical air, whereas gasification takes place with excess carbon. The gasification of solid fuels containing carbon is accomplished in an air sealed, closed chamber under slight vacuum or pressure relative to ambient pressure. The fuel column is ignited at one point and exposed to the air blast. The gas is drawn off at another location. Depending upon the positions of air inlet and gas withdrawal with reference to the fuel bed movement, three broad types of gasifiers have been designed and operated to date. They are: (*a*) updraft, (*b*) downdraft and (*c*) cross draft gasifiers, which are schematically shown in Figs. 7.26, 7.27 and 7.28.

The *advantages* of a gasifier are:

(*i*) It is very easy to operate the gasifier.

(*ii*) Its maintenance is easy.

(*iii*) It is sturdy in construction and,

(*iv*) reliable in operation.

Gasification—The conversion alternative

A diverse range of routes are available for conversion of biomass to various forms of energy. Most of these processes or technologies are suitable for either one or few of the biomass resources and have limitations in terms of total available energy and the form of energy.

The direct use as fuel, densification and subsequently use as fuel, digestion to generate biogas, gasification, incineration and steam generation are some of the available processes/technologies for conversion of biomass to energy, especially electrical and/or heat energy.

However, among all these alternatives, gasification is the best suitable alternative in view of the following:

(*i*) Gasification offers high flexibility in terms of various biomass materials as the feedstock.

(*ii*) Gasification has thermo-chemical conversion efficiencies in the range of 70% to 90% which is highest among various alternative.

(*iii*) Gasification output capacity, especially in the high output ranges, is controlled only by availability of adequate feed materials rather than other technical consideration.

(*iv*) The area requirement for gasification equipment is lowest per unit output of energy in the form of heat and/or electricity.

(*v*) The gasification equipment has high turn down ratios comparable to biogas and higher than steam turbine systems.

(*vi*) Gasification outputs ore suitable as a fuel to all types of internal combustion engines with capacity derating in the range of 15% to 30%.

7.24.2. Classification of Biomass Gasifiers

Gasifiers are classified as per:

(*i*) the direction of the gas flow, and

(*ii*) output or capacity of the gasifiers.

(1) Gasifier is a vertical flow packed bed reactor through which oxygen or air for combustion is passed through it. It may pass downward or upward or across the bed. The direction of the gas flow defines it as a *down draught, up draught,* or *cross-draught* producer gas generator.

(2) As per the *output power,* gasifiers are classified as

(*i*) *Small size gasifiers* with output upto 10 kW.

(*ii*) *Medium size gasifiers* with outputs in the range of 10 kW to 50 kW.

(*iii*) *Large size gasifiers* with output in the range of 50 kW to 300 kW.

(*iv*) *Very large gasifiers* with outputs of 300 kW and above.

(*v*) Gasification for direct thermal energy applications.

(3) Gasifiers may also be classified as per the type of the bed *i.e.,* (1) Fixed bed and (2) Fluidised bed.

(1) Fixed Bed Gasifiers. The three main designs of fixed bed gasifiers are (*i*) Up draught, (*ii*) Down draught and (*iii*) Cross draught.

(*i*) *Up Draught Gasifiers.* In updraught gasifier air enters below the combustion zone and the producer gas leaves near the top of the gasifier. This type of gasifier is easy to build and operate. The gas produced has practically no ash but contains tar and water vapour because of passing of gas through the unburnt fuel. Hence updraught gasifiers are suitable for tar free fuels like charcoal, especially in stationary engines (Fig. 7.26).

(*ii*) *Down-draught Gasifier.* In down draught gasifier air enters at the combustion zone and the gas produced leaves near the bottom of the gasifier. In this type of gasifiers, the volatiles and the tars produced from the descending fixed [illegible] have to pass through the reaction zone where mostly they are racked and [illegible] constriction provided in the hearth, *i.e.,* the throat ensures that the gaseous products pass through the hottest zone, the gas produced contains less of tar and more of ash, These gasifiers are suitable for fuels like wood and agricultural wastes. They may be used for power generation upto above 150 kW, and beyond that there may be geometrical limitations upon gas quality. They are cheap and easy to make (Fig. 7.27).

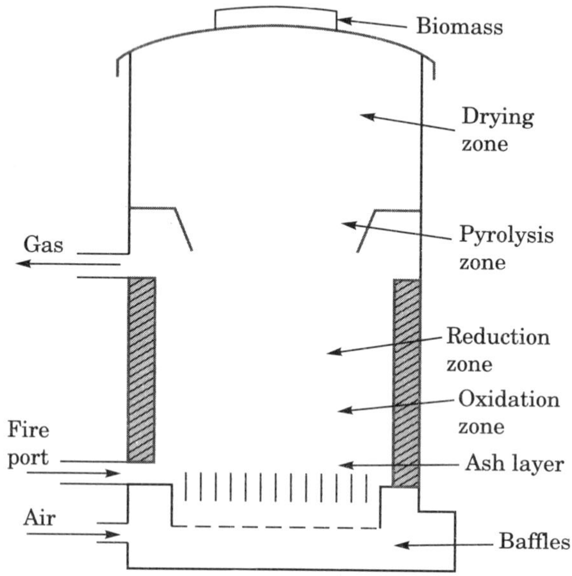

Fig. 7.26. Schematic diagram of an updraft gasifier.

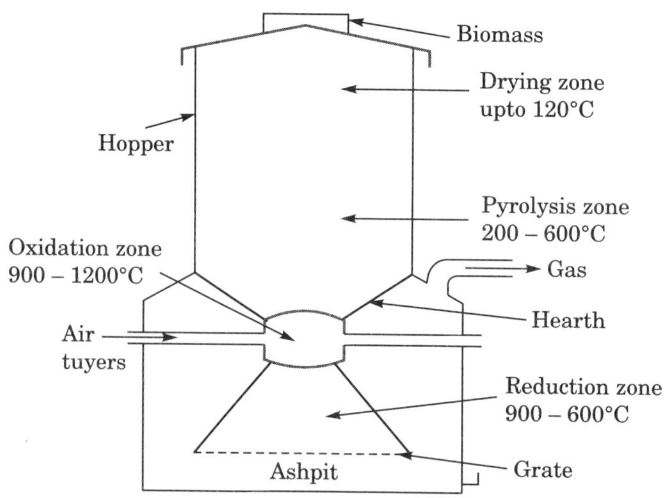

Fig. 7.27. Schematic diagram of a downdraft gasifier.

In the downdraft gasifier, air enters through radial tuyers near the top of the fire box. The partial combustion takes place in the zone in-front of the tuyere openings. The heat generated pyrolyzes the fuel immediately above and the hot gases proceed downward through the firebox constriction, and exit through the grate. The throat (constriction) forces the raw gases to pass through a high temperature zone so that most of the unburnt pyrolysis products (tars) can be cracked into gaseous hydrocarbons, thus producing a relatively clean gas.

The downdraft gasifier, also called the cocurrent moving bed gasifier, is most commonly used for engine, applications because of its ability to produce a relatively clean gas. However, the presence of the throat poses problems of fuel movement. In the recent years, so called throatless gasifiers have been developed at Solar Energy Research Institute, Colorado, USA, Indian Institute of Science Bangaluru; Nimbkar Agricultural Research Institute, Phaltan; and Tata Energy Research Institute, New Delhi. For large sizes, fluidized bed gasifiers and entrained bed gasifiers have also been considered.

(iii) Cross Draught Gasifiers. In this type of gasifiers, the gas produced passes upwards in the annular space around the gasifier that is filled with charcoal. The charcoal acts as an insulator and a dust filter. They are usually suitable for power generation upto 50 kW.

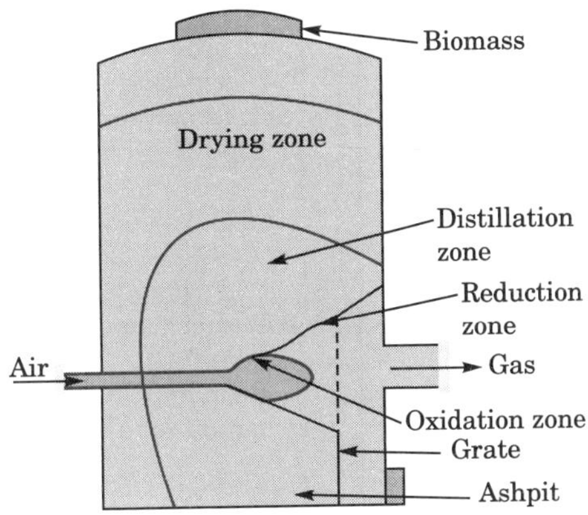

Fig. 7.28. Schematic diagram of a crossdraft gasifier.

Air enters the cross draft gasifier through a water cooled nozzle mounted on one side of the firehox. The gas is produced in the horizontal zone in front of the nozzle and passes through a vertical grate into the hot gas port on the opposite side. Because of the short path-length for the gasification reactions, this type of gas producer responds most rapidly for changes in gas production. However, the cross draft gasifier is not commnonly used.

Understanding of the various complex reactions occurring in a gasifier has so far been incomplete. However, the following reactions explain fairly well the process of gasification.

$$C + O_2 \longrightarrow CO_2 + 393800 \text{ kJ/kg mol} \qquad \text{(Combustion)}$$

$$C + H_2O \longrightarrow CO + H_2 - 131400 \text{ kJ/kg mol} \qquad \text{(Water gas)}$$

$$CO + H_2O \rightleftharpoons CO_2 + H_2 + 41200 \text{ kJ/kg mol} \qquad \text{(Water shift reaction)}$$

$$C + CO_2 \longrightarrow 2CO - 172600 \text{ kJ/kg mol} \qquad \text{(Boudouard reaction)}$$

$$C + 2H_2O \longrightarrow CO_2 + 2H_2 - 78700 \text{ kJ/kg mol}$$

$$C + 2H_2 \longrightarrow CH_4 + 75000 \text{ kJ/kg mol} \qquad \text{(Methane reaction)}$$

The sequence of these reactions in a downdraft gasifier is shown in Fig. 7.29.

With the help of the above reaction scheme and also with the knowledge of the equilibrium constants, it is possible to predict the equilibrium composition of the gaseous composition of the gaseous products. The equilibrium composition for a given solid fuel depends upon the air supply per unit weight of the biomass. A dimensionless parameter known as the equivalence ratio (*ER*) is usually defined to characterize the air supply conditions as follows:

$$ER = \frac{\text{(Weight of oxygen/Weight of dry fuel)}}{\text{(Weight of oxygen/Weight of dry fuel)}}$$

The denometer in above equation is the oxygen stoichiometric required for complete combustion of the fuel and varies from fuel to fuel. It is generally observed that for effective gasification, *ER* should be in the range of 0.2—0.4. Below an *ER* value of 0.2 pyrolysis predominates the process and above an *ER* value of 0.4, combustion predominates.

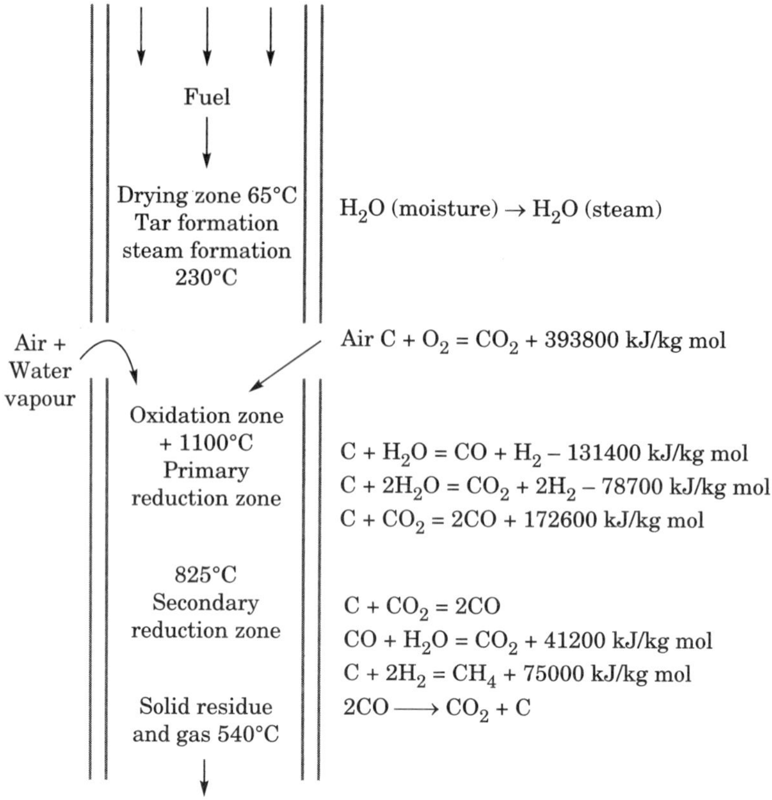

Fig. 7.29. Sequence of reactions in a downdraft gasifier.

In a real gasifier, however, the composition of the gas is far from equilibrium value and is generally dependent on the gasifier design. Factors that affect the gas composition are temperature distribution in the fuel bed, average gas residence time, and the residence time distribution. These are in turn dependent upon the mode of air entry, dimensions of the gasifier, and heat losses to surroundings. Modelling of the various processes occurring in the gasifier requires not only a knowledge of kinetics but also an understanding of the heat and mass transfer processes occurring in the various zones of the gasifier.

In all the existing gasifier designs, the raw producer gas generated contains varying amounts of moisture, dust char particles, and tar; the gases are also at a high temperature. Admission of hot gases into an engine results in loss of power and, hence, the gas has to be cooled. Also, the impurities present in the gas are detrimental to the operation, of IC engines and hence the gas has to be cleaned before admitting into the engine. No standard has yet been evolved to determine the degree of purity required for engine operation, but a tar, content of less than 100 mg/Nm^3 is generally acceptable. The conditioning of raw producer gas for use in IC engines usually consists of: (*i*) removal of larger char particles, (*ii*) scrubbing of gas to remove tar, (*iii*) cooling the gas, (*iv*) removal of moisture, and (*v*) removal of fine dust particles. These processes are accomplished in a gas cooling/cleaning train such as the one shown in Fig. 7.30.

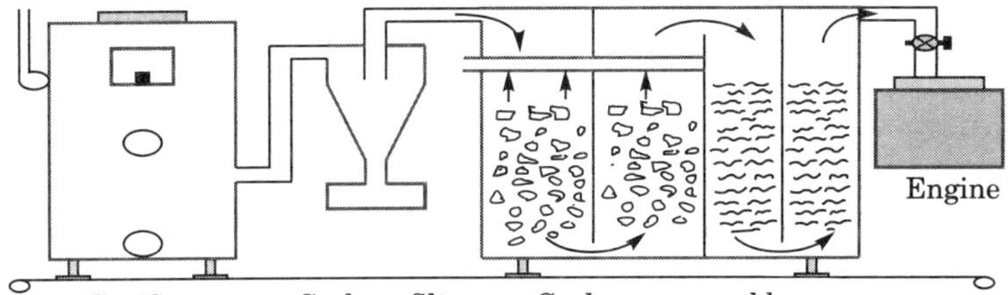

Fig. 7.30. Schematic diagram of a complete biomass gasifier-engine system.

Any carbonaseous matter can be gasified in principle. But from the point of view of operational reliability, some fuels are better than others. In general, fuels with low ash content such as charcoal, wood and coconut shells are extremely suitable for continuous and reliable operation. Commercial gasifiers have generally avoided the use of high ash or so called *slagging fuels*. However, successful operation of rice husk gasifiers for years in Italy and China, for example, show that it is technically feasible to utilize even problematic biomass residues for gasification.

(2) Fluidised Bed Gasifier. The selection of the design of the gasifier either fix bed or fluidised bed will depend on the actual application. However at this stage it is worth noting that the performance of a fixed bed gasifier will depend on the fuel properties, such as chemical analysis, volatile contents, calorific value, size distribution and ash characteristics. On the contrast, a fluidised bed gasifier is most versatile and any biomass (including sewage sludge; pulping effluents etc.) can be gasified using this type of gasifier. It provides a means of burning any combustible material from wet sewage sludge to refuse, with high efficiency and with minimal pollution. At the heart of a fluidised bed combustor is a hot bed of inert particles which are held in suspension— 'fluidised' by an upward current of air (refer Fig. 7.30). The calorific value of biomass is not a constraint. Besides being highly efficient because of high heat release rates as well as effective heat transfer resulting from rapid mixing and turbulence within the fluidised bed, fluidised bed gasifier can handle biomass with high ash content (for example rice husk).

Fluidised bed generally contain either inert material (sand) or reactive material (limestone or catalyst). These aid heat transfer and provide catalytic or gas clearing action. The bed material is kept in fluid state by the rising column of the gas. Normally the operating temperature of the bed is maintained within the range of 750–950°C, so that the ash zones do not get heated to its initial deformation temperature and this prevents clinkering or slagging.

Advantages of fluidised bed gasifier are:

(*i*) Fuel flexibility and type of fuel with calorific value ranging from 800 to 8000 kcal/kg can be used.

(*ii*) *Good heat storage capacity.* The bed has a very high heat storage capacity to always ensure combustion. In fact, when due to power failure if the gasifier had to cut off, it would be possible to restart the gasifier even after 5 hrs time without going through the initial starting procedure. Such a high heat storage capacity helps in elimination of flame sensing device and the associated problems connected with the same.

(*iii*) Quick start up.

(*iv*) High combustion efficiency.

(*v*) High output rate.

(*vi*) Consistent rate of combustion.

(*vii*) Usage of fuel with high moisture content.

(*viii*) Rapid, response to fuel input changes.

(*ix*) Because of the low temperature combustion, corrosion caused by alkali compounds in ash significantly reduced.

(*x*) Require much less boiler plan area than a stoker.

(*xi*) Uniform temperature throughout the furnace volume.

(*xii*) Reduced emission of harmful nitrous oxide.

(*xiii*) Sulphur dioxide emission can be reduced to acceptable level with less expense.

(*xiv*) Operation is as simple as that of an oil fired boiler.

7.24.3. Chemistry of the Gasification Process

The producer gas is formed by the partial combustion of solid biomass in a vertical flow packed bed reactor. In conventional producer gas theory, the reactions take place in three zones of a deepfuel bed, namely the oxidation, reduction and distillation zones. In the oxidation zone the oxygen in the air-steam blast reacts with the carbon in the fuel to reduce carbon to form hydrogen and carbon monoxide. The CO_2 coming from the oxidation zone is also reduced to carbon monoxide in the reduction zone. The final gas composition relies on the water-gas shift reaction

$$CO + H_2O \longrightarrow CO_2 + H_2 \qquad ...(7.8)$$

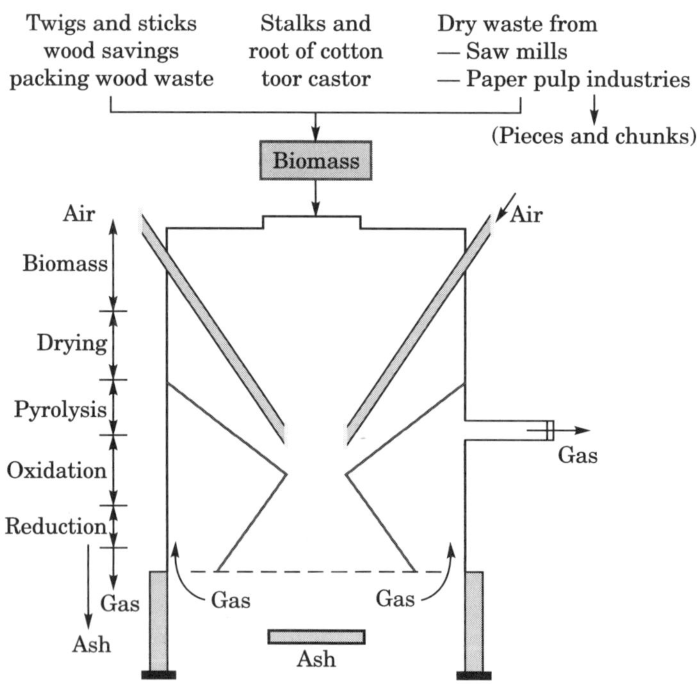

Fig. 7.31. Schematic diagram of a gasifier.

which approaches chemical equilibrium corresponding to the prevailing exit gas temperature. In the distillation zone the raw fuel is preheated and carbonised giving of condensable and noncondensable gases.

The process is called "gasification" as it transfers the majority of the chemically bound energy of the solid fuel into the gas phase.

As already stated, *Pyrolysis* (destructive distillation) converts organic wastes to char, tar and oils, and gas. Here gases produced are CO, CO_2 and H_2. The organic plants which are used in operation are called pyrolysis plants and vessel in which this takes place is called pyrolyzer.

In the Fig. 7.31 four separate zones are shown *i.e.,* Drying, *pyrolysis,* oxidation and reduction.

Thermo-chemical Reactions

The conversion process revolves around the combustion reaction Equation (7.9) which supplies heat to the neighbouring zones in the reactor. Carbon constitutes the major proportion of the feed and the main reactions are:

$$C + O_2 + 3.79\ N_2 = 3.79\ N_2 + CO_2 \quad \text{...(7.9)}$$

Exothermic reaction 395,000 kJ/kg atom of carbon. (Heat is evolved in the reaction). The carbon dioxide formed is reduced in the presence of flowing carbon, over 90% of CO_2 is reduced to CO at temperatures above 900°C,

$$C + CO_2 + 3.79\ N_2 = 3.79\ N_2 + 2CO \quad \text{...(7.10)}$$

It is endothermic reaction 172,000 kJ/kg atom of carbon (Heat is absorbed in the reaction).

The main reaction which ideally should occur in the reactor bed is

$$2C + O_2 + 3.79\ N_2 = 3.79\ N_2 + 2CO \quad \text{...(7.11)}$$

Exothermic reaction 223,300 kJ/kg atom of carbon.

The composition of producer gas formed under 100% conversion conditions assuming a carbon food based on the above equations would be 34.5% carbon monoxide and 65.5% nitrogen by volume. The calorific value of the gas thus formed is 4359 kJ/m^3 at STP. Based on the above, the potential heat available in 2 kg atoms of carbon is 790,800 kJ and of this 223,300 kJ are evolved as sensible heat (see equation 7.10), consequently, 567,500 kJ are available in the producer gas as potential heat in 5.79 kg moles of producer gas. The potential cold gas efficiency is therefore approximately 72%.

The hot gas reserves the heat evolved in equation (7.11) and therefore in practice the hot gas efficiency is much higher. The process is particularly efficient therefore when the gas is to be used as a heat source.

$$\text{Hot gas efficiency} = \frac{\text{Potential heat in the gas} + \text{Sensible heat in gas}}{\text{Total heat available in the solid fuel}}$$

The injection of water into the reactor by introduction in the feed or by direct injection of steam into the reaction zone and/or increasing the humidity of the combustion air can improve the calorific value of the resultant gases formed by shifting the gasification in favour of equation (7.12) below.

An important function of the combustion zone is to raise the reduction zone temperature to promote the carbon/steam gasification reaction which has a higher activation energy. This reaction requires temperature of 900°C and above.

$$C + H_2O = CO + H_2 \quad \text{...(7.12)}$$

It is endothermic reaction 130,000 kJ/kg atom of carbon. This practice is termed *"wet-gasification"* and has the added advantage of dropping the temperatures in the immediate vicinity of the hearth and reducing the formation of clinker.

This reaction (7.12) is very important in producer gas generation as it can enrich the gas manufactured with hydrogen, thus enhancing its calorific value. The other reaction with carbon steam occurs at lower temperature and predominates between 500–600°C

$$C + 2H_2O = CO_2 + 2H_2 \quad ...(7.13)$$

It is endothermic reaction 88,000 kJ/kg atom of carbon.

A further steam reaction which occurs in the gas producer with an excess of steam is the water gas shift reaction. This should be avoided in the producer gas generation as it reduces the cold gas efficiency, though, on the other hand, it reduces the carbon monooxide which is highly toxic in fuel gas distribution for public use.

$$CO + H_2O = CO_2 + H_2 \quad ...(7.14)$$

Exothermic reaction 42,000 kJ/kg mole of CO.

As it can be seen from the above equations, the presence of water in the biomass-feed effects the quantity of the gas produced and the efficiency of the conversion process. This applies also the hydroxyl groups present in other compounds in the biomass.

For low capital cost units it is preferred, however, to employ 'dry gasification' and not to introduce water as such into the reactor bed, as its addition in metered amounts is difficult to control and creats other practical problems. When using wood, peat or other uncarbonised residues sufficient steam is generated internally by chemical break-down. Excess moisture can produce poorer quality gas as shown in Equation.

Further factors affecting gasification are *air velocity, zone thickness, particle size and chemical composition of the feed.* In process design the after treatment of the gas is also very important that the carbon monoxide leaving the reduction zone should be cooled rapidly to below 600°C as its breakdown, to soot predominates at high temperatures. This cooling down of the hot gases reduces the deposit of soot and is called "Freezing the equilibria".

Other reactions which occur in the *char* at temperature of about 500°C are:

$$C + 2H_2 = CH_4 \quad ...(7.15)$$

Endothermic reaction 75,000 kJ/kgmole of carbon

$$CO + H_2O = CO_2 + H_2 \quad ...(7.16)$$

Endothermic reaction 42,000 kJ/kg mole of CO.

7.24.4. Applications of the Gasifier

The gasification technology has tremendous potential in terms of its applications as the process outputs can be converted into electrical, mechanical and/or heat energy.

The potential applications of the gasifier can be analysed based on the total energy requirements in a particular application.

Small Size Gasifiers. (upto 10 kW)

Normally this category of gasifiers shall find applications in rural areas, especially for providing shaft line power to agricultural pumps, processing machinery and agricultural—processing machineries like Thrashers, straw choppers, etc.

For example, a gasifier can be used to provide shaft line-power to a 10 HP pump used for irrigation. Normally the operating period for such pumps ranges from 600 to 1200 hours per year which can irrigate 10 to 15 hectares of land. In such application, the annual feed requirement of the gasifier shall range from 5.5 to 11.0 MT per year. This requirement is very nominal as this quantity of biomass can easily be available from one or two hectares only.

In fact, a high yielding crop like maize can produce as large as 6 to 8 tonnes of residue per hectare and if this residue is gasified, it can easily fuel a 15 to 20 hectares irrigation system.

Medium Size Gasifiers (10 kW—50 kW)

This category of gasifiers can easily meet the shaft line power requirements of various rural industries like saw mills, carpentary workshops, mechanical fabrication shops as well as small rice mills. They can also find extensive applications as a decentralised source of electrical energy in milk chilling centres, primary health coverage centres and for rural electrification.

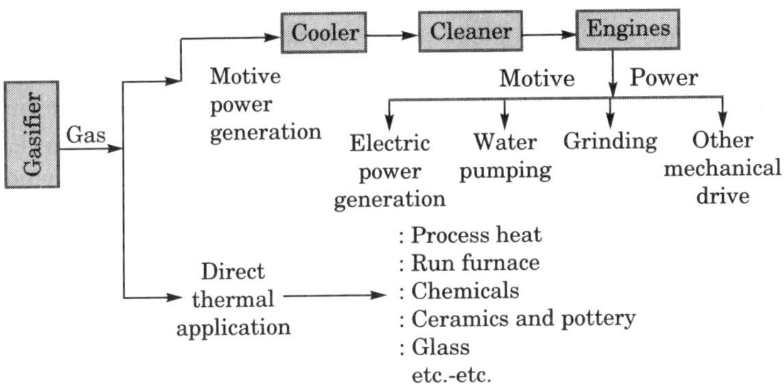

Fig. 7.32. Applications of the gasifier.

This category gasifiers can also be used to provide vehicular power or for combinations applications like vehicular power or for shaft line power to vehicle mounted equipment.

A 40 to 50 HP tractor/van carry a load of 5 to 7 tonnes. On this vehicle a straw chopping cum bailing machine with power requirement of 30 kW and weighing almost 1 tonnes can easily be mounted.

Now assuming that this mobile unit travels 50 kms in to and fro journey and operates the equipment for 6 hours in a day, the requirement of biomass as fuel to gasifier shall be 0.4 to 0.5 tonnes per day. The van can easily carry this quantity at fuel besides the mounted equipment. In fact, at the end of the day, the van carry back 4 to 5 tonnes of chopped and baled straw. This can give almost 75% capacity utilization for the vehicle mounted equipment.

In most of the applications, the gasifiers under this category can be installed without any heat recovery system. However, in special cases, the gasifier systems can have heat recovery in the form of hot water.

For example, a small size chilling centre with a storage capacity of 5000 litres of milk shall have a shaft line power requirement of 30 to 40 kW and hot water requirement of 1000 to 2000 litres per day. Both these requirements can be met through the gasifier and heat recovery system respectively.

Besides this, either mobile or stationary type of gasifier in this category can be used for decentralised electrification in small size villages or temporary establishments like camps, etc. A 40 kW system can meet the street light and in house electrification requirements for a camp with 200 to 300 tents.

In fact the decentralised electrification application has tremendous potential in remote and hilly areas where the cost of electricity transmission through normal channels is prohibitive or the establishments in such areas are of a temporary nature.

Large Size Gasifiers (50 kW and above)

This category of gasifiers can find applications in rural as well as urban industries, besides being a source of decentralised electrification. This category of gasifiers can meet the shaft power requirements directly or indirectly of various industries like dairy, oil mill, mineral processing, brick manufacturing, ceramics and pottery inductries etc. These gasifiers can also be used in mining operations, forest based wood processing units, well drilling etc. These gasifiers can be used for total electrification of small and medium size villages.

These gasifiers can be installed with or without heat recovery system depending upon the total energy requirements and its forms. For example, a gasifier system installed to generate 200 kW electrical energy for the dairy industry may have a heat recovery system to meet a vast of the hot water requirements in the dairy.

Direct Thermal Applications

The primary application for direct heat gasifiers is in industries where fuel oil or coal/lignite is being used to generate process heat or run furnaces and kilns. This opens a large number of possible uses, as fuel oil or coal/lignite is used extensively by industrial sector in most developing countries. Industries that commonly use fuel oil or coal/lignite for supplying heat include the following:

Cement manufacture,
Glass making,
Brick making,
Ceramics and pottery,
Rubber manufacture,
Food processing,
Brewing,
Crop drying,
Fertilizer production and
Chemicals.

In short all the categories have tremendous potential for varied applications. However the viability, both in terms of technical and economics, needs to be assessed in each case individually.

7.24.5. Problems in Developments of Gasifiers

Some of the major problems, technical and commercial, which are the barriers in the development of gasifier technology, are discussed briefly as follows.

(1) *Fuel Supply and Cost.* The fuel requirement for gasifiers vary from 1.0 to 1.3 kgs per kW depending on the feed properties like moisture content, calorific value, ash content, etc. and the conversion efficiencies of the system.

Accordingly, a 100 kW and 300 kW systems can annually require approximately 850 MT and 2550 MT of feed respectively. Normally such quantities, especially, 1000 MT and above, are not easily available at a particular location from a single source except in case of large agricultural or forestry processing centres. Against this, the feed requirements, for small and medium size gasifiers *i.e.*, upto 50 kW output can be marginal and be easily available from a single source of supply.

In view of this, the large system needs to be designed to have flexibility in terms of feed materials. The biomass materials, especially agricultural residues are not available throughout the year and hence need to be transported and stored for continuous availability. In practice, this can cause problems due to scattered and perishable nature, high cost of transportation, and inventory costs ensure continuous and ready availability of biomass feed materials.

The cost of fuel supply is also important, biomass resources may become costlier in future.

(2) *Capital Costs and Economics.* It is important, to evaluate how gasifiers compete against other technologies. The task is difficult since both the capital costs as well as economics of not only gasifiers but also of other competing systems can be highly location and application specific.

In considering economics, the capital, present charges for energy, operating costs, capacity utilization, etc. become important factors.

Depending on the cost of present source of energy supply various alternatives can become economically favourable or unfavourable. Besides this, the economics of other competing technologies also needs to be considered before an appropriate decision can be taken in favour of gasification.

(3) *Technology Development.* In gasifiers, the following three areas need further technical development.

(1) Bridging and jamming associated with handling of feed and/or ash either inside or outside gasifiers.

(2) Lack of proven gas clean up systems, which results in tar build up on the valves and fast wearing of engine cylinder and pistons. This, in turn, results in frequent overhaul of the system.

(3) Lack of data to carryout proper modifications in air fuel mixing systems as well as the engines.

In fact, the ready availability of different types of engines and their characteristics, especially with special reference to excess air provision, mixing systems, compression ratio, etc. need to be considered while planning the development of the technology.

(4) *Safety.* Producer gas is a low heating value gas which is poisonous and explosive. Hence, proper carbon monooxide metering arrangement as well as safety devices need to be provided with the systems, especially when gasifiers are installed in buildings. The systems also need provision to divert the gas to stack in an emergency shut down.

The larger systems also need proper control loops to ensure adequate fuel in the gasifier and matching combustion air flow as well as to carry out technical shut down for the entire system under abnormal operating conditions. This may include various conditions like over-crank, over speed, low lube oil pressure, high gas intake, temperature, low cooling water level, etc. In short, adequate safety precautions need to be provided depending on size and application of the system.

(5) *Market Barriers.* The known technologies and conventional fuels can pose problems for gasifier marketing. The potential users recognize the risk in the new technology that has been fully demonstrated to be significantly less expensive or more reliable. Hence the gasification system suppliers must convince the potential users that the cost is worthy and there is no risk of operating a new technology.

7.25. Advantages and Disadvantages of Bio-logical Conversion of Solar Energy

Advantages

(*i*) The initial investment is comparatively low, compared to other methods of solar energy utilization, since sophisticated or expensive equipment is not needed.

(*ii*) Processing biomass for fuel from sewage and wastes reduces the environmental hazards.

(*iii*) The techniques for raising such plants are simple and similar to conventional agriculture. Hence they can be readily used in the rural areas, of developing countries.

(*iv*) Small units can be located near to the accumulation of waste in the forming region.

(*v*) The by products of such a system can be fully recycled.

(*vi*) The methods of power production are relatively low polluting.

(*vii*) Bio gas can be distributed through GI pipes for domestic use, and

(*viii*) It can be stored in container which can be transported to consumers.

(*ix*) In rural areas hygenic and sanitation conditions are improved.

(*x*) The production of biological fuels may be coupled to the synthesis of protein.

Disadvantages or Limitations

(*i*) The land area required is relatively large and hence this method is ruled out for areas where land cost is high.

(*ii*) Photosynthesis efficiency or a small percentage (of the order of 0.1%) of solar radiations are converted into biomass by plants.

(*iii*) The cost of energy produced is higher than in urban areas with centralised power generation stations.

(*iv*) Relatively low concentration of biomass per unit area of land.

(*v*) Collection and transportation of biomass become expensive due to high moisture dontents (of the order of 50—95%).

(*vi*) Since the gas cannot be liquified ordinarily there is storage problems, it can not be bottled like LPG into cylinder.

7.26. Pyrolysis (destructive distillation)

We use *'pyrolysis'* as a general term for all processes whereby organic material is heated or partially combusted to produce secondary fuels and chemical products. The input may be wood, biomass residues, municipal waste or indeed coal. The products are gases, condensed vapours as liquids, tars and oils and solid residue as char (charcoal) and ash. Traditional *charcoal making* is pyrolysis with the vapours and gases are not collected. *Gasification* is pyrolysis adapted to produce a maximum amount of secondary fuel gases. Various pyrolysis units are shown in Fig. 7.33. Vertical top loading devices are usually considered the best. The fuel products are more convenient clean or transportable than the original biomass. The chemical products are important as chemical feedstock for further processes, or as directly marketable goods. Partial combustion devices, which are designed to maximize the amount of combustible gas rather than char or volatiles, are usually called *gasifiers.* The process is essentially pyrolysis, but may not be described as such.

Efficiency is measured as the heat of combustion of the secondary fuels produced, divided by the heat of combustion of the input biomass as used. Very high efficiencies of 80 to 90% can be reached. For instance gasifiers from wood can produce 80% of the initial energy in the form of combustible gas (largely H_2 and CO), suitable for operation in converted petroleum fueled engines. In this way, the overall efficiency of electricity generation (say 80% × 30% = 24%) could be higher than that obtained with a steam boiler. Such gasifiers are potentially useful for power generation less than 150 kW.

The chemical processes in pyrolysis are much related to similar distillations of coal to produce synthetic gases, tars, oils, and coke. For instance the large scale use of piped town gas (H_2 + CO) in Europe, before the recent change to natural gas (mainly CH_4), was possible from the reaction of water on heated coal with reduced air supply:

$$H_2O + C \longrightarrow H_2 + CO \quad ...(7.17)$$
$$C + O_2 \longrightarrow CO_2; \quad ...(7.18)$$
$$CO_2 + C \longrightarrow 2CO \quad ...(7.19)$$

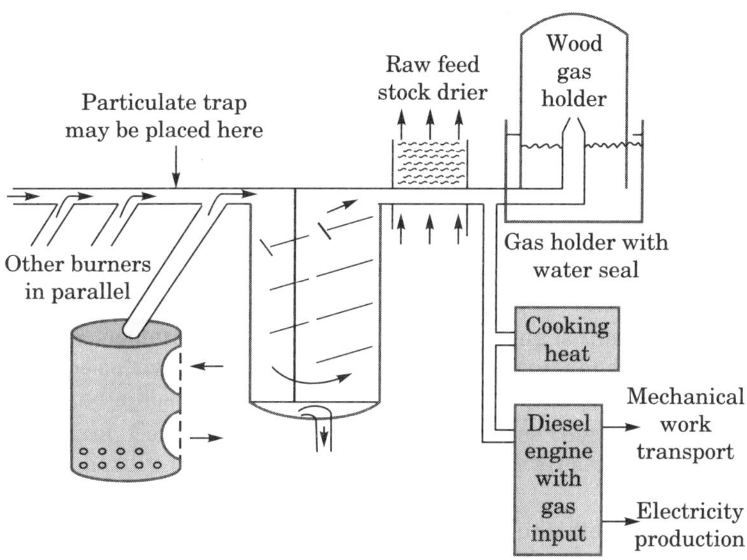

Fig. 7.33. Small scale pyrolysis unit.

The following is given as a summary of the wide range of conditions and products of pyrolysis. The input material needs to be graded to remove excessive non-combustible material *(e.g.,* soil, metal), dried if necessary (usually completely dry material is avoided with gasifiers), chopped or shredded, and then stored for use. The air/fuel ratio during combustion is a critical parameter affecting both the temperature and the type of product. Pyrolysis units are most easily operated below 600°C. Higher temperatures of 600 to 1000°C, need more sophistication, but more hydrogen will be produced in the gas. Below 600°C there are generally four stages in the distillation process:

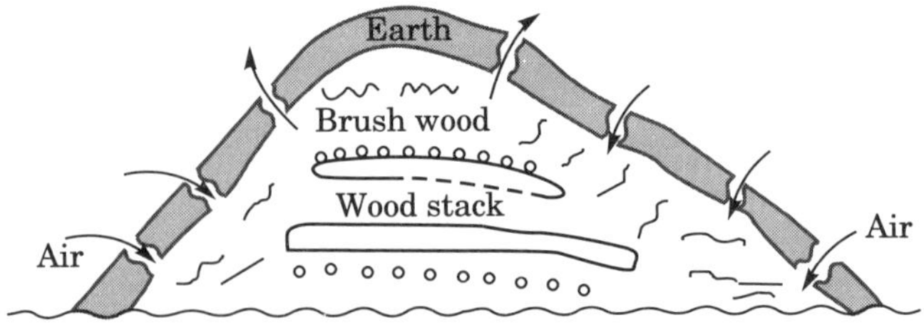

Traditional charcoal kiln.

Fig. 7.34. Pyrolysis systems.

(1) **100** to **120°C.** The input material dries with moisture passing up through the bed.

(2) **275°C.** The output gases are mainly N_2, CO and CO_2; acetic acid and methanol distill off.

(3) **280—350°C.** Exothermic reactions occur, driving off complex mixtures of chemicals (ketones, aldehydes, phenol, esters), CO_2, CO, CH_4, C_2H_6 and H_2. Certain catalysts, *e.g.* ZnCl, enable these reactions to occur at power temperature.

(4) **Above 350°C.** All volatiles are driven off, a higher proportion of H_2 is formed with CO, and carbon remains as charcoal with ash residues.

The condensed liquids, called tars and pyroligneous acid, may be separated and treated to give identifiable chemical products (*e.g.* methanol, CH_3OH, a liquid fuel). Table (7.10) gives examples and further details.

The secondary fuels from pyrolysis have less total energy of combustion than the original biomass, but are far more convenient to use. Some of the products have significantly greater energy density (*e.g.,* CH_4 at 55 MJ/kg) than the average input. Convenience includes better control of combustion, easier handling and transport, greater variety of end-use devices, and less air pollution. The following general comments apply:

Table 7.10. Pyrolysis yields from the dry wood

Yields per 1000 kg dry wood	
Charcoal	300 kg
Gas (combustion 10465 kJ/m^3)	140 m^3 (NTP)
Methyl alcohol	141
Acetic acid	531
Esters	81
Acetone	31
Wood oil and light tar	761
Creosote oil	121
Pitch	30 kg

1. **Solid char (mass yield 25—35% maximum)**

Modern charcoal retorts operating at about 600°C produce 25 to 35% of the dry matter biomass as charcoal. Traditional earthen kilns usually give yields closer to 10% since there is less control. Charcoal is 75 to 85% carbon, unless great care is taken to improve quality (as for chemical grade charcoal), and the heat of combustion is about 30 MJ/kg. Thus if charcoal alone is produced from wood, between 15 and 50% of the original chemical energy of combustion remains. Charcoal is useful as a clean controllable fuel. Chemical grade charcoal has many uses in laboratory and industrial chemical processes. Charcoal is superior to coal products for making high quality steel.

2. **Liquids (condensed vapours, mass yield ~ 30% maximum)**

These divide between (*i*) a sticky phenolic tar (creosote) and (*ii*) an aqueous liquid, pyroligneous acid, of mainly acetic acid, methanol (maximum 2%) and acetone. The liquids may be either separated or used together as a crude fuel with a heat of combustion of about 22 MJ/kg. The maximum yield corresponds to about 400 litres of combustible liquid per tonne of dry biomass.

3. **Gases (mass yield ~ 80% maximum in gasifiers)**

The mixed gas output with nitrogen is known no wood gas, synthesis gas, producer gas or water gas, and has a heat of combustion in air of 5 to 10 MJ/kg (4 to 8 MJ/m^3 at STP). It may be used directly in diesel cycle or spark ignition engines with adjustment of the fuel injector or carburettor, but extreme care has to be taken to avoid intake of ash and condensable vapors. The gas is mainly N_2, H_2 and CO, with perhaps small amounts of CH_4 and CO_2. The gas may be stored in gas holders near atmospheric pressure, but is not conveniently compressed. A much cleaner and more uniform gas may be obtained by gasification of wet charcoal rather than wood, since the majority of the tars from the original wood have been removed.

7.27. Further Thermochemical Processes

In the previous sections biomass has been used directly after preliminary sorting and cutting for combustion or pyrolysis. However the biomass may be treated chemically, (1) to produce material suitable for alcoholic fermentation (section 7.27) or (2) to produce secondary or improved fuels.

Consider the following few important examples from the great number of possibilities.

(1) Hydrogen reduction. Dispersed, shredded or digested biomass, *e.g.*, manure, is heated in hydrogen to about 600°C under pressure of about 50 atmospheres. Combustible gases, mostly methane and ethane, are produced that may be burnt to give about 6 MJ per kg of initial diy material.

(2) Hydrogenation with CO and steam. As above, but heated with CO and steam to about 400°C and 50 atmospheres. A synthetic oil is extracted from the resulting products that may be used as a fuel. A catalyst is needed to produce reactions of the following form:

$$CO + H_2O \longrightarrow CO_2 + H_2$$

$$C_n(H_2O)_n + (n + 1)\, H_2 \longrightarrow nH_2O + H(CH_2)_nH \qquad ...(7.20)$$

where the latter reaction implies the conversion of carbohydrate material to hydrocarbon oils. The energy conversion efficiency is about 65%.

(3) Acid and enzyme hydrolysis. Cellulose is the major constituent (30% to 50%) of plant dry biomass and is very resistant to hydrolysis and hence fermentation by micro-organisms. Conversion to sugars, which can be fermented, is possible by heating in sulfuric acid or by the action of enzymes (celluloses) of certain micro organisms. The products may also be used as cattle feed.

(4) Conversion of coconut oils to esters as fuel. This is an example of many chemical processes becoming available to improve bio fuels. About 50% of the white 'meat' of coconuts, copra, is oil that is extracted by squeezing and rolling. Substantial trade exists in this oil, *e.g.* the 'Philippines' annual production of coconut oil is 10^6 t/year. 1The oil may be used directly in diesel engines with adjustment to the injection system, but toxic fumes are produced, deposits occur and the fuel solidifies below 23°C. Reaction with 20% by volume of ethanol or methanol produces volatile easters and glycerol. The easters are better diesel engine fuels than the oil, and the glycerol is a valuable byproduct. The ethanol can be made by fermentation from natural sugars, and the methanol *via.* gasification from wood wastes. The energy density of the easters is about 38 MJ/kg, which is higher than the oil and near to petroleum at about 46 MJ/kg. Other vegetable oils can be used similarly in diesel cycle engines.

(5) Methanol liquid fuel. Methanol, is a toxic liquid, is made from the catalytic reaction of H_2 and CO at 330°C and at 150 atmospheric pressure:

$$2H_2 + CO \longrightarrow CH_3OH$$

The gases are the components of synthesis gas and may be obtained from gasification of biomass. Methanol may be used as a liquid fuel in petroleum engines with an energy density of 23 MJ per kg.

7.28. Alternative Liquid Fuels (Alcohol fuels)

Ethanol, C_2H_5OH, is produced naturally by certain micro-organisms from sugars under acidic conditions, pH 4 to 5. This alcoholic termentation process is used world wide to produce alcoholic drinks. The most common micro organism, the yeast *Saccharomyces cerevisiae,* is poisoned by C_2H_5OH concentrator greater than 10%, and so higher concentrations upto 95% are produced by distilling and fractionating (Fig. 7.35). When distilled, the remaining constant boiling point mixture is 95% ethanol, 5% water. Anhydrous ethanol is produced commercially with azeotropic removal of water by codistillation with solvents such as benzene. Only about 0.5% of the energy potential of the sugars is lost during fermentation, but significant amounts of process heat are required for the concentration and separation processes. This process heat may be provided by the combustion or gasification of otherwise, waste biomass.

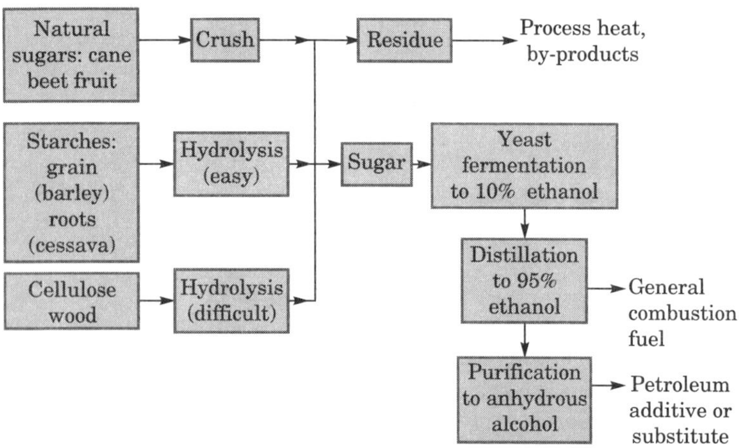

Fig. 7.35. Ethanol production.

The sugars may be obtained by the following routes:

(*i*) Directly from sugarcane

(*ii*) From sugar beet

(*iii*) From starch crops *e.g.*, grain and cassava and

(*iv*) From cellulose

The improved production of *ethanol* has been viewed with particular interest since it can be used in existing motor vehicles as a gasoline extender (gasohol) or octane booster. Alcohol—gasoline blends, although they require strict elimination of water from the alcohol, have continued to gain importances.

Alcohols can also be used for heating and lighting, in simple heaters on wick lamps, or in pressure stoves and lanterns. They are clean, comparatively safe fuels, and have the advantage over petroleum fuels of being mixable with water, which makes accidental fires easier to control. Use of ethanol as a boiler fuel could help consume the increased production. Electrical generation is another use that avoids the necessity of new storage facilities for temporary surpluses. Further, costs of boiler conversion from oil to ethanol are modest compared with those required for coal or charcoal, and ethanol has a low atmospheric pollution factor.

Methanol has been used successfully in gas turbines methanol has also been used developmentally in fuel cells, devices in which the chemical energy of the fuel is converted directly to electrical power. If fuel cells prove practicable, this could provide a significant market for methanol.

Butanol can also be produced from non-petroleum source and used as a fuel. It is, however, comparatively expensive.

Some characteristics of these alcohols are compared with those of gasoline and diesel oil in Table 7.11. Differences include the fact that the alcohols, being essentially pure chemicals, boil at one temperature while the petroleum fuels are mixtures of several chemicals with a wide range of boiling temperatures. Alcohols contain oxygen, whereas the petroleum fuels do not. The heating values for the alcohols are significantly lower than for the petroleum fuels with that of ethanol being about two-thirds that of gasoline.

Table 7.11. Liquid fuel characteristics of alcohols, diesel and gasoline

Characteristic	*Gasoline*	*Diesel*	*Methanol*	*Elthanol*	*Butanol*
Boiling point °C	32-210	204-343	65	78	118
Lower heating Value (MJ/kg)	44.5	43	19.6	26.9	33.1
MJ/lit.	31.9	38.9	15.4	21.1	26.7

The wide boiling range of gasoline offers a special advantage; the lower boiling components are sufficiently volatile to allow engine start-up at low temperatures. Unmodified engines powered by pure methanol or ethanol will not start below 10-15°C. The presence of oxygen in the alcohols gives them compatibility with water, gasoline or diesel fuels and this means that less air is required for combustion. The alcohols have an affinity for water sufficient to draw it from the air, a characteristic that can cause problems in their unprotected use and storage.

Ethanol has greater potential for use as an *industrial solvent* and *chemical* than as a liquid fuel. For example, about half the ethanol sold in the U.S.A. is used as a solvent. Printing inks, shellacs, varnishes, nitrogen cellulose coating, cosmetics, pharmaceuticals and some foods, all use ethanol in varying amounts. Paints, brake fluids, lubricants, herbicides, pesticides and explosives, all ethanol at some point in their manufacture.

For use as a chemical feedstock, ethanol is dehydrated to ethylene. By using a fixed or fluidised catalyst bed at about 300°C, a yield of 85—90% can be obtained (about 1.70—2.0 *t* ethanol produces 1 *t* of ethylene). Ethanol has also been used to produce acetaldehyde which is a precursor for acetic acid, acetic anhydride, butanol, DDT ethyl hexanol and butadience. The bulk of these chemicals is now based on ethanol.

Fermentation ethanol may have the best future as a substrate for single-cell protein production. Its advantage in this use include: purity, acceptability; ease of storage and handling; non-toxicity; versatility as a substrate or CO substrate; miscibility with water; relatively low oxygen demand; relatively low temperature fermentation; and high cell yields.

(1) **Ethanol production.** Ethanol production is based on two major techniques—*chemical synthesis* and *fermentation.* The primary synthetic route is the catalytic hydrolysis of ethylene derived from petroleum:

$$CH_2 = CH_2 + H_2O \longrightarrow CH_3CH_2OH \qquad ...(7.21)$$

Other raw materials that may be used by this method include natural gas, coal, oil shales and tar sands. Synthetic ethanol plants range in annual production from 40 ml to 450 ml.

For the fermentation sources, there are four categories of feedstocks: sugar-containing materials, carbohydrates, lignocellulosics, and urban and industrial wastes.

Raw materials. Materials containing sugar require the least costly preparation, as the sugars are already available in degradable form. Sucrose occurs in free form in sugarcane, and clarified can juice may be used directly for ethanol production. However, molasses is used rather than the corresponding juices for various reasons. This class of raw materials is relatively the most expensive to obtain since it has other markets. Sugar beets and sweet sorghum also have potential for ethanol production, but sweet sorghum juice contains some starch and aconitic acid that cause difficulties in sugar crystallization.

Starch or carbohydrate is an attractive feedstock for many fermentation processes. High yields and starch content (50–60%) give corn an advantage. Spoiled and low quality starches, not suitable for food or animal feed, can be effectively used for production of alcohol. Even if food grade starch is used, the resulting biomass byproduct represents an excellent, protein—fortified human or animal food.

Potatoes and cassava offers a high yield of starch per hectare of cropland, but the problem of storage to allow year—round ethanol production has not been overcome. Cassava, one of the most efficient photosynthetic plants, contains 20-35% starch and 1-2 % protein.

Lignocellulose is the major renewable from the carbohydrate in the world. It has been estimated that about 100 Gt cellulose is produced worldwide each year, a significant fraction of this occurs as municipal and agricultural waste. The energy crisis of the 1970s created renewed interest in the use of cellulose, but it is still difficult to convert it efficiently to its degradable manomer sugar units. A substantial improvement in cellulose hydrolysis technology is required before this resource material can be used on a large scale as a biomaas feed stock for ethanol production. While there are great quantities of urban and industrial wastes which require processing for environmental reasons, these materials are so widely distributed that only a few isolated sources have economic potential.

The collection, transportation, concentration and storage of these materials present technological difficulties. As well, lignocel-lulosic materials included in urban and industrial wastes require pre treatment to yield degradable carbohydrates.

Pretreatment. The fibre component of cellulosic biomass, hemi cellulose and cellulose, can be hydrolyse, but the structural features of cellulose, such as the degree of crystallinity, the degree of polymerization, the degree of water swelling and surface area, make it less amendable to hydrolysis. Some form of pretreatment is, therefore, necessary to obtain high yields of monosaccharides. Pretreatment methods, which disrupt the highly ordered cellulose structure and the lignin carbohydrate complex, remove lignin and increase the surface area available to enzymes (or acids), not only promote hydrolysis, but also enhance the rate and extent of cellulose sachharification.

Several techniques have been proposed to enhance the degree of hydrolysis of lignocellulosic to sugars. These techniques are generally classified as physical, chemical, biological or a combination of these, according to their mode of action.

The physical pretreatment involves particle reduction to very small mesh sizes. The shearing and compressive forces involved in milling cause a reduction in crystallinity, a decrease in the mean degree of polymerization and an increase in bulk density. The milled material also allows for a high slurry concentration, thereby reducing the reactor volume and hence, capital cost. In general, the power requirements for these grinding or milling methods are so large as to make them quite expensive chemical treatments with strong acids or bases, such as sulphuric acid or sodium hydroxide, or with other cellulose swelling or dissolving agents also effectively increase hydrolysis. These chemicals are generally quite corrosive and expensive and must be recovered for reuse. Furthermore, they are often toxic or inhibitory to micro organisms (or their enzymes) so that their removal from the pretreated cellulosic materials must be almost complete. All these factors combine to increase the expense and difficulty of such chemical treatment methods.

Another technique of current interest is the steam explosion process (Muzzy et al. 1982). Lignocellulosic materials (usually wood chips) are saturated with water under pressure of 2000–3400 KPa (21–35 kg/cm^2) at elevated temperatures (215–260°C). When the pressure is released, the water evaporates rapidly and the wood fibres tend to separate. In addition, the moisture and high temperature liberate acetic acid which catalyses the hydrolysis of the hemicellulose.

Although this process is very effective, it requires considerable thermal energy and some of the sugars are inevitably degraded by the high temperatures involved.

Conversion Processes

Acid hydrolysis can be carried out with dilute or concentrated acids. The important advantage of concentrated acid hydrolysis is that, since it is carried out at low temperatures, the sugars formed are not degraded. However, the large quantity of acid needed to wet the feedstock can not be discarded. Acid recovery is thus necessary, and this is a every expensive operation. As a result, virtually all recent development work has been with dilute acids.

Cellulose hydrolysis in dilute acids requires higher temperatures (at least 180°C) which cause the product sugars to decompose. The reaction kinetics have been extensively studied increasing either the acid concentration (within the dilute range) or the temperature causes more net production of the desired product, glucose. Further more, for a given acid concentration, the maximum glucose yield that can be achieved at a given temperature occurs at an optimum reaction time. This optimum reaction time decreases as the reaction temperature increases.

The basic dilute acid hydrolysis process consists of upto six steps : feed stock preparation, prehydrolysis, hydrolysis of cellulose, neutralization and clean up of the hydrostate, fermentation of hexose sugars to ethanol and distillation.

Pretreatment of the feedstock is not as critical for acid hydrolysis as it is for other conversion methods. With wood, size reduction to chips is generally sufficient. Further size reduction may be required for plug flow reactors.

Prehydrolysis of the hemicellulose is often carried out so that hemicellulose sugars (pentose and hexose) can be recovered before they are degraded by the more severe cellulose hydrolysis step. The prehydrolysis step also improves the yields of the main hydrolysis, particularly for substrates such as wood which contain a significant amount of lignin.

Atleast four reactors types have been investigated—percolator semi continuous, plug flow, cocurrent and counter current moving beds and simple batch. Percolator, cocurrent and counter current reactors have intrinsic yield advantage over plug flow and batch reactors. However, another important reactor consideration is the product sugar concentration. Dilute solution such as those obtained in the percolator are much more expensive to ferment and distil. Higher product concentrations may be achievable in plug flow reactors, which also offer potentially higher through puts. After hydrolysis, lignin is removed by filtration or centrifugation and sent to the boiler.

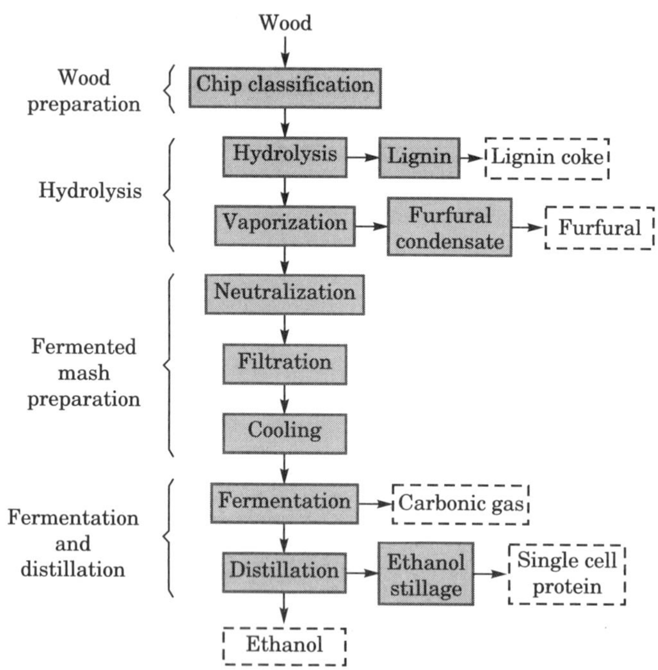

Fig. 7.36. Ethanol from wood by acid hydrolysis.

The by products from an acid hydrolysis process depend primarily on the pretreatment and prehydrolysis options chosen. The lignin residue from acid hydrolysis is not very reactive and probably suitable only for use as fuel. However, if markets for lignin are available that justify the cost, it can be recovered ahead of acid hydrolysis by steam explosion and caustic or other solvent extraction techniques.

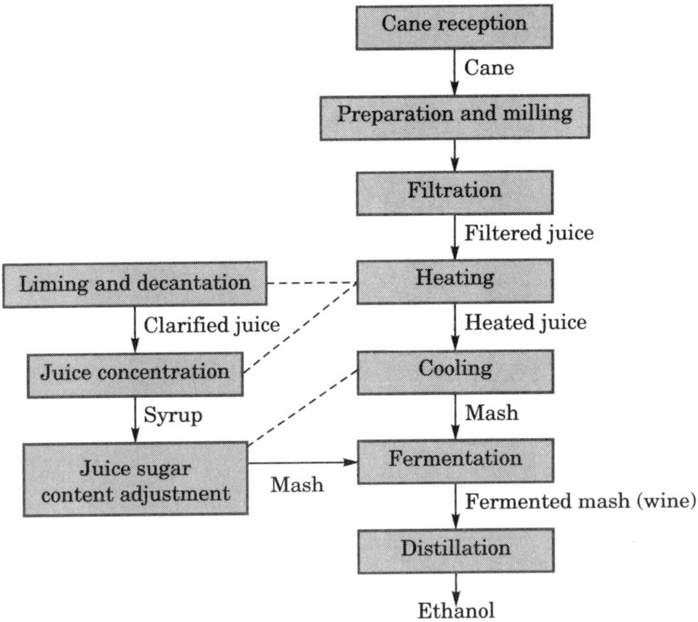

Fig. 7.37. Ethanol from sugar cane.

If a prehydrolysis step is included the hemicellulose sugars can be recovered with little degradation, and the hexose can be fermented with the glucose produced from the cellulose. Technologies are also being developed to convert the pentose sugars to energy products; otherwise, they may be marketable in other forms, such as animal feed molasses. If there is no prehydrolysis, potentially recoverable amounts of furfural, acetic acid and other products are formed.

Acid hydrolysis is suitable for any cellulosic feedstock, although it is most often used with wood. For the batch or semi continuous hydrolysis processes, the permeability of the bulk substrate is important. Some cellulosic feedstocks tend to compact and cause plugging of in the beds. Thus feedstocks like bagasse, corn stover and municipal wastes may be difficult to hydrolyse in a batch reactor. The moisture content of the feedstock, determines the amount of acid needed to achieve the desired concentration. It is undesirable to dry the feedstock before hydrolysis, as this reduces the ability of the acid to diffuse into the substrate.

When compared with enzymatic hydrolysis processes and direct conversion technologies, acid hydrolysis has many *advantages*. It has been much more extensively studied and practised at all levels, from the laboratory to commercial operations. Also, the use of wood as feedstock introduces no technical problems and does not require extensive pretreatment. Another advantage is that asepsis is not an important issue for the acid hydrolysis unit operation. Finally, it has much shorter reaction time than enzymatic hydrolysis. The acid catalyst is cheap and readily available. No expensive enzyme production facilities are necessary.

Offsetting these advantages are certain limitations. For example, acid hydrolysis has a lower potential yield of ethanol than enzymatic hydrolysis or direct conversion. Because of lower yields, the potential sugar concentration is lower than that of form enzymatic hydrolysis. Furthermore, some of the degradation products formed are toxic to the yeast used in fermentation. Also, the by-product, lignin, is of lower potential market value than that from enzymatic hydrolysis.

Another problem is that the more severe operating conditions— acid combined with high temperatures—cause corrosion and require expensive construction materials. Metals from the reactor can have an inhibitory effect on the hydrolysis reaction, thereby decreasing the fermentable sugar yield.

Unlike acid hydrolysis of cellulosic materials, enzymatic conver-sion has not been used on a commercial scale. The basic enzymatic hydrolysis process consists of these essential operations: pretreatment of the feedstock, production and recovery of cellulose enzymes, sacharification of the pretreated cellulose, recovery of the fermentable sugars, fermentation of the sugar syrups, recovery of the product (alcohol) and processing of by products and waste.

The enzymes are produced by growing a cellulase producing organism in a submerged culture, with cellulose or another suitable carbohydrate as the primary source of carbon. Because of higher yields and superior cellulase qualities, *Trichoderma* is almost always preferred.

For a typical batch fermentation with *trichoderma,* the level of carbon source in the medium ranges from 2-8%, depending on the feedstock and its degree of pretreatment. The medium also consists of mineral salts, a source of complex nitrogen and trace elements. The fermentation takes place at 28-32°C, and requires 2-13 days. Enzyme concentration in the spent medium as high as 40 mg protein per ml broth has been reported, but this was achieved in a laboratory under very controlled conditions with specially pretreated cellulose.

The degree of hydrolysis is a function of time and temperature. If the hydrolysis remains below 40°C, the sachharification of any concentration of cellulose can theoretically be complete, given sufficient reaction time. At temperatures above 40°C, the reaction proceeds more quickly; but the enzyme begins to degrade or inactive. Most processes rely on fairly high sachharification rates at the expense of enzyme activation. Temperatures of 45-50°C with holding times of 24-48 hours are most common.

Undigested insolubles, mostly lignin and cellulose, can be removed by filtration. The insolubles are generally easily dewatered by mechanical means to 50% solid. At this stage, they represent a reasonable source of solid fuel.

Nearly any cellulosic—based material can be enzymatically hydrolysed if it has been adequately pretreated. Wood chips require extensive pretreatment, such as explosive decompression using high-pressure steam. Most hardwoods, particularly those of the poplar species are excellent substrates after this type of treatment. Softwoods and some hardwoods, such as *eucalyptus,* when exposed to this type of process produce undesirable by products that inhibit enzymatic hydrolysis and produce sugar solutions containing materials toxic to yeast. Therefore, more, expensive pretreatments that produce a substantially comminuted product and possibly a partially pulped material would be necessary to convert softwood species.

One of the major advantage of enzymatic hydrolysis is that the enzyme reactions are specific. No secondary reactions with the products or, other components are catalysed under the relatively mild temperatures and pH levels. This result in potentially higher yields and higher sugar concentration as well as a lignin rich by product that is still highly reactive. Secondly, the mild operating conditions do not require the use of expensive corrosion resistant alloys and pressure vessel construction.

But the asepsis requirement for enzyme production may impose severe limitations. The hydrolysis reaction is also highly susceptible to microbial contamination. Todate, there has been limited experience in the operating of hydrolysis reactors for extended periods. The dilute sugar concentrations, long holding times, mild temperatures, carryover of essential salts and relatively high concentrations of complex nitrogen are ideal conditions for the growth of many organisms.

Fermentation Systems

The conversion of cellulose, starch and sugars to ethanol (C_2H_5OH) is:

$$\underset{\text{(Cellulose)}}{(C_6H_{10}O_5)_n} + nH_2O \longrightarrow \underset{\text{(Glucose)}}{nC_6H_{12}O_6} \quad ...(7.22)$$

$$C_{12}H_{22}O_{11} + H_2O \longrightarrow \underset{\text{glucose}}{2C_6H_{12}O_6} \quad ...(7.23)$$

$$C_6H_{12}O_6 \longrightarrow \underset{\text{ethanol}}{2C_6H_5OH} + 2CO_2 \quad ...(7.24)$$

The conventional batch process for the production of ethanol from molasses was developed in the days of inexpensive energy and must be considered obsolete. The concept of continuous tower fermentation was developed for the brewing industry in Britain in the 1960s. However, many factors limit the use of tower fementers for the fuel-grade ethanol production. For example, as the feed sugar concentration is increased (to 15-20%) to raise the product ethanol concentration, so too do the fluid specific gravity and viscosity, both factors leading to increased expectation of yeast floc washout in the tower with an increase in the hydraulic loading. Also, it has been suggested the yeast will flocculate properly in the presence of significant concentrations of glucose, again leading to washout of the yeast.

Significant improvements in ethanol production technology are necessary in order to reduce substantially production costs. To minimize costs, a balance must be struck among the three principal goals of ethanol fermentation:

(*i*) High substrate utilization,

(*ii*) High ethanol productivity, and

(*iii*) High ethanol concentration in the product stream to the recovery section.

These goals are to some extent incompatible. In batch fermentations, complete sugar consumption means long fermentation times and, hence, low average ethanol productivity by volume with continuous stirred tank fementers, it is common to have the point of maximum ethanol productivity at a dilution rate as near to maximum as possible, leaving some residual sugar in the fermenter effluent. The choice of the best strategy is therefore a function of factors such as the type of raw material, cost and availability of distillation energy, available waste disposal options etc.

Because of these factors, cost is minimized by streams using high initial substrate concentrations in rapid fermentations. The rate of fermentation and, hence the size of the fermentation equipment, are affected by the use of high concentrations of yeast to speed fermentation and the removal of the rate inhibitory ethanol from the fermentor as fast as it is formed.

Methanol Production

Methanol is a clear, colourless liquid that freezes at – 73°C and boils at 65°C. It is an excellent antifreeze, as it is completely miscible with water. Its octane rating is high and it burns cleaner than gasoline in an internal combustion engine.

Current processes for producing methanol use, synthesis gas, a mixture of carbon monoxide and hydrogen usually derived from natural gas. This mixture is reacted under pressure of 10–20 MPa (100-200 atmospheres) at high temperatures (250–330°C) to yield methanol in the presence of a suitable catalyst:

$$CO + 2H_2 \longrightarrow CH_3OH \qquad ...(7.25)$$

The methanol-containing gases leaving the converter after the reaction are cooled to condense the methanol, and the unreacted gases are recycled. The methanol is then purified by distillation. Only a small amount of by product is obtained, including dimethyl ether and higher alcohols. Although any carbon source can be used to make synthesis gas, natural gas has been preferred because of cost, convenience and efficiency. It is possible to generate mixture of CO and H_2 from coal, wood (including wood wastes) peat or urban solid wastes, although no commercial plants use this process. Several countries, however including Australia, Brazil, Canada, the U.S.A. and New Zealand, have done extensive engineering and economic studies on wood to methanol processes.

A drawback to any biomass-to-methanol scheme is the size of the plant required and the amount of raw materials needed. The minimum practical plant size (about 200 t/day of methanol) required about 500 t/day of oven dry wood or equivalent biomass. Such volumes of biomass are not usually available at a given site without extensive gathering and transportation costs. In addition to its possible energy use, methanol is currently used as a chemical intermediate. Nearly half of the methanol produced in the U.S.A. is converted to formaldehyde; some is used in solvents and the remainder is used for a variety of syntheses.

Raw Materials

Methanol production from wood involves two stages of thermochemical transformation. The first stage involves gasification of the wood to produce syngas. The second stage is the actual synthesis step which converts syngas to methanol. This step is proven commercially available process. Gasification has been the bottleneck to commercialization. Considerable research and development is required to make the use of wood more economic than the use of natural gas. About 2.25 kg of natural gas is required to produce 4 l of methanol, as opposed to about 9 kg of dry wood.

Gasification of wood

Wood is charged at the top of the reactor and ash is discharged from the bottom. Air and steam are charged near the base of the reactor.

The sequence of reactions occurring in wood gasification is:

Drying (100°—200°C)

Moist wood and heat ⟶ Dry wood and water vapour

Pyrolysis (200—500°C)

$$\text{Dry wood and heat} \longrightarrow \text{char} + CO + CO_2 + H_2 + CH_4 + \text{tar and pyroligneous acids} \qquad ...(7.26)$$

Gasification (500°C +)

$$\text{Char} + O_2 + H_2O \longrightarrow CO + H_2 + CO_2 \qquad ...(7.27)$$

The raw gas typically contains hydrogen (18 %), carbon monoxide (22.8%), carbon dioxide (9.2%), methane (2.5%), other hydrocarbons (0.9%), oxygen (0.5%) and nitrogen (45.8%).

Gas purification and Shift conversion

The raw gas is then purified to remove all gases but hydrogen and carbon monoxide. This mixture is reacted with water so that the final mixture contains a 2 : 1 ratio of H_2 to CO. In this conversion, additional CO_2 is formed and must be removed before methanol synthesis. The raw gas from the reactor then passes through a scrubber, cooling the gas to about 32°C and removing tars and acid. The gas is then compressed to about 700 KPa (7 kg/cm^2) and treated in two stages to remove carbon dioxide. In the first stage, a hot potassium carbonate solution reduces the CO_2 content to about 300 ppm. In the second stage, monoethanoloamine (MEA) is used to reduce the CO_2 content to about 50 ppm. The gas then passes through a cryogenic system, which removes the residual CO_2 and water vapour, plus methane and other hydrocarbons and finally nitrogen.

The purified gas is a mixture of hydrogen (approximately 44%) and carbon monoxide (approximately 56%). Further processing is required to produce the 2 : 1 ratio of H_2 to CO needed to produce methanol.

Following cryogenic purification, the gas is compressed to about 2800 KPa (28 kg/cm^2), for shift conversion. Part of the CO reacts with water vapour in the presence of an iron catalyst to form additional hydrogen so that the exit gas contains a 2 : 1 ratio of H_2 to CO.

Since the shift reaction also produces CO_2, it is necessary to rescrub the gas with a potassium carbonate absorption system.

Synthesis

The syngas is then compressed to 14000–28000 KPa (140–280 kg/cm^2) and passed into the methanol synthesis reactor. In the reactor, approximately 95% of the gas is converted to methanol over a zinc-chromium catalyst. The unreacted gases are separated and recycled and the methanol purified by distillation.

Gasification equipment

Wood conversion technologies currently under consideration for methanol production can be divided into two broad categories; oxygen gasifiers and pyrolytic gasifiers. Oxygen gasifiers allow combustion of a fraction of the wood feedstock to provide the process heat. The use of oxygen reduces capital costs by increasing reactor throughout and eliminates the problem of nitrogen dilution of the gas product. (This problem arises from the use of air). It is more economical to separate nitrogen from air (to produce oxygen) prior to gasification, then to separate nitrogen from the gaseous products after gasification. Oxygen processes may be carried out in any one of the following reactors for methanol production from wood.

(*i*) *Updraft reactor*—a fixed bed where wood flow is counter current to the oxygen flow (offered commercially, but no recent successful demonstration using wood feedstocks; gas product is tarry when compared to fluidized bed gas product).

(*ii*) *Downdraft reactor*—a fixed bed where wood flow is cocurrent with the oxygen flow (not commercially available).

(*iii*) *Fluidized bed reactor*—the bed is composed of inert solids and wood and resembles a boiling liquid as oxygen passes up through it (offered commercially; highest efficiency of all oxyges units; yet to be demonstrated at elevated pressures).

Pyrolytic gasifiers employ either an internal non-combustion heat source or an external heat source (transferred through the reator walls) to drive the process. As there are no combustion products such as water or carbon dioxide, the yields of synthesis gas, and ultimately merhanol are higher.

7.29. Conclusion

Photobiological effects could be used as a source of fuel. A wholly different approach for methane production (bio gas) through solar energy plantations. The main idea is to use solar energy for photosynthetic plant method which is then into biogas generator and then to produce methane. In these solar plantations or energy farming, the plant method is grown with specific objective of bioconversion to CH_4, and therefore one seeks the highest yiels. Grasses, Algae, water hyacinth (floating water plants) are examples of such vegetations. Algae could be grown in nutrient-loaded municipal sewage to produce methane as a result of their fermentation. Electricity could thus be generated on a large scale at a low cost. This method would in turn solve the problem of sewage treatment in large cities. The grown algae are first harvested, then treated to yield about 5% of solids, the rest being water. The solid material is subjected to anaerobic fermentation and after about three weeks, that the organic material is converted into methane. This methane could be used as a regular fuel to produce thermal energy and eventually electricity. Biomass processes can be designed to produce solids (wood and charcoal), liquids (oils and alcohols), gases (methane and hydrogen), or electricity. The various processes for conversion of bio mass into useful fuels are *direct combustion, anaerobic digestion, hydrolysis, pyrolysis, hydrogasification* and *hydrogenation.* Of course much of the energy (1/3 to 2/3) bound in biomass is lost during its conversion to useful fules. Methanol and ethanol are useful biomass fuels which can be blended with gasoline.

Solar energy can also be used to increase the output of Gobar gas plants, by glass green house over the digester to trap the solar energy for heating. On absolute basis, the cost of biomass energy is found to be generally higher at present than that from the best coal based plants. However several other factors have to kept in consideration. Firstly the cost of biomass energy can be reduced by the development of vertices with higher yields whereas with coal the cost will inevitable increase in the future. Secondly, for the rural areas, especially in developing countries, biomass energy could be boon. Such areas are rarely served adequately by conventional networks and even then the true cost of energy supplied to them is very high. Thirdly biomass is a perennial and low polluting type of energy will all its by products capable of being recycled.

It is evident that biogas can give hand to an extremely great extent in the substitution of fossil fuels and a thorough investigations in many angles, the generation and utilization are to be explored. The value of fertilizers for the yield of crops after the extraction of biogas must be popularized among farmers, thereby making them use the by product biogas for their energy needs. Digesters design must be perfected and must be made popular in rural areas. Ban should be made in burning cow dung cakes. With such a potential energy source, India will not suffer due to depletion of fossil fuels if these potentials are properly tapped and used.

QUESTIONS

1. How biomass conversion takes place?
2. What is the difference between biomass and biogas?
3. Explain the process "photosynthesis." What are the conditions, which are necessary for it?
4. What is meant by anaerobic digestion? What are the factors, which affect biodigestion? Explain briefly.
5. How are biogas plants classified? Explain them briefly.
6. What are the advantages and disadvantages of floating drum plant?
7. Name the various models of biogas plant.
8. Explain the constructional detail and working of KVIC digester.
9. What is meant by wet fermentation and dry fermentation?
10. What are the main problems in straw fermentation? How they are minimised?
11. What is a community biogas plant? What are the main problems encountered in its operation?
12. Give a list of the materials used for biogas generation.
13. What are the factors, which affect the size of the biogas plants?
14. What are the techniques suggested for maintaining the biogas production? Explain.
15. Write the main applications of biogas.
16. What is meant by 'energy plantation'? What are its advantages and disadvantages?
17. What are the main plants proposed for energy plantation especially in India?
18. How are Gasifiers classified? What is pyrolysis?
19. What are the potential applications of the gasifier?
20. What are the advantages and disadvantages of biological conversion of solar energy?
21. Calculate: (*i*) the volume of a biogas digester suitable for the output of four cows, and (*ii*) the power available from the digester. Retention time is 20 days, temperature 30°C, dry matter consumed 2 kg/day, biogas yield 0.24 m^3 per kg. Burner efficiency is 60%, methane proportion is 0.8. Hm the heat of combustion of methane may be assumed to be 28 MJ/m^3 at STP. **[Ans.** (*i*) 3.2 m^3, (*ii*) 300 W]

UNIT—4

GEOTHERMAL ENERGY

- **Geothermal Sources**
- **Prime Movers for Geothermal Energy Conversion**
- **Applications, Operational & Environmental Problems**

CHAPTER 8

Geothermal Energy

(Energy from the Earth)

8.1. Introduction

Energy present as heat (*i.e.,* thermal energy) in the earth's crust; the more readily accessible heat in the upper most (10 km) or so, of the crust constitutes a potentially useful and almost inexhaustible source of energy. This heat is apparent from the increase in temperature of the earth with increasing depth below the surface. Although higher and lower temperatures occur, the average temperature at a depth of 10 km is about 200°C.

U.S. Geological survey defines geothermal source as "all of the heat stored in the earth's crust above 15°C to a depth of 10 km." It occurs when the immense heat energy in the core of earth rises closer to the surface of the earth due to cracks or faults as accounted by the "Plate Techonics Theory", in the crust and heats the surrounded rock. These hot spots can be liquid dominated, vapour dominated, petro-thermal or geopressure system, depending upon several geological and hydrological factors. These inturn are tapped artificially to use the vast stored heat energy, for power generation and several other uses, depending upon the temperature of occurrence and other parameters.

Hot molten (or partially molten) rock, called "Magma" is commonly present at depths greater than 24 to 40 km. In some places, however, anomalous geologic conditions cause the magma to be pushed up toward the surface, in an active volcano, the magma actually reaches the surface, where heat of the magma is being conducted upward through an overlying rock layer.

Fig. 8.1 shows a typical geothermal field. The hot magma (molten mass) near the surface (A) solidifies into igneous rock (B). Igneous is Latin word, igneous meaning "of fire" specially formed by volcanic action or great heat. (Igneous rock found at the surface is called volcanic action rock). The heat of the magma is conducted upward to this igneous rock. Ground water that finds its way down to this rock through fissures in it, will be heated by the heat of the rock or by mixing with hot gases and steam emanating from the magma. The heated water will then rise convectively upward and into a porous and permeable reservoir (C) above the igneous rock. The reservoir is capped by a layer of impermeable solid rock (D) that traps the hot water in the reservoir. The solid rock, however, has fissures (E) that act as vents of the giant underground boiler. The vents show up at the surface as geysers fumarols (F) (steam is continuously vented through fissures in the ground, these vents are called fumarols) or hot spring (G). A well (H) taps steam from the fissures for use in a geothermal power plant.

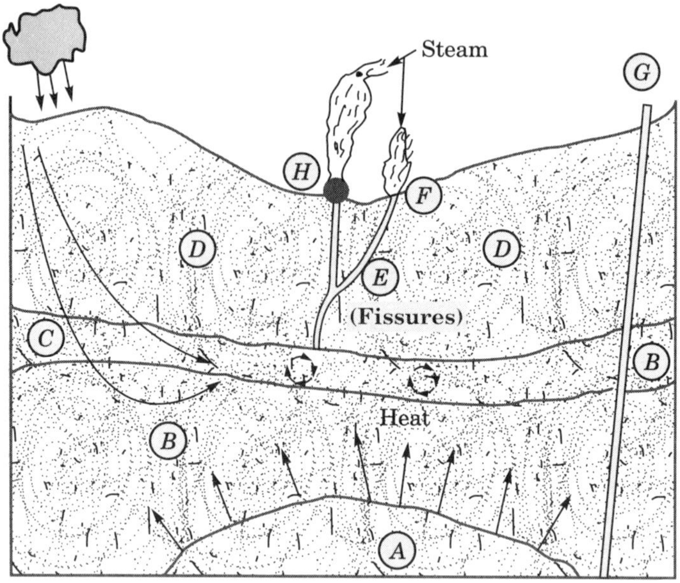

Fig. 8.1. A typical geothermal field.

It can be seen that geothermal steam is of two kinds that originating from the magma itself, called *magmetic steam,* and that from ground water heated by the magma called *meteoritic steam.* The latter is the largest source of geothermal steam. Not all geothermal sources produce steam as described above. Some are lower in temperature so that there is only hot water. Some receive no ground water at all and contain only hot rock. Geothermal sources are therefore of three basic kinds: (1) *hydrothermal* (2) *geopressured* and (3) *petrothermal.* These will be explained later.

The total amount of energy in the outer 10 km of the earth's crust exceeds greatly that obtainable by the combustion of coal, oil and natural gas. At present, however, only the relatively small proportion of the geothermal energy in wet reservoirs (a geothermal reservoir is defined as a region where there is a concentration of extractable heat), may be regarded as economically useful. Nevertheless, this amount is large enough to make a significant contribution to the energy resources. With advances in technology, a portion of the much larger dry geothermal energy resources may also become available.

Historically, the first applications of geothermal energy were for space heating, cooking, and medicinal purposes. The earliest record of space heating dates back to 1300 in iceland. In the early 1800s, geothermal energy was used on what was then a large scale by the conte Franceso de Laderel to recover boric acid. The first mechanical conversion was in 1897 when the steam of the field at Larderallo, Italy, was used to heat a boiler producing steam which drove a small steam engine. The first attempt to produce electricity also took place at Larderello in 1904 with an electric generator that powered four light bulbs. This was followed in 1912 by a condensing turbine; and by 1914, 8.5 MW of electricity was being produced. By 1944 Larderello was producing 127 MW. The plant was destroyed near the end of World War II, but was fortunately rebuilt and expanded and eventually reached 360 MW in 1981.

In the United States, the first attempt at developing the Geysers field was made in 1922. Steam was successfully, taped, but the pipes and turbines of the time were unable to cope with the corrosive and abrasive steam. The effort was not revived until 1956, till that time stainless steel alloys were developed that could withstand the corrosive steam, and the first electric generating unit of 11-MW capacity began operation in 1960. Since then 13 generally progressively larger units have been added to the system. After-ward number of units were planned which brought the total capacity to about 1500 MW by the late 1980. Other electric-generating fields of note are in New Zealand, Japan, Mexico, the Phillipines, the Soviet Union and Ice land (a large space heating programme).

In 1979, the global electricity production from geothermal resources was 1872 MW and units under construction were of 1650 MW capacity. For space and process heat, total power utilization was 1281 MW. So that total power production was about 4850 MW. This power consumption is likely to grow up upto 100,000 MW by the turn of the next century.

It has been analysed that about 40 to 50% of total heat required by the society is a heat below 200°C. 30% heat is required at about 150°C. 20% heat is required at about 100°C.

Thus there is a vast scope to use geothermal energy for low temperature applications. There is a ample scope to develop geothermal power in India, but still development in geothermal field is in initial stage. There are about 340 known thermal areas in India, each represented by hot/warm spring. About 113 spring area, discovered so far where geothermal power is available. 46 of these systems are of high temperature type (above 150°C), which could generate 1838 MW for a period of 30 years. 59 of these are of intermediate temperature type (90° to 150°C), which could be considered for power generation using binary vapour cycle and other are of low temperature type below 90°C.

Till now only one pilot plant is in operation in puga valley, in Jammu and Kashmir, having 20 MW capacity. Another plant is at Parvati Valley, Himachal Pradesh is under construction. A 7.5 tonne capacity cold storage pilot plant based on geothermal energy was installed at Manikaran, Himachal Pradesh. A 5 kW pilot power plant is under fabrication at the National Aeronautical Laboratory, Bangaluru. This plant will run on geothermal energy which will be recovered from the hot springs at Manikaran in Himachal Pradesh. Plans are being made to undertake further research and development studies in the area of geothermal energy. Exploitation of this energy source for non power sectors like poultry farming, mushroom cultivation, space heating is possible in this Country on fairly large scale.

It can be seen that while geothermal energy is not the sought after sole and long range solution to our energy problems, it nevertheless represents a not insignificant factor if its resources are developed in a careful and efficient manner.

Geothermal energy is also the one of the renewable energy sources, which are defined as those resources that draw on the natural energy flows of the earth (another term "the *alternative energy sources*" is also common in use). Renewable energy sources are so named because they recur, are seemingly inexhaustible, and are free for the taking. Geothermal energy has practically no intermittency, has the highest energy density, and is economically not far removed from conventional technologies. Geothermal energy is classified as renewable because the earth's interior is and will continue in the process of cooling for the indefinite future. Hence, geothermal energy from the earth's interior is almost as inexhaustible as solar or wind energy, so long as its sources are actively sought and economically tapped.

8.2. Estimates of Geothermal Power

The estimates vary very widely. However the following give a rough estimate. For a depth of 3 kms, the total stored energy of known fields is approximately 8×10^{21} Joules and for a depth of 10 km the total stored energy is estimated to be about 4×10^{22} Joules. The energy stored in hot springs is about 10% of the above quantities. If the above energy is extracted from a 3 km belt with 1% thermal energy recovery factor at a uniform rate of over a 50 year period, thermal power of 50 GW is obtained. With a thermal electric conversion efficiency of 20% will yield only 10 GW of electric power. For the estimate based on a 10 km depth on electric power of 50 GW is predicted U.S.A. has 5—10% geothermal fields and India much less.

8.3. Nature of Geothermal Fields

It is convenient to classify earth's surface into three broad groups.

(1) *Nonthermal areas* having a temperature gradient of 10—40°C per km depth.

(2) Semithermal areas having a temperature gradient of 70°C per km depth.

(3) *Hyperthermal areas* where the temperature gradients are many times greater than in non thermal areas.

Geothermal fields may further be classified into three types:

(A) Hyperthermal Fields

(1) *Wet fields.* Where the water is pressurized and temperatures are above 100°C. When they are led to the surface a fraction will be splashed into steam and a major part remains as the boiling water.

(2) *Dry fields.* They produce dry saturated steam or superheated steam at pressure above atmospheric.

(B) Semithermal Fields

These are capable of producing hot water at temperatures above 100°C.

8.4. Geothermal Sources

Five general categories (or kinds) of geothermal resources have been identified:

(1) Hydrothermal convective systems.

These are again subclassified as:

(*a*) Vapour dominated or dry steam fields.

(*b*) Liquid dominated system or wet steam fields, and

(*c*) Hot water fields.

(2) Geopressure resources.

(3) Petro-thermal or Hot dry rocks (HDR).

(4) Magma resources.

(5) Valcanoes.

The hydrothermal convective systems are best resources for geothermal energy exploitation at present. Hot dry rock is also being considered.

Hydrothermal Systems. Hydrothermal systems are those in which water is heated by contact with the hot rock, as explained earlier.

Vapor dominated Systems. In these systems the water is vaporized into steam that reaches the surface in a relatively dry condition at about 200°C and rarely above 7 kg/cm^3 (8 bar). This steam is the most suitable for use in turbo electric power plants, with the least cost. It does, however, suffer problems similar to those encountered by all geothermal systems, namely, the presence of corrosive gases and erosive material and environmental problems. These type of systems are very less in numbers; they are only five known sites in the world. The Geysers plant in the United States, the largest in the world today, and Larderello in Italy, are both vapor-dominated systems.

Liquid dominated Systems. In these systems the hot water circulating and trapped underground is at a temperature range of 175 to 315°C. When tapped by wells drilled in the right places and to the right depths, the water flows naturally to the surface or is pumped upto it. The drop in pressure, usually to 7 kg/cm^2 (8 bar) or less, causes it to partially flash to a two—phase mixture of low quality *i.e.,* liquid-dominated. It contains relatively large concentration of dissolved solids ranging between 3000 to 25,000 ppm and sometimes higher. Power production is adversely effected by these solids because they precipitate and cause scaling in pipes and heat exchanger surfaces, thus reducing flow and heat transfer. Liquid dominated systems, however, are much more plentiful than vapour dominated systems, and, next to them, require the least extension of technology.

(2) **Geopressured Systems.** These resources occur in large, deep sedimentary basins. The reservoirs contain moderately high temperature water (or brine) under very high pressure. They are of special interest because substantial amounts of methane CH_4 (natural gas) are dissolved in the pressurized water (or brine) and are released when the pressure is reduced. Geopressured water is tapped in much deeper underground acquifers (it is a water-bearing stratum of permeable rock, gravel or sand), at depths between about 2400 to 9000 m. This water is thought to be at the relatively low temperature of about 160°C and is under very high pressure, from the overlying formation above, of about 1050 kg/cm^2 (more than 1000 bar). It has a relatively high salinity of 4 to 10 per cent and is often referred to as brine. The geopressured resources are quite large: they could be used for the generation of electric power and the recovery of natural gas if suitable technology could be developed and if individual reservoir productivity and longevity prove to be adequate.

(3) **Hot Dry Rocks** (or Petrothermal Systems). These are very hot solid rocks occurring at moderate depths but to which water does not have access, either because of the absence of ground-water or the low permeability of the rock (or both). In order to utilize this resource, means must be found for breaking up impermeable rock at depth, introducing cold water, and recovering the resulting hot water (or steam) for use at the surface. The known temperatures of HDR vary between ~ 150 to 290°C. This energy, called petrothermal energy, represents by far the largest resource of geothermal energy of any type, as it accounts for large percent of the geothermal resource. Much of the HDR occurs at moderate depths, but it is largely impermeable as stated above in order to extract thermal energy out of it, water will have to be pumped into it and back out to the surface. It is necessary for the heat transport mechanism that a way be found to render the impermeable rock into a permeable structure with a large heat transfer surface. A large surface is particularly necessary because of the low thermal conductivity of the rock. Rendering the rock permeable is to be done by fracturing it. Fracturing methods that have been considered involve drilling wells into the rock and then

fracturing by (1) high-pressure water or (2) nuclear explosives. Efforts in this direction are in progress.

(4) **Magma Resources.** These consist of partially or completely molten rock, with temperatures in excess of 650°C, which may be encountered at moderate depths, especially in recently active volcanic regions. These resources have a large geothermal energy content, but they are restricted to a relatively few locations. Furthermore, the very high temperatures will make extraction of the energy a difficult technological problem.

The geothermal resources outlined above and means used or proposed for their development are described more fully in the following sections.

8.5. Hydrothermal (Convective) Resources

These are wet reservoirs at moderate depths containing steam and/or hot water under pressure at temperatures upto about 350°C. These systems are further subdivided, depending upon whether steam or hot water is the dominant product. Hydrothermal resources represent only a small fraction of the potential geothermal resources, but they are the only ones that have been utilized commercially so far. If the temperature is high enough, the water or steam can be used to generate electricity, otherwise the geothermal energy is best supplied to process and space heating.

Hydrothermal resources arise when water has access to high temperature rocks, this accounts for the description as "hydrothermal". The heat is transported from the hot rocks by circulating movement (*i.e.*, by convection of the water in a porous medium). The general geological structure of a hydrothermal convective region is shown in simplified form in Fig. 8.2. The molten rock (magma), raised by internal earth forces is overpaid by an impervious rock formation, through which heat is conducted upward. Above this is a permeable layer into which water has penetrated, often from a considerable distance. The permeability could result from fractures or intergranular pores. The heat taken up by the water from the rocks below, is transferred by a convection to a layer of impervious rocks above, (In convection, the heated water rises, because of its lower density, and then descends when it is cooled by transferring heat to the colder rocks above).

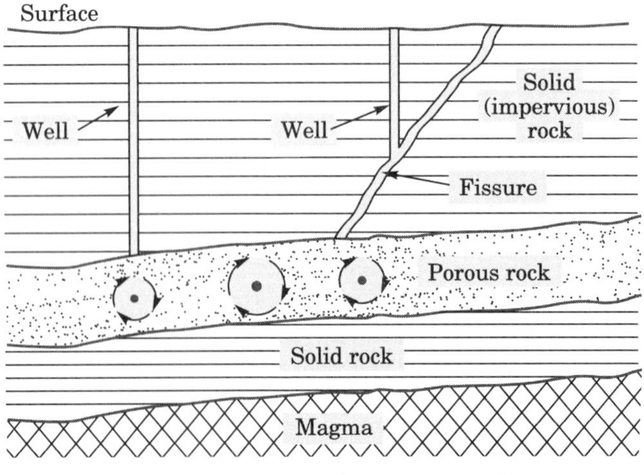

Fig. 8.2. Hydrothermal convective region.

Hot water or steam often escapes through fissures in the rock, thus forming hot springs, geysers fumarols, etc. In order to utilize the hydrothermal energy, wells are drilled either to intercept a fissure or, more commonly, into the formation containing the water (*i.e.*, the hydrothermal reservoir). Most hydrothermal wells range in depth from about 600 to 2100 m, although there are some shallower and deeper production wells. As already mentioned for practical purposes hydrothermal resources are further subdivided into vapor-dominated (Dry steam) and liquid-dominated (wet steam) types. In vapour dominated systems, the wells deliver steam, with little or no liquid water, usually at temperatures of about 150°C to 250°C but occasionally higher. Liquid dominated systems, on the other hand produce a mixture of steam and hot water.

8.5.1. Vapour Dominated Systems

As indicated previously these are the most attractive geothermal resources because they are the most easily developed. They have the lowest cost and least number of serious problems. However they constitute only a few per cent of hydrothermal resources and a much smaller proportion of the accessible geothermal energy resources.

Fig. 8.2 (*a*) and 8.2 (*b*) show a schematic and T.S. diagram of a vapour-dominated power system. Dry steam from the wells is collected, filtered to remove

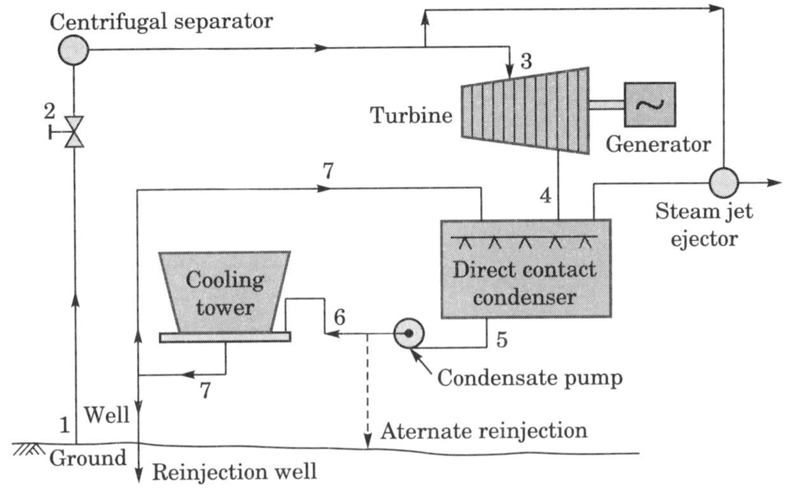

(*a*) Scheme of a vapour dominated power plant

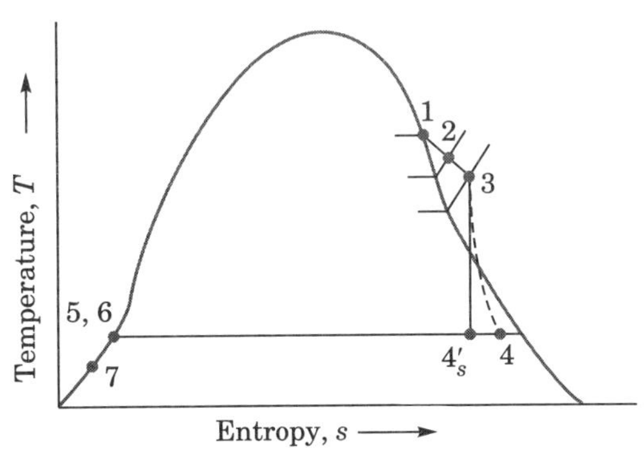

(*b*)

Fig. 8.2. Vapour dominated system on T.S. diagram.

abrasive particles and passed through turbines, which drive electric generators in the usual manner. The essential difference between this system and a conventional steam turbine generator system, using fossil or unclear fuel, is that geothermal steam is supplied at a much lower temperature and pressure. The dry steam from the well (1) at perhaps 200°C is used. It is nearly saturated at the bottom of the well and may have a shut off pressure upto about 35 kg/cm^2 is ($\simeq$ 35 bar). Pressure drops through the well causes it to slightly super heat at the well head (2). The pressure there rarely exceeds 7 kg/cm^2 (~ 7 bar). It then goes through a centrifugal separation and then enters turbine after additional pressure drop. Processes between well and well head, and centrifugal separator and turbine are essentially throttling processes with constant enthalpy. The steam after expansion in the turbine (3) entres the condenser at 4.

Because turbine flow is not returned to the cycle but reinjected back into the earth, a direct contact condenser of barometric or low level type may be used. Direct contact condensers are more effective and less expensive than surface type condensers. As with steam (or any vapour) turbines in general, the efficiency is improved if the back (or exhaust) pressure is maintained at a low level by condensing the steam. In a conventional plant, the pure water produced in the condenser is required as feed water for the steam boilers ; consequently, the steam and impure condenser cooling water are kept separate. In a hydrothermal plant, however, there is no need for feed water. Hence, a direct contact system is used at the Geysers in which the turbine exhaust steam is condensed by direct contact with cooling water. The resulting warm water is circulated through a mechanical draft *cooling tower* and returned to the condenser.

The condensation of steam continuously increases the volume of the cooling water. Part of this is lost by evaporation in the cooling towers (6), and the remainder is injected deep into the ground (7) for disposal. The turbine exhaust steam at 4 mixes, with the cooling water (7) that comes from cooling tower. The mixture of cooling water coming from the cooling tower and turbine exhaust is saturated vapour at (5) that is pumped to the cooling tower (6).

As stated above condensation of the steam continuously increases the volume of the cooling water. Part of this is lost by evaporation in the cooling towers, and the remainder is injected deep into the ground for disposal. The ability to produce their own cooling water is a unique feature of many hydrothermal plants; conventional steam turbine generating plants require an external source of condenser (make up) cooling water.

The surface condensers are also, however, used in some units with H_2S removal system.

Geothermal plants use much lower temperature and pressure steam and higher condenser pressures than conventional plants, and hence they are much lower in efficiency. At the Geysers installation, the more recent deep (roughly 2500 m) wells supply steam to the turbines at a temperature of about 180°C and a gauge pressure of 7.8 atm. (0.78 M Pa). (The temperatures and pressures in the reservoir are higher). For comparison, the steam temperature in a modern fossil fuel plant may be about 550°C and the pressure 240 atm (24 M Pa). Hence, the geothermal electric plants require special low pressure turbines. Furthermore, because of the low steam temperature, the *thermal efficiency* (*i.e.*, the proportion of heat supplied to the turbine that is converted into electric energy) is only about 15 per cent, compared with 40% (or so) in a modern fossil fuel plant.

The Geysers installation in California (U.S.A.), Larderello in Italy and Motsukawa, Japan are: good examples of dry steam fields. In U.S.A., the Geysers installation started commercially operation in 1960, its electric capacity has been steadily increased and reached about 1000 MW in 1980. This capacity is increasing further, as reservoir conditions are not affected.

Characteristics of geothermal steam electric plants. A number of environmental effects are characteristics of geothermal steam electric facilities. The steam may contain 0.5 to 5% by weight of non condensable gases which appear in the turbine exhaust. These gases consist mainly of carbon dioxide with small amounts of methane and ammonia, which are largely harmless in the quantities present. In addition, the gases may contain upto 4 or 5% of hydrogen sulfide; not only does this gas have an unpleasant (rotten-eggs) odor, it can also be harmful to plant and animal life if it should accumulate. At the geysers, hydrogen sulfide is considered to be a nuisance rather than a hazard.

In the past the non condensable gases have been released to the atmosphere where the hydrogen sulfide is gradually destroyed by oxidation. The products, however, are oxides of sulphur which can themselves be harmful at appreciable concentrations. Plans are therefore under way to remove most of the hydrogen sulfide from the gases before they are discharged.

The geothermal steam may also contain boron, arsenic; mercury, and other potentially poisonous elements which, together with some of the ammonia, are found in the turbine condensate. These substances must be disposed of in a safe manner. This is achieved at the Geysers by the reinjection of excess condensate into the ground at a considerable depth.

The withdrawal of large amounts of steam (or water) from a hydrothermal reservoir may result in surface subsidence. Such subsidences have sometimes occurred in oil fields and are dealt with by injecting water into the ground. The reinjection of excess water from a geothermal steam electric plant may serve the same purpose. Care must be taken, however, that the water is reinjected at some distance from a ground fault. Steam-electric plants of all types discharge to the atmosphere much of the heat present in the turbine exhaust steam. In the Geysers facilities this discharge occurs in the cooling towers. Because of the low thermal efficiency, a larger proportion of the heat supplied is discharged than from fossil-fuel or nuclear plants. Vaporization water in the cooling towers also result in the addition of large amounts of moisture in the atmosphere. Hitherto, geothermal steam has been used on a relatively moderate scale and the environmental effects of the heat and moisture have not been significant.

Example 8.1. *A 250 MW vapour-dominated hydro-thermal power plant uses saturated steam at 31 kg / cm^2 at shut-off. The steam is throttled to a turbine at inlet pressure of 9.55 kg / cm^2. A direct contact condenser operates at a pressure of 0.35 kg / cm^2 with the cooling-tower exist is at 25°C. The turbine polytropic efficiency is 0.80 and the turbine-generator combined mechanical and electrical efficiency is 0.9.*

Calculate:

(*i*) *Steam flow rate in kg / hr and m^3 / kg;*

(*ii*) *The cooling water flow kg / hr;*

(*iii*) *Plant efficiency and*

(*iv*) *Heat rate.*

Reinjection occurs prior to the cooling tower.

Solution. Refer Fig. 8.2 (*a*) and (*b*).

From steam tables.

Enthalpy at point 1 at (31 kg/cm^2) = 669.7 kcal/kg

$H_1 = H_2 = H_3$, enthalpy remains constant during throttling.

$\therefore \quad H_3 = 669.7$ kcal/kg.

At point 3, $\quad P_3 = 9.55$ kg/cm^2

Specific volume $\quad v_{s3} = 0.22$ m^3/kg

Entropy $\quad S_3 = 1.580$

$T_3 = 190°$C (degree of superheat = 13°C)

Point 2 and 3 are in superheated region.

At point 1 steam is saturated, but at points 2 and 3, it becomes superheated. Here Mollier diagram can also be used to find the above parameters.

S_{4-s} at 0.34 kg/cm^2 = S_3

$$x_{4-s} = 0.838$$

and $$H_{4-s} = h_s + xL = 72 + 0.838 \times 556 = 538 \text{ kcal/kg}$$

(H_{4-s} can be read directly from Mollier's diagram)

Isentropic turbine work

$$= H_3 - H_{4-s} = 669.7 - 538 = 131.7 \text{ kcal/kg.}$$

Actual turbine work = 0.80 × 131.7 = 105.36 kcal/kg

$$H_4 = 669.7 - 105.36 = 564.34 \text{ kcal/kg}$$

h_{5-6} (ignoring pump work) = 72 kcal/kg

$$h_7 = h_5 = \text{sensible heat} = 25 \text{ kcal/kg}$$

$$\text{Turbine steam flow} = \frac{250 \times 0.860 \times 10^6}{105.36 \times 0.9}$$

$$= 1.838 \times 10^6 \text{ kg/hr.} \quad \textbf{Ans.}$$

$$= \dot{m}_4 \text{ say}$$

$$\text{Turbine volume flow} = \frac{1.836 \times 10^6}{60} \times vs_3$$

$$= \frac{1.836 \times 10^6}{60} \times 0.22$$

$$= 6.732 \times 10^3 \text{ m}^3\text{/min.}$$

Cooling water flow $\dot{m}_7$:

$$\dot{m}_7 (h_5 - h_7) = \dot{m}_4 (H_4 - h_5)$$

$$\dot{m}_7 = \frac{H_4 - h_5}{h_5 - h_7} \dot{m}_4$$

$$= \frac{564.34 - 72}{72 - 25} \times 1.836 \times 10^6$$

$$= 19.21 \times 10^6 \text{ kg/hr.} \quad \textbf{Ans.}$$

Heat added $$= H_1 - h_6$$

$$= 669.7 - 72 = 597.7 \text{ kcal/kg.}$$

Plant efficiency $$\eta_p = \frac{\text{Actual turbine work} \times \eta_{mg}}{\text{Heat added}}$$

$(n_{mg}$ = combined mechanical and electrical efficiency of turbine generator)

$$= \frac{105.36}{597.7} \times 0.90 = 0.1586$$

$\therefore \quad n_{plant} = 15.86\%$. **Ans.**

$$\text{Plant heat rate} = \frac{860 \times \text{heat added}}{\text{net work}}$$

$$= 860 \times \frac{597.7}{105.36 \times 0.90} = \frac{860}{n_{plant}}$$

$$= \frac{860}{0.1586} = 5422 \text{ kcal/kWH.} \quad \textbf{Ans.}$$

8.5.2. Liquid-Dominated Systems (Wet steam fields)

In the liquid dominated reservoir, the water temperature is above the normal boiling point (100°C). However because the water in the reservoir is under pressure, it does not boil but remains in the liquid state. When the water comes to the surface the pressure is reduced; rapid boiling then occurs and the liquid water "flashes" into a mixture of hot water and steam. The steam can be separated and used to generate electric power in the usual manner. The remaining hot water can be utilized to generate electric power or to provide space and process heat, or it may be distilled to yield purified water.

The water comes with various degrees of salinity, ranging from 3000 to 280,000 ppm of dissolved solids, and at various temperatures. There are, therefore, various systems for converting liquid dominated system into useful work that depend upon these variables. For liquid dominated (*high temperature*) systems two methods which will be covered are:

(*a*) The *flashed-steam system,* suitable for water in the higher temperature range, and

(*b*) The *binary-cycle system,* suitable for water at moderate temperatures.

(*c*) A third method, called the *total flow system,* concepts of it is also covered, but this approach awaits further development.

Liquid dominated (*low temperature*) system is also described in this section.

Liquid dominated (high temperature) *system:* Steam from liquid dominated, high temperature reservoirs is being used in several countries to generate electric power. The most extensive development has been in the volcanic Wairakei field in Newzealand. The water in the hydrothermal reservoir is at the temperature of about 230°C and pressure of 40 atm (4 MP*a*). The liquid originates mainly from depths of 600 to 1400 m and is flashed into a mixture of steam and water at the surface.

After passage through a *cyclone separator,* to remove the water, the steam is supplied to the turbine connected to electric generators. The maximum initial steam temperature is about 175°C and the gauge pressure is 3.5 atm (0.35 MP*a*). The exhaust steam is condensed by direct contact with cold water from the nearby Waikaro river; the warm condenser cooling water is then discharged to the river. This simple procedure, which does not require, cooling towers, is possible only because of the ample flow of river water.

(*a*) **The Flashed-Steam System.** This is illustrated by the flow and *T-S* diagram of Figs. 8.3 (*a*) and (*b*). Water from the underground reservoir at 1 reaches

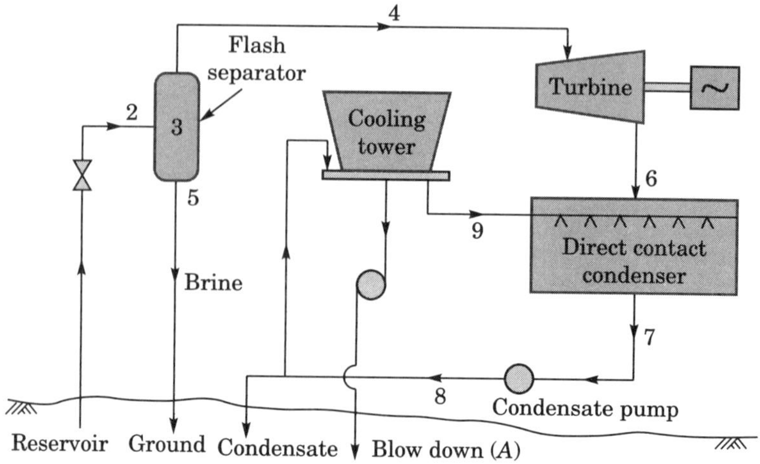

(*a*) Schematic of a liquid dominated single flash steam system.

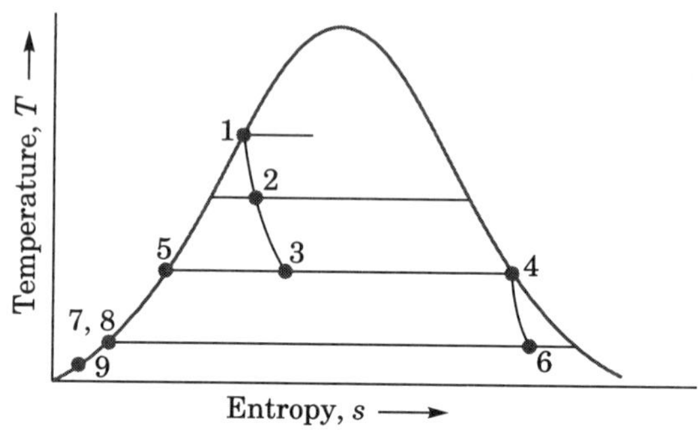

(*b*) *T-S* diagram of a liquid dominated single flash steam system.

Fig. 8.3

the well head at 2 at a lower pressure. Process 1-2 is essentially a constant enthalpy throttling process that results in a two-phase mixture of low quality at 2. This is throttled further in a flash separator resulting in a still low but slightly higher quality at 3. This mixture is now separated into dry saturated steam at 4 and saturated brine at 5. The latter is reinjected into the ground.

The dry steam, a small friction of the total well discharge (because of low quality at 3), and usually at pressures below 7 kg/cm^2 gauge (0.7 MP*a*), is expanded in a turbine to 6 and mixed with cooling water in a direct-contact condenser with the mixture at 7 going to a cooling tower in the same fashion as the vapor dominated system. The balance of the condensate after the cooling water is recirculated to the condenser is reinjected in to the ground.

The power generation from such system can be made more economical by associating chemical industry with power plant to make use of the brine and the gaseous effluent.

Because of the high water temperature of more than 300°C and a pressure of about 120 atm. (12 MP*a*) at a depth at 1200 m, the lower (and larger) reservoir is of special interest as a source of energy. However, the mineral content (mainly silica and sodium, potassium and calcium chlorides) of the water ranges upto 20 or 30% by weight ; this may be compared with the 3.3 percent (average) of salts in sea water. The hot, highly saline water referred to as *geothermal brine*, is very corrosive.

Moreover when the temperature is lowered, as it is when the heat is utilized as an energy source, dissolved silica and salts and corrosion products from solid deposits on pipes pump surface etc. The hot-water withdrawn from the steam water separator after flashing is also discharged to the well or river. Because heat present in the water is discarded in this manner, the overall thermal efficiency for electric power generation in the above case is about 8 per cent. The hot water could however be used for space heating before discharge.

The flashed steam system described above has the following *limitations* as compared with the vapour dominated system.

(1) This system requires much larger total massflow rates through the well.

(2) Due to large amount of flows, there is a greater degree of ground surface subsidence.

(3) The system provides a greater degree of precipitation of minerals from the brine, resulting in the necessity for design of valves, pumps, separator internals, and after equipment for operation under scaling conditions.

(4) Greater corrosion of piping, well casing, and other conduits.

(5) Many times temperature and pressure of the water may not be sufficient to produce the flash steam.

Flashed steam systems have been widely used in Japan, New zealand, Italy, Mexico, and U.S.A.

The spent brine leaving the separator at 5 Fig. 8.3 has a large mass-flow rate and a large total energy compared with that in the steam used to drive the turbine at 4. The ratio of the brine enthalpy to the steam enthalpy was found to be 1.31 : 1. Improvement in the cycle would therefore use some of this otherwise lost energy in the cycle. *Double flash cycle* can give more power than the single flash cycle under the same condition.

Figs. 8.4 (*a*) and (*b*) show a schematic flow and *T-S* diagram of a double flash steam system. Depending upon the original water conditions, the brine at 5 is admitted to a second, lower pressure separator, where it flashes to a lower pressure steam (6) that would be admitted to a low-pressure stage in the turbine. The remaining spent brine at 8 is reinjected in to the ground. An example of the double flash system is the 50 MW Hatchobaru plant build on the island of Kyushu in Japan. It uses an innovative steam condenser and gas extraction system and a dual admission double flow steam turbine.

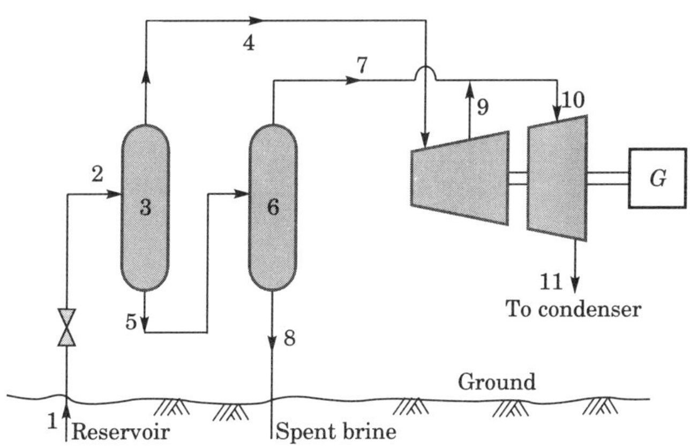

(*a*) Schematic of liquid dominated double flash steam system.

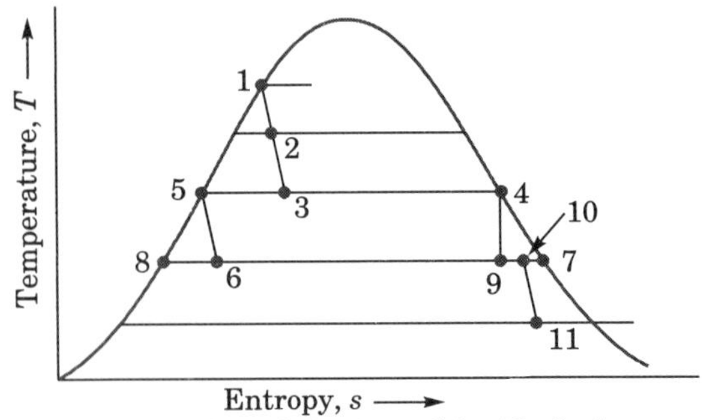

(b) T-S diagram of the liquid dominated double flash system.

Fig. 8.4

(b) **Liquid Dominated Systems: Binary Cycle.** In order to isolate the turbine from corrosive or erosive materials and/or to accommodate higher concentration of noncondensable gases, the binary cycle concept is now receiving considerable attention as an alternate power cycle concept. This is basically a Rankine cycle with an organic working fluid. A heat exchanger system is used to transfer a fraction of the brine enthalpy to vaporize the secondary working fluid. Expansion through a turbine to a lower pressure, fixed by the heat rejection temperature, provides the means for power generation.

About 50% of hydrothermal water is in the moderate temperature range of 153°C to 205°C. This water which is available in lower temperature ranges is unsuitable for power production. It is however suitable for direct utilization for domestic and industrial process heating. If this water is used in a flashed steam system, it would have to be throttled down to such a low pressures that results in excessively large specific volume flows as well as even poor cycle efficiencies. Instead this water is used as a heat source for a closed cycle that uses another working fluid that has suitable pressure temperature volume characteristics. As stated above the binary fluid (or two fluid) system, is being investigated to overcome these limitations of flashed steam system. In the binary system an organic fluid with a low boiling point, such as *isobutane* (2 methal propane) C_4H_{10} (normal boiling point at one atm. pressure = 10°C) and Freon—12 (normal boiling point – 29.8°C) are usually recommended. Ammonia and propane may also be used. The working fluid would operate at higher pressures, corresponding to the source water and heat sink temperatures.

Flow diagram of a binary cycle system is shown schematically in Fig. 8.5. Hot water or brine from the underground reservoir, either as unflashed liquid or

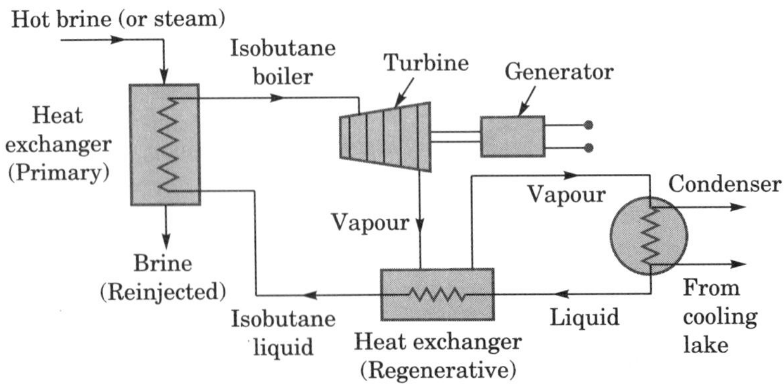

Fig. 8.5. Binary fluid geothermal power system.

as steam producing by flashing is circulated through a primary heat exchanger. In the heat exchanger the hot brine transfer its heat to the organic fluid thus converting it to a superheated vapour that is used in a standard closed Rankine cycle. The vapour drives the turbine generator. The exhaust vapour from the turbine is cooled in the regenerative heat exchanger and then condensed, using either and air cooled condenser or a water cooled condenser and cooling tower. The condensed liquid organic fluid is returned to the primary heat exchanger by way of the regenerative heat exchanger. The hot geothermal fluid and the organic fluid, constitute the two fluids of the binary fluid system.

The condenser is cooled by water from a natural source, if available, or a cooling tower circulation system. The blow down from the tower may be reinjected to the ground with the cooled brine. Make up of the cooling tower must be provided however.

In the binary cycle there are no problems of corrosion or scaling in the working cycle components, such as the turbine and condenser. Such problems are confined only to the well casing and the heat exchanger. The heat exchanger is a shell and tube type so that no contact between brine and working fluid takes place.

If the temperature and salinity of the geothermal brine are not high, the tendency for solids (scale) to deposit on surfaces in the heat exchanger is not too great. The liquid brine under pressure may then be pumped directly through the heat exchanger and reinjected into the geothermal reservoir.

However, where the temperature and salinity of the brine are high, this procedure may not be practical because the heat exchangers may soon be rendered ineffective by scale deposition. Methods for scale control and removal are being developed, but in the meantime the high temperature brine is flashed into steam, which is scrubbed and passed on to the heat exchanger. Nearly all the salt content remains in the residual brine in the flash tank. To achieve maximum utilisation of the heat content of the brine, it may be flashed in two more stages, at successively lower temperature and pressure, before it reinjected into ground.

The first binary cycle was installed in the Soviet Union on the Kamchatka Peninsula in 1967. It has a gross output of 680 kW using a low temperature water reservoir at 80°C and Freon 12 as working fluid. The first binary cycle to be built in the United States is an 11 MW plant built by the Magma Company in California. The site has a potential of 10,000 MW. Isobutane was chosen in preference to Freon 12 primarily because of its lower cost. A second U.S. binary cycle plant is being built at Raft River Idaho. It is a 10 MW plant.

(*c*) **Total Flow Concept.** A third approach called the *total-flow concept* would utilize both the kinetic energy and the heat energy of the steam liquid mixture produced by flashing the geothermal brine. The overall efficiency for conversion into electrical energy should be greater than in the methods described above in which only the heat content of the brine is utilized. One proposed total flow system utilizes the principle of the Lysholm machine known in this connection as the helical rotary (or screw) expander (or mixed phase expander). The expander is simply a compressor in reverse.

The principle of the total flow concept is simple. The flow and *T-S* diagrams are shown in Fig. 8.6 (*a*) and (*b*). The hot brine from geothermal well at 1 is throttled to 2, where it becomes a two phase mixture of low quality. The two phases at this point are not separated, the full flow is expanded to 3, condensed to 4. Then the brine is reinjected into the ground at 5. Comparing the *T-S* diagrams of Figs. 8.6 (*a*) and 8.6 (*b*) show that the throttling process 2-3 that occurs in the flash separator in the former is no longer necessary and that, considering equal pressures at 2, the full available energy at 2 is used in the latter, while part of it is destroyed in the former as a result of throttling. In addition the flow in the flashed steam turbine and hence the work per unit flow from the well head is only a small fraction, equal to the quality in the flash separator, x_3, of the total flow and work that would occur in the latter.

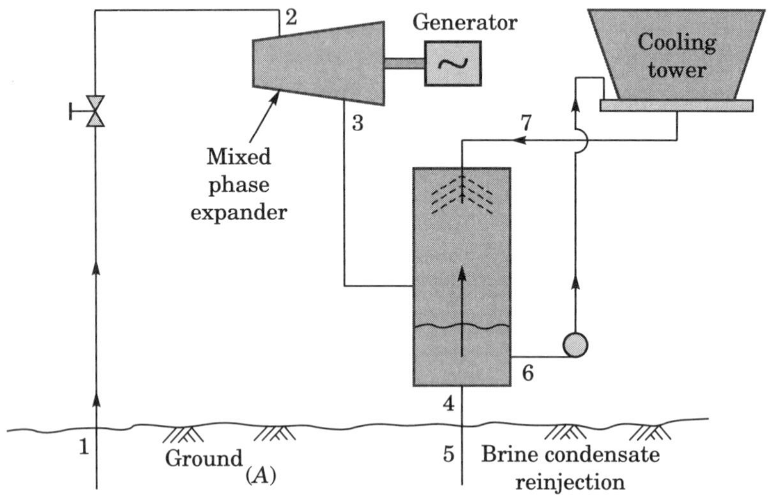

(*a*) Schematic of a liquid dominated total flow concept.

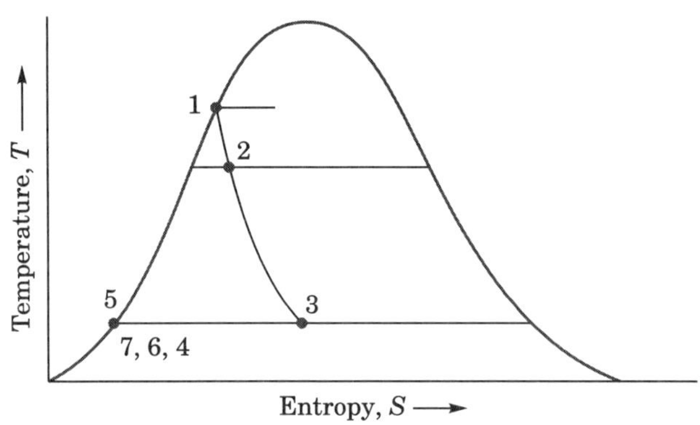

(*b*) *T-S* diagram for liquid dominated total flow system.

Fig. 8.6

The other characteristic of total flow concept system is that it requires the use of a *mixed phase expander* powered by a two phase mixture of low quality (as indicated by point 2 in figure). Whereas flashed steam systems and vapour dominated systems rely on axial flow multistage steam turbines similar to those used in conventional power plants, only difference is that they are designed for much power pressures. These turbines use relatively clean high quality or even superheated steam.

The *requirements* of mixed phase expanders are that they should be able to overcome the losses associated with the impingement of liquid droplets on blades (turbine operates less efficiently as the quality decreases). They must also be able to withstand the corrosive and erosive effects of the significant quantities of dissolved solids in the brine.

The experimental and analytical work is going in this field to develop the required expander.

Example 8.2. *A hot water geothermal plant of the total flow type receives water at 225°C. The pressure at turbine inlet is 10.5 kg/cm²a. The plant uses a direct contact condenser that operates at 0.35 kg/cm²a. The turbine has a polytropic efficiency of 0.65. For a cycle net output of 10 MW, calculate:*

(a) the hot water flow, in kg/hr;

(b) the condenser cooling water flow, in kg/hr, if such water is available at 27°C;

(c) the cycle efficiency and

(d) plant heat rate.

Solution.

Refer Figs. 8.6 (*a*) and (*b*).

From steam tables and Mollier charts,

We have $H_1 = H_2 = 669.6$ kcal/kg

Pressure at point 2, is 10.5 kg/cm^2

Thus, $T_2 = 195°C$ (14°C, degree of superheat)

$s_2 = 1.567$

$v_{sup} = 0.27$

$x_{3s} = 0.832$

$H_{3s} = 535$ kcal/kg

Isentropic turbine work

$= H_2 - H_{3s} = 669.6 - 535$

$= 134.6$ kcal/kg

Actual turbine work

$= 0.65 \times 134.6 = 87.49$ kcal/kg

$H_3 = 669.6 - 87.49 = 582.11$ kcal/kg

h_{4-5} (ignore pump work)

$h_4 = 72.4$ kcal/kg

h_6 at 27°C = 27.0 kcal/kg

Turbine steam flow or Hot water flow,

$$= \frac{\text{Power output}}{\text{Actual turbine work}} = \frac{10 \times 10^6 \times 0.86}{87.49}$$

$= 0.983 \times 10^5$ kg/hr. **Ans.**

Condenser cooling water flow $\dot{m}_4$,

$$\dot{m}_3 \times (H_3 - h_4) = \dot{m}_4 (h_4 - h_6)$$

or $$\dot{m}_3 = \frac{582.11 - 72.4}{72.4 - 27} \times 0.983 \times 10^5$$

$= 14.03 \times 10^5$ kg/hr. **Ans.**

Heat added $= H_1 - h_4$

$= 669.6 - 72.4 = 597.2$ kcal/kg

$$\text{Plant efficiency} = \frac{\text{Turbine work}}{\text{Heat added}}$$

$$= \frac{87.49}{597.2} = 0.1465 = 14.65\%. \quad \textbf{Ans.}$$

$$\text{Plant heat rate} = \frac{860}{\text{Plant efficiency}}$$

$$= \frac{860}{0.1465} = 5870 \text{ kcal/kWh.} \quad \textbf{Ans.}$$

Liquid Dominated (Low temperature) **Systems** (Hot water resources). The low temperature below 150°C, liquid dominated reservoirs are the most numerous hydrothermal source in the world. The main use would be to provide heat for homes, commercial, industrial, and agricultural buildings, including greenhouses and animal shelters, and for food and industrial processes. Hot water can also be utilised for air conditioning and refrigeration. The general principles of possible applications are the same as those considered in connection with solar energy thermal applications.

Because the water temperature is not very high, little mineral matter is extracted from the rock medium; hence, the mineral salt content of the water is usually small. The geothermal water is then not very corrosive, and it can often be circulated directly through a heat distribution system and reinjected into the ground. If there is danger of corrosion, however, a heat exchanger would be used to transfer heat from the natural hot water to ordinary service water which is then distributed. The relatively small geothermal water circuit could then be made of more expensive corrosion resistant materials. Economic use of the hot water referred to above depends on the proximity of utilization centres. It does not appear to be feasible at present to transport hot water for distances greater than about 24 km. This is so partly because of the cost of the pumps, piping etc., and partly because heat losses increase with distance. The cost of installing a distribution system in an existing city would be prohibitive, but in a new housing development the hot-water piping could be laid at the same time as that for service water. If the hot water can not be used directly in the vicinity of its source, the heat may be converted into electrical energy which can be transmitted to more distant points. A binary cycle could be used in which the hot water serves to vaporize and heat a liquid of low boiling point, as described earlier. Whether or not such a system for generating electricity will prove to economic depends on circumstances (*e.g.,* cost of drilling wells, temperature of the hot water, availability of condenser cooling water, etc.).

8.6. Geopressured Resources

Drilling for oil and gas has revealed the existence of reservoirs containing salt water at moderately high temperatures and very high pressures in a belt some 1200 km in length. Because of the abnormally high pressure of the water, up-to 1350 atm. (137 MPa) in the deepest layers, the reservoirs are referred to as geopressured. This was observed along the Texas and Louisiana coasts of the Gulf of Mexico (U.S.). The geopressured hot water (or brine) reservoirs were apparently formed by accumulation of geothermal heat stored over several million years, in water trapped in a porous sedimentary medium by the overlying impervious layers. The upward loss of heat is relatively small and there are no obvious surface indications of the deep, high temperature reservoirs. In typical geopressured systems in Texas, the pressures are from 680 to 950 atm. (68 to 95 MPa) and temperatures from 160 to 200°C at depths from 4 to 5 km. Higher pressures and temperatures have been measured at greater depths. The amount of dissolved salt in the water varies with the location and depth of the reservoir, ranging from very small to about three times that in sea water.

A special feature of geopressured waters (or brines) is their content of methane (natural gas). The energy value of the brines thus depends on their temperature. The solubility of methane in water at normal pressure is quite low, but it is increased at the high pressures of the geothermal reservoirs. When the water is brought to the surface and its pressure reduced, the methane gas is released from solution. The gas content of geopressured brine is usually about 1.9 to 3.8 cu m. gas per cu m. water but higher values have been reported in brief tests. However, the amount of natural gas recoverable economically from geopressured reservoirs is presently unknown.

8.7. Hot Dry Rock Resources or Petrothermal Systems

Petrothermal systems are those that are composed of hot dry rock (HDR) but no underground water. They represent by far the largest geothermal resource available. The rock, occurring at moderate depths, has very low permeability and needs to be fractured to increase its heat transfer surface. The thermal energy of the HDR is extracted by pumping water (or other fluid) through a well that has been drilled to the lower part of the fractured rock. The water moves through the fractures, picking up heat. It is then travels up a second well that has been drilled to the upper part of the rock and finally back to the surface. There, it is used in a power plant to produce electricity.

These types are not associated with hydrothermal activity. Such resources, with rock temperatures exceeding about 200°C at depths upto 5 km, are estimated to be significant and worthy of development as a source of energy. Hot dry rocks exist because they are impermeable to water or because water does not have access to them. Quite often both conditions occur; that is to say, the rocks are impermeable and there is little or no surface water in the vicinity. In principle, the recovery of heat from such hot dry rocks involves breaking up or cracking the rock to make it permeable and then introducing water from the surface. The water is heated up by the rock and is returned to the surface where the heat is utilized as stated above.

There are two methods to tap this geothermal energy one possible method is to detonate a high explosive at the bottom of a well drilled into the rock. This may be a nuclear explosive. Water would be injected into this well, circulated through the cavities so formed to extract heat from the rock. The water or water steam mixture is withdrawn through another well Fig. 8.7.

Another method is to use hydraulic fracturing to produce the heat transfer surface and permeability required to extract energy at a high rate from hot dry rock. Hydraulic fracturing, which is performed by pumping water at high pressure into the rock formation, is commonly used in oil and gas fields to improve the flow. The crack produced in this manner is roughly in the from of a large, thin vertical disk, resembling a pancake, possibly a few hundred meters in diameter but less than a centimeter thick. Heat can then be extracted from the hot rock by circulating water through the crack.

The hot water or water steam mixture is utilized for generation of electricity with a binary liquid system using Freon (R—114) as the turbine working fluid.

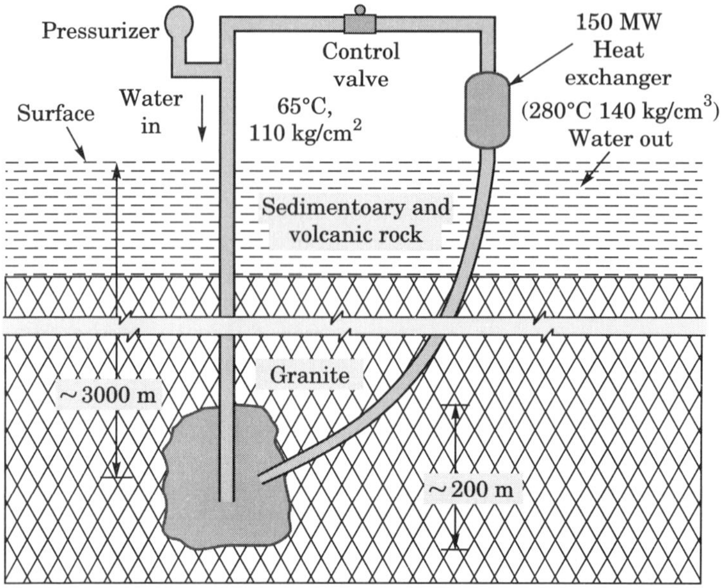

Fig. 8.7. Heat extraction from hot dry rocks.

A feature of this scheme is that, as the reservoir heat is depleted with time, temperature differences within the rock result in stresses that cause the original fractures to propagate, thereby unlocking more HDR surface to the water and resulting in a pancake shaped fracture zone.

It is believed that HDR systems offer more flexibility in operation and design than other geothermal systems. For example, the designer can have a choice of water flow rates and temperatures by drilling to various depths, and the operator can change pumping pressure and hence flow rates to suit load conditions.

One of the uncertainties associated with the hydraulic fracturing and heat extraction procedure is the useful life time of the cracked region. This can be determined only be continued testing. However computer modelling suggests that, as heat is extracted, the cracking should be increase with time and make more hot rock accessible to water. The effective life time of the fractured system should thus be extended.

There are problems which are faced by developers include leakage of water (or other fluid) underground and the necessity of make up for it from resources above ground, the effect of the water or fluid on rock composition; material carry over with the fluid, and cost. It should be noted that two wells are to be drilled instead of the one for the hydrothermal energy and that these wells are drilled deeper and in much harder rock. This is expected to make petrothermal exploitation very costly, unless the underground rock being developed is very hot.

Many more studies on the mechanical, thermodynamic; and economic aspects of petrothermal systems are necessary before commercial exploitation becomes feasible.

8.8. Magma Resources (Molten rock chamber systems)

In some cases, especially in the vicinity of relatively recent volcanic activity molten or partically molten rock (*i.e.*, magma) occurs at moderate depths (*e.g.*, less than 5 km). The very high temperature above 650°C and the large volume make magma a substantial geothermal resource. However, extraction of the heat from the molten rock will be difficult and may not be feasible for some time.

A concept being studied for the U.S. Department of energy by the Sandia National Laboratories is to place a heat exchanger within the magma. Heat would be transferred to a suitable liquid and brought to the surface. The hot liquid could be used to produce a working fluid, possibly steam, to operate a turbine and electric generator. The liquid would then be recirculated through the heat exchanger in the magma. One problem would be to construct a heat exchanger that will withstand the high temperatures of the magma. Another is to maintain flow of the viscous magma around the heat exchanger to provide a steady supply of heat and to prevent deposition of solid.

The capital cost of discovery, development and installation of geothermal heating systems have tended to discourage their use for power generation or process heating. Of course exceptions are the areas where geothermal resource was found by chance or was known to exist at shallow depths. For the production of electric power from geothermal sources however, the costs are comparable to or lower than those of all other methods except hydroelectric. Although the geothermal energy sources that are now used for electric power generation are of the shallow-steam reservoir type but in near future with technological developments, deeper reservoirs, fractured rocks and other schemes like molten magma will probably be used and may become economical in future.

8.9. Comparison of Flashed Steam and Total Flow Concept

A cycle diagram comparing the various liquid, dominated (the flashed steam) and total flow methods on the *T-S* diagram is shown in Fig. 8.8. From the comparison it is very much obvious that direct expansion of the total well head product is thermodynamically the simplest, and provides an upper bound on cycle efficiency regardless of the number of stages of separation used in the flashed steam system.

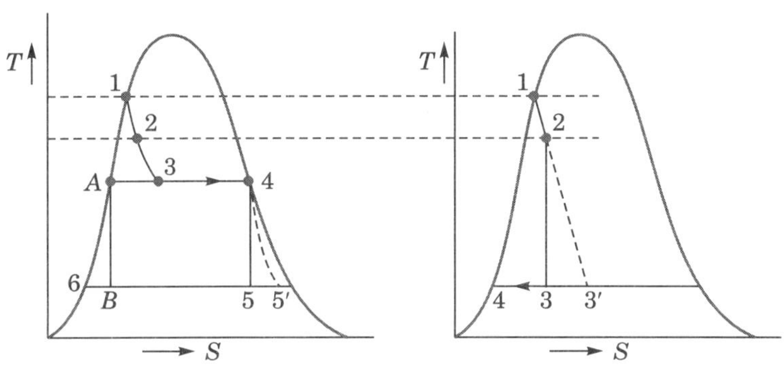

Fig. 8.8. Flashed steam process and total flow process on *T-S* diagrams.

In both the systems condensing is accomplished with a barometric direct-contact condenser. For high salinity water, however, the condenser for the total flow system will require modification to separate brine fraction from the vapour-fraction in order to prevent fouling of the cooling system.

Fig. 8.9 shows the results of an analytical comparison of various liquid dominated systems. The curves indicating that the total flow concept produces the highest specific power per unit mass flow rate at the well head. The calculations were based on the same condenser temperatures of 50°C (120°F), the same turbine efficiencies of 35 percent, and the same in plant power requirements of 30% of the gross. Differences in these would naturally change the differences between the curves.

In the binary cycle system the net power output is the gross turbo-generator output less the parasitic power for pumping and other plant requirements. An upper and lower bound calculated for the binary cycle performance is given in Fig. 8.9.

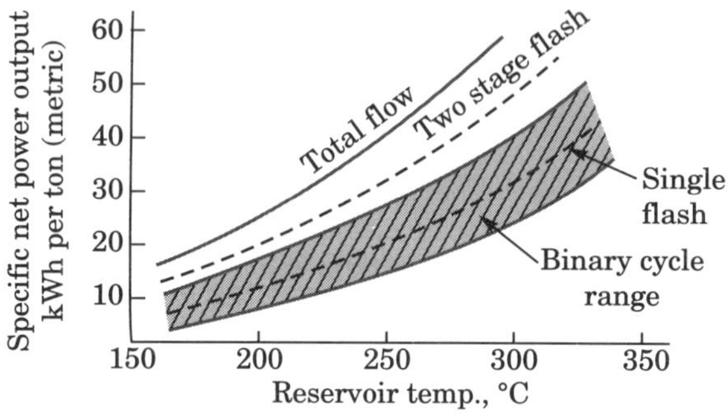

Fig. 8.9. Comparison of various liquid dominated systems.

The above curve was obtained by optimizing the system relative to heat exchanger performance by minimizing exchanger area per unit of heat transferred and per unit of net power output. The upper curve then represents the condition for a minimization of the cost of both well and exchanger.

8.10. Interconnection of Geothermal Fossil Systems (Hybrid Systems)

The concept of hybrid geothermal fossil fuel systems utilizes the relatively low temperature heat of geothermal sources in the low temperature end of a conventional cycle and the high temperature heat from fossil-fuel combustion in the high temperature end of that cycle. The concept thus combines the high-efficiency of a high temperature cycle with a natural source of heat for part of the heat addition, thus reducing the consumption of the expensive and non renewable fossil fuel.

Arrangement for hybrid plants: There are two arrangements:

(*i*) geothermal preheat, and (*ii*) fossil superheat.

Geothermal preheat arrangement is suitable for low-temperature liquid dominated systems whereas fossil superheat system is suitable for vapour dominated and high temperature liquid-dominated systems.

1. Geothermal Preheat Hybrid Systems. In these systems, the feed water of a conventional fossil fueled steam plant is heated by low temperature geothermal energy. Geothermal heat replaces some or all, of the feedwater heaters, depending upon its temperature. A cycle operating on this principle is illustrated schematically in Fig. 8.10. In it, geothermal heat heats the feed water throughout the low-temperature end prior to an open-type deaerating heater. It is followed by a boiler feed pump and three closed-type feedwater heaters with drains cascaded, backward. These receive heat from steam bled from higher-pressure stages of the turbine. No steam is bled from the lower pressure stages because geothermal brine fulfills this function.

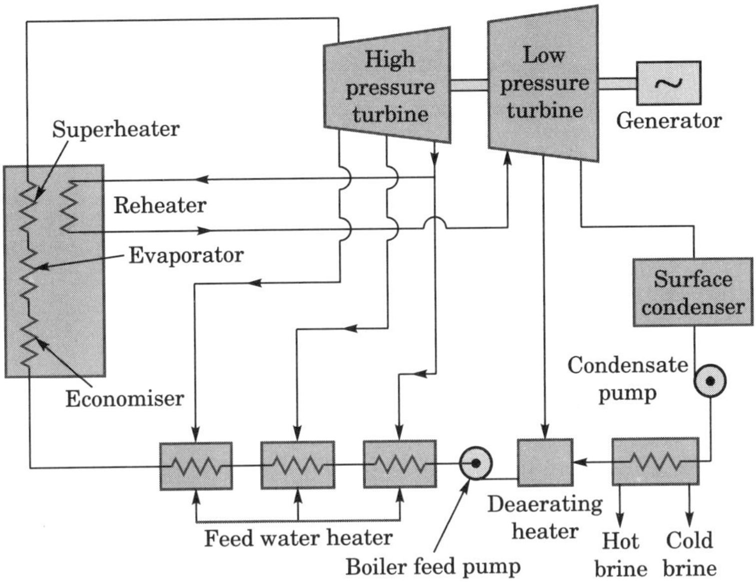

Fig. 8.10. Schematic of a geothermal preheat hybrid system.

2. Fossil Superheat Hybrid Systems. In these systems, the vapour dominated steam or the vapour obtained from a flash separator in a high temperature liquid-dominated system is super heated in a fossil fired superheater. Based on this principle, a cycle is illustrated in Fig. 8.11 and *T-S* diagram of the system is shown in Fig. 8.12. The system consists of a double flash geothermal steam system. Steam produced at 4 in the first stage flash separator is preheated from 4 to 5 in a regenerator by exhaust steam from the high pressure turbine at 7. It is then superheated by a fossil fuel fired superheater to 6, and expands in the high pressure turbine to 7 at a pressure near that of the second stage steam separator. It then enters the regenerator, leaves it at 8, where it mixes with the

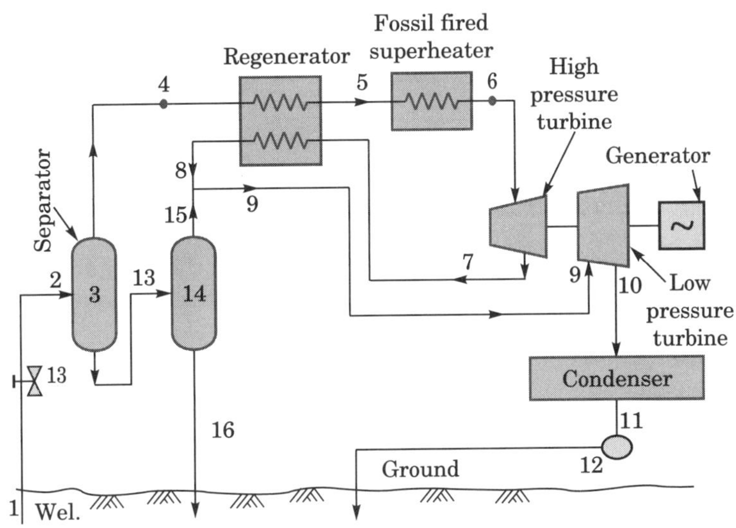

Fig. 8.11. Schematic of a fossil superheat hybrid system with two stage flash separator, regenerative and fossil fired superheater.

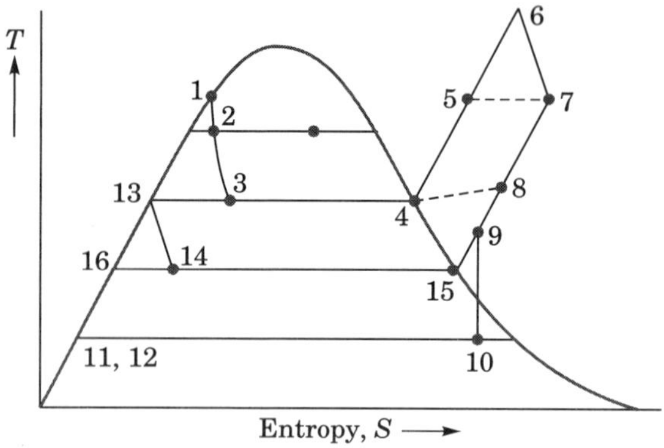

Fig. 8.12. *T-S* diagram for Fossil superheat systems in Fig. (8.11).

lower pressure steam produced in the second stage flash separator at 15, and produces steam at 9, which expands in the lower pressure turbine to 10. The condensate at 11 is pumped and reinjected into the ground at 12. The spent brine from the second stage evaporator is also reinjected into the ground at 16.

8.11. Prime-Movers for Geothermal Energy Conversion

The prime movers can be classified as below:

1. Impulse/Reaction machines

(*a*) Axial flow—curtis, Rateau steam turbine.

(*b*) Radial inflow—Francis turbine, multiple disc drag turbine.

(*c*) Radial outflow—rotating nozzle (pure reaction), Hero's turbine.

(*d*) Multiple disc turbine—bladeless impulse or reaction drag turbine.

2. Positive displacement machines

(*a*) Helical, screw expander.

(*b*) Rotating oscillating vane machine.

3. Impulse machines

(*a*) Tangential flow pelton wheel, Re-entry type turbine.

(*b*) Axial flow, De-Laval turbine, Curtis turbine.

8.11.1. Impulse Reaction Machines

Generally reaction machines will likely find limited applications. Since the pressure and temperature drop is required in the rotating component, scale formation if it occurs will be difficult to remove. Further, maximum power output requires a high ratio of tip speed to absolute speed of the fluid at the nozzle exit, that leads to high speeds and stresses. There are practical problems of maintaining rotating seals, modification of geometry as the withheld pressure drops with time and lowered nozzle efficiency due to the segregation of liquid and vapour in the rotating passage.

(*a*) **Axial flow Impulse turbine.** The idea of axial flow impulse turbine is tried in USSR, and the research on the performance of an axial flow impulse turbine to which water/steam mixtures are admitted through expansion nozzle is conducted at the Lawrence Livermore laboratory, it was found that the power developed from the water/steam mixtures from a text bore was virtually the same as that produced from the steam phase alone from the same well, but that the optimal rotational speed was nearly twice as much with the steam as with the wet mixture.

(*b*) **Radial outflow turbines.** Two phase outflow nozzle driven reaction machine is shown in Fig. 8.13. The hot water at high pressure is expanded through a nozzle to a pressure below the saturation pressure of hot water, water vapours are produced. As the specific volume of water vapour is less than that of the liquid water only a small fraction of liquid needs to vapourise to give a mixture primarily rich in gas phase with water droplets entrained in it. If the liquid droplets are assumed to be at the same velocity as the gas then

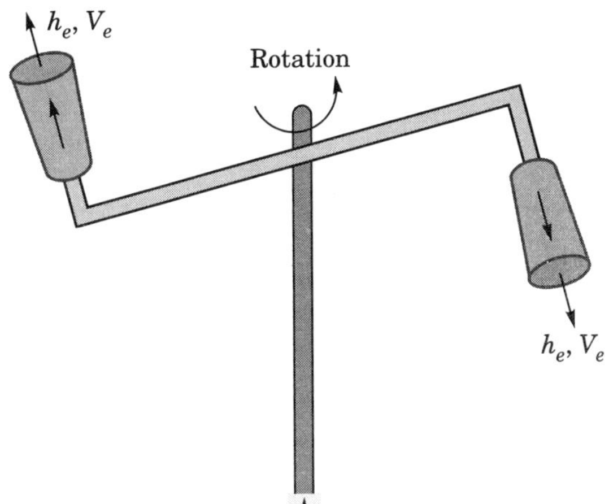

Fig. 8.13. Radial outflow turbine.

$$\frac{U^2}{2gJ} = h_i - h_e = \text{heat drop.}$$

If the nozzle is stationary then workdone = 0, but if say two nozzles are attached to shaft then a rotating couple will act and make it rotate.

Advantages

(*i*) The principle is simple.

(*ii*) Construction too is easy.

Disadvantages:

(1) For reasonable outputs the exhaustages velocity relative to nozzle has to be 330 m/sec (1000 ft/sec).

(2) To accelerate the liquid droplets to the same velocity as that of the gas large nozzles are required.

(3) Fluid friction on the droplets made the process irreverssible.

(4) There is a gas drag on the nozzles as a result of rotational speed in the exhaust chamber.

(*c*) **Hero's Turbine** (Armstead-Hero turbine). The intended purpose of this concept is to provide a cheap turbine capable of handling wet or dry geothermal fluids at the well heads of newly sunk productive bores, so as to put to immediate use the fluid that would normally be blown to waste. It would thus serve as a pilot plant. As a pilot plant it could be moved from field to field as new fields are explored. It is a reaction turbine of the simplest type. For use with dry steam the arrangement shown in Fig. 8.14 (*a*) is used, steam being admitted through the hollow shaft and discharge through tangential peripheral

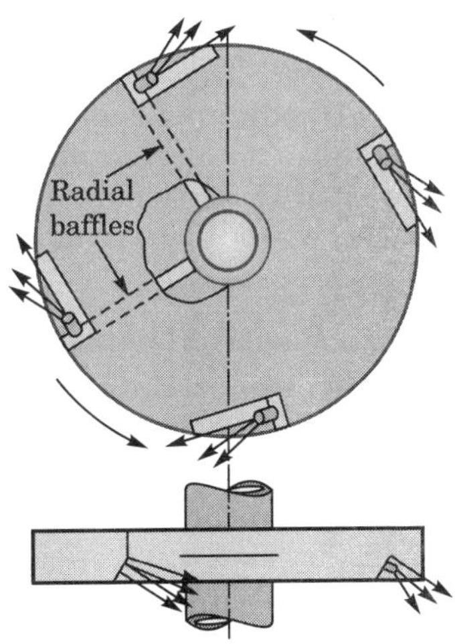

Fig. 8.14 (*a*)

nozzles provided with small axial rake to clear the exhaust away from the flat cylindrical rotor. For wet steam, the two phases would be separated within the rotor itself by centrifugal action and the use of scoop Fig. 8.14 (*b*) that enable the water to be discharged at a velocity appropriate to its intrinsic heat energy (which would be less than that of the steam discharged at the periphery). The water jet quantity would be self adjusting owing to the behaviour of boiling water when passing through bell mouthed orifices. Advantage is taken of the moderate sacrifice of efficiency resulting from a large drop in rotational speed to limit the speed to a practicable maximum value.

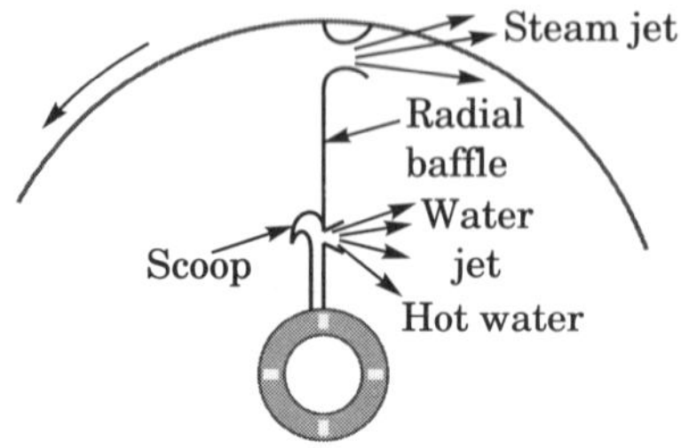

Fig. 8.14 (*b*)

Advantages:

(1) As already mentioned it is the cheap type of turbine that can handle all quality geothermal fluids (by changing fixed nozzles). Absence of blades make it simple.

(2) It can serve as a pilot plant.

(3) It could also be used as an emergency stand by unit for which high efficiency is unimportant.

Disadvantages. Steam friction would limit its high efficiency.

(*d*) **Multiple Disk Turbine** (Bladeless turbine). This consists of closely spaced thin disk mounted side by side on a shaft. Geothermal fluid (hot water or/and steam) enters through the nozzles acquires high velocity and is admitted into the spaces between the discs at the periphery and directed to follow an inward spiral path towards a central exhaust port. The driving force is derived from the frictional boundary layer drag between the fluid and disc surfaces (*i.e.,* during its passage the fluid exerts frictional shear stress on disks resulting in net torque on the shaft).

Advantages:

(1) Such a machine as described would be quite easy and cheap to construct and to maintain.

(2) Discs could be easily dismantled and cleaned of chemical deposits and reassembled.

Disadvantages. No detailed analysis has been made so far. As the basic mechanism for momentum transfer is friction drag and the much of energy would be wasted in frictional turbulence and the possibility that the entrained liquid droplets will be thrown radially outward by centrifugal forces, the turbine efficiency is low.

8.11.2. Positive Displacement Machines

(A) Keller Rotor Oscillating Vane Machine. This is a positive displacement machine designed to handle all quality geothermal fluids. Its somewhat fanciful action is self evident from Fig. 8.15.

Advantages. The advantage is that it can handle variety of geothermal fluids, (dry steam and or/geothermal hot water, it can handle very wet steam too).

Disadvantages: (1) It is mechanically quite complex and dependent upon the contact between the rotor piston and vane for sealing during expansion. Even then it offers high expansion ratios.

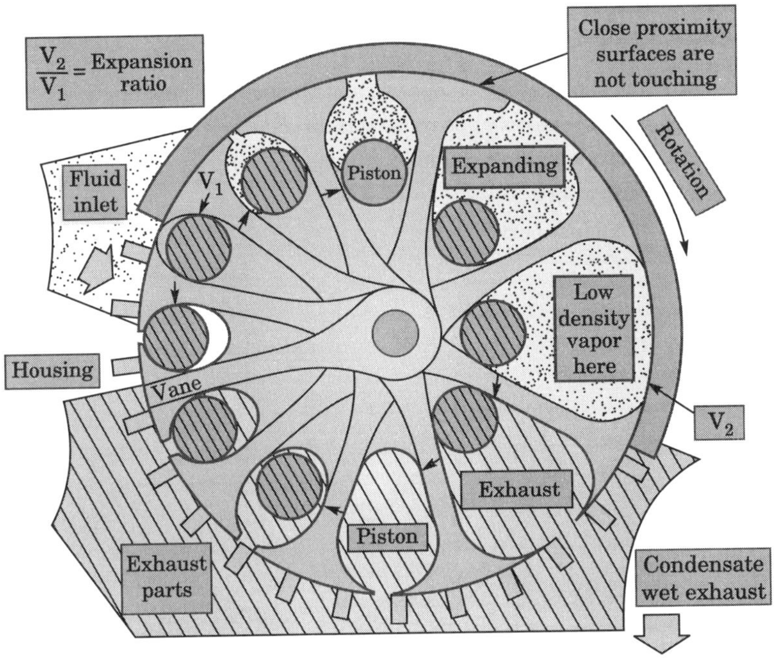

Fig. 8.15. Diagrammatic representation of the KROV (Keller Rotor Oscillating Vane) positive displacement machine.

(2) The efficiency of the device is thus low due to the above mentioned reason.

(B) The Helical Rotor Expander (Sprankle HPC prime movers Hydrothermal Power Co.):

A prototype of this machine is actually working in the Salten Sea area, California. It is illustrated in Fig. 8.15 (*a*) and (*b*). It consists of two intermeshed helical gears. Such a machine when driven by external power can work as a

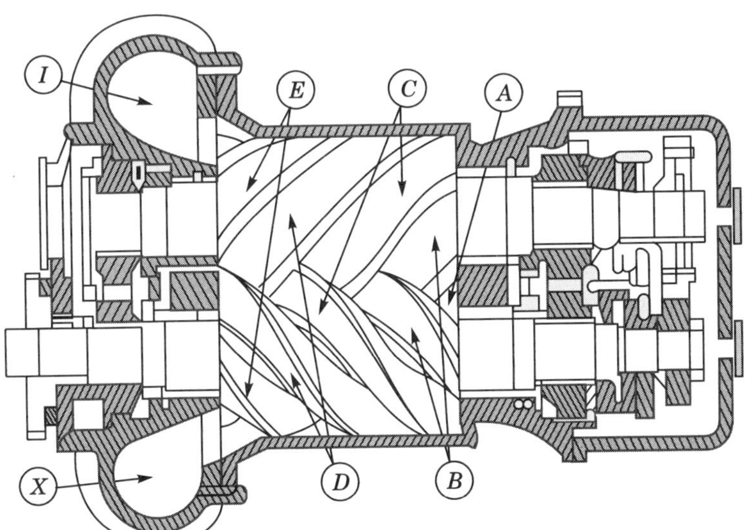

(*a*) Plan section view

A	High pressure entry pocket
B	Successive regions occupied by expanding fluid
C	as the passage elongates, and the point of meshing
D	of the screws appears to retreat axially from
E	the expanding prime.
I	Inlet port
X	Exhaust port

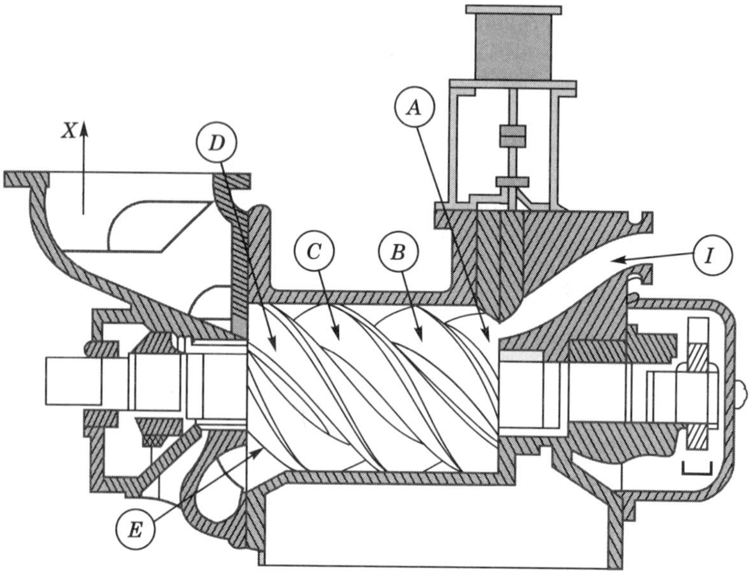

(*b*) Side section view

Fig. 8.15. The Sprankle (Hydrothermal Power Co.) double helical hot brine expander.

compressor, when used in reverse it can act as a positive displacement prime mover. It is principally meant for hot brines. The hot brine is admitted to the rotors at point A, and passes through the helical passages B, C and D falling in pressure and flashing off vapour as it proceeds, being finally exhausted from E. The pressure exerted by the fluid upon the helical vanes produces a rotary driving force. It is thus a two phase flow machines.

Advantages: (1) It is mechanically simple and self clearing as a result of rotor-to-rotor case relative motion.

(2) The makers are offering units up to 5 MW capacity and claim that the process approximates to an isentropic expansion from a saturated liquid. Thus the efficiency of the unit is high.

(3) No scale deposition problem arises with 1000 hrs of continuous operation with brines.

Disadvantages: (1) If the brine concentration is at the saturated salts are likely to be deposited on the helical and stator surfaces and the rubbing action is alleged to form a hard salt surface, sealing the machine against leakage and short circuiting of fluids past their intended path through the machine.

(2) The machine weight per kW of output is quite high.

(3) It will also be necessary to establish whether the long term effects of flashing will be damaging to the machine surfaces in a way that cavitation damages the marine propellers.

(4) Positive displacement machines are limited in volume flow rate capacities relative to impulse reaction or pure impulse ones. Fundamentally this is due to the internal losses related to sonic velocity of fluid. So these expanders must be physically large enough to produce significant power output.

8.11.3. Impulse Machine

Looking at the necessities of requiring inherently simple, compact machines it can be concluded that pure impulse devices have significant advantages. Because of geometric considerations the axial flow machine currently appears the best for total flow expansion. Refer Fig. 8.16 (*a*) and (*b*). Expansion of two phase well product through the convergent-divergent nozzle converts the brine thermal energy at high well head pressure at 2 to kinetic energy in the form of high velocity fluid stream at back pressure 3. The wheel efficiency is a measure of the wheel's ability to convert fluid kinetic energy in to shaft work. It is a function of blade geometry, turbulence, fluid friction, exist losses fanning losses etc. Advantages are already given above (like simplicity, compactness, high mass flow handling capacity, high efficiency.).

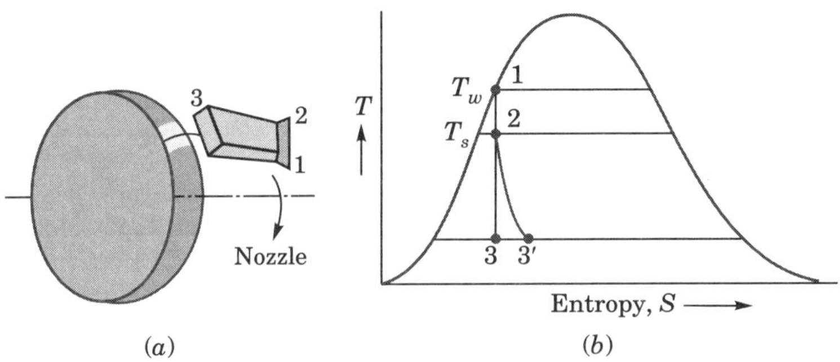

Fig. 8.16. Principle of Impulse Machine for Geothermal Power Plant.

The disadvantage is that the intersection of liquid droplets with the turbine blade during passage of the two phase supersonic flow lessens the efficiency.

8.12. Advantages and Disadvantages of Geothermal Energy over other Energy Forms

This energy form possess distinct *advantages* over conventional and new energy sources:

(1) Geothermal energy is versatile in its use.

(2) It is cheaper, compared to the energies obtained from other sources both zero fuels and fossil fuels.

(3) Geothermal energy delivers greater amount of net energy from its system then other alternative or conventional systems.

(4) Geothermal power plants have the highest annual load factors of 85% to 90% compared to 45% to 50% for fossil fuel plants.

(5) Geothermal energy is the least polluting compared to the other conventional energy sources.

(6) The greatest attraction of geothermal energy is its amenability for multiple uses from a single resource.

(7) Geothermal energy is the renewable resource that has practically no intermittency, has the highest energy density, and is economically not far removed from conventional technologies, geothermal energy is classified as renewable because the earth's interior is and will continue in the process of cooling for the indefinite future. Hence, geothermal energy from the earth's interior is almost as inexhaustible as solar or wind energy, so long as its sources are actively sought and economically tapped.

This energy has also got the following *disadvantages*:

(1) Overall efficiency for power production is low, about 15 per cent, compared to 35–40% for fossil fuel plants.

(2) The withdrawal of large amounts of steam or water from a hydrothermal reservoir may result in surface subsidence (or settlement).

(3) The steam and hot water gushing out of the earth may contain H_2S, CO_2, NH_3 and radon gas etc. If these gases are vented into the air, air pollution will be a real hazard. These gases are to be removed by chemical action, before they are discharged.

(4) Drilling operation is noisy.

(5) Large areas are needed for exploitation of geothermal energy as much of it is diffused.

8.13. Applications of Geothermal Energy

There are three main applications of the steam and hot water from the wet geothermal reservoirs:

(1) Generation of electric power,

(2) Industrial process heat and

(3) Space heating for various kinds of buildings.

One of the greatest benefits of geothermal energy is its varied application and versatility. Though electric power generation has been demonstrated and has taken a place of pride, still there are many other important applications. The following list is the indicative of the industries that would benefit from low grade heat between 200–250°C. *The list does not include power generation, space heating, farming and balneology.*

Desalination

Many chemical industries including extraction of valuable minerals from geothermal fluids

Salt production from sea water

Heavy water production

Protein and vitamin production

Sulphur and sulphuric acid production

Textile industry

The dye industry

Paper manufacture

Mining, including: recovery acid upgrading of minerals

Mining in areas of intense frost (*e.g.*, Siberia)

Peak dying

Plastic manufacture

Alumina from bauxite

Rayon production

Total gasification of coal (Lurgi process)

Timber seasoning

Refrigeration and air conditioning

Mushroom culture

Powdered coffee production

Dried milk production

Fruit and juice canning and bottling

Cattle feed from Bermuda grass

Sugar processing (with the use of bagasse as an industrial material, and not as a fuel)

Sugar beet industry

The bottling of mineral water

Fish drying and fish meal production

Other food processing and corning

Crop drying (Seawood, grass etc.)

Molasses fermentation

Road heating in frosty climates

Anti freeze for fire fighting

Sewerage heat treatment.

Table 8.1. Applications of Geothermal Energy at different Temperatures

Temperature	Application
210	
200	
190	
180	Evaporation of highly concentrated solutions Refrigeration by ammonia absorption Digestion of paper pulp (kraft)
170	Heavy water *viz*. Hydrogen sulphide process Drying of diotomacious earth
160	Drying of fishmeal, drying of timber
150	Alumina *via* Bayer's process
140	Drying farm products at high rates: Canning of food
130	Evaporation in sugar refining: extraction of salts by evaporation and crystallisation. Fresh water by distillation.
120	Most multi effect evaporation concentration of saline solution.
110	Drying and curing of light aggregate cement slabs.
100	Drying of organic materials, seaweeds, grass, vegetables, etc. washing and drying of wood.
90	Drying of stock fish: Intense de-icing operation.
80	Space heating (building and greenhouses)
70	Refrigeration (lower temperature limit)
60	Animal husbandry. Green houses by combined space and hot bed heating.
50	Mushroom growing balneology
40	Soil warming
30	Swimming pools, biodegradation, fermentations, warm water for year round, mining in cold areas, deicing.
20	Hatching of fish: Fish farming.

Sat. Steam: 210 to 100

Hot water: 120 to 20

Conventional power production: Refrigeration by ammonia absorption to Canning of food (180–140)

High grade heat is generally more versatile in applications including power generation. Whenever low grade heat is more suited to heating purposes and certain process industries.

The non-electrical application of geothermal energy is an important part of the total geothermal energy consumption. The total installed capacity of the geothermal power is only few MW compared to non electrical consumption. The non-electrical uses of geothermal energy is about 50% of that exploited for the better known electrical use. As already mentioned that the electrical conversion efficiency is of the order of 10–12%.

Low enthalpy geothermal fluids are used in the following sectors:

(1) Balneology (2) Agriculture

(3) Aquaculture (4) Climate control

(5) Industrial applications.

Current industrial applications of geothermal energy are given in the Table 8.1.

Space Heating. Geothermal energy for space heating is of great importance in some countries in Iceland, for instance where about 50% of the population enjoys such heating for its homes. The geothermal fluids for such applications usually come from geothermal reservoirs at temperatures ranging from 60 to as high as 150°C. Thermal fluids within the temperature range occur at economically acceptable depths in many parts of the world, and extensive areas are known. Thus wide spread use of this form of energy is possible.

Although geothermal space heating may serve a single house in the rural area, the most usual approach is distinct heating surfaces which serve whole areas, towns, or a city. As a rule, space heating by geothermal energy causes minimal pollution problems, in as much as there is no smoke and warm effluents are distributed widely to the sewage system in many areas where such systems have been installed the cost of energy provided is very low when compared with fossil fuels. A depreciation time for equipment of 20 to 30 years is generally used for economic evaluations.

Most distribution systems where hot water is the heating medium are single pipe system which involves the discharge of the water to the sewage system after use. The distribution temperature of the water is preferably in the range 80 to 90°C, and will cooldown to around 40°C upon use. The supply mains to the distribution system will ordinarily discharge into the storage tanks which help in taking care of daily fluctuations in hot water load. Booster pumping is usually necessary in order to maintain sufficient pressure in the distribution system.

In *house heating* application, generally central heating systems are used for houses. The hot water usually is admitted directly to these systems and discharged to sewage after use.

Agricultural and Related Applications. Geothermal energy for heating green houses is important in a number of countries. Since the temperature of the heat source will vary greatly from one location to the next, as well as variations in heating requirements, the surface area of the radiator system (often consisting of bare pipes) must be carefully tailored to local conditions. Heating fluid temperatures well above 100°C are rarely practical. Small green houses may take advantage of heat in the effluent from ordinary space heating systems. The most important crops of heated green houses of this type include cut flowers, tomatoes, cucumbers, and seedlings of many varieties. Animal husbandry, fish farming, and hatching stations also frequently take advantage of available geothermal hot water.

Climatic Variations. Generally geothermal energy to date has been exploited for those most obvious of applications: (1) for the generation of electric power, particularly more recently with distribution and shortage problems associated with fossil fuels ; and (2) dating back many more years, for space heating and some secondary heating applications in colder climates, as has been representative of the situation in Iceland. Only modest attention has been given thus far by engineers and planners to the exploitation of geothermal energy sources which may be located in warmer climates as for cooling and for processing and manufacturing applications. In certain climatic situations, since refrigeration may be achieved through heating, it is suggested that the load factor may be improved by supplying heat for refrigeration as needed, as well as for heating when needed.

Process Heating. Since geothermal energy resources exit in many countries, it is of interest to point out some common factors which affect the viability of exploitation. Since the application for heating in cold climates and the generation of electric power are obvious uses and are receiving considerable attention, it may be well to concentrate on the possibilities for process heating uses in any climate.

Examples of application of geothermal energy for process use are given in Table 8.2.

Table 8.2. Geothermal Energy for Process Use

Product	*Application*	*Form of geothermal energy*
Pulp and paper	Evaporating, digesting drying	Steam
Timber drying and seasoning	Drying, seasoning	Steam, hot water
Diatomic processing	Drying, heating deicing	Steam
Sea wood drying	Drying	Hot water
Washing of wool	Heating and drying	Steam, hot water
Curing and drying of building material	Heating and drying	Steam, hot water
Salt recovery from sea water	Evaporation	Steam
Brewing distillation	Heating and evaporation	Steam

8.14. Material Selection for Geothermal Power Plants

Most geothermal waters contain dissolved solids. The total amount of dissolved solids is typically in the range of 300–1500 ppm (parts per million), of which silica amounts to some 25–50%. The dominant ions are sodium, chloride, and sulphate. In some areas saline geothermal waters occur resulting from sea water perculating, into the bedrock and mixing with fresh water of meteoric origin in various portions. Where seawater does not enter geothermal systems, variations in the thermal fluids composition can be related to reservoir temperature and be correlated with mineral solubilities and ion-exchange equilibria with hydrothermal minerals. Geothermal water also contains some dissolved gases such as carbon dioxide, hydrogen sulfide, and to a lesser extent, hydrogen and ammonia. The transfer of carbon dioxide and hydrogen sulfide into the steam phase upon flashing may cause some harmful effects to some materials. When the steam condenses, substantial amounts of the carbon dioxide and hydrogen sulfide are dissolved in the condensate and render it quite acidic (pH 3–5); the acid condensate is highly corrosive.

Initially, when geothermal power plants were built, the same materials were used as for conventional steam power plants and experience was obtained through failures. The general principle is that the better the alloy in regard to high strength properties, the worse it behaves. A good deal could be deducted from careful inspection of parts, such as casing and valves after exposure for some time to geothermal fluids. Experience showed severe cutting of valve seats and faces where leakage of wet steam took place. Many valves adequate in other industries were found to be incapable of making a tight seal or closing against the flow under pressure. Stellating of the faces was the remedy and use of stainless steel trim seemed to remedy the problem.

By the time large geothermal turbines started to be built, evolution of blade material had already settled on 12–13% chrome steel as the best all around material. It was found to standup to hydrogen sulfide as long as it was not in the hardened state. Stellating of the blades has also been found necessary because of the wetness of the steam (approximately 12–14%). So far only limited number of manufacturers produce geothermal turbines.

To summarize, materials which have been found, through expe rience and testing, to perform well in various geothermal environments throughout the world and which are least corrosive in nature are:

1. *Carbon steel,* to be used for dry and wet steam transmission pipes and separators.

2. *Stainless steel* for nozzles and diaphragms, 12–13% chrome stainless steel for rotors.

3. Austenitic stainless steel for most metal components in the condensate cooling system.

4. Aluminium or stainless steel for most structures in atmospheric exposures.

5. Platinum or gold-rhodium platings for surfaces of electrical contacts and tin plating of all insulated copper.

6. Redwood and Douglas fir for cooling tower fill, polyvinyl chloride also preferred because of its strength and fire retardant qualities.

8.15. Geothermal Exploration

The existence of a geothermal field is often apparent from steam or hot water at the surface. Because of high costs of geothermal drilling, however, something needs to be known about the energy potential of a reservoir before drilling is undertaken. In some geothermal fields, the surface indications provide meager (or misleading) information as to the reservoir capacity.

At present, relatively little is known about methods for predicting geothermal reservoir characteristics procedures under study include the following:

(i) rate of upward heat flow in the ground;

(ii) chemical compositions of surface and ground waters;

(iii) electrical resistivity of the ground at depth; and

(iv) passive and active seismic measurements.

These procedures are discussed briefly below. In addition, gravity and magnetic surveys can sometimes provide useful information.

Heat Flow Rate. The heat flow (or flux) in the ground is measured in bore holes to a depth of about 100 m or more. It is commonly expressed in *heat flow units* (HFU), where 1 HFU is equal to 10^{-6} calorie per sq cm per sec (0.0418 watt per sq m). The observation of substantially higher value would indicate a possible geothermal reservoir.

Water Chemistry. Analysis of hot well or spring water can often indicate the probable origin of the water and the reservoir temperature. Concentration of mineral salts in the water may be used to distinguish between geothermal reservoirs of different types.

Electrical Resistivity. Measurements are made of the electrical resistivity of the ground at a surface depth at many points and also at several depths at a fixed location over a potential geothermal reservoir. The electrical resistivity depends largely on the salinity and temperature of the ground water and the porosity of the rocks. A low resistivity may indicate the presence of hot and/or saline water. Hot water containing substantial amounts of dissolved mineral salts is often present in wet geothermal reservoirs.

Seismic Monitoring. Geothermal reservoirs may be associated with persistent microseismic (minor earthquake) activity which can be readily directed by a *seismometer*. This is referred to as a passive seismic technique. There are some indications that the microseismic signals may be related to reservoir depth and temperature gradient.

Active seismic surveys are similar to those used in exploring for oil. A shock wave is generated in the ground by means of an explosion or impact with a heavy mass. The signals defected at a distance can provide information about the subsurface structure. More needs to be known, however, concerning the response of geothermal reservoirs in order to interpret the results.

8.16. Geothermal Well Drilling

Drilling into geothermal reservoirs is usually conducted with rotary bits in a manner similar to that used in drilling for oil or gas. Geothermal drilling is, however, more difficult and expensive. The temperatures, upto 350°C, encountered in geothermal drilling are higher than in oil (or gas) well drilling; moreover, it may be necessary to penetrate hard, abrasive volcanic rocks. Even the best tungsten carbide bits suffer excessive wear and must be renewed frequently. The drilling muds used to lubricate and cool the drill bit deteriorate rapidly at temperatures above about 175°C. The deterioration results in an increase in viscosity, thus making circulation of the mud more difficult. The mud may form a hard, caked lining in the bell bore and ruin the production.

In dry formations or where water-bearing strata have been ceased off, a stream of air can be used instead of drilling mud. The high air velocity required to remove rock cuttings can, however, weaken the drill pipe string by abrasion; the drill pipe must therefore be replaced more often than in oil well drilling. Where water is encountered, there is no choice at present other than to use ordinary drilling mud.

Efforts are being made to improve the techniques and reduce the costs of drilling wells into geothermal reservoirs. These efforts include the development of: (1) unproved drill bits including special synthetic diamond bits, (2) drilling techniques that do not require rotary bits or which permit changing the cutting face while the bit is down-hole, (3) high temperature drilling fluids and well cementing materials, and (4) well logging instruments that can operate at high temperatures. (Well logging involves the lowering of various instruments down a well hole to provide indications of the characteristics of the formation being penetrated).

8.17. Operational and Environmental Problems

Geothermal process is not altogether a very clean process unless certain precautions are taken; otherwise there may be pollution and relative effects. Fortunately all these precautions can be taken easily and without spending large amounts of money. For all this it is necessary first to be aware of the possible sources of geothermal pollution and to avoid these. All such are given below briefly.

1. *Solid particles and noncondensable gases.* Steam and water from both hydrothermal systems contain, besides the dissolved solids in the water, entrained solid particles and noncondensable gases. The entrained solids are removed, usually be centrifugal separators at the well head. They are further removed by strainers, usually before turbine entry. The noncondensables gases are mostly CO_2 (about 80 percent), plus varying amount of CH_4, (methane), H_2, N_2, NH_3 and H_2S (Hydrogen sulphide). In new well, the percentage of these gases is higher ($\simeq$ 4%). Besides finding their way with the fluid into the plant equipment, the noncondensables also partly escape the atmosphere *via* the particle centrifugal separators, the condenser ejectors, and in some cases the cooling towers.

The presence of the noncondensable gases has several effects. First, the large quantity of these gases, relative to noncondensables in conventional steam systems, necessitates the careful design of adequate gas ejectors to maintain vacuum in the condenser. Second, although the presence of acid forming gases cause no particular problems in dry steam lines that are made of ordinary carbon steels, their corrosive effect in wet conditions necessitates the use of stainless steel in all equipment exposed to wet steam or condensate. Such equipment includes turbine erosion shields and shaft seals, exhaust duct lining, condenser lining, condensate lines and pumps, and metal parts in cooling towers, (condensers in geothermal plants may be of the direct contact type, and hence cooling towers are exposed to geothermal condensate). In the turbine, steam nozzles and blades subjected to dry or high quality steam are usually made of 11 to 13% chrome steel. The nozzles are usually designed with large throat areas on a wide pitch to minimize scaling. Because nickel is particularly sensitive to H_2S corrosion, it is not recommended for use in the rotor. The cooling towers are usually designed with plastic fill and concrete shells, the later coated with coaltar epoxy. Aluminium is recommended for condensate pipes and valves that are made large enough to allow low velocities and hence erosion. Aluminium is also recommended for switchyard structures that are in the open but in a generally corrosive atmosphere.

H_2S is almost always present in geothermal field. This if present in excess quantities may cause harmful effects on the bearings. This also attacks electrical equipments and it may have adverse effects on crops and on river life.

At power plants it occurs in high concentration at the gas ejector discharge points. At some places, all regular escape path of H_2S are being trapped and the gas is being burned to form SO_2 which is then scrubbed with cooling tower water. Elemental sulphur is then precipitated and filtered out, while the contaminated water is reinjected into the ground. Where river cooling is adopted, some H_2S would escape in solution in the cooling water. With large river flow this is not of importance but in small streams, the only solution could be the conversion to cooling towers.

A further effect of the noncondensables is that they are environmentally undesirable because they partly escape into the atmosphere. Most are corrosive in the normally damp atmosphere of the plant site and are noxious and toxic and hence major air pollutants. The most objectionable are H_2S and, to a lesser extent, NH_3. The incondensable gases that accompany the bore fluids consists of CO_2. With high CO_2 content in water discharged into the rivers can aggravate weed growth.

(2) *Re-injection.* Re-injection would avoid discharging large quantities of heat into rivers, with consequent hazards to fisheries, and would also avoid infecting rivers and stream with toxic substances emitted from the bores which would endanger down stream drinking water supplies, fisheries and farming activities. Since the removal of huge quantities of underground water is known to cause land subsidence reinjection could at least mitigate this. Reinjection minimizes surface pollution.

(3) *Land-Erosion.* Closer control, replanting of shrubs and trees, more careful site selection and improved construction methods are helping to solve this problem. The close spacing of several walls within a single levelled area, combined with directional drilling could help in this respect.

(4) *Noise.* Noise pollution is another problem. Exhausts, blow downs and centrifugal separation are some of the sources of noise that necessitate the installation of silencers on some equipment. The noise cause a serious health hazard. Workers on new well sites have to wear ear-plugs or muffs lest their hearing be damaged. Reinjection could eliminate noise due to flashing steam where hot water is now being discharged to waste.

(5) *Water-borne poisons.* The water phase in wet fields sometimes contain toxic mercury, arsenic, ammonia etc., which if discharged could contaminate water downstream. Highly saline bore water too may be harmful. Possible solution to this include reinjection, disposal into sea through ducts and channels and the use of evaporator ponds.

(6) *Air-borne poisons.* From various points harmful substances may escape into the air at thermal sites. These may contain radioactive materials also. Horizontal well discharge in a controlled direction could be a solution to this problem. Systematic monitoring is advisable in this case.

(7) *Heat pollution.* The necessary adoption of relatively low temperature for geothermal power production results in low efficiencies and emission of huge quantities of waste heat. In wet fields, enormous source of heat wastage can arise from the reinjection of very hot unwanted bore water into rivers and streams or into storage ponds and then into atmosphere. Heat pollution in river water can damage fisheries and encourage growth of unwanted water weeds. Possible remedies include reinjection of surplus cooling tower water and unwanted bore water, the generation of additional power by means of binary fluid cycles etc. The escape of heat from cooling towers may affect local climate.

(8) *Silica.* Silica can be particularly troublesome with distinct heating systems. Re-injection of silica-loaded waters could flow up the permeability of the substrate thus necessitating the constant changing of bore re-injection sites. Settlement ponds could mitigate this trouble.

(9) *Subsidence.* The withdrawal of huge quantities of under-ground fluids from a wet field can cause substantial ground subsidence, which could cause fitting and stressing of pipelines and surface structures. Re-injection is really the best solution to this problem. As stated earlier reinjection also minimizes surface

pollution. Large extractions and reinjection also pose the possibility of seismic disturbances.

(10) *Seismicity*. Some fears have been expressed that prolonged geothermal exploitation could trigger off earthquakes especially if reinjection is practised in zones of high shear stress where fairly large temperature differentiation occur.

(11) *Escaping steam*. Huge volumes of flash steam escaping into the air could cause dense fog to occur, which may drift across to nearly roads and cause traffic hazards. The proportion of escaping steam should be small.

(12) Geopressured water, in addition to the above problems, is thought to carry large quantities of sand, especially at the high flows required. The result is increased erosion and scaling problems.

8.18. Geothermal Energy in India: Prospects

Brief History of Investigation. Between 1963 and 1965 the National Geophysical Research institute, the Geological survey of India and the Jadavpur University Kolkata, spearheaded respectively, heat flow, geotectonic and geochemical studies encompassing the field of research and development relating to geothermal resource evaluation.

In 1968, 1972, and 1973, the hot springs committee of the ministry of irrigation and power and separate sub committees of the National Committee on Science and Technology classified all the 250 known hot springs of India on a Geotectonic basis, by grouping them into geothermal provinces and by identifying the pattern and the required priorities of exploration and exploitation under short term and long term plans.

In 1971, encouraged by the finding of the hot springs committee's report the Government of India launched a Geothermal Resource Development project in Manikaran, Himachal Pradesh, with the assistance and collaboration of the United Nations Development Programme.

Also in 1973, the first large scale and systematic field investigations in the country for the exploration and utilization of geothermal resources was taken up under the lead agency of the Geological Survey of India, in one of the remote and relatively backward regions of the country *i.e.*, Puga in Ladakh, Jammu and Kashmir state, where fossil fuels were non-existent and hydro power resources were difficult and costly to develop. Many national agencies such as Atomic Energy Commission, National Geophysical Research Institute and Geological Survey of India cooperated to integrate and execute geological, geophysical, geochemical and drilling activities in Puga area.

This integrated activity was continued in the Puga and chumathang geothermal fields of Ladakh in 1974 and was also extended to cover two other geothermal fields at Manikaran and Kasol in the state of Himachal Pradesh and Sona in Haryana State.

Geothermal Occurrence in India. Important geothermal provinces and the places where hot springs occur in India are:

(*i*) N.W. Himalayan geothermal province
(*ii*) North India precambrian province
(*iii*) Cambay Graben province
(*iv*) East Indian Archaean province
(*v*) N.E. Himalayan Geothermal sub province
(*vi*) Naga-Lushai Geothermal province

(*vii*) Damodar valley Graben province

(*viii*) Mahanadi valley Graben province

(*ix*) Narbada Tapti Graben province

(*x*) Godavary valley Graben province

(*xi*) South Indian Archaean precambrian Geothermal province

(*xii*) West coast Geothermal province, and

(*xiii*) Andaman Nicobar Geothermal province.

Potential of Geothermal Resources in India. There are about 340 known thermal areas in India, each represented by hot or warm springs. Many more areas are being discovered and reported, in the 12 well defined geothermal provinces.

The total stored heat potential of the 93 systems considered is 36.87×10^{18} calories, which is equivalent to the combustion energy of 5160 million tonnes of coal or 25440 million barrels of oil. 38 of these systems are of high temperature type whose heat energy could be considered for electrical power generation. Their estimated cumulative potential for power generation is of the order of about 500 MW for 100 years or 1650 MW for a 30 years period of utilization.

Of the remaining thermal areas, 49 are of intermediate temperature and 6 are of low temperature geothermal resources type which could best be used for non electrical applications. Their cumulative stored heat potential is 19.37×10^8 calories out of which only 1.135×10^8 calories could be beneficially put to practical utilization. Since most of the non transportation energy needs could be met at temperature below 150–200°C and if the potentials of all 93 systems is considered for non electrical applications the cumulative beneficial heat will be of the order of 2.185×10^{18} calories. If this heat is to be supplied from electrical power, 10,000 MW of electricity could be required for 30 years period.

Thus these springs have a potential to substitute about 10,000 MW of electricity could be required for 30 yrs period. Thus these springs have a potential to substitute about 10,000 MW of electricity if utilization potentials are available near each site. This is roughly 10% of the total anticipated power production in India by the turn of the century.

Utilization Studies. Several pilot projects were undertaken by Geological Survey of India in collaboration with other agencies such as N.A.L. Bangaluru, IIT Delhi in the geothermal area of North-West Himalayan province, which have conclusively proved the vast potentialities for exploitation.

A pilot project for "space heating" at puga, in Ladakh in 1975 at an altitude of 4500 M, involved construction of a shed and using steam at 125°C, at 2.5 kg/cm^2 from a nearby geothermal well for heating the space, with the help of an aluminium finned, copper, tube radiator converter. A difference of 25°C was achieved with respect to the ambient temperature.

The another project named *"Green house pilot project"* at Chumathang, (Ladakh, 4400 m altitude) in 1974, was commissioned. It utilizes hot water from a nearby geothermal well to heat the soil and environment of a green house separately. Heating of soil was done by laying zig-zag pipes below the complete area of soil and allowing geothermal fluid to pass through them. This project proved that 41 varieties of plant including flowers, creepers, vegetables, and trees can be grown even at the peak of winter using geothermal energy and constructing green house, whereas normally no germination is possible for 10 months in a year in this area. This project was undertaken by Geological Survey of India in collaboration with "Field Research Laboratory" at Leh.

A third pilot project "cold storage plant" has been recently commissioned at Manikaran (Himachal Pradesh) by the joint collaboration of Geological Survey of India, IIT Delhi, and Himachal Pradesh Government. This plant avails the hot water at 90°C from a nearby geothermal well and cold water at 10°C from the Parbati river, flowing nearby, and is capable of removing 400,000 kcal/hr of heat to obtain a permanent cold storage temperature of 5°C. This can if successful help the local farmers to store nearly 15,000 tonnes of fruits and potatoes after the harvest and enable them to sell them throughout the year at a uniform price.

A 5 kW pilot power plant is under fabrication at National Aeronautical Laboratory, Bangaluru. This plant will run on geothermal energy which will be recovered from the hot springs at Manikaran in Himachal Pradesh. This will utilize a binary cycle process using R-113 as the working fluid.

All these studies have confirmed the suitability of the resources for utilization for various purposes. Plans are being made to undertake further research and development studies in the area of geothermal energy.

QUESTIONS

1. (*a*) Define a Geothermal source.
   (*b*) Classify Geothermal sources.
2. What are the sub classification of hydrothermal convective systems? Describe a vapour dominated or dry steam field.
3. A 100 MW vapour dominated system as shown in Fig. 8.2 uses saturated steam from a well with a shut-off pressure of 28 kg/cm^2 a steam enters the turbine at 5.6 kg/cm^2 a and condenses at 0.14 kg/cm^2 a. The turbine polytropic efficiency is 0.82 and the turbine—generator combined mechanical and electrical efficiency is 0.90. The cooling tower exist is at 21°C. Calculate the necessary steam, kg/hr, and m^3 min; the cooling water flow, kg/hr, and the plant efficiency and heat rate, kcal/kwh, if reinjection occurs prior to the cooling tower.
   [**Ans.** 0.844 × 10^6 kg/hr, 5.2 × 10^3 m^3/min, 13.72 × 10^6 kg/hr, 16.49%, 5213 kcal/kwh]
4. What are the limitations of a flashed steam system? What are the advantages of double flash system?
5. Describe a Binary cycle system for liquid dominated system.
6. Describe the principle of total flow concept. Compare it with other system.
7. Describe the main types of turbines in brief, which may be used for Geothermal energy conversion.
8. What are the advantages and disadvantages of Geothermal energy forms?
9. What are the main applications of Geothermal energy?
10. What are the possible sources of Geothermal pollution? How these are avoided?
11. Give a brief note on prospects of Geothermal energy in context to India.

UNIT—5

ENERGY FROM THE OCEANS

- Ocean Thermal Energy Conversion (OTEC)
- Tidal Energy
- Ocean Waves
- Small Scale Hydroelectric

CHAPTER 9

Energy from the Oceans

(OTEC, Tides, Waves, and Small Hydroelectric)

9.1. Introduction

Broadly the ocean sources of energy are Ocean Thermal Energy Conversion (OTEC) and the Tidal energy, wave energy and fourth form of the energy that emanates from the sun-ocean system stems from the mechanism of surface water evaporation by solar heating *i.e.,* hydrological cycle. These energy sources (except tidal) are the result of the absorption by the seas and oceans of solar radiation, which causes, like the wind, ocean currents and moderate temperature gradients from the water surface downward, especially in tropical waters. The oceans and seas constitute some 70% of the earth's surface area, so they represent a rather large storage reservoir of the solar input.

The conversion of solar energy stored as heat in the ocean into electrical energy by making use of the temperature difference between the warm surface water and the colder deep water. The facilities proposed for achieving this conversion are commonly referred to as OTEC plants or sometimes as solar sea power plants (SSPP). Since the ocean waters are heated by the sun, they constitute a virtually in-exhaustible source of energy. However, unlike direct solar energy, the ocean energy is available continuously rather than only in the daytime.

The operation of the OTEC plant is based on a well established physical (thermodynamic) principle. If a heat source, is available at a higher temperature and a heat sink at a lower temperature, it is possible in principle, to utilize the temperature difference in a machine or *prime mover* (*e.g.,* a turbine) that can convert part of the heat taken up from the source into mechanical energy and hence into electrical energy. The residual heat is discharged to the sink at the lower temperature. In the OTEC system, the warm ocean surface water is the heat source and the deep colder water provides the sink.

The temperature gradient can be utilized in a heat engine to generate, power. This is *called ocean thermal energy conversion* (OTEC). OTEC may be considered solar energy once removed. This energy form has very low efficiency and has very high capital cost, because the temperature difference is small even in tropics.

Tide is a periodic rise and fall of the water level o7f the sea which are carried by the action of the sun and moon on the water of the earth. Tides are primarily caused by the lunar, and only secondarily by solar, gravitational forces acting together with those of the earth on the ocean waters to create tidal flows. The tidal range varies from few cms to about 8 to 10 m in some parts of the world. The potential energy of the tides can be trapped to generate power, but at extremely high capital costs. Tidal energy can furnish a significant portion of all such energies which are renewable in nature. It has been estimated that about a billion kW of tidal power is dissipated by friction and eddies alone. This is slightly less than the economically exploitable power potential of all the rivers of the world. It is only indication of the magnitude of the tidal power available, all of it is not economically feasible also. The first attempt to utilize energy of the ocean was in the form of tidal mills in the eleventh century in Great Britain and later in France and Spain.

The power in the *ocean waves* has been part of the human experience for thousands of years. However, this history of attempts to exploit this power for human purposes has been extensive. Patents have been used on a variety of devices, and many schemes have been described conceptually. Some small scale prototype devices have been tested. Upto now no major development programme has been carried through in any country. Small devices are available, however, and are in limited use as power supplies for buoys and navigational aids. From the engineering development point of view, wave energy development is not nearly as far along as wind and tidal energy.

The hydrological cycle results in rain fall, which causes river flows, that can be trapped behind dams to even out the variations in the river flows and thus become the source of either low head or high-head (dam) *hydroelectric energy*. Hydroelectricity accounts for some 25% of the world's total electricity capacity. Some countries, such as Norway, Switzerland, and Canada, because of favourable tapographics and rainfall, for exceed this average. Hydroelectricity used to account for much more than these percentages in the nineteenth century, but the advent of cheap coal and steam power plants and the lack of additional favourable *hydro-siies,* caused a decline in its use. Many small river hydroelectric plants were abandoned. With the coming of the energy crisis of the 1970s, interest was renewed in these small plants to supplement input to power grids. *Small scale hydroelectric* facilities can supply in principle significant amounts of electricity for irrigation or potable water pumping, lighting, or health or educational purposes. The total potential amount of such a resource is very poorly documented but is apt to be large.

In this chapter, power production from the energies associated with the ocean temperature difference (OTEC), the tides and the waves will be discussed. Hydroelectric power is a major topic and needs a separate complete details, the readers are advised to consult the book on *'power plant technology"* by the author.

9.2. Ocean Thermal Electric Conversion (OTEC)

9.2.1. Introduction

Then ocean thermal energy concept was proposed as early as 1881 by the French physicist Jacques *d'* Arsonval. In this indirect form of solar energy at sea, collection and storage are free. The surface of the water acts as the collector for solar heat while the upper layer of the sea constitutes infinite heat storage reservoir. Thus heat contained in the oceans, which is solar in origin could be converted into electricity by utilizing the fact that the temperature difference between the warm surface water of the tropical oceans and the colder waters in the depths is about 20-25°K. Warm surface water could be used to heat some low boiling organic fluid, the vapour of which would run a heat engine. The exit vapour would condensed by pumping cold water from the deeper regions. The amount of energy available for ocean thermal power generation is enormous, and is replenished continuously.

Solar energy absorption by the water takes place according to *Lambert's law* of absorption, which states that each layer of equal thickness absorbs the same fraction of light that passes through it. Mathematically

$$-\frac{dI_{(x)}}{dx} = kI$$

or $$I_{(x)} = I_0 e^{-kx} \qquad ...(9.1)$$

where, I_0 and $I_{(x)}$ are the intensities of radiation at the surface ($x = 0$) and at a distance x below the surface. K is an *extinction coefficient* (or absorption coefficient) that has the unit L^{-1}, K has values of 0.05 m^{-1} for very clear fresh water, 0.27 for turbid fresh water and 0.50 m^{-1} for very salty water. Thus the intensity decreases exponentially with depth and, depending upon K, almost all of the absorption occurs very close to the surface of deep waters. Because of heat and mass transfer at the surface itself, the maximum temperatures occur just below the surface.

Considering deep water in general, the high temperatures are at the surface, whereas deed water remains cool. In the tropics, the ocean surface temperature often exceeds 25°C, while 1 km below, the temperature is usually not higher than 10°C. Water density decreases with an increase in temperature (above 4°C where pure water's density is maximum, decreasing again below this temperature, the reason ice floats). Thus there will be no thermal convection currents between the warmer, lighter water at the top and deep cooler, heavier water. Thermal conduction heat transfer between them across the large depths, is too low to after this picture, and thus mixing the retarded, so the warm water stays at the top and the cool water stays at the bottom. It is said, therefore, that in tropical waters there are two essentially infinite heat reservoirs, a heat source at the surface at about 27°C and a heat sink, some 1 km directly below, at about 4°C ; both reservoirs are maintained annually by solar incidence. The concept of ocean thermal energy conversion (OTEC) is based on the utilization of this temperature difference in a heat engine to generate power.

The surface temperatures (and temperature differences) vary both with latitude and season, both being maximum in tropical, subtropical, and equatorial waters *i.e.*, between the two tropics, making these waters the most suitable for OTEC systems.

Several such plants are built in France after World War II (the largest of which has a capacity of 7.5 MW). With a 22°K temperature difference between surface and depths, such as exists in warmer ocean areas than in north sea, the Carnot efficiency is around 7%. This is obviously very low, and comparable to that expected from a flat plate collector. In fact, by the time the overall efficiency has been reduced by using a practical engine (operating on a Rankine cycle say) together with heat exchangers, the propositions might seem hopeless. One major difference between these two heat sources is that solar energy arrives with a low power density, and requires a large acreage of flat plate collector. Whereas on ocean thermal gradient source can operate with a small area collector by pumping sufficient water through the heat collector. Indeed the attraction of the solar sea power plant lies in its present day engineering feasibility and possible competitive cost with fossil fuel power stations. As stated the idea of ocean thermal energy conversion with a suitable working fluid was originated by *d'* Arsonval, but the technical feasibility of the open cycle system was demonstrated by Claude with an installation on the south coast of Cuba in 1929. It was a remarkable achievement at the time. The electric power generated was 22 kilowatts with an overall efficiency more than 1 per cent. The hot and cold water were conducted through the long pipes to the machinery ashore. With the limited technology and cheap fuel at that time, there was then little prospects for economic feasibility. A larger installation with two units totalling 7 megawatts was constructed on the Ivory coast by the French in 1956, but encountered troubles and was abandoned.

The process of OTEC, requires that the warm surface water and cold water from depth (about 1000-1500 m), be brought into proximity so they act as the heat source and the heat sink, respectively for a heat engine. In other words, solar energy collected and stored as heat by the world's major oceans, can be converted into electricity through a generation process similar to that of conventional power plants, except that in the case of OTEC, no depletable fuel is required. Furthermore, although there is some seasonal variation in the ocean thermal resource at a given OTEC power plant location, there is little diurnal variation. Accordingly OTEC power plants are analogous to solar hydropower plants in that they smooth out the diurnal intermittance of the solar radiation, in contrast to other electric power options. Thus OTEC power plants provide a potentially substantial renewable source of base load electricity, albeit located mainly at sea.

Although it is possible to find good land sites where OTEC power plants can be located, by bringing the warm and cold water onto shore *via* aqueducts (artificial cannal/conduit), it is clear that such opportunities will be much limited on a global basis than the ample opportunities for generating substantial amounts of OTEC electricity abroad floating OTEC platforms. This is both because of the special technical requirements for on shore OTEC plants and because of the limited market potential (atleast in the near term) for OTEC electricity at such sites. On shore OTEC power plants will be viable mainly at locations where three requirements are all simultaneously satisfied with satisfactory economics:

(*i*) Coastal zone land must be available,

(*ii*) Sea floor must descend sufficiently rapidly from the shore based plant location; and

(*iii*) The seasonal availability of warm and cold water without undue gradation by the warm and cold water effluents from the OTEC plant must meet certain criteria. In any event, it is probable that available and attractive on shore and near shore OTEC power plant, location will be populated early in the development and implementation of the OTEC concept, both as convenient locations for pilot and demonstration plants and because they will constitute attractive inter mediate markets for OTEC electricity and by products.

OTEC power generation system gives less efficiency, as stated above. However, because of the OTEC requirement for parasitic power (such as for pumping up the cold water supply) and other losses, the achievable net conversion efficiency is only about 2.5% (Carnot efficiency 7%). This compares a net efficiencies of 30 to 40% associated with conventional power plants.

Some engineers question whether such as an extremely low net efficiency will ever allow OTEC to become economically viable. However, it is important to consider the matter in more sophisticated terms than net efficiency; since in the case of OTEC there is no fuel cost, only the requirement to pay for circulating much more warm and cold water, than is normally associated with power generation. This means that extensive areas heat exchangers will be required for "closed cycle" OTEC plants (which would employ a working fluid such as ammonia) or that degasifiers (to remove gases dissolved in the sea water) and tremendous turbines would be required for "open cycle" OTEC plants that would operate by the flash evaporation of sea water. Thus, although the net efficiency of the OTEC plant must certainly be positive and as high, as is readily attainable, the key economics question is the resulting cost of OTEC electrical energy, not the actual value of the net efficiency.

9.2.2. Methods of ocean thermal electric power generation

There are two rather different methods for harnessing ocean thermal differences. One is the *open cycle,* also known as the *Claude cycle,* and other is the *closed cycle* system, also known as the *Anderson cycle.* These are covered in the next two sections.

In the *closed cycle* system, a liquid working fluid, such as ammonia or propane, is vaporized in an evaporator (or boiler); the heat required for vaporization is transferred from the warm ocean surface to the liquid by means of a *heat exchanger.* The high pressure vapour leaving the evaporator drives an expansion turbine, similar to a steam turbine that it is designed to operate at a lower inlet pressure. The turbine is connected to an electric generator in the usual manner. The low pressure exhaust from the turbine is cooled and converted back into liquid in the condenser. The cooling is achieved by passing cold, deep ocean water, from a depth of 700 to 900 m or more, through a heat exchanger. The liquid working fluid is then pumped back as high pressure liquid to the evaporator, thus closing the cycle.

In the *open cycle* turbine system, water is the working fluid. The warm surface water is caused to boil by lowering the pressure, without supplying any additional heat. The low pressure steam produced then drives a turbine, and the exhaust steam is condensed by the deep colder water and is discarded. A heat exchanger is not required in the evaporator, and direct contact between the exhaust steam and a cold water spray makes a heat exchanger as necessary in the condenser. On the other hand, because of the low energy content of the low pressure steam, very large turbines or several smaller units operating in parallel would be required to achieve a useful *electric power* output.

The *Claude cycle* or *open cycle* which is older one, utilizes the vapour pressure of sea water itself as the working medium and has been demonstrated to be practicable. The other method, *a. closed cycle* known as the Rankine cycle, uses a working fluid with higher vapour pressure (such as ammonia, hrydrocarbon or halocarbon) at the temperature available. This cycle is favoured for the future development in expectation of higher efficiency. The first published work on OTEC by *d'*. Arsonvol in 1881, suggests a closed cycle, and that article proposed sulfer dioxide (SO_2) as the working fluid. However, the first OTEC experiments by Claude in the 1920s utilized on open cycle where sea water was evaporated under a partial vacuum.

9.2.3. Open cycle OTEC System (Claude cycle)

'Open cycle' refers to the utilization of sea water as the working fluid, wherein sea water is flash evaporated under a partial vacuum. The low pressure steam is passed through a turbine, which extracts energy from it, and then the spent vapour is cooled in a condenser. This cycle drives the name 'open' from the fact that the condensate need not be returned to the evaporator, as in the case of the 'closed cycle'. Instead, the condensate, can be utilized as desalinated water if a surface con-denser is used, or if a spray (direct-contact) condenser is used, the condensate is mixed with the cooling water and the mixture is discharged back into the ocean. A schematic diagram of the open cycle system is shown in Fig. 9.1. Its corresponding T-S diagram is also

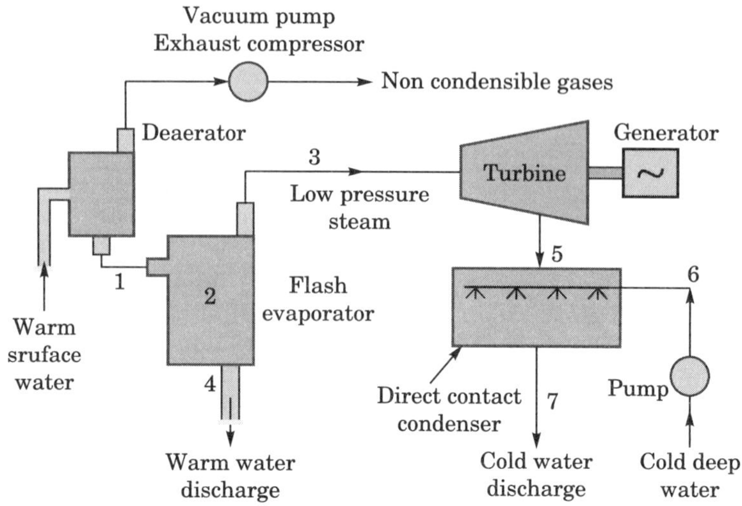

Fig. 9.1. Schematic of the OTEC open cycle.

shown in the Fig. 9.2. In the cycle shown warm surface water at say 27°C is admitted into an evaporator in which the pressure is maintained at a value slightly below the saturation pressure corresponding to that water temperature. At the new pressure, water which is entering the evaporator gets 'superheated'. As shown in Fig. 9.2 the warm water which is at 27°C, has a saturation pressure of 0.03619 kg/cm^2 (0.0356 bar) (point 1). The evaporator pressure is 0.03213 (0.0317 bar), which corresponds to 25°C saturation temperature. This temporarily superheated water undergoes *volume boiling* (as opposed to pool boiling which taked place in conventional boilers due to an immersed heating surface), causing that water to partially flash to steam to an equilibrium two phase condition at the new pressure and temperature of 0.03213 kg/km^2 and 25°C (point 2). Process 1-2 is a throttling and hence constant, enthalpy process. The low pressure in the evaporator is maintained by a vacuum pump that also removes the dissolved non condensable gases from the evaporator.

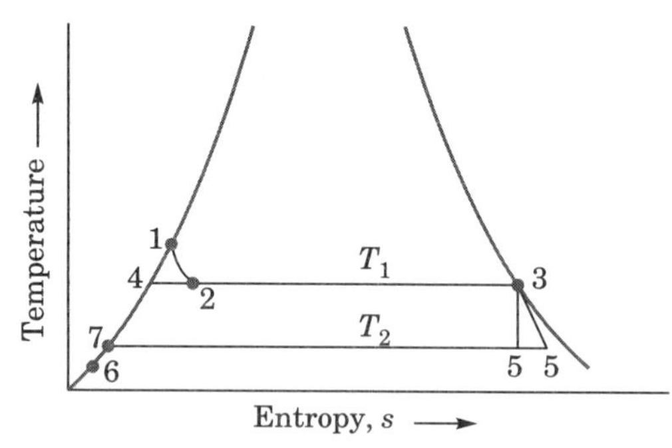

Fig. 9.2. *T-S* diagram corresponding to Fig. (9.1).

At point 2, the evaporator contains a mixture of water and steam of very low quality. The steam is separated from the water as saturated vapour at 3. The remaining water is saturated at 4 and is discharged as brine back to the ocean. The steam at 3, has a very low pressure and high specific volume (0.03213 kg/cm^2, 43.40 m^3/kg), as compared to conventional fossil power plant, which has about 160 kg/cm^2 pressure and 0.021 m^3/kg specific volume. The steam expands in a specially designed turbine that can handle such conditions to 5. The condenser pressure and temperature at point 5 are of the order of 0.01729 kg/cm^2 (0.017 bar) and 15°C. A direct contact condenser is used as the turbine exhaust steam will be discharged back to the ocean in the open cycle system. In the condenser, the exhaust steam is mixed with cold water from the deep cold water pipe at 6, which results in a near saturated water at 7. This water is allowed to discharged to the ocean. The cooling water from the deep ocean which is at about 11°C, on reaching the condenser, its temperature rises to about 15°C, due to heat transfer between the progressively warmer outside water and cooling water inside the pipe, as it ascends towards the top.

It can be seen that very large ocean water mass and volume flow rates are used in open OTEC systems and that the turbine is a very low pressure until that receives steam with specific volumes more than 2000 times that in a modern fossil power plant. Thus the turbine resembles the few lost exhaust stages of a conventional turbine and is thus physically large.

Because of the need in the open cycle to harness the energy in low pressure steam, extremely large turbines (compared to wind turbines) must be utilized. Further more degasifiers (deaerators) must be used to remove the gases dissolved in the sea water unless one is willing to accept large losses in efficiency. On the other hand, since there are no heat transfer problems in the evaporator, the problem of bio-fouling control is minimized.

The cost of an open cycle system for providing substantial number of megawatts is presently regarded by most OTEC workers as being significantly greater than for closed cycle system. The turbine cost constituted almost half the cost of the power system, but may be amenable to reductions that could result from design innovations.

9.2.4. The Closed or Anderson, OTEC Cycle

A schematic of a closed cycle OTEC power plant is shown in Fig. 9.3. Heat exchanger known as evaporators and condensers are a key ingredient, since extensive areas of material are needed to transfer significant amounts of low quality heat of the low temperature differences being exploited. In other words, large

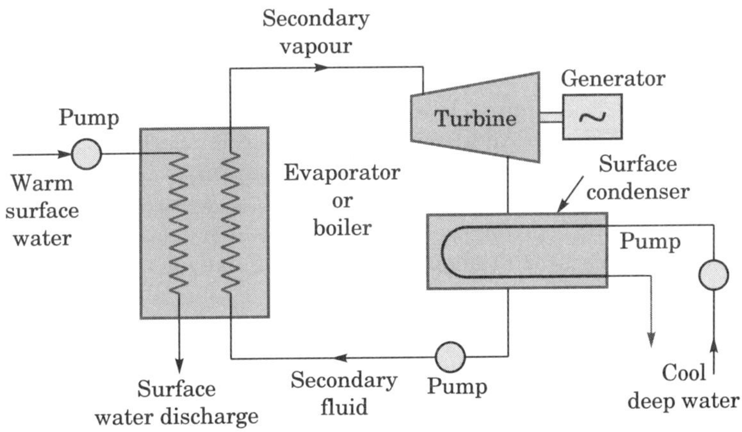

Fig. 9.3. Schematic of an OTEC closed cycle system.

volumes of water must be circulated through the OTEC power plant, requiring commensurately large heat exchangers. The actual components employed in an OTEC closed cycle system would appear more like the hardware illustrated in Fig. 9.4, another closed cycle schematic. This cycle requires a separate working

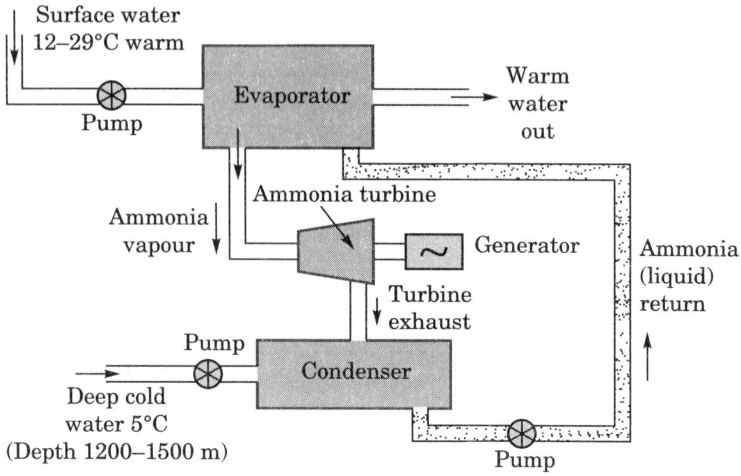

Fig. 9.4. Schematic of a closed OTEC ammonia cycle.

fluid that receives and rejects heat to the source and sink *via* heat exchangers (boiler or evaporator and surface condenser). The working fluid may be ammonia, propane, or a Freon. The operating (saturation) pressures of such fluids at the boiler and condenser temperatures are much higher than those of water, being roughly 10 kg/cm^2 ($\simeq$ 10 bar) at the boiler, and their specific volumes are much lower, being comparable to those of steam in conventional power plants.

Such pressures and specific volumes result in turbines that are much smaller and hence less costly than those that use the low pressure steam of the open cycle. The closed cycle also avoids the problems of the evaporator. It however, requires the use of very large heat exchangers (boiler and condenser) because, for an efficiency of about 2%, the amounts of heat added and rejected are about 50 times the output of the plant. In addition, the temperature differences in the boiler and condenser must be kept as low as possible to allow for the maximum possible temperature difference across the turbine, which also contributes to the large surfaces of these units.

The closed cycle approach was first proposed by Barjot in 1926, but the most recent design was by Anderson and Anderson in the 1960s. The closed cycle is sometimes referred to as the *Anderson. Cycle*. In the cycle propane was chosen as the working fluid. The temperature difference between warm surface and cool surface was 20°C. The cool surface was at about 600 m deep. Propane is vaporized in the boiler or evaporator at about 10 kg/cm^2 (10 bar) or more and exhausted in the condenser at about 5 bar.

Instead of usual heavier and more expensive shell and tube heat exchangers, the Anderson OTEC system employs thin plate type heat exchangers, which minimizes the mass and the amount of material and hence cost. The heat exchangers are placed at depths where the static pressure of the water in either heat exchanger roughly equals the pressure of the working fluid, this helps in reducing the thickness of the plates.

A fundamental requirement in closed cycle systems is to transfer heat efficiently across the heat exchanger surfaces constituting the evaporators and condensers, so as to achieve a high value of overall heat transfer coefficient (U) measured in watts per kelvin per square meter or W/°K/m^2. For the evaporation, this overall heat transfer coefficient is a measure of how effectively heat is transferred sequentially from sea water through the heat exchanger materially (a metallic alloy) and hence to the working fluid *(e.g.,* ammonia). For the condenser, an overall U characterized the reverse heat transfer process.

In an ocean environment, it is likely that a layer of slime known as "bio fouling" will *eventually* accumulate on the water side of the heat exchangers. Such slime is first comprised or microorganisms, at which stage, the bio fouling is called "micro fouling". Subsequently, if the slime is not removed, additional bio-fouling in the form of micro organisms will become attached, augmenting the slime layer. The occurrence of micro-fouling seems to be a pre-requisite for the attachment of macro organisms. A film of corrosion and possibly of calcareous (*e.g.,* minerals) deposits can also accumulate on the water side (and conceivably through leakage—even on the working fluid side of the heat transfer surfaces). The total formulation of bio-fouling, corrosion, and so on, is referred to a "fouling" (or scaling) and will tend to inhibit heat transfer through it. The "fouling factor" is a measure of the thermal resistance 'R_f' of a fouling film. This thermal resistance is the reciprocal of the corresponding heat transfer coefficient 'h_f' of the fouling film. To maintain viable OTEC heat exchangers, provisions must be made to inhibit the formation of fouling layers and to remove any significant fouling that forms. Removal can be accomplished by periodically cleaning the heat exchanger surfaces through mechanical, chemical or other means.

Although both closed-and open cycle turbine systems are being explored, it appears that closed-cycle systems offer the most promise for the near future. Each of the possible working fluids (*i.e.,* ammonia and propane) has advantages and disadvantages. Ammonia has better operating characteristics than propane and it is much less flammable. On the other hand ammonia forms a noxious vapour and probably could not be used with copper heat exchanger (see below). Propane is compatible with most heat exchanger materials, but it is highly flammable and forms an explosive mixture with air. Ammonia has been used as the working fluid in successful tests of the OTEC concept with closed cycle systems.

9.2.5. Heat Exchangers (Evaporators)

The maximum (or ideal) efficiency for the conversion of heat into mechanical work (or electricity) in a turbine depends on the drop in temperature of the working fluid in its passage through the turbine and the turbine inlet temperature. In Fig. 9.4, the temperature drop in the turbine is 10°C (10 K) and the inlet temperature is 20°C (20 + 273 = 293 K), hence the maximum thermal efficiency is 10/273 = 0.034 or 3.4%. In an OTEC system, departure from ideal behaviour in the turbine and allowance for the energy required to pump the cold water from great depths would reduce the net efficiency for electric power generation to 2 to 2.5%. This may be compared with the almost 40 percent efficiency of a modern coal fired power plant. In the OTEC system, the low conversion efficiency would be compensated by the enormous amounts of heat available in ocean surface waters. But in order to utilize this heat to generate electrical energy in economically useful amounts, water must be pumped through the heat exchangers in both evaporator and condenser at very high rates. In a facility designed to produce 100 mega watts of electrical power, for example, the total flow of water might be more than 500 million gallons (2.2 million cu. m.) per hour. The area of the heat exchange surfaces for both evaporator and condenser would be about 1 million sq. m. (Note that 100 megawatts is only about one tenth of the electric power generated by a single modern coal fired or nuclear plant.).

The effectiveness and cost of the heat exchangers are regarded as critical for the OTEC concept. The electric power that can be generated depends, in the first place, on the rate of heat transfer from the warm ocean water to the working fluid in the evaporator. Further more, conversion of this heat into electrical energy with maximum efficiency requires that the temperature of the working fluid entering the turbine should be as high as possible and that of the fluid leaving the turbine as low as possible. All these requirements can be met only if there is effective heat transfer in the heat exchangers.

Special efforts are being made to improve the engineering design of heat exchangers suitable for OTEC use. In addition, the constructional materials, must have good heat conductivity and be resistance to corrosion and erosion by rapidly flowing ocean water. Among the materials being considered, the prime candidates are : (1) titanium (2), aluminium (or an alloy) (3) an alloy of copper (90%) and nickel, and (4) plastic.

(1) **Titanium** is resistant to corrosion and erosion by ocean water and it has good mechanical strength. At present, however, it is an expensive metal of limited availability. Nevertheless, since titanium is not a rare element in nature, the supply could probably be increased and the cost of the metal decreased if there were a sufficient demand.

(2) **Aluminium** is considerably cheaper than titanium, but the common form of the metal is more susceptible to corrosion by ocean water. However, some alloys have been found to be suitable for use with salt water. Even if the aluminium alloy heat exchangers had to be replaced more often than those of titanium, they might be more cost effective in the long run.

(3) A **90/10 copper-nickel** alloy has been used extensively in both land-based and shipboard power plant condensers with ocean water as coolant. It is less expensive than titanium although more than aluminium alloys. Although the copper nickel alloy is resistant to corrosion by sea water, the copper is readily attacked if ammonia is present. Hence, copper heat exchangers could probably not be used with ammonia as the working fluid. In normal operation, the ocean water and ammonia would not be in contact, but small leaks in the heat exchanger surface, which are almost impossible to avoid, would result in severe deterioration of the copper.

(4) The possibility of fabricating heat exchangers from relatively inexpensive plastic material has attracted interest. The heat conductivity of plastics is normally too low for efficient heat exchange, but it could be increased by inclusion of graphite. But even the graphite filled plastic is inferior to most metals in this respect and correspondingly large heat exchange areas would be necessary for a specified electrical output. Plastics are expected to be resistant to corrosion by ocean water and ammonia, but some may be affected by propane. Materials with sufficient mechanical strength and resistance to erosion by high velocity sea water remain to be developed.

9.2.6. Bio-fouling

The deposition and growlh of micro-organisms, called biological fouling (or bio-fouling), on the cooling-water side of the condenser heat exchanger, is a problem encountered in most power plants. It would also be expected to arise in both the evaporator and condenser heat exchangers of an OTEC plant. Bio-fouling is less with copper (or copper alloy) heat exchangers because traces of dissolved copper act as a biocide. Biofouling is important because it reduces the heat transfer efficiency and is usually dealt with by chemical (chlorination) or mechanical (brushes or rubber balls) means. Increasing the flow rate of the water is advantageous because the organisms are less likely to become attached to the heat-exchanger surfaces. However, the flow rate must not be high enough to cause erosion.

Bio-fouling effects and ways of dealing with them are being studied in connection with the design and location of OTEC plants. Such effects are expected to be especially significant for the evaporator heat exchanger where the warmer water would be conductive to the growth of marine organisms.

9.2.7. Site Selection

In selecting a site for an OTEC facility, the primary considera-tion is, of course, a significant temperature difference—at least about 20°C—between surface and deep ocean waters (for 700-900 m depth or more) that will permit year round operation. The greater the difference, the lower will be the cost of generating electricity. The best sites are in the tropical belt between about 20°N and 20°S latitude. There are, however, several locations outside this area, that might be suitable for OTEC plants. In choosing a site, consideration should be given to the potential for bio-fouling effects as noted earlier.

As a general rule, an OTEC plant would be located offshore in order to provide access to the deep colder water. However, an ideal situation might be one where the shoreline dropped steeply to a considerable depth. Most of the installation could then be more conveniently build on land.

9.2.8. Energy Utilization

If possible an OTEC plant should be less than about 30 km from shore. The electricity generated could then be transmitted inexpensively to land by submarine cable. At somewhat greater distances, the transmission costs would be increased but might be tolerable. If the plant is so far from shore that these costs become prohibitive, the electricity generated can be utilized at the plant site to produce energy intensive materials.

One suggestion is to use direct electric current to decompose sea water by the process of electrolysis; the main products would be hydrogen and oxygen. The hydrogen could be liquified and transported by tanker to a point where it could be used as fuel. Alternatively, the hydrogen could be combined with atmospheric nitrogen to form ammonia for use as fertilizer, thereby saving natural gas which is presently the main source of hydrogen for this purpose.

9.2.9. Hybrid Cycle

There are several variations on the standard OTEC open cycle system. One variation is the "hybrid cycle" which is an attempt to combine the best features and avoid the worst features of the open and closed cycles. First, as shown in Fig. 9.5, sea water is flash evaporated to steam, as in the open cycle. The heat in the resulting steam is then transferred to ammonia in an otherwise conventional closed Rankine cycle system.

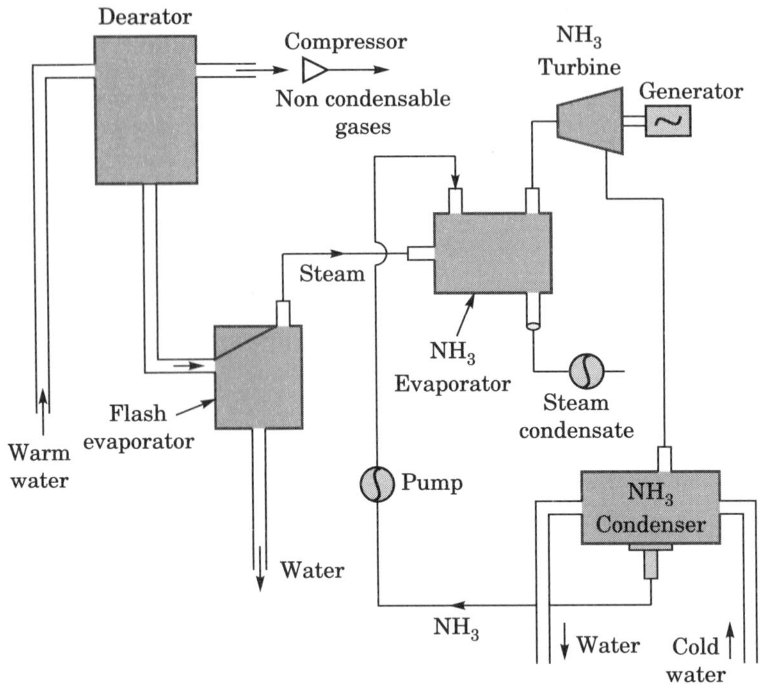

Fig. 9.5. Hybrid cycle.

9.2.10. Conclusions

A comparative study of the closed (ammonia), open (steam), and hybrid cycles showed the closed cycle system to be most economical in cost and to require the least parasitic power.

Closed cycle is favoured for the future development in expectation of higher efficiency but does not yet have the advantage of having been put in practice. Compared with the prospects of wind power for large scale development to help meet out energy needs soon, it is this lack of experience that puts oceans thermal difference development at a disadvantage and leads to the emphasis on wind power. In the case of wind power the favoured method has been demonstrated as practicable in several countries over the past half century and has failed only the economic test in the era of cheap fuel whereas the favoured method of ocean thermal energy conversion has not been so demonstrated. It is to be hoped that it soon will be ; there are reports that Japan plans to lead the way in actual development. Research and actual development particularly of materials, fouling problems, and design are in progress in this country.

9.2.11. Prospects of Ocean Thermal Energy Conversion in India

The OTEC project cell established at IIT, Chennai has completed the preliminary feasibility study for establishing *a*, 1 MW OTEC plant in Lakshadeep Island at Minicoy. The OTEC works on the principle of utilizing the temperature difference of sea water at depth and that at the surface. The surface sea water is used to vaporise a low boiling chemical which drives, a turbogenerator. The vaporised chemical is then compressed, it is condensed by using cold sea water from depth. Preliminary oceanographic studies on eastern side of Lakshadeep Island suggest the possibility of the establishment of shore based OTEC plant at the island with a cold water pipe line running down the slope to a depth of 800 to

1000 m. Both the island have large lagoons on the western side. The lagoons are very shallow with hardly any nutrient in the sea water. The proposed OTEC plant will bring up the water from 1000 m depth which has high nutrient value. After providing cooling effect in the condenser, a part of deep sea water is proposed to be diverted to the lagoons for the development of aqua culture. A hydrographic survey of the proposed site was undertaken by National Hydrographic Office, Dehradun. The preliminary assessment of survey indicates the availability of suitable conditions for establishment of OTEC plant.

9.3. Energy from Tides

9.3.1. Introduction

Tide is a periodic rise and fall of the water level of sea which are carried by the action of the sun and moon on the water of the earth. Tide energy can furnish a significant portion of all such energies which are renewable in nature. It has been estimated that about a billion kW of tidal power is dissipated by friction and eddies alone. This is slightly less than the economically exploitable power potential of all the rivers of the World. It is only indication of the magnitude of tidal power available; all of it is not economically feasible also. The first attempt to utilize energy of the ocean was in the form of tidal "mills" in the eleventh century in Great Britain and latter in France and Spain.

The large scale up and down movement of sea water represents an unlimited source of energy. If some part of this vast energy can be converted into electrical energy it would be an important source of hydro power. The main feature of the tidal cycle is the difference in water surface elevations at the high tide and at the low tide. If this differential head could be utilized in operating a hydraulic turbine, the tidal energy could be converted into electrical energy by means of an attached generator. In principle, this is not very difficult as water, at the time of high tide, is at a high level and can be let into a basin to be stored at a high level there. The same water can be let back into the sea during the low tide through the turbines, thus producing power. Since the basin water level is high and sea water is low, there is a differential head comparable to the tidal range, that can be utilized for the running of the turbines. Basically it appears to be a simple proposition, the problems involved in it, are many. The Tides, as we see, although free, were inconvenient because they come at varying times from day to day, have varying ranges (heads) and, for large outputs required large capital expenditures. Their early use declined and eventually came to a half with the coming of the age of steam and cheap coal. With the beginning of the energy crisis in the 1970s, the tidal energy, like other renewable energy sources, received renewed attention.

The first tidal power plant was commissioned by General De-Gaulle at La Rance in 1966 which marked a break through. The average tidal range is 8.4 m

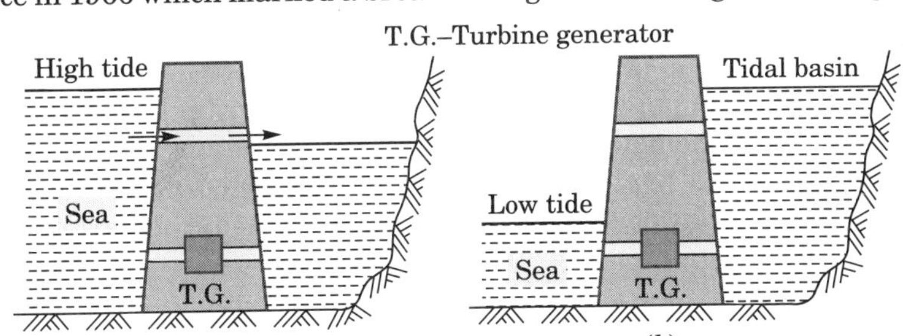

Fig. 9.6. Principle of Tidal power generation.

(± 4.2 m), and the maximum is 13.5 m. Effective basin surface is 22 km^2 ; basin volume is 184,000,000 m^3. There are no special problems with this site, and it was a very sensible choice for the world's *first* tidal power station. It has used a single basin and submerged reversible propeller type turbine generators that could generate power with the water flowing in either direction through the turbine runner. Bulb turbines have been provided in this project, they are operating satisfactorily for the last 25 years.

The dam contains a small lock for commercial vessels, six movable sluice gates, and 24 turbo generators units of 10 MW each. The movable gates are used to accelerate filling and emptying of the storage basin at small differential water levels. Installed capacity of 240 MW, but of course the average power generating capacity is less, because power can be generated only intermittently. Maximum electric energy production capacity is 544,000 MW h/yr., which gives a plant capacity factor, without any allowance for maintenance or down time, of 0.26. Maximum utilization of the stored hydraulic potential energy is 18%, which may be increased to 24% in future by additional of 80 MW more generating capacity.

There are presently two tidal power stations operating, one in France and one in the USSR, The one in France is a full commercial station. The one in the USSR is more in the nature of a pilot plant engineering experiment, and is rather small in size.

The French project at Rance near St. Malo in Brittany, uses a dam that goes straight across the estuary of the Rance river. The dam is not too long—750 m, shore to shore. The depth is never more than 12 m below mean sea level, there is an above surface rock part way across, the climate is moderate, Kisoloya Guba on the Barents sea, 70 km north of Murmansk. It is small and was intended as a pilot project. It has one generator of 400 MW rating and delivers 700 to 800 MW h/yr, to the local electrical grid operation began in 1968.

Nearly for 17 years there was a dull period in the tidal power development due to the problems involved in it. The main disadvantage of tidal power plant is that the production of power from tidal station occurs at times and in magnitudes which are dependent on the relative positions of earth, moon and sun to one another and not on the electrical demand of consumers.

The success of La Rance stimulated interest in Canada, where on the eastern coast, in the Bay of Fundy, the largest, tidal ranges in the World occur—being as high as 16 m on spring tides. The pioneering project in Nova Scotia, Canada at Annapolis Royal was commissioned in 1983. For the first, time conventional bulb turbines were replaced by large rim generator units of straflo turbines with runner diameter of 7.6 m and capacity of 10 MW. Though this type of turbine has been in use for many years in Europe, its runner diameter was limited to about 2 m by sealing problems at the runner periphery. If these prove successful, it will provide an attractive alternative to the bulb unit, since the straflo unit has much higher inertia, thus enhancing system stability, and access to the generator is easier, thus facilitating maintenance.

In India, there are possible tidal projects in the Gulf of Kutch and Cambay, and on a smaller scale, in the Sunderbans regions of the Bay of Bengal.

In Korea, there have been a series of studies by Canadian and French firms, the latest one being of Garolim Bay. A comparative statistics of some recently proposed tidal schemes are given in the table 9.1.

Table 9.1. Comparative Statistics of Some Recently Proposed Schemes

Country site	*Installed capacity (MW)*	*Annual energy (GWh)*	*No. of 7.6 m diameter turbines*	*Mean tidal range (m)*
UK Seven Estuary	7200	13000	230	9.3
Mersy Estuary	525	1090	21	6.7
Strangford Lough	210	530	30	3.1
Eire Shannon Estuary	318	715	30	3.8
India Gulf of Kutch : Kandla	600	1600	43	5.2
Korea Garolim Bay	480	1200 approx.	32	4.6
Brazil Bacanga	30	55	2	4.1
U.S.A. Knoik Arm	2220	5500	80	7.8
Canada Cumber land basin	1147	3420	37	10.5
Cobequid Bay	4028	12600	106	12.4
Annapolis Royal	20	50	1	6.7
China Liangxia	3	11	6 (2.5 m)	– 6
USSR Lumbousky	400	—	—	– 6
Mezenskaya	10000	—	—	– 9

9.3.2. Basic Principal of Tidal Power

Tides are produced mainly by the gravitational attraction of the moon and the sun on the water of solid earth and the oceans. About 70% of the tide producing force is due to the moon and 30% to the sun. The moon is thus the major factor in the tide formation.

Surface water is pulled away from the earth on the side facing the moon, and at the same time the solid earth is pulled away from the water on the opposite side. Thus high tides occur in these two areas with low tides at intermediate points. As the earth rotates, the position of a given area relative to the moon changes, and so also do the tides. There are thus a periodic succession of high and low tides.

Although there are exceptions, two tidal cycles (*i.e.*, two high tides and two low tides) occur during a lunar day of 24 hours and 50 minutes (The lunar day is the apparent time of revolution of the moon about the earth). That is to say, the time between high tides and low tide at any given location is a little over 6 hours. A high tide will be experienced at a point which is directly under the moon. At the same time, a diametrically opposite point on the earth's surface also experiences a high tide due to dynamic balancing. Thus a full moon as well as a no moon produce a high tide. In a period of 24 hrs 50 minutes, there are therefore, two high tides and two low tides. Fig. 9.7. These are called semidiurnal tides. The rise and fall of the water level follows a sinusoidal curve, shown with point *A* indicating the high tide point and point *B* indicating the low tide point. The average time for the water level to fall from *A* to *B* and then rise to *C* is approximately 6 hours 12.5 min.

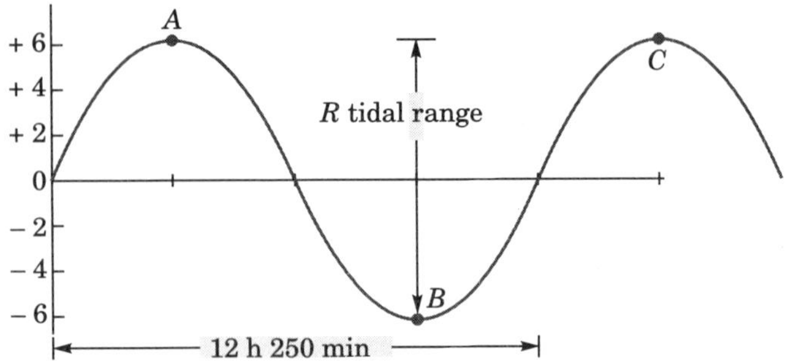

Fig. 9.7. The tides of sea.

The difference between high and low water levels is called the *range* of the tide. The *tidal range R* is defined as :

R = water elevation at high tide—water elevation at low tide.

Because of the changing positions of the moon and sun relative to the earth, the range varies continuously. There are however, some characteristic features of this variation.

At times near full or new moon, when sun, moon and earth are approximately in a line, the gravitational forces of sun and moon enhance each other. The tidal range is then exceptionally large, the high tides are higher and low tides are lower than the average. These high tides are called *spring tides*, on the other hand, near the first and third quarters of the moon, when the sun and moon are at right angles with respect to the earth, *neap tides* occur. The tidal range is then exceptionally small; the high tides are lower and the low tides higher than the average. Hence the range is not constant. It varies during the 29.5 day lunar month Fig. 9.8 being maximum at the time of new and full moons, called the spring tides, and minimum at the time of the first and third quarter moons, called the neap tides. The spring neap tidal cycle lasts one-half of a lunar month. A typical *mean range* is roughly one third of the spring range. The actual variations in range are some what complicated by seasonal variations caused by the ellipticity of the earth's orbit around the sun.

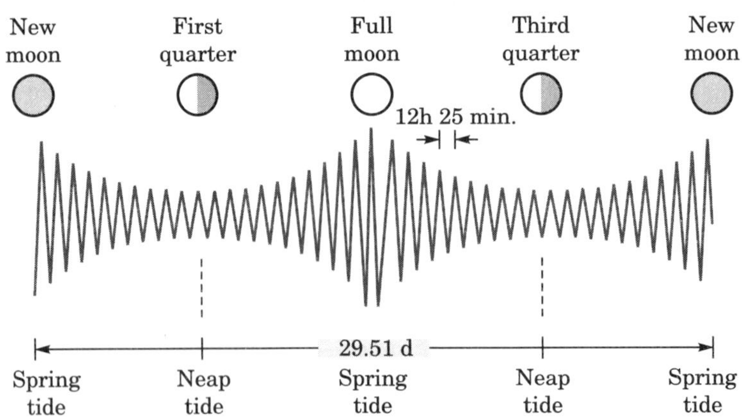

Fig. 9.8. Relative high and low tides showing variation in range during lunar month.

The variations in the periodicity and monthly and seasonal ranges must, of course, by taken into account in the design and operation of tidal power plants. The tides however, are usually predictable, and fairly accurate tide tables are usually available.

Tidal ranges vary from one earth location to another. They are influenced by such conditions as the profile of the local shoreline and water depth. When these are favourable, a resonance like effect causes very large tidal ranges. Ranges have to be very large to justify the huge costs of building dams and associated hydroelectric power plants. Such tides occur only in a few locations in the world.

Following *points* have to be specially noted in connection with the tidal phenomenon.

(1) The tides are a periodical phenomenon but no two tides in any cycle are alike. Since the relative positions of sun and moon and their distances from earth are continuously changing, the tides are also influenced accordingly. Of the two high tides in a single day, one tide is higher than the other. In any month, the tides on the full moon and no moon days are particularly higher than the rest, as on these days sun's and moon's attraction acts in a directly additive manner. These are termed as the spring tides. In any year, the tides that occur at the time of versnal and autumnal equinoxes will we even higher due to the relative location of the sun and earth. Thus the tidal range R shown in the figure varies from time to time. Generally, a long time mean value of R is designated as mean tidal range at any particular place.

(2) The mean tidal range varies from place to place. The shape of the tidal cycle depends upon the interaction of the sea with the coast line. Where the coast line offers a resonating influence, the tidal range gets accentuated, at the other places, the land may produce a dampening effect on the tidal phenomenon. For instance, in land locked seas, the tidal phenomenon is always much subdued. Because of this interacting effect, the tidal range (as well as the tidal period) varies from place to place. For example; the mean tidal range on the west coast of India is as high as 7 to 8 meters near the gulf of Kutch, whereas it is only one meter or so near Kerala, down south. Bay of Fundy (Canada) has one of the greatest tidal ranges in the world *i.e.,* of 20 m, whereas the Adiatic sea at the Zara has virtually static water with the range being only of a few cm. Thus the tidal phenomenon is a unique feature of every coast line.

(3) In spite of their complexity, the tides are amenable to mathematical analysis. As a result the exact time and the water level for a high tide as low tide can be forecast with great accuracy.

9.3.3. Components of Tidal Power Plants

There are *three* main components of a tidal power plant *i.e.*

(*i*) The power house.

(*ii*) The dam or barrage (low wall) to form pool or basin.

(*iii*) Sluice ways from the basins to the sea and *vice versa*.

The turbines, electric generators and other auxiliary equipments are the main equipments of a power house. The function of *dam* to form a barrier between the sea and the basin or between one basin and the other in case of multiple basins.

The sluice ways are used either to fill the basin during the high tide or empty the basin during the low tide, as per operational requirement. These are gate controlled devices.

It is generally convenient to have the power house as well as the sluice ways in alignment with the dam.

The design cycle may also provide for pumping between the basin and the sea in either direction. If reversible pump turbines are provided, the pumping operation can be taken over at any time by the same machine. The modern tubular turbines are so versatile that they can be used either as turbines or as pumps in either direction of flow. In addition, the tubular passages can also be used as sluice ways by locking the machine to a stand still. As compared to conventional plants, this, however, imposes a great number of operations in tidal power plants. For instance, the periodic opening and closing of the sluice-ways of a tidal plant are about 730 times in a year.

Dam (Barrage). Dam and barrage are synonymous terms. Barrage has been suggested as a more accurate term for tidal power schemes, because it has only to withstand heads a fraction of the structure's height, and stability problems are far more modest. However, the literature does not always make the distinction, even though heads are small with tidal power cut offs.

Tidal power barrages have to resist waves whose shock can be severe and where pressure changes sides continuously.

Since barrage length adds also to the price tag of the plant, short barrages are preferred even if basin size may have to be smaller as a result of site choice. Up to a height of 20 m, cost remains proportional to length as it is not changed by the need to build a dam wall to withstand high hydrostatic pressure. When, the elevation exceeds the 20 m limit, costs increase faster with length. Most tidal power plants do not have heads exceeding 20 m.

The barrage needs to provide channels for the turbines in prestressed or reinforced concrete. To build these channels a temporary cofferdam is necessary, but it is now possible to built them on land, float them to the site, and sink them into place.

Flatness is required for the sea bottom ; sandy bottom usually necessiates piling. There, where sand or rock can bear the weight of the structure to be implanted, the bottom can he prepared, the structure placed on it and then anchored. Prefabricated concrete blocks can be used as the core for large barrages and voids filled with rocks or concrete remaining holes with sand, and the entire construction then asphalted.

Construction of a barrage usually will influence the tidal amplitude. Indeed, such a construction modifies the effective length of the embayment or basin and its shape as well, particularly if the scheme involves supplementary spur dams, or brings about relocation or disappearance of natural obstruction as is foreseen for the severn plant and has occurred in the Rance estuary.

The construction influence the resonance of the bay, and most bays are less than the resonant length of the tidal wave. If resonance is reduced, the range will decrease ; if measures are taken to augment the resonance, tidal amplitude may be increased.

Tidal barrages require sites where there is a sufficiently high tidal range to give a good head of water—the minimum useful range is around three meters. The best sites are bays and estuaries, but water can also be impounded behind bunded reservoirs built between two points on the same shore line.

The precise design of barrage and its mode of operation depend critically on the requirements for power and on a careful analysis of the economics. The simplest and cheapest schemes would normally involve a single barrage designed to trap water in a basin at high tide and to generate on the ebb. More complex schemes could involve generation on both the ebb and flood tides, or the construction of a secondary basin which would permit water to be stored and discharged whenever desired. This would provide more firm power on a flexible basis. The expertise necessary to design and build such structures is available.

The location of the barrage is important, because the energy available is related to the size of the trapped basin and to the *square* of the tidal range. The nearer it is built to the mouth of an estuary or bay, the larger the basin, but the smaller the tidal range. A balance must also be struck between increased output and increased material requirements and construction costs.

Gates and Locks. Tidal power basins have to be filled and emptied. Gates are opened regularly and frequently but heads very in height and on the side where they occur, which is not the case with conventional river projects. The gates must be opened and closed rapidly and this operation should use a minimum of power. Leakage, is tolerable for gates and barrage. Since we are dealing with sea water, corrosion problems are actute, they have been very successfully solved by the cathodic protection and where not possible by paint. Gate structures can be floated as modular units into place.

Though, in existing plants, vertical lift gates have been used, technology is about ready to substitute a series of flaps that operate by water pressure. Flap gates are gates that are positioned so as to allow water into the holding basin and require no mechanical means of operation. If used they, are positioned, in the case of modular construction, in the caissons. A caisson is then floated into place. They have to be built so that they will be adequate for the maximum tidal amplitude. Top hung on a gate-hoisting beam, a gate would transfer its hydraulic load to the concrete structure. If operation is to be rapid and efficient, gates must open under the maximum differential head; this favours as flat as possible the tidal basin face of the sluice.

The flap gates allow only in the direction of the sea to basin. Hence, the basin level rises well above to sea level as ebb flow area is far less than flood flow area.

Power House. Because small heads only are available, large size turbines are needed; hence, the power house is also a large structure. Both the French and Soviet operating plants use the bulb type of turbine. Of the propeller type, with reversible blades, bulbs have horizontal shafts coupled to a single generator. The cost per installed kilowatt drops with turbine size, and perhaps larger turbines might be installed in a future major tidal power plant.

The Bulb Group (Rance Example). A bulb type turbine is an axial flow turbine the bulb set, resembling in appearance a small submarine, is made up of an ogive shaped steel shell containing an alternator and a Kaplan turbine. It is placed in a horizontal hydraulic duct and entirely surrounded by water; a shaft provides communication with the engine room of the power plant. The set function as a turbine and as a pump, and regulates the flow in both directions of flow, tide to reservoir (basin) and reservoir to tide.

The alternator is directly coupled to the turbine and turns it at some pressure. The turbine is a kaplan wheel with four mobile blades and guide vanes. The group, functioning as a turbine in the direct sense basin to sea, and furnishing power in reverse, sea to basin direction, functioning as a pump in direct sense. (Fig. 9.9).

Rim Type Turbines. Different types of turbines are under study; usually mentioned are included shaft turbines, rim type turbines, or straight flow turbines, where the generator is attached peripherally on the turbine blades, an arrangement that couples two turbines of conventional type to one generator, and a hydraulic system in which upto six turbines are coupled to hydrostatic pumps and to drive a pelton-wheel, which, in turn, drives a high speed generator. The main problem in rim type turbine, in which the rotor surrounds the turbine runner as a rim carried by the runner blades, is the seals between the stationary parts and the rotating rim. Engineers who favour straight flow generators against the bulb turbine generators point out the lower inertia characteristic of the bulb type, claiming this could lead to problems during power system disturbances. The designers of a new type of straight flow units put forth savings in civil works and in generator and auxiliary electrical equipment because large unit capacities than with the bulb type unit, for the same head, would be possible.

9.3.4. Operation Methods of Utilization of Tidal Energy

The generation of electricity from water power requires that there should be a difference in levels (or heads) between which water flows. A number of concepts have been proposed for generating electricity by utilizing the head that can be produced by the rise and fall of the tides to operate a hydraulic turbine. The power generation from tides involves flow between an artificially developed basin and the sea. However in order to have a more or less continuous generation, this basic scheme can be elaborated by having two or more basins. Accordingly we can distinguish the following types of arrangements:

(1) Single basin arrangement, (2) Double basin arrangement.

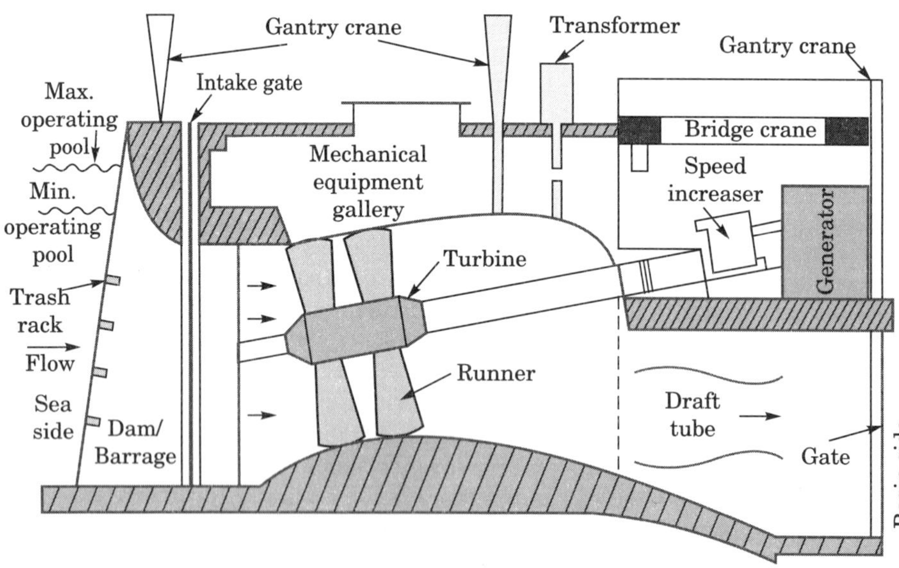

Fig. 9.9. Schematic layout of Tidal power house (Typical main unit way).

Single basin schemes can generate power only intermittently, but a double basin scheme can provide power continuously, or on demand, which is a tremendous advantage. The drawback is that the civil works become more extensive. In the simplest double basin scheme there must be a dam between each basin and the sea, and also a dam between the basins, containing the power house. One basin is maintained always at a lower level than the other. The lower reservoir empties at low tide, and the upper reservoir is replenished at high tide. If the generating capacity is to be large, the reservoirs must be large, which usually means that the dams will be long.

(1) Single Basin Arrangement

The simplest way to generate tidal power is to use a single basin with a retaining dam in the following manner :

In a single basin arrangement there is only one basin interacting with the sea. The two are separated by a dam (or barrage) and the flow between them is through sluice ways located conveniently along the dam. Potential head is provided by rise and fall of tidal water levels, this is usually accomplished by blocking the mouth of a long narrow estuary with a dam across it, thereby creating a reservoir. The dam or barrage embodies a number of sluice gates and low head turbine sets. The generation of power can be achieved in a single basin arrangement either as a

(*a*) Single ebb cycle system, or (*b*) Single tide cycle system, or

(*c*) Double cycle system.

(a) Single ebb cycle system. When the flood tide (high tide) comes in, the sluice gates are opened to permit sea-water to enter the basin or reservoir, while the turbine sets are shut. The reservoir thus starts filling while its level rises, till the maximum tide level is reached. At the beginning of the ebb tide the sluice gates are closed. Then the generation of power takes place when the sea is ebbing (flowing back of tide) and the water from the basin flows over the turbines into the lower level sea water. After two or three hours when there is sufficient difference between the full reservoir level and the falling tide level, to run the turbines, they are started and keep working until the rising level of the next flood tide and the falling reservoir level together reduce the effective head on the turbines to the extent where it can no longer work safely and efficiently. The turbines are then closed and the sluice gates opened again; to repeat the cycle of operations. Since in an estuary, the ebb tide has a long duration than the flood tide, the ebb operation provides an increased period of actual work.

(b) Single tide cycle system. In single tide cycle system, the generation is affected when the sea is at flood tide. The water of the sea is admitted into the basin over the turbines. As the flood tide period is over and the sea level starts falling again, the generation is stopped. The basin is drained into the sea through the sluice ways.

Flood operation scheme needs larger size plant, operating, for shorter period and hence less efficient as compared to ebb tide operation. The ebb operation plant will be of smaller size, but will operate over a large period. The aim and effort should be to obtain as long a period of operation as possible at the beginning and finishing the work at the minimum operating head. Adjustable blade bulb turbines area specially suited for such operations under low and variable heads. All these turbines may be reversible which can work as a pumps after the high tide, to increase the amount of water in the storage lake, until the generation can start. This way the turbine operation period may be prolonged to ten or more hours, out of the twelve hours twenty five minutes of the tidal cycle.

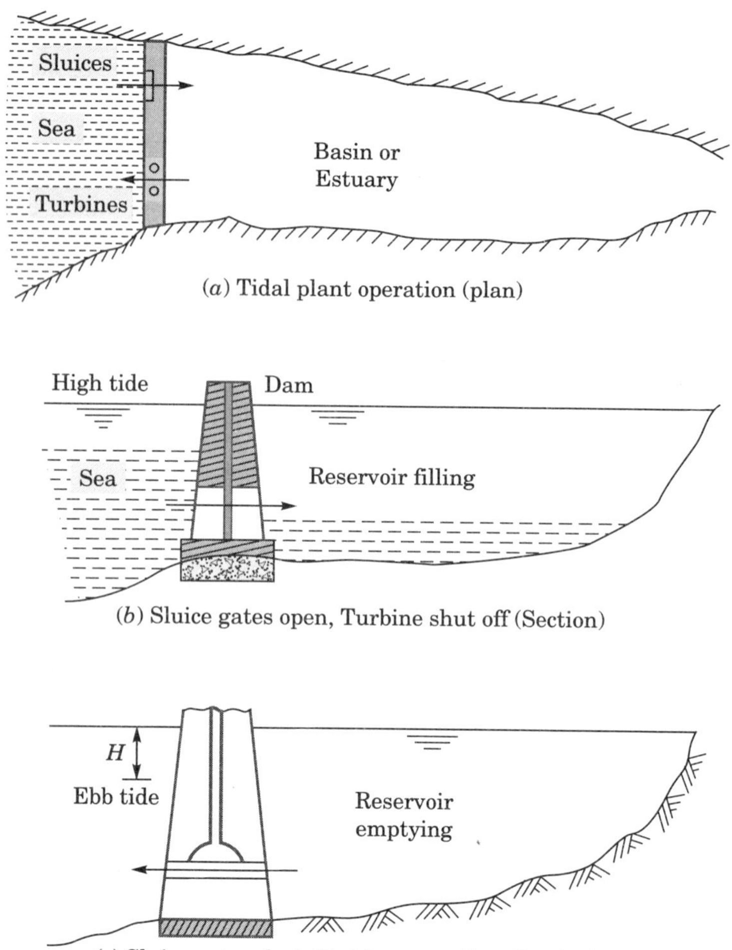

(*a*) Tidal plant operation (plan)

(*b*) Sluice gates open, Turbine shut off (Section)

(*c*) Sluice gates shut, Turbine operation (Section)

Fig. 9.10. Tidal power plant (single basin operation).

The tide cycle requires a deeper reservoir so as to locate the sills of the sluice gates deeper and, thus, requires greater construction costs. It has been estimated that the energy produced by an ebb cycle system can be as much as 1.5 times that by a tide cycle system.

The *main disadvantage* in both the ebb cycle as well as the tide cycle systems is the intermittent nature of their operation. However, since the intermissions occur at regular intervals, there is possibility of connecting another supplementary system, so as to balance the discontinuity. Such possibilities can regulate the output. The system can be so geared as to generate power, both during the ebb and flood tides with the help of single basin only. This system is known as the *double cycle system.*

(*c*) *Double cycle system.* As stated above in double cycle system, the power generation is affected during the ebb as well as in flood tides. The direction of flow through the turbines during the ebb and flood tides alternates, but the machine acts as a turbine for either direction of flow. In this method, the generation of power is accomplished both during emptying and filling cycles. Both filling and emptying processes take place during short periods of time, the filling when the ocean is at high tide while the water in the basin is at low tide level, the emptying when the ocean is at low tide and the basin at high-tide level.

The flow of water in both directions is used to drive a number of reversible water turbines, each driving an electrical generator. Electric power would thus be generated during two short, period during each tidal period of 12 *h*, 25 min. or once every 6*h*, 12.5 min.

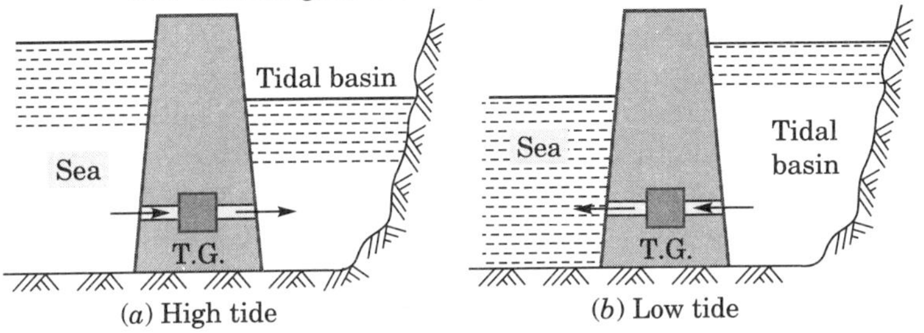

Fig. 9.11. Double cycle system.

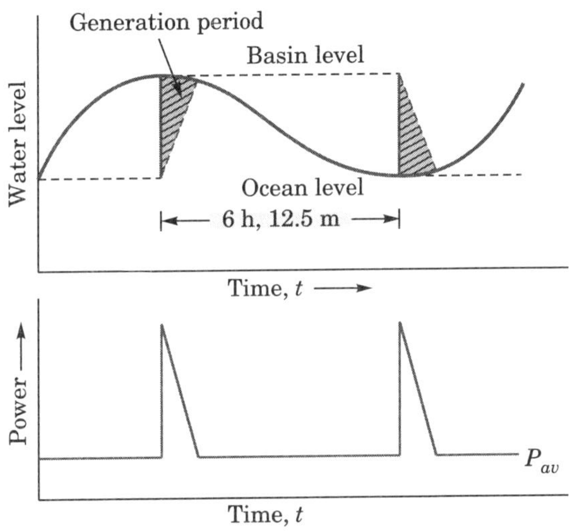

Fig. 9.12. Ocean and pool levels and power generated in a single basin tidal system.

Though the double cycle system has only short duration interruptions in the turbine operation, yet a continuous generation of power is still not possible. Furthermore the periods of power generation coincide only occasionally with periods of peak demand. These problems are solved to some extent in the two basin scheme described below. However, a fundamental drawback to all methods for generating tidal power is the variability in output caused by the variations in the tidal range.

(2) Double Basin Arrangement

It requires two separate but adjacent basins. In one basin called "upper basin" (or high pool), the water level is maintained above that in the other, the low basin (or low pool). Because there is always a head between upper and lower basins, electricity can be generated continuously, although at a variable rate.

In this system the turbines are located in between the two adjacent basins, while the sluice gates are as usual embodied in the dam across the mouths of the two estuaries. At the beginning of the flood tide, the turbines are shut down, the gates of upper basin A are opened and those of the lower basin B are closed. The basin A is thus filled up while the basin B remains empty. As soon as the rising water level in A provides sufficient difference of head between the two basins, the turbines are started. The water flows from A to B through the turbines, generating power. The power generation thus continues simultaneously with the filling up the basin A. At the end of the flood tide when A is full and the water level in it is the maximum, its sluice gates are closed. When the ebb tide level gets lower than the water level in B, its sluice gates are opened whereby the water level in B, which was arising and reducing the operating head, starts falling with the ebb. This continues until the head and water level in A is sufficient to run the turbines. With the next flood tide the cycle repeats itself. With this twin basin system, a longer and more continuous period of generation per day is possible. The small gaps in the operation of such stations can be filled up by thermal power.

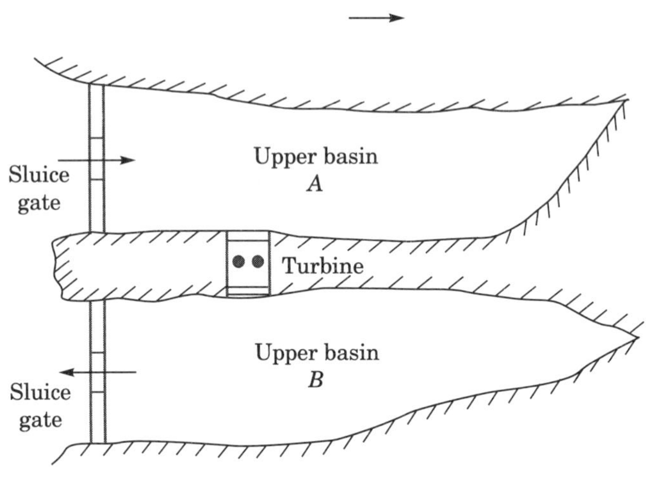

Fig. 9.13. Tidal power plant Double Basin Operation.

The operation of the two basin scheme can be controlled so that there is a continuous water flow from upper to lower basin. However since the water head between the basins varies during each tidal cycle, as well as from day to day, so also does the power generated. As in the case with single basin scheme, the peak power generation does not often correspond in time with the peak demand. One way of improving the situation is to use off-peak power, from the tidal power generators or from an alternative system, to pump water from the low basin to the high basin. An increased head would then be available for tidal power generation at times of peak demand. This is very similar to pumped storage system in hydro-electric power stations.

9.3.5. Estimate of Energy and Power in Simple Single Basin Tidal System

The expression of maximum energy that can be generated during one generation period can be derived with the help of Fig. 9.14, which shows the case of the basin beginning at high tide level, emptying through the turbine to the ocean, which is at low tide. (The identical energy will be generated in reverse process).

During the *emptying process,* the differential work done by the water is equal to its potential energy at the time. Considering a tidal range R, and intermediate head, at a given time, the amount of work is calculated, considering a small head dh, for a intermediate head h, as shown in fig.

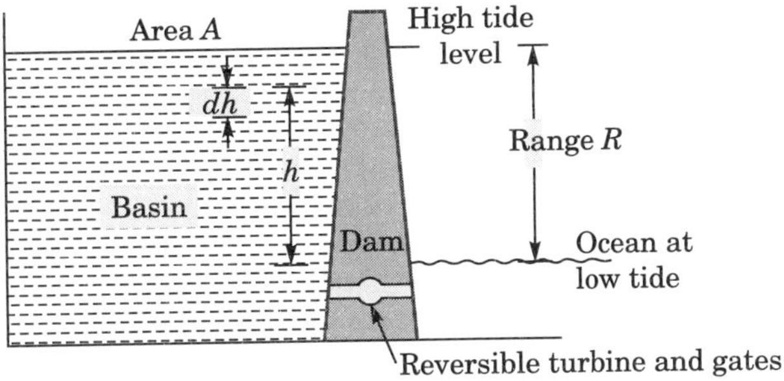

Fig. 9.14. Single Basin Tidal System.

We can write

$$dw = dm.g.h \qquad ...(9.2)$$

but, $$dm = -\,p.A.dh \qquad ...(9.3)$$

so that $$dw = -\,p.A.dh\; g\; h \qquad ...(9.4)$$

where, W = workdone by water, kcal/kg or Joule.

g = gravitational constant

m = mass flowing through turbine, kg

h = head m

ρ = water density, kg/m^3

A = basin surface area, considered constant m^2

The total theoretical work during a full emptying (or filling) period is obtained by integrating the expression (9.4) *i.e.*

$$W = \int_R^0 dw = -\,g\,\rho\,A \int_R^0 h\,dh$$

$$= \frac{1}{2}\,g\,\rho\,A\,R^2 \qquad ...(9.5)$$

The work is proportional to square of the tidal range.

The power is the rate of doing work.

The power is generated during emptying (or filling) and no power is generated during rest of the time. The *average theoretical power delivered* by the water is W divided by the total time it takes each period to repeat itself. Duration is 6*h*, 12.5 minutes as shown in Fig. 9.10. (6*h*, 12.5 minutes = 22,350 seconds).

Thus average theoretical power in watts:

$$\text{Pav.} = \frac{W}{\text{time}} = \frac{g\,\rho\,A\,R^2}{2\times 22350} = \frac{1}{44{,}700}\,g\,\rho\,A\,R^2 \qquad ...(9.6)$$

Assuming an average sea water density = 1025 kg/m^3, the average power per unit basin area is given by

$$\frac{\text{Pav.}}{A} = \frac{1}{44{,}700} \times 9.80 \times 1025\,R^2$$

$$= 0.225\,R^2 \text{ watts/m}^2 \text{ (MW/km}^2\text{)} \qquad ...(9.7)$$

The actual power generated by a real tidal system would be less than the average theoretical power obtained by the above expressions due to frictional losses and conversion efficiencies of turbine and electric generators. The actual power generated may be about 25 to 30% of the theoretic power.

Example 9.1. *A tidal power plant of the simple single basin type, has a basin area of 30 × 10⁶ m². The tide has a range of 12 m. The turbine, however, stops operating when the head on it falls below 3 m. Calculate the energy generated in one filling (or emptying) process, in kilowatt, hours if the turbine generator efficiency is 0.73.*

Solution. The total theoretical work W

$$= \int_R^r dw$$

R is the range = 12 m.

r the head below turbine stops operating = 3 m.

$$W = \int_R^r - g\,\rho\,A\,h\,dh = -\,g\,\rho\,A \int_R^r h\,dh$$

$$= \frac{1}{2} g\,\rho\,A\,(R^2 - r^2) \qquad \text{...}(i)$$

Thus the average power

$$\text{Pav}, = \frac{W}{\text{time}} = \frac{g\,\rho\,A\,(R^2 - r^2)}{44{,}700} \qquad \text{...}(ii)$$

The average power generated

$$= \frac{1}{44{,}700} \times 9.80 \times 1025 \times 30 \times 10^6\,(12^2 - 3^2) \text{ Watts}$$

$$= \frac{1}{44{,}700} \times 9.80 \times 1025 \times 30 \times 10^6 \times 135 \text{ Watts}$$

$$= 911.25 \times 10^6 \text{ Watts} = \frac{911.25}{1000} \times 3600 \times 10^6 \text{ kWh}$$

$$= 3280.5 \times 10^6 \text{ kWh}$$

Considering turbine generator efficiency, the energy generated

$= 3280.5 \times 10^6 \times 0.73 = 2395 \times 10^6$ kWh. **Ans.**

9.3.6. Estimation of Energy and Power in a Double Cycle System

In a double cycle system, the estimate of energy and power can he made as per usual calculation of power in a hydropower plant, *i.e.*, considering the average discharge and available head at any instant. Then total energy can be obtained by integrating the value of instantaneous power. The energy that is available for a tidal plant depends upon the range of the tide and volumetric capacity of the basin. The power from the plant is generally obtained in parts, in duration of 3 to 3½ hrs in a flood tide cycle and for equal duration in an ebb cycle. There are approximately 705 full tidal cycle in a year. The power or total energy is first calculated per tidal cycle and then yearly total energy generated can be obtained as explained below.

Let V be the volume of the basin

$$V = Ah_0 \qquad \text{...(9.8)}$$

where, A is the average cross sectional area of the basin in m^2, and h_0 is the difference between maximum and minimum water levels.

$$\therefore \quad \text{Average discharge } Q = \frac{Ah_0}{t} \qquad \text{...(9.9)}$$

t is the total duration of generation in one filling/emptying operation.

Now power generated at any instant

$$P = \frac{\rho Q h}{75} \times \eta_0, \text{ H.P.} \qquad \text{...(9.10)}$$

$$= \frac{\rho Q h}{75} \times \eta_0 \times 0.736, \text{ kW} \qquad \text{...(9.11)}$$

where, ρ is the average density of sea water = 1025 kg/m^3,

h is the available heat at the instant,

Then total energy

$$= \int_0^t P\,dt = \int_0^t \frac{\rho Q h}{75}\,dt \times \eta_0 \times 0.736 \text{ kW per tidal cycle} \qquad \text{...(9.12)}$$

Then yearly power generation

$$= \int_0^t \frac{\rho Q h}{75}\,dt \times \eta_0 \times 0.736 \times 705 \text{ kWh/year.} \qquad \text{...(9.13)}$$

The available head, h, is a function of time and $\int_0^t h\,dt$ can be obtained graphically.

Example 9.2. *The observed difference between the high and low water tide is 8.5 m, for a proposed tidal site. The basin area is about 0.5 sq km which can generate power for 3 hours in each cycle. The average available head is assumed to be 8 m, and the overall efficiency of the generation to be 70%. Calculate the power in h.p. at any instant and the yearly power output. Average specific weight of sea water is assumed to be 1025 kg/m^3.*

Solution. Volume of the basin $= Ah_0$

$$= 0.5 \times 10^6 \times 8.5 = 4.25 \times 10^6 \text{ m}^3$$

$$\text{Average discharge } Q = \frac{\text{Volume}}{\text{Time period}} = \frac{Ah_0}{t} = \frac{4.25 \times 10^6}{3 \times 3600}$$

$$= 0.03704 \times 10^6 = 393.5 \text{ m}^3\text{/s}$$

Power at any instant

$$P = \frac{Q\rho h}{75}\eta_0 \text{ h.p.} = \frac{393.5 \times 1025 \times 8}{75} \times 0.70$$

$$= 310.15 \times 10^2 \text{ h.p.} \quad \textbf{Ans.}$$

The total energy in kWh/tidal cycle

$$E = 301.15 \times 10^2 \times 0.736 \times 3 = 664.93 \times 10^2$$

Total number of tidal cycle in a year = 705

Therefore total output per annum $= 664.93 \times 10^2 \times 705$

$$= 468.78 \times 10^5 \text{ kWh/year.} \quad \textbf{Ans.}$$

9.3.7. Site Requirements

The utilization of tidal energy requires construction of a barrier (or barriers) across a narrow inlet to an estuary or bay, thus forming an enclosure (or basin) in which ocean water can be impounded. Electricity can be generated by allowing water to flow through a turbine from the basin filled at high tide to the open ocean during falling tides and also as the basin is being filled from the ocean during rising tides.

In each case, the maximum amount of electrical energy that can be generated depends on the product of the tidal range (R) and the mass (or volume) of water flowing through the turbine. The volume is equal to the range multiplied by the area of the impounded water. Hence, the electrical energy is proportional to the square of the range and the area of the enclosed basin. A favourable site for a tidal power plant should then have a large tidal range, and the geographic features should permit enclosure of large areas with reasonably short dams or other barrages. Sluice gates in the dams permit water to pass to or from the enclosed basin (or basins).

9.3.8. Storage

Storage is necessary when alternative electricity production schemes are not, or can not be, connected with the electrical grid. Tidal power plants are not an exception to this rule; there is a strong case for associating them with storage to provide for the varying needs of the consumer, also taking into account the eventual presence close by of other plants. The tidal power plant has, by its very nature, many of the components required of a pumped storage scheme. It has also the advantage that, with such a scheme, added tidal power becomes a firm day time source of power.

This is important because the principal saving provided by tidal energy is that of fuel, and probably also of pollution abatement costs. The fuel saved depends on the type of plant whose load factor is reduced and how such plants perform with the addition of tidal power or without it. Evidently, this argument applies only to highly developed countries where there are numerous other plants, and for less to developing countries.

The alternate to storage, when energy is available (but power is not needed because of timing), is not to generate, but the potential thus left untapped is not recoverable at a later time. It is pretty much a case of "store it or you will never have it when needed and no potential is available". Hence the extra capital investment may well be justified.

For tidal power being produced in 'mechanical form' only *batteries, compressed air,* and *hydraulic storage* can be considered. *Fly where* could be used, but its technology is not yet ready.

Flywheel Storage. By connecting a flyhweel to a motor/generator, it is spinned upto high speed and absorbs energy. This energy can be returned to the motor—generator by decreasing the flywheel's speed. However only small quantities can be stored, and they are limited to short periods of time.

Batteries Storage. Perhaps most flexible and practical are electrochemical batteries. These batteries are the most common, but research is being conducted with less expensive and lighter types. Also presently available are aqueous solution nickel and nickel cadmium types, but they are much more costly. Again, this type of storage is suited to small plants, particularly self contained ones.

Hydraulic Storage. Hydraulic (or pumped—water) storage is most frequently mentioned in connection with tidal power plants. Two reservoirs, at different elevations, are linked by pumps and turbines and their motors and generators. They constitute the accumulator (a reversible pump turbine, connected to motor-generator achieves the same result). Off peak energy, which could be used, is put to work to pump water from the lower into the upper reservoir, which can be run through the turbine and generator and provide energy when needed at a later time.

Power can be provided immediately on demand; storage stabilizes an otherwise intermittent source and constitutes an efficiency improvement, Reservoirs must be found or built, of course, and a certain amount of energy is lost, to activate and work the pump.

Compressed Air Storage. Air is compressed during off peak periods and thus stored underground. It can then be called upon to work gas turbines and provide electricity. Tidal energy can be used directly to drive air turbo compressors.

9.3.9. Advantages and Limitations of Tidal Power Generation

Advantages. (1) The biggest advantage of the tidal power is besides being inexhaustible, it is completely independent of the precipitation (rain) and its uncertainty. Even a continuous dry spell of any number of years can have no effect whatsoever on the tidal power generation.

(2) Tidal power generation is free from pollution, as it does not use any fuel and also does not produce any unhealthy waste like gases, ash, atomic refuse.

(3) These power plants do not demand large area of valuable land because they are on the bays (sea shore).

(4) Peak power demand can be effectively met when it works in combination with thermal or hydroelectric system.

Limitations. There are a number of reasons, why the tidal power generation is still a novelty, rather than a normal source of energy. The reasons can be enumerated as below

(1) The fundamental drawback to all methods of generating tidal power is the variability in output caused by the variations in the tidal range.

(2) The tidal ranges is highly variable and thus the turbines have to work on a wide range of head variation. This affects the efficiency of the plant.

(3) Since the tidal power generation depends upon the level difference in the sea and an inland basin, it has to be a intermittent operation, feasible only at a certain stage of the tidal cycle. This intermittent pattern could be improved to some extent by using multiple basins and a double cycle system.

(4) The tidal range is limited to a few meters. Thus the bulb turbine technology was not well developed, use of conventional Kaplan runners was the only alternative. This was found to be unsuitable. Now with the development of reversible flow bulb turbines, this difficulty is overcome.

(5) The duration of power cycle may be reasonably constant but its time of occurance keeps in changing, introducing difficulties in the planning of the load sharing every day in a grid. This handicap can be removed now with the help of computerised programming.

(6) Sea water is corrosive and it was feared that the machinery may get corroded.

Stainless steel with a high chromium content and a small amount of molybednum and the aluminium bronzes proved to be good corrosion resistant at La Rance project. The vinyl paint exhibited good results.

(7) Construction in sea or in estuaries is found difficult.

(8) Cost is not favourable compared to the other sources of energy.

(9) It is feared that the tidal power plant would hamper the other natural uses of estuaries such as fishing, or navigation.

9.3.10. Prospects of Tidal Energy in India

The possible sites for tidal power generation in India are obviously those where high tidal ranges occur *e.g.,* Gulf of Cambay (Bhavnagar Sonrai), Gulf of Kutch (Kandla, Navalakhi) and of Houghly river. The maximum tidal range in the Gulf of Cambay (10.8 m) and is attractive for a tidal plant. There are two possible sites on the western bank namely, Sonrai creek and Bhavnagar creek which have the essential requirements for locating probable plants. However, the silt change of the Gulf of Cambay is about 5000 ppm which is thought to be high and needs a closer study for future development. Gulf of Kutch has a maximum spring tide range of 7.5 m. The silt change here (near Navalakhi in the Gulf of Kutch) is much lower (nearly 4000 ppm).

The tidal ranges and power potential of these sites are indicated in the Table 9.2.

Table 9.2. Tidal Power Potential in India

Sites	*Spring Tidal range in metres*	*Assumed area in sq. kms.*	*Total maximum potential energy 10^6 kWh/yr.*	*Single Basin Cycle MW10^6 kW/yr.*		*Two basins*			
						Alternatively operating MW10^6 kWh/yr.		*Cooperating MW10^6 kWh/yr.*	
Gulf of Kutch Navalakhi	7.5	10	1110	43	376	48	419	16.4	143
Gulf of Cambay	10.8	10	2300	89.4	784	100	880	34.2	300
Sagar	4.85	10	464	18	157	20.1	176	6.9	60
Diamond Harbour	3.9	10	686	26.6	233	29.7	262	10.15	89

There is at present no indication regarding the cost of generation from tidal power. Preliminary studies already carried out by the CPWD and for tidal station in the Gulf of Cambay indicated higher cost of generation from conventional sources. However, the cost of coal and other allied materials is increasing which may open up the possibility of exploitation of this source of power. Adequate data will have to be collected for any realistic assessment of tidal power potential and possible impact on the environment, current patterns, tidal reflections, sedimentation, erosion etc. Detailed feasibility reports based on full technology assessment are called for before venturing into this field.

9.4. Ocean Waves

9.4.1. Introduction

Ocean and sea waves are caused indirectly by solar energy like the wind and OTEC. Wave energy derives from wind energy, which drives in turn from solar energy. As stated earlier, the wind energy is caused by the uneven solar heating and subsequent cooling of the earth's crust and the rotation of the earth. Wave energy at its most active, however, can be much more concentrated than the solar energy. Devices that convert energy from waves can therefore produce much higher power densities that solar devices.

The power in the ocean waves has been part of the human experience for thousands of years. However, the history of attempts to exploit this power for human purposes has been extensive. It was the recent energy crisis that prompted serious attempts at harnessing the waves for the production of electricity. During the decade of the 1970s dozens of patents were filed to do this, though most are complicated and rather fragile in the face of the gigantic power of the ocean storms. Patents have been issued on a variety of devices and many schemes have been described conceptually. Some small scale prototype devices have been tested. Upto now no major development programme has been carried out through in any country. Small devices are available, however, and are in limited use as power supplies for buoys and navigational aids. From the engineering development point of view, waves energy development is not nearly as far along as wind and tidal energy.

World sites that may be suitable for 'harvesting' of energy from waves include the Molakai and Alenuihaha channels in the Hawalian islands, where 2 to 3 m high (crest to trough) waves are typical during normal trade wind periods, the Pacific coast of North America, the Arabian sea of India and Pakistan, the North Atlantic coast of Scotland, the coast of New England etc.

9.4.2. Advantages and Disadvantages of Wave Energy

Advantages : (1) The wave energy has the advantage over wind or solar that the energy has been naturally concentrated by accumulation overtime and space and transported from the point at which it was originally present in the winds.

(2) The degree of power concentration effected by waves is quite substantial whereas the power density in wind may, at a good site, average some hundreds of watts per square meter, wave from being transported across a plane perpendicular to the wave propagation direction at a good site is from 10 to 100 times at large.

(3) It is a free and renewable energy source.

(4) Wave power devices do not use up large land masses unlike solar or wind.

(5) These devices are relatively pollution free and, because they remove energy from the waves, leave the water in a relatively placid (calm) state in their wakes.

Disadvantages : (1) The major disadvantage of wave energy, as compared to wind, is that the energy is available on the ocean. The extraction equipment must operated in a marine environment with all that implies in terms of maintenance, construction cost, life time, and reliability. The energy may have to be transported a greater distance to shore.

(2) Wave, energy converters must be capable of withstanding very severe peak stresses in storms.

(3) There is relatively scarcity of accessible sites of large wave activity.

(4) Wave energy conversion devices that have been proposed are relatively complicated.

(5) Economic factors such as the capital investment, costs of the maintenance, repair and replacement, as well as problems of biological growth of marine organisms, are all relatively unknown and seem to be on the large side.

9.4.3. Energy and Power from the Waves

A two dimensional progressive wave as shown in Fig. (9.15), is represented by the sinusoidal simple harmonic wave shown at time $t = 0$, and at time t. The wave may be expressed by the following relation involving some parameters:

$$y = a \sin\left(\frac{2\pi}{\lambda}x - \frac{2\pi}{\tau}t\right) \quad \text{...(9.14)}$$

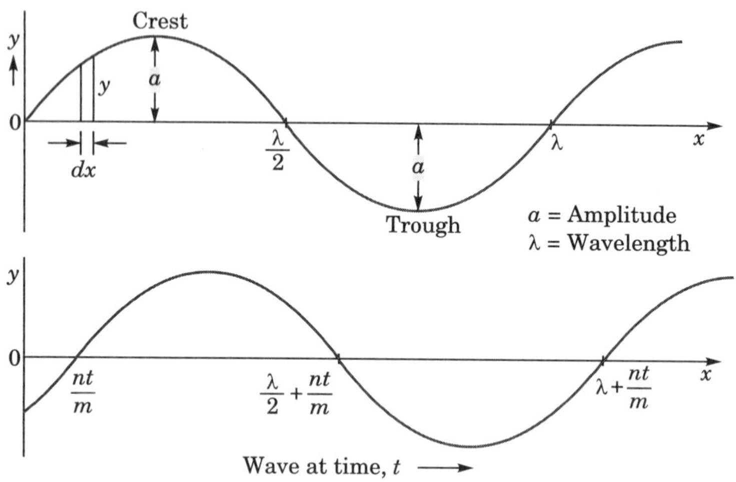

Fig. 9.15. A typical progressive wave.

where, y = height above its mean level in m. a = amplitude in m
λ = wave length in m t = time in seconds
τ = period in seconds

$$2\pi = \left(\frac{x}{\lambda} - \frac{t}{\tau}\right) = \text{phase angle (dimensionless)}$$

The relationship between wavelength and period is approximately

$$\lambda = 1.56\ \tau^2 \quad \text{...(9.15)}$$

The expression (9.14) can be expressed as

$$y = a \sin(mx - n\,t) \quad \text{...(9.16)}$$

where, $m = \dfrac{2\pi}{\lambda}$ and $n = \dfrac{2\pi}{\tau}$ = phase rate

$2a$ = height (from crest to trough).

Potential Energy. The potential energy arises from the elevation of water above mean level (*i.e.*, $y = o$) considering a differential volume $y \cdot dx$, it will have a mean height $\lambda/2$. Thus its potential energy is

$$dPE = m\,g\,y/2 = (\rho\,y\,dx\,L)\,g\,y/2$$

$$= \frac{g\rho y^2\,L\,dx}{2} \quad \text{...(9.17)}$$

where, m = mass of the liquid in $y \cdot dx$, kg

g = gravitational constant

ρ = water density, kg/m^3,

L = arbitrary width of the two dimensional wave, perpendicular to the direction or wave propagation x, m.

Combining (9.16) and (9.17), we obtain:

$$PE = \frac{\rho L a^2 g}{2} \int_0^\lambda \sin^2 (mx - nt)\, dx \qquad ...(9.18)$$

$$= \frac{\rho L a^2 g}{2} \left(\frac{1}{2} mx - \frac{1}{4} \sin 2\, mx \right)_0^\lambda$$

$$= \frac{1}{4} g \rho a^2 \lambda L \qquad ...(9.19)$$

The potential energy density per unit area is PE/A, where, $A = \lambda L$, in J/m^2.

$$\therefore \qquad \frac{PE}{A} = \frac{1}{4} g \rho a^2 \qquad ...(9.20)$$

Kinetic Energy. The derivation of the K.E. is rather complex and beyond the scope of the book. From hydrodynamic theory this can be expressed as

$$\text{K.E.} = \frac{1}{4} g \rho a^2 \lambda L \qquad ...(9.21)$$

and the K.E. density is

$$\frac{\text{K.E.}}{A} = \frac{1}{4} g \rho a^2 \qquad ...(9.22)$$

Total energy and power density can be written as

$$\frac{E}{A} = \frac{1}{2} g \rho a^2 \qquad ...(9.23)$$

$$\frac{P}{A} = \frac{1}{2} g \rho a^2 . f \qquad ...(9.24)$$

where, f is the frequency. (The Power P = energy × frequency).

9.4.4. Wave Energy Conversion Devices

The mechanical energy in waves takes different forms. There is the energy of forward motion of the wave the highly noticeable energy that slams into ships and cliffs. Some of the proposed schemes are oriented toward this forward motion kinetic energy. Any geometric arrangement that absorbs energy by converting the forward momentum of the wave into motions of its internal parts, without reemitting as much energy as it absorbs, can extract this forward motion energy.

Also very promising from the energy extraction point of view is the potential energy of the raised water at the wave crest. The gravity head between crest and trough is not large enough to permit practical power generation by driving a hydro-turbine directly, but the potential energy in the wave is very considerable. The rate at which work is done on a large ship by the ocean as the ship is lifted up by the swell is typically several times the power being delivered by the ship's engine.

It is not difficult to invent a variety of ratchet, valve, or intermittent pump mechanisms that can convert wave energy to hydraulic, pneumatic, or electrical form. The engineering challenge is to find a cost effective way to do it on a large scale. Since waves come in a wide range of wave lengths and amplitudes, any effective device will either have to be broad band that is, non resonant—or it will have to have its resonance frequency continuously adjusted. Many mechanisms have been proposed, and a number are being investigated, For example, waves can be made to compress air in the top of *floating tank,* using one way air valves. Electricity is generated as the air is bled out through a pneumatic or air turbine.

In a different devices, waves passing a pipe standing vertically in the water and equipped with an internal flop valve that allows water to move upward in the pipe but not downward, cause a water column to rise in the pipe, to a height many times the wave height. The water can be released from an elevated part in a controlled manner, to drive a conventional hydro turbine. Another class of mechanism employs the floats with different dynamic response characteristics. The differential motion is used to operate a pump or mechanical engine. For example, a long vertical cylinder extending far below the surface will remain nearly stationary as the waves pass by. A toroidal float surrounding it will rise and fall with the valves, and can be made to push and pull on pump plungers attached to the cylinder.

Some of the main concepts for converting wave energy into mechanical or electrical are described briefly as follows :

Wave Energy Conversion by Floats

Wave motion is primarily horizontal, but the motion of the water is primarily vertical. Mechanical power is obtained by floats making use of the motion of water. The concept visualizes a large float that is driven up and down by the water within relatively stationary guides. This reciprocating motion is converted to mechanical and then electrical power is generated.

A system based on this principle is shown in Fig. 9.16, in which a square float moves up and down with the water. It is guided by four vertical manifolds that are part of a platform. There are four large under water floatation tanks which stabilizes the platform. Platform is supported by buoyancy forces and no vertical or horizontal displacement occurs due to wave action. Thus the platform is made stationary in space. A piston which is attached to float as shown in figure moves up and down inside a cylinder. The cylinder is attached to platform and is therefore relatively stationary. The piston and cylinder arrangement is used as a reciprocating compressor. The downward motion of the piston draws air into the cylinder *via* an inlet check valve. This air is compressed by upward motion of the piston and is supplied to the four underwater floatation tanks, through an outlet check valve *via* the four manifolds. In this way, the four floatation tanks serve the dual purpose of buoyancy and air storage, and also the four vertical manifolds and float guides. An air turbine is run by the compressed air which is stored in the buoyancy-storage tanks, which in turn drives an electrical generator, producing electricity which is then transmitted to the shore *via* an under water cable.

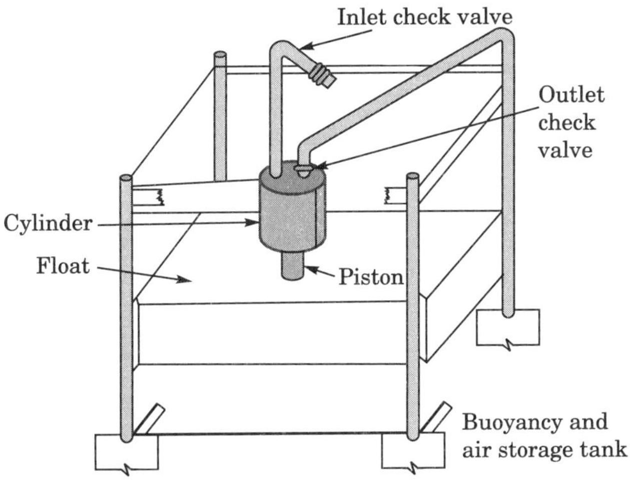

Fig. 9.16. Schematic of a float wave power conversion device.

High-Level Reservoir Wave Machine

The concept of this device is illustrated with reference to Fig. 9.17, in which a magnification piston is used. The pressurized water is elevated to a natural reservoir above the wave generator, which would have to be near a shore line, or to an artificial water reservoir. The water in the reservoir is made to flow through a turbine coupled to an electric generator, and then back to sea level. Calculations made shown that a 20 m diameter generator can produce 1 MW power.

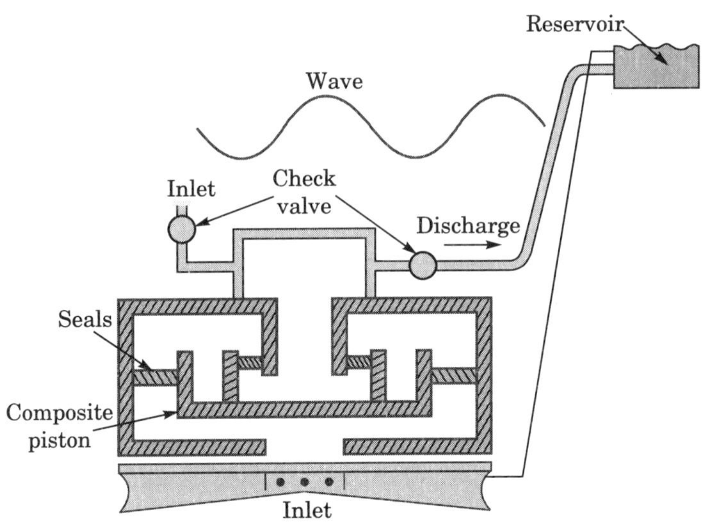

Fig. 9.17. Schematic of a high level reservoir wave machine.

The Dolphin Type Wave Power Machine

This type of wave generator, which is designed by Tsu Research laboratories in Japan is shown in Fig. 9.18. The system consists of following major components:

(*i*) a dolphin, (*ii*) a float, (*iii*) a connecting rod, and (*iv*) two electrical generators.

This device uses the float which has two motions. The first is a rolling motion about its own fulcrum with the connecting rod. Revolving movements are caused between the float and the connecting rod. The other is a nearly vertical or heaving motion about the connecting rod fulcrum. It causes relative revolving movements between the connecting rod and the stationary dolphin. In both the cases, the movements are amplified and converted by gears into continuous rotary motions that drive the two electrical generators.

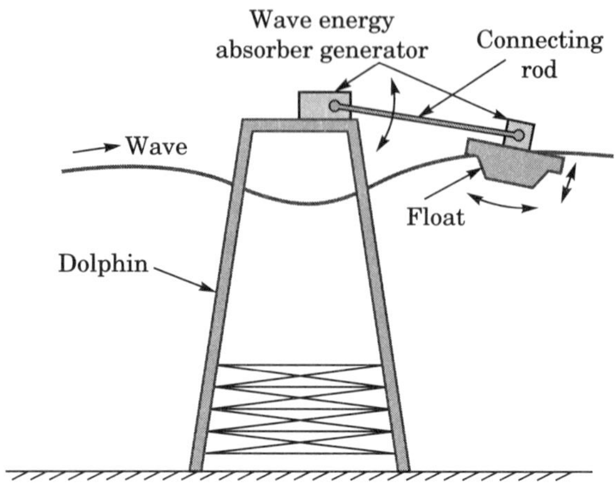

Fig. 9.18. Schematic of the Dolphin type wave generator.

Other Wave Machines

Hydraulic accumulator wave machines are also used, which instead of compressing air, the water itself is pressurized and stored in a high pressure accumulator or pumped to a high level reservoir, from which it flows through a water turbine electric generator. This is done by transforming large volumes of low pressure water at wave crest into small volumes of high-pressure water by the use of a composite piston. The piston is composed of a large-diameter main piston and a small-diameter piston at its centre. The main piston moves up and down with the entering of the wave water through the opening. A closed water loop exists above the small piston. The pressure on the main piston is magnified on the small piston during the upstroke. The high pressure water is conducted through a one-way valve to a hydraulic accumulator at the top of the generator. Two air (or other gas) volumes counter balance and act as cushions in a chamber above the main piston and in a sealed compartment in the hydraulic accumulator. The latter also maintains the high water pressure. Part of the high-pressure water flows through a Pelton wheel or Francis hydraulic turbine that drives an electrical generator and is then discharged to a storage chamber below the turbine.

In another differential motion concept, called the "*nodding duck*" employs large cam shaped "ducks" mounted on a very long, floating frame. The 'ducks' oscillate or 'nod' relative to the frame, driving hydraulic pumps. This 'nodding duck' scheme though only in the model testing stage, has been investigated more than most of the other concepts, and is currently one of the leading contenders in the British wave energy programme. It is a rather serious criticism it was claimed bya wind power proponent that the energy pay back time for wave power systems in general and the nodding duck scheme in particular may be excessively long, owing to the large amount of steel and concrete required. This claim was disputed by the inventor of the nodding duct concept.

It is interesting to recall that S. J. Savonius used a rotor of the type generally known as a Savonius rotor and generally used as a windmill to extract energy from the waves of ocean. The rotor was fixed with its axis horizontal and perpendicular to the direction of wave propagation (near the shore, to which it was anchored rigidly, in this case). Water passing the rotor flowed both forward and back at different times and depths ; the asymmetric Savonius rotor can take power from both the forward and back flows. Power outputs of a few kilowatts per square meter were obtained with equipment that was far from perfect.

9.5. Small Scale Hydroelectric

9.5.1. Introduction

Small scale hydroelectric facilities can supply in principle significant amounts of electricity for irrigation or potable water pumping, lighting, or health or educational purposes. The total potential amount of such a resource is very poorly documented but is apt to be large.

Upto 1972, hydro engineers concentrated on developing the larger sites, where the economy of scale enabled the production of energy at a cost low enough to compete thermal power, fueled with low cost oil. However with the prospect of rapidly depleting fossil fuels coupled with steady rise in oil prices, attention has returned to the smaller sites previously regarded as uneconomic. Moreover, the remarkable advancement in the technology of development uf turbines suitable for utilising small falls and small discharges efficiently has increased the chances of development of *small* (*mini* and *micro* included) hydel installations to a large extent. Manufacturers have been quick enough to develop packages designs for small units. For very small hydroplants of less than 500 kW capacity, *electronic load controllers* have been developed to replace the governor. These controllers maintain a constant load on the turbine and hence constant flow, surplus power is diverted to a resistor and either wasted or used to heat water.

Many countries now have active small hydro development and rural electrification programmes, due to the several advantages offered by these plants.

The advantage of their operation in hilly and remote areas and the elimination of long transmission system, and lesser gestation periods have lent added attraction. It has little or no adverse environmental impact, effects on stream ecology are minor.

China has concentrated on small hydro, has over 1 lakh sites developed and is now exporting low cost units. These mini hydro schemes which are spreaded all over the country, are supplying about 10% of China's total installed capacity which is in the vicinity of 7,000 MW. Other countries such as Malaysia, Indonesia and Phillipins are actively pursuing small and micro-hydro, developing sites using a minimum of elaborate equipment.

In India, the potential of small hydropower is estimated to be 5,000 MW at present, while further investigations and surveys are expected to indicate a higher potential. *Small hydropower is covered in the renewable energy programme.* The alternate hydro-energy centre at Roorkee works on the development of solar hydro-power systems as well as *Hybrid*, hydro systems. If small hydro power stations are set up all over the country, decentralized availability of power will be become possible.

Sites for low-head installations can be found everywhere in the mountain region, plains or even at the sea level. They may range from micro sets (less than a megawatt) to the largest axial flow turbines (upto 50 MW). Also, low head hydro-power sites can be close to power consumption areas, which is an advantage. In water logged areas like the Sundarbans in Bengal or remote hilly areas such as Ladakh, low head installations are often the only source of energy.

It is in this context that the Bulb-Turbine, the youngest member in the family of hydroturbines came into picture. Bulb turbine including its generator is enclosed in a shell known as bulb which can be conveniently installed horizontally in a low head stream. It has more or less a straight flow path. It can work under a head from as low as half a metre to ahead of 95 meters. It can generate power from as low as 5 kW to as high as 50 MW. Bulb turbines can easily be standardised in the step of 100 kW in the micro-range, in the step of 1 MW in the mini range and in the step of 5 MW in the range of small turbines.

The small hydro power potential which remains largely untapped so far because of an impression that:

1. Small hydro electric projects entail higher capital costs on per kW installation basis.

2. Higher managerial and administrative costs.

3. Relatively low utilization.

4. Unstable operation of isolated systems due to low inertia.

5. Certain problems caused by the disbursed nature of projects in site identification, preparation of reports, proper construction planning and management and operation etc.

Yet one can not deny a place for small hydel in the larger energy frame work primarily because:

1. Small or mini hydro power (SHP) is a non consumptive generator of electrical energy utilizing a renewable resource which is made continually available through the hydrologic cycle of environment.

2. Small hydro power (SHP) is essentially non polluting and release no heat. Adverse environmental impacts are negligible and for small installation may be totally eliminated.

3. With the development of compact efficient machines the investment per kW installed is not very high. SHP projects do not require large capital investment.

4. Compared to other conventional energy generation schemes, these projects have low gestation period ranging from 8 to 24 months.

5. Operating costs are low and the equipment does not need trained and skilled personnel. With the introduction of microprocessor the SHP station may run virtually unattented. Further freedom from fuel dependence together with long life of SHP plant make installation resistant to inflation.

6. SHP is ideal decentralized energy generation source. It can supply energy to rural feeders thereby cutting distribution losses to a large extent.

7. SHP can be synchronised with grid and has been demonstrated in national demonstration projects in H.P., Haryana and Tamilnadu. SHP synchronisation with utility grid improves voltage profile.

8. With the interconnection and synchronisation with grid, plant utilization factor becomes very high. Thus SHP becomes economically attractive.

9. Due to their small size and local availability of utility grid, SHP stations generally use induction generators. Thus induction generators are not only inexpensive but also the problem of synchronization and interconnection is also vastly simplified.

10. SHP can be a catalyst in remote areas using relative simple technology in mobilizing productive resources and cutting enhanced economic opportunities for local residents.

11. Most of these schemes, in India, can be constructed on existing canals and irrigation system thereby necessitating minor modification and result in minimal disturbances to surrounding environment.

12. SHP can be developed to augment hydro power capacity at existing irrigation dams and power houses. The possibility of retrofits and additional turbines and generators makes the upgrading of present installations attractive. The determinants in the economic feasibility of alternative proposals of power generation may not necessarily depend upon the variables of the highest gain, from any individual scheme, but variables on the basis of the impact on economy of the system as a whole. From the considerations of higher efficiency and lower cost per kW capacity, a larger set may be individually more economical but this individual economic gain might be out weighted by the adverse economy as a consequence of:

(*a*) Low operation efficiency due to partial loading.

(*b*) Sudden loss of unit.

(*c*) Longer gestation period due to sizeable difference between the generation capacity of the unit and the increment rate of load developed in the region served by the plant.

(*d*) Transport difficulties in hilly areas.

(*e*) Transmission losses.

(*f*) Transmission costs.

(*g*) High cost of fossil fuels when compared to thermal power.

In coming years the small hydel (mini, micro and small) shall be more feasible and economical than other sources like thermal power because cost of fossil fuels.

9.5.2. What is Small Hydel Development ?

There is no formal definition of a small hydro plant but this may generally be taken as power station/plant having output upto 5000 kW. Some associate the concept of small hydro with low head say upto 15 m. This may not generally be true as there is no restriction on head for small hydro development. Stations upto output of 1000 kW are called micro and upto 5000 kW as mini hydroplant. Concepts of small hydro are naturally relative and vary from country to country depending on the size of its existing development. In U.S.A. even 15000 kW is considered small hydro development while in France and some other countries it is upto 3000 kW. The hydro electric installations having plants of higher unit rating can generally be made by utilising conventional technology of hydro equipment and tailor made designs both for equipment and civil works for power stations. It may be mentioned that source of energy from sea waves, coastal tides and ocean water itself, attempts in development of which also are gaining momentum, are not included in small hydro development. These are distinct types of development requiring all together different type of technology. Conceptually small hydro development can be categorised in two types *one utilizing small discharges but having high head* and the second utilising *large discharges under comparatively very small head.* These are also distinguished by the nature of the generating plant required in these two types. For high head discharges being small, the physical size of the plant required is small. For the second type, as the discharges handled are high, the physical size of the generating unit and the power station consequently is quite big compared to the general notions of the rating. Also for the latter type proper arrangement for entry of water and its discharge is required to be made.

9.5.3. Nature of Small Hydro Development

Equipment technology is available to utilise discharge as small as 200 litres/sec. (0.2 cumec), head ranging from even 1 metre and producing an output of 1 kW with reasonable cost. Development of potential, relatively small is of course dependent on the pressing need and the resources.

The small hydro development of the first type which is confined mainly to hilly areas is characterised by relatively very simple features of works. The civil work involved comprise a small structure to divert the flow of the hill stream/river, small water conductor system such as a channel, flume or buried conduits, power house building and a small length of transmission line.

There is no need for substantial storage and generally run of river is utilised. The power generally consumed in the local area eliminating requirement of long transmission lines. The grid line would normally be far away and hence the need of local development.

Features of development of the second type which normally belongs to the plains are somewhat different. The head available is rather low and discharges have to be comparatively large to be economically viable. The development thus can take place on small river, irrigation outlet, canal falls etc. The difference usually comes in because of regulating arrangement for inlet and discharge of water and the type of the generating equipment that may be bulb or tube type requiring larger size of power house. The grid transmission line would generally be available nearby and the power station also be connected to it.

9.5.4. Classifications of Small Hydro Power Stations

As per central Electricity Authority and Bureau of Indian standards, the SHP stations are classified as follows:

1. **Depending on Capacity**

Size	*unit size*	*Installation*
Micro	upto 100 kW	100 kW
Mini	101 to 1000 kW	2,000 kW
Small	1001 to 6000 kW	15,000 kW

2. **Depending on Head**

Ultra low head	Below 3 metres
Low head	less than 30 metres
Medium head	Between 30 to 75 metres
High head	Above 75 metres.

The above mentioned definitions vary from country to country.

On the basis of heads available small hydel schemes (mini, micro and small) may be further categories as

(*i*) Independent scheme (*ii*) Sub ordinate scheme.

Independent Scheme. Independent schemes are those schemes where a stream flow is captured, regulated and developed for the principle objective of power generation only, in these schemes high head scheme and medium head schemes are considered but medium head schemes are successful. The low head schemes are not found economical for using as independent generation. These schemes are successful in Himalayan territories states like, Himachal Pradesh, Jammu and Kashmir, Nagaland and Manipur etc.

Sub ordinate Schemes. The subordinate schemes are those where the principle objective is other than power generation. The region in which there is extensive network of irrigation canals and water is stored in the form of reservoir are used for subordinate scheme where primary aim is irrigation or drinking water and secondary aim is power generation. The feasible sites for subordinates schemes is indogagetic plains (canals) and Peninsular India (reservoirs).

9.5.5. Components of a Hydroelectric Scheme

The basic and common components of a hydroelectric schemes are given below: (refer Fig. 9.19)

(*a*) Diversion and intake (*b*) Desilting chamber

(*c*) Water conductor system (*d*) Forebay/balancing reservoir

(*e*) Surge tank (if necessary) (*f*) Penstock

(*g*) Power house—Turbine, Generator, Protection and control equipment, Dewatering, Drainage system, Auxiliary, Power system, Grounding, Emergency and stand by Power system, Lighting and Ventilation.

(*h*) Tail race channel.

Dam, Barrages, solid boulder structure and trench type weir are usually employed to divert, the required flow from the river bed/streams to the intake structure Fig. 9.19.

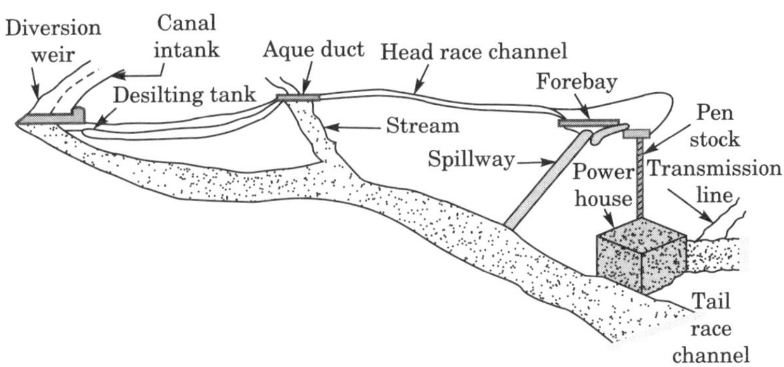

Fig. 9.19. Typical arrangement of small Hydro Power Station.

For small hydro stations usually solid boulder structure and trench type weirs are employed. Trench type is suitable where rock is not available in the river bed. Boulder type could be preferred if boulders are available in the river and rock is encountered in the river bed within one meter depth. If solid boulder structure type weir is provided in nonrocky foundations, maintenance and repair of damages due to scour may create problems. The diversion weir should be capable of diverting all the lean season flows and the structure should be reasonably safe during monsoon floods. In hill streams, diversion structures face a recurring problems due to choking of intake. Further underground structures are vulnerable to damage from heavy boulders and bed loads. Trench type weirs are preferred from this consideration.

Densilting Chamber is necessary where the water contains large quantities of course silt to minimise erosion damage to the turbine runner etc. The extent of desilting requirements would depend on the quantum and type of silt carried by the stream and the runner material. Abrasion effect becomes more pronounced with increasing head. The desilting chamber may be designed to exclude the particles coarser than the sizes mentioned in table for various heads of water to achieve a power draft free of abrasion effects.

Head	*Size of the Silt particles to be removed*
Medium head	0.2 to 0.5 mm
High head	0.1 to 0.2 mm.

Water Conductor system should be designed to ensure least loss of head and loss of water due to seepage. Further flow velocity should be adequate enough to prevent reduction of discharging capacity due to settling of silt. The canal can be lined with tiles. For small power stations, it may be economical to adopt lining with low density poly ethylene (LDPE), film with single tile cover especially where it traverse porous strata.

If the forebay is to be used as a balancing reservoirs storage for about 4–6 hours could be provided if justifiable on economic reasons. If the forebay is just used as a transit point, a storage of about 2 minutes may be adequate creation of pondage is preferred in cutting from economic reasons.

When the length of the water conductor conduit is more than 5 times the head on the machines, surge tank is necessary. In case of mini/micro hydel schemes, where impulse turbines are used, the pressure surges can be avoided with deflection of jet. This would eliminate the need of surge tank though this would involve some wastage of water, in reaction type turbine, water wasting relief valves or air vessel can be provided to avoid governor instability and hunting due to oscillations in the water column.

Penstocks can be made of steel pipes (ERW or Spiral Welded MS pipe), Hume pipes and PVC pipes depending on the design pressure, Commonly a bell mouth entry is provided to reduce the head losses and to ensure a smooth entry of water from the forebay tank into the penstock. Intakes are provided with trash racks to prevent entry of trash, debries and ice (in cold climate). Flow velocity through trash rack is kept at 0.6 to 0.9 metres per sec, so that head losses are not significant. Clogging to an extent of 30% of the trash rack area may be considered in the design without materially decreasing the head on the turbine. Trash rack for small intakes should be slopped to fascilitate cleaning of the debris. The bar spacing should be close enough to restrict the passage of any object that should not pass through the turbine.

Tailrace is a simple water channel or a cut and cover conduit with a maximum water velocity of 1 meter/second transporting the water from the turbine outlet (draft tube) to the river.

9.5.6. Civil works design considerations for Mini and Micro Hydel Projects

Introduction. A large potential of energy remains untapped in India in hilly streams and canal falls in the range of mini and micro hydro. The rough estimates show that the total potential in small hydro is of the order of 5000–6000 MW. At present only about 3 to 4% of this potential has been utilised. The construction for generation of about 200. MW is in progress in different parts of the country. This clearly shows that the country has embarked upon the programme of development of small hydro on a very large scale. But the main deterrents in taking up small hydro is their larger per kW costs as compared to the conventional large hydro which is also available in abundance. The immediate step required to be taken for making mini/micro hydro economically viable is to make efforts to reduce the cost of equipment and civil works.

Civil works form a major part of the total cost of the project. Therefore, the reduction in cost of civilworks will play a major role in reducing the overall cost of the project. The main components of civil works are diversion structure, water conductor system, desilting tank, Forebay, spillway, penstock, power house, tailrace channel and protection works. Some of these are discussed as follows:

Diversion Structure

The diversion structure provided should be simple in construction as well as economical. It should involve minimum maintenance. Depending upon the type of river bed the diversion structure may be of two types *viz.,* Boulder weir and Trench type weir. Boulder weir is quite cheap and is suitable where rock is encountered just below the river bed. Where the streams carry big boulders during flood season, it may destroy the boulder type weir constructed across the river. This problem can be solved by providing trench type weir. The trench type weir consists of a trapezoidal trough located below the bed of the river with top kept at the bed level of the river. The intake is located at the end of the trench weir.

It is usually constructed in reinforced concrete or masonry. A gate is also provided with the intake to release the water to the desired extent. The horizontal trash rack is provided at the bed level with slope in the down stream direction so that the water is tapped and boulders may roll down. The spacing between bars of trash racks should be so selected that it is adequate to draw that entire flow during lean period even if half of the effective area is clogged. The length of the trench weir which is equal to the water way, should be adequate to pass the designed flood. The longitudinal-slope provided in the trench weir should be adequate to develop such velocity of water that the bed load entering the trench is flushed.

Water Conductor System

Water conductor system is a very important, component of the hydro power scheme. The type of water conductor system depends on the site conditions and the materials available. The design of the water conductor system should ensure minimum head loss, adequate velocity of flow so that silt does not settle down. The materials of construction should be such that the loss due to seepage is also minimized. The most commonly used channel section is trapezoidal. To ensure minimum seepage, the lining of channel can be done with LDPE film. In steep hill slope, RCC hume pipes are most suitable as they can be hurried in a cut and cover trench.

Desilting Tank

Desilting tank is provided usually in the initial reaches of water conductor to trap the suspended silt load and pebbles etc., so as to minimize the erosion damages to the turbine runner. The size of the silt particles to be trapped for medium head power stations is from 0.2 to 0.5 mm and for high head it is from 0.1 to 0.2 mm. The depth of tank may be kept between 1.5 to 4 m. The horizontal flow velocity should not exceed 0.4 to 0.6 m/s.

Forebay

Forebay is usually constructed in reinforced concrete of stone masonry. It is provided so that a minimum head over the penstock is always there. Storage capacity of forebay is usually kept 2 minutes. In case where load fluctuation is much, balancing reservoirs may be provided to take care of peak demand.

Penstock

Water is fed to the machines through penstocks. Design of penstocks involves two aspects *viz,* hydraulic design and structural design. The hydraulic design consists of determining diameter of penstock. The diameter of penstock should also involve economic considerations. In structural design, thickness of penstock is determined. Generally the penstock are fabricated of mild steel. However PVC and RC hume pipes may also be used to reduce the cost.

Spillway

In the case of load rejection, the water level may rise and flood the area. Therefore, a spillway is provided keeping its crest at the permissible water level so that water level may not rise above the maximum permissible level. Spillway can be provided in the form of channel or pipe.

Power House Building

Power house building accommodates turbine, generator, control panels, auxiliary equipment etc. It must be a simple structure constructed either in RCC or stone masonry. The plane size of power house building depends upon the size and number of machines and centre to centre distance between them. This spacing should be kept minimum permissible so that size is reduced. The height of building is kept about 3–5 m. Where ever possible crane should be replaced by pulley block to make the scheme more economical. In cases where crane is provided, RCC columns may be provided to support it otherwise ordinary masonry may be used for the construction of walls. The roof may be provided on tubular trusses to reduce the cost.

Tail Race Channel

Water after flow through machine is fed to the stream downstream of power house. It may be trapezoidal or rectangular channel constructed in stone masonry or brick masonry depending upon the material available locally.

Materials for Construction

Efforts should be made to use the locally available material. The material generally used are Reinforced concrete, structural steel, stone masonry Brick masonry etc. Research is going on for the development of new materials *e.g.* ferrocement and steel fibre reinforced concrete. These are discussed briefly as follows:

Ferro Cement. Ferrocement a composite material, usually brittle, is reinforced with fibres dispersed throughout the composite. In ferro cement structures, the reinforcement consists of small diameter wiremesh in which the proportion and distribution of the reinforcement are made uniform by spreading out the wire meshes throughout the thickness of element. This dispersion of the fibres in the brittle matrix offer not only convenience and practical means of achieving improvement in many of the engineering properties of the materials such as fracture, tensile and flexural strength, toughness, fatigue resistance and impact resistance but also provides advantages in terms of fabrication of products and components.

The use of ferrocement may be made in gates, penstocks slabs etc.

Steel Fibre Reinforced Concrete

Steel fibre reinforced concrete (SFRC) is a new concrete for which gradients are hydraulic cement, fine and coarse aggregate and discontinuous discrete steel fibres. In recent years, it has been recognised that the additions of small closely spaced and uniformly dispersed discrete steel fibres in uncrete would substantially improve its static and dynamic properties. The SFRC has better characteristics as compared to ordinary concrete as regards compressive strength, tensile strength, flexural strength, shear strength, fatigue strength and impact resistance.

SFRC may be used for Forebay, gates and penstocks.

9.5.7. Turbines and Generators for Small Scale Hydro Electric

(1) *Bulb or Tubular Turbine.* This is rather a recent development in which complete turbine and the generator is enclosed inside the conduit which carries water from headrace to tail race. This enclosure is a water tight housing (or bulb), which is supported centrally within a horizontal water channels. The turbine is essentially of Kaplan type or a propeller type. The generator may be directly coupled to the turbine shaft or driven through a set of gears for increasing the speed to reduce the generator size. The generator is usually on the upstream side of the turbine now a days. The turbine, located at the downstream end, is connected to the generator, by way of a sealed shaft. Water flows in the axial direction around the bulb and through guide vanes and wicket gates to the turbine runner with propeller type blades. Fig. 9.20.

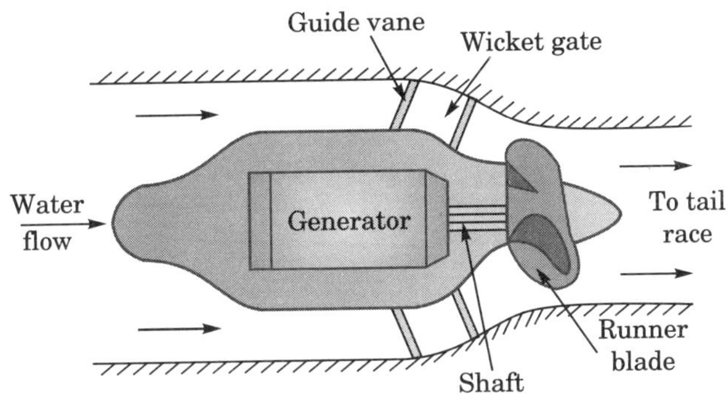

Fig. 9.20. Bulb turbine.

This development is most suitable for utilisation of comparatively low heads, the range being from 3 to 18 m. The output rating of the bulb type design are available right from 200 MW to 40 MW, though the bigger one may not come under small hydroelectric development.

Arrangement of the bulb unit (See Fig. 9.20) indicates that this is rather complicated. But the design has been almost perfected and there are a large number of units of various sizes in operation all over the world. The efficiencies of the generating set of this type are very high compared to the size and are comparable with those of conventional units as hydraulic losses are kept minimum. The units are available as package units which facilitates erection and dismentling.

(2) *Tube Turbine.* Modification of the Kaplan—Type turbine have been developed for water heads below some 15 m (50 ft). In one such type, called the *tube turbine,* the turbine was a horizontal (or almost horizontal) shaft is located in a tubular channel; the water flows in an axial direction through the channel with the runner. The water is directed between the propeller blades by fixed guide varies, and the flow is controlled by wicket (or similar) gates.

If the shaft is horizontal the channel slopes downward beyond the turbine so that the attached generator is above the water level in the tail race. Fig. 9.21. Alternatively, the turbine shaft may be sloped upward to achieve the same objective. An advantage claimed for the tube turbine is that the water does not change direction and hence lose energy before it enters the turbine.

This type of turbine is a varient of bulb turbine. In this type of management, only the turbine is housed inside the conduit and the generator is mounted outside in a pit by bringing the turbine shaft out of the turbine casing. This type is also known as "pit type". The arrangement can be either with horizontal shaft or inclined shaft as explained above. The arrangement is having the advantage of easy accessibility of the generator.

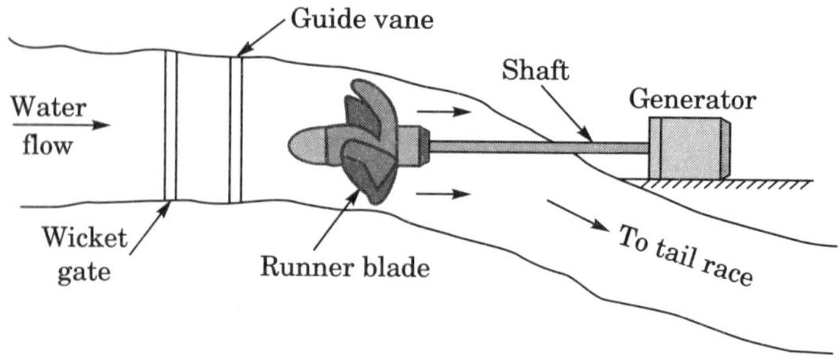

Fig. 9.21. Tube Turbine.

(3) *Straflo Turbine.* This is another design rather marvellous, the turbine of which is similar to that in bulb type. In this arrangement there is no separate rotor provided for the generator as the poles are mounted on the periphery of the turbine runner. There is an elaborate specialised arrangement of seals for preventing water getting into poles. As the name suggests the shaft is horizontal in this design and the water flows is always axial. Such a design requires smaller size of the power house. Another advantage in this type is a large range of output and head that can be utilized.

Generators

Conventionally hydro electric A.C. synchronous generators are used. The generator (rotor) should withstand turbine runaway speed *i.e.,* speed attained on sudden load through off. Brushless excitation system generators are used in small hydro power stations which reduces maintenance costs and time. Induction generators need excitation source and as such can only operate when connected to an existing electrical system.

Speed Regulation

The speed regulation of turbine is an important and complicated problem. The magnitude of the problem varies with size, type of machines, type of electrical load and whether or not the plant is tied into an electrical grid. It should also be kept in mind that runaway or no load speed can be higher than the design speed by factors as high as 2.5. This is an important design consideration for all rotating parts, including the generator. The cost of standard speed governor is disproportionately high in the smaller sizes. Regulation of speed is normally accomplished through flow control. Adequate control requires sufficient rotational inertia of the rotating parts.

The precision of governing necessary will depend on whether the electrical generator is synchronous or induction type. There are advantages to the induction type of generator. It is less complex and therefore, less expensive. Its frequency is controlled by the frequency of the grid it is feeding into, thereby eliminating the need of an expensive conventional governor. It can not operate independently but can only feed into a network and does so with lagging power factor which may or may not be disadvantage, depending on the nature of the load. Long transmission lines, for example, have a high capacitance and in this case the lagging power factor may be an advantage.

9.5.8. Protection, Control and Management of Equipments

Small hydro power plant controls are required for unit start, unit shut down, unit synchronising and unit loading and control (speed and voltage), unit electrical and mechanical protection and emergency shut-down, hydraulic control.

A major cost in small hydros is operation and management man power. These costs can be prohibitive in microhydel. Low cost automation is therefore, of importance if mini and micro hydel is to survive unit starting and synchronising can be manual. But emergency shut down must be fool proof to avoid damage. It is considered that even if man power is provided for management, the level of expertise will be of low level. Accordingly, dependable and sample control/remote control of unit is required. Microprocessors are being used.

Control of induction generator grid connected units is easy synchronisation, speed and voltage control are automatically taken care of Generator are less in cost. Automation costs are very low. Large scale installation of these units on our existing irrigation works like canals etc., can be highly economical.

In isolated areas local grids can be formed and smaller units can be induction generator type and can be controlled from a manned control mini hydro station where synchronous generator are installed.

9.5.9. Advantages and Limitations of Small Scale Hydroelectric

Advantages. (*i*) Smaller hydro projects takes the shortest time for developing a unit.

(*ii*) Once it is built the running expenditure is almost negligible.

(*iii*) The operation and maintenance of such power station is the simplest.

(*iv*) These energy sources are free from hazards of pollution.

(*v*) Unlike the big hydrodevelopments, these have no environmental problems, no submergence of land, no loss of agricultural land, no dislocation of habitation, no interstate problems and ecology of the region remains unaffected.

(*vi*) The construction of the various elements of the works, which are always simple, can generate employment for local people. The small hydro projects can will be operated and maintained by the local people. This factor alone can be the greatest incentive for the development which has been exploited most beneficially by China. In China, small communes have been encouraged to construct, own, operate and maintain such power stations.

Limitations. There are numerous reasons for comparatively smaller rate of development of small hydroelectric projects. Most of the factors responsible for this state of affairs have been quite formidable. These are not intended to be elaborated here. The main reasons are

(*i*) Prohibitive economics compared to that of conventional hydro power which is still abundantly available in the country.

(*ii*) Non availability of indigeneous equipment for generating plant (except for a particular type in a restricted range) and the import procedures are time consuming.

(*iii*) Remoteness of sites especially, in the hilly areas and adverse geological conditions in Himalayan region.

(*iv*) General lack of a awareness of benefits from small development.

With concentrated efforts it would be possible to improve the conditions and create proper climate for development. The present sources of small energy development which is mainly diesel, would become too scarce a commodity to be used for power generation in very near future.

9.5.10. Hybrid Systems

Small hydro can be extensively developed to exploit the available resources to meet rural energy requirements. Further electrical energy being the highest form of energy can be used to upgrade other systems. These hybrid systems may be used to harness/exploit and upgrade other available renewable resources which are presently unutilised/under utilised in rural areas. In addition to developing various hybrid systems, biomass gasification based cogeneration systems are proposed for peak load demands when microhydel ayafems are operated in decentralised manner. The required biomass will be obtained from high yielding and high density plantation.

In decentralised system, the power from micro-hydel system is preferably fed to the cluster of villages for meeting their requirements of power for domestic and street lighting, irrigation and rural industries. Depending upon the availability of power, hybrid systems in consultation with District Industries Officer could also be put by private enterpreneurs. The peak demand exceeding the micro-hydel capacity will be met by biomass gasification based cogeneration system.

A major cause of low benefit cost ratio (reported to be 1.68) of isolated small hydroplants is due to poor utilisation of power (low head factor). This can be overcome by using surplus off-peak high grade energy in hybrid systems.

The peak load carries a special significance in utility generation industry. Higher the peak load, greater will be capacity of power generation. If the total load of the system is taken by a single power plant. This induces higher initial investment, low efficiency of generation and higher generating costs. To overcome above problems, the micro-hydel system is proposed to operate as base load plant and biomass gasification based electrical generation system as peak load plant. The first combined system (300 kW micro-hydel and 200 kW biomass gasification based co-generation) is coming up at *Kakroi* (Haryana). The production cost per unit of electricity in micro-hydel system having annual plant factor of 0.40 comes to ₹ 1.25 and integration with biomass gasification system this reduces to ₹ 0.66. The integrated system also meets the entire energy requirement. However this requires prior information regarding the load curve before selecting the capacities of micro-hydel and gasifier systems.

9.5.11. Conclusion

It has been concluded a compaign should be made to give proper publicity for finding out the maximum mini, micro and small hydro electric generation sites on the following grounds:

(*i*) that mini, micro and small hydro electric power is a new source of renewable energy.

(*ii*) that mini, micro and small hydro electric is an alternative way of developing hydro electric resources of a country *i.e.,* it can develop the same potential as would be available for normal schemes.

(*iii*) that mini, micro and small hydropotential is more economical.

(*iv*) that it saves fossil fuels,

(*v*) that it reduces the transmission losses,

(*vi*) that it is suitable for isolated loads,

(*vii*) that it is easy to operate,

(*viii*) that it has low gastation period,

(*ix*) that its equipments can be standardized easily.

QUESTIONS

1. What is the basic principle of ocean thermal energy conversion (OTEC)?
2. What are the main types of OTEC power plants? Describe their working in brief.
3. Describe the 'closed cycle' OTEC system, with its advantages over 'open cycle' system.
4. Explain with sketches the various methods of tidal power generation. What are the limitations of each method?
5. What are the difficulties in tidal power developments?
6. The basin area of a tidal power plant is 20×10^6 m^2. The tidal range is 8 m, calculate the energy generated in kwh. [**Ans.** 10.368×10^8 kWh]
7. In Gulf of Cambay, which is being considered for possible tidal power generation, during the tide cycle, the observed difference between the high and low water of the tide was 10.8 m. It has been estimated that this estuary having an area of 10 sq. km can generate power for 3 hours in each cycle. Assuming the average available head to be 10 m, and the overall efficiency of the generation system to be 7 5%. Calculate:
   (*i*) The power (in h-p) at any instant and
   (*ii*) The total energy generated in the year. Take specific weight of sea water = 1025 kg/m^3. [**Ans.** (*i*) 102.5×10^4 h-p, (*ii*) $E = 15955.56 \times 10^5$ kWh/yr.]
8. In an estuary, which is being developed for tidal power generation during the tide cycle the observed difference between the high and low water of the tide was 5.5 m. It is estimated that the estuary's area is 0.5 sq km which can generate power for 3 hours in each cycle. Assuming the average available head to be 5 m, and the overall efficiency of generation to be 75%, calculate,
   (*i*) The power in hp at any instant and
   (*ii*) The total energy in the year.
   Sea water specific gravity can be taken equal to 1025 kg/m^3.
   [**Ans.** (*i*) 130.5×10^2 hp ; (*ii*) 203×10^5 kWh/yr.]
9. What are the advantages and limitations of wave energy conversion ?
10. Write a short note on wave energy conversion machines.
11. Write a short note on small head hydro power development.
12. What type of turbine is best suited for micro hydel plant? Describe it.
13. Describe the different types of turbines are in use for small scale hydroelectric plants.
14. What are the advantages and limitations of small scale hydroelectric power generation?

UNIT—6

CHEMICAL ENERGY SOURCES

— Fuel Cells and Batteries
— Hydrogen Energy

CHAPTER 10

Chemical Energy Sources

(Fuel Cells and Batteries)

10.1. General Introduction

This energy source includes augmentation of efforts on the development of fuel cells and batteries. The development of particular type of fuel cells and batteries holds promise for (*i*) running small and light commercial vehicles in the beginning followed by for heavy vehicles ; (*ii*) rural electrification of remote small villages having a population of 100—500 through rechargable power batteries, (the batteries can be charged with the help of wind mills). This would also help achieving the social objectives of the Government to provide electricity for lighting rural areas where laying of transmission faces hazards. The work on batteries and fuel cells, *viz.* sodium sulphur batteries, metal air rechargable cells, hydrogen oxygen and lithium fuel cells etc. is already in progress in a number of places in our country.

The Department of Non-conventional Energy Sources (DNES), India, has been promoting R&D efforts relating to chemical and electrochemical sources of energy keeping in view their potential role in the long term.

The Department extended support for studies relating to Beta, Alumina for Sodium Sulphur batteries, Lithiumion conductors for solid state batteries, fuel cell materials and development of high energy density storage batteries suitable for photo voltaic and other applications.

10.2. Fuel Cells

10.2.1. Introduction

A cell (or combination of cells) capable of generating an electric current by converting the chemical energy of a fuel directly into electrical energy. The fuel cell is similar to other electric cells in the respect that it consists of positive and negative electrodes with an electrolyte between them. Fuel in a suitable form is supplied to the negative electrode and oxygen, often from air, to the positive electrode. When the cell operates, the fuel is oxidised and the chemical reaction provides the energy that, is converted into electricity. Fuel cells differ from conventional electric cells in the respect that the active material (*i.e.*, fuel and oxygen) are not contained within the cell but are supplied from outside.

But for its costs, pure (or fairly pure) hydrogen gas would be preferred fuel for fuel cells. Alternatively impure hydrogen obtained from *hydrocarbon fuels*, such as natural gas or substitute natural gas (methane), *liquified petroleum gas* (propane and butane) or liquid *petroleum products*, can be used in fuel cells. Efforts are being made to develop cells that can use carbon monoxide as the fuel; if they are successful, it should be possible to utilise coal as the primary energy source. *Main uses of fuel cells are in power production, automobile vehicles and in special military use.*

10.2.2. Design and Principle of Operation of a Fuel Cell (with Special Reference to H_2, O_2 cell)

As stated these are electrochemical devices in which the chemical energy of fuel is converted directly into electric energy. The chemical energy is the free energy of the reactants used. This conversion takes place at constant temperature and pressure. The basic feature of the fuel cell is that the fuel and its oxidant are combined in the form of ions rather than neutral molecules.

The first practical fuel cells was demonstrated by Francis T. Bacin and J.C. Frost of Cambridge University in 1959. As per the fuel used the main types of fuel cells are:

(*i*) Hydrogen (H_2) fuel cell, (*ii*) Hydrazine (N_2H_4) fuel cell,
(*iii*) Hydrocarbon fuel cell, and (*iv*) Alcohol (methanol) fuel cell.

The operation of the fuel cell can best be described with reference to a specific device. Fuel cell can be adopted to a variety of fuels by changing the catalyst. Here hydrogen, oxygen (Hydrox) cell is described for example. These types are the most efficient and the most highly developed.

The main components of a fuel cell are:

(*i*) a fuel electrode (anode),
(*ii*) an oxidant or air electrode (cathode), and
(*iii*) an electrolyte.

In most fuel cells, hydrogen (pure or impure) is the active material at the negative electrode and oxygen (from the oxygen or air) is active at the positive electrode. Since hydrogen and oxygen are gases, a fuel cell requires a solid electrical conductor to serve as a current collector and to provide a terminal at each electrode. The solid electrode material is generally porous.

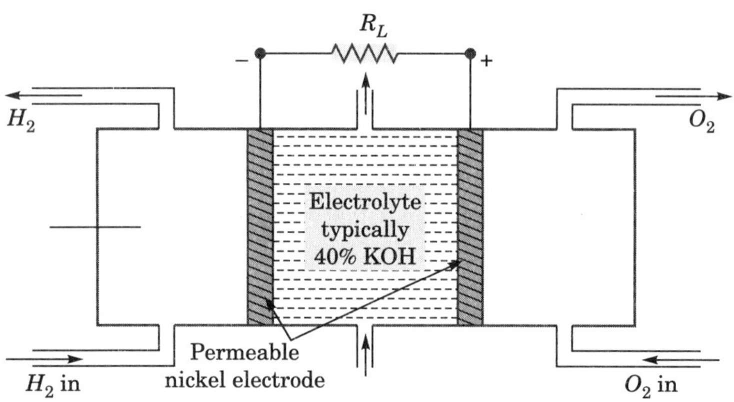

Fig. 10.1. A hydrox (H_2, O_2 cell).

Porous nickel electrodes and porous carbon electrodes are generally used in fuel cells for commercial applications. Platinum and other precious metals are being used in certain fuel cells which have potential utility in military and space applications. The porous electrode has a larger number of sites, where the gas, electrolyte and electrode are in contact; the electro chemical reactions occur at these sites. The reactions are normally very slow, and catalyst is included in the electrode to expedite them. *The best electro chemical catalysts are finely divided platinum or platinum like metal deposited on or incorporated with the porous electrode material.* Since the platinum metals are expensive, other catalysts, such as nickel (for hydrogen) and silver (for oxygen), are used where possible. The very small catalyst particles provide a large number of active sites at which the electrochemical reactions can take place at a fairly rapid rate.

Although practical fuel cells differ in design details, the essential principles are the same, as indicated by the schematic illustration in Fig. 10.1. Hydrogen gas is supplied to one electrode and oxygen gas (or air) to the other. Between the electrodes is a layer of electrolyte. Most existing fuel cells operate at temperature below about 200°C; the electrolyte is then usually and aqueous solution of an alkali or acid. The liquid electrolyte is generally retained in a porous membrane; but it may be free flowing in some cells. Different electric current is drawn from the cell in the usual manner by connecting a load between the electrode terminals.

The electrochemical reactions occurring at the electrodes of a hydrogen-oxygen cell may vary with the nature of the electrolyte, but basically they are as follows. At the negative electrode, hydrogen gas (H_2) is converted into hydrogen ions (H^+) *i.e.,* hydrogen with a positive electric change, plus an equivalent number of electrons (*i.e.,* $\bar{e}$) ; thus

$$H_2 \longrightarrow 2H^+ + 2\bar{e} \qquad ...(10.1)$$

At this electrode, hydrogen is diffused through the permeable nickel in which is embedded a catalyst. The catalyst enables the hydrogen molecules, H_2 to be absorbed, on the electrode surface as hydrogen atoms, which react with the hydroxyl ions (OH^-) in the electrolyte to form water.

When the cell is operating and producing current, the electrons flow through the external load to the positive electrode; here they interact with oxygen (O_2) and water (H_2O) from the electrolyte to form negatively charged hydroxyl (OH^-) ions ; thus

$$\frac{1}{2} O_2 + H_2O + 2\bar{e} \longrightarrow 2OH^- \qquad ...(10.2)$$

The hydrogen and hydroxyl ions then combine in the electrolyte to produce water

$$H^+ + OH^- \longrightarrow H_2O \qquad ...(10.3)$$

The electrolyte is typically 40% KOH solution because of its high electrical conductivity and it is less corrosive than acids.

The above equations show that hydroxyl ions produced at one electrode are involved in the reaction at the other. Also electrons are absorbed from the oxygen electrode and released to the hydrogen electrode. Addition of the three forgoing reactions show that when the cell is operating, the overall process is the chemical combination of hydrogen and oxygen (gases) to form water that is

$$H_2 + \frac{1}{2} O_2 \longrightarrow H_2O \qquad ...(10.4)$$

The oxygen and hydrogen are converted to water, which is the waste product of the cell. The reactants are stored outside the cell (note difference from storage battery), and the electrodes and electrolyte are not consumed in the overall process. These properties load to the design of convenient small size and long life power units.

If the electrodes are on open circuit (Fig. 10.2), the hydrogen electrode accumulates a surface layer of negative charges. These attract potassium ions,

K^+, of the electrolyte, providing an electrical double layer. Similarly, the loss of electrons from the oxygen electrode results in a layer of positive charges, which in turn attracts hydroxyl ions, OH^-, from the electrolyte. These electrical double layer at the electrodes build up until the potentials are such that they inhibit only other further reactions between the electrolyte and the fuel gases. This situation is illustrated in the figure showing that an open circuit voltage is developed between the electrodes. The magnitude of this emf is 1.23 volt at 1 atm and 25°C.

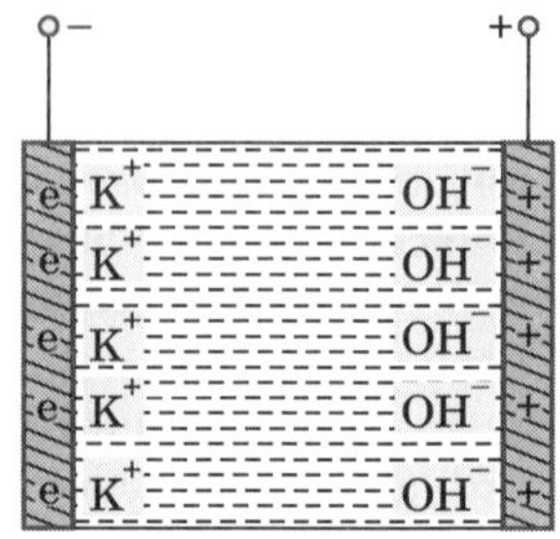

Fig. 10.2. Electrodes on open circuit.

If the circuit is closed (Fig. 10.3), the electrons can now leave the electrodes pass through the connecting circuit to the oxygen electrodes, and take part in the reaction of equation (10.2) above. This movement of electrons constitutes a current passing through an external load. In this way useful electrical work is obtained directly from the chemical process. Note that the electrons flow is from the hydrogen to the oxygen electrode.

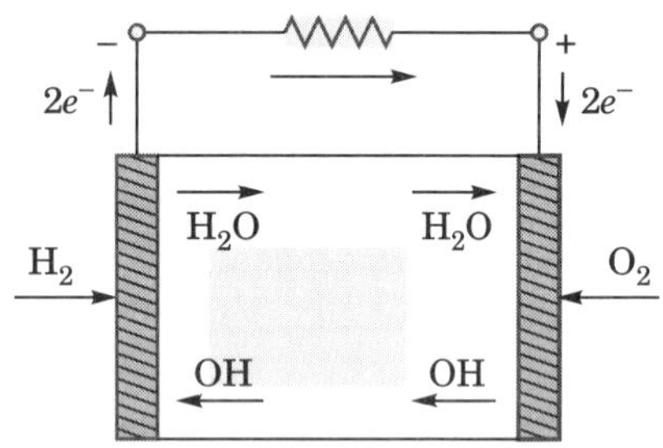

Fig. 10.3. Electrodes on closed circuit.

Hydrogen fuel cells (Hydrox) are of two types:

1. *Low temperature cell.* The electrolyte temperature is 90°C. It is sometimes pressurised, but not by a great amount, usually say upto 4 atmospheres.

2. *High pressure cell.* Pressure is upto about 45 atmospheres and temperatures upto 300°C say. A single "Hydrox" fuel cell can produce an e.m.f. of 1.23 volts at 1 atm and 25°C. By connecting a number of cells, it is possible to create useful potential of 100 to 1000 volts and power levels of 1 kW to 100 MW nearly. The current depends upon the physical size of the cell. The output of the fuel cell varies directly with the pressure, so as to increase the cell output, the gas pressure is raised. The optimum size of the cell at present is about 0.27 cu. m, per kW.

The gases in the hydrogen oxygen cell must be free from carbon-dioxide, because this gas can combine with the potassium hydroxide electrolyte to form potassium carbonate. If this occurs, the electrical resistance of the cell is increased and its output voltage is decreased. Consequently, when air is used to supply the required oxygen, carbon dioxide must first be removed by scrubbing with an alkaline medium (*i.e.*, lime).

Fuel cells are particularly suited for low voltage and high current applications.

Hydrogen oxygen fuel cells have been proposed for propulsion of electric vehicles, with the hydrogen provided by a metal hydride. However this use is limited by the heavy weight of the hydride and the cost of the relatively pure hydrogen required. It appears that for the present, at least the use of hydrogen oxygen cells will be restricted mainly to manned space vehicles. Such cells with porous nickel electrodes and potassium hydroxide electrolyte have been used to provide electric power for the Apollo and shuttle space craft. The hydrogen and oxygen for operating the cell are stored in liquid form to minimize the volume occupied.

10.2.3. Classification of Fuel Cells

Before describing the different types of fuel cells, it is necessary to have some method of classification of various types of fuel cells, which are either in existence or are being invented. Several methods of classification of fuel cells have appeared in the literature. One of the difficulties in arriving at a systematic classification is that several operational variables exist.

For example, fuel cells may be classified according to the temperature range in which they operate,

Low temperature	25—100°C
Medium temperature	100—500°C
High temperature	500—1000°C
Very high temperature	above—1000°C

Another method would be according to the type of *electrolyte, e.g.*, aqueous, non-aqueous, molten or solid. One could also classify the fuel cells according to the physical state of the fuel:

Gas—hydrogen, lower hydrocarbons

Liquid—alcohols, hydrazine, higher hydrocarbons

Solid—metals etc.

In the present discussion, a broad division is first made according to weather the fuel cell system is a primary or secondary one. A primary fuel cell may be defined as one in which the reactants are passed through the cell only once, the products of the reaction being discarded. *e.g.*, H_2—O_2 fuel cell. A secondary fuel cell is one in which the reactants are passed through the cell many times because they are regenerated from the products by thermal, electrical, photochemical methods, *e.g.*, Nitric oxide-chlorine fuel cells.

10.2.4. Types of Fuel Cells

Following fuels are mostly used in fuel cells:

(1) Hydrogen, (2) Fossil fuel, (3) Hydrocarbon fuel, (4) Alcohol fuel, (5) Hydrazine fuel.

There are discussed in the following sections.

(1) Hydrogen, oxygen (H_2, O_2) cell of primary systems is already described using 40% KOH solution as electrolyte. *Here now Ion exchange membrane cell will be described which uses membrane electrolyte.*

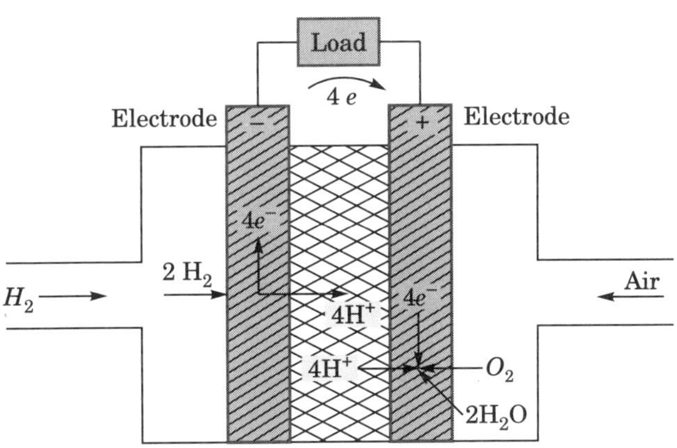

Fig. 10.4. Ion Exchange Membrane cell.

Ion Exchange Membrane cell. The basic design of the cell, which consists of a solid electrolyte Ion exchange membrane, electrocatalysts and gas feed tubes is represented in Fig. 10.4. The distinctive feature of this cell is that it uses a solid electrolyte in the form of an ion exchange membrane. The membrane is non-permeable to the reactant gases, hydrogen and oxygen, which thus prevents them from coming into contact. The membrane is however, permeable to hydrogen ions which are the current carriers in the electrolyte.

The desired properties of an ideal ion exchange membrane electrolyte are:

(*i*) High ionic conductivity.

(*ii*) Zero electronic conductivity.

(*iii*) Low permeability of fuel and oxidant.

(*iv*) Low degree of electro-osmosis.

(*v*) High resistance to dehydration.

(*vi*) High resistance to its oxidation or hydrolysis and,

(*vii*) Mechanical stability.

A considerable amount of research has been carried out in a search for the ideal membrane. Interpolymers of polyflurocarbon and polystyrene sulfonic acids have been found to be quite satisfactory. In order that electrolyte resistance be as low as possible, a thin sheet of this material (0.076 cm thickness) is used as the electrolyte. The use of even thinner electrolyte sheets is prevented by problems of gas permeability, mechanical stability etc. An advantageous feature of this electrolyte is that it retains only a limited quantity of water and rejects excess water produced in the fuel cell.

The two electrodes, which consist of the electrocatalyst and a plastic material for water proofing the electrode are in the form of fine metallic wire screens. They are bonded on either side of the electrolyte layer. The wire screen material is titanium or platinum. Metallic current collectors are ribbed onto each electrode. The hydrogen compartment of the cell is enclosed; the hydrogen gas enters this compartment through a small inlet and circulates throughout the ribbed current collectors and distributes itself evenly over the electrode. On the opposite side oxygen or air enters the compartment, coolant tubes run through the ribs of the current collectors. On the oxygen side, the current collectors, also hold wicks which absorb water, the product of fuel cell reaction and carry it over by capillary action. The water leaves the cell through an exit from the oxygen compartment. Oxygen is prevented from leaving its compartment by the inclusion of a differential pressure water separation system.

The ion exchange membrane electrolyte is acidic and the current carriers in solution are hydrogen ions. The hydrogen ions are produced by the reaction at anode according to,

$$2H_2 \longrightarrow 4H^+ + 4\bar{e} \quad ...(10.5)$$

These ions are then transported to the cathode through the electrolyte and the electrons reach the cathode *via* the external circuit. At the cathode, oxygen is reduced producing water as represented by:

$$O_2 + 4H^+ + 4\bar{e} \longrightarrow 2H_2O \quad ...(10.6)$$

Thus the overall cell reaction is,

$$2H_2 + O_2 \longrightarrow 2H_2O \quad ...(10.7)$$

This cell operates at about 40–60°C. The thermodynamic reversible potential for the reaction is 1.23 volts at 25°C.

(2) *Fossil Fuels Cells.* The most interesting fuel cells for the near future are modified hydrogen oxygen (or air) cells, in which a gaseous or liquid hydrocarbon is the source of hydrogen. Eventually, coal may serve as the primary energy source for fuel cells. Cells based on *fossil fuels* have three main components:

(1) The fuel processor which converts the fossil fuel into a hydrogen rich gas,

(2) The power section consisting of the actual fuel cell (or combination of cells), and

(3) The inverter for changing the direct current generated by the fuel cell into alternating current to be transmitted to user.

Main components of fuel cell system are shown schematically in Fig. 10.5. The most highly developed fossil fuel cells are phosphoric acid cells, molten carbonate cells, solid oxide electrolyte cell.

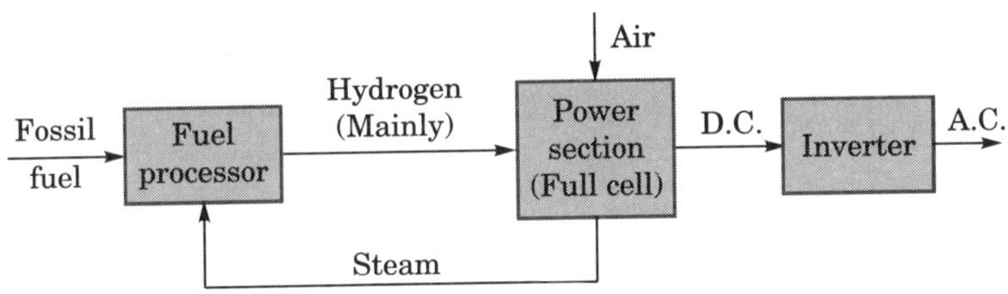

Fig. 10.5. Main components of fuel cell system.

In phosphoric acid cell utilizes a concentrated aqueous solution of phosphoric acid as the electrolyte. The primary fuel is light hydrocar bon, such natural gas or naphtha. The operating temperature is 150 to 200°C and the discharge voltage is 0.7 to 0.8 volt. Each cell unit is only a few millimeters thick so that a large number can be stacked in a package of reasonable size to produce the desired voltage and power.

Molten Carbonate Cells. These are high temperature fuel cells, with a molten carbonate mixture as the electrolyte, offer the prospect for use with a variety of fossil fuels, including coal. A special feature of these cells is that during operation they can oxidize carbonmonoxide to carbondioxide as well as hydrogen to water. Hence gaseous mixture of hydrogen and carbon monoxide, which are relatively inexpensive to manufacture, can be used in the cell, the presence of carbon dioxide would have only a minor effect.

Several methods are available for fuel processing. These methods are essentially the same as those used for the commercial production, of mixtures of hydrogen and carbonmonooxide, known as *synthesis gas*. Synthesis gas can be made from coal by reactions with air and steam or from various petroleum products. In order to prevent poisoning of the electrodes in the fuel cell, the gas mixture must be desulfurized before being supplied to the power section.

The electrolyte in the high temperature fuel cells under development is a molten mixture of alkali metal (lithium, sodium and potassium) carbonates at a temperature of 600-700°C. This is retained in an inert porous matrix sandwiched between two porous nickel electrodes.

Because of the high temperature, electrochemical catalysts are apparently not necessary on the electrodes. The mixture of hydrogen and carbon monoxide is supplied to the negative electrode and oxygen (from the air) to the positive electrode. The discharge e.m.f. of the cell is about 0.8 volt.

The electrolyte is usually held in a sponge—like ceramic matrix, Metallic electrodes are placed in direct contact with this solid electrolyte. A hydrocarbon fuel, such as methane or kerosene is used. The fuel is reacted inside the cell to produce H_2 and CO. At the fuel electrode, H_2 and CO react with CO_3 ions in the electrolyte, releasing electrons to the electrode, and forming H_2O and CO_2 (Fig. 10.6). At the oxygen electrode O_2 reacts with the returning electrons and CO_2 diverted from the fuel electrode to form CO_3 ions. These CO_3 ions then migrate through the electrolyte to the fuel electrode. Thus the reactions are as follows.

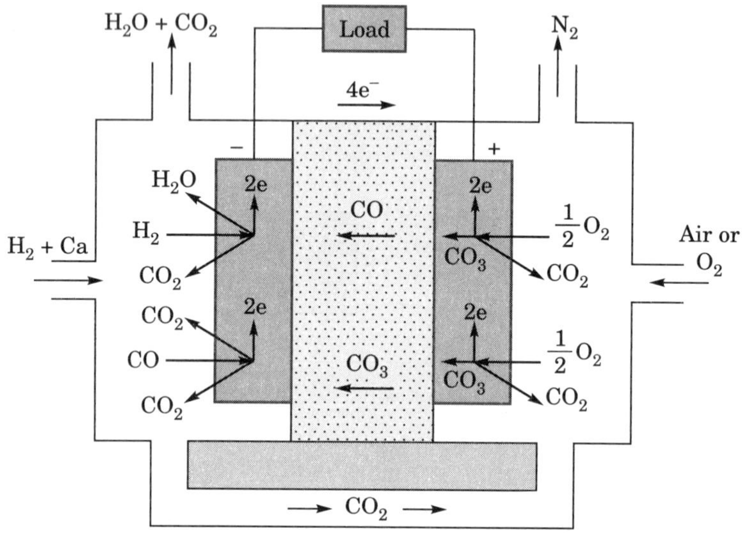

Fig. 10.6. Molten carbonate fuel cell.

At the fuel electrode:

$$H_2 + CO_3^{--} = H_2O + CO_2 + 2e$$
$$CO + CO_3^{--} = 2CO_2 + 2e$$

At oxygen electrode:

$$O_2 + 2\,CO_2 + 4e = 2\,CO_3^{--}$$

Overall cell reaction:

$$H_2 + CO + O_2 = H_2O + CO_2$$

An important, aspect of the molten carbonate fuel cells is that the discharged gases, consisting mainly of the steam and carbondioxide products and nitrogen from the air, are at a temperature exceeding 540°C. The hot gases could be used to provide industrial process heat, to operate a gas turbine or to produce steam in a waste-heat (heat exchanger) boiler to drive a steam turbine. The turbine would be attached to a generator to produce additional electric power. The overall efficiency for fuel use would thus be substantially increased.

Solid Oxide Electrolyte Cells. Certain solid, ceramic oxides are able to conduct electricity at high temperatures and can serve as electrolytes for fuel cells. These cells could utilize the same fossil fuels as the molten carbonate cells. The processing operation would then be the same as described above. Solid-oxide cells are in the early stages of development. A possible electrolyte is zirconium dioxide containing a small amount of another oxide to stabilize the crystal structure; this material is able to conduct oxygen ions (O_2^-) at high temperatures. The electrode material might be porous nickel and the operating temperature in the range of 600—1000°C. Electrochemical catalysts would not be required.

Other energy sources or fuels, that can be conveniently stored and transported in liquid form, such as *methanol, ammonia* and *hydrazine*, have been proposed for the fuel cells. Methanol can be catalytically reformed with steam at about 200°C to yield a mixture of hydrogen (75 volume %) and carbondioxide. This gas can be supplied to the negative electrode of a fuel cell with air at the positive electrode. The cell with aqueous phosphoric acid solution as the electrolyte, might be similar to those already described.

In the ammonia (NH_3) oxygen (air) fuel cell, ammonia gas, obtained from the stored liquid, is decomposed catalytically into hydrogen (75 volume %) and nitrogen. Part of the hydrogen is burned in air to provide the heat required for the decomposition. The bulk of the hydrogen is then supplied to the negative electrode of a hydrogen air fuel cell. The most suitable electrolyte would probably be potassium hydroxide solution. The nitrogen formed in the decomposition of ammonia is inert and plays no role in the cell. The advantage of ammonia as fuel is that it can be stored in the liquid form. Disadvantages include higher vapour pressure and cell does not operate satisfactorily at low temperatures.

A compact fuel cell for a mobile source, possibly for vehicle propulsion, utilizes the liquid hydrazine (N_2H_4) and hydrogen peroxide (H_2O_2) or air are the energy source. Hydrazine is injected as required into the aqueous potassium hydroxide electrolyte to provide the active material at the negative electrode. The oxygen for the positive electrode is obtained either by the catalytic decomposition of hydrogen peroxide at ordinary temperatures or from the ambient air. Each electrode may consist of a nickel screen matrix with nickel (negative) or silver (positive) as the electrochemical catalyst. The overall cell reaction is the oxidation of hydrazine to water and nitrogen, but the discharge emf is similar to that of the hydrogen oxygen cell.

Basic reactions are:

$$NH_2\,NH_2 + 4\,OH^- \longrightarrow N_2 + 4H_2O + 4e \qquad ...(10.7)$$

$$O_2 + 2H_2O + 4e \longrightarrow 4OH^- \qquad ...(10.8)$$

Overall cell reaction:

$$NH_2\,NH_2 + O_2 \longrightarrow N_2 + 2H_2O \qquad ...(10.9)$$

Basic drawback of H_2—O_2 cell is that Hydrogen and Oxygen are gases, so there is storage difficulty. Liquid fuel hydrazine is highly reactive. Power output is more as open circuit voltage (OCV) for this case is 1.56 as compared to 1.23 volt of H_2—O_2 cell.

Main drawback of hydrazine fuel is that it is highly toxic as well as costly.

Aluminium—Oxygen (Air) Cell (or battery). It is being developed by the Lawrence Livermore National Laboratory (U.S.A), mainly for electric vehicle propulsion. This cell its unusual in the respect that the metal aluminium is effectively the fuel which is consumed during operation and replaced as required. The aluminium (Al) forms the negative electrode of the cell, and oxygen (from air) is the positive electrode; the electrolyte is an aqueous solution of sodium hydroxide. The overall cell reaction is symbolically

$$\underset{(-)}{Al} + \underset{(+)}{\frac{3}{4}\,O_2\,(air)} + \frac{3}{4}\,H_2O \longrightarrow Al\,(OH)_3. \qquad ...(10.10)$$

so that aluminium, oxygen (from air), and water (from the electrolyte) combine to form aluminium hydroxide (Al) $(OH)_3$. The aluminium (negative) electrodes are made of the metal containing a small amount of gallium, and the air (positive) electrodes are carbon coated with an electro chemical catalyst, possibly silver. Before entering the battery, the air is scrubbed to remove CO_2. The operating temperature of the battery is about 50 to 60°C.

Regenerative Systems. A regenerative fuel cell is one in which the fuel cell product (*e.g.*, water in the hydrogen-oxygen fuel cell) is recovered into its reactants (*e.g.*, hydrogen and oxygen) by one of several possible methods—thermal, chemical, photochemical, electrical or radio chemical.

Since there are two stages in a regenerative fuel cell:

(1) Conversion of fuel cell reactants into products while producing electrical energy, and

(2) Reconversion of fuel cell products into reactants, —it is clear that the overall efficiency of a regenerative fuel cell is the product of the efficiencies of these two stages. In the following section, the principle of the regenerative systems is explained by considering the photochemical mode of regeneration.

Photochemically regenerative fuel cells. In this method, the products of the fuel cell reaction are transformed into its reactants by light. Because of the ample availability of *solar energy*, this method should be valuable, provided that there are suitable substances which can undergo photo chemical dissociation.

The sequence of reactions which are taking place in this fuel cell can be represented as follows:

Electrochemical : $A + B \longrightarrow AB$ + electricity

Photochemical : AB + light $\longrightarrow A + B$

Overall : Light + electricity

The nitric oxide—chlorine fuel cell, in which the overall reaction is,

$$2NO + Cl_2 \longrightarrow 2NOCl \qquad ...(10.11)$$

the product nitrosyl chloride is decomposed photochemically to chlorine and nitrous oxide. The system is schematically represented in Fig. 10.7.

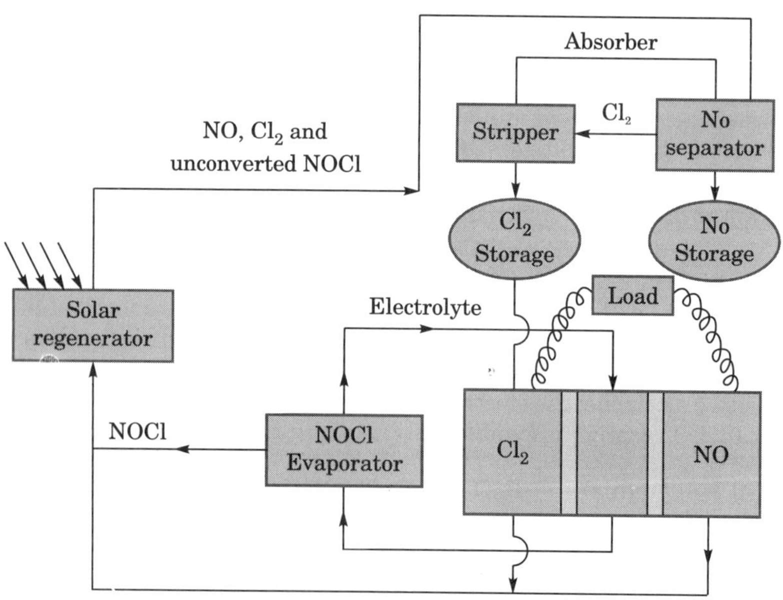

Fig. 10.7. Regenerative fuel cell.

The cell has a standard reversible cell potential of 0.21 volt. The reactants may be regretted from NOCl, in the liquid phase by light. Under these conditions, the quantum yield is low because the back reaction is also rapid. In the gas phase regeneration is easier although there is some problem of separating the NO and Cl_2. The currents obtainable are low, probably as a result of the low value of the reversible potential.

10.2.5. Advantages and Disadvantages of Fuel Cell

Advantages. (1) It has very high conversion, efficiencies as high as 70% have been observed, since it is a direct conversion process and does not involve a thermal process. In the conventional thermal process for generating electricity, heat energy produced by combustion of the fuel is converted partially into mechanical energy in a steam turbine and then into electricity by means of a generator. The efficiency of a heat engine is limited by the operating temperatures, and in the large modern steam-electric plants above 40% of the heat energy of the fuel is converted into electrical energy. Fuel cells, on the other hand, are not heat engines and are not subjected to their temperature limitations.

(2) Fuel cells can be installed near the use point, thus reducing electrical transmission requirements and accompanying losses. Consequently considerably higher efficiencies are possible.

(3) They have few mechanical components; hence, they operate fairly quietly and require little attention and less maintenance.

(4) Atmospheric pollution is small if the primary energy source is hydrogen, the only waste product is water; if the source is a hydrocarbon, carbon dioxide is also produced, nitrogen oxides, such as accompany combustion of fossil fuels in the air, are not formed in the fuel cell. Some, heat is generated by a fuel cell, but it can be dissipated to the atmosphere or possibles used locally.

(5) There is no requirement for large volumes of cooling water such as are necessary to condense exhaust system from a turbine in conventional power plant.

(6) As fuel cells do not make noise, they can be readily accepted in residential areas.

(7) The fuel cell takes little time to go into operation.

(8) The space requirement for fuel cell power plant is considerably lean as compared to conventional power plants.

Disadvantages. The main disadvantages of fuel cells are their high, initial costs and low service life.

10.2.6. Conversion Efficiency of Fuel Cells

The electrical energy generated by a fuel cell depends on what is called the "free" energy rather than on the heat energy, of the overall cell reaction. The free energy of formation of 1 mole (18 grams) of liquid water from hydrogen and oxygen gases at atmospheric pressure is 56.67 kcal or 237 kJ at 25°C. The heat energy (or enthalpy) of the reaction under the same conditions is 68.26 kcal (286 kJ). The theoretical efficiency of the conversion of heat energy into electrical energy in a hydrogen-oxygen fuel cell is thus $(56.67/68.26) \times 100 = 83\%$.

Efficiencies as high as 70% have been observed, but the practical cells using pure hydrogen and oxygen generally have conversion efficiencies in the range of 50 to 60%. The efficiencies are some what lower when air is the source of oxygen. The overall thermal to electrical conversion efficiencies are also lower when the hydrogen is derived from hydrocarbon sources. Nevertheless, they should be higher than those obtainable from the same fuels is most steam electric plants.

The theoretical emf (or voltage) of a fuel cell can be calculated from the reaction free energy. For the hydrogen-oxygen cell at 25°C, with the gases at atmospheric pressure, the ideal emf. is 1.23 volts; at 200°C, it is about 1.15 volts. The discharge voltages observed in actual cells are always below the theoretical value, the difference increasing with increasing strength of the current drawn from the cell (Refer Fig. 10.8).

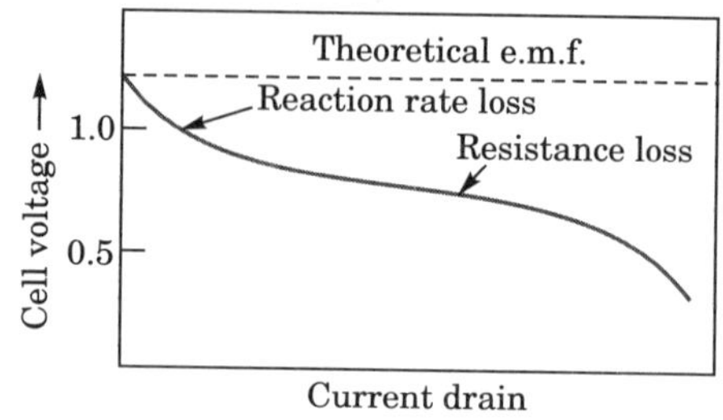

Fig. 10.8. Current and voltage in discharge of fuel cell.

For the moderate currents at which fuel cells normally operate the emf is 0.7 to 0.8 volts. This deviation from the theoretical emf accounts for the conversion efficiency of a fuel cell being below the ideal maximum value.

The departure of a fuel cell from ideal behaviour arises from several factors one is the inherent slowness of the electrode reactions; this is dominant at low current drains. It can be reduced by an effective electrochemical catalyst and by increasing the operating temperature. At larger currents there is an additional contribution from the electrical resistance of the electrolyte (multiplied by the current strength). A low resistance (*i.e.*, high conductivity) electrolyte is therefore desirable.

Even in an ideal hydrogen-oxygen cell, 100 – 83 = 17% of the chemical reaction energy (enthalpy) would be liberated as heat. The proportion is increased in an actual cell because the conversion efficiency is less than the maximum of 83%. In order to avoid an excessive temperature rise, heat is removed from the fuel cell during operation. Possible ways of doing this are by the flow of excess air past the positive electrode or by circulating the electrolyte through an external cooler.

In some fuel cell designs it is proposed to utilize the heat to provide space heating of a building and to supply hot water. The heat released in high temperature cells might be used for industrial purposes (*i.e.,* process heat) or to generate steam.

Polarization in Fuel Cells. In a practical fuel cell the theoretical output voltage is not attained for several reasons, as outline below. The difference between the theoretical voltage and the actual voltage is known as the *polarisation*. Some authors call this the overvoltage, by analogy with the electrolysers. The effect of polarisation is to reduce the efficiency of the cell from the theoretical maximum. The significant drop in voltage and hence energy loss takes place as the current density is increased as shown in Fig. 10.9.

Three main types can be distinguished:

(1) Activation polarisation,

(2) Resistance or ohmic polarisation, and

(3) Concentration polarisation.

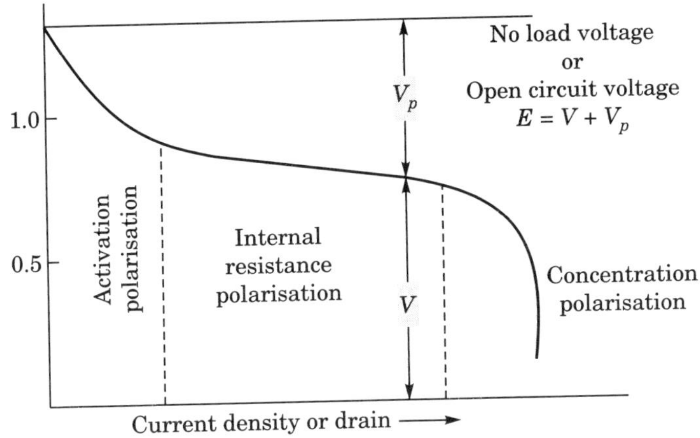

Fig. 10.9. Polarisation in fuel cells.

(1) **Activation Polarisation** (chemical polarisation). This is related to the activation energy barrier for the electron transfer process at the electrode. In fuel cells electrons are liberated and reaction is chemisorption reaction. At low current densities significant number of electrons are not emitted, which results in such a potential loss. This, process requires that certain minimum activation energy supplied so that sufficient number of electrons are emitted, this energy is supplied by output of the cell. This loss is known as activation or chemical polarisation.

This polarisation may be reduced by using better electrode catalysts, increasing surface area, and by raising the operating temperature.

(2) **Resistance Polarisation.** The voltage drop is linearly related to the current flow according to ohm's law. The internal resistance is composed of the electrode resistance, the bulk electrolyte resistance, and interface contact resistance between electrode and electrolyte. The loss due to resistance polarisation is significant when current density is quite large. The reduction in internal resistance is the main design criterion for low resistance polarisation losses. With large electrodes the resistance may be quite significant, particularly when these are made thin or consists of extremely thin coating on a non conducting support or when the electrode catalyst is a poor electric conductor.

The electrolyte resistance can be decreased usually by using a more concentrated electrolyte by closer spacing of electrodes and by increased temperature. Hence loss due to resistance polarisation can be reduced by:

(*i*) Selecting proper shape of the electrode to have minimum contact between electrode and electrolyte,

(*ii*) Reducing the gap between electrodes,

(*iii*) By using concentrated electrolyte.

(3) **Concentration Polarisation.** This type of polarisation tends to limit the current drawn, as shown in the figure. This broad heading covers a slow step in any of the mass transport processes in the cell, and is generally divided into two categories *i.e.*,

(1) Electrolyte side polarisation and

(2) Gas side polarisation.

The electrolyte-side polarisation is due to slow diffusion in the electrolyte causing a change in concentration at the electrode. This effect can be minimized by increasing the electrolyte concentration or by stirring or circulating the electrolyte.

Gas side polarisation is caused from slow diffusion of reactants through a porous electrode to the reaction site, or of slow diffusion of products away from the reaction site. The loss in voltage due to gas side polarisation is reduced by using electrodes of smaller pore size. Concentration polarisation is also reduced by increasing temperature.

We have observed that all the three losses in a fuel cell are decreased by increasing temperature, due to this, a given cell is usually operated in practice at the higher end of its temperature range.

All the losses in a fuel cell may by included under *voltage efficiency* and it can be expressed as:

$$\eta_v = \frac{\text{Operating voltage}}{\text{Theoretical voltage}}$$

$$= \frac{\text{On load voltage}}{\text{Open circuit voltage}} = \frac{V}{E} \quad ...(10.12)$$

where, V = operating voltage at a given current density, and

E = the theoretical open circuit voltage (emf).

In the case of a fuel cell which acts ideally, the terminal cell potential is constant and is equal to the thermodynamic reversible potential of the cell at any value of the current density drawn from the cell. In order that the efficiency of electrochemical energy converters may be compared with those of other energy conversion devices, it is necessary to have a common base. Most of the energy converters convert heat energy into electricity. The heat energy in many cases is provided by the heat of a chemical reaction. The efficiency of these cases is defined as the work output divided by the heat input.

Thus, in an analogus manner, the heat input for the electrochemical energy converter should be taken as the enthalpy change of the reaction (ΔH). The work output in an electrochemical energy converter which operates at the thermodynamic reversible potential of the cell is the free energy change of the reaction ($\Delta W_{max} = -\Delta G$), Eqn. (10.20). Thus, the ideal efficiency of an electrochemical energy converter is defined by the equation

$$\eta_i = \frac{\Delta G}{\Delta H} = -\frac{\eta F E}{\Delta H} \quad ...(10.13)$$

where, η = number of electrons transferred per molecule of the reactant.

F = Faraday's constant

E = e.m.f. of cell.

Generally, ΔG is quite close to ΔH and hence the efficiency of fuel cells ideally would be close to unity.

In the case of practically all fuel cells, the terminal cell potential decreases with increasing current density drawn from the cell. The main reasons for this decrease as shown above are:

(*i*) the slowness of one or more of the intermediate steps of the reactions occuring at either or both of these electrodes.

(*ii*) the slowness of mass-transport processes (*e.g.*, reactants to and/or products from the electrode) and

(*iii*) ohmic losses through the electrolyte.

The first two losses occur at the anode and cathode and the third type generally in the electrolyte. The potentials expressing these losses are termed overpotentials. The three types of over potentials are given the names activation, concentration, and ohmic respectively. When the terminal potential is E, the energy output during the formation of 1 mole of products in the cell becomes μFE. Under these conditions, the efficiency is expressed by

$$\eta_m = \frac{\eta FE}{\Delta H} \quad \text{...(10.14)}$$

The magnitude of each type of over potential increases with current density drawn from the cell. Hence, the terminal cell potential and the efficiency of the cell decreases with increasing current density.

There is also another reason for the loss in efficiency of an electrochemical energy converter. This loss is owing to the fact that either there is an incomplete conversion of the reactants at each electrode to their corresponding products, or sometimes the reactant from one electrode diffuses through the electrolyte and reaches the other electrode, where it reacts with the reactant at this electrode. The overall efficiency taking into account this loss, is obtained by multiplying the efficiency expressed by Eqn. (10.13), by the Faradaic efficiency r, which represents the amount of product formed to that expected theoretically on the basis of complete conversion and is often equal to unity.

10.2.7. Types of Electrodes

These are of two types:

(*i*) Porous electrodes and; (*ii*) Nonporous electrodes.

(i) Porous electrodes. The design of these electrodes has been based on the assumption that it is necessary to maintain a threephase boundary between gas, electrode and electrolyte. Two main methods have been used to maintain such a stable interface. In the first method, *dual-porosity electrodes* are used, the interface being maintained at the junction of the fine pore and coarse pore layers by means of a differential pressure across the electrode. In a dual pore electrode the fine pore need not be a conductor and porous plastics have been used for this purpose. The second method uses a trapped electrolyte held in a porous nonconducting diaphragm whose pores are smaller than those of the contacting electrodes. This form has been widely used for high temperature cells. The ion exchange membrane cell is a special example of this type. *Porous nickel* electrodes and *Porous carbon* electrodes impregnated with catalysts are generally used in fuel cells for commercial applications. Platinum and other precious metals are being used in certain fuel cells which have potential utility in military and space applications.

(ii) Non-porous electrodes. The concept of non porous diffusion membrane electrodes, an example of which is the hydrogen diffusing palladium alloy membrane allows the complete separation of the reactant gases from the electrolyte. This obviates, the problems of interface control and of direct chemical reactions with the electrolyte.

10.2.8. Work Output and EMF of Fuel Cell

In a fuel cell a chemical reaction takes place whereby the reactants are converted into products in a steady flow process. It is to be noted that fuel and air supplied to the fuel cell at nearly atmospheric temperature and pressure. Also the products of combustion leave at atmospheric pressure and may be cooled, to atmospheric temperature before leaving the fuel cell. Here we are interested to find the maximum useful work that can be obtained from a steady flow stream which undergoes a process at the beginning and end of which, it is in pressure and temperature equilibrium with the atmosphere. This work is obtained by combining the first and second laws of thermodynamics as follows:

From the first law of thermodynamics for a steady flow process,

$$\Delta Q - \Delta W = \Delta H + \Delta(KE) + \Delta(PE) \quad ...(10.15)$$

where, ΔQ = heat transferred to the steady flow stream from the surroundings,

ΔW = workdone by the flow stream on the surroundings,

ΔH = change in enthalpy of the flow stream from entrance to exist and

$\Delta(KE)$ and $\Delta(PE)$ denote the changes in kinetic and potential energy of the stream, which in this case are usually negligible. Hence equation (10.15) can be written as:

$$\Delta W = \Delta Q - \Delta H \quad ...(10.16)$$

If ΔW is to be maximum, the process must be reversible. For a reversible process, from second law of thermodynamics we have

$$\Delta Q = \int T\, ds \quad ...(10.17)$$

But the surroundings have a constant temperature T. Hence, reversible transfers of heat can only be at this temperature T which is also the temperature of the end states of the process. The equation (10.17) becomes,

$$\Delta Q_{revers} = T\,\Delta S \quad ...(10.18)$$

Combining equations (10.16) and (10.18)

$$\Delta W_{max} = T\,\Delta S - \Delta H = -\,(\Delta H - T\,\Delta S) \quad ...(10.19)$$

Gibbs free energy G is given by

$$G = H - TS$$

where, H = enthalpy

$$\therefore \quad \Delta G = \Delta H - Tds - SdT$$

Since the temperature of the flow stream at both entrance and exist is T,

$$\Delta G = \Delta H - T\Delta S$$

Hence $\quad \Delta W_{max} = -\,\Delta G \quad ...(10.20)$

The equation (10.20) is a very important equation of thermodynamics and it states that the maximum useful work done by a steady flow system during a process, the end states of which are at p and T, equals the negative of the change in Gibbs free energy, of the system. It may be noted that equation (10.20) is also valid for a closed system provided that at the beginning and end of the process, the system is at pressure p and temperature T. Also this equation does not require that the system be at p and T throughout the process, the intermediate states of the system may change but end states must be at p and T.

The efficiency of energy conversion of a fuel cell at 25°C is the ratio of the useful work to the heat of combustion of the fuel, *i.e.*,

$$\eta_c = \frac{\Delta W}{-\,(\Delta H^\circ)_{25°\text{C}}}$$

or

$$\eta_{max} = \frac{\Delta W_{max}}{-\,(\Delta H^\circ)_{25°\text{C}}} \quad ...(10.21)$$

where, ΔW_{max} is the smaller of the two values $-\,(\Delta G)_T$ and $(-\,\Delta H)_T$.

The values of $\Delta G°$ (ΔG at 1 atm and temperature T) and η_{max} for various fuels are given in table (10.15).

In order to determine the reversible voltage or e.m.f. of the cell recall that the reversible electrical work is given by

$$dW_{rev} = E\,dq$$

or $$\Delta W_{rcv} = E\,\Delta q \quad ...(10.22)$$

where, Δq is the charge transported. For a chemical reaction the charge transported Δq is given by

$$\Delta q = n\,F \quad ...(10.23)$$

where, n = number of electrons transferred per molecule of the reactant

F = Faraday's constant,

= (96,500 coloumbs per grammole).

Hence equation (10.22) and (10.23) yields

$$\Delta W_{rev} = nFE \quad ...(10.24)$$

or $$E = \frac{\Delta W_{rev}}{nF} = \frac{-\Delta G}{nF} \quad ...(10.25)$$

A relationship between ΔH and ΔE may be derived as follows : Since Gibbs free energy G is a function of temperature and pressure

then $$DG = \left(\frac{\delta G}{\delta T}\right)_p dT + \left(\frac{\delta G}{\delta p}\right)_T dp \quad ...(10.26)$$

Table 10.1. The Values of ΔG° and ηmax for Various Fuels at 25°C (MKS Unit)

Reaction	$\Delta H°$ *kcal/kg mole*	$\Delta S°$ *kcal/kg mole °C*	$\Delta G°$ *kcal/kg mole*	η_{max}	*E* (*Volt*)
$H_2 + \frac{1}{2}O_2 = H_2O$ (liquid)	– 68317	– 38.997	– 56690	0.825	1.230
$H_2 + \frac{1}{2}O_2 = H_2O$ (gas)	– 57798	– 10.607	– 54635	0.800	1.185
$C + O_2 = CO_2$	– 94052	+ 0.697	– 94260	1.0	1.020
$CH_4 + 2O_2 = CO_2 + 2H_2O(l)$	– 212800	– 58.013	– 195500	0.92	1.063
$CH_4 + 2O_2 = CO_2 + 2H_2O(g)$	– 191760	– 1.233	– 191390	0.90	1.038
$C_3H_8 + 5O_2 = 3CO_2 + 5H_2O(l)$	– 530600	– 89.40	– 503926	0.952	1.094
$C_3H_8 + 5O_2 = 3CO_2 + 5H_2O(g)$	– 488530	+ 24.08	– 495706	0.920	1.060

Table 10.2. ηmax for Various Fuels at 298 K

Reaction	$\Delta H°$ *kJ/kg mole*	$\Delta S°$ *kJ/kg mole °C*	$\Delta G°$ *kJ/kg mole*	η_{max}	*E Volt*
$H_2 + \frac{1}{2}O_2 = H_2O(l)$	– 285838	– 163.16	– 237191	0.825	1.230
$H_2 + \frac{1}{2}O_2 = H_2O(g)$	– 241827	– 44.38	– 228593	0.800	1.185
$C + O_2 = CO_2$	– 393514	+ 2.916	– 394384	1.0	1.020
$CH_4 + 2O_2 = CO_2 + 2H_2O(l)$	– 890355	– 242.73	– 817972	0.92	1.063
$CH_4 + 2O_2 = CO_2 + 2H_2O(g)$	– 802323	– 5.159	– 800776	0.90	1.038
$C_8H_8 + 5O_2 = 3CO_2 + 5H_2O(l)$	– 2220030	– 374.05	– 2108426	0.952	1.094
$C_3H_8 + 5O_2 = 2CO_2 + 5H_2O(g)$	– 2044010	+ 100.75	– 2074034	0.920	1.060

From equation (10.18), we have

$$\Delta Q_{revers} = T\Delta S$$

or $$T \,.\, \Delta S = \Delta U + P\,\Delta V$$

or $$T \,.\, ds = DU + PdV \qquad ...(10.27)$$

Also we have

$$G = H - TS$$

$$dG = dH - Tds - SdT = D(U + PV) - Tds - SdT$$

$$= DU + PdV + Vdp - Tds - SdT$$

$$= Tds + Vdp - Tds - SdT = Vdp - SdT \qquad ...(10.28)$$

Comparing coefficient of Eqns. (10.26) and (10.28), we get

$$\left(\frac{\delta G}{\delta T}\right)_p = -S \quad \text{and} \quad \left(\frac{\delta G}{\delta P}\right)_T = V$$

We know, $\Delta G = \Delta H - T\Delta S$ (for constant temp. T)

Substituting the value of S, from above, we have

$$\Delta G = \Delta H - T\Delta\left(-\frac{\delta G}{\delta T}\right)_p$$

$$= \Delta H + T\left(\frac{\delta \Delta G}{\delta T}\right)_p \qquad ...(10.29)$$

From equation (10.25),

$$\Delta G = -nFE$$

Therefore equation (10.29) becomes

$$-nFE = \Delta H + T\left(\frac{\delta \Delta G}{\delta T}\right)_p$$

and $$\Delta H = -nFE - T\left(\frac{\delta \Delta G}{\delta T}\right)_p$$

or $$\Delta H + nFE + T\left(\frac{\delta \Delta G}{\delta T}\right)_p = 0$$

or $$\Delta H = -nFE + T\,nF\left(\frac{\delta E}{\delta T}\right)_p \qquad (\because \ \Delta G = -nFE)$$

$$= -nF\left[E - T\left(\frac{\delta E}{\delta T}\right)_p\right]$$

or $$\Delta H_p.T = nF\left[T\left(\frac{\delta E}{\delta T}\right)_p - E\right] \qquad ...(10.30)$$

Equation (10.30) is called the *Gibbs-Helmholtz* equation and may be used to compute the enthalpy of reaction from measurement of E.

Example 10.1. *Find the reversible voltage for Hydrogen oxygen fuel cell, having the reaction;*

$$H_2 + \tfrac{1}{2}O_2 = H_2O(l)$$

Given. $(\Delta G°)_{25} = -$ *56,690 cal/gm-mole of* H_2 *or* -237.3×10^3 *Joules/gm-mole of* H_2

Solution. Reversible voltage E of a cell is given by

$$E = \frac{\Delta W_{rev}}{nF} = -\frac{\Delta G}{nF}$$

Since 2 electrons are transferred per molecule of H_2. Thus

$$E = \frac{237.3 \times 10^3}{2 \times 96{,}500} = \mathbf{1.23\ volt.}$$

Example 10.2. *A hydrogen oxygen fuel cell in which the following reactions occur:*

At the anode : $H_2 \longrightarrow 2H^+ + 2e^-$

At the cathode: $2e^- + \tfrac{1}{2}O_2 + H_2O \longrightarrow 2OH^-$

operates at 25°C. Calculate

(i) the voltage output of the cell *(ii) the efficiency*

(iii) the electrical work output per mole of H_2 *consumed and per mole of* H_2O *produced.*

(iv) the heat transfer to the surroundings.

Take $\Delta H^\circ_{298\,K} = -\,68{,}317$ *kcal/kg mole*
$\Delta G^\circ_{298\,K} = -\,56{,}690$ *kcal/kg mole.*

Solution. At the cathode the one oxygen atom combine with the two electrons arriving by the external circuit. Here n = 2 or 2 electrons are transferred per molecule of H_2.

(*i*)
$$E = -\frac{\Delta G}{nF} = \frac{237.3 \times 10^3}{2 \times 96500} = 1.23 \text{ volt.} \quad \textbf{Ans.}$$

(*ii*)
$$\eta_{max} = \frac{\Delta W_{max}}{-(\Delta H)_{25^\circ C}} = \frac{-(\Delta G)T}{(-\Delta H^\circ)_{25}} = \frac{56690}{68317}$$
$$= 0.825. \quad \textbf{Ans.}$$

(*iii*) The electrical work output per mole

$$\Delta W_{rever} = nFE = 2 \times \frac{96500}{4.184} \times 1.23$$
$$= \textbf{56690 kcal/kg mole.} \quad \textbf{Ans.}$$

(*iv*) Since there are 1 mole of H_2O produced for each mole of fuel as evident from the equation:

$$H_2 + \frac{1}{2} O_2 \longrightarrow H_2O$$

There is also a work output of 56690 kcal/kg mole.

The heat transfer is

$$Q = T\,\Delta S = \Delta H^\circ_{298} - \Delta G^\circ\ \Delta G^\circ_{298}$$
$$= -\,68{,}317 + 56690 = -\,\textbf{11627 kcal/kg mole.}$$

The negative sign indicates that the heat is removed from the cell and transferred to the surroundings.

Example 10.3. (SI unit). *A hydrogen-oxygen fuel cell operates at 25°C. Calculate the voltage output of the cell, the efficiency, and the electric work output per mole of* H_2 *consumed and per mole of* H_2O *produced. Also compute the heat transferred to the surroundings.*

Given : $\Delta H^\circ_{298^\circ k} = -\,285838$ *kJ/kg mole*
$\Delta G^\circ_{298^\circ k} = -\,237191$ *kJ/kg mole.*

Sol. According to equation $H_2 + 2OH^- \longrightarrow 2\,H_2O + 2e^-$, there are two electrons transferred per molecule of H_2, so that η_e = 2. The cell voltage is given as:

$$E = -\frac{\Delta G^\circ_{298\,k}}{\eta_e F} = \frac{237.191 \times 10^3}{2 \times 96.500 \times 10^3} = 1.229 \text{ V} \quad \textbf{Ans.}$$

The efficiency η_{max} is $= \dfrac{\Delta G^\circ_{298}}{\Delta H^\circ_{298}}$

$$= \frac{-\,237.191 \times 10^3 \text{ kJ/kg mole}}{-\,285.838 \times 10^3 \text{ kJ/kg mole}} = 0.83 = 83\% \quad \textbf{Ans.}$$

The electrical work output per mole of the fuel is

$$W_e = -\,\Delta G^\circ = -\,237.191 \text{ kJ/kg mol}$$

Since there is 1 mol of H_2O produced for each mole of fuel, there is also a work output of 237191 kJ/kg mol H_2O.

The heat transferred is

$$Q = T.\, \Delta S = \Delta H^\circ_{298} - \Delta G^\circ_{298}$$
$$= -28584 + 23719 = -4865 \text{ kJ/kg mole}$$

The negative sign indicates that the heat is removed from the cell and transferred to the surroundings.

Example 10.4. *A galvanic cell has the following reaction,*

$$Ag(s) + HgCl(s) = Ag\,Cl(s) + Hg(l)$$

if $E^\circ_{25°C} = 0.0455$ volt, and $(dE^\circ/dT) = 0.000338$ volt/degree, calculate ΔG°, ΔS°, and ΔH°.

Solution. We have the relation $\Delta G^\circ = -nFE$

where, ΔG° = Gibbs free energy of the system at 1 atm and temperature (T)

n = number of electrons transferred per molecule of reactant

F = Faraday's constant (96,500 coloumbs per gram-mole)

E = e.m.f. of the cell

$$\Delta G^\circ = -(1 \times 96{,}500 \times 0.0455)$$
$$= -4{,}400 \text{ joules.}$$ **Ans.**

Again we have $$\Delta S^\circ = nF\left(\frac{\delta E}{\delta T}\right)_p = nF\,\frac{dE^\circ}{dt}$$

$\Delta^\circ S$ = Entropy change of the system at temperature T and press p = 1 atm in this case

$$= 1.96{,}500 \times 0.000338$$
$$= 32.6 \text{ joule/deg.}$$ **Ans.**

And entropy change is given by the relation

$$\Delta H^\circ = nF\left[T\,\frac{dE^\circ}{dT} - E^\circ\right]$$
$$= 1 \times 96{,}500\,[298 \times 0.000338 - 0.0455]$$
$$= 5{,}300 \text{ joule.}$$ **Ans.**

Example 10.5. *Perform the necessary calculations to show that the maximum efficiency for the methane fuel cell is 92% and the ideal cell voltage is 1.15 volts. What flow rate in kg/hr of methane and oxygen would be required to produce a power output of 100 kW. What heat transfer rates would be involved under these circumstances?*

Given that, $$\Delta G^\circ_{25°C} = -195{,}500 \text{ cal/gm mole}$$
$$\Delta H^\circ_{25°C} = -212{,}800 \text{ cal/gm mole.}$$

Solution. For methane cell, the cell reaction is

$$CH_4 + 2H_2O \longrightarrow CO_2 + 8H^+ + 8e^-$$

Here, $n = 8$

$$E^\circ_{25°C} = \frac{(-\Delta^\circ G)_{25°C}}{nF}$$

$$= \frac{-195,500}{8 \times 96,500} \times 4.184 \text{ Joules/coulomb}$$

($\because$ F = 96,500 coulombs/gm-mole)

$$= 1.0595 \text{ volts} \simeq 1.060 \text{ volts.}$$

Max. efficiency,

$$\eta_m = \frac{\Delta W_{max}}{(-\Delta H)_{25°C}} = \frac{195,500}{212800} = 0.92. \quad \textbf{Ans.}$$

Voltage efficiency,

$$\eta = \frac{\text{On load voltage}}{\text{Open circuit voltage}} = \frac{\text{Operating voltage}}{\text{Theoretical voltage}}$$

$$\text{Theoretical voltage} = \frac{1.060}{0.92} = 1.15 \text{ volts.} \quad \textbf{Ans.}$$

Power developed

$$= 100 \text{ kW} = 100 \times 10^{-3} \text{ W}$$

$$= 100 \times 10^{-3} \times 0.86 \text{ kcal/hr}$$

[$\because$ 1 Watt = 1 Joule/sec = 0.86 kcal/hr]

$$= 86,000 \text{ kcal/hr.}$$

$$\Delta G = -195,500 \text{ cal/gm mole}$$ [given]

$$\text{Required flow rate of Methane} = \frac{86,000}{\frac{195,500}{16}} \text{ kg/hr}$$

(Methane moles) = 16 = 6.547 kg/hr. **Ans.**

The heat transfer,

$$Q = T\Delta S = \Delta H + \Delta W$$

$$= \Delta H°_{25} - \Delta G°_{25} = -212800 + 195500$$

$$= -116,700 \text{ kcal/kg mole.} \quad \textbf{Ans.}$$

10.2.9. Applications of Fuel Cells

The applications of fuel cell may be discussed with reference to the following:

(*i*) Domestic use

(*ii*) Central power stations

(*iii*) Automotive vehicles

(*iv*) Special applications.

The e.m.f. or voltage of a fuel cell depends to some extent on the discharge current strength. The average voltage per cell is 0.75 volt. By joining a number of cells in series and parallel can provide any reasonable voltage and current. Fuel cells generate direct current which can be used for electric lamps and some small applications such as heat pumps, motors etc., conversion into alternating current by means of an inverter might be necessary.

Fuel cells can be made in modules of different size that are readily transportable. They can then be assembled at any location to provide a specified voltage and power output. The modular design should make it possible to construct plants of various capacities for different requirements.

If fuel cells of reasonably low cost and long life can be produced, a major use might be by electric utilities for load levelling as explained below. A long term possibility is a centralstation power plant in which coat is gasified and the gas is used to generate electricity directly by means of fuel cells. Such an installation is expected to have a higher efficiency for fuel utilization than a conventional steam-electric plant.

Portable generating sets seem to be a favourable field for fuel cells. Here, already, fuel cells appear to be competitive as compared with conventional sources. Low temperature fuel cells have a favourable position for operating times of 3,000 to 4,000 hours per year, using methanol as a fuel.

Large generating stations operate most efficiently at a steady (rated) power output, but the demand for power is variable. When the demand or load is less than the rated output, the excess would be used to generate hydrogen by electrolysis of water. At times when the load is greater than the power supply, the hydrogen would be used in fuel cells to satisfy the additional demand. By siting fuel cells near load centres where the demand exists, electrical transmission and distribution costs would be reduced, although there would be some cost for transporting the hydrogen. Sometimes new load centres are formed as a result of housing and industrial developments. To satisfy the power demand utilities, will either built additional large plants, which require considerable capital expenditures, or utilized diesel engines or gas turbines operated by natural gas or a petroleum fuel. The same fuel might be utilized more economically in fuel cells located near the new load centre.

Fuel sources have been proposed for remote or rural areas or unattended locations, for mobile and emergency power sources, and for vehicle propulsion. The high temperature batteries may be the best candidates for vehicle propulsion in the long term, but certain fuel cells are potential alternatives to storage batteries for electric vehicles. Such vehicles may have longer travel ranges than those with the most advanced storage batteries. The fuel cells of special interest are the aluminiumair, methanolair, and the hydrogenoxygen cell. The aluminium-air cell is of special interest for electric vehicle propulsion because of the high specific energy that is possible. An aluminiumair battery, and associated equipment may weigh roughly the same as the gasoline engine and fuel in a medium size automobile. A five passenger electric vehicle is expected to have a travel range of atleast 1600 km before replacement of the aluminium electrodes is necessary. However, the aluminium hydroxide produced when the cell operates. It must be removed every 400 to 600 km, and water must be added at similar intervals.

The range of a vehicle with a methanolair battery depends on the volume of the fuel tank, that is, on the quantity of methanol that can be carried. If fairly pure hydrogen gas were available at a reasonable price, the hydrogenoxygen (air) fuel cell might be used for electric vehicle propulsion. The major problem would be storage of the fuel on the vehicle. However, the high efficiency of the fuel cell would compensate for the weight of the hydrogen storage system.

Many of the fuel cells currently under development are for *special applications* where convenience is of paramount importance, cost is secondary. For these applications hydrogen is the superior fuel from the view point of reactivity and availability of invariant electrolyte, although it is relatively costly. It seems likely that hydrogenoxygen and hydro carbonoxygen cells will be used to an increasing extent in special military and space projects.

10.3. Batteries

10.3.1. Introduction

A battery is defined as a combination of individual cells. A cell is the elemental combination of materials and electrolyte constituting the basic electrochemical energy stores. A battery can also be thought of as a black box into which electrical energy is put, stored electro chemically, and later recovered as electrical energy. A storage cell is one that can be recharged after discharge by passing a direct current through the cell in the opposite direction to the discharge current. The chemical reaction taking place in the cell when it is charged is reversed when the cell is discharged. Thus, in the charged cell, electrical energy is stored as chemical energy which can be recovered as electrical energy when the cell is discharged.

There are two principal divisions of a battery:

(1) *Primary Batteries* nonchargable, *e.g.*, "drycell" flash light batteries. In primary batteries the chemical reactions are nonreversible.

(2) *Secondary Batteries* rechargeable, *e.g.*, a lead acid battery. There are many types of secondary batteries. The chemical reactions are reversible in secondary batteries.

Secondary batteries are of chief interest for solar electrics (Solar and wind energy, electrical storage), and we therefore discuss here about secondary batteries or storage batteries.

In a storage battery, individual cells are connected in various ways to provide the desired power output. Since storage batteries are generally portable, they are common mobile sources of energy. A major use, for example, is the starting-lighting ignition (SLI) system of automobiles. Among the many other applications are those in the operation of mine locomotives, forklift trucks, golf carts, road vehicles, and submarines and other underwater craft. If maintained in a charged state, storage batteries are a dependable source of electrical energy; they are consequently used in stationary emergency power systems. Furthermore, instruments of many industrial plants are operated by batteries.

Storage batteries are expected to play an important role in energy resources conservation. Among the potential uses are: (1) Storage of electrical energy produced from the solar energy, (photovoltaic conversion) and wind energy, (2) Propulsion of road vehicles and (3) load leveling and peak saving in electrical utilities.

10.3.2. Basic Battery Theory

A generalized cell consists of two electrodes called the anode and cathode immersed in a suitable electrolyte. When an electrical load is connected between the electrodes charge separation occurs at the interface between one electrode and the electrolyte, freeing both an electron and ion. The electron flows through the external load and ion through the electrolyte recombining at the other electrode. Advanced treatment of the electrochemistry is discussed in proceeding article *i.e.*, in classes of batteries.

The polarity and magnitude of the cell terminal voltage is, in general, a function of the electrode materials, electrolyte, cell temperature, and other factors.

Many generic versions of cell exist. All have arisen from the electrochemistry field.

10.3.3. Definitions of Fundamental Quantities

Open circuit voltage (OCV). The battery or cell terminal voltage with $R_{load} = \infty$, *i.e.*, no load.

Energy capacity of a battery is of 3 types:

(*a*) Energy stored (watt hours or kilowatt hours)

(*b*) Energy stored weight (watt hours/kg)

(*c*) Energy stored/volume (watt hours/m^3).

Power capacity is the rate at which stored energy can safely be taken out of a battery and restored.

Specific power is the maximum rated power output per kg, the battery can supply. Rating various batteries by specific power permits rapid performance comparisons of different kinds of batteries.

Energy Efficiency of a battery is, defined as:

$$\text{Energy Efficiency} = \frac{\text{Useful energy out (watt hrs)}}{\text{(Recharge energy (watt hrs)}}$$

$$= \frac{\int_0^{t_1} I_1 E_1 \, dt}{\int_0^{t_2} I_2 E_2 \, dt} \qquad \text{...(10.31)}$$

where, I_1 = Battery discharge current

E_1 = Battery discharge terminal voltage

I_2 = Battery charging current

E_2 = Battery charging terminal voltage

t_1 = Battery discharging time

t_2 = battery charging time

Two special cases of equation (10.31) are of interest to the system engineer. If the discharge and charge occurs at constant voltages, then equation (10.31) reduces to:

$$\text{Energy efficiency} = \frac{E_1 \int_0^{t_1} I_1 \, dt}{E_2 \int_0^{t_2} I_2 \, dt} = \frac{E_1 \times \text{Amp. hour output}}{E_2 \times \text{Amp. hour input}} \qquad \text{...(10.32)}$$

The integrals might be evaluated graphically. Once found, multiplication by the respective terminal voltage gives the watt-hours of input or output. Equation (10.32) suggests why for some types of batteries with nearly constant terminal voltage the chief interest might be in the current I v_s time t curve for discharge or charge. The ampere hours output of a battery is a very common way of rating.

If the discharge and charge occurs at constant currents, then Equation (10.31) reduces to:

$$\text{Energy Efficiency} = \frac{I_1 \int_0^{t_1} E_1 \, dt}{I_2 \int_0^{t_2} E_2 \, dt} \qquad \text{...(10.33)}$$

If neither current or voltage is held constant, then we must resort to equation (10.31) directly, and either graphically or electronically evaluate the integrals to determine the energy efficiency for the battery.

Cycle life is the number of times the battery can be charged and discharged under specified conditions. The cycle life of a battery may vary greatly with the depth of discharge, deep discharge tending to result, in short cycle life.

10.3.4.1. Battery Fundamental Characteristics

These are shown in a generalised form in Fig. 10.10. These are the terminal characteristics of the battery. Clearly at infinite resistive load we have the open circuit voltage at point *O*; as resistance decreases to R_1 the battery load system operates at point. A delivering P_1 power to the load. A further decrease in load resistance to .R_2 moves the operating point to *B*, the maximum power output being p_3. Further decrease in load resistance to R_3 results in operation at *C* and a decrease in load power to P_2. If the terminals are short circuited ($R = 0$) then we arrive at point *D*.

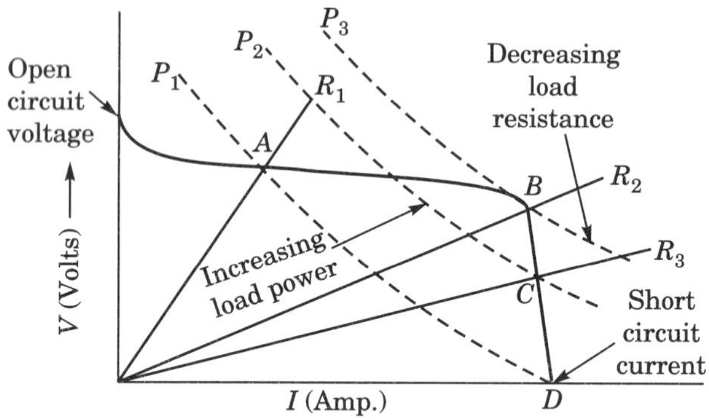

Fig. 10.10. The *I-V* characteristics of 0 generalized battery and its relationship to a d.c. load.

The shape of the battery is characteristic *I-V* curve will vary with the type and size of the battery. The *O-A-B* portion of the curve is generally not flat, though for some cell types it may be approximately so. It usually will have a portion which is approximately straight. The battery's curve *O-A-B-C-D* is generally found experimentally under fixed known set of temperature, pressure, or other conditions.

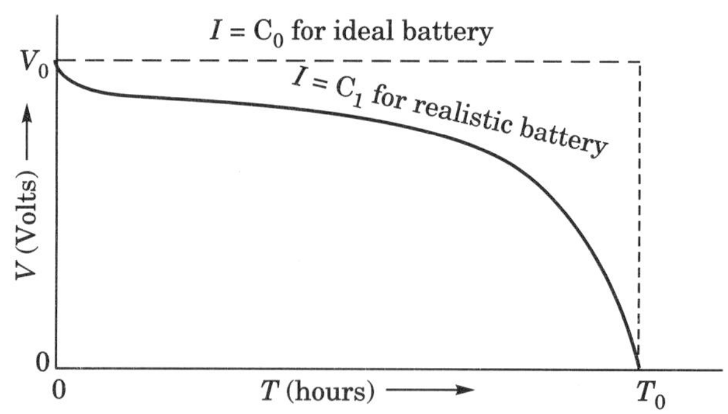

Fig. 10.11. The voltage time curve for a battery.

Another fundamental characteristics of a battery is shown in Fig. 10.11, the voltage v_s time curve under the constraint of constant discharge current.

The electrical energy recovery is:

$$= \int_0^{T_0} I\, V(t)\, dt = I \int_0^{T_0} V(t)\, dt \qquad \text{...(10.34)}$$

$= I \times$ area under curve (watt hours)

For the ideal battery shown maintaining a constant terminal voltage, energy recovered:

$= V_0 . I . T_0 = V_0 \times$ "amp-hour capacity".

In general each kind of battery has a different $V(t)$ curve. The curves are obtained generally for a constant discharge current. The curves are generally found experimentally. There may be marked differences in these curves for different kinds of batteries, *e.g.* a nickel-cadmium battery is well known for its very sudden exhaustion in contrast to the gradual fall-off shown in Fig. 10.11.

I-V characteristic and the $V(t)$ curve, both are affected significantly by the following factors:

(*a*) *Battery temperature.* Low temperature generally result in decreased battery performance and *vice versa*.

(*b*) Whether the battery is used continuously or only part time.

(*c*) *Battery environment, e.g.,* altitude, vibration, and other factors.

(*d*) Type, size and manufacture of the battery.

(*e*) The internal resistance of the battery.

(*f*) Battery previous history.

Battery Equivalent Circuit models are shown in Fig. 10.12. They are:

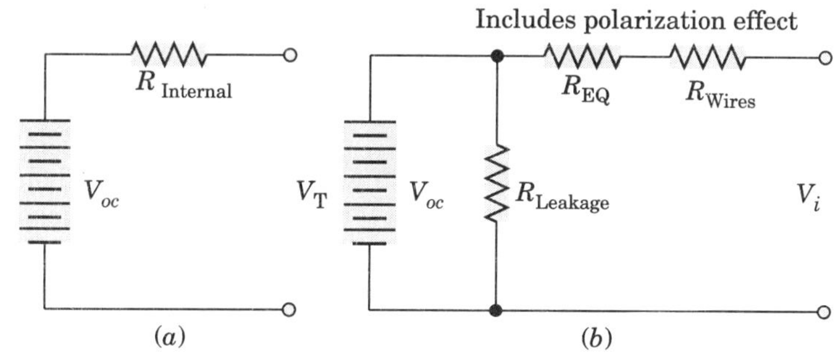

Fig. 10.12. Battery equivalent circuit models.

(*a*) First order model, in which leakage effects are negligible, and it is the one which is most widely used.

(*b*) Second order model showing leakage as the cause of "run down" and short shelf life on secondary batteries.

10.3.5. Different Types of Battery Arrangement

In a storage battery, individual cells are connected in various ways. Like all electric cells, a storage cell has positive and negative electrodes to which the respective terminals are attached. Individual cells may be combined in *series* with the positive electrode of each cell connected to the negative electrode of the adjacent cell Fig. 10.13. The total electromotive force (emf) or voltage of the battery is then the sum of the separate voltages. In most automobile batteries, for example, six cells each with an emf of close to 2 volts are connected in series to provide a 12 volt output. The total current (in amperes) drawn from the series of cells, is however, the same as that drawn from each cell.

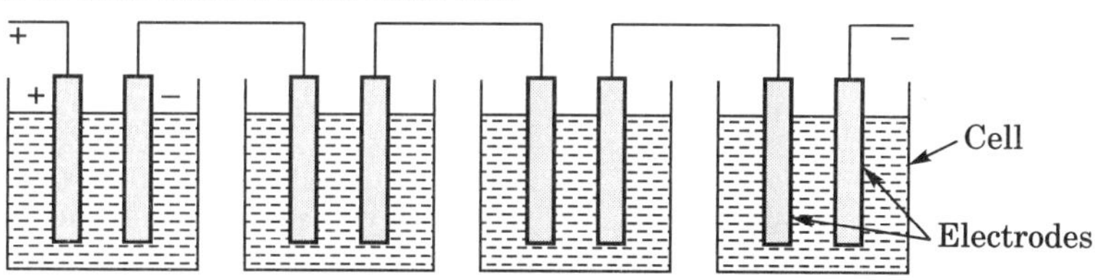

Fig. 10.13. Electric cells connected in series.

Storage cells in a battery can also be combined in *parallel*; all the positive electrodes of the individual cells are connected together and so also are all the negative electrodes. Fig. 10.14. The battery voltage is now the same as that of a single cell, but the current is the sum of the currents supplied by the individual

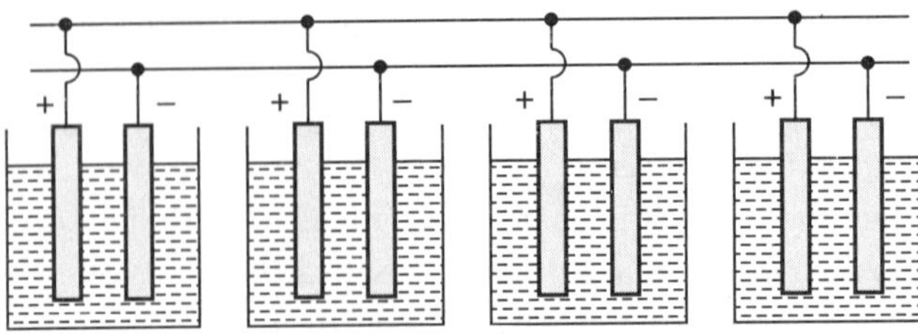

Fig. 10.14. Electric cells connected in parallel.

cells. By combining appropriate numbers of cells in both series and parallel (Fig. 10.15), the battery can deliver the desired voltage and current. The power of a battery (*i.e.*, the rate at which stored energy is withdrawn) in *watts* is

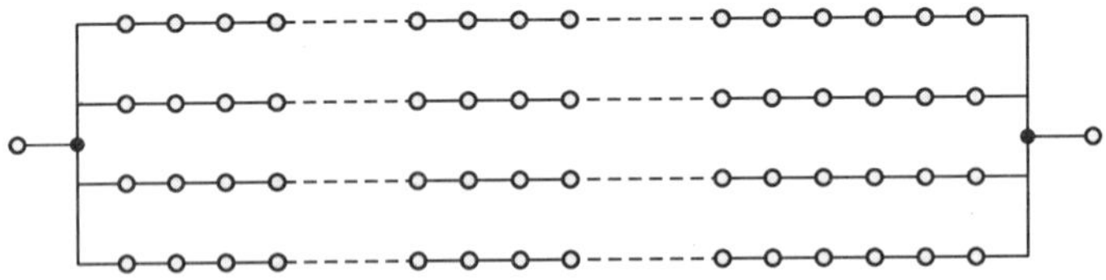

Fig. 10.15. Cell arrangements in series and parallel.

equal to the product of the emf in volts and the current in amperes; thus

Power (watts) = EMF (volts) × current (amperes)

Hence, an appropriate combination of cells can provide the desired power output.

The energy storage capacity of a battery is often expressed in ampere hours; this is the product of the current (in amperes) the battery can deliver multiplied by the delivery time (in hours). The usuable energy (in watt hours) stored in the ampere hour capacity multiplied by the emf of the battery in volts. An automobile battery, for example with an emf of 12 volts and 50 ampere hour capacity, delivers 12 × 50 = 600 watt-hours (or 0.6 Kilowatt-hours) of energy.

The practical storage capacity of a battery depends on the discharge rate (or discharge time). Increase in the discharge rate (or decrease in discharge time) for a given battery results in a decrease in the amount of electrical energy that can be delivered. This effects is very marked in the leadacid battery commonly used in automobiles. For example, a battery with an energy capacity of 0.6 Kilowatt hour (kW hr) at 2.5 amperes discharge rate might have a capacity of only 0.24 kW hr at 150 amperes. However, upon standing after a rapid discharge, the battery may experience a partial recovery and deliver more of its stored energy. In a lead acid cell, the recovery arises mainly from local readjustments in the acid concentration which decreases when the cell is discharged.

10.3.6. Classifications of Batteries

There are many kinds of batteries which have been created and studied in the past. These are classified into three types:

(*a*) *Conventional Batteries*, examples are lead acid, Nickel cadmium; Nickel iron, Nickel zinc. Silver Zinc, Silver Cadmium, Zinc-Bromine etc.

(b) *Metal Gas Batteries*, examples, are Iron Air, Zinc Air, Zinc Oxygen, Zinc chlorine, Nickel hydrogen, Cadmium oxygen, Cadmium air, Aluminium air, Lithium Sulfur Dioxide, Sodium-air, Magnesium air etc.

(c) *Alkali metal-High temperature Batteries*, examples are Sodium sulfur, Sodium Chlorine, Lithium Sulfur, Lithium Iron Sulfide, Lithium Chlorine, Lithium Copper, Lithium Nickel halide Aluminium Chlorine, etc.

We shall be discussing some of these in the following paragraphs.

Lead Acid Battery. The lead acid battery, invented in Gaston Plante in France in 1859, was the first practical storage battery; it is still, in spite of its defects, the one in widest use, especially for the SLI (starting-lighting ignition) system of road automobiles and farm vehicles. The name arises from the chemical nature of the electrodes (lead) and the electrolyte (acid) in which the electrodes are immersed in the charged battery, the active material of the positive electrode is lead dioxide (PbO_2) whereas that of the negative electrode is metallic lead (Pb). The electrolyte is an aqueous solution of sulphuric acid.

When the cell is discharged, the active materials of both, electrodes are converted into lead sulfate ($PbSO_4$). The process in reversed when the cell is charged; lead dioxide is regenerated at the positive electrode and lead at the negative electrode. The net chemical reactions taking place upon discharging and charging the battery are as follows:

$$\underset{(+)}{PbO_2} + \underset{(-)}{Pb} + 2H_2SO_4 \underset{\text{charge}}{\overset{\text{discharge}}{\rightleftharpoons}} \underset{(+)}{PbSO_4} + \underset{(-)}{PbSO_4} + 2H_2O \quad ...(10.35)$$

The (+) and (–) signs below the symbols refer to positive and negative electrodes, respectively. The emf of a lead acid cell is independent of the amounts of active material present on the electrode, but it does depend on the concentration of the sulfuric acid in practice, the concentration is indicated by the specific gravity of the electrolyte. In a fully charged cell, the specific gravity is usually about 1.26 to 1.28 at 25°C; the emf of a lead acid cell is then close to 2.1 volts (at a low discharge rate). An increase or decrease in the specific gravity (*i.e.*, in the acid concentration) is accompanied by an increase or decrease, respectively in the emf.

When the lead acid cell is discharged, sulfuric acid is consumed at both electrodes as the lead dioxide (positive) and lead (negative) are converted into lead sulfate. Furthermore, as seen in the equation given above, water (H_2O), which dilutes the acid, is produced in the discharge; this occurs at the positive electrode. As a result of the decrease in the acid concentration (and specific gravity), the emf of a lead acid cell decreases steadily as it is discharged. The cell is generally regarded as being fully discharged when the specific gravity is about 1.08, at 25°C and the emf is roughly 1.7 volts. Both acid concentration and emf are restored when the cell (or battery) is charged.

The electrodes in a lead acid cell are commonly in the form of a plates upon (or within) which the active materials are deposited. Although the amounts of these materials do not affect the cell emf, they determine the energy storage capacity of the cell/battery. In order to increase the storage capacity, it is necessary therefore, to increase the accessible amount of active material. However, for efficient utilization of this material, the plates must not be too thick. Hence, the storage capacity is increased by increasing the area of active material exposed to the electrolyte. This is achieved by connecting in parallel a substantial number (ten or more) of positive plates alternating with a set of connected negative plates Fig. 10.16. A large area (and mass) of accessible active material can thus be obtained within a cell of reasonable size.

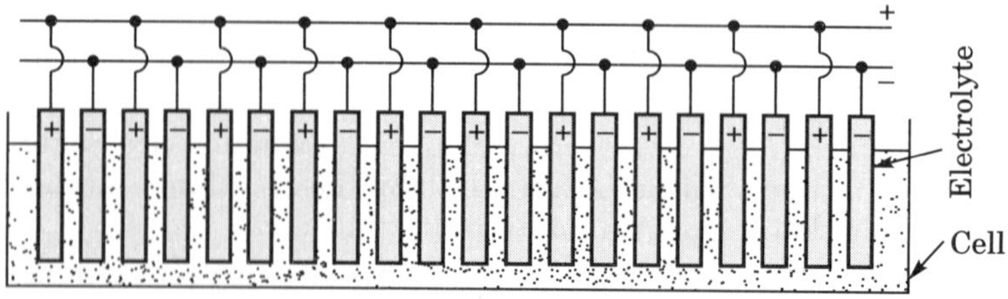

Fig. 10.16. Single cell with positive and negative plates in parallel.

For automobile batteries, the plates are commonly made from a cast grid (or lattice) of lead, alloyed with small amounts of other elements (*e.g.*, antimony, calcium, etc.) to improve corrosion resistance and mechanical properties. The spaces in the grid are then filled with a paste made from lead monoxide (PbO), some finely divided lead, and aqueous sulfuric acid. These are called *pasted plates*.

The plates are now "formed" by treating them as if they are being charged at a low rate in dilute acid. The material on the positive plates is oxidized to active lead dioxide and that on the negative plates is reduced to spongy (active) lead. The plates are assembled with porous separators between them, to form a unit cell of the desired storage capacity. Sulfuric acid electrolyte of the proper concentration (specific gravity) is then added. Finally, the unit cells, each with an emf of about 2 volts, are connected to form a battery of the required voltage (*e.g.*, six cells in a standard 1.2 volt automobile battery).

A disadvantage of the lead acid battery, especially for vehicle propulsion, is its low *specific energy*, frequently called the energy density, this is the energy stored (and recovered) per unit mass. It is commonly expressed in watt hours of stored energy per Kilogram (W-hr/kg).

There are several possible ways in which the energy densities of lead-acid batteries might be increased. They include : (1) decreasing the weight of the lead grid used in the making of the plates, (2) improving the efficiency of utilization of the active materials, and (3) increasing the concentration of the acid electrolyte.

1. *Grids,* required to support the active material, must be strong enough to maintain their integrity throughout the service life of the battery, in spite of inevitable corrosion by the concentrated sulphuric acid (about 38 weight percent in a charged cell). The grids must also provide electrical conduction throughout the mass of active material. So far grids have been made of lead (or lead alloys) because it is the only metal of reasonable cost that is not severely attacked by the acid. Even lead is corroded to some extent, so that the grids must be fairly massive.

Attempts are being made to design a light weight grid for lead acid batteries. The most promising approach appears to be the use of lead coated aluminium. An under coat of a resistant material would retard corrosion of the aluminium even if the outer lead coating is not perfect. A grid of this type would make possible a marked decrease in battery weight for the same storage capacity. There would also be a decrease in the electrical resistance, thereby reducing energy lost as heat.

2. When a lead acid cell is completely discharged for all practical purposes, a substantial proportion, often about half, of the active material remains on the plates. The fraction of the active material actually consumed in the production of energy, called the *utilization coefficient*, depends on the design of the cell, the rate of discharge, and the temperature.

Since sulfuric acid is required for the electrochemical discharge reactions at both positive and negative electrodes, an important factor is the accessibility of the acid electrolyte to the active material. In the course of the cell discharge, the formation of lead, sulfate tends to block the pores in the active material and thus prevents access of the acid electrolyte to the interior. Consequently, the active material is not utilized completely. Low acid concentration, high discharge rate, and low temperature enhance sulfate blockage and reduce the utilization coefficient. In addition, repeated deep charge discharge cycles and standing in the discharged state cause physical changes in the electrode materials that decrease the utilization.

Any improvement in the utilization coefficient and an equivalent increase in the specific energy should be attainable. This would probably require new concepts in plate design and a better understanding of the details of the electrode processes. Although the overall electrochemical reactions are well known, there are many complex physical changes in the materials that occur when a cell is charged and discharged.

3. The emf of a lead acid battery can be increased by increasing the acid concentration (or specific gravity); there is consequently an increase in the specific energy for a given discharge (in ampere-hours). Because the corrosion of lead increases with the acid concentration, the maximum specific gravity in a fully charged cell has been taken to be 1.30 at 25°C. However it is reported that acid with a specific gravity of 1.4 has been used in Japan, with an accompanying increase in specific energy. The specific energy of a lead-acid cell can be increased by operating at higher temperatures, upto about 40°C, but this would also result in greater corrosion.

Nickel-Iron cell. The nickel iron alkaline cell was invented by Thomas A. Edison in the early 1900s. The electrochemical reactions occurring when discharged and charged are represented (approximately) by,

$$\underset{(+)}{2NiOOH.\,H_2O} + \underset{(-)}{Fe} \underset{\text{charge}}{\overset{\text{discharge}}{\rightleftharpoons}} \underset{(+)}{2Ni(OH)_2} + \underset{(-)}{Fe(OH)_2} \qquad ...(10.36)$$

When $Ni(OH)_2$ is hydrated form of a lower nickel oxide (NiO) and $Fe(OH)_2$ is hydrated ferrous oxide (FeO) formed when the cell is discharged. As in all electrical cells when producing current, reduction (in this case from a higher to a lower oxide) takes place at the positive electrode and oxidation (from metal to the oxide) at the negative electrode.

Unlike the lead acid cell, the electrolyte, potassium hydroxide solution, is not involved in the electrochemical reactions. Hence concentration does not change appreciably when the cell is discharged (or charged). The specific gravity of the electrolyte can not then be used to indicate the extent of discharge, as it can in the lead-acid battery. (The role of the electrolyte in an alkaline cell is to carry electrical charges from one electrode to the other, through the cell).

The positive plates of the Edison cell are made of a nickel plated steel grid holding perforated tubes of the same material, the tubes are packed with alternating layers of nickel oxide and metallic nickel flakes to improve the electrical conductivity. The negative plates are also nickel plated steel grids with long, rectangular perforated pockets containing iron oxide powder, this is reduced electrolytically to finely divided iron metal. Alternating positive and negative plates are connected in parallel as in Fig. 10.16 to form a single cell.

Advantages of the Nickel Iron Battery. The Edison battery has been used for rail road car and mine lighting, for mine locomotives, and for industrial trucks, largely because of its rugged construction and moderately high specific energy (about 53 W hr/kg). The weight of nickel iron batteries is less, although the volume is greater, than for leadacid batteries of equal energy storage capacity. Another advantage of the Edison batteries is that they can be stored for long periods in the uncharged state without suffering damage, because of physical changes in the lead sulfate, this is not the case with lead acid batteries.

Limitations of the Battery. Major drawbacks of the nickel-iron battery are that, upon standing, it loses its charge (*i.e.,* moderately rapid self discharge) and requires addition of water from time to time. Both effects are due to the chemical reaction of the active iron with the electrolyte solution, the water is decomposed with the liberation of hydrogen gas and conversion of the iron into inactive ferrous hydroxide. Liberation of hydrogen gas may also occur when the cell is being charged. Another drawback of the Edition cell is that its performance falls rapidly at temperatures below about 10°C; some form of heating would thus be required in cold weather.

Nickel Cadmium Battery

This is similar to the Edison cell and was developed by W. Junjner in Sweden. The main difference is that the active negative electrode material is cadmium metal rather than iron. The electrochemical reactions are:

$$\underset{(+)}{2NiOOH.\ H_2O} + \underset{(-)}{Cd} \underset{\text{charge}}{\overset{\text{discharge}}{\rightleftharpoons}} \underset{(+)}{2Ni(OH)_2} + \underset{(-)}{Cd(OH)_2} \qquad ...(10.37)$$

where, $Cd(OH)_2$ is a hydrated cadmium oxide (CdO). The potassium hydroxide electrolyte is again not involved directly in the cell reactions.

In the early nickel cadmium cells, both positive and negative plates were of the same type as the negative plates of the Edison cell, that is with relatively long, rectangular pockets containing the active materials. These are called "pocket- type" cells.

A later development is the "sintered plate" electrode. Fine metallic nickel power is molded around a nickel (or nickel plated) screen and sintered (*i.e.,* heated to cause particle agglomeration) to form a nickel matrix.

Because a large part of the active material in an electrode does not participate in the charge discharge processes, the nickel powder may be mixed with graphite before sintering. The porous matrix is then impregnated with a solution of a nickel salt (for the positive electrode) or of a cadmium salt (for the negative electrode) and treated to form the respective hydroxides.

Sintered plate cells are more expensive than the pocket type, but they have a smaller internal resistance, a higher specific energy and better utilization of the active electrode materials. Sintered plates are being proposed for use with other cells having nickel (hydroxide) electrodes.

Advantages of the nickel cadmium cell over the Edison cell are:

(*i*) a much lower rate of self discharge and longer life;

(*ii*) unlike iron, cadmium does not react with the KOH electrolyte;

(*iii*) there is little loss of water and no hydrogen evaluation either when the cell is standing in the charged state or when it is being charged, consequently, nickel cadmium cells are often sealed.

Because of their high initial cost, nickel cadmium batteries are used where the long life and reliability compensate for high cost small sealed cells are used for hearing aids, power tools, pocket calculators, etc.; whereas larger batteries are employed in space craft, air craft, and in land based auxiliary power systems. There appears to be little prospect for the use of nickel-cadmium cells for road vehicle propulsion or electric utility applications.

Zinc Air Battery

The cell with a positive oxygen (air) electrode and a zinc negative in an alkaline electrolyte was initially developed as a primary cell. It is rechargeable, however, and the air-zinc storage battery is being considered for vehicle propulsion because of its low cost and potentially high specific energy.

When the cell is discharged, the oxygen at the positive electrode is reduced to water and the zinc on the negative electrode is oxidized to zinc hydroxide. The reverse reactions occur when the cell is charged; thus,

$$\underset{(+)}{\frac{1}{2}O_2(\text{air})} + \underset{(-)}{Zn} + 2H_2O \underset{\text{charge}}{\overset{\text{discharge}}{\rightleftharpoons}} \underset{(+)}{H_2O} + \underset{(-)}{Zn(OH)_2} \quad ...(10.38)$$

Part of the zinc hydroxide dissolves in the alkaline electrolyte to form zincate ions, as noted earlier. The cell emf for low discharge current is close to 1 volt, but it decreases as the discharge rate is increased.

The chief drawback to the zinc-air battery has been its short, life, due mainly to deleterious changes in the zinc electrode. Solution to this problem have been proposed based on a unique feature of storage cells with air electrodes. When other cells are charged, both positive and negative electrodes must be restored; in air-electrode cells, however, the positive material (*i.e.*, oxygen) is always available from the ambient air and only the negative electrode needs restoring.

Chlorine Electrode Cells (chlorine zinc battery)

The chlorine (gas) positive electrode offers the prospect of storage cells with high emf and specific energy using relatively low cost materials. A battery with a chlorine positive and a zinc negative electrode in an aqueous zinc chloride solution as electrolyte has an emf of about 2 volts. A specific energy of 80 to 100 W-hr/kg has been predicted, but this apparently does not allow for the weight of auxiliary equipment for refrigeration and pumping. A zinc chlorine battery has been used successfully to propel a demonstration road vehicle.

The chlorine electrode consists of chlorine gas dissolved in the electrolyte, with a suitable conductor, such as graphite (carbon), to provide electrical contact. The negative electrode is also of graphite on which metallic zinc is deposited. The charge and discharge reactions are then

$$\underset{(+)}{Cl_2} + \underset{(-)}{Zn} \underset{\text{charge}}{\overset{\text{discharge}}{\rightleftharpoons}} \underset{(+)}{2Cl^-} + \underset{(-)}{Zn^{2+}} \rightleftharpoons ZnCl_2 \quad ...(10.39)$$

where the zinc (Zn^{2+}) and chlorine (Cl^-) ions form zinc chloride ($ZnCl_2$) in solution. Formation of dendritic zinc growths is much less in zinc chloride solution than in the alkaline cells described earlier.

The problem with the zinc chlorine cell is the storage and supply of chlorine gas one proposed solution is based on the use of two compartments; one compartment contains the electrodes and the zinc chloride electrolyte solution, whereas the other, which can be refrigerated, provides storage of the chlorine as the solid hydrate, usually Cl_2, $6H_2O$, at a temperature below (9°C). When the cell is charged, metallic zinc is deposited on the negative electrode and chlorine gas is released at the positive electrode. The chlorine passes to the cooled storage compartment where it comes in contact with a small quantity of zinc chloride solution, and the solid hydrate is formed and retained.

To discharge the cell and draw current from it, the positive electrode must be kept supplied with chlorine. This is done by pumping warm electrolyte solution from the electrode compartment to the storage compartment. As the temperature in the latter increases, chlorine is released from the hydrate and is carried by the electrolyte back to the electrode compartment.

The highly corrosive nature of wet chlorine (*e.g.*, in solution) is a serious drawback to the zinc-chlorine cell. Titanium is one of few metals that is resistant to attack by wet chlorine, but it is expensive. Possible plastic materials may be found that can withstand the action of chlorine.

High Temperature Batteries

Storage cells (and batteries) operating at high temperatures, with non-aqueous electrolytes, are being studied in many countries for possible use in electric vehicles and for utility energy storage. For a variety of reasons, including the absence of water from the electrolyte, the low densities of the active materials, and the fairly high emfs, some of the high temperature cells have specific energies as high as 150 to 200 kW-hr/kg. The most promising cells of this type are the lithium alloy metal sulfide and sodium sulphur systems. Other high temperature cells have been proposed but they have received less attention.

Lithium (Aluminium) Alloy Metal Sulphide Cells. A study of the lithium sulphur system for a high temperature cell was initiated at the Argonne National Laboratory (ANL) U.S.A in 1968. In the earliest cells, the negative electrode was liquid lithium retained in a porous metallic support; the positive electrode was molten sulphur. The electrolyte was a liquid eutectic (*i.e.*, minimum melting point) mixture of lithium chloride (about 60 molecular percent) and potassium chloride melting at 352°C. The operating temperature of the cell was about 400°C. However, significant losses were experienced from both lithium and sulphur electrodes.

The Liquid lithium negative electrode was later on replaced by a solid lithium-aluminium allow, and a solid metallic sulfide, iron sulphide (FeS or FeS_2) in a particular is the positive electrode material. The electrolyte is the lithium chloride potassium chloride eutectic mixture at a temperature of 400°C to 450°C. The emf of the cell is less than that of the original lithium sulphur cell, but, the advantages of the lithium (aluminium)—iron sulphide cell are sufficient to compensate for the lower voltage (and specific energy).

The electrochemical reactions in the cell are very complex and are given here in the simplified terms. When the cell is discharged, the lithium at the negative electrode forms lithium ions (Li^+). These ions migrate through the molten electrolyte to the positive electrode where they react with the iron sulphur to form lithium sulfide (Li_2S) and iron. The reverse processes take place when the cell is charged. It is to be noted that with the high temperature cells described here, it is more convenient to consider the negative electrode material before the positive material, this is the reverse of the procedure used in low temperature cells.

In the active positive electrode material is iron disulfide (FeS_2), the electrochemical reactions occur in two stages; one is

$$\underset{(-)}{2Li} + \underset{(+)}{FeS_2} \underset{\text{charge}}{\overset{\text{discharge}}{\rightleftharpoons}} \underset{(-)}{Li_2S} + \underset{(+)}{FeS} \quad ...(10.40)$$

for which the cell emf is close to 1.6 volts; the other, involving the ferrous (FeS), is

$$\underset{(-)}{2Li} + \underset{(+)}{FeS} \underset{\text{charge}}{\overset{\text{discharge}}{\rightleftharpoons}} \underset{(-)}{Li_2S} + \underset{(+)}{FeS} \quad ...(10.41)$$

at an emf of 1.3 volts. Thus, a cell with iron disulfide as the positive electrode material has discharge levels (or plateaus) at 1.6 and 1.3 volts.

The high operating temperature of the battery is attained initially or after standing idle by means of an external electrical resistance heater. During operation, either when discharging or charging, heat generated within the cell as a result of its internal resistance serves to maintain the temperature. A large battery operating at high power may then require an air blower to prevent over-heating.

Sodium Sulphur Cell. This type of battery was initiated by the Ford Motor Company. Because of its inherent simplicity, the cell has attracted wide interest, especially for vehicle propulsion. The sodium-sulphur battery could be also utilized for stationary energy storage systems.

The operating temperature of the cell is between 300 and 400°C, which is the well above the melting point of sodium and sulphur; hence both active electrode materials are in the liquid state. The sodium sulphur cell is unusual in this respect as well as in having a solid electrolyte that also serves to separate the liquid electrode materials.

In most designs, the electrolyte is a ceramic called beta alumina, consisting mainly of sodium oxide (Na_2O) and alumina (Al_2O_3). The general formula of beta alumina is Na_2O. x Al_2O_3, where x is in the range of about 5 to 11. Sometimes small amounts of lithium oxide (Li_2O) or other oxide may be added to improve the strength and the electrical conductivity. At the cell operating temperature, sodium ions can move through the beta-alumina in the same direction as the current. Borate glass which is also a sodium ion conductor, is used as an electrolyte.

When a sodium sulphur cell is discharged, the sodium forms sodium ions (Na^+) at the negative electrode, these ions travel through the solid electrolyte to the positive electrode where they react with the sulphur. The process is reversed when the cell is charged. In the simplest terms, the electrochemical reactions may be represented by:

$$\underset{(-)}{2Na} + \underset{(+)}{S} \underset{\text{charge}}{\overset{\text{discharge}}{\rightleftharpoons}} Na_2S \quad ...(10.42)$$

In practice, however, the sodium sulphide (Na_2S) interacts with excess sulphur at the positive electrode to form poly sulphides having the general formula Na_2S_x, where x usually ranges from 5 (fully charged) to 3 (discharged). Within the range, the sodium poly sulphide is soluble in the liquid sulphur when x is less than 3, a solid poly sulphide separates and interferes with the cell operation. The cell emf is about 2 volts for very small discharge currents, but it is less for larger currents. Overall specific energies of more than 100 W hr/kg should be possible for sodium sulphur batteries.

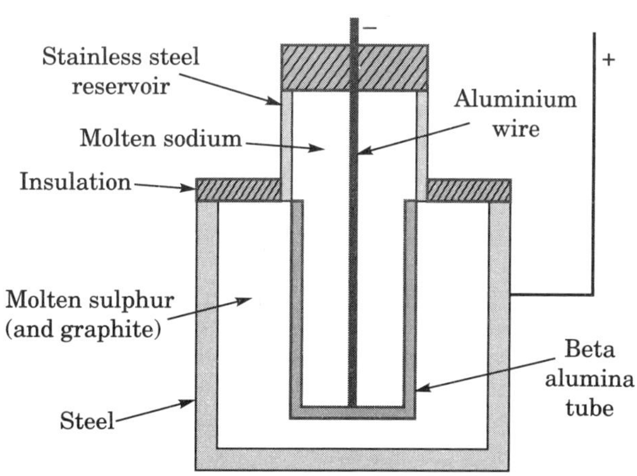

Fig. 10.17. Sodium sulphur cell with beta alumina electrolyte separator.

In a common cell design, a tube mode of beta alumina is sealed to the open bottom of a stainless steel reservoir, both tube and reservoir contain molten sodium. Since sodium is a good electrical conductor, it serves as its own current collector. An aluminium wire inserted into the sodium, which does not attack aluminium, provides the negative electrode connection.

The beta alumina tube is surrounded by molten sulphur in a suitable metal possibly stainless steel, container. Sulphur is a very poor electrical conductor, and hence an effective current collector is required. Carbon (graphite) in the form of felt, foam or specially shaped structures, immersed in the sulphur has been generally used for this purpose. The metal container then provides the positive electrode connection (See Fig. 10.17). The high operating temperatures are attained in the same way as described earlier for lithium (aluminium)—iron sulphide cells.

The major *characteristics* of the sodium sulphur cell are:

(*i*) High open circuit and operating voltage: 1.8-2.1 V.

(*ii*) High specific energy: 200 to 300 Wh/kg.

(*iii*) High energy density: 150 to 250 kWh/m^3.

(*iv*) High power density or discharge rate: 1 to 3 hour rate.

(*v*) Hermetically scaled construction.

(*vi*) No gas evolution.

(*vii*) Zero-self discharge or excellent charge retention.

(*viii*) High recharge efficiency; 85% at three hour rate.

(*ix*) Liquid reactants: No aging of active materials.

(*x*) Potentially five to ten year life.

Limitations of the sodium-sulphur battery are:

(*i*) Operating temperatures 300—400°C.

(*ii*) Possible aging of solid electrolyte deterioration of mechanical and electrical properties on cycling.

(*iii*) High reactive components safety considerations, corrosion.

(*iv*) No overcharge capability: Need for voltage control when charging.

The corrosion problem, which is caused by the hot, molten polysulfide on the metal container, can be minimized by the use of refractory liners with good electrical conductivity *e.g.* graphite or chromium oxide, Cr_2O_3. Another solution to the problem is to reverse the electrode positions, the molten sulfur (poly sulfide) is held in the interior of the beta-alumina electrolyte tube with the sodium on the outside in stainless steel container.

The future of the sodium-sulphur battery depends mainly on materials development. In particle, commercial production at a reasonable cost of beta alumina with adequate electrical conductivity, mechanical strength, and life time is essential. In addition, a low-cost, corrosion-resistant material is required for sealing the stainless steel sodium reservoir to the beta-alumina tube.

10.3.7. Advantages of Batteries for Bulk Energy Storage

Advancement in batteries development has brought the battery field for utility load levelling and electric vehicles. The benefits and advantages of batteries for bulk energy storage can be summarized as follows:

(*i*) Mitigation of oil shortage and import problems.

(*ii*) Lower cost of electric energy—as the result of saving in fuel cost and improved utilization in capital equipment.

(*iii*) Modulability and lower capital cost of batteries.

(*iv*) Environmental pollution abatement of air and noise, and aesthetic conservation.

(*v*) Substantive savings in power transmission.

(*vi*) Shorter lead time for construction.

(*vii*) Reliability in power emergency and regulation.

QUESTIONS

1. What is a fuel cell? Describe the principle of working of a fuel cell with reference to H_2—O_2 cell.
2. Derive an expression for emf of a fuel cell.
3. Derive the expressions for free energy and potential of the fuel.
4. (*a*) Describe with the help of a neat sketch the various steps in reaction mechanism of a $H_2.O_2$ fuel cell. What are the different polarization that occur in such cells.

   (*b*) Derive the following relations for fuel cell.

$$\Delta S = nF\left(\frac{\delta E}{\delta T}\right)_p, \; T\left(\frac{\delta E}{\delta T}\right)_p = \frac{\Delta H}{nF} + E.$$

5. Find the reversible voltage for the methanol fuel cell having the reaction.

$$CH_3OH + 1.5\ O_2 = CO_2 + 2\ H_2O\ (l) \quad \ldots(i)$$

$$CH_3OH + H_2O \rightleftharpoons CO_2 + 6H^+ + 6e^- \quad \ldots(ii)$$

   Given ΔG_{25} °C = – 167,340 Cal/gm-mole. **[Ans.** 1.21 volt]

6. (*a*) What is the principle of fuel cell. Discuss problems associated with operation of fuel cell.

   (*b*) Find the output voltage and theoretical conversion efficiency of an oxygen hydrogen fuel cell operating at 600°C. Assume that the hydrogen is supplied at 1.1 atm and oxygen is supplied from air at 1.2 atm and assume that the steam produced is at 1 atm. Take:

$$\Delta G° \text{ (for steam)} = 228 \times 10^6 \text{ J/kg mole } H_2O$$

$$\Delta H° \text{ (for steam)} = 241 \times 10^6 \text{ J/kg mole } H_2O$$

$$\Delta H° \text{ (for water)} = 286 \times 10^s \text{ J/kg mole } H_2O$$

$$F = 9.65 \times 10^6 \text{ C/kg mole electrons.}$$

7. What is polarisation? List different types of polarisations that occur in fuel cells. Show how does the electrode structure help in reducing the polarisation?
8. What are the advantages and disadvantages of a fuel cell?
9. Write short notes on:

   (*a*) Applications of fuel cell.

   (*b*) Types of electrodes for a fuel cell.

   (*c*) Advantages of batteries for bulk energy storage.
10. What is the difference between a fuel cell and a battery?
11. Explain the principle of working of a battery. Describe a lead acid battery.
12. What are the different arrangements of a storage battery, describe briefly.
13. What are the different classes of a batteries?

    Describe briefly a high temperature battery for road vehicles.

CHAPTER 11

Hydrogen Energy

11.1. Introduction

Hydrogen as an energy carrier can play an important role as an alternative to conventional fuels, provided its technical problems of production, storage and transportation can be resolved satisfactorily and the cost could be brought down to acceptable limits. One of the most attractive features of hydrogen as an energy carrier is that it can be produced from water which is abundantly available in nature. Hydrogen has the highest energy content per unit of mass of any chemical fuel and can be substituted for hydrocarbons in a broad range of applications, often with the increased combustion efficiency. Its burning process is non-polluting and it can be used in the fuel cells to produce both electricity and useful heat. Use of hydrogen as an energy source involves five basic issues: (1) Production, (2) Storage and transportation, (3) Utilization, (4) Safety and management, (5) Economy. These are discussed in the following articles.

The hydrogen can be used as a fuel directly, or it might be used as a raw material to produce methanol, ammonia, or hydrocarbons by using either carbon dioxide or nitrogen from the atmosphere.

The combination of hydrogen with oxygen (*e.g.,* from air) results in the liberation of energy, with water as the sole material product; thus,

$$H_2 + \frac{1}{2} O_2 \longrightarrow H_2O + \text{energy} \qquad ...(11.1)$$

The reaction can be carried out and the energy made available in several different ways, so that hydrogen is a versatile fuel material. Hydrogen is chemically very reactive and hence it is not found in its free state on the earth. However, combined chemically with other elements, it is present in water, fossil hydrocarbons, biological materials such as cellulose, starch etc., minerals such as biocarbonate rocks. Energy must be supplied to those compounds to break the chemical bonds to release hydrogen. Unlike *fossil* (coal, natural gas and petroleum) and *nuclear fuels* and solar radiation, which are primary energy sources, hydrogen is a secondary fuel that is produced by utilizing energy from a primary source. Much of this energy can be recovered by recombination of the hydrogen with oxygen in the ways indicated above.

The current and projected needs of hydrogen are growing rapidly, applications include nitrogeneous fertilizers, coal liquefaction basic chemicals etc. Longer term applications are for electric power producing fuel cells. It can be used as a fuel for *a gas turbine* or for a *spark-ignition engine.* Hydrogen is ecologically an ideal fuel for transportation purposes. Its heating value of 28000 kcal/kg is almost three times that of hydrocarbon fuels. So when used in air crafts, it reduces the take-off load considerably.

The possible areas of use for hydrogen in the near future are as follows:

(*i*) production of electrolytic hydrogen, for full load exploitation of nuclear power stations;

(*ii*) direct addition of hydrogen to the existing natural gas distribution network;

(*iii*) use of hydrogen in the processing of heavy oil;

(*iv*) use of hydrogen for the manufacture of synthetic liquid or gaseous fuels;

(*v*) direct use of hydrogen as a motor vehicle fuel in urban transport, particularly where air pollution problems are already critical;

(*vi*) reduction of iron oxides by means of hydrogen in the steel industry;

(*vii*) direct use of hydrogen as an aircraft fuel in air transport.

While hydrogen can be produced from fossil hydrocarbons it is desirable to use processes which recover hydrogen by a closed cycle water splitting without the use of much hydrocarbons, Such an approach has many advantages because hydrocarbons are not a renewable source and their supply is becoming increasingly unreliable. The cost of crude oil is escalating day by day and in the near future alternative technologies will be competitive.

Several cycles have been suggested for closed cycle water splitting. The basic Westing houso Electro-chemical thermal sulfur cycle is one such promising cycle which has received some attention in production of hydrogen.

The simplest practical way to obtain hydrogen from water is means of electrical energy. Hence electrical energy can be conveniently stored and transmitted by way of hydrogen in some situations. Electricity is generated from distributed primary sources, such as geothermal energy, wind energy and some other forms of solar energy, often at a distance from a load centre or an electric utility grid. Conversion into hydrogen fuel would then be a practical means of energy transmission.

One of the potential advantages of hydrogen as a secondary fuel is that it can be transmitted and distributed by pipe line in much the same way as natural gas. Alternatively, the hydrogen gas could be converted into liquid form by cooling to a very low temperature and transported in insulated tanks by high way railroad. On a smaller scale, hydrogen gas could be compressed into transportable cylinders. Another possibility is to form a solid compound of a metal with hydrogen from which the gas can be recovered, when required, by heating. Thus, hydrogen can serve as a means of carrying energy from the place where a primary source is available to a distant load centre where the energy is used.

An important aspect of hydrogen utilization is that it is accompanied by little or no atmospheric pollution. In many cases, such when hydrogen is burned in pure oxygen, in a flameless (catalytic) burner in air, or in a fuel cell with air or oxygen, water is the sole product. However, if hydrogen is burned normally in air or in a spark-ignition engine or a gas turbine, *nitrogen* oxides may be produced by combination at high temperature of the nitrogen and oxygen in the air. By using an excess of air relative to the hydrogen consumed or by introducing water vapour, the flame temperature can be reduced to the point where nitrogen oxide formation is small.

Properties of Hydrogen

Hydrogen at ordinary temperature and pressure is a light gas with a density only 1/14th that of air and 1/9th that of natural gas under the same conditions. By cooling to the extremely low temperature of – 253°C at atmospheric pressure, the gas is condensed to a liquid with a specific gravity of 0.07, roughly 1/10th that of gasoline.

The standard *heating value* of hydrogen gas is 12.1 MJ/cu m compared with an average of 38.3 MJ/cu m for natural gas. The heating value of liquid hydrogen is 120 MJ/kg or 8400 MJ/cum; the corresponding value for gasoline (or approximately for jet fuel) is 44 MJ/kg or 32,000 MJ/cum. Hence, for producing a specific amount of energy, liquid hydrogen is superior to gasoline (or jet fuel) on a weight basis but inferior on a volume basis.

The flame speed of hydrogen burning in air is much greater than for natural gas, and the energy required to initiate combustion (*i.e.,* the ignition energy) is less. One consequence of the low ignition energy is that flameless combustion on a catalytic (finely divided metal) surface is possible with hydrogen at much lower temperatures than flame burning.

Mixture of hydrogen and air are combustible over an exceptionally wide range of compositions; thus, the flammability limits at ordinary temperatures extend from 4 to 74% by volume of hydrogen in air. (Detonation can occur between 18 and 59 per cent). This wide range has an important bearing on the use of hydrogen fuel in internal combustion engines. The engine will operate, although not necessarily with the same efficiency, from very rich (excess fuel) to very lean (excess air) mixtures. The adjustment of air to fuel ratio is thus much less critical than in a gasoline engine.

11.2. Hydrogen Production

11.2.1. Introduction

Commercial production of hydrogen today is carried out by the steam reformation or partial oxidation of hydrocarbons (natural gas, naphthas, or crude oil) depending on supplies and economics. Clearly, with gas and oil in diminishing supply, these sources for hydrogen will become less attractive as a basis for an energy delivery system.

Hydrogen made from coal is of questionable importance as a fuel gas. On one hand, it is relatively inexpensive (compared with the sources we discuss next), but on the other hand, it would not be as conveniently transported or utilized in transmission and combustion systems as would be a synthetic conventional fuel such as methane or liquid hydrocarbons (which are probably as easy to make from coal as is pure hydrogen). Hydrogen made from coal can probably only be justified as a fuel for special applications where the unique characteristics of hydrogen can be put to advantage, such as its weight or its non-polluting characteristics.

Small quantities of hydrogen are commercially produced today by the *electrolysis* of water usually in situations where electric power is cheap, or where reliable unattended operations is required. Electrolysis is the only presently available technology by which hydrogen can be made from non-fossil energy (nuclear, solar, geothermal, or wind), and it is this aspect that makes electrolysis so important in the context of a hydrogen energy system.

Electrolysis suffers from a disadvantage as an energy conversion system because its raw energy requirement must be supplied as a high quality energy form: electricity. Considerable research work is underway to develop a thermal cycle that would utilise heat to achieve the chemical splitting of water to its elements without the need for intermediate electricity generation, and without the need to use the extremity high temperature of 2500°C or more, (which would be required to dissociate water directly). This so called *thermochemical* hydrogen production. Concept is an important building block of a hydrogen energy system because it offers the potential for non-fossil hydrogen production at a higher efficiency and lower cost than electricity generation by electrolysis. At present, research is at the stage where these potentials have not been realized in practice, but progress is sufficiently encouraging to continue.

The methods of producing hydrogen may be classified according to the immediate source of addition of energy to decompose, thus *electrical energy* (in electrolysis), *heat energy* (in thermo-chemical methods), *fossil fuels,* and *solar energy.*

11.2.2. Electrolysis or the Electrolytic Production of Hydrogen

The process of splitting water into hydrogen and oxygen by means of a direct electric current is known as *electrolysis*; this is the simplest method of hydrogen production. In principle, an electrolysis cell consists of two electrodes, commonly flat metal or carbon plates, immersed in an aqueous conducting solution called the electrolyte. A source of direct current voltage is connected to the electrodes so that an electric current flows through the electrolyte from the positive electrode (or anode) to the negative electrode (or cathode). As a result, the water in the electrolyte solution is decomposed into hydrogen gas (H_2) which is released at the cathode, and oxygen gas (O_2); released at the anode. Although only the water is split, an electrolyte *(e.g.,* KOH solution) is required because water itself is a very poor conductor of electricity.

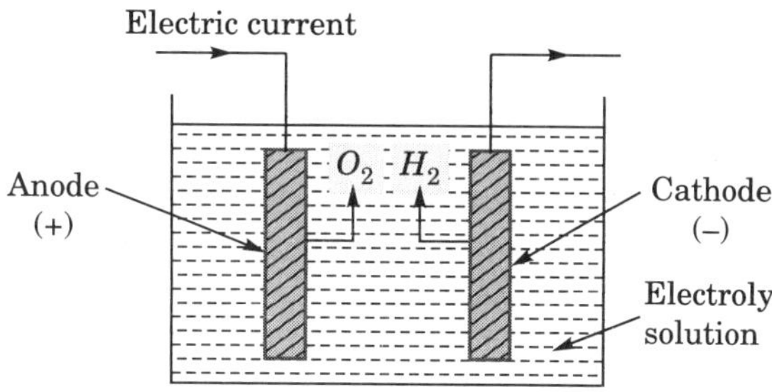

Fig. 11.1. Simple electrolytic cell.

Ideally, a voltage of 1.23 volts should be sufficient for the electrolysis of water at normal temperature and pressure. For various reasons, especially the slowness of the electrode processes that lead to the liberation of hydrogen and oxygen gases, higher voltages are required to decompose water. The decomposition voltage increases with the current density (*i.e.,* the current per unit (area of electrode). Since the rate of hydrogen production is proportional to the current strength, a high operating current density is necessary for economic reasons. Hence, in practices the decomposition voltage (per cell) is usually around 2 volts.

Theoretically, 2.8 kW-hr of electrical energy should produce one cu.m. of hydrogen gas. Because of the higher than ideal decomposition voltage, however the actual electrical energy requirement is generally from 3.9 to 4.6 kW-hr per cu.m.). This means that the efficiency of electrolysis (*i.e.,* the proportion of the energy supplied that is used in electrolysis) is roughly 60 to 70%.

The electrolysis efficiency can be increased by decreasing the decomposition voltage for a given current density. To achieve this, the electrode surface must be able to catalyze (*i.e.,* expedite) the electrode processes. One of the best catalyst is platinum in a finely divided form, deposited on a metal base. However, because of the high cost of platinum, other electrode surface materials are used commercially. For practical water electrolysis, the electrodes are generally of nickel-plated steel. The effective electrode surface area (and hence the rate of the electrode process) is increased by depositing porous nickel on a wire gauge, or a highly corrugated steel base. Research is being directed at the development of improved electrodes that will give better electrolysis efficiency at a reasonable cost.

Diaphragms prevent electronic contact between adjacent electrodes and passage of dissolved gas or gas bubble; from one electrode compartment to another (leading to a decrease in current efficiency and possible to explosions), without themselves offering an appreciable resistance to the passage of current within the electrolyte. Dissolved gas crossover is serious only in pressure operations; to prevent the passage of gas bubbles, the diaphragm must consist of small pores whose capillary pressure is greater than the maximum differential pressure applied across the cell.

Asbestos is the most common material for cell diaphragms. At atmospheric pressure, woven asbestos cloth is used, sometimes with fine nickel wire to support the structure. Pressure electrolyzers usually have a mat made of woven or felted asbestos fibres that produces a fine pore structure, giving a higher resistance to the generation of gases. This mat is sometimes supported by the electrodes.

Three major factors determine the usefulness of an electrochemi cal cell for hydrogen production. One is the energy efficiency, related to the cell's operating voltage: another is the capital cost of the plant, related to the rate of hydrogen production from a cell of a given size. These two factors are closely inter related. The third factor is the life time of the cell and its maintenance requirements, which involve the materials used in its construction and the operating conditions selected.

A number of advantages can be gained from operating an electrolyzer at higher pressures, including (*a*) reduction in specific power consumption, (*b*) delivery of gas at pressure, thus reducing or eliminating the cost of gas compressors; and (*c*) reduction in the size of electrolysis cells. It can be shown theoretically that the reversible cell voltage increases with pressure. However, as decreased gas volume and higher operating pressures result in a reduced over potentially there is usually a small overall reduction in the cell voltage. This real gain in efficiency is offset by increase in the cost of pressure vessels or stronger components. Operating voltages as stated above can be lowered for a given current by using electrodes that carry precious metal catalysts or incorporate sophisticated metallurgical structures, both of which are expensive but increase efficiency.

Several large electrolytic hydrogen plants, consuming over 100 MW, have been operated successfully, while thousands of smaller units are in use for special applications.

Two types of electrode arrangements are used by industry for the electrolysis of water. On this basis they are classified as:

(*i*) Tank type electrolyzer, and

(*ii*) Filter press or bipolar electrolyzer.

(*i*) Tank Type Electrolyzer. The first applications of water electrolysis in industry used the tank electrolyzer, in which a series of electrodes, alternating anodes (+) and cathode (–) are suspended vertically and parallel to one another in a tank filled (most commonly) with a 20-30% solution of potassium hydroxide in demineralized water. Alternate electrodes, usually the cathodes, are surrounded

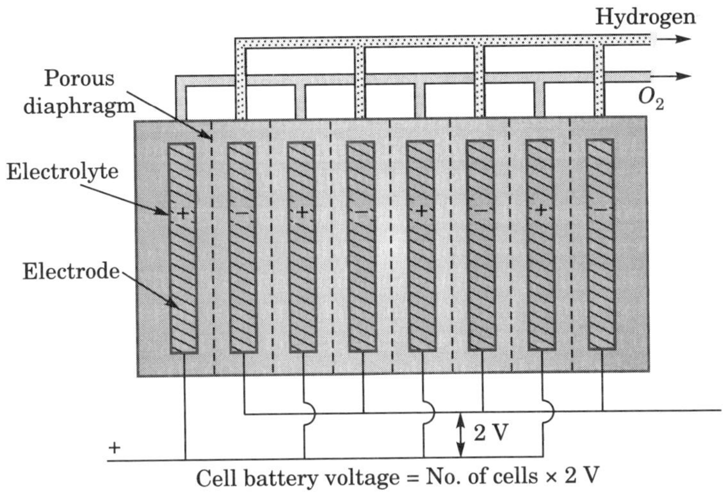

Fig. 11.2. Schematic diagram of tank type (unipolar) electrolyzer.

by porous diaphragms (*e.g.*, asbestos) impermeable to gas but permeable to the cell's electrolyte, that prevent the passage of gas free from one electrode compartment to another. The whole assembly is hung from a series of gas collectors. All the anodes in the tank are connected to the same positive terminal of the direct current voltage source, and all the cathodes are connected to the same negative terminal (See Fig. 11.2).

By connecting the electrodes in parallel in this manner the voltage required for a tank of several electrode pairs, regardless of the number is little more than for a single pair (or cell), that is, about 2 volts. As a rule, a number of tanks are connected in series: the operating voltage is then roughly $2n$ volts, where n is the number of tanks so connected.

The major *advantages* of tank type electrolyzers are two-fold relatively few parts are required and those needed are relatively inexpensive; and individual cells may be isolated for repair or replacement simply by short circuiting the two adjacent cells with a temporary busbar connection. Some *disadvantages* of the tank electrolyzers are:

(*a*) Inability to handle high current densities because of cheaper component parts and

(*b*) Inability to operate at high temperatures because of heat losses from the large surface areas of connected cells.

(*ii*) Filter-Press Electrolyzer (Bipolar Electrolyzer).

The alternative and more widely used water electrolysis system is called the filter-press electrolyzer because of its superficial resemblance to a filter press. Except at the ends of the cell, the electrodes are bipolar that is, one face of each plate electrode is an anode and the other face is the cathode. (Refer Fig. 11.3). As in the tank system, porous diaphragms between adjacent electrodes prevent mixing of the hydrogen and oxygen gases.

In the filter press electrolyzer the cells are connected in series with an anode at one end and a cathode at the other end of the series. The total voltage required is then approximately 2 C volts, where C is the number of cells in the series. The cells which are connected in series, their individual cell voltages are additive within a battery. Because the cells can be made relatively thin, a large gas output is achieved from a relatively small volumes.

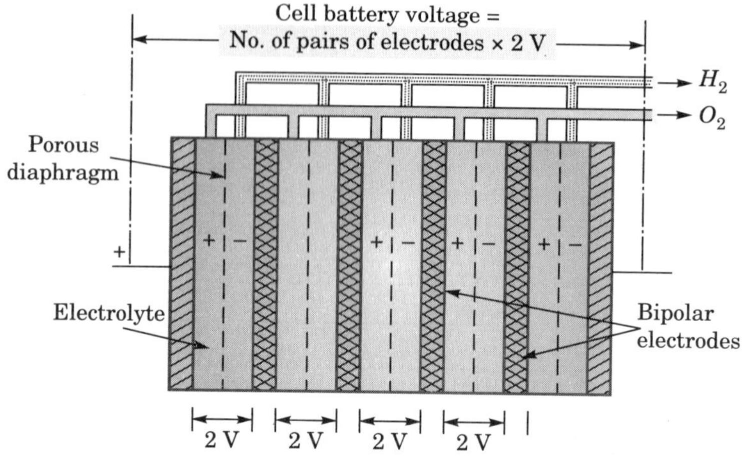

Fig. 11.3. Filter press (Bipolar) cell construction.

It is usually desirable to circulate electrolyte through the cell, thereby separating the gas and the electrolyte, and in many designs, this is accomplished in a separating drum mounted on top of the electrolyzer. The electrolyte, free of gas, is recirculated through the cells, and the circulation is maintained by gas lift of the generated oxygen and hydrogen.

The filter press electrolyzer is generally preferred, because it occupies less space and can be operated at a higher current density than the tank type. The economics are thus more favourable, since a larger hydrogen production is possible in a plant size.

Although filter press electrolyzers can operate at higher current densities and appear to occupy relatively less space than tank types, they require a much closer tolerance in construction and are more difficult to maintain. Breakdowns in filter press electrolyzers are rare, but when they occur, rejuvenation is difficult and repair may take considerable time. If an individual asbestos diaphragm is damaged, the entire battery must be dismantled. The greater capital costs of bipolar electrolyzers are offset because these electrolyzers can operate at higher current densities (producing more hydrogen per unit area of electrode) with virtually the same operating voltages as the tank type unit.

Electrolyzer Modifications. In addition to the search for improved electrodes, efforts are being made to increase electrolyzer efficiencies. For example, operation at temperatures and pressures above normal is reported to decrease the decomposition voltage. In a commercial German design, for which a high efficiency is claimed the pressure in the cell is maintained at some 50 atm. (5 MPa) and the electrolyte temperature is 90°C.

General Electric Company of Lynn. Massachusetts (U.S.A.), has been developing a water electrolysis system based on solid polymer electrolyte (SPE) fuel cell technology. SPE fuel cells were first used in space exploration during the Gemini program, where they provided primary on board power for seven of the space craft flights.

The SPE is a thin, solid, plastic sheet of perfluorinated sulfonic acid polymer having many of the physical characteristics of Teflon. Unlike Teflon, however, when a thin sheet of this material is saturated with water, it is an excellent ionic conductor, providing low electrical resistance. Used in an electrolysis cell, it is the only electrolyte required; there are no liquid acids or alkaline in the system. Hydrated hydrogen ions ($H^+ . xH_2O$) move through the sheet of the electrolyte.

A cell consists of a thin sheet of SPE with a finely divided precious metal (*e.g.*, platinum) coated on each side to catalyze the electrode processes. Metal plates pressed against the surfaces of the SPE provide electrical contact with the coatings and serves as electrodes for connection with voltage source.

The electrolysis of water is of interest as an economic industrial scale method of producing hydrogen in cases where (*i*) fossil fuels such as petroleum, natural gas and coal (from which hydrogen can also be obtained) are scarce or expensive, or both, and (*ii*) electrical energy from sources other than fossil fuels is available in sufficient quantity and relatively cheaply.

Nuclear power stations have to be operated with a load which is constant in time; hence, during the seasons or times of day when there is a decreased demand for electricity the energy produced is particularly cheap and could be used to produce hydrogen.

So far, water electrolysis technology has been developed to the point where extremely robust and reliable electrolysers are available which operate without excess pressure or under a pressure of 30 bar. However, most electrolysers operate with relatively high cell voltages (between 1.8 and 2.2 V). Furthermore these electrolysers are operated with relatively modest current densities of 0.15 and 0.30 A/cm^2.

High cell voltages imply high energy costs, and low current densities imply that the electrolysers give small yields per unit time, leading to relatively high capital costs. For these reasons hydrogen produced electrolytically is at present twice as expensive as hydrogen produced from fossil fuels.

11.2.3. Thermo-chemical Methods

The overall efficiency for the conversion of primary energy from fossil and nuclear fuels into hydrogen by electrolysis is dependent, in the first place on the net efficiency of generating electricity. This efficiency may be upto about 38% for modern fossil fuel plants and 32% for nuclear installations. Assuming that electrolyzer efficiency can be increased to 80%, the overall efficiency for hydrogen production would be only 25 to 30%.

A higher conversion efficiency might be possible if the heat produced by the primary fuel could be used directly to decompose water, without the intermediary of electrical energy. Such direct decomposition into hydrogen and oxygen is possible, but it requires a temperature of atleast 2500°C. Because of the temperature limitations and conversion process equipment, direct single step water decomposition can not be achieved. However a sequential chemical reaction series can be devised in which hydrogen and oxygen are produced, water is consumed and all other chemical intermediates are recycled. The operation is called a *thermochemical cycle,* it is so called because energy is supplied as heat at one or more of the chemical stages. In the reaction series, water is taken up at one stage, and hydrogen and oxygen are produced separately in different stages. The net result is the decomposition of water into hydrogen and oxygen. This multistep thermochemical method offers the potential for processes that could use high temperature nuclear heat and be contained in conventional or developable chemical equipment.

For practical reasons, primarily the availability structural and containment materials, the maximum temperature in a thermochemical cycle is considered to be about 950°C. Heat energy should then be convertible into hydrogen energy with an efficiency around 50%; this is a marked improvement over what is possible by electrolysis. Unless the upper temperature of the thermochemical cycle is above 700°C, the efficiency is little better than for electrolysis.

At present, no commercial process for the thermal splitting of water to hydrogen and oxygen is in operation. Several workers have proposed many multistep reaction sequence that thermally decompose water at lower overall temperatures. An example of such a reaction sequence is as follows:

$$2CrCl_2 + 2HCl \longrightarrow 2CrCl_3 + H_2 \quad (325°C) \quad ...(11.2)$$

$$2CrCl_3 \longrightarrow 2CrCl_2 + Cl_2 \quad (875°C) \quad ...(11.3)$$

$$H_2O + Cl_2 \longrightarrow 2HCl + \frac{1}{2}O_2 \quad (850°C) \quad ...(11.4)$$

As can be seen, in this reaction sequences, only water is split, all other materials are completely recycled.

The maximum temperatures available from newly designed high temperature gas cooled reactors (HTGRs) are about 800- 900°C. Some what higher temperatures may be available in the future from the conceptual, ultra-high temperature gas cooled reactors.

A very large number of thermochemical cycles have been proposed, based on chemical and thermodynamic considerations. Several of the more promising cycles are being investigated in many countries. Although the various stages in a cycle may be thermodynamically possible, there is no assurance, without experimental studies, that they will take place at a useful rate. Also several cycles involve highly corrosive substances, such as sulphur dioxide and trioxide, chlorine, and hydrogen bromide, containment problems may then be dominant on the whole, the thermochemical production of hydrogen is regarded as a long term project.

Some of the possible thermochemical cycles are given in Table 11.1.

Table 11.1. Possible Cycle for Hydrogen Production

S. No.	*Chemical Reaction*	*Temperature °C*
1.	$C + H_2O \longrightarrow CO + H_2$	725°C
	$CO + 2Fe_3O_4 \longrightarrow 3Fe_2O_3 + C$	225°C
	$3Fe_2O_3 \longrightarrow 2Fe_3O_4 + \frac{1}{2}O_2$	1425°C
2.	$Cd + H_2O \longrightarrow CdO + H_2$	125°C
	$CdO \longrightarrow Cd + \frac{1}{2}O_2$	1225°C
3.	$3FeCl_2 + 4H_2O \longrightarrow Fe_3O_4 + 6HCl + H_2$	600°C
	$Fe_3O_4 + \frac{1}{4}O_2 \longrightarrow \frac{3}{2}Fe_2O_3$	400°C
	$\frac{3}{2}Fe_2O_3 + 9HCl \longrightarrow 3FeCl_3 + \frac{9}{2}H_2O$	100°C
	$\frac{3}{2}FeCl_3 \longrightarrow 3FeCl_2 + \frac{3}{2}Cl_2$	400°C
	$\frac{3}{2}H_2O + \frac{3}{2}Cl_2 \longrightarrow 3HCl + \frac{3}{2}O_2$	800°C

4.	$Fe_3O_4 + 2H_2O + 3SO_2 \longrightarrow 3FeSO_4 + 2H_2$		125°C
	$3FeSO_4 \longrightarrow \frac{3}{2}Fe_2O_3 + \frac{3}{2}SO_2 + \frac{3}{2}SO_3$		725°C
	$\frac{3}{2}Fe_2O_3 + \frac{1}{2}SO_2 \longrightarrow Fe_3O_4 + \frac{1}{2}SO_3$		925°C
	$2SO_3 \longrightarrow SO_2 + O_2$		925°C
5.	$2Cu + 2HCl \longrightarrow 2CuCl + H_2$		100°C
	$4CuCl \longrightarrow 2CuCl_2 + 2Cu$		100°C
	$2CuCl_2 \longrightarrow 2CuCl + Cl_2$		600°C
	$Cl_2 + Mg(OH)_2 \longrightarrow MgCl_2 + H_2 + \frac{1}{2}O_2$		80°C
	$MgCl_2 + 2H_2O \longrightarrow Mg(OH)_2 + 2HCl$		350°C

11.2.4. Some Thermochemical cyclic processes

Numerous candidate cycles have been suggested during past few years. Here, we shall consider only three cyclic processes which are being worked on at the present time, and are at an advanced stage of development, or are of special importance. These cyclic processes should be mentioned particularly, for which demonstration models are already available, constructed mainly of glass or quartz and giving a continuous production of about 100 *l* hydrogen per hour. These are the Westinghouse sulphur cycle, the Ispra Mark 13 bromine—sulphur cycle and the General Atomic Co. iodine—sulphur process of these, the first two are hybrid processes.

(1) **Westinghouse Electrochemical Thermal Sulphur Cycle**

This is a two-step process, where in hydrogen and sulphur acid are produced electrolytically by the reaction of sulphurous acid and water. SO_2 is recovered by reducing SO_3 obtained from sulphuric acid at high temperature, oxides of sulphur serve as recycling intermediates within the process. The use of sulfur compounds results in several process advantages.

The process in the most general form consists of two chemical reactions: one for producing oxygen and the other for producing hydrogen. The production of oxygen occurs *via* thermal reduction of sulphur trioxide obtained from sulphuric acid.

$$H_2SO_4 \longrightarrow H_2O + \frac{1}{2}O_2 + SO_2 \qquad ...(11.5)$$

The equilibrium for reaction (11.5) lies to the right at temperatures above 1000°K. Catalysts are available for accelerating the rate of sulphur trioxide reduction to sulphur dioxide and oxygen. The process is completed by using sulphur dioxide from thermal reduction step to depolarise the anode of a water electrolyzer. The overall reaction occurring electro chemically is:

$$2H_2O + SO_2 \longrightarrow H_2 + H_2SO_4 \qquad ...(11.6)$$

Reaction (11.6) comprises of two individual reactions.

Cathode $2H^+ + 2e^- \longrightarrow H_2$ ...(11.7)

Anode $H_2SO_3 + H_2O \longrightarrow 2H^+ H_2SO_4 + 2e^-$ ...(11.8)

$(E° = -0.17 \text{ volt})$

It is clear by summing reactions (11.5) and (11.6) that the overall process decomposes water into hydrogen and oxygen and involves only sulphur oxides as recycling intermediates. Although electrical power is required in the electrolyzer, much smaller quantities than those necessary for conventional electrolysis are needed. The theoretical voltage to decompose water is 1.23 volts with many commercial electrolyzers requiring over 2 volts. The power requirements (0.17 volts at unit activity of reagents and products) are seen to be less than 15% of those required in conventional electrolysis. This aspect changes dramatically the heat and work requirements for water decomposition and leads to high thermal efficiencies.

Basic Process Description

The process is shown in Fig. 11.4. Hydrogen is generated electrolytically in an electrolysis cell which anodically oxidises sulphurous acid to sulphuric acid

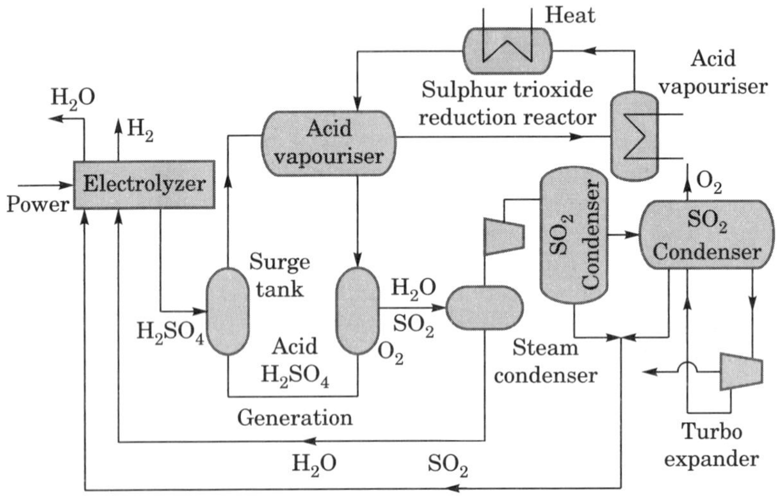

Fig. 11.4. Westinghouse sulphur cycle flowsheet schematic.

while simultaneously generating hydrogen at cathode. Sulphuric acid formed in the electrolyzer is sent to a tank from where it is fed to two vaporizers in series. The first of these is a recuperative heat exchanger heated by the effluent from the high temperature sulphur trioxide reduction reactor. The second is heated by helium from an ultra high temperature reactor. The sulphur trioxide steam mixture from the second vaporizer flows to the helium heated reduction reactor where sulphur dioxide and oxygen are formed. These gases are subsequently cooled against the incoming acid and unreacted sulphur trioxide is recovornd on sulphuric acid in knockout system. Steam is first condensed, following with the SO_2/O_2 mixture is compressed and sulphur dioxide recovered.

Bulk sulphur dioxide removal is accomplished by condensation against cooling water. Final removal is achieved by condensation against low temperature oxygen. This refrigeration and some auxiliary power production is generated by expansion of the oxygen stream prior to its venting.

Hence in this process sulphur dioxide in sulphuric acid solution is oxidized electrochemically to sulphuric acid, hydrogen being formed at cathode. The sulphuric acid is highly concentrated, decomposed at a high temperature, and the sulphur dioxide and water formed are used again in the process.

$$SO_2 + 2H_2O \xrightarrow{298-373\text{ K}} H_2SO_4 + H_2 \textbf{ electrolysis} \quad ...(11.9)$$

$$H_2SO_4 \xrightarrow{925-1125\text{ K}} H_2O + SO_2 + \frac{1}{2}O_2 \textbf{ thermolysis} \quad ...(11.10)$$

However, although at first sight the process appears to be simple, there are serious difficulties in carrying it out in practice. The electrochemical stage functions best in dilute sulphuric acid an optimum of 30 wt% has recently been cited whereas highly concentrated acid is most favourable for the high temperature stage. If, however, higher concentrations are used in the electrolytic cell both the resistance of the electrolyte and the anode potential increase considerably. Also, from a concentration of about 70-75% onwards there is reduction of the acid at the cathode to sulphur and hydrogen sulphide. All this makes it necessary to optimize the acid concentration for the two stages and to use considerable amounts of heat in producing highly concentrated acid.

There is a further problem connected with the electrolytic cell. In order to avoid losses of potential and of hydrogen at the cathode the cathode compartment must be kept free from sulphur dioxide. This can be achieved either by a higher static pressure of the electrolyte on the cathode side or by a specially constructed cell containing two cation exchange membranes, between which there flows an auxiliary electrolyte. The electrodes are made of porous graphite, graphite gauge or graphite felt. Both the anodic and the cathodic reactions must be catalysed.

(2) Ispra Mark 13 cycle

$$2HBr \xrightarrow{353-473\text{ K}} H_2 + Br_2 \textbf{ electrolysis} \qquad ...(11.11)$$

$$Br_2 + SO_2 + 2H_2O \xrightarrow{293-373\text{ K}} 2HBr + H_2SO_4\,(aq) \qquad ...(11.12)$$

$$H_2SO_4 \xrightarrow{925-1125\text{ K}} H_2O + SO_2 + \frac{1}{2}O_2 \textbf{ thermolysis} \qquad ...(11.13)$$

In the first, stage concentrated hydro bromine acid is electrolysed at about 373 K, giving hydrogen and bromine. The latter is separated by distillation, and in the second stage reacts with sulphur dioxide giving an aqueous solution of hydrogen bromide and sulphuric acid. One thus obtains gaseous hydrogen bromide and fairly concentrated sulphuric acid (about 75-80%), which can then be decomposed at high temperatures, either directly or after further concentration. It is well known that hydrogen bromide cannot be appreciably decomposed thermally in a single stage below 1300 K. As in the case of water, direct decomposition is possible only with the aid of electrical energy. If one compares the standard cell potentials for the two electrolyses, the small difference is rather discouraging (1.23 – 1.07 = 0.16 V). However, for hydrobromic acid it is possible to lower the reversible cell voltage markedly (by several 100 mV), notably by increasing the concentration of hydrogen bromide. Industrial realization of the electrolysis of hydrobromic acid has been well investigated.

Graphite is used as the electrode material. The over voltage for the discharge of hydrogen can be kept low by the addition of small amounts of noble metal salts. No catalyst is needed at the anode. The over voltage for the discharge of bromine is very small (in contrast to that for chlorine), and can be neglected in practice.

At 373 K cell voltages of 0.75—0.8 V at 2 kA/m^2 are obtained. Long-term experiments have shown that there are no obstacles to the industrial realization of the electrolysis of hydrobromic acid. The second stage raises no particular problems on a laboratory scale.

At the present stage it is difficult to make a comparison with the Westinghouse process. The somewhat lower cell voltage in the Westinghouse sulphur process is counterbalanced by the advantage of the higher concentration of sulphuric acid obtained in the Mark 13 process. It should also be mentioned that the incorporation of an additional element—bromine in the Mark 13 process may also be advantageous, since it offers more possibilities in the industrial operation of the process. For example, the cycle can be modified so that no free bromine is formed. Purification of the oxygen from sulphur dioxide or separation of the mixture of oxygen and sulphur dioxide can be achieved by the Bunsen reaction. Electrolysis can be carried out with either one or two cell cycles. A further reduction of the cell voltage can be obtained by carrying it out under pressure. Without departing from the conventional pressure range, the use of a precipitation technique for the sulphuric acid makes it possible to obtain at 373 K a cell voltage of 0.7 V at about 5 kA/m^2.

These considerations show that further work is necessary to determine which is the optimum process.

(3) Iodine-Sulphur Cycle

Among the purely thermochemical cyclic processes, those belonging to the iodine-sulphur family are of most interest at the present time. The following three stage process has been developed by the General Atomic Co. in particular:

$$2HI \xrightarrow{393-800\text{ K}} H_2 + I_2 \quad ...(11.14)$$

$$I_2 + SO_2 + 2H_2O \xrightarrow{370\text{ K}} H_2SO_4 + 2HI \quad ...(11.15)$$

$$H_2SO_4 \xrightarrow{1150\text{ K}} H_2O + SO_2 + \frac{1}{2}O_2 \quad ...(11.16)$$

The main difficulty in this process lies in the fact that, if side reactions are to be avoided, the two acids can be obtained only in dilute solution in the second stage, and are difficult to separate even on a laboratory scale. Separation by distillation is impossible without decomposition. The General Atomic Co. has developed a continuous process in which an excess of iodine is used to effect separation into approximately 56% sulphuric acid and hydriodic acid rich in iodine. After phosphoric acid has been added to the latter, treatment at a high temperature yield's liquid iodine and gaseous hydrogen iodide. The further working up of the solutions consumes a considerable amount of heat. Nevertheless, an overall efficiency of 47% is attained.

The difficulties associated with this process have led in recent years to numerous investigations and suggestions for modifying the process. For instance, scientists have attempted to carryout the second stage electrochemically. On short circuiting the cell the following electrode reactions occur:

$$\text{Cathode:} \quad I_2 + 2H^+ + 2e^- \longrightarrow 2HI \quad ...(11.17)$$

$$\text{Anode:} \quad SO_2 + 2H_2O \longrightarrow H_2SO_4 + 2H^+ + 2e^- \quad ...(11.18)$$

The anode and cathode compartments are separated by an ion exchange membrane. However, the current densities so far obtained have been too low, and the properties of both the ion exchange membranes used and the electrode materials have been satisfactory.

11.2.5. Fossil Fuel Methods

Mostly a gaseous mixture of carbon monoxide and hydrogen is formed in the first stage, in the production of hydrogen by using a fossil fuel (*i.e.,* natural gas, petroleum product, or coal). Such a mixture can be made by any method used for an intermediate heat value fuel gas, synthesis gas or water gas. The procedures in common use are steam reforming of methane or other hydrocarbon gas or light liquid hydrocarbon and partial oxidation of a heavier hydrocarbon in the presence of steam at a high temperature. In all these cases part of the hydrogen produced originates in the hydrocarbon.

To remove the carbon monoxide, the mixture is submitted to the water gas shift reaction with steam. The carbon monoxide is thereby converted into carbon dioxide with the formation of additional hydrogen.

$$CO + H_2O = CO_2 + H_2 + 1440 \text{ kJ/kg.} \qquad ...(11.19)$$

The carbon dioxide is an acid gas that can be absorbed in an alkaline medium. If the small amounts of carbon monoxide and dioxide remaining are undesirable, they can be converted into methane which can be separated as a liquid by cooling to a moderately low temperature.

Several processes proposed for converting coal into gaseous and liquid hydrocarbon fuels require a hydrogen rich gas. The hydrogen would then be made by reacting coal or *Char* obtained in the early stages of coal treatment, with steam and a limited amount of oxygen. The heat generated when the carbon in the coal (or char) reacts with oxygen produces the high temperature required for the carbon steam reaction. The product is a mixture of hydrogen and carbon monoxide. The carbon monoxide may be removed in the manner described above.

Air can be used instead of oxygen to supply the heat for the carbon steam reaction, but about half (by volume) of the product gas is inert nitrogen from the air. Although this procedure is more economical, the high proportion of nitrogen which can not be removed easily, is often a drawback. However, the iron steam process, described below is designed to use air, steam, and coal char to make hydrogen essentially free from nitrogen and also from carbon monoxide without, the need for water gas shifts.

This method, which depends on the reaction of steam with iron at a temperature of about 815°C at a pressure of 70 atm (7 MPa). The products are fairly pure hydrogen gas and a solid iron oxide; thus

$$\underset{\text{Iron + Steam}}{Fe + H_2O} \longrightarrow \underset{\text{Iron oxide + Hydrogen}}{FeO + H_2} \qquad ...(11.20)$$

The iron is recovered from the oxide in a separate vessel and returned for further reaction with steam. The conversion (reduction) of iron oxide to iron is achieved by means of a reducing-gas mixture of carbon monoxide, hydrogen, and nitrogen at a temperature of 1095°C made by the air-steam

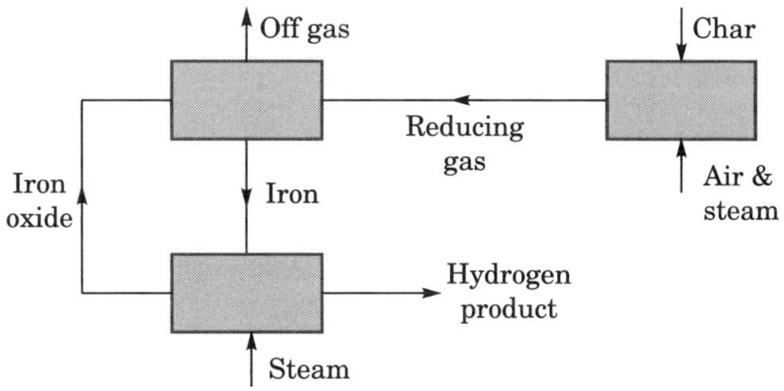

Fig. 11.5. Hydrogen production by the iron process.

chat process (refer Fig. 11.5). Carbon monoxide and nitrogen are absent from the product, as the iron-steam reaction occurs in separate vessel.

Coal gasification for the production of Hydrogen

Introduction. In the gasification of coal, there complete conversion of the organic part of the coal into gas, so that ash alone remains. This is done by reacting the coal with a gasifying agent, *e.g.,* steam above 700°C. The following account is concerned solely with coal gasification, the fundamental physical and chemical principles are explained first, end this is then followed by a discussion of the state of the art and also new developments.

Fundamental physical and chemical principles

The basic reactions are set out in Table 11.2. The main reaction is the heterogeneous water gas reaction (1), in which carbon is converted with steam to hydrogen and carbon monoxide. The reaction products can then react either with the gasifying agent or with each other in two further reactions the homogeneous water gas shift reaction, (2) and the methanation reaction, (3) with formation of carbon dioxide and methane. Processes (2) and (3) are exothermic, but the endothermic main reaction (1) is dominant, so that overall the gross reaction consisting of the sum of the three reactions listed under "steam gasification" is heat consuming, or endothermic.

Hydrogasiflcation (4), *i.e.,* the direct reaction of hydrogen with the carbon of the coal, is of great importance for the production of methane as substitute natural gas from coal. Lastely, the reaction with oxygen (5) and (6) also play a part in gasification techniques, on the one hand serving as heat producing processes in association with endothermic steam gasification but on the otherhand also resulting as in partial combustion (6) in CO. Reaction (2) is of great importance for the manufacture of pure hydrogen, in the subsequent gas conversion.

Table 11.2. Basic reaction of coal gasification

	Steam gasification	*ΔH KJ / mol*
(1)	$C + H_2O \rightleftharpoons H_2 + CO$	+ 118.5
(2)	$CO + H_2O \rightleftharpoons H_2 + CO_2$	– 42.3
(3)	$CO + 3H_2 \rightleftharpoons CH_4 + H_2O$	– 206.0
	Hydrogasification	
(4)	$C + 2H_2 \longrightarrow CH_4$	– 87.5
	Combustion	
(5)	$C + O_2 \longrightarrow CO_2$	– 406.0
(6)	$C + \frac{1}{2}O_2 \longrightarrow CO$	– 123.0

The steam gasification process and its potential applications

Fig. 11.6 gives a schematic review of the steam gasification process and its potential applications. Coal and steam are fed into a gas generator, which can be one of the three different types (two autothermic and one allothermic) depending on the heat production technology used. Gas generators classed as autothermic are those in which the heat is produced in the generator itself by an *exothermic* reaction (generally the combustion reaction). If air/oxygen is used for this, the nitrogen introduced along with it causes considerable dilution of the raw gas; the end product is therefore a lean gas with a calorific value of the order of 4000-5000 kJ/m^3 (NTP), which can be used in industrial firing for heating purposes provided that no lengthy transport distances are involved. Examples are smelting or metal working plants, glassworks, brickworks, limeworks and the like, which at one time operated gas generators in Germany itself. This technique is used now a days in countries where natural gas is not available or can not be distributed because there are no gas networks. There are numerous plants in operation in developing countries, *e.g.,* India. The use of a lean gas in power stations provides the opportunity of first removing, in an intermediate purification stage, all the noxious substances particularly sulphur, that were originally contained in the coal, and thus producing environmentally clean electricity from coal in addition, converting coal into gas opens up the possibility of using gas turbines, which greatly improves the efficiency. Work is on this technology is therefore in progress in R & D projects in a number of industrialized countries.

Autothermic gasification using pure oxygen gives an undiluted gas consisting essentially of hydrogen, carbonmonoxide, methane and carbondioxide. After purification and conversion, one area of application is to use this to obtain synthesis gas for the manufacture of ammonia, methanol and liquid hydrocarbons. It is mainly for this purpose that coal gasification plants are being operated today.

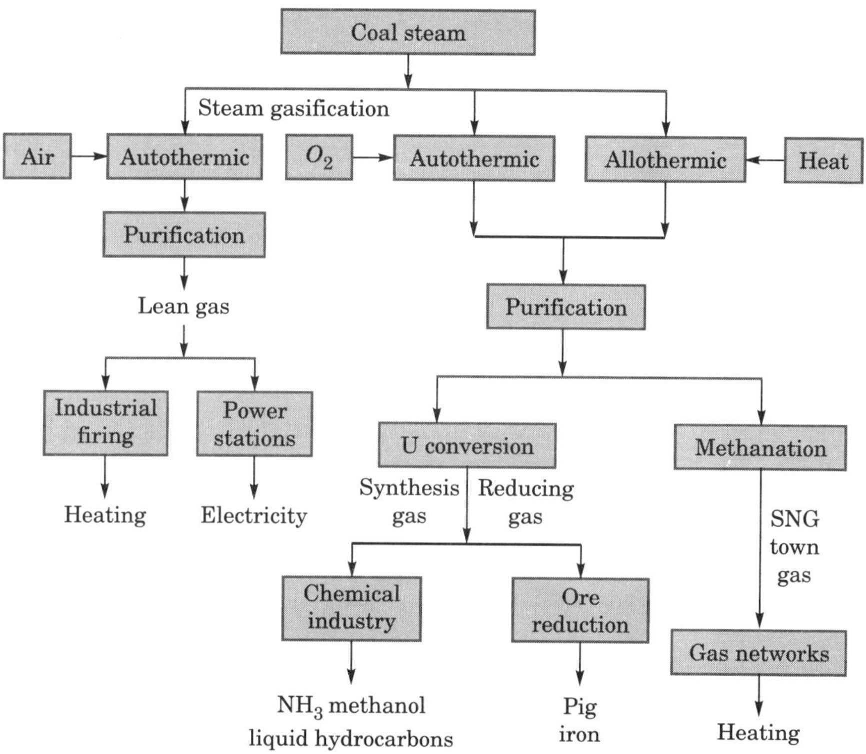

Fig. 11.6. Steam gasification process and its applications.

Another area of use for the gas produced is the direct reduction of ores, particularly iron ores. This technology is primarily of interest for countries that have a poorly cokable coal and can employ the coal gasification route to use such coal for smelting iron ore. Where coal with very good coking properties is available, this route does not really make any sense either technically or economically. Lastly it is also possible to convert the CO and H_2 produced in an autothermic process into methane in a catalytic reaction in accordance with equation (3), and to distribute this as substitute natural gas or town gas *via* the gas networks.

Raw gases with lower CO_2 contents but of otherwise similar quality are also produced by allothermic gas generators, which differ in that coal, coal products or gasification products are burned out-side the generator and the heat is then introduced into the generator using appropriate methods of heat transfer. This obviates the need to produce pure oxygen. However, these processes have not yet won acceptance on a large industrial scale because of their complexity. They will regain importance in the fairly near future by virtue of the fact that it is intended to utilize heat from nuclear high temperature reactors for gasification.

Coal gasification plants

The coal gasification plants used in industry throughout the world today have gas generators of the types illustrated in Fig. 11.7. They are based exclusively on autothermic process in which coal is gasified with steam and oxygen.

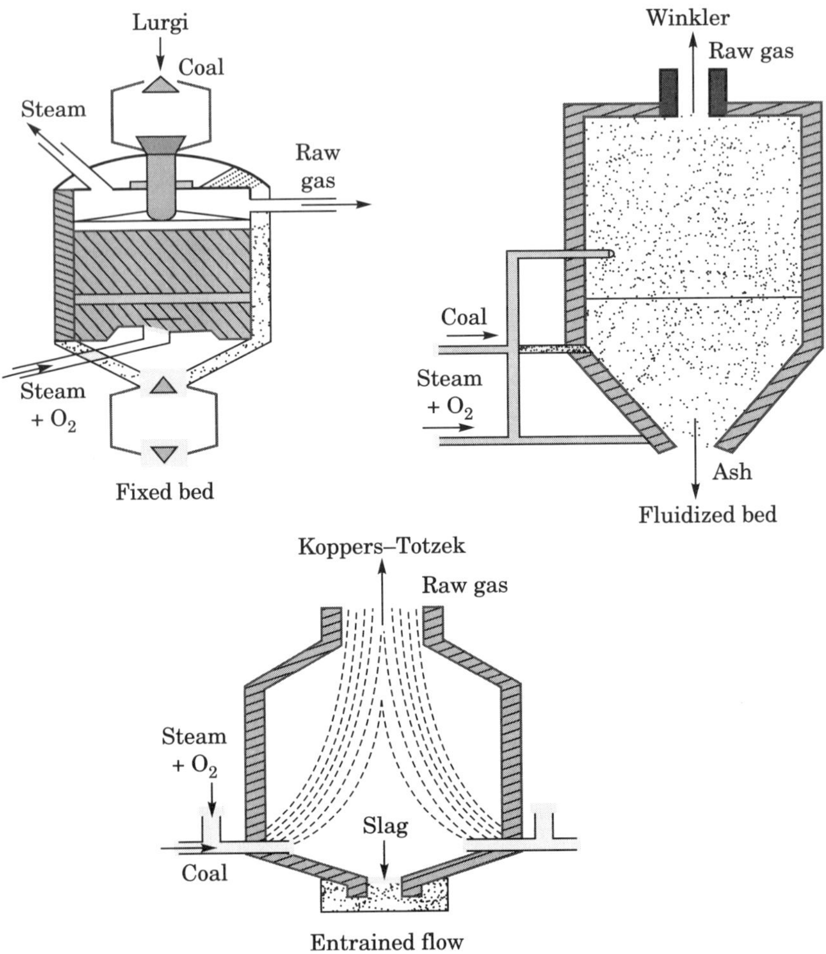

Fig. 11.7. Gas generators for present day industrial coal gasification.

In the *Lurgi gas generator,* gasification is carried out in a fixed bed in which the coal and gasifying agent are introduced in counter current flow. This results in the formation in the generator of a temperature profile and separate zones, in which with increasing temperature, the coal is dried and devolatilized and then gasified and combusted. The process takes place under pressures of between 25 and 35 bar. The coal is therefore fed into the generator *via.* a lockhopper system and the ash is removed *via* a similar system. The generator is a water-cooled pressurized vessel 3-4 m in diameter.

The fluidized bed process, called the *Winkler process* after its inventor, was developed for coal gasification on a large industrial scale as long ago as the 1920s. Here, coal is gasified in a fluidized bed with steam and oxygen in a cylindrical gas generator that tapers to a conical-shape at the bottom. The coal is fed in near the centre of the bed *via* a screw conveyor. The gasifying agent is introduced firstly at the bottom and secondly also above the fluidized bed. This ensures that entrained coal particles being carried out with the gas can also be gasified, resulting in a high carbon burn-up and because of the high temperatures that build up above the fluidized bed considerable reduction of the by products. Uptill now, the Winkler process has been operated solely at atmospheric pressure.

In the *Koppers-Totzek* entrained-flow reactors pulverized coal is gasified with oxygen and steam in an entrained flow of coal dust. Because of the short residence times, the introduction of relatively large quantities of oxygen produces very high temperatures and high conversion rates are therefore achieved. The ash runs down the walls as liquid slag and is removed from beneath the generator. Directly above the generator there is a waste heat boiler for heat recovery.

Table 11.3 makes it possible to compare these three currently used coal gasification processes. The most important technical data on the generators have already been given in the text. As regards the feed coal for the process, it should be noted that the Lurgi and Winkler processes are suitable at best for moderately caking coals, whereas caking properties are immaterial with the Koppers-Totzek process. Also the fixed-bed process does not tolerate fine grained material: only lump coal can be used, and fines have to be diverted to some other use such as a power station for the generation of electricity and steam for coal gasification. However, it is only fair to record that future developments are aimed at broadening the feed coal tolerance. As regards the gas composition, it should be pointed out that in the fixedbed-process there is markedly higher methane content in the raw gas, which is attributable both to the separate zones that form inside the gasifier and to the elevated pressure. Consequently, the Lurgiprocess is particularly well suited to the production of a high-heating value gas and was in fact originally developed for the production of town gas. The raw gas from the Winkler and Koppers-Totzek processes, in contrast, exhibits a very low methane content and is therefore very suitable for conversion into pure hydrogen. The table also lists a number of criteria relevant to the choice of process according to the rank and cost of the coal and the target product.

A major field of application of coal gasification is in the manufacture of ammonia for fertilizer production. For this purpose in various developing countries such plants are based on hard coal, and the process most widely used is the Koppers-Totzek process. The Lurgi process is used, in South Africa for the gasification of hard coal with a wide range of products. A large number of Lurgi gasifiers are in operation in which hard coal is gasified from which tonnes of liquid hydrocarbons are obtained by Fischer Tropsch synthesis.

11.2.6. Solar Energy Methods

The following two approaches are under consideration *i.e.*

(*i*) Bio photolysis, and (*ii*) Photo electrolysis.

Table 11.3. Comparative data on coal gasification processes in Industrial use

Gas generator		*Lurgi*	*Winkler*	*Koppers-Totzek*
	Type	Fixed-bed	Fluidized-bed	Entrained flow
	Temp. (°C)	700–1000	Approx. 1000	>1300
	Pressure (bar)	35	1	1
	Gas production (m^3/h)	35–50000	17–20000	20000 (2 nozzles) 50000 (4 nozzles)
Coal	Rank	Lignite and hard coal upto moderately caking	Lignite and hard coal upto moderately caking	All coals
	Grain size (mm)	6–40	0–8	< 0.1
Raw gas	Vol. % H_2	37–39	35–46	31
	CO	20–23	30–40	58
	CO_2	27–30	13–25	10
	CH_4	10–12	1–2	0.1
Other criteria		• Good heat utilization • Little discharge • High carbon content • Tar and oil • High steam consumption	• No tar and oil • High carbon content • High discharge • Low degree of gasification	• No by products • High steam decomposition • High coal and oxygen consumption • High discharge

Biophotolysis. This method utilizes living systems (or material derived from such systems) to split water into its constituents hydrogen and oxygen. In normal photosynthesis in green plants the green pigment chlorophyll takes up energy from sunlight and in a complex series of reactions breaks up water molecules into oxygen gas, hydrogen ions (*i.e.*, hydrogen with a positive electric charge), and electrons (*i.e.*, particles with a negative charge). The oxygen is evolved from the green plant, but the hydrogen ions and electrons are removed by interaction with carbon dioxide (from the air) to produce simple sugars.

Certain single cell green algae are able to make the enzyme (*i.e.*, biological catalyst) hydrogenase. In these algae, the second stage of photosynthesis can be circumvented by eliminating carbon dioxide. The hydrogen ions and electrons then combine in the presence of hydrogenase to form hydrogen gas. Thus exposure of these algae to sunlight and water (plus essential mineral salts) yields a mixture of oxygen and hydrogen gases that can be separated in various ways. The formation and activity of hydrogenase are inhibited by accumulated oxygen gas; consequently, the gases produced are removed by a flow of nitrogen.

Blue-green algae differ from green algae in several respects. In particular, in addition to normal photosynthesis cells in which reaction with carbon dioxide occurs, they contain some larger cells (heterocysts) where hydrogen can be formed. In the presence of nitrogen *(e.g.*, from the atmosphere), however, the nitrogen combines with the hydrogen ions and electrons to produce ammonia. By preventing access of nitrogen (*e.g.*, in an inert argon atmosphere), blue-green algae decompose water in *sun light* to yield hydrogen and oxygen.

Instead of using living algae to obtain hydrogen from water, a more convenient approach is to utilize biological materials obtained from plants or bacteria. One advantage is the ability to vary the conditions to optimize hydrogen production.

Chloroplasts, the small bodies containing the chlorophyll in green plants, retain their photosynthetic activity when extracted from the plant. Hydrogen and oxygen can then be obtained from water by exposing chloroplasts to sunlight together with the enzyme hydrogenase and ferredoxin, an electron carrier, also of biological origin. It is possible, although less efficient to replace the ferredoxin by a synthetic electron carrier and the hydrogenase by an inorganic (platinum) catalyst. An ultimate objective of research on the decomposition of water by sunlight is the efficient simulation of biological processes without using biological materials.

Photoelectrolysis. In ordinary electrolysis, water is decomposed into hydrogen and oxygen by passing an electric current, from an outside source, between two electrodes in an electrolyte solution. In photoelectrolysis, a current is generated by exposing on or both electrodes to sunlight. Hydrogen and oxygen gases are liberated at the respective electrodes by the decomposition of water, just as an ordinary electrolysis. Atleast one of the electrodes in photoelectrolysis is usually a *semiconductor*; a catalyst may be included to facilitate the electrode process. In the cells studied so far, the efficiency for the conversion of solar energy into hydrogen oxygen energy has been very low. Research is being directed at increasing this efficiency by selection of electrode materials, electrolyte solutions, and electrode catalysts.

Electrolysis is a more attractive way of producing hydrogen with solar radiation since it can be operated intermittently and therefore needs no storage.

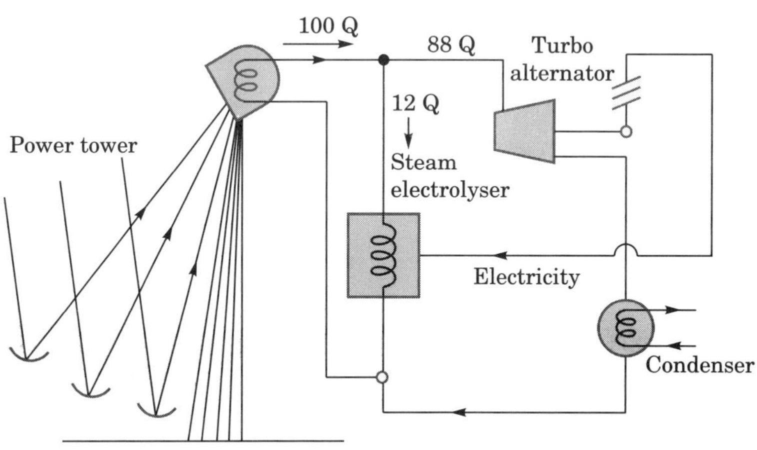

Fig. 11.8. Hybrid electrolysis.

The solar electricity needed for electrolysis can be produced either photoelectrically or thermochemically. Both technologies are available today. Solar electrolytic hydrogen production is therefore a question of the cost of component and of development. Solar thermal power plant (Tower concept *Fig. 11.8*) is an example of a solar electricity generation station. When the break through in high-temperature electrolysis comes it could lead to an interesting application for thermomechanical solar electricity generation, only part of the coolant heated in the receiver is released into the turbo generator; the remainder is used to supply the heat to electrolyser. Saving can be made with such hybrid electrolysis systems.

11.3. Hydrogen Storage

In an energy system there is a need to be able to store energy somewhere between the production point and the utilization point. The need for storage is due to the almost inevitable mismatch between the optimum production rate of energy and the fluctuations in demand for energy by the users.

In the electric energy system storage, presents considerable difficulty because electricity itself is not readily storable. In contrast to the electrical energy system, both the gas energy and the oil energy systems are endowed with the capability store very large quantities of energy. In the oil industry, liquid fuels are stored in above and below ground tanks at the refinery, at the distribution centres and even at the consumers premises. In the gas industry, storage is provided to a small extent in the pipeline themselves and to a large extent in the underground systems which can be either depleted gas or oil fields or so called aquifer systems: rock formations resembling oil fields but containing water.

The location of energy storage systems is very important. If a large storage system can be installed very close to customer, the load factor on the transmission systems is automatically raised, and therefore the transmission cost becomes lower.

One of the advantages often claimed for a hydrogen energy systems is that hydrogen is storable. However, it must be realized that storage of hydrogen is not an easy problem compared with storage of liquid fuels such as gasoline or oil. It is only when it is compared with electricity that storage of energy as hydrogen seems relatively easy. It is when hydrogen is considered as a replacement fuel in applications currently met by natural gas and oil, that bulk energy storage becomes very important.

There are five principle methods that have been considered for hydrogen storage, these are:

(1) Compressed gas storage.

(2) Liquid storage (cryogenic storage in vacuum insulated or superinsulated storage tank).

(3) Line pack system (allowing the pressure in the transmission or distribution system to vary).

(4) Underground storage (in depleted oil and gas fields or in aquifer systems).

(5) Storage as metal hydrides.

Hydrogen can best be stored in different ways, depending on the various field of applications.

Storage can be effected in a gaseous or liquid state, or in the structure of solids. According to the field of application, it is necessary to distinguish between storage on a large scale and on a small scale. The former applies particularly to stationary applications, and the latter also to mobile ones.

1. *Compressed Gas Storage.* Hydrogen is conveniently stored for many applications in high pressure cylinders. This method of storage is rather expensive and very bulky because very large quantities of steel are needed to contain quite small amounts of hydrogen. In the conventional industrial hydrogen system, compressed gas is used to supply relatively small amounts of hydrogen, but when hydrogen is considered as a fuel, it is soon realized that tank storage of hydrogen is not really a practical proposition.

2. *Liquid Storage.* On a small scale or moderate scale, hydrogen is frequently stored under high pressure in strong steel cylinders, as stated above; this type of storage would be too costly for a large scale applications. A more practical approach is to store the hydrogen as liquid at a low temperature, (*i.e.,* cryogenic storage). For example, the liquid hydrogen fuel used as rocket propellant in the space program is stored in large tanks. Very large facilities for hydrogen liquifaction have been designed and built, and large storage tanks have also been constructed. One major difference exists between handling liquid natural gas and liquid hydrogen the storage temperature. Liquid hydrogen boils at – 253°C and therefore must be maintained at or below this temperature in storage unless pressure build up can be tolerated. It is commonly regarded as necessary to use vacuum, insulated storage vessels, where liquid natural gas can be maintained in the liquid state at a considerable higher temperature by using superinsulation, but without the need for vacuum jackets. The principal need for vacuum jacketed containers is that the liquid hydrogen is below the temperature at which air condenses on the surface, and thus any air in the contact with the cold walls of the hydrogen container. There is also a flammability danger from the fact that liquefied atmospheric gases (rich in oxygen) would concentrate in the vicinity of the hydrogen tank. Another problem concerning storage of liquid hydrogen is the considerable amount of energy required to convert hydrogen gas into the liquid phase. Not only must the latent heat of the phase change be removed from the system, but some additional precooling is required to bring the gaseous hydrogen down below the Joule-Thompson temperature above which liquid hydrogen heats up on expansion. Thus, a liquid hydrogen plant normally requires some kind of primary refrigeration, such as a liquid nitrogen plant, to precool hydrogen. The net result is that about 25 to 30% of the heating value of hydrogen is required to liquefy hydrogen.

(3) *Line Packing.* The use of line pack storage in the natural gas industry provides a relatively small capacity storage system, but one with a very fast response time that can take care of minute by minute or hour by hour variations in demand. A hydrogen transmission and distribution system running on hydrogen would have a similar capability although the capacity would be reduced by a factor of about 3 because of the reduced heating value of hydrogen, compared with natural gas.

(4) *Underground Storage.* The cheapest way to store large amounts of hydrogen for subsequent distribution would probably be in underground facilities similar to those used for natural gas; these facilities would include depleted oil and gas reservoirs and aquifers. More expensive alternatives would be caverns produced by conventional mining or by dissolving out salt with water. Since hydrogen gas tends to escape readily through a porous material, some geologic formations that may be suitable for storing natural gas may not be suitable for hydrogen.

(5) *Metal Hydrides (Storage in chemically bound form).* Considerable interest has been shown recently in the possibility of storage of hydrogen in the form of a metal hydride. A number of metals and alloys form solid compounds, called metal hydrides, by direct reaction with hydrogen gas. When the hydride is heated, the hydrogen is released and the original metal (or alloy) is recovered for further use. Thus, metal hydrides provide a possible means for hydrogen storage. An important property of metal hydrides is that the pressure of the gas released by heating a particular hydride depends mainly on the temperature and not the composition. At a fixed temperature, the gas pressure remains essentially constant until the hydrogen content almost exhausted.

Several studies are being made to find a metal hydride that would satisfy the requirements for hydrogen storage. These requirements include the following:

(*i*) the metal (or alloy) should be fairly inexpensive,

(*ii*) the hydride should contain a large amount of hydrogen per unit volume and per unit mass,

(*iii*) the hydride should be formed without difficulty by reaction of the metal with hydrogen gas, and it should be stable at room temperature, and

(*iv*) the gas should be released at a significant pressure from the hydride at a moderately high temperature (preferably below 100°C).

Three of the more promising hydrides are those of lanthanum nickel ($LaNi_5$), iron titanium (FeTi), and magnesium nickel) (Mg_2Ni) alloys. The maximum hydrogen contents are represented approximately by the formulae $(LaNi_5)H_6$, $(FeTi)H_2$ and (Mg_2NiH_4) respectively. These hydrides contain somewhat more hydrogen than an equal volume, but much less than an equal weight, of liquid. Thus, in theory $(LaNi_5)$ Hg contains 1.35% by weight, $(Fe\ Ti)H_2$ contains 1.9%, and $(Mg_2Ni)H_6$ contains 3.6% of hydrogen. It appears that hydrides would be of interest for stationary storage of hydrogen, when the small volume is advantageous. They might be less useful, however, for mobile storage on a light vehicle, where their weight would be a drawback *i.e.,* the weight of the hydride is more relative to its hydrogen content.

Hence as cited above it has been shown that certain metals and their alloys can reversibly store hydrogen chemically or electrochemically by forming metal hydrides. The formation of various exothermic hydrides depending on pressure and temperature and their stoichiometry can be determined by means of concentration/pressure isotherms. If gaseous hydrogen is to react with an alloy, the molecules first have to dissociate into hydrogen atoms which can then be absorbed by the metal. For physical reasons, hydride formation always takes place at constant pressure.

The dissociation of the molecular hydrogen into atomic hydrogen and the bonding in the metal phases means that

- the gas pressure of the hydrogen above the hydride is one to three orders of magnitude below the values for gas storage in high-pressure cylinders,
- the hydrogen density in the hydride is greater than that of hydrogen in liquid form, and
- because of hydrogen's calorific value of 33,000 Wh/kg and the weights of metal alloys used for storage, the energy densities of the low temperature hydride can be about 600 Wh/kg ($TiFeH_2$, $Ti\ Fe_{0.8}\ Mn_{0.2}\ H_2$) and those of the high-temperature hydrides between 1200 Wh/kg ($Mg_2\ Ni\ H_4$) and 2400 Wh/kg ($Mg\ H_2$).

In practice, energy densities of from 500 Wh/kg (already achieved) to a maximum of 1000 Wh/kg can be attained.

If this is compared with the energy density of gasoline (including tank), *i.e.*, some 10,000 Wh/kg and that of a lead battery, *i.e.*, 25—30 Wh/kg (account being taken of the efficiency of electric propulsion, which is 2-3 times higher than that of the combustion engine), as a rough approximation it can be said that for the same energy content a hydride storage unit (with cladding) will always be 10-20 times heavier than a gasoline tank but 5-10 times heavier than a conventional lead battery with the same range. Thus, reversible chemical storage of hydrogen in metals gives technically acceptable figures in terms of weight, volume, pressure safety and energy consumption.

The chemical reaction equation for exothermic formation of the hydride from hydrogen and metal is:

$$H_2 + Me \underset{\text{discharge}}{\overset{\text{charge}}{\rightleftharpoons}} \text{hydride} + \text{heat} \begin{cases} \rightarrow \text{free} \\ \leftarrow \text{added} \end{cases}$$

This means that during charging up with hydrogen heat is always produced at the same time, and in order to withdraw the hydrogen from the hydride it is always necessary to add heat.

For the technical application of hydrides the temperature at which their dissociation pressures attain values above 1 bar is of particular interest. Of the many that exist, this group of hydrides are called low temperature hydrides (LTH), because even at temperatures below freezing point their dissociation pressure is above 1 bar; all the other hydrides, for which the pressure limit of 1 bar is above the boiling point of water, are known as high temperature hydrides (HTH).

If the reaction equation for hydride formation is viewed not as a hydrogen reaction but as a thermal reaction, the basic equation for the storage of heat in hydrides is:

$$Q + MeH_2 \longrightarrow Me + H_2$$

Heat is stored and hydrogen is given off by the dissociation of the hydrides.

The back reaction is accordingly:

$$H_2 + Me \longrightarrow MeH_2 + Q$$

Now the hydrogen is absorbed and (the formerly stored) heat released. This means that metal hydrides fulfil a dual role. They store both fuel (hydrogen) and heat.

Low temperature hydrides (LTH) have a heat storage density of upto 0.3 MJ/kg($TiFeH_2$, $CaNi_5H_6$) or 1.5×10^3 MJ/m^3 in a temperature range from – 20°C to 100°C at a hydrogen pressure of ~ 10 bar.

High temperature hydrides (HTH) have a heat density of upto 3 MJ/kg (MgH_2, Mg_2NiH_4) or 6×10^3 MJ/m^3 in a temperature range from 150°C to 550°C (TiH_2 : 3 MJ/kg, 20×10^3 MJ/m^3, T = 500°C) and at a hydrogen pressure of ~ 10 bar.

Thus any combustion process (in industry, the home or vehicles) which uses hydrogen from a hydride storage unit can be combined with partial or total storage of the waste heat of combustion of the hydrogen.

In combustion processes (*e.g.*, H_2 engines or H_2 domestic heating etc.) waste heat is produced and delivered to the hydride to release hydrogen, it is therefore stored in the metal and not given off into the atmosphere simultaneous with combustion. To ensure continuous release of hydrogen, the waste heat from the combustion process has, of course, to be greater than the energy needed for the release of the hydrogen from the hydride.

If the quantity of heat available, together with all the transmission losses, is the same as the energy needed for the dissociation of the hydride, the combustion process produces no waste heat to the outside, since it is then a closed system of hydrogen combustion/waste heat storage. Only when recharging the metal with hydrogen (and only then) does the hydride formation lead to the release of the previously stored heat of combustion. The waste heat energy can be made available for practical use, *e.g.*, for space heating, in various temperature ranges (by varying the pressure, in recharging).

This interaction of fuel and heat storage can be used to particular advantage in connection with combustion engines.

11.4. Hydrogen Transportation

Pipe lines. At present, the long distant pipelining of hydrogen is an operation that is carried out by only a few specialized companies in different parts of the world. It is of interest to compare the design requirements of a pipeline for hydrogen with those of a pipeline for natural gas. Heating value of hydrogen in only 12.1 MJ/cum, as compared to about 38.3 MJ/cum for natural gas. This implies that to deliver the same quantity of energy, three times the volumes of hydrogen must be transmitted. On closer inspection, however, one finds that the capacity of a pipeline depends upon the square root of the density of the gas, and because the density of hydrogen is about one ninth (1/9) that of natural gas, there is a compensating factor of the one third that results in the given pipe having essentially the same energy carrying capacity for natural gas as for hydrogen. This is true at atmospheric pressure. As the pressure increases to typical pipeline operating pressures of 50 kg/cm^2 (5 MPa) or so, the compressibility factor for hydrogen is somewhat different that for natural gas, and this results in a slightly unfavourable carrying capacity for hydrogen. At 50 kg/cm^2 (5 MPa), the ratio of heating values for a given compressed volume of hydrogen and natural gas has changed from 3 : 1 to 3.83 : 1. Long distances gas transmission lines of lengths greater than about 90 km must be supplied with pipeline compressors at fairly regular intervals. Hydrogen compressors must handle a considerably greater volume of the gas, somewhere between three to four times the number of cu m for the same energy capacity. Secondly, the horse power required to drive a hydrogen compressor is considerably greater than that needed to drive a natural gas compressor for the same gas energy throughout. Thirdly, the design of rotory compressors commonly used for natural gas lines appears to be inadequate for hydrogen operation.

It is possible to estimate the cost of transmitting hydrogen by pipeline from a knowledge of the required pipeline diameter, compressor capacity and horse power, and energy throughout required. In the case of design of a pipeline of hydrogen transmission the cost of fuels used to drive the engines for compressors is also a highly deciding factor, because the compression energy is so much higher than that of natural gas.

One of the principal concerns about hydrogen transmission is the fear of hydrogen *embrittlement of the* pipeline materials. A number of metals lose their mechanical strength on exposure to hydrogen; the phenomenon called hydrogen embrittlement, is specially significant for steel in hydrogen under pressure. Operating experience with common pipeline steels at pressures up to about 35 kg/cm^2 (3.5 MPa) has shown no problems of consequence. Normal pipeline steels are satisfactory for pressures up to the typical operating values of 50 atm (5 MPa). However the behaviour at higher pressure is uncertain, and more experimental work needs to be carried out to get some definite data.

Liquid Hydrogen Transportation. Hydrogen in bulk can be transported and distributed as the liquid. Double walled, insulated tanks of liquid hydrogen with capacities of 7000 gal (26.5 cu m) or more are carried by road vehicles and upto 34,000 gal (129 cu m) by rail road cars. Distribution of liquid hydrogen by pipelines, jackted with liquid nitrogen, has been proposed. The costs would be substantially greater than for gas pipelines, but it might be justifiable for certain fuel applications where the liquid is required.

Metal Hydride Transportation. Hydrogen can also be transported as a solid metal hydride. The main drawback, as stated earlier, is the weight of the hydride relalive to its hydrogen content.

11.5. Utilization of Hydrogen Gas

Hydrogen gas can be utilised:

(*i*) For residential uses

(*ii*) For industrial uses

(*iii*) For as an alternative transport fuel

(*iv*) For as an alternative fuel for aircraft

(*v*) For electric power generation (utilities).

(*i*) Residential uses. Electricity for *lighting* and for operating domestic appliances (*e.g.*, refrigerators) could be generated by means of fuel cells, with hydrogen gas at one electrode and air at the other.

Hydrogen can be used in *domestic cooking*. The burners of domestic appliances (*e.g.*, stoves) would have to be modified if hydrogen were to replace natural gas. In the first place, since the heating value per unit volume of hydrogen gas is less than that of natural gas, a larger volume would have to reach the burners to achieve the same heating effect. Furthermore, the greater flame speed when hydrogen burns in air would facilitate backward propagation of the flame (*e.g.*, backfiring) in a natural gas burner. These problems can be overcome by changing the design of the burner, including the hole size and the air supply system.

Direct combustion product of hydrogen in air, is water only. Nitrogen oxides may also be formed at the flame temperature but their amounts should be quite small. The stove may not require venting to the outside atmosphere unless the humidity level from the water vapour were objectionable.

Hydrogen would be useful in *radiant space heaters,* because of the possibility of flameless combustion on a catalytic surface. Such devices would operate spontaneously when the gas was turned on and no pilot light or other ignition system would be required. Because of the low combustion temperature, nitrogen oxide formation would be negligible and venting would be unnecessary.

(*ii*) Industrial uses. There are many potential uses for hydrogen in industry, either as a fuel or a chemical reducing (*i.e.*, oxygen removal) agent if the economics were favourable. For example, in several industrial processes natural gas has been the most satisfactory source of heat in a hydrogen energy economy, hydrogen could replace natural gas in these operations. Hydrogen gas could also be used with advantage, instead of coal or coal derived gases, to reduce oxide ores (*e.g.*, iron ore) to the metal (iron).

***(iii)* Road vehicles.** The use of hydrogen fuel in internal combus- tion engines for automobiles, buses, trucks, and farm machinery has attracted interest as a means of conserving petroleum products and of reducing atmospheric pollution. Because of the fuel is a gas, the conventional carburetor of a spark-ignition engine, in which liquid gasoline is vapourized in air, must be modified for use with hydrogen. The methods of using hydrogen as a fuel in CI engines are as follows:

(*i*) A mixture of fuel gas and air, with an approximately constant fuel to air ratio, is introduced into the cylinder intake manifold. The engine power (*i.e.*, vehicle speed) is controlled by varying the quantity of mixture entering the cylinder by means of a throttle value. (The same procedures were used in the otto and ol.lier early I.C. engines with coal gas as the fuel). Stable operation, especially at higher speeds, may require addition of water vapour to the fuel air mixture; this can be achieved by returning part of the exhaust gas to the manifold.

(*ii*) The hydrogen gas under pressure is injected through a valve directly into the engine cylinder, and the air is admitted through another intake valve. Since the hydrogen and air are supplied separately, an explosive mixture does not occur except in the cylinder. This scheme is considered to be safer than the previous one, in which such a mixture is formed in the manifold. The engine power output is controlled by varying the pressure of hydrogen gas from about 14 atm at low power to 70 atm at high power. The hydrogen is required to be stored as a compressed gas.

(*iii*) The hydrogen gas at normal or moderate pressure is drawn through a throttle value into the engine cylinder during the intake stroke. At the same time, unthrottled air is drawn in through the intake port. Here, also, there is no explosive mixture,except in the cylinders. The engine power is varied by adjusting the hydrogen inlet throttle. Since the air supply is unthrottled, there is a change in the proportion of fuel in the cylinder and consequently a change in the power developed. This scheme of power variation is possible because of the wide composition range over which hydrogen air mixture can be ignited.

Another modification arises from the high speed of the hydrogen flame in air; this require that the ignition time be retarded (*i.e.*, less spark advance) compared with a gasoline engine.

Advantages claimed for hydrogen fuel engines are:

(*i*) They can higher efficiencies *i.e.*, utilize a higher proportion of the energy in the fuel than gasoline engine.

(*ii*) The amount carbon monoxide and hydrocarbons in the exhaust would be very small since they would originate only from the cylinder lubricating oil.

However the nitrogen oxide levels due to high combustion temperature may be high, it may be reduced by reducing the combustion temperature by injecting water vapour into cylinder from the exhaust as described earlier.

The other way to utilize hydrogen as a vehicle fuel is the use of fuel cells. Hydrogen is utilized in funl cells and then electricity is generated, which could be utilised to operate electric motors to propel the vehicle.

For vehicles the storage of hydrogen can be as *compressed gas, liquid,* and *metal hydrides.*

(iv) **Air Craft Applications.** The earliest application of liquid hydrogen fuel is expected to be in a jet air craft; this possibility was demonstrated in a subsonic air craft in 1957. The main advantage is the much lower overall weight of the fuel and storage tank than for ordinary jet fuel. The volume of liquid hydrogen would be greater than for regular fuel, but this could be accommodated on a large aircraft. The cold liquid hydrogen could be used directly or in directly to cool the engine and the air frame surfaces of a high speed aircraft. If a hypersonic aircraft (*i.e.,* speed more than five times the local speed of sound) is ever developed, liquid hydrogen may be the only practical fuel. Even in the subsonic range liquid hydrogen makes it possible to reduce the starting weight of high capacity air craft by half, compared with hydrocarbon fuelled aircraft of the same range and pay load. Because of the smaller total weight it is possible to achieve shorter take off runs, steeper climbing paths (less nuisance from noise) and/or smaller engine thrust (less noise production).

It may also be possible to decrease the size and weight of the engines.

Hydrogen's favourable diffusion properties and high thermal conductivity lead to better mixing than with liquid hydrocarbons, even with shorter combustion chambers. The wide range of ignition for hydrogen air mixtures (5% to 75% by volume) makes the engine more readily controllable, especially under partial loads, and reduces the emission of noxious substances. Since it is possible to use engines of shorter construction, thus reducing the time for which the fuel is at a high temperature, the emissions of nitrogen oxides (in any case small with hydrogen) can be still further reduced. The problems, in addition to that of the economic production of Hydrogen, lie in the provision of the necessary infrastructure at international air ports and the bridging of the transitional period, which will necessarily be a long one. A further problem concerns the placing of the fuel tanks, which are more bulky than those for kerosene. It is not possible to site them in the wings, since the surface/volume ratio of cryogenic containers must be kept as low as possible in order to keep down evaporation losses. Foam insulation can be used, in accordance with current technical practice, but the optimum construction requires further investigation with present techniques double walled tanks with an intervening vacuum for insulation would be too heavy. External tanks located on the wings would increase the aerodynamic drag considerably. Placing this tanks in the extended fuselage represents a reasonable compromise between space utilization, aerodynamic drag and safety. The insertion of suitable bulk heads would provide adequate safety, even in the case of accidents.

The heat required to vaporize and heat up the hydrogen for the engines can be obtained through certain sections of the outer skin of the wings and fuselage. In this way the boundary layer is cooled so as to produce laminar flow, resulting in a lowering of the aerodynamic drag and hence of the fuel consumption (laminar flow control), this could not be achieved to the same extent in any other way.

(v) **Electric Power Generation.** It is unlikely that hydrogen would serve as a major fuel for electric power generation by a utility. However, its substitution for natural gas in peak shaving turbines is possible. Hydrogen could also be used as a means for storing and distributing electrical energy.

This also comprises the production of electricity by using hydrogen in fuel-cell systems. The modular construction of fuel-cell systems makes it possible to obtain favourable efficiencies even with small assemblies. A further advantage of the modular structure of fuel-cell systems lies in the extremely high potential for reducing manufacturing costs when producing a large number of units. However, in the present state of fuel cell technology such applications are feasible only for pilot plants. A number of developments are still necessary, especially in the fields of electrodes and catalysts, to make fuel cell systems applicable on a large scale.

It is also of importance that the conversion efficiency of fuel cells is independent of the load factor over a wide range, so that a high efficiency can be obtained even with partial loads. Furthermore, with fuel cells which at high temperatures high-grade waste heat can be used for thermal energy production.

On the basis of experience of electricity generation with small hydrogen-oxygen assemblies, research and development work is now in progress (especially in U.S.A.), with the object of developing fuel-cell power stations for the centralized and local generation of electricity. The present position with regard to various fuel cells can be described as follows.

Low and medium temperature fuel cells with operating temperature between 80 and 250°C have been equally for developed in three directions:

(*i*) alkaline fuel cells with aqueous caustic potash as electrolyte;

(*ii*) acid fuel cells with concentrated phosphoric acid solutions as electrolyte;

(*iii*) solid electrolyte fuel cells with acidic ion exchange groups in a polymer matrix.

Alkaline fuel cells can be used advantageously for producing electricity from pure hydrogen and oxygen, since high efficiencies can be obtained here with cheaper catalysts. Fuel cells with acid electrolytes can also be operated with fossil derived impure hydrogen or hydrocarbons, but as yet require the use of catalysts of platinum or other noble metals.

Along side the use of fuel-cells operating at low or medium temperatures and pressures, the development of optimum configuration of large fuel cell modules therefore depends on the choice of suitable inexpensive materials for electrodes of sufficient long term stability and enhanced procedures for manufacturing the individual components and modules.

High temperature fuel cells operate in the temperature range from 600 to 1000°C. Because of their high operating temperature they have the advantage that no special catalysts are needed to activate the reactions, so that crude gas or CO/H_2 mixtures can be used directly for producing electricity, and high grade waste heat is available for use.

High temperature fuel cells with solid electrolytes based on conduction by oxide ions represent the reverse of the high temperature electrolysis of steam, and involve the same problems as those described in that connection. In particular, it will be necessary to solve the fundamental problems connected with constructional materials arising from the high operating temperatures (around 1000°C) before development to an industrial scale is possible.

A more advanced stage of development has been reached with high-temperature cells with molten carbonate electrolytes, operating at about 600°C. Laboratory cells have been undergoing tests for sometime now. Further development into larger units depends essentially on solving basic constructional problems and developing suitable materials. If these development problems could be solved fuel cell power stations could compete seriously with H_2/O_2 or H_2/air steam turbine processes.

11.6. Hydrogen as an Alternative Fuel for Motor Vehicles

The consequences of using hydrogen in motor vehicles are best seen by comparing it with other chemical energy vectors. **Table 11.4** shows the required tank volume and weight of fuel for typical non-conventional fuels compared with iso-octane, which is typical of the various hydrocarbons that constitute gasoline. Comparison of isooctane with methane (liquid natural gas), methanol and hydrogen shows that, although liquid hydrogen requires a larger tank volume, it requires the lowest weight of fuel on account of its high minimum calorific value. Except for the tank volume liquid methane corresponds most closely with gasoline, while methanol requires an appreciably larger tank volume and also (on account of its lower calorific value) a considerable heavier weight, of fuel **(Table 11.5).**

Table 11.4. Comparison of Various Alternative Fuels and Storage

	Storage capacity wt. %	*Density of Hydrogen g/l*	*Energy density*		*Operating temperature (at 2 bar)*
			kWh/kg	*k/Wh/l*	
Magnesium hydride	7	101	2.33	3.36	350°C
Magnesium nickel hydride	3.16	81	1.05	2.69	250°C
Iron titanium hydride	1.75	96	0.58	3.18	– 10°C
Liquid hydrogen	100	71	33.3	2.36	– 250°C
Methanol			5.6	4.42	
Liquid methane			13.8	5.8	– 161°C
Ammonia			5.14	4.21	
Hydrazine			4.6	4.65	
Iso-octane (It is taken here as representative of gasoline, which is a mixture of low boiling hydrocarbons)			12.7	8.76	

Table 11.5. Weight Required For the Storage of Fuels

	Hydrogen (kg)	*Hydride (kg)*	*Storage container, etc. (kg)*	*Total weight (kg)*
Magnesium hydride	8.5	121		193
Magnesium nickel hydride	8.5	269		430
Iron titanium hydride	8.5	485		775
Liquid hydrogen	8.5	—	50	58.5
Equivalent quantity of gasoline	22.3	—	15	37
Equivalent quantity of methanol	50.5	—	20	70

In the use of hydrogen as an alternative fuel, the storage of the fuel on board the vehicle constitutes an important problem. For motor vehicles hydrogen could in principle be stored as compressed gas, chemically as a metal hydride, or as a liquid. The storage of compressed gas would be difficult on grounds of safety because of the high pressures required. Furthermore, even at pressures around 200 bar the quantity of hydrogen stored amounts to only about 1% of the weight of the container, while the volume of the container is only 20 times greater than that of an equivalent gasoline tank. In metal hydrides hydrogen atoms are incorporated in the crystal lattice of metals or alloys (inter metallic compounds), and each lattice atom can absorb up to two hydrogen atoms, or in individual cases between two and three. When using alloys like iron titanium the storage material can contain only upto 1.75 wt% of hydrogen, since in iron titanium hydride (Fe Ti H 1.95) the iron and titanium constitute about 98.25 wt% of the substance. The amounts of hydrogen stored are more favourable when the lattice consists of lighter atoms, for example magnesium.

However, other problems then arise, since the formation of the hydride and its decomposition to give hydrogen occur rapidly enough to produce a sufficiently high hydrogen pressure only at fairly high temperatures (250-350°C).

This temperature can be lowered by suitable additions to the alloy, particularly of nickel. However, this causes the heat of formation of the hydride, which must be removed when the storage unit is being charged and supplied when the hydrogen is being liberated, to assume very high values (about 30% of the minimum calorific value of hydrogen see Table 11.4). This can cause problems in running the vehicle, since on account of the high operating temperature of these hydrides the heat available from the exhaust gases is insufficient to liberate from the hydride the hydrogen required by the engine. Since hydride vehicles have already been investigated intensively, a considerable amount of information is how available about the performance and potential of mobile hydride storage units.

For a given amount of energy stored in the vehicle, hydride storage requires about the same volume as liquid hydrogen, but 10-20 times the weight. Hydride vehicles must therefore be regarded as alternatives to electric vehicles, and would be used mainly for urban transport within a moderate range. However, it is precisely in urban transport, that carrying the additional weight of the storage system is a disadvantage; for example, it may around to about 300 kg for a maximum range of 150 km. The consequent increases in fuel consumption in urban transport makes the. advantages that hydride storage offers over storage as liquid hydrogen, such as the possibility of loss-free storage at ambient temperatures and the absence of energy consumption for the liquefaction of hydrogen, seen less attractive. Thus, although the consumption of primary energy for liquefaction may be regarded as a particular disadvantage, it does not represent a limitation in comparison with hydride storage, since on the one hand liquid and gasoline are comparable as regards weight and range even when the storage tank is taken into account and on the otherhand a considerable proportion of the energy used in liquefaction can be regained as mechanical energy when the cold hydrogen is warmed up. It is also of importance for engine operation that a pump for liquid hydrogen can be used to obtain the extra pressure needed for internal mixing in the spark ignition engine or indeed for diesel operation, with only a small expenditure of energy. This would be difficult to achieve with hydride storage, either hydrogen at ambient temperature would have to be compressed with a compressor, which because of the large volume of gaseous hydrogen would require an appreciable proportion of the output of the vehicle engine, or the injection pressure of the hydrogen into the engine would have to be reduced so much that the fuel could no longer be supplied at the most favourable moment.

A problem could arise with evaporation of hydrogen gas from the storage tank of the vehicle, especially when it is parked in an enclosed space such as a garage. However, according to the most recent observations the rate of evaporation from the latest vacuum insulated storage containers can be kept down to a very low level by taking suitable precautions. Laboratory storage containers are now available for liquid hydrogen with capacities of 100-200 litres which can be kept for up to one year before all the liquid has evaporated. Storage tanks for motor vehicles giving storage periods of weeks or months are technically quite feasible today at reasonable cost. The small amount of hydrogen with evaporates can be used to generate electricity in a fuel cell, thus making it possible to dispense with a rotary electric generator. In this way the car battery can be charged efficiently and with little consumption of fuel when the car is stationary and during town driving, and the emission of hydrogen to the surroundings, which is particularly dangerous in enclosed spaces, is avoided completely.

11.7. Safety and Management

Since hydrogen is a highly flammable gas, it must be handled with care in special equipment designed for safety. The extensive use of H_2 on the grounds is dangerous. Natural gas and gasoline are also hazardous materials, yet they are in common daily use. The danger of fire or explosion can be minimized by taking proper precautions with hydrogen. Larger quantities of the gas and liquid have has been carried by pipeline, road and rail for use in industry and in space vehicles in many respects, hydrogen gas is no more, and possible less, hazardous than natural gas. The lower flammability limit of hydrogen is 4% (by volume) in air, whereas that of methane in natural gas is 5%. However, hydrogen will escape from a leak of fixed size about three times as fast as natural gas hence, in a closed space, the flammability will be reached soon with hydrogen. On the other hand, the energy contained in the space at the lower flammability limit is about one fourth that for natural gas. Because of its low density, hydrogen disperses more readily in an open (or ventilated) space, and it would therefore take a much longer time for the flammability limit to be reached.

The chief danger from hydrogen gas is associated with the very low ignition energy, which is less than a tenth than of natural gas. Consequently, a spark that is too weak to ignite a flammable air natural gas mixture may ignite an air hydrogen mixture. In handling hydrogen, special care is to be taken to avoid flames and to prevent spark formation.

Liquid hydrogen presents another safety problem. The low temperature of the liquid may cause air in the vicinity to liquefy; since oxygen is more readily liquefied than nitrogen, an oxygen rich liquid may form. This would greatly increase the flammability danger. Such a situation could arise if a liquid hydrogen tank or pipeline were not properly insulated so that the exterior temperature fell low enough to liquefy oxygen from the air. The special double walled insulation as stated earlier is thus desirable.

Liquid oxygen could also be formed from the air present in an "empty" tank or pipe when it is filled with liquid hydrogen. The air must therefore be previously removed by purging with hydrogen gas or preferably helium. If hydrogen is used, care must be taken in disposing of the purged gas mixture.

Hydrogen is used on a large scale today in industry, especially the chemical industry, where the safe handling of hydrogen is an established technique. Nor are any essential problems to be expected in this area of use in the future. However, in considering the possible future public use of hydrogen as an energy vector, for example as an alternative fuel, it has to be borne in mind that at present hydrogen (and especially liquid hydrogen) is generally considered as extremely dangerouo substance to handle, although completely realistic assessments can be made on the basis of extensive experience in space travel. This is because such experience is not sufficiently widely known, and because the public has been accustomed for nearly a hundred years to dealing with gasoline and kerosene, although these too are by no means free from danger. Natural gas is also generally accepted today as an energy vector, although in the 1940s it gave rise to a number of extremely serious accidents with liquid gas tankers and large scale storage in the U.S.A. In the case of hydrogen this 'familiarization effect' has not yet developed, although in the past hydrogen has been used in considerable quantities in town gas, which contains between 40% and 70% of hydrogen.

Although safety problems with hydrogen are somewhat different, in the present state of knowledge no greater potential dangers are to be envisaged than in dealing with natural gas, while the unequalled environmental compatibility of hydrogen must be counted an additional advantage.

The hazards—both real and supposed—presented by hydrogen must be evaluated before it can be used as a fuel and energy vector in any particular system. The public's questions about safety with have to be answered if hydrogen is to be introduced for use as a transport fuel, for heating in the home and in industry, for new industrial processes such as generating electricity at periods of peak demand or, to a greater extent than present, for industrial processes such as ammonia, fertilizer and methanol production and coal gasification or liquefaction. This enlarged hydrogen market will be accompanied by a corresponding growth in the production of hydrogen in new techniques for its production, in storage capacity and in transportation across or beneath public territory. All these elements of a 'hydrogen energy system' are involved in any discussion of safety.

It must be stressed that hydrogen has been used as a fuel and as a chemical for more than two centuries. Hydrogen is used mainly as an intermediate product in the chemical and petroleum industries, the three largest consumers being the production of ammonia and methanol and the refining of crude oil.

Almost all the hydrogen in the petrochemical industry is produced and consumed in one place. Only a few small pipeline systems carrying hydrogen between several plants are in use. The largest hydrogen pipeline system is in Germany and has a total length of approximately 210 km and an annual throughout of approximately 3×10^8 m^3. Total world production of hydrogen in 1973, was 2.1×10^{11} m^3, which is equivalent to 1.8×10^7 tonnes. These figures show that considerable experience exists in the handling of hydrogen.

Causes of files and preventive measures:

Leaks from seals at flanges and valves are responsible for 65% of the fires. Fittings that resulted in the most leaks that in turn led to fires were located in the heat exchangers upstream or downstream of the high temperature methanation conversion process, in waste heat boilers and in ammonia conveters. All leaks at valve flanges that led to fires generally are in the synthesis circuit.

The analysis of causes indicates the characteristic tendency of hydrogen to penetrate normally "air-tight" seals. The volume that leaks through a seal is determined either by the density of the gas (molecular weight) or by its viscosity, for a given pressure difference and leakage area. Like methane, hydrogen has a lower viscosity and molecular weight than air; hence a fitting or component that is normally air tight is by no mean a necessarily "gas tight" to hydrogen or methane.

When leaking hydrogen has a temperature near or above room temperature it rises upwards rapidly owing to the difference in density. The density of bydrogen is only 1/8 that of methane; hydrogen therefore has a corresponding faster rate of rising. Hydrogen's relatively high diffusion coefficient also promotes dispersion even under still atmospheric conditions.

In the event of an escape of hydrogen in an enclosed space, its tendency to rise imposes a limit on the degree to which it is diluted by the air. As recommended for this reason hydrogen should not be handled in enclosed spaces but always in the open air waste gas systems and furnaces (heat treatment furnaces and reformer furnaces) are more often affected by faults than any other components, Compressors and heat exchangers are likewise frequently affected. The relatively high accident rate for compressors, furnaces, heat exchangers and reformers is understandable when it is remembered that these usually contain hydrogen at temperatures close to its spontaneous ignition point (585°C).

The relatively high accident rate for systems in which the operating personnel are'not specifically familiar with the hazards presented by hydrogen should also be mentioned. This in especially noticeable in, for example, nuclear power stations, chlorine plants and semi conductor factories, where hydrogen is only an unwanted side product that has to be handled. Leaks from flanges and failure of waste gas collection systems and in the case of chlorine, the gas collection system are the fittings and components mostly affected by the faults. In view of the possible use of water electrolysis as a method of producing hydrogen in the future, it is important to point out that electrolysis has a high accident rate, most of the accidents today occurring in chlorine/alkali plants. It is therefore particularly important to ensure that the hydrogen and the oxygen are kept separate in the electrolysis cell and in the gas collection system.

Traditional fittings and components like pressurized tanks, distillation columns and heat exchangers as such appears to be affected to a negligible extent by hydrogen faults.

Preventive measures. In discussing preventive measures it is necessary to distinguish between measures intended to prevent accidents inside the system and outside the system. Three conditions must be fulfilled simultaneously for a fault inside a system: for example, hydrogen as fuel, oxygen/chlorine as oxidising agent and, lastely a source of ignition. The elimination of sources of ignition may be an adequate preventive measure for avoiding hydrogen explosions because of the extremely small amount of energy required to ignite hydrogen. Consequently, to avoid explosions in hydrogen systems, it is absolutely essential to *prevent oxygen from entering the system.* This can be accomplished either by making the pressure in the system higher than that of the oxidizing atmosphere or by using an inert gas within the system. It should be pointed out that only 4% volume of hydrogen is needed to cause an explosion provided conditions are suitable in other respects. Since the lower explosion limit of hydrogen is also dependent on pressure, particular attention has to be given to determining the lower ignition point for the particular gas composition and pressure concerned.

The inside of the *system can be rendered inert* not only by using nitrogen but also by using carbondioxide, halons or extinguishing powders such as ammonium phosphate or potassium chloride.

The latter can be used for temporarily rendering mixtures of hydrogen and air inert in order to prevent explosions progressing through the system in this case they are therefore actually *explosion suppressors.*

Another way of preventing explosions from propagating through the system is to install *flame traps.* Here too, reliable flame traps for hydrogen are difficult to design because hydrogen requires a very narrow gap to prevent a flame from passing.

Whilst a high internal pressure and an inert internal atmosphere efficiently and reliably prevent the formation of explosions in pure hydrogen systems, other measures are needed in systems containing hydrogen and other products to prevent the dangerous consequences of explosions.

The *measures* are essentially as follows:

1. systems that withstand the force of an explosion,
2. systems that withstant the force of pressure surges,
3. explosion-relief systems,
4. explosion-suppression systems and, for pipeline system alone,
5. flame traps, flame suppressors, explosion relief devices and rapid closing devices.

Measures for preventing hydrogen explosions outside the system are greatly limited in effectiveness because an oxidizing agent, air, is always present and the presence of a source of ignition always has to be assumed even though it may have previously been demonstrated that numerous leaks, and therefore the presence of hydrogen in the free atmosphere, had occurred without an explosion and that the hydrogen had diffused away harmlessly. Consequently a fundamental preventive measure in the case of hydrogen plants is that they be built in the open air. Further, it is very important to detect any hydrogen leaks as early as possible. Because of hydrogen's low density, standard analysers like gas detectors can be used to only a limited extent and can be made effective only in open air plant. The other preventive measures are largely secondary measures giving protection solely against further effects resulting from fires and explosions. Examples of such secondary measures are automatic or manual shut down of the process, sprinklers, flooding, water spraying, extinguishing systems, halon systems, and explosion relief devices for entire buildings.

All these systems have proved their effectiveness in the past. Automatic sprinklers, flooding systems and water spraying for cooling have a success ratio of 80%. This result seems to confirm the advantages of the qualitative recommendation made in published guidelines on explosion prevention. Powder extinguishing systems have a successive ratio of 86% in extinguishing hydrogen fires. Shutting down the entire hydrogen plant is the most natural method of preventing fire and has a success ratio of 81%. Explosion relief devices have a success ratio of 90% and have contributed greatly to preventing damage to apparatus and buildings.

11.8. Hydrogen Technology Development in India

The Department of Non-conventional Energy Sources (DNES), ministry of energy. India has so far supported 22 Research and development projects of repute and universities on various aspects of Hydrogen technology namely:

(*i*) Production of hydrogen by photo electrolysis of water using solar energy.

(*ii*) Production of hydrogen by blue green algae and by certain bacterial species.

(*iii*) Storage of hydrogen through metal hydrides/non metal hydrides (rice husk and misch metal).

(*iv*) Problems relating to utilization of hydrogen as a fuel, that is, development of suitable engines and burners etc.

(*v*) Liquid hydrogen production, storage and utilisation etc.

Satisfactory progress has been reported on the various funded projects advancing the state of art one step further.

In comparison with other important secondary energy vectors, such as electricity, natural gas, synthesis gas, or remote heating, hydrogen satisfies a number of necessary conditions for a potential future secondary energy vector. These include availability, manufacture without harm to the environment, good storage properties and very ready conversion into other forms of secondary energy.

A coupling of hydrogen and electricity in a future energy economy is technically feasible. Hydrogen can thus successively replace petroleum based products, including natural gas, and also synthesis gas.

Hydrogen can also be mixed with natural gas or synthesis gas, or can be used in pure state: it can therefore play an important part in ensuring gas supplies when fossil energy vectors become scarce.

Hydrogen can play a specially significant role when coupled with regenerative primary energy sources such as the sun and wind: since it can be stored, it can serve here to even out a fluctuating supply of energy and to transport energy to areas of high demand.

Hydrogen is also a candidate among the various alternative transport fuels such as synthetic hydrocarbons, methanol and liquid natural gas. Because of its favourable environmental characteristics it can be used both in air transport and in surface transport.

In view of the requisite order of magnitude of a future hydrogen market, however, enormous capital investment will be required for production, distribution and storage, and this factor represents the essential obstacle to a rapid introduction of hydrogen.

QUESTIONS

1. What are the different methods for hydrogen production? Explain in brief.
2. What is an electrolysis? Describe the more popular method of hydrogen production.
3. Describe thermo chemical method, for hydrogen production.
4. What is the Westinghouse electro chemical Thermal Sulphur cycle? Describe it.
5. Write short notes on:
   (*a*) Solar energy method of H_2 production
   (*b*) Fossil fuel method of H_2 production
   (*c*) Metal hydrides
   (*d*) Safety precautions in hydrogen utilization.
6. What are the different methods for hydrogen storage? Describe their advantages and disadvantages.
7. Write a brief note on hydrogen transportation.
8. Write the main applications of hydrogen gas.

UNIT—7

ADDITIONAL ALTERNATE ENERGY SOURCES AND IMPROVED ENERGY UTILIZATION

- Magneto Hydro Dynamic (MHD) Power Generation
- Thermo Electric Power
- Thermionic Generation
- Thermo Nuclear Fusion Energy

CHAPTER 12

Magneto Hydro Dynamic (MHD) Power Generation

12.1. Introduction

Eighty % of total electricity produced in the world, is hydal, while remaining 20% is produced from nuclear, thermal, solar, Geothermal energy and from magneto hydrodynamics (MHD) generators.

MHD power generation is a new system of electric power generation which is said to be of high efficiency and low pollution. In advanced countries MHD generators are widely used but in developing countries like India it is still under construction. This construction work is in progress at Trichi in Tamilnadu under joint efforts of BARC (Bhabha Atomic Research Centre) BHEL, Associated Cement Corporation (ACC) and Russian technologists.

As its name implies, magneto hydro dynamics (MHD) is concerned with the flow of a conducting fluid in the presence of magnetic and electric field. The fluid may be gas at elevated temperature or liquid metal like sodium or potassium. An MHD generator is a device for converting heat energy of a fuel directly into electrical energy without a conventional electric generator. In this system, an MHD converter system is a *heat engine,* in which heat taken up at a higher temperature is partly converted into useful (electrical) work and the remainder is rejected at a lower temperature. Like all heat engines, the thermal efficiency of an MHD converter (*i.e.,* the proportion of the heat taken up that is converted into useful work) is increased by supplying the heat at the highest practical temperature and rejecting it at the lowest practical temperature.

MHD generation looks the most promising of the direct conversion techniques for the larger scale production of electric power. It is observed that economic and physical factors will lead to design outputs of the order of 1000 MW, In fact MHD is really of interest only for central power generation, its potentialities for a propulsion unit are remote.

12.2. Principles of MHD Power Generation

The principle of MHD generation is simply that discovered by Faraday: when an electric conductor moves across a magnetic field, a voltage is induced in it which produces an electric current. This is the principle of the conventional generator also, where the conductors consist of copper strips. In MHD generator, the solid conductors are replaced by a gaseous conductor; an ionized gas. If such a gas is passed at a high velocity through a powerful magnetic field, a current is generated and can extracted by placing electrodes in a suitable position in the stream. This arrangement as illustrated in Fig. 12.1 provides d.c. power directly.

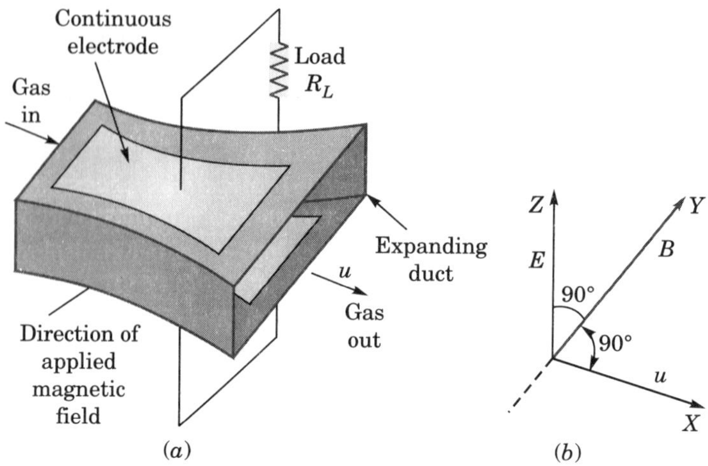

Fig. 12.1. Principle of MHD Power Generation.

The principle can be explained as follows. An electric conductor moving through a magnetic field experiences a *retarding force* as well as an induced electric field and current. This effect is a result of Faraday's law of electromagnetic induction. The induced emf is given by

$$E_{ind} = \vec{u} \times \vec{B} \qquad ...(12.1)$$

where, u is the velocity of the conductor and B is the magnetic, field intensity as shown in Fig. 12.1 (*b*). The induced current density is

$$\vec{J}_{ind} = \sigma\, \vec{E}_{ind} \qquad ...(12.2)$$

where σ is electric conductivity.

The retarding force on the conductor is the Lorentz force given by

$$\vec{F}_{ind} = \vec{J}_{ind} \times \vec{B} \qquad ...(12.3)$$

From an energy point of view, the movement of the force through a displacement (mechanical work) is converted to electrical work (current flow against potential difference) by means of the electromagnetic induction principle. *This is a work energy conversion and is not limited by the Carnot principle, of course, electromagnetic induction forms the basis of operation for conventional electric generators.*

The electromagnetic induction principle need not be limited to solid conductors; the movement of a conducing fluid through a magnetic field may also be employed for electric energy conversion when a fluid is used, the energy conversion technique is called the *magneto hydrodynamic* (MHD), *energy conversion.*

The direct conversion of kinetic (or motion) energy into electrical energy by the flow of an electrically conducting fluid, usually a gas or a gas liquid combination, through a stationary magnetic field. If the flow direction is at right angles to the magnetic field direction, an electromotive force (or electric voltage) is induced in the direction at right angles to both flow and field directions, as depicted in figure. This is the basic principle of MHD conversion (refer Fig. 12.2).

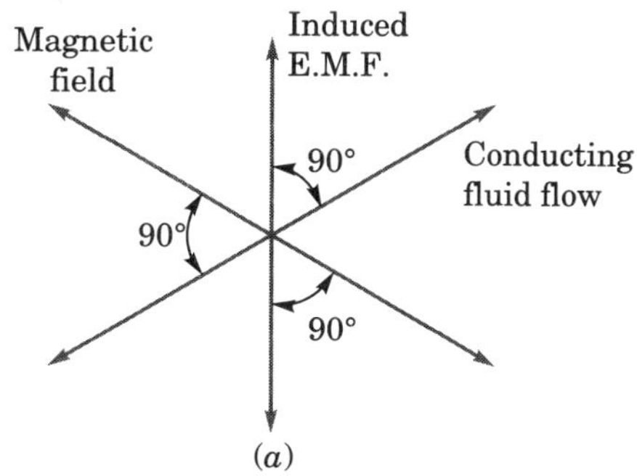

Fig. 12.2. Principle of magneto hydrodynamic conversion.

A schematic of MHD generator is shown in Fig. 12.3. The conducting flow fluid is forced between the plates with a kinetic energy and pressure differential sufficient to overcome the magnetic induction force F_{ind}. The end view drawing illustrates the construction of the flow channel. An ionized gas is employed as the conducting fluid. Ionization is produced either by *thermal means i.e.,* by an elevated temperature or by *seeding* with substance like cesium or potassium vapours which ionize at relatively low temperatures. The atoms of the seed element split off electrons. The presence of the negatively charged electrons makes the carrier gas an electrical conductor. The other way is to incorporate a liquid metal into a flowing carrier gas. Since the metal is a good electrical conductor, the gas metal mixture can be used as the working fluid in an MHD generator.

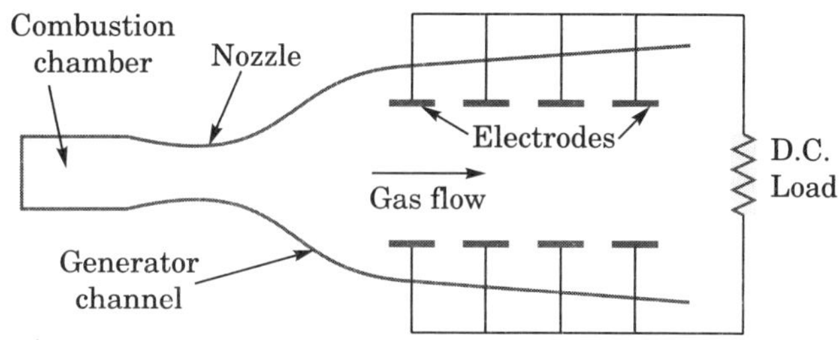

Fig. 12.3. Simple MHD Generator (magntic field perpendicular to the plant of the page).

In the overall power cycle, the MHD converter takes the place of a turbine in a conventional vapour or gas turbine cycle. Still, a compressor must be used to elevate the pressure, heat is added at high pressure and the flow is accelerated before entering the converter. A schematic of the MHD power cycle is shown in Fig. (12.4 *a* and *b*), along with a T-S diagram for the ideal processes.

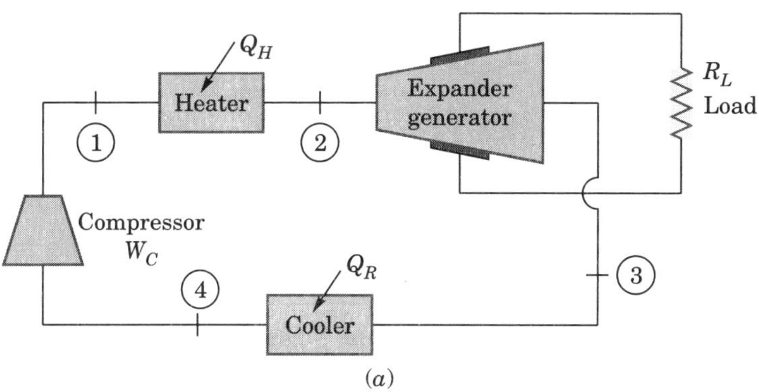

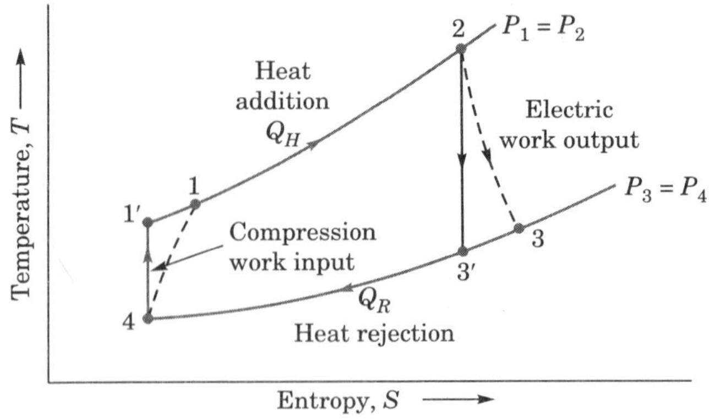

(b)

Fig. 12.4. (*a*) Schematic of MHD Power cycle (closed system)
(*b*) T-S diagram for cycle

In addition to the induced electric field E_{ind}, there is also some applied field E at the electrodes. The total current and force are, then

$$J = \sigma\,(\vec{E} + \vec{u} \times \vec{B}) \qquad \text{...(12.4)}$$

$$F = \vec{J} \times \vec{B} = \sigma\,(\vec{E} + \vec{u} \times \vec{B}) \times \vec{B} \qquad \text{...(12.5)}$$

Thermal efficiency. Clearly the overall MHD cycle is thermal power cycle and as such is limited by the Cornot efficiency. In terms of the T-S diagram of Fig. 12.4, the cycle thermal efficiency may be written as

$$\eta_{th} = \frac{\text{work output}}{\text{heat input}} = \frac{(h_{20} - h_{30}) - (h_{10} - h_{40})}{(h_{20} - h_{10})} \qquad \text{...(12.6)}$$

Where the indicated enthalpies are *stagnation* values which take into account the K.E. of the flow. The stagnation enthalpy is defined as

$$h_o = h + \frac{u^2}{2g_c} \qquad \text{...(12.7)}$$

where, u is the flow velocity. In practical MHD converters, high velocity ionized gases are usually employed for the conversion process, so that the K.E. of the flow represents a substantial portion of the total energy. The ideal process 2-3′ is shown as an isentropic which would assume perfect conversion of the flow pressure and kinetic energy into electric energy.

Several factors can reduce the efficiency of the converter. These are:

(*i*) Dissipation of energy in the internal resistance of the ionized gas.

(*ii*) A space charge barrier at the electrode surface.

(*iii*) Heat transfer through the electrode and insulator walls.

(*iv*) Various losses associated with fluid friction etc.

(*v*) Hall effect losses resulting in current induction in the direction of the flow. Such losses can be substantial when the seeded combustion gases are employed as the working fluid. The Hall effect causes a voltage gradient in a direction perpendicular to the applied magnetic field and current flow.

The gas, experiences a breaking force due to electromagnetic interaction, which can be compared to the retarding force produced by the turbine blades in the conventional system. As mentioned above, the gas must be ionized in order that it may conduct sufficiently for successful operation. In practice the degree of ionization required is very small (0.1%) so the gas is still composed mostly of neutral particles. It is these neutral particles that carry nearly all the K.E. of the stream, and, of course, they are unaffected by electromagnetic forces. The exact mechanism of the breaking force has not yet been explained, but it is thought that the applied magnetic field manifests itself through the force that it exerts on the electrons in the gas. This force is then coupled to the neutral particles by the electron-ion coulomb forces and the ion-neutral collisions. The retarding force will thus be a complex function of collision cross section and magnetic field density.

From the analysis it is evident that to achieve a large power output, the gas must have a high velocity, 10^3 m/s, and the applied magnetic field density must be as large as possible. (See equation 12.25). One of the major problems to achieve adequate conductivity in the gas (> 10 mhos/m). To achieve equilibrium conductivities in the pure gas, thermal ionization, temperatures of tens of thousands of degrees are required. By seeding the gas with elements which have low ionisation potentials, such as alkali metals caesium and potassium as stated above, it is possible to achieve reasonable conductivities at gas temperatures in the region of 2,000°C. It is because this temperature is just within the limit of materials technology that MHD generation has aroused so much interest in the last, few years.

In a practical MHD converter (or generator), the energy of motion of the conducting fluid is derived from heat obtained by burning a *fossil fuel*. Hence, an MHD generator is a device for converting heat energy directly into electrical energy without a conventional electric generator. The major reason for the interest in MHD conversion is that it can take better advantage than other heat engines of the high temperatures attained in the combustion of a fossil fuel. As a result high thermal efficiencies should be possible. For example, using MHD conversion as a *topping cycle* for a conventional steam turbine, an overall efficiency of 50 to 60% should be possible. This may be compared with about 40% for the most efficient steam turbine.

12.3. MHD Systems

12.3.1. Introduction

Magneto hydrodynamic conversion systems can operate in either *open* or *closed* cycles. In an open cycle system, the working fluid is used on the once through basis. The working fluid after generating electrical energy is discharged to the atmosphere through a stack. In the closed cycle, system the working fluid is continuously recirculated; the discharged working fluid is reheated and returned to the converter. In an open cycle system the working fluid is air. In closed cycle systems helium or argon is used as the working fluid. In open-cycle systems, the hot combustion gases offer seeding, can be used directly as the working fluid, in closed cycle systems, however, heat is transferred from the combustion gases to the working fluid by means of a *heat exchanger*. A higher working temperature and a better thermal efficiency are thus possible in open cycles, provided suitable construction materials are available.

Thus the MHD systems can be classified broadly as follows:

(1) Open cycle systems.

(2) Closed cycle systems. This may be further sub classified as:

(*i*) Seeded inert gas systems,

(*ii*) Liquid metal system.

These are described in the following sections.

12.3.2. Open-Cycle Systems

The arrangement of the system is shown schematically in Fig. 12.5. In this system, fuel used may be oil through an oil tank or gasified coal through a *coal gasification plant.* The fuel (coal, oil or natural gas) is burnt in the combustor or combustion chamber. The hot gases from combustor is then seeded with a small amount of an ionized alkali metal (cesium or potassium), to increase the electrical conductivity of the gas. The seed material, generally potassium carbonate, is injected into the combustion chamber, the potassium is then ionized by the hot combustion gases at temperatures of roughly (2300 to 2700°C).

To attain such high temperatures, the compressed air used to burn the coal (or other fuel) in the combustion chamber, must be preheated to atleast 1100°C. A lower preheat temperature would be adequate if the air were enriched in oxygen. An alternative is to use compressed oxygen alone for combustion of the fuel, little or no preheating is then required. The additional cost of the oxygen might be balanced by the savings on the preheater. The hot, pressurized working fluid leaving the combustor flows through a *convergent-divergent nozzle* similar to a rocket nozzle. In passing through the nozzle, the random motion energy of the molecules in the hot gas is largely converted into directed, mass motion energy. Thus, the gas emerges from the nozzle and enters the MHD generator unit at a high velocity.

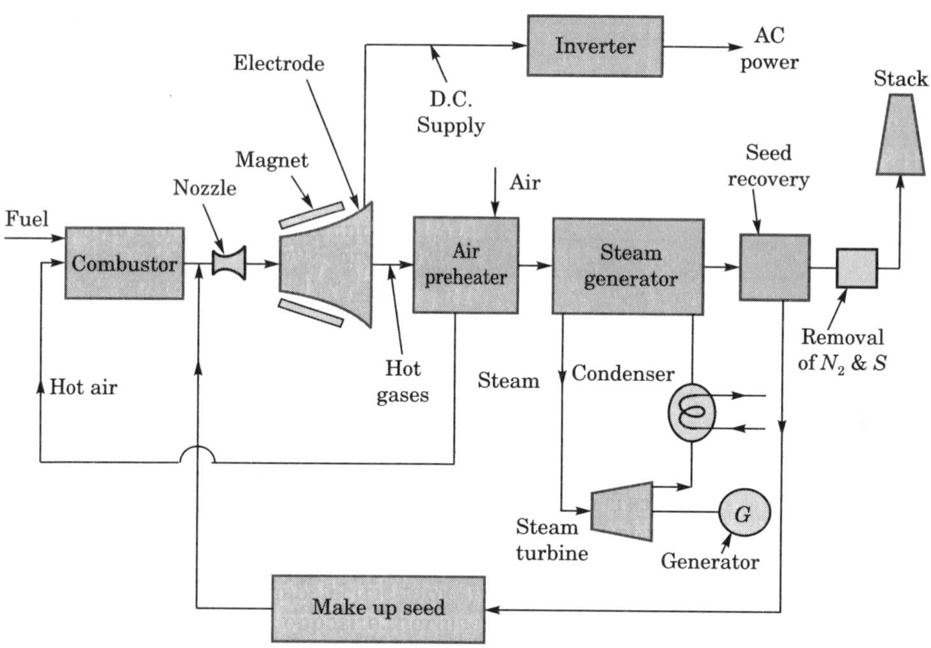

Fig. 12.5. Schematic of an open cycle MHD Generators.

The *MHD generator* is a divergent channel (or duct) made of a heat-resistant alloy (*e.g.*, Inconel) with external *water cooling*. The hot gas expands through the rocket like generator surrounded by powerful magnet. During the motion of the gas the positive and negative ions move to the electrodes and constitute an electric current. The magnetic field direction, which is at right angles to the fluid flow, would be perpendicular to the plane of paper. A number of oppositely located electrode pairs are inserted in the channel to conduct the electric current generated to an external load. The electrodes pair may be connected in various ways one of which is shown in Fig. 12.3. An MHD generator, unlike a conventional generator, produced direct current, this can be converted into commonly used alternating current by means of an inverter.

The arrangement of the electrode connections is determined by the need to reduce losses arising from the *Hall effect*. By this effect, the magnetic field acts on the MHD—generated (Faraday) current and produces a voltage in the flow direction of the working fluid rather than at right angles to it. The resulting current in an external load is then called the Hall current.

Various electrode connection schemes have been proposed to utilize the Faraday current while minimizing the Hall current. A simple way, although not the best is shown in Fig. 12.3. A better, but more complicated, alternative is to connect each electrode pair across a separate load, as in Fig. 12.6. Another possibility is to utilize the Hall current only; each electrode pair is short-circuited outside the generator, and the load is connected between the electrodes at the two ends of the MHD generator (Fig. 12.7).

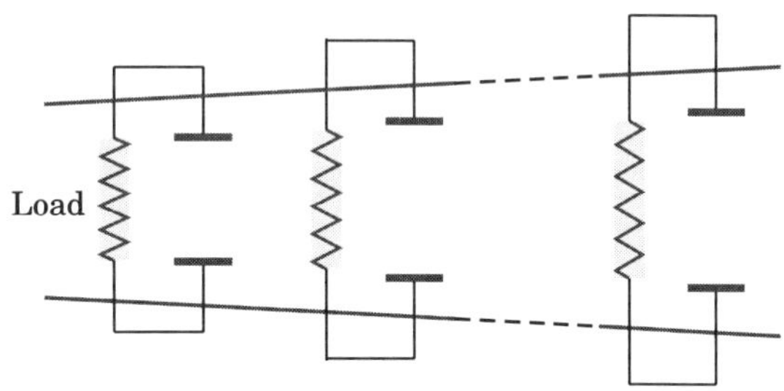

Fig. 12.6. MHD electrode connections to minimize Hall current.

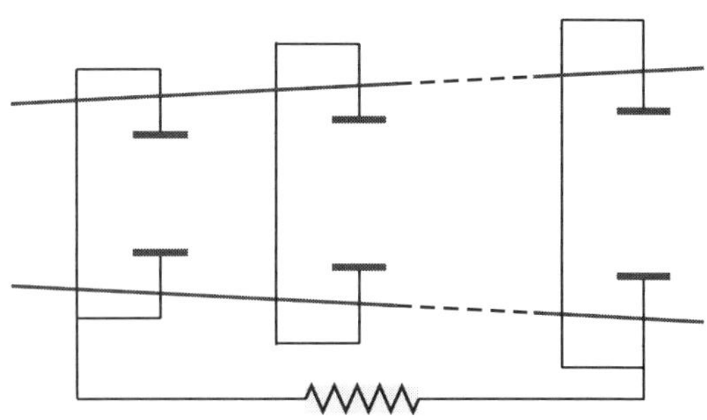

Fig. 12.7. Use of Hall current in MHD generator.

As the working fluid travels along the MHD generator and its energy is converted into electricity, its temperature falls. When the gas temperature reaches about 1900°C, the extent of ionization of the potassium is insufficient to maintain an adequate electrical conductivity. This places a lower limit on the useful operating temperature of the MHD system. The large residual heat available from the hot discharge working gas can then be utilized in several ways. For example, it conserve to preheat the combustion air by way of a heat exchanger similar to the regenerator in a gas turbine.

At this stage, some 25 to 35% of the heat energy in the working fluid should have been converted into electrical energy. The still hot gas leaving the air preheater would be used in waste heat (heat exchanger) boiler to produce steam for operating a turbine generator. In this way, another 25 to 30% of the initial heat should be recovered as electrical energy in a *combined cycle system*.

The seed material is recovered for successive use in seed recovery apparatus. Prior to the discharge of the working gas (as flue gas) from the steam boiler to the atmosphere the flyash from the coal fuel must be removed. However, instead of discharging the flyash, as is usually done, it may have to be treated for recovery of the seed material which is mixed with ash. Unless the sulphur in the coal has been removed (*e.g.* in a fludized bed combustor), the original potassium carbonate seed will have been converted into potassium sulphate. This must be extracted from the flyash and reconverted by chemical reactions into potassium carbonate.

The removal of residual sulphur by the potassium carbonate seed eliminates need for desulfurization of the flue gas, but *nitrogen oxides* are not removed. When oxygen alone is used for combustion of the coal (or other fossil fuel), the problem of nitrogen oxide formation does not arise. However, if nitrogen (from air) is present, the nitrogen oxide content of the combustion gases will be high because of the required high temperature of the working fluid. Consequently, a controlled combustion procedure is used to reduce the nitrogen oxide level in the discharge flue gas. The air supplied to the combustion chamber is not sufficient to permit complete fuel burning; combustion of the unburned fuel gases is then completed by introducing additional air at a later stage, beyond the MHD generator. The lower combustion temperature is accompanied by a decrease in the nitrogen oxide concentration.

For efficient practical realization an MHD system must have the following features:

1. Air superheating arrangement to heat the gas to around 2500°C, (the inlet temperature of MHD is about 2500°C), so that the electrical conductivity of the gas is increased.
2. The combustion chamber must have low heat losses.
3. Arrangement to add a low ionization potential seed material to the gas to increase its conductivity.
4. A water cooled but electrically insulating expanding duct with long life electrodes.
5. Seed recovery apparatus—necessary for both environmental and economic reasons.

12.3.3. Closed cycle Systems

Two general types of closed cycle MHD generators are being investigated. In one type, electrical conductivity is maintained in the working fluid by ionization of a seed material, as in open cycle systems; and in the other, a liquid metal provides the conductivity. The carrier is usually a chemical inert gas, although a liquid carrier has been used with a liquid metal conductor. The working fluid is circulated in a closed loop and is heated by the combustion gases using a heat exchanger. Hence, the heat source and the working fluid are independent. The working fluid is helium or argon with cesium seeding.

1. Seeded inert gas system. In a closed cycle system the carrier gas (argon/helium) operates in a form of *Brayton cycle*. The gas is compressed and heat is supplied by the source, at essentially constant pressure; the compressed gas then expands in the MHD generator, heat is removed from the gas by a cooler, this is the heat rejection stage of the cycle. Finally the gas is recompressed and returned for reheating.

A closed cycle MHD system is shown in Fig. 12.8. The complete system has three distinct but interloking loops. On the left is the external heating loop. Coal is gasified and the gas (having a high peak value of about 5.35 MJ/kg and at a temperature of about 525°C) is burnt in a combustor to provide heat. In the primary heat exchanger, this heat is transferred to a carrier gas argon/helium (working fluid) of the MHD cycle. The combustion products after passing through the air preheater (to recover a part of the heat of combustion products) and purifier (to remove harmful emissions) are discharged to atmosphere. Because the combustion system is separate from the working fluid, so also are the ash and flue gases. Hence, the problem of extracting the seed material from flyash does not arise. The flue gases are used to preheat the incoming combustion air and then treated for flyash and sulphur-dioxide removal, if necessary, prior to discharge through a stack to the atmosphere.

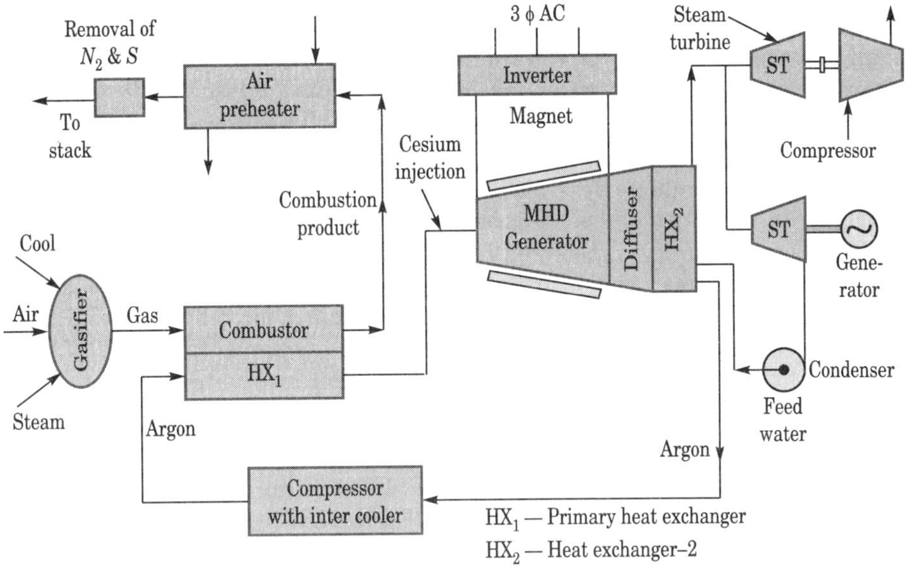

Fig. 12.8. A closed cycle MHD system.

The loop in the centre is the MHD loop. The hot argon gas is seeded with cesium and resulting working fluid is passed through the MHD generator at high speeds. The d.c. power out of MHD generator is converted to A.C. by the inverter and is then fed into the grid.

The loop shown on the right hand side in fig. is the steam loop for further recovery of the heat of working fluid and converting this heat into electrical energy in the diffuser the working fluid is slowed down to a low subsonic speed. Then hot fluid enters a secondary heat exchanger, which serves as a waste heat boiler to generate steam. This steam is partly utilized to drive a turbine generator and for driving a turbine which runs the argon (or helium) compressor. The output of the generator is also fed to the main grid. The working fluid is returned back to primary heat exchanger after passing through compressor and inter cooler.

A closed cycle system can provide more useful power conversion at lower temperature (around 1900°K as compared to 2500° K for open cycle system). The somewhat lower operating temperature of a closed cycle, MHD converter than of an open cycle system have an advantage in permitting a wider choice of materials. On the other hand, the lower temperature of the working fluid also means a lower thermal efficiency. Furthermore temperatures in the combustion chamber are still high, and special construction materials are required for the primary heat exchanger. Moreover the working fluid must be kept absolutely pure. The electrical stability of the flow in the generator poses problems because the gas is subjected to electrical fields approaching breakdown conditions.

The closed cycle MHD using rare gases as working fluid is developed and it is the most promising system among all. Generally, the heat source used is gas cooled nuclear reactor.

2. Liquid Metal System. When a liquid metal provides the electrical conductivity, an inert gas (*e.g.*, argon or helium) is a convenient carrier. The carrier gas is pressurized and heated by passage through a (primary) heat exchanger within he combustion chamber. The hot gas is then incorporated into the liquid metal, usually hot sodium, to form the working fluid. The latter then consists of gas bubbles uniformly dispersed in an approximately equal volume of liquid sodium.

The working fluid is introduced into the MHD generator through a nozzle in the usual ways; the carrier gas then provide the required high directed velocity of the electrical conductor (*i.e.,* the liquid metal). After passage through the generator, the liquid metal is separated from the carrier gas. Part of the heat remaining in the gas is transferred to water in a (secondary) heat exchanger to produce steam for operating a turbine generator. Finally the carrier gas is cooled, compressed, and returned to the combustion chamber for reheating and mixing with the recovered liquid metal. The working fluid temperature is usually around 800°C, as the boiling point of the sodium, even under moderate pressure, is below 900°C.

The lower operating temperature then in other MHD conversion systems may be advantageous from the material standpoint, but the maximum thermal efficiency is lower. A possible compromise might be to use liquid lithium, with a boiling point near 1300°C, as the electrical conductor. Lithium is much more expensive than sodium, but losses in a closed system be small.

The closed cycle liquid metal system has the basic advantage of high electrical conductivity. However, liquids being practically incompressible, high velocities can not be produced by expansion as in the case of gas. In order to achieve reasonable velocities, either in low efficiency jet pump or a two-phase system with vapour bubbles can be used. Even though liquid metal are excellent conductors, their vapours are poor electrical conductors. Thus there are several practical problems which have to be solved before liquid metal MHD systems can be commercially exploited. These systems are also appropriate for being coupled to nuclear reactors and the temperature range of interest to the liquid metal MHD (800°K—1300°K) is similar to liquid metal fast breeder reactors.

A liquid metal MHD cycle is illustrated in Fig. 12.10, in which liquid potassium after being heated in breeder reactor is passed through a nozzle to increase its velocity. The vapour formed due to nozzle action are separated in the separator and condensed and then pumped back to the reactor as shown. Then the liquid metal with high velocity attained is passed through MHD generator to produce d.c. power. The liquid potassium coming out of MHD generator is passed through the conventional steam plant, where in the heat exchanger the heat of liquid potassium is utilized to generate steam to run steam turbine generator.

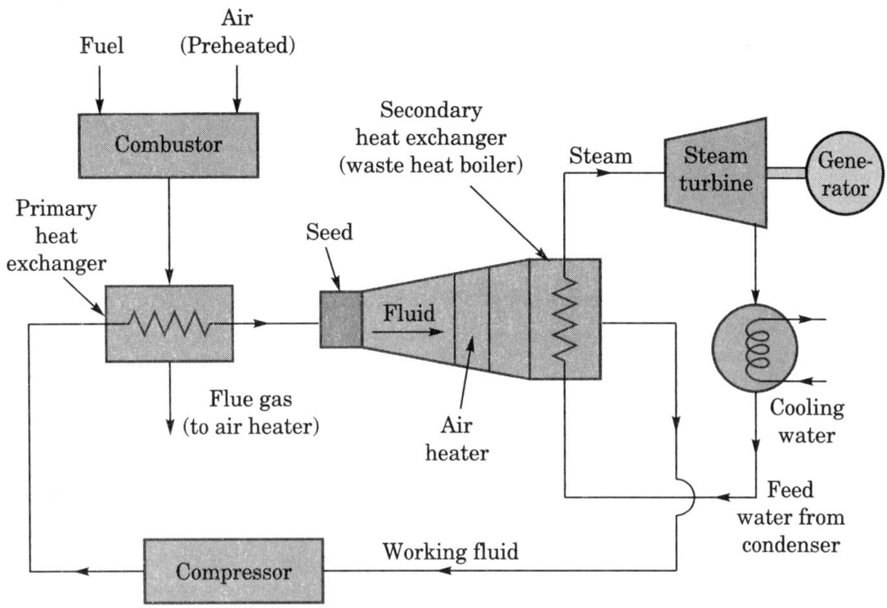

Fig. 12.9. Closed-cycle MHD Generator.

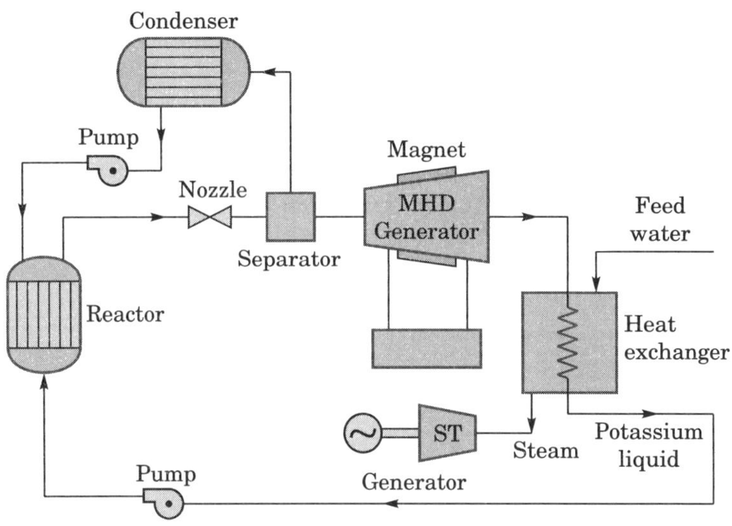

Fig. 12.10. Closed cycle MHD generator using liquid metal as working fluid coupled with steam genarator.

12.4. MHD Design Problems and Developments

The experiments have been demonstrated for the scientific feasibility of MHD generation, but the efficiencies attained so far have been relatively low and the life of the equipment has been short. Some of the matters requiring resolution before MHD generation can become economically practical are outlined below.

The combustor, MHD-generator channel, electrodes, and air preheater are exposed to corrosive combustion gases at very high temperatures, materials must be developed to permit an adequate operating life for the components. The ash (or slag) residue from the burning coal is carried over with the combustion gases and tends to cause erosion of exposed surfaces. However, deposition of the slag on such surfaces may provide some protection. Another problem is separation of the seed material (as potassium sulfate) from the flyash and its reconversion into its original (carbonate) form.

The difficulties associated with slag and seed recovery can be eliminated by using a fuel gas derived from coal rather than coal itself, in the combustor (open cycle system). An ash free low heat value fuel gas, made from coal at a moderate cost and treated for sulphur removal, would make a suitable fuel for an MHD conversion combustion. Burning of the gas in preheated air should provide adequate working fluid temperatures.

A more advanced concept is to use hydrogen gas made from coal and water. When this is burned in (compressed) oxygen, the product would be high-temperature steam. After seeding and passage as working fluid through the MHD generator, the steam would be used to drive a turbine generator, thus avoiding the need for a waste-heat boiler.

The power output of an MHD generator is theoretically propor-tional to the square of the magnetic field strength (Eqn. 12.25) hence a strong magnetic field is desirable. Conventional electro magnets, in which the field is generated by direct current passing through water cooled copper coils have been used in MHD studies. Such magnets would not be practical for large scale MHD power generation because they would require very large currents. Consequently superconducting magnets, which use very little current, will be necessary. Magnets of this type are being developed for use in MHD generators.

12.5. Advantages of MHD Systems

(1) The conversion efficiency of an MHD system can be around 50% as compared to less than 40% for the most efficient steam plants. Still higher thermal efficiencies (60-65%) are expected in future, with the improvements in experience and technology.

(2) Large amount of power is generated.

(3) It has no moving parts, so more reliable.

(4) The closed cycle system produces power free of pollution.

(5) It has ability to reach the full power level as soon as started.

(6) The size of the plant (m^2/kW) is considerably smaller than conventional fossil fuel plants.

(7) Although the costs can not be predicted very accurately, yet it has been reported that capital costs of MHD plants will be competitive with those of conventional steam plants.

(8) It has been estimated that the overall operational, costs in an MHD plant would be about 20% less than in conventional steam plants.

(9) Direct conversion of heat into electricity permits to eliminate the gas turbine (compared with a gas turbine power plant) or both the boiler and the turbine (compared with a steam power plant). This elimination reduces losses of energy.

(10) These systems permit better fuel utilization. The reduced fuel consumption would offer additional economic and special benefits and would also lead to conservation of energy resources.

(11) It is possible to utilize MHD for peak power generations and emergency service (upto 100 hours per year). It has been estimated the MHD equipment for such duties is simpler, has the capability of generating in large units and has the ability to make rapid start to full load.

12.6. Electrical Conditions: Voltage and Power Output of MHD Generator

The MHD power generation technology is the production of electric power utilizing a high temperature conducting plasma moving through an intense magnetic field.

Let us assume that a charge particle having charge q (a positive charge), moving with a velocity v' towards right and a perpendicular magnetic field of flux density B, pointing into the paper is applied. A magnetic force F acts on the charged particle. This force is given by

$$\vec{F} = q\,(\vec{v}' \times \vec{B}) \qquad ...(12.8)$$

Fig. 12.11 gives the force on the charge particle moving in the magnetic field.

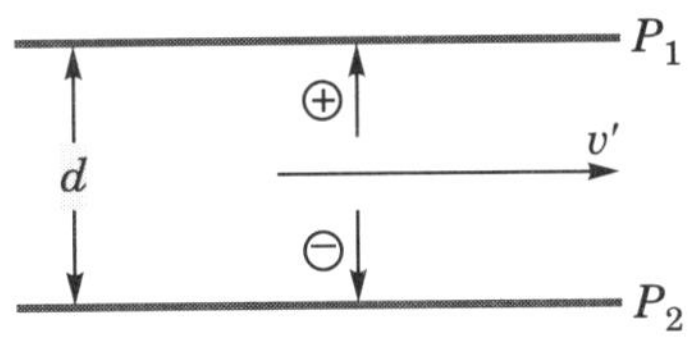

Fig. 12.11. Force on the charge particle.

Force F, velocity v', and magnetic flux density B are vectors. A positively charge particle is forced upwards and negatively charge particle is forced downward. If the particle is replaced by a mass of ionised gas molecules, moving at a velocity v', the positive ions would be accelerated towards the upper plate P_1 and the negative ions would be accelerated towards the plate P_2. If the plates P_1 and P_2 are externally connected through a resistance, a current would flow through the resistance. Thus mechanical energy is extracted from the gas and converted into electrical energy. This is the principle of MHD generation. It is a reversible process. If an emf is applied, in place of resistance, in a direction opposite to the direction of above current, energy would be supplied to gas and the gas particles would be accelerated.

If an electric field of strength E, is also operate in the region, the total force experienced as

$$\vec{F} = q\vec{E} + q(\vec{v}' \times \vec{B})$$

$$= q(\vec{E} + \vec{v}' \times \vec{B}) \qquad ...(12.9)$$

Throughout the following analysis, the symbol q denotes a positive charge. If in the final expressions, the solutions for electrons are required, then all that is needed is the substitution $q = -e$, where e is equal to 1.602×10^{-19} Coulomb. Equation (12.8) defines the direction of force. If the directions of $\vec{v}'$ and $\vec{B}$ are known, then turning from v_1' towards $\vec{B}$ a corkscrew will advance in the direction of $\vec{F}$.

In the case of the MHD generator, the velocity v' in equation (12.9) is the vector sum of the gas velocity u, and the particle drift velocity (being driven by current) v *i.e.*

$$v' = \vec{u} + \vec{v} \quad \text{...(12.10)}$$

Equation (12.9) now becomes

$$\vec{F} = q\vec{E} + (q\vec{u} \times \vec{B}) + (q\vec{v'} \times B)$$

$$= q(\vec{E'} + \vec{u} \times \vec{B} + \vec{v} \times \vec{B}) \quad \text{...(12.11)}$$

Equation (12.11.) can be written as

$$\vec{F} = q\vec{E'} + q(\vec{v} \times \vec{B}) \quad \text{...(12.12)}$$

where, $E' = (E + \vec{u} \times \vec{B})$, is the effective field seen by the moving particle. The drift velocity is related to the force on the particles through the mobility, μ *i.e.,*

$$F = \frac{qv}{\mu} \quad \text{...(12.13)}$$

Equation (12.11) then becomes

$$\vec{v} = \mu\vec{E} + \mu(\vec{u} + \vec{B}) + \mu(\vec{v} \times \vec{B}) \quad \text{...(12.14)}$$

This is the general equation. If we confine the scalar quantities to the case shown in Fig. (12.12), where the gas flows in the positive x direction across a magnetic field applied in the positive z direction (*i.e.,* $u_x = u$, $u_y = u_z = 0$; $B_y = B$, $B_x = B_z = 0$, and replace the terms $\mu B = \beta$ equation (12.14) becomes

$$\left.\begin{aligned} \vec{v_x} &= \mu E_x - \beta v_z \\ v_y &= \mu E_y \\ v_z &= \mu E_z + \beta u + \beta v_x \end{aligned}\right\} \quad \text{...(12.15)}$$

From these equations we can solve for v_x and v_z explicity. The y terms are omitted below because, in the example given, there will be no forces acting in the y-direction.

From equation (12.15)

$$\left.\begin{aligned} v_x &= \frac{\mu(E_x - \beta E_z) - \beta^2\mu}{1+\beta^2} \\ v_y &= \frac{\mu(E + \beta E_x) + \beta u}{1+\beta^2} \end{aligned}\right\} \quad \text{...(12.16)}$$

The current density J, is related to the particle drift velocity by the simple expression

$$J = \frac{\sigma v}{\mu} \quad \text{...(12.17)}$$

where, σ is the gas conductivity.

Therefore

$$\left.\begin{aligned} J_x &= \frac{\sigma(E_x - \beta E_z) - \sigma\beta B u}{1+\beta^2} \\ J_z &= \frac{\sigma(E_z + \beta E_x) + \sigma B u}{1+\beta^2} \end{aligned}\right\} \quad \text{...(12.18)}$$

These are the basic equations required to investigate the MHD generator. As seen there are currents produced in both the X and Z directions, the former being in the direction of gas flow. This leads to two basic designs of MHD generator that are possible, for current can be extracted and passed through a load from either J_z or J_x. These two basic designs have come to be called the *Faraday generator* and *Hall generator* respectively. The reason for this latter name is that the current density, J_x in the flow direction is due to the Hall effect. Here Faraday generator is discussed in following analysis.

Referring to Fig. 12.12, the motion of the gas is *in* x direction, magnetic field B is in y direction and the force on the particle is in z direction. When the current I is flowing across load resistance R_L, with voltage V across the load, then electric intensity between the plates will be given by

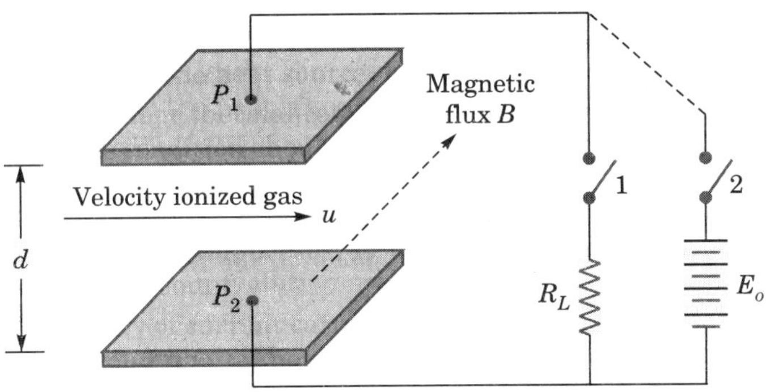

Fig. 12.12. Direction of magnetic field, gas velocity and force in MHD system.

$$\vec{E}_z = -\frac{V}{d} \quad ...(12.19)$$

where, d is the distance between plates. The total electric field is

$$E'_z = \vec{E}_z + \vec{B}u \quad ...(12.20)$$

$$= -\frac{V}{d} + \vec{B}\,u$$

$$= \frac{1}{d}\,(Bud - V) \quad ...(12.21)$$

The electromagnetic field E_z and B acting on the moving gas produce the same force on the ions as the electromagnetic fields. E_z' and B produce on a gas with zero average velocity. Evidently the open circuit voltage

$$E_0 = Bud \quad ...(12.22)$$

Let, R_g = internal resistance of the generator

$$= d/\sigma A \quad ...(12.23)$$

where, σ = conductivity of the gas A = plate area

Maximum power output (P_{max}) is obtained when internal resistance and total resistance are equal *i.e.*,

$$R_g = R_L$$

$$P_{max} = E_0 I$$

$$= I R_g I = I^2 R_g$$

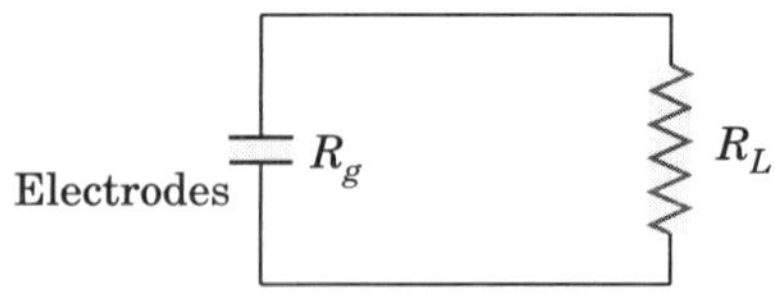

Fig. 12.13. Electrical circuit of generator of Fig. 12.1.

$$= \left(\frac{E_0}{R_L + R_g}\right)^2 R_g = \frac{E_0^{\ 2}}{4R_g} \qquad \text{...(12.24)}$$

$$= \frac{B^2 u^2 d^2}{4R_g} \qquad (\because \quad E_0 = Bud)$$

Putting the value of R_g from equation (12.23)

$$P_{max} = \frac{B^2 u^2 d^2}{4d} \sigma A = \frac{1}{4} \sigma u^2 B^2 \ dA$$

$\therefore$ Maximum Power per unit volume

$$= \frac{1}{4} \sigma u^2 B^2 \qquad \text{...(12.25)}$$

It can be shown with the help of Fig. 12.12 that if an external voltage greater than Bud (or E_0) be connected instead of Load by putting the switch to position 2, then the directions of current and force acting on ions are reversed. The system tends to speed up the gas velocity. The velocity of the ejected gas is higher than the velocity of inlet gas. The reaction force acting on the magnet tends to push the MHD generator in the negative x direction. Electrical energy is converted in mechanical energy.

Example 12.1. *With the following specifications for an MHD generator, calculate the open circuit voltage and maximum power output.*

Plate area	$= 0.25\ m^2$
Distance between plates	$= 0.50\ m$
Flux density	$= 2\ Wb/m^2$
Average gas velocity	$= 10^3\ m/second$
Gaseous conductivity	$= 10\ Mho/m.$

Solution. We have open circuit voltage

$$E_0 = B\ u\ d$$

where, B = flux density

u = Average gas velocity

d = distance between plates.

$E_0 = 2 \times 1000 \times 0.50 = 1000$ volts. **Ans.**

Generator resistance

$$R_g = \frac{d}{\sigma A} = \frac{0.50}{10 \times 0.25} = 0.20 \text{ ohm}$$

$$\text{Maximum power} = \frac{E_0^{\,2}}{4R_g} \text{ watts}$$

$$= \frac{(1000)^2}{4 \times 0.20} = \frac{1000 \times 1000}{0.80}$$

$$= \frac{10}{0.80} 10^5 \text{ watts} = 1250 \text{ kW. } \textbf{Ans.}$$

12.7. Gas Conductivity

The gas conductivity (σ) is one of the parameters in the expression for the power generated. It is this parameter that has impeded the progress of the MHD generation. For to attain reasonable conductivity in a gas is a difficult task. In this context a reasonable conductivity refers to a value between about 10 and 100 mhos/m. Such a conductivity is very small in comparison with those of materials with which we are more familiar. For instance, most metals have conductivities of the order of 10^7 mhos/m and even transistor materials have conductivities exceeding 100 mhos/m.

The gas has to be made sufficiently conducting, in order that it may be utilised as a working fluid. In other words, it has to be *ionised*—a process in which electrons are removed from an atom. Conduction is due to the free electrons and positive ions which move under the effect of a magnetic field. There are several types of ionization, *viz.*,

(*i*) Thermal ionization;

(*ii*) Magnetically induced ionization;

(*iii*) Radio-frequency wave induced ionisation;

(*iv*) Radioactivity;

(*v*) Photoionisation;

(*vi*) Electron-beam ionization; and

(*vii*) Flames.

Thermal ionization is the most important method of ionising a plasma. Ionization is obtained by imparting enough thermal energy to the gas. This method of ionization is directly dependent upon temperature. As the temperature of a gas is raised, the kinetic energy of its constituents is increased. When the temperature has reached a sufficiently high value than an electron can ionize a molecule with which it colloides. Such electron-molecule collisions thus produce more electrons, giving rise to what is called *thermal ionisation*.

In *magnetically induced ionisation,* a d.c. voltage is applied across a gas at reduced pressure to create an electric field which will supply energy to electrons.

Radio frequency waves can be used to produce a low amount of ionization in a gas. This is not effectively enough for MHD generation.

The other methods are yet to be developed to be of practical use. The gas does not need to be very highly ionized for the MHD generator. A minimum conductivity of about 10 mho/m is in order. It may be noted that this compared with the values of 10^2 mho/m for transistor materials, 10^5 mho/m for graphite and 10^7 mho/m for metals. In practice the gas is 'seeded' with materials that can easily be ionized so that a reasonable conductivity may be obtained at a temperature within the bonds of material technology. It is found that seeded gas operating at about 2000°C forms fairly a good working media for MHD generators. Elements with low ionisation potential, such as alkali metals, are used for seeding. The best element is found to be caesium which has an ionisation potential of 3.87 eV. This element is, however expensive and salt of potassium which is cheaper is being considered.

The working gas, although ionized is still composed mainly of neutral atoms, which carry nearly all of the kinetic energy of the gas stream. The magnetic field applied exerts a force on the free electrons in the working medium so that they move in tight circles in the magnetic field and are effectively trapped. These electrons in turn act upon the positive ions due to the coulomb electrostatic attractive forces. Ion-neutral collisions then provide the last link in the chain which provides an effectively coupling between the magnetic field and the neutral atoms of the gas. The magnitude of the "braking force" is a complex function of collision cross-section and magnetic field strength. It is this braking force that slows down the gas, thus converting the kinetic energy directly to electrical energy.

Simplified Analysis

The gas must be made a conductor by ionizing some fraction of it. If the temperature of the gas is raised, the constituent atoms have increased K.E., with the result that ionizing collisions become more probable. The major ionization mechanism is electron atom collisions, represented by the formula

$$A + e^- \rightleftharpoons A^+ + 2e^- \qquad ...(12.26)$$

Consider a gas consisting of charged and neutral particles of mass m. Each charged particle carries a charge q. If an electric field E is applied, the particle will experience a force qE. Such particles will drift in the field, and will travel a distance ξ in the direction of the field. If the particle is considered to lose all its velocity on collision then, from elementary kinetics,

$$\xi = \frac{1}{2}\left(\frac{qE}{m}\right)T^2 \qquad ...(12.27)$$

where, T = mean free time between collisions.

Thus the average drift velocity $\overline{v}$ in the field direction is given by

$$\overline{v} = \frac{1}{2}\frac{qE}{m}T \qquad ...(12.28)$$

If there are n charged particles per unit volume, the current density is

$$\delta = nq\,\overline{v} = \frac{nq^2}{2}\frac{ET}{m} \qquad ...(12.29)$$

and the electrical conductivity is,

$$\sigma = \frac{J}{E} = \frac{nq^2T}{2m} \qquad ...(12.30)$$

The mean free time is equal to the mean free path λ, divided by the mean thermal speed of the particles, $\overline{C}$. From the kinetic theory,

$$\overline{C} = 0.92\left(\frac{3KT}{m}\right)^{1/2} \qquad ...(12.31)$$

and

$$\lambda = \frac{1}{\Sigma n_i \alpha_i} \qquad ...(12.32)$$

where, K is Boltzmann's constant,

n_i is the density of the particles of species i in the gas,

α_i is the collision cross-section between them and charged particles, and T is the absolute temperature coming equations (12.31) and (12.32), we get

$$T \simeq \frac{1}{\Sigma n_i \alpha_i}\left(\frac{m}{3\,KT}\right)^{1/2} \qquad ...(12.33)$$

Thus the conductivity is given by

$$\sigma \simeq \frac{nq^2}{\Sigma n_i \alpha_i}\frac{1}{(mKT)^{1/2}\,.\,2\sqrt{3}} \qquad ...(12.34)$$

As the conductivity depends on $m^{-1/2}$, electrons in the gas will be atleast 43 ($= 1840^{-1/2}$) times as effective as ions, and hence it is usually only the electronic conductivity that is significant. If the gas is only slightly ionized (as it is in MHD generator system), the gas a whole is effectively electrically neutral. This means that the number of positive iona present, n_p is equal to the number of electrons n, assuming there are no negative ions, *i.e.*,

$$n_p = n \qquad ...(12.35)$$

Consider a gas containing a total of n_n particles per unit volume of which n_p are positive ions. Also let α_p and α_n be the collision cross-sections for the ions and neutral particles respectively. Expanding the summation in equation (12.34) to cover the ions and the neutral particles, the conductivity is

$$\sigma \sim \frac{e^2}{(mKT)^{1/2}}\frac{n}{n\alpha_P + (n_n - n)\alpha_n} \qquad ...(12.36)$$

where, $-e$ is now used instead of to denote the charge of an electron. The conductivity thus depends upon the factor

$$\frac{1}{(\alpha_p - \alpha_n) + \frac{\alpha_n}{N}} \qquad ...(12.37)$$

where, $$N = \frac{n}{n_n}$$

i.e., N is fractional degree of ionisation.

Consider equation (12.37) it implies to regims:

(*i*) $N < \dfrac{a_n}{\alpha_P - \alpha_n}$ ...(12.38)

In this case the conductivity is directly proportional to N *i.e.*, the degree of ionization. In this region electron neutral collisions are more important than electron ion collisions.

(*ii*) $N > \dfrac{\alpha_n}{\alpha_P - \alpha_n}$ ...(12.39)

This occurs when electron-ion collisions are the dominant process. Here the conductivity is independent of N, and hence of the degree of ionization.

For a gas of interest in MHD power generation the factor $\dfrac{\alpha_n}{(\alpha_p - \alpha_n)}$ is typically of the order 10^{-3}. Therefore equation (12.40) shows that there is no point in trying to achieve a degree of ionization higher than 0.1% for beyond this figure. It has little effect on the electrical conductivity. This is illustrated in Fig. (12.14) with the assumption that $\frac{\alpha_n}{\alpha_P - \alpha_n} = 10^{-3}$.

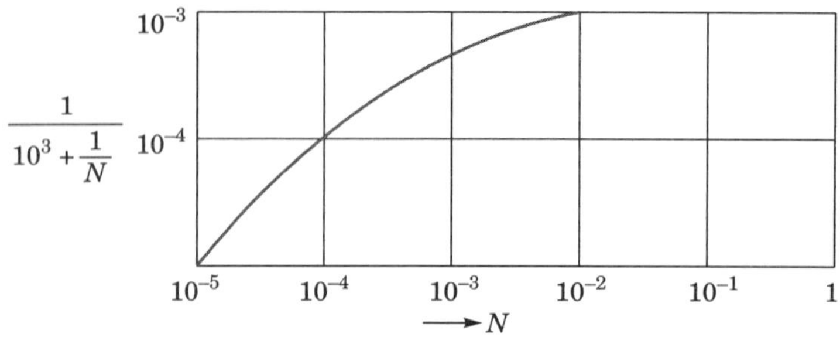

Fig. 12.14. Fractional degree of ionization.

It may be pertinent here to note that the gas in the generator can be termed a plasma *i.e.,* an ionized gas in which the positive ions and electrons are present in equal numbers, and where the shielding distance is small compared with the other distance of interest. The gas in the MHD system satisfies this criterion. It is to be noted, however, that it is very different from the fully ionized plasmas associated with thermonuclear fusion.

Thermal Ionization. A thermodynamic analysis by Saha leads to an equation relating the electron density to temperature.

This is

$$n^2 = n_n G \left(\frac{2\pi KTm}{h^2}\right)^{3/2} \exp. \frac{-eV_i}{KT} \qquad ...(12.40)$$

where, h is Plancks constant,

V_i is the ionization potential of the gas atoms, and

G is a statistical weight factor.

Equation (12.40) is actually an approximation to the case where the electron and ion densities are very much smaller than the density of the neutral atoms *i.e.,* it holds for low level of ionization.

Elementary kinetic theory provides the relationship

$$p = n_n KT \qquad ...(12.41)$$

where p, is the pressure of the gas.

Using this expression, together with equations (12.36) and (12.40), we obtain for the case appropriate to condition Eqn. (12.38)

$$\sigma \simeq \frac{e^2}{(mp)^{1/2}} \frac{G^{1/2}}{\alpha_n} \left(\frac{2\pi KTm}{h^2}\right)^{3/2} . \exp. \frac{-eV_i}{KT} \qquad ...(12.42)$$

This shows that the conductivity increases as the pressure is reduced, and that the ionization potential of the gas atoms has a great influence. If V_i is halved, the conductivity is increased by a factor of 7.4.

Equation (12.42) also shows that, to achieve conductivities above 10 mhos/m in the pure gas by thermal ionization alone would require temperatures of several tens of thousands of degrees. Such temperatures are far beyond the scope of present day materials technology, and consequently it must be concluded that thermal ionization in the pure gas can not be used to obtain the required conductivities.

It is necessary either to reduce the temperature requirement in some way or to increase the electron concentration by a method not based on temperature. Such methods are described below.

Seeding. One way of increasing the conductivity of gas, without the need of exceedingly high temperatures, is to introduce into the gas a material whose ionization potential is lower than that of the gas atoms themselves. This means that the addition; or seeding agent, will ionize more readily than the gas itself, and will thus enhance the electrical conductivity.

As stated earlier, main seeding materials are alkali metals like caesium, potassium etc. The addition of the seeding material allows the attainment of suitable conductivities at very much lower temperatures. It is seen that conductivities between 10 and 100 mhos/m can be achieved in the temperature range 2000—2400°K, according to the percentage of seed material added and the gas pressure. These temperatures are still very high, but materials technology has now advanced sufficiently to make working at such temperatures atleast feasible. However owing to the difficulties incurred, alternative methods of increasing the conductivity are intensively studied.

Non-thermal Ionization. An alternative approach to the problem of achieving adequate conductivity is to obtain the required electron density by a non equilibrium process. This non-equilibrium state must then be maintained for a time sufficiently long to allow the gas to pass through the generator duct in non-equilibrium ionization systems, the carrier is a monoatomic gas (*i.e.,* a gas with any one atom in the molecule) this is usually the chemically inert gas argon, although helium is possible alternative. The seed is cesium metal, which volatilizes and ionizes more readily than potassium in potassium carbonate. Caesium is more expensive than potassium, but in a closed system the losses should be small. In the monoatomic gas carrier, non-equilibrium ionization can occur; the extent of ionization then corresponds to that which would normally exist at a much higher temperature (*e.g.,* as in combustion gases). Adequate electrical conductivity for MHD applications can thus be attained in a monoatomic gas with a cesium seed at temperatures below 2000°C.

Recommendation methods, as electron heating and seeding can be employed.

12.8. Materials for MHD Generators

Owing to the high temperature of plasma (2700°C), refracting materials are required in several parts of generators like electrodes, channel or duct wall. Due to stringent environmental conditions selection of materials is very limited. Following are the important factors to be considered for selection of materials.

(*i*) Thermal shock resistance (*ii*) Electrical conductivity

(*iii*) Corrosion resistance (*iv*) Erosion resistance

(*v*) Oxidation reduction resistance

(*vi*) Melting point of materials and (*vii*) Density.

Requirements for materials is MHD generators:

1. *Electrodes*—For high conductivity, high temperature is required
   —Electrically conductive and structurally stable at high temperature.

2. *Duct liner*—Cooler than electrodes
   Must be electrical insulator
   Must be thermal insulator.
3. *Magnet*—stronger than permanent magnets or Ferromagnetic cores can provide
   Few materials have high melting points
   For Insulators—oxides, nitrides, zirconates etc.
   For conductors—Refractory metals, nitrides borides and carbides.

Field coils

Permanent magnet can produce only 0.4 Wb/m^2. Electromagnets can produce only 1 Wb/m^2 but massive and lossy.

In magnetic field below a certain temperature metals, compounds and alloys exhibit reduction in resistance to *zero-superconductivity*. (Current carrying capacity at a given field, at 10 Wb/m^2—NbTi, Nb_3Sn can take 10^4 amp/cm^2 and 2×10^4 amp/cm^2). Increase in magnetic field above a certain temp. (Critical field) *destroys* superconductivity.

Electrode materials. Electrodes with higher electrical conductivity contribute towards better charge transfer across the electrodes/plasma interface. Most of the initial channel wall designs involved water cooled metallic electrodes due to the oxidising condition. But the cold surface creats cold boundary layer which generally promote high thermal losses, low conductivity and leads to considerable voltage drop in this region. The alkali seed condensation and cold slag condensation degrade the properties of electrode.

Important aspects of any practical electrode system for MHD generator are high reliability and long life of operation (10,000 hrs) and efficient in transferring current densities of 1—5 amp/cm^2. Electrode material should be stable chemically, thermally and mechanically.

There are several potential ceramic materials for the electrode system that operate near 1500—2000°C. The carbides (SiC, ZrC, MbC), borides (ZrB_2, TiB_6, LaB_6), silicides (MO Si_2, Ws_1), chromites (Sr, Mg, doped), oxides and their solid solutions (ZrO_2—CeO_2, CeO_2) and graphite as electrically conducting rafractory materials. Most promising of these, however stabilized zirconia.

Zirconia based materials. Zirconia is a good refractory oxide (Melting point 2800°C). However, pure zirconia can not be used as it undergoes phase transitions at different temperatures. During these phase transitions, non-linear thermal expansion takes place. Hence elimination of the disruptive phase transition is necessary. This can be achieved by the addition of CaO (calcium oxide) Y_2O_3 (yitrium oxide), MgO, which results in the formation of cubic phase that is stable over a wide range of temperature and composition 15 to 27% CaO and 7 to 52% Y_2O_3 are required for stabilization of ZrO_2. This stabilization process results in formation of oxygen vacancies which lead to oxygen ion conduction at elevated temperatures. Due to all these problems simply stabilized zirconia CSZ (calcium stabilized zirconia) or YSZ (yitrium stabilized zirconia) is not preferred as MHD material.

On the other hand substitution of teravalent metal ions for Zr^{4+} will not lead to anion vacancy formation and at the same time undesirable phase transitions may be supressed. In the process it will enhance electronic conduction. Transition and rare earth metal oxides are potential dopant materials for this purpose.

Multivalency of these ions gives rise to dropping conduction *e.g.*, Ceria doped ZrO_2, maximum conductivity is obtained at 18—19 M/O, CeO_2 substitution.

Chromite based materials. Lanthanum Chromite $LaCrO_3$ is one of the materials which has been thoroughly investigated for this application. Incorporation of substantial amount of MgO improves its thermochemical properties without significantly changing its thermochemical properties and electrical properties as compared to SrO doped chromite (5-20 M/O). It is a good electronic conductor with little variation in conductivity with temperature. Materials of an optimum composition LC 20 M (20% MgO doped $LaCrO_3$) is found to be highly promising for use in MHD. In chromite based materials major problem is the volatility in oxidizing environment. It shows considerable weight loss at 1700°C which deteriorates the electrodes. However sintering of LC 20 m is not significantly dependent on the partial pressure of oxygen unlike in case of Sr doped $LaCrO_3$.

Solid solution of La AlO_3 and $SnZrO_3$ retards vaporization but electrical conductivities of these materials are relatively low.

From electrical conductivity point of view, $LaCrO_3$—Cr Cermet (40—60 M/O) has been considered. This cermet has high thermal conductivity more than that of $LaCrO_3$ and also high electrical conductivity. However, oxidation of chromium metal is a main problem.

Aluminate based materials. Spinel (Mg Al_2O_4) is a high resistivity insulator and hence used for interelectrode insulation on addition of Fe_3O_4 it becomes conducing electrodes material. As its melting point is 1860°C, it cannot be used safely above 1750°C. It is suitable for coal MHD generators.

Nickel oxide based materials. Lithium doped NiO is a good electronic conductor but does not passes high enough melting point. Hence the most suitable materials are CeO_2 and ZrO_2—CeO_2 solid solution.

Materials for Channel or Duct Wall

Construction of duct through which to pass the very hot, sometimes very corrosive gas is second major problem. Materials to withstand temperatures of the order of 3000°C are hard to find. Search of suitable ceramics or coatings of ceramics is continuing. Steam cooled walls which are made of insulated laminations of very thin nickel or stainless steel have been tried.

Apart from high temperature, the duct wall has to withstand high temperature gradients as well as withstand the stresses encountered in use and also the thermal shock in the event of accidental shut down of the gas supply. The duct material has to be electrically insulating at the operating temperatures.

The electrodes which are required to emit a cupious stream of electrons throughout their life time are required to withstand more regorous environment than the duct. Water cooled copper electrodes have been used with some success.

High temperature coupled with corrosion by alkali seeded plasma and erosion by high velocity plasma constitute severe environments experienced by structural materials in MHD generators. Ceramic materials, inherently, offer superior chemical stability towards oxidation and corrosion by alkali seed materials. However, thermal shock of ceramic is necessary to be taken into account. The density of ceramic should be high in order to reduce the penetration of slag, seed and gas diffusion which enhance degradation of insulation and electrode material. At the same time density should not be high to jeopardize the thermal shock resistance.

Resistivity and dielectric breakdown strength of ceramic should also be taken into account. Dielectric breakdown strength of Al_2O_3 is 10^5—10^6 V/cm at room temperature. This value decreases at high temperature, it drops from 10^6 V/cm to 10^4 V/cm at 1400°C. Al_2O_3 MgO, $MgAl_2O_3$ are potential channel wall materials. However, Al_2O_3 reacts with potassium seed at high temperature to give potassium Beta alumina ($K Al_{11}O_{17}$), on the other hand, MgO, stands seed and slag corrosion. It also has good thermal shock resistance. Hence use of MgO as duct wall material is favourable.

12.9. Magnetic Field

As the output power delivered by an MHD generator (refer Eq. 12.25) is proportional to B^2 (magnetic flux density^2), hence the magnet should be as large as possible. The magnet being one of the most expensive items of an MHD generator, it is imperative that the production of large magnetic fields is optimised, constant research in this direction has been in progress. In this connection four types of magnets are identified which are:

(*i*) Permanent magnets,

(*ii*) Water-cooled electromagnets,

(*iii*) Cryogenically cooled electromagnets,

(*iv*) Superconduction magnets.

At the outset, the iron core of a magnetic circuits which is used to reduce the reluctance of the circuit is no longer of any consequence where field strength of over 2 Wb/m^2 are required. Thus air cooled electromagnets are used.

The first two types of magnets mentioned above are self explanatory. The cryogenically cooled magnet employs refrigerator principle for cooling. In a superconducting magnets there is no Jouleon heating. Superconductivity is exhibited by some metals compounds, and alloys when they are in a magnetic field, and below a certain temperature usually a few degrees Kelvin, their resistance decreases to zero. Each material has a critical field strength, beyond which if the field is increased, the superconductivity is destroyed. The transition temperature and critical field for some superconducting materials are given in Table 12.1.

Table 12.1

Properties of Superconductivity Materials

Material	*Critical field* Wb/m^2	*Transition temp.* °K
Tin	0.08	7.18
Niobium	0.25	8.0
Niobium 33 Zirconium	8.00	10.50
Niobium-titanium	12.00	9.80
Niobium 3 tin	22.10	18.00

12.10. Superconductivity

The ability of certain metallic elements, alloys, or compounds at extremely low temperatures to carry electric current with essentially no resistance. Once an electric current is established in a superconducting loop, it will continue almost indefinitely.

As the superconducting material is cooled, its resistance falls to zero at certain *transition* (or *critical) temperature.* For nearly all superconducting metallic elements, the transition temperature is below 10 K, that is, within 10°C of the absolute zero, the lowest conceivable temperature; some metallic alloys and compounds, however, have transition temperatures upto about 22 K. [A temperature of T Kelvins (K) is equivalent to 273.15 + T in °C].

The recorded transition temperatures for superconductors are those measured at low (essentially zero) current and magnetic field. An increase in strength of the current or of the surface magnetic field results in a decrease in the transition temperature. Above a certain *critical current* or *critical magnetic field* strength, the normal conducting state is restored and persists no matter how low the temperature. Hence superconductivity can occur only if the current and surface magnetic field are both less than the respective critical values.

Superconducting materials are classified as Type—I and Type—II. Those of Type—I are mostly pure metals; they have relatively low critical currents and magnetic fields and are of little practical value. Type II superconductors, on the other hand, which are either compounds or solid-solution alloys of two or more metals, have certain characteristics that make them potentially useful materials. As a general rule, Type II superconductors retain their superconductivity at high magnetic fields and high currents. However, the critical temperature above which superconductivity ceases is decreased as the/ magnetic field and the current are decreased.

Of the high current Type—II superconductors known are: (*i*) one is a solid-solution alloy of 30 to 40 atom % of niobium (Nb) with titanium (Ti), represented by Nb-Ti; its transition temperature is about 10 K. (*ii*) The other is a compound of niobium and tin (Sn), formula Nb_3Sn, with a transition temperature of 18 K. Not only does the transition to the superconducting state occur at a higher temperature, which is advantageous from the cooling stand point, but the critical current and magnetic field are higher for Nb_3Sn than for Nb-Ti. On the other hand Nb_3Sn is a brittle material that has proved difficult to fabricate.

Applications. Several energy-related applications of high current type II superconductors are being studied. In an electromagnet, a magnetic field is produced by passing a current through a conductor, usually a copper coil wound around an iron core. One factor determining the field strength is the current strength, and this is limited by heating of the coils. The heating rate is equal to I^2R, where, I is the current strength and R is the resistance of the coil. In a superconducting coil, R is essentially zero and large currents can flow with little or no heating.

A thin wire (or film) of superconductor can thus carry more current than a much thicker normal conductor without over heating. Hence if a sufficiently low temperature can be maintained, electromagnets with compact superconducting coils can be designed to produce strong magnetic fields. Such electromagnets have potential applications in *electric generators, magnetohydrodynamic conversion* and *magnetic-confinement fusion.*

The amount of electric power that can be transmitted by an underground line of normal conductor is limited by resistance heating. The virtual absence of resistance heating in superconductors thus makes these materials of interest for underground *elective power transmission.* Another possible use of superconductors in the power industry is for electrical *energy storage.* The stored energy is equal to

V. I, where *V* is the voltage and *I* is the current strength. Since large values of *I* are possible, substantial amounts of energy can be stored.

Fabrication. Production of superconductors for practical use is complicated by the significant changes in their properties that accompany apparently minor variations in fabricating conditions. Nevertheless, both Nb-Ti and Nb_3Sn have been produced in useful forms as thin filaments or layers in a matrix of a normal conductor (*e.g.*, copper). The latter serves as a heat sink to prevent undesirable local temperature increases that would cause the superconductor to revert to the normal state and to provide an alternative temporary path for the current in the event of a loss of superconductivity.

Niobium titanium superconductor composites are fabricated by embedding thin rods of this material in copper and drawing out into wires containing fine superconducting filaments. This is possible because Nb-Ti is ductile. Several wires may be twisted or braided together to form a multifilament superconductor. Flat taps or ribbons can be produced in similar manner.

Since Nb_3Sn is brittle, direct wire drawing is impracticable. The problem has been solved by producing filaments of niobium, which is ductile metal, and then converting them into Nb_3Sn. For example, rods of niobium are embedded in a copper-tin alloy and drawn into wires; upon heating, tin diffuses into the niobium filaments to form thin surface layers of Nb_3Sn. Alternatively, filaments of niobium in copper are coated with tin and heated; the tin diffuses through the copper and converts the niobium into Nb_3Sn. The outer layer of tin, which may be undesirable, is then removed by acid. Superconducting taps (or ribbons) required for certain applications are fabricated in a somewhat analogous manner by rolling sheets rather than drawing wires.

Low Temperatures. For cooling to the superconducting state, both Nb-Ti and Nb_3Sn require the use of liquid helium as refrigerant; at normal atmospheric pressure the temperature is 4.2 K. This temperature may be necessary for Nb-Ti when carrying a substantial current, but it is probably lower than required by Nb_3Sn. Since helium is an expensive and relatively rare material, a more readily available alternative refrigerant would be desirable. The only possibility is liquid hydrogen, but since its temperature is 20.3 K at atmospheric pressure, it can not be used with either Nb-Ti or Nb_3Sn.

Consequently, a search is being made for superconductors with transition points above 20 K. The compound of niobium and germanium, Nb_3Ge, has a zero-current transition temperature of 22 to 23 K. However, this compound is not formed as readily as the analogous Nb_3Sn, and, in any event a temperature below 20 K would be required to maintain superconductivity with large currents. This might be possible by using a refrigerant consisting of a slurry of solid and liquid hydrogen, which forms at 14°K.

12.11. International Status of MHD Power Generation and its Future Prospects

International co-operation has been prominent in the development of MHD power generation over the past 3 decades, research and development in Magneto hydrodynamic (MHD) electric power generation have been accompanied by a considerable upswing in international cooperation. The first demonstration of the feasibility of MHD was made in the USA in 1959. Beginning in 1960, related scientific research and engineering development was undertaken by USA, USSR, Japan, UK, France, Poland and other countries.

A 25 MW MHD pilot plant was commissioned in USSR in 1970, which has successfully supplied electricity to the Moscow grid. Encouraged by the successful operation of the U-25 facility, the USSR has designed a 500 MW power station with a 250 MW MHD component and it is, located near Moscow. A pilot 250 MW Engineering Test Facility (ETF), which is planned to be fully integrated MHD/steam power plant, is commissioned in USA. Further a Commercial Demonstration Plant (CDP) for demonstrate the commercial feasibility of an MHD system for base load application is likely to be commissioned by the year 2000. In Japan the work on a 20 MW demonstration plant near Tokyo is in progress. Poland and China are also believed to be developing MHD systems. USA has been collaborating with USSR, Poland, Japan and Netherlands in this field.

The Indian R and D programme on MHD started in 1977. It concentrates on developing necessary competence in this area and associated technology which will form the basis of creating commercially viable equipment for coal based MHD power generation. The Depart-ment of Non-conventional Sources of Energy (DNES) has sponsored research and development programme on coal based MHD power generation. Bhabha Atomic Research Centre (BARC) in collaboration with BHEL and Institute of High Temperature (USSR) is executing this research and development programme. The layout for a 5–15 MW pilot MHD plant has been prepared and the construction work is taken up at the plant site at Tiruchirapalli. All the proposed MHD installation throughout the world are hybrid MHD steam open cycle systems. In March 1985, an experimental 5 MW (Thermal input) plant was commissioned, at Tiruchirapally. It is an open cycle MHD pilot plant using blue water gas as fuel, preheated 40% oxygen enriched air as oxidiser and Aq. Sol. K_2CO_3 as seed material. An iron core magnet of 2 tesla lined intensity gives the required magnetic field. A ceramic lined water cooled combustor is used to obtain desired plasma parameters. The main duct is lined with high temperature material capable of withstanding temperature in excess of 2500°C. Faraday type MHD generator is used to extract the MHD power. The hot exhaust gases from MHD channel outlet are cooled in a heat recovery system which is made of spray cooling modules. The seed is recovered through wet recovery method and the spent combustion products are exhausted out to the atmosphere through chimney. The experimental run for power generation on this plant was carried out successfully during August 1985. The experimental studies of individual component performance are in progress. There will help in developing scientific data and design capabilities for large size plants.

The benefits occurring from a full fledged MHD programme are two fold. Apart from the cheap generation of electricity and conservation of known resources of fossil fuel, the programme undertakes development efforts in advanced fields of technology. These have far reaching influence in nuclear, aerospace, defence, chemical and metallurgical fields. These benefit include the development and improvement in properties of various refractory materials, metals and alloys; a better understanding of super conducting technology, construction and fabrication of efficient high temperature, high pressure heat exchangers and reactors.

REVIEW QUESTIONS

1. Discuss the principle of MHD generation.
2. Derive the equations for the voltages and power output of an MHD generator.
3. An MHD generators has the following parameters:
   Plate area = 0.20 m^2
   Distance between plates = 0.4 m^2
   Flux density = 2 Wb/m^2
   Average gas velocity = 1000 m/s
   Conductivity of the gas = 10 mho/m
   Calculate the open circuit voltage and maximum power output.
   [**Ans.** 800 volts, 800 kW]
4. With the following specifications for an MHD generator calculate:
   (*i*) open circuit voltage and (*ii*)maximum power output.
   Plate area = 0.1 m^2
   Distance between plates = 0.5 m
   Flux density = 3 Wb/m^2
   Average gas velocity = 10^3 m/s
   Gaseous conductivity = 10 mho/m [**Ans.** 1500 volts, 1125 kW]
5. How MHD systems are classified? Describe them in brief.
6. Describe an MHD open cycle system. What are the main advantages of an MHD power generation?
7. Describe an MHD closed cycle system, with its advantages and disadvantages.
8. What are important factors to be considered while selecting materials for an MHD generators? Give names of the materials for electrodes and generator duct.
9. What are the various losses associated with operation of MHD generator?
10. Derive the expression for maximum power per unit volume for a Faraday generator.
11. Write short notes on:
    (*i*) Materials for MHD generator (*ii*) Seeding
    (*iii*) Superconductivity (*iv*) Gas-conductivity.

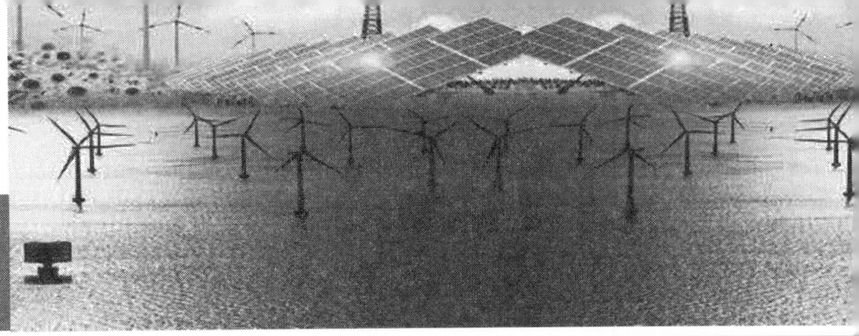

CHAPTER 13

Thermoelectric Power

13.1. Introduction

In 1822 the German Scientist Seebeck discovered that a loop of two dissimilar metals developed an e.m.f. when the two junctions were kept at different temperatures. This effect has long been used in thermo-couples to measure temperatures. This phenomenon offers one method of producing electrical energy directly from the heat of combustion, but its thermal efficiency is very low, of the order of 1 to 3%. It looks not very attractive except for very special small power applications. Efficiency of thermoelectric generator depends upon the temperatures of hot and cold junctions in any heat engine.

The basic simplicity of this method of electric power production has encouraged research to find materials that might work more efficiently. Much of the study has centred on the semiconductor class of materials that stand midway between electrical insulators and conductors. The material in addition to working with reasonable efficiency must also to withstand high temperature service. Encouraging results have been obtained with ceramic type materials made of semiconductors 'doped' with materials to increase their conductivity to a level about halfway between semiconductors and metal conductors. These materials, known as manganese, iron, cobalt and nickel treated with oxygen, sulphur, selinium, and tellurium. Researches report efficiencies of the order of 8% and confidently expect to develop materials that would work at 20% thermal efficiency. They envision the probability of developing peak load power stations with capacities of the order of 100 MW. Where fuels are very cheap, this device can generate power for stand by or even base load plants.

13.2. Basic Principles of Thermoelectric Power Generation

Thermoelectric generator is a device which converts heat energy (thermal energy) into electrical energy through semiconductor or conductor. The direct conversion of heat energy into electric energy (*i.e.*, without a conventional electric generator) based on *the Seebeck Thermo*

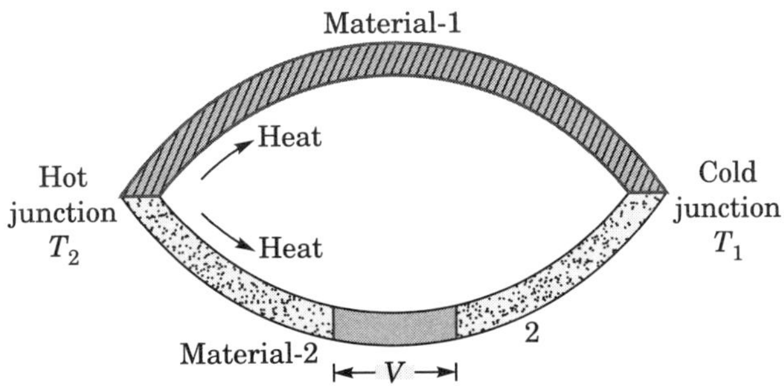

Fig. 13.1. Seebeck effect principle of Thermal couple.

electric effect. Consider two dissimilar materials joined together in the form of a loop so that there are two junctions. Such a system is illustrated in Fig. 13.1. If a temperature difference is maintained between these two junctions, an electric current will flow round the loop.

The magnitude of the current will depend on both the materials used and the temperature difference of the junction ($\Delta T = T_2 - T_1$). If the circuit is broken an open circuit voltage 'V' appears across the terminals of the break as shown in fig. The thermo emf, V produced by the device is given by

$$V = \alpha_{S_{1-2}} \Delta T \quad \text{...(13.1)}$$

where, $\alpha_{S_{1-2}}$ is the seebeck coefficient. The subscript $1-2$ indicates that the value is particular to the materials composing the loop. In most cases $\alpha_{S_{1-2}}$ is depend on temperature, so that for larger temperature difference it is more accurate to express equation (13.1) in the form

$$V = \int_{T_1}^{T_2} \alpha_{S_{1-2}} \, dT \quad \text{...(13.2)}$$

where, T_2 will be used throughout to denote the temperature of the hot junction and T_1 that of the cold one. From equation (13.2) it will be seen that the Seebeck coefficient is the temperature coefficient of thermo e.m.f. or the rate of change of thermo-emf with temperature.

$$\alpha_{S_{1-2}} = \underset{\Delta T \to 0}{\text{Lt}} \frac{\Delta V}{\Delta T} = \frac{dV}{dT} \quad \text{...(13.3)}$$

The Seebeck coefficient depends upon the choice of the materials and can not be attributed to either material above. It has unit of volt per degree. Depending upon the choice of materials, the drop in the potential may be either positive or negative in the direction of the drop of temperature. Thus the sign as well as magnitude of the Seebeck coefficient of significant. Microscopically, the Seebeck effect arise because the concentration of the charge carriers in a conductor depends upon the temperature. The presence of a temperature gradient in a material causes a carrier concentration gradient and an electric field is established, which causes the net flow of charge carriers under open circuit conditions to be zero. Clearly two dissimilar materials are necessary in order to notice this effect. We can select the materials in such a way that these potential differences may add up so that an electric current flows through the circuit with heat being converted into the electrical current.

The Seebeck effect has been used for a great many years in the thermocouple, which is simple and robust instrument for the measurement of temperature. By placing the hot junction at the point where a knowledge of temperature is required, and maintaining the other at a known fixed temperature, the output open circuit voltage, or the circulating current produced, is then proportional to the temperature.

A qualitative idea of this phenomenon is as follows. The materials of which the loop illustrated is constructed contain certain densities of free electrons. When the two materials are placed in contact the free electron gas will diffuse throughout them. In general the materials will first contain different densities of free electrons. The result will be that to reach equilibrium electrons will have to diffuse from the regions of high density to those of low density. Consider Fig. 13.2, material 2 has a higher density of free electrons than materials 1, and also $T_2 = T_1$. When the materials are in contact, the free electrons of the electron gas will diffuse (just as molecules in real gas), to equate the density throughout the region. As the density was originally greater in material 2, electrons will leave this material, so that it becomes positively charged, and will enter material 1, thus charging it negatively. As more electrons flow across, the potentials will increase. Eventually an electric field build up across the junction just sufficient to oppose the net diffusion of any further electrons. The differential diffusion results in the appearance of a potential difference across the junction; which is known as the contact potential. In the loop of Fig. 13.2 (*a, b*) considering the junction to be at the same temperature, each will develop a contact potential in equilibrium these are equal and opposite, so that there is no net current round the loop. If the temperature is increased but the incremental increase in one material will be greater than in other. Consider that one junction of the couple shown is raised in temperature resulting in greater increase of electrons in material 2 that in material 1. To maintain equilibrium more electrons will have to leave material 2 than previously, resulting in an increased contact potential. As the other junction has been kept at the same temperature its contact potential has remained the same. The result of increasing the temperature at one junction is thus to produce an imbalance of contact potentials. The difference produces a resulting emf that will derive a current round the closed loop. The sequence of events is illustrated in Fig. (13.2. *a* and *b*).

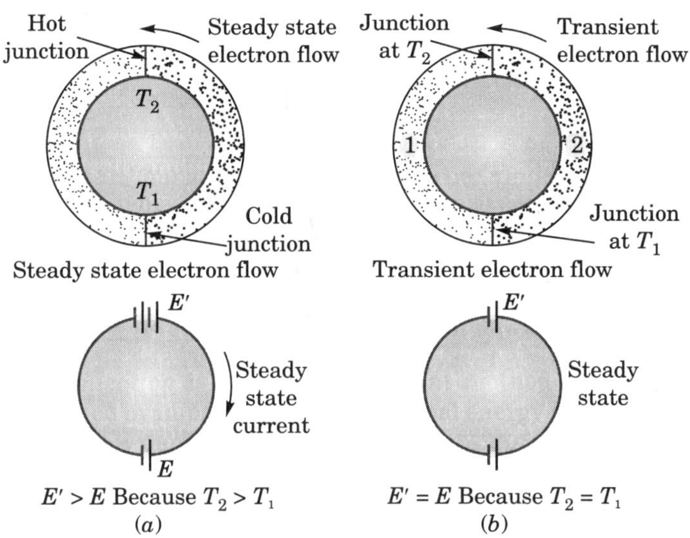

(*a*) Two-metal loop with junctions at same temperature.

(*b*) Two-metal loop with junctions at different temperature.

Fig. 13.2. Production of a thermo emf

Equation (13.1) shows that the Seebeck coefficient $\alpha_{S_{1-2}}$, is dependent on both of the materials which make up the complete element. If α_S is the absolute coefficient particular to one material. With this idea $\alpha_{S_{1-2}}$ can be expressed as

$$\alpha_{S_{1-2}} = \alpha_{S_1} - \alpha_{S_2} \qquad ...(13.4)$$

The coefficient $\alpha_{S_{1-2}}$ is defined as positive, if the conventional current tends to flow from material 1 to material 2 at the hot junction. The magnitude of $\alpha_{S_{1-2}}$ is found to decrease with an increase in carrier density.

The Fig. (13.3) shows the variation with the temperature of the Seebeck coefficients of metals, and p and n type semiconductors. With the above sign convention of $\alpha_{S_{1-2}}$, the coefficient is positive for p type materials and negative for n type. Thus if a thermoelectric device is made up from a pair of such materials equation (13.4) becomes

$$\alpha_{S_{1-2}} = \alpha_{S_1} + \alpha_{S_2} \quad ...(13.5)$$

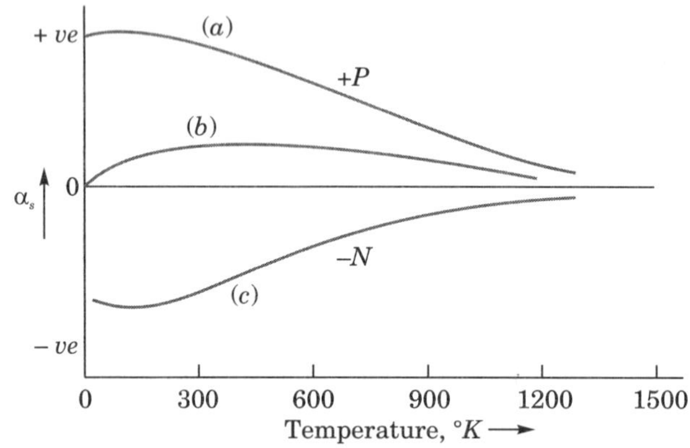

(a) p-type semiconductor (b) metal
(c) n-type semiconductor.

Fig. 13.3. Variation of Seebeck Coefficient with temperature.

Showing immediately the great enhancement of emf resulting from this combination. Note that also by definition.

$$\alpha_{S_{1-2}} = - \alpha_{S_{2-1}} \quad ...(13.6)$$

The parameter $\alpha_{S_{1-2}}$ is also called the relative thermoelectric power of the wires. In many thermocouple systems, $\alpha_{S_{1-2}}$ varies slowly with temperature so that the Seebeck voltage in Equation (13.2) may be written:

$$V_{1-2} \simeq \alpha_{S_{1-2}} (T_2 - T_1) \quad ...(13.7)$$

provided that $(T_2 — T_1)$ is not too large. For example, the Seebeck coefficient for an iron-constantan thermocouple is = 53 μ V/°C for 0°C ≤ 100°C.

A thermocouple performs useful work only if it is supplying a current to an external device. In short circuit operation the electrical energy is dissipated as resistive heat within the wires. In the open circuit no work is done by the thermocouple because no current is flowing. When power is supplied to an external device (Fig. 13.4) heat is absorbed at certain points along the thermocouple wire and rejected at others. The difference is converted to electrical work and applied to the device.

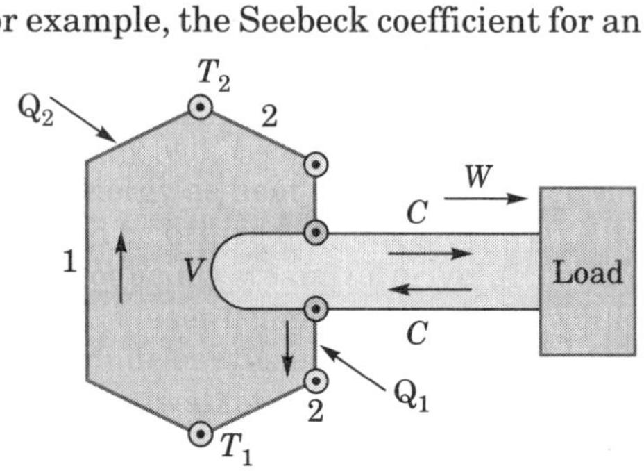

Fig. 13.4. A thermocouple of meterial 1 and 2 extracting heat Q_2, rejecting heat Q_1, and supplying electrical work to a load. The power supplied is the product of the terminal voltage V, and the current, I delivered to the load.

The thermodynamic conversion of heat to work actually involves four distinct processes, only two of which are reversible. The reversible processes are the *Peltier* and *Thomson* effects. The irreversible ones are the *Fourier* effect (heat conduction) and the *Joule* effect (electrical resistance heating).

Before applying elementary thermodynamics to the thermocouple system, the different *thermoelectric effects i.e.,* Joule effect, Seebeck effect, Peltier effect, and Thomson effect should be defined.

Joule Effect. This refers to the *irreversible conversion* of electrical energy into heat when a current I flows through a resistance R, an amount of heat equal to I^2R is generated per unit time. This heat is called the *Joulean heat.*

$$Q_j = I^2R \quad \text{...(13.8)}$$

As already described the *Seebeck effect* involves the conversion of heat supplied at the *hot junction* into electrical current. The electrical power output due to the Seebeck effect is however reduced on account of heat conduction and electric power loss in the leads. Since a good electrical conductor is also a good heat conductor, much of the heat transferred to the hot junction is carried away by the wires. The selection of proper materials is therefore important to reduce the above loss in output.

Peltier Effect. When an electric current flows across an *isothermal* junction of two dissimilar materials, there is either an evolution or absorption of heat at the junction. This effect is called the *peltier heat* (Fig. 13.5). The Peltier effect is a reversible effect because if heat is evolved when current flows in one direction, the same amount of heat is absorbed at the junction if the current flow is reversed.

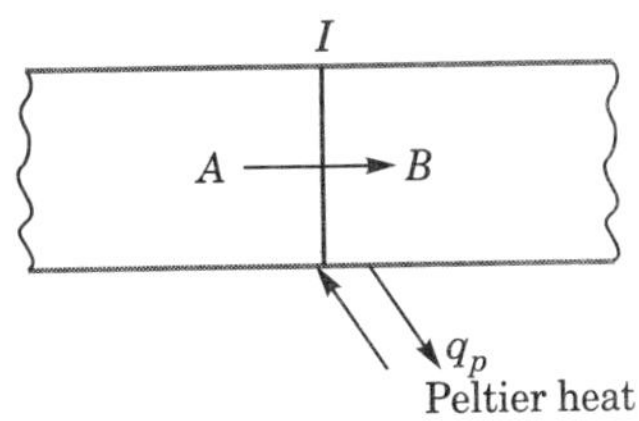

Fig. 13.5. Peltier effect.

The *Peltier Coefficient* $\alpha_{P_{1-2}}$ is defined as the heat evolved or absorbed at the junction per unit current flow, per unit time

$$\alpha_{P_{1-2}} = \alpha_{P_1} - \alpha_{P_2} = \frac{q_P}{I} \quad \text{...(13.9)}$$

and $$q_P = \alpha_{P_{1-2}} . I \quad \text{...(13.10)}$$

I = Peltier heat per unit time.

Thomson Effect. When an electric current flows through on material having a temperature gradient, there is an evolution (or absorption) of heat. This phenomenon is called the *Thomson effect* (Fig. 13.6). The Thomson heat is reversible because reversing the direction of current flow reverses the direction of heat transfer without change in magnitude. The *Thomson Coefficient* σ is defined as the Thomson heat absorbed (or evolved) per unit time per unit electric current and per unit temperature gradient;

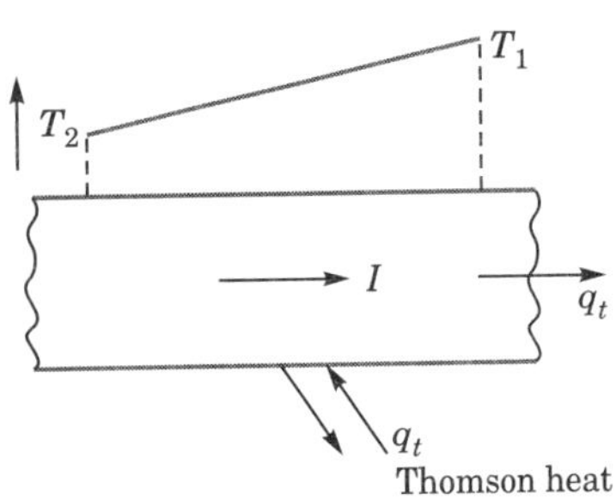

Fig. 13.6. Thomson effect.

$$\sigma = \frac{dq_T/dx}{d_T/dx} \quad \text{...(13.11)}$$

where, $\frac{dq_T}{dx}$ = heat interchange per unit time per unit length of conductor.

$\frac{d_T}{dx}$ = temperature gradient.

Hence, the Thomson heat per unit time is given by

$$dq_T/dx = \sigma I \frac{d_T}{dx}$$

The Seebeck, Peltier and Thomson effects are more pronounced in *semiconductors* as compared to metals.

The relations between the Seebeck, Peltier and Thomson Coefficients as derived by Kelvin are as follows:

$$\left.\begin{aligned} \alpha_p &= \alpha_S \cdot T \\ \text{or} \quad \alpha_{P_{1-2}} &= \alpha_{S_{1-2}} \cdot T \end{aligned}\right\} \quad ...(13.12)$$

$$\left.\begin{aligned} \sigma &= T \frac{d\alpha_S}{d_T} \\ \sigma_{1-2} &= \frac{T\, d\alpha_{S_{1-2}}}{d_T} \end{aligned}\right\} \quad ...(13.13)$$

In order to apply elementary thermodynamics of the thermoelectric system, we must ignore the irreversible processes. If the wires are assumed to have zero thermal conductivity, irreversible conduction of heat between the hot and cold reservoirs can not occur. Furthermore, if the wires also have zero electrical resistivity, the irreversible process of Joule heating is absent. We are thus left with the reversible *Peltier* and *Thomson* effects, which will be described below.

Equation (13.2), can be integrated by parts as

$$V_{1-2} = \int_{T_1}^{T_2} (\alpha_{S_1} - \alpha_{S_2})\, d_T$$

$$= (\alpha_{S_1} - \alpha_{S_2}) T \Big|_{T_1}^{T_2} - \int_{T_1}^{T_2} \left\{ \frac{d}{d_T} \left(\alpha_{S_1} - \alpha_{S_2} \right) \right\} d_T \quad ...(13.14)$$

We will define the following functions of the Kelvin temperature:

$$\alpha_{P_{1-2}} = (\alpha_{S_1} - \alpha_{S_2}) T$$

$$= \alpha_{S_{1-2}} \cdot T = \text{Peltier Coefficient at junction } 1-2 \quad ...(13.15)$$

$$\left.\begin{aligned} \sigma_1 &= \frac{-T\, d\alpha_{S_1}}{d_T} = \text{Thomson Coefficient along wire 1} \\ \sigma_2 &= \frac{-T\, d\alpha_{S_2}}{d_T} = \text{Thomson Coefficient along wire 2} \end{aligned}\right\} \quad ...(13.16)$$

In terms of these Coefficients, Equation (13.16) can be expressed as

$$V_{1-2} = \alpha_{P_{1-2}} \cdot T_2 - \alpha_{P_{1-2}} \cdot T_1 + \int_{T_1}^{T_2} \sigma_1\, d_T - \int_{T_1}^{T_2} \sigma_2 d_T \quad ...(13.17)$$

The electrical power supplied to a device is found by multiplying equation (13.17), by the current flowing I, giving

$$P = I \,.\, V_{1-2}$$

$$= I\,\alpha_{p_{1-2}}\,T_2 - I\,\alpha_{P_{1-2}}\,T_1 + I\int_{T_1}^{T_2}\sigma_1\,d_T - I\int_{T_1}^{T_2}\sigma_2 d_T \qquad ...(13.18)$$

From Equation (13.18) it can be seen that the electrical power produced by the thermocouple is equal to the difference between two Peltier effects and two Thomson effects.

Whenever an electrical current traverses a junction immersed in a heat reservoir at temperature *T*, heat is either absorbed or rejected according to the direction of current. This heat is called *Peltier heat* as stated above, is given by

$$q_p = \alpha_{p_{1-2}}\,I\left(\frac{\text{Peltier heat per unit}}{\text{time}}\right)$$

In addition to Peltier heat transferred at the junctions, there is also *Thomson heat* transferred to and from the surroundings along the wires. This heat rate per unit length of wire is

$$\frac{dq_{thomson}}{dx} = I\sigma_1\,\frac{d_T}{dx}$$

So that the total heat transferred along wire 1, is

$$q_{thom} = I\int_{T_1}^{T_2}\sigma_1\,d_T \ \text{(Thomson heat)} \qquad ...(13.19)$$

Returning to the equation (13.18) we may write the electrical power produced as

$$p = q_{p2} - q_{p1} + q_{t1} - q_{t2} \qquad ...(13.20)$$

where, q_{p2} is the Peltier heat absorbed at the hot junction.

q_{p1} is the Peltier heat rejected at the cold junction.

q_{t1} is the Thomson heat absorbed along wire 1.

q_{t2} is the Thomson heat rejected along wire 2.

Since Thomson heat is reversibly transferred along a series of reservoirs varying in temperature from T_1 to T_2, the system will not necessarily have the *Carnot* efficiency. However consider the highly idealized case in which the thermoelectric power of each wire is *independent* of temperature so that, according to Equation (13.16), the Thomson coefficient is zero and consequently the Thomson heat vanishes. In that case the system becomes a reversible heat engine operating between two reservoirs-one at each junction. The efficiency of this engine is, using Eqn. (13.7) and (13.14)

$$\eta = \frac{P}{q_p}$$

$$= \frac{I\,V_{1-2}}{I\,\alpha_{p_{1-2}}}$$

$$= \frac{I\,.\,\alpha_{S1-2}\,(T_2 - T_1)}{I\,\alpha_{S_{1-2}}\,T_2} = \frac{T_2 - T_1}{T_2} \qquad ...(13.21)$$

The equation (13.21), as expected, leads to Carnot efficiency.

Example 13.1. *The Seebeck coefficient of a junction is 55 μV/°K at 373 K and 50 μV/°K at 373 K. Find the Peltier heats absorbed and rejected when the thermocouple is operating between these heat reservoirs and is supplying a current of 10 mA.*

Solution. Using equation (13.12), the Peltier coefficients at these junctions are $\alpha_{P_{1-2}} = \alpha_{S_{1-2}} T$

$$\therefore \quad \alpha_{P_{1-2}} \text{ at } 273\text{ K} = (55 \times 10^{-6})(373) = 2.05 \times 10^{-2} \text{ W/amp}$$

and
$$\alpha_{P_{1-2}}(273) = 50 \times 10^{-6} \times (273) = 1.37 \times 10^{-2} \text{ W/amp}$$

From equation (13.9) we find the Peltier heats absorbed and rejected to be

$$q_{2\ Peltier} = \alpha_{P_{1-2}} \cdot I$$
$$= 2.05 \times 10^{-2} \times 10 \times 10^{-3} = 205\ \mu\text{W}$$

and
$$q_{1\ Peltier} = (1.37 \times 10^{-2}) \times 10 \times 10^{-3} = 137\ \mu\text{W}.$$

If no other heat transfer were involved, the difference between these values, 205 – 137 = 68 μW, would be supplied as electric power.

Example 13.2. *Find the Thomson heat transferred to the surroundings from a wire whose end points are maintained at 373 and 273 K. A current of 10 mA is flowing in the wire and its absolute thermoelectric power increases linearly with temperature at a rate* $\frac{d\alpha S_1}{dT} = 5.4 \times 10^{-9}$ V/°K².

Solution. The Thomson coefficient σ, varies with temperature and is given by equation (13.16) *i.e.*,

$$\sigma_1(T) = -T\frac{d\alpha S_1}{dT} = -(5.4 \times 10^{-9} \times T)\ \text{V/°K}$$

The Thomson heat is given by equation

$$q_{th} = I \int_{T_1}^{T_2} \sigma_1\, dT$$

$$= 10 \times 10^{-3}(5.4 \times 10^{-9}) \int_{273}^{373} T dT$$

$$= 1.74\ \mu\text{W}. \textbf{ Ans.}$$

13.3. Thermoelectric Power Generator

A simple arrangement for utilizing the Seebeck coefficient is shown in Fig. 13.7. The thermocouple material *A* and *B* are joined at the hot end, but the other ends are kept cold; an electric voltage (or electromotive force) is then generated between the cold ends. A direct current will flow in a circuit or load (*e.g.*, a motor resistance, etc.) connected between these ends. The current will continue to flow as long as the heat is supplied at the hot junction and removed from the cold ends. For a given thermocouple, the voltage and electric power output are increased by increasing the temperature difference between the hot and cold

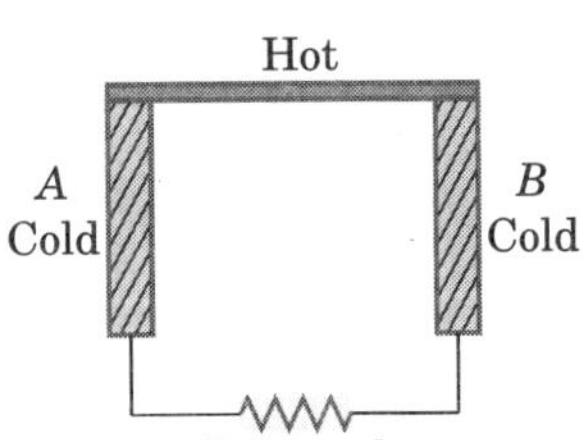

Fig. 13.7. Simple thermocouple arrangement.

ends. In a practical thermoelectric converter, several couples are connected in series to increase both voltage and power, Fig. 13.8. If the output voltage is not sufficient to operate a particular device or equipment, it can be increased, with little loss of power, by an inverter transformers combination. The direct current generated by the thermocouples is first changed into alternating current of essentially the same average voltage by means of an inverter. The alternating current and voltage is then increased to the desired value with a transformer. The high voltage alternating current can be reconverted into direct current if required, by a rectifier.

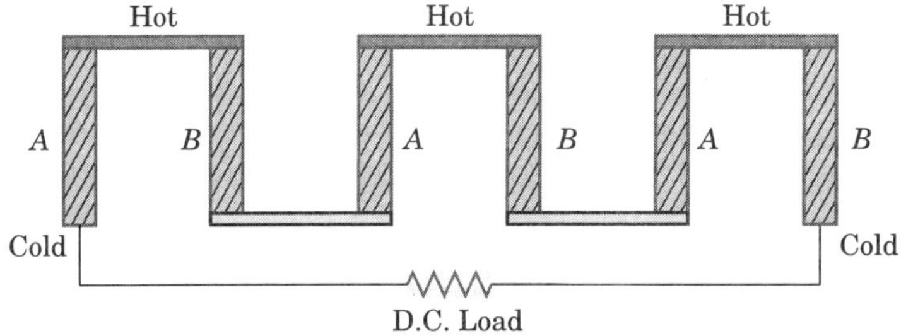

Fig. 13.8. Thermocouples in series to increase voltage.

A thermoelectric converter is a form of *heat engine.* Heat is taken up at an upper temperature (*i.e.,* the hot junction) and part is converted into electrical energy; the remainder is discharged (or removed) at a lower temperature (*i.e.,* the cold ends). As with other heat engines, the thermal efficiency of a given thermocouple for conversion of heat into work (electrical energy is increased by increasing the upper temperature and/or decreasing the lower temperature. The latter is usually that of the environment and is roughly constant; hence, the conversion efficiency of a given thermocouple is determined in practice mainly by the temperature of the hot junction.

Thermoelectric generators have been built with power outputs ranging from a few watts to kilowatts. The source of heat is immaterial and any available fuel may be used consequently, thermoelectric generators have been employed in areas outside the regular electric power distribution system. An important application is the use of radioactive decay heat to generate power in space and other remote locations. If the efficiency could be increased substantially thermoelectric conversion might become practical for generating electricity using heat supplied by *solar energy.*

13.4. Performance Analysis of Thermo electric Power Generator

We have seen above that if a resistance is placed across the output terminals of a thermocouple, current will flow and the thermocouple becomes a power producing device. (Fig. 13.9) shows a thermoelectric converter schematically. It consists of two blocks A and B of semiconductor materials connected together by a conductor. The conductor receives heat from a thermal source and the lower open (not conducted) ends of the blocks reject heat to a low temperature sink. The two junctions 1 and 2 are kept at a temperature difference ΔT. The sides of the blocks are insulated. Hence, heat flow occurs along the length of the blocks only, supplying heat to the hot junction causes the *electrons* in the n type block and *holes* in the p type block to flow away from the hot junction, thereby producing a potential difference between two open ends. If the circuit is completed at the cold junction, an electric current will flow through the load.

The operation of the system can be reversed by supplying electrical energy in place of heat. The current will flow now serve to remove heat from one junction and to deliver heat to the other junction. The device can then be represented by an emf α_{SAB}. ΔT with an internal resistance R, which is the total series resistance of the two parts A and B of the thermoelectric circuit. If an external load resistance R_L is applied, the current is α_{SAB}. $\Delta T/(R + R_L)$, and the power fed into external load is

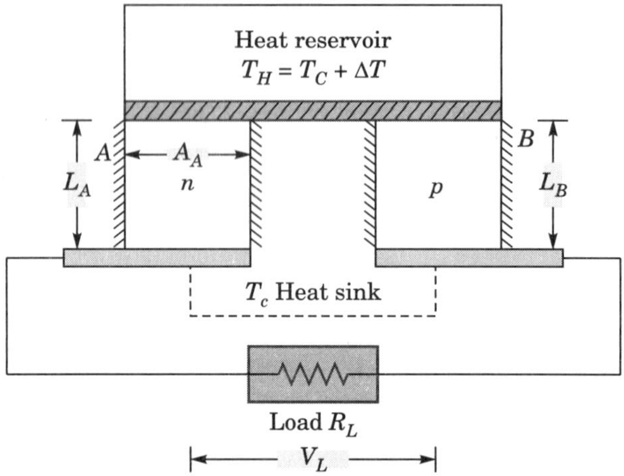

Fig. 13.9. Schematic of Thermoelectric converter.

$$P = \left[\frac{\alpha_{SAB}\ \Delta T}{(R + R_L)}\right]^2 .R_L \qquad ...(13.22)$$

which has a maximum value, when $R = R_L$

i.e.,

$$P_{max} = \frac{\alpha^2{}_{SAB}\ \Delta T^2}{4R} \qquad ...(13.23)$$

The factor $\alpha^2{}_{SAB}/R$ is called the *figure of merit* for this application. The condition for a large maximum power and therefore large ΔT and a large value of $\alpha^2{}_{SAB}/R$. The *first* is easily accomplished, though for large values of ΔT the expression α_{SAB} ΔT must be replaced by $\overline{\alpha}_{SAB}$ ΔT, where α_{SAB} is the average value of α_{SAB} over the temperature interval ΔT;

$$\overline{a}_{SAB}\ \Delta T = \int_T^{T+\Delta T} \alpha_{SAB}\ dT \qquad ...(13.24)$$

The second condition requires more careful consideration. First of all, one should make the length of the rod small and the diameters of the rods large, since that decrease the resistance ($R = \rho l/A$). One can not go too far in that direction, however, because the heat reservoir then loses a large amount of power by heat conduction, diminishing the overall efficiency of the system. Next increases the impurity concentration of the rods. This tends to decrease α_{SAB} and decrease R, but since at first R decreases faster than α_{SAB}, the optimum design lies at high impurity concentration (around $10^{19}/cm^3$). Finally, it is important to give the rods 'A' and 'B' opposite thermoelectric forces per degree. If the two rods have equal resistance and equal and opposite thermoelectric forces per degree, α_{SAB} increases by a factor 2, R increases by a factor '2' and $\dfrac{\alpha^2{}_{SAB}}{R}$ increases by a factor '2'.

Rather than optimizing the power, it is often more meaningful to optimize the efficiency 'η' defined as the ratio of the power. P_L developed into the load resistance R_L to the heat flow q_H from the source at the hot junction.

$$\eta = \frac{P_L}{q_H} \quad ...(13.25)$$

Now, $$P_L = I^2 R_L,\ I = \frac{\alpha_{SAB} \cdot \Delta T}{R + R_L} \quad ...(13.26)$$

Let us now apply the first law energy balance to the hot plate for which during a unit time the energy effects are:

(*a*) An amount of energy q_H is supplied to the plate from the source.

(*b*) Since a current flows through the circuit, an amount of heat q_p (Peltier effect) is absorbed at the junction.

(*c*) An amount of heat q_c is conducted away from the plate through the two blocks.

(*d*) An amount of heat $q_j/2$ is transferred to the plate. This amount represents one half of the Joule heat generated by the current flow. The remaining one half of the Joule heat is transferred to the cold junction.

Hence, $$q_H + \frac{1}{2} q_j = q_p + q_c$$

or $$q_H = q_c + q_p - \frac{1}{2} q_j$$

$$= K\,\Delta T + (T + \Delta T)\,\alpha_{SAB} I - \frac{1}{2} I^2 R \quad ...(13.27)$$

where, I is the current flowing through the circuit

R is the resistance of the thermocouple = $(R_A + R_B)$

K is the thermal conductance = $(K_A + K_B)$.

$T + \Delta T$ is the temperature of the hot junction.

$(T + \Delta T)\ \alpha_{SAB}$—The Peltier Coefficient.

As per Peltier, the thermal power evolved or absorbed at a junction is given by

$$q_p = \alpha_{p_{1-2}} I, \quad ...(13.28)$$

where, $\alpha_{p_{1-2}}$ *is* the Peltier coefficient. The Peltier coefficient $\alpha_{p_{1-2}}$ is related to Seebeck coefficient αs_{1-2} by the following expression:

$$\alpha_{p_{1-2}} = \alpha_{S_{1-2}} T \quad ...(13.29)$$

I^2R—Heat developed in the thermocouple, of this, half the heat flows to the heat sink at the cold junction (temperature T_1) and half flows back to the heat source (hence the term = $1/2\ I^2R$), it is called joule heat, generated by current flow through a resistance. Joule heat $q_j = I^2R$.

If $$T + \Delta T = T_H$$

then, $$\eta = \frac{I^2 R_L}{K\Delta T + \alpha_{SAB} T_H I - \frac{1}{2} I^2 R} \quad ...(13.30)$$

Substituting for I and introducing $m = \frac{R_L}{R}$ as a new variable, one obtains

$$\eta = \frac{\left(\frac{\alpha_{SAB}\,\Delta T}{R+R_L}\right)^2 .RL}{K\Delta T + \alpha_{SAB}\,T_H \frac{\alpha_{SAB}\,\Delta T}{R+R_L} - \frac{1}{2}\left(\frac{\alpha_{SAB}\,\Delta T}{R+R_L}\right)^2 R} \quad ...(13.31)$$

$$= \frac{\frac{\alpha^2{}_{SAB}\,\Delta T^2}{\left(\frac{R+R_L}{R}\right)^2} \cdot \frac{R_L}{R^2}}{K\Delta T + \frac{\alpha^2{}_{SAB}\,T_H \Delta T}{\frac{R+R_L}{R} \times R} - \frac{1}{2}\frac{\alpha^2{}_{SAB}\,\Delta T^2}{\left(\frac{R+R_L}{R}\right)^2} \times \frac{1}{R^2} \times R}$$

$$= \frac{\frac{m}{R}\,\frac{\alpha^2{}_{SAB}\,\Delta T^2}{(1+m)^2}}{K\Delta T + \frac{\alpha^2{}_{SAB}\,T_H\,\Delta T}{(1+m)\,R} - \frac{1}{2}\,\frac{\alpha^2{}_{SAB}\,\Delta T^2}{(1+m)^2}\,\frac{1}{R}}$$

$$= \frac{\frac{m}{R}\,\frac{\alpha^2{}_{SAB}\,\Delta T^2}{(1+m)^2}}{\frac{K\Delta T\,.\,R(1+m^2) + 2(1+m)\,\alpha^2{}_{SAB}\,T_H\,\Delta T - \alpha^2{}_{SAB}\,\Delta T^2}{2R\,(1+m)^2}}$$

$$= \frac{m\,\alpha^2{}_{SAB}\,\Delta T^2/\alpha^2{}_{SAB}\,\Delta T\,T_H}{\frac{K\Delta T\,.\,R(1+m^2)}{\alpha^2{}_{SAB}\,\Delta T\,.\,T_H} + \frac{(1+m)\,\alpha^2{}_{SAB}\,T_H\,\Delta T}{\alpha^2{}_{SAB}\,\Delta T\,.\,T_H} - \frac{1}{2}\,\frac{\alpha^2{}_{SAB}\,\Delta T}{\alpha^2{}_{SAB}\,\Delta T\,.\,T_H}}$$

$$= \frac{m\left(\frac{\Delta T}{T_H}\right)}{\frac{(1+m^2)}{T_H}\,\frac{RK}{\alpha^2{}_{SAB}} + (1+m) - \frac{1}{2}\,\frac{\Delta T}{T_H}} \quad ...(13.32)$$

Equation (13.32) can be written as

$$\eta = \frac{\left(\frac{T_1 - T_2}{T_1}\right) \times m}{\left[\frac{(1+m)^2}{T_1}\,\frac{(K_A + K_B)\,(R_A + R_B)}{\alpha^2{}_{SAB}} + (1+m) - \frac{1}{2}\left(\frac{T_1 - T_2}{T_1}\right)\right]} \quad ...(13.33)$$

where, T_1 = higher temperature *i.e.*, $T + \Delta H$
= source temperature (T_H)
T_2 = lower temperature *i.e.*, sink temperature
$= T_C$

$$K_A = \frac{k_A A_A}{L_A}, K_B = \frac{k_B A_B}{L_B}$$

k_A and k_B are thermal conductivities of the materials A and B respectively

$R_A = R_B$ = electrical resistance

$$R_A = \frac{\rho_A L_A}{A_A}, R_B = \frac{\rho_B L_B}{A_B}$$

In equation (13.33) the first term equals the Carnot efficiency for an engine working between T_1 and T_2 and the value of the second term (which is always less than unity) depends upon the material properties, geometry and operating temperatures of the converter.

An index used in rating thermoelectric converters is called the figure of merit, Z, defined for a single material as

$$= \frac{\alpha_s^2}{\rho k} (°K^{-1}) \frac{\alpha_s^2}{\rho k} (°K^{-1}) \qquad ...(13.34)$$

$$Z = \left[\frac{\alpha_{SA} - \alpha_{SB}}{\sqrt{\rho_A k_A} + \sqrt{\rho_B k_B}} \right]^2 \qquad ...(13.35)$$

The *figure of merit* depends on the properties of thermoelectric materials used. A *high value of Z is obtained by using materials of large Seebeck coefficient, small thermal conductivity and small electrical resistivity.*

The larger the value of Z is, the higher the quality of the thermocouple. The value of Z approaches infinity as k and ρ each approaches zero (See Equation (13.34). The effectiveness of a thermocouple depends on its figure of merit as defined in Equation (13.32). In order to have a large Z, a thermoelectric material must have a large Seebeck coefficient, a low electrical resistivity, and a low thermal conductivity. Although most metals do have a low electrical resistivity, their thermal conductivity is high and the Seebeck coefficient is only moderate. These are specially prepared semiconductor junctions that have high electrical conductivities and yet have relatively low thermal conductivities. In addition they have larger Seebeck coefficients than metals, semiconductor thermocouples are, however, more difficult to fabricate and more fragile than metallic thermocouples. The highest values of Z obtained are typically $Z \simeq 0.002$ °K^{-1}. For reservoirs of 373 and 273 K, even such a large figure of merit leads, to a thermodynamic efficiency of only 3%. The corresponding Carnot limit is 27%.

To produce a usable voltage, a series of thermocouple junctions known as *thermopile* are required as stated earlier. Such a solar powered thermopile has overall efficiency even smaller, than its thermodynamic efficiency, because much of the solar energy intercepted by the absorber plate is lost to the surroundings, only a part is supplied to the thermoelectric system. A typical overall efficiency is only a few per cent.

It can be shown that the ideal (maximum) efficiency of a thermoelectric converter is given by

$$\eta_{ideal} = \left(\frac{T_H - T_C}{T_H} \right) \left[\frac{M - 1}{M + \frac{T_C}{T_H}} \right] \qquad ...(13.36)$$

where, $$M = \left[1 + \frac{Z}{2}(T_H + T_C)\right]^{1/2} \quad \text{...(13.37)}$$

Example 13.3. *For a thermoelectric power generator following parameters are given:*

Temperature of the hot reservoir of source	*= 600°K*
Temperature of the sink	*= 300°K*
Figure of merit for the material	$Z = 2 \times 10^{-3}$ *°K*$^{-1}$

Determine the efficiency of the thermoelectric generation. What will be its Carnot efficiency?

Solution.

$$M_{optimum} = \left[1 + \frac{Z}{2}(T_H + T_C)\right]^{1/2}$$

$$= \left[1 + \frac{2 \times 10^{-3}}{2}(600 + 300)\right]^{1/2}$$

$$= (1.9)^{1/2} = 1.38$$

Efficiency of the thermoelectric generator is

$$\eta = \left(\frac{T_H - T_C}{T_H}\right)\left[\frac{M_{opt} - 1}{M_{opt} + (T_C/T_H)}\right] \times 100$$

$$= \frac{600 - 300}{600} \cdot \frac{1.38 - 1}{1.38 + (300/600)} \times 100$$

$$= \frac{1}{2} \times \frac{0.38}{1.88} \times 100$$

$$= 0.101 \times 100 = 10.1\%. \textbf{ Ans.}$$

whereas efficiency of the Carnot cycle (reversible)

$$\eta_C = \frac{T_H - T_C}{T_H} \times 100$$

$$= \frac{300}{600} \times 100 = 50\%. \textbf{ Ans.}$$

Example 13.4. *A thermoelectric generator operates between the temperature limits of 923 K and 323 K. The cross-sectional areas and lengths of n and p type elements are:* $A_1 = 2.3$ *cm*^{2}, $l_1 = 1.5$ *cm, and* $A_2 = 1.308$ *cm*^{2}, $l_2 = 0.653$ *cm, respectively. Using the optimum value for the product of internal resistance and overall thermal conductance, calculate the maximum generator efficiency and the efficiency for maximum power. Also calculate the power output for both the cases.*

The following properties may be used.

	n type	*p type*
Seebeck coefficient (α_s)	-190×10^{-6},	190×10^{-6} *volt/°C*
Specific resistivity (ρ)	1.45×10^{-3},	1.8×10^{-3} *ohm-cm*
Fig. of merit (z)	2.0×10^{-3},	1.7×10^{-3} *°K*$^{-1}$

Solution. Conductivity (*n type*) $k_1 = \dfrac{\alpha_{S_1^2}}{\rho_1 Z_1}$

$$= \frac{(190 \times 10^{-6})^2}{1.45 \times 10^{-3} \times 2 \times 10^{-3}}$$

$= 124 \times 10^{-3} = 0.0124$ W/cm °C.

Similarly, $$k_2 = \frac{\alpha_{S_2^2}}{\rho_2 Z_2} = \frac{190^2 \times 10^{-12}}{1.8 \times 10^{-3} \times 1.7 \times 10^{-3}}$$

$= 1.18 \times 10^{-3} = 0.0118$ W/cm °C

$$Z_{opt} = \frac{(\alpha_{S_1} - \alpha_{S_2})^2}{\left[\sqrt{\rho_1 k_1} + \sqrt{\rho_2 k_2}\right]^2}$$

$$= \frac{(-190 - 190)\, 10^{-12}}{\left[\sqrt{.45 \times 10^{-3} \times 1.24 \times 10^{-2}} + \sqrt{1.18 \times 10^{-2} \times 1.8 \times 10^{-3}}\right]^2}$$

$= 1.85 \times 10^{-3}$ °K

Thermal conductance $$K = \frac{k_1 A_1}{l_1} + \frac{k_2 A_2}{l_2}$$

$$= 1.24 \times 10^{-2} \frac{2.3}{1.5} + 1.18 \times 10^{-2} \frac{1.303}{0.653}$$

$= 0.0452$ W/°C.

R = Resistance of the generator $= R_1 + R_2$

$$= \frac{\rho_1 l_1}{A_1} + \frac{\rho_2 l_2}{A_2}$$

$$= \left[1.45 \times \frac{1.5}{2.3} + 1.8 \times \frac{0.653}{1.303}\right] \times 10^{-3}$$

$= 1.840 \times 10^{-3}$ ohm.

$$M_{opt} = \left[1 + \frac{Z}{2}(T_H + T_C)\right]^{1/2}$$

$$= \left[1 + \frac{1.85 \times 10^{-3}}{2}(923 + 323)\right]^{-1/2} = 1.43$$

$R_L = M_{opt} \times R$

$= 1.43 \times 1.840 \times 10^{-3} = 2.63 \times 10^{-3}$ ohms.

Optimum efficiency η_{opt}

$$= \frac{T_H - T_C}{T_H} \times \frac{M_{opt} - 1}{M_{opt} + (T_C/T_H)}$$

$$= \frac{600}{923} \times \left[\frac{1.43 - 1}{1.43 + (323/923)}\right] = 0.65 \times 0.24 = 0.1570$$

= 15.70%. **Ans.**

Efficiency for maximum power output η

$$= \frac{\frac{T_H - T_C}{T_H} \times m}{\left[\frac{(1+m)^2}{T_H} \times \frac{KR}{\alpha_{S12}^2} + (1+m) - \frac{1}{2}\left(\frac{T_H - T_C}{T_H}\right)\right]}$$

For maximum power output

$R_L = R$, *i.e.*, $m = 1$.

$$n_{max} = \frac{\frac{600}{923} \times 1}{\frac{(1+1)^2}{923} \times \frac{1}{Z_{opt}} + (1+1) - \frac{1}{2}\frac{600}{923}}$$

$$\left[\because \; Z_{opt} = \left\{\frac{(\alpha_{S_1} - \alpha_{S_2})}{\sqrt{\rho_1 k_1 + \rho_2 k_2}}\right\}^2\right]$$

$$\therefore \qquad \eta_{max} = \frac{0.65}{\frac{4}{923} \times \frac{1}{1.85 \times 10^{-3}} + 2 - 0.325}$$

$$= \frac{0.65}{4.342 - 0.325} - 0.1618 = 16.18\%. \text{ **Ans.**}$$

Power output $P_{opt} = I^2 R_L$

$$= \frac{\alpha_{S_{1-2}^2} (T_H - T_C) R_L}{(R + R_L)^2} = \frac{\alpha_{S_{1-2}^2} (T_H - T_C)^2}{(1 + M_{opt})^2 R_{L_{opt}}}$$

$$= \frac{(-190 - 190)^2 \times 10^{-12} \times 600^2}{(1 + 1.43)^2 \times 2.63 \times 10^{-3}}$$

= **3.318 Watts. Ans.**

For *Maximum power* P_{max} $(R_L = R)$

$$= \frac{\alpha_{S_{1-2}^2} (T_H - T_C)^2 RL}{(r + R_L)^2} = \frac{\alpha_{S_{1-2}^2} (T_H - T_C)^2 R_L}{4 R_L^2}$$

$$= \frac{(-190 - 190)^2 \, 10^{-2} \, 600^2}{4 \times 1.84 \times 10^{-3}} = \textbf{7.12 Watts. Ans.}$$

Example 13.5. *A 100 kW, 115 volt, thermoelectric generator operates between1500 and 1000 °K. The material properties are:*

a_{SAB} *at 1250 °K = 0.0012 volt / °K*

$k_A = 0.02$ *watt / cm °K,* $k_B = 0.03$ *watt / cm °K*

$\rho_A = 0.01$ *ohm cm,* $\rho_B = 0.12$ *ohm cm.*

For an optimum design

$A_A = 43.5 \text{ cm}^2$

$A_B = 48.6\ cm^2$

$L_A = L_B = 0.49\ cm.$

and the current density in the elements is limited to 20 amp/cm², calculate:

(i) Maximum efficiency of the thermoelectric generator.

(ii) Number or thermocouples in series.

(iii) Open circuit voltage.

(iv) Heat input and rejected at full load.

Solution. Maximum or ideal efficiency of a thermoelectric converter is given by

$$\eta_{max} = \left(\frac{T_H - T_C}{T_H}\right)\left[\frac{(M-1)}{\left(M + \frac{T_C}{T_H}\right)}\right]$$

where, $M = \left[1 + \frac{Z}{2}(T_H + T_C)\right]^{1/2}$

We shall calculate Z (figure of merit) first.

$$\because \qquad Z = \left[\frac{(\alpha_{SA} - \alpha_{SB})}{\sqrt{\rho_A k_A} + \sqrt{\rho_B k_B}}\right]^2$$

$$= \left[\frac{0.0012}{\sqrt{0.02 \times 0.01} + \sqrt{0.03 \times 0.012}}\right]^2 = \left[\frac{12 \times 10^{-4}}{3.314 \times 10^{-2}}\right]^2$$

$$= 13 \times 10^{-4}\ °K^{-1}$$

Now $M = \left[1 + \frac{13 \times 10^{-4}}{2}(1500 + 1000)\right]^{1/2} = 1.635$

$$(i)\ \therefore \quad \eta_{max} = \left(\frac{1500 - 1000}{1500}\right)\left(\frac{1.635 - 1}{1.635 + (1000/1500)}\right)$$

$= 0.091 =$ **9.1%. Ans.**

(ii) Open circuit voltage per couple

$V = \alpha_{SAB}(T_H - T_C)$

$= 0.0012 \times (1500 - 1000) = 0.6$ volt

$$R = R_A + R_B = \frac{\rho_A L_A}{A_A} + \frac{\rho_B L_B}{A_B}$$

$$= \frac{0.01 \times 0.49}{43.5} + \frac{0.012 \times 0.49}{48.6} = 2.34 \times 10^{-4}\ \text{ohm}.$$

Now, $V_L = V - RI$

$= V - (R_A + R_B)I = 0.6 - (2.34 \times 10^{-4})(20 \times 48.6)$

$= 0.374$ volt

Hence number of couples in series

$$= \frac{\text{Total voltage required}}{\text{Voltage required by one couple}} = \frac{115}{V_L} = \frac{115}{0.374} = 309. \textbf{ Ans.}$$

(*iii*) Open circuit voltage = $V \times 309$

$= 0.6 \times 309 = 185.4$ volt. **Ans.**

(*iv*) Heat input at full load;

$$= \frac{\text{Output}}{\text{Efficiency}} = \frac{100}{0.091} = 1100 \text{ kW. } \textbf{Ans.}$$

Heat rejected at full load

= Heat input – work output

= 1100 – 100 = 1000 kW. **Ans.**

13.5. Thermoelectric Materials

The following materials have been found suitable for use in thermoelectric elements: Lead telluride (Pb Te) in *n* and *p*-type forms, bismuth telluride ($Bi_2 Te_3$), bismuch sulfide ($Bi_2 S_3$), antimony telluride ($Sb_2 Te_3$), tin telluride (Sn Te), indium arsenide, germanium telluride (Ge Te), cesium sulfide (CeS) and zinc antimonide (Zn Sb).

Lead telluride (Pb Te), a compound of lead and tellurium, containing small amounts of either bismuth (*n*-type) or sodium (*p*-type), has been commonly used in recent times for thermoelectric converters. However, the proportion of heat supplied that is converted into electrical energy is only about 5 to 7%. Efforts are being made to provide more efficient thermocouple materials; semiconductors based on silicon-germanium and compounds of selenium (*i.e.*, selenides) appear to be promising for high temperature applications. The figure of merit of some thermoelectric materials are given in Table (13.1).

Table 13.1. Fig. of Merit for Thermoelectric Materials

Materials	$Z\ (°K^{-1})$
Bismuth telluride (doped with Sb or Se)	4×10^{-8}
Lead telluride	1.5×10^{-3}
Germanium telluride (with bismuth)	1.5×10^{-3}
Zinc antimonide (doped with silver)	1.5×10^{-3}
Cesium sulfide	1.0×10^{-3}

13.6. Selection of Materials

We shall now consider the selection of the most suitable materials for thermoelements. First of all, it should be noted that semiconductors, owing their much higher values of 'α_S' are definitely preferable to metals. In fact for most metals have $a_S < 10 \times 10^{-6}$ volt/degree. The ratio of electrical conductivity ($\sigma = 1/\rho$) to thermal conductivity *k* for all metals is close to the theoretical value predicted by quantum mechanics.

$$\frac{k}{\sigma} = \frac{\pi^2}{3}\left(\frac{k}{e}\right)^2 T = 2.44 \times 10^{-8}\, T$$

At an average temperature $T = 500°$ K, $Z = 8.2 \times 10^{-6}$ and at $T_H = 700°$K and $T_C = 300°$K, $\eta = 8.5 \times 10^{-4} = 0.085\%$. For the best pair of metals Bi – Sb, we can put,

$$\alpha_S = 100 \times 10^{-6}, K = 9 \times 10^{-6},$$
$$Z = 1.1 \times 10^{-3}, T_H < 450°K, T_e = 300 °K$$

and $$\eta = 3.1\%$$

In several semiconductors, Z reaches, and sometimes exceeds, the value of 1×10^{-3} and η approaches 10%. Furthermore, several semiconductors possess a high melting point thus permitting operation at high values of T_H. Therefore, efficiencies of the order of 10-15% and more can not be regarded as unattainable with the use of semiconductor. So semiconductor is most suitable thermo element compared with the metal. Let us now formulate the conditions for achieving high efficiency. It has been shown that material should be characterized by as high a value of $Z = \frac{\alpha_S^2}{\rho k}$ °K^{-1} $= \frac{\alpha_S^2 \sigma}{K}$ as possible, and be capable of operation upto very high temperature T_H of the heat source. For this purpose.

(1) The component thermal conductivity of the semiconductor should be as low as possible. This condition should be satisfied by substances with a low Debye temperature, consisting of heavy molecules weekly bound to each other. The value k can be further reduced by introducing of impurities and the formation of solid solutions, provided these impurities do not simultaneously reduce the electron mobility by a similar amount.

(2) The mobility of current carriers (electrons or holes) should be as high as in compatible with condition 1. The mobility should preferably not be reduced much by the introduction of impurities which raises the electrical conductivity, σ. This condition will be best satisfied by atomic or molecular lattices. With the predominance of acoustic vibrations in the thermal spectrum, the mobility can be slightly increased by through annealing and in some cases by the introduction of impurity atoms tending to occupy empty lattice sites and reduce any lattice distortion.

(3) One of the arms should consists of a purely hole type and the other of a purely electronic type semiconductor. For this purpose semiconductor should be selected which have low ionisation energy and forbidden band of such a width that the concentration should be determination solely by impurity atoms upto the hot junction temperature. Only at high ratios electrons-to-hole mobilities is it possible to employ semiconductors with a narrow forbidden band.

(4) The electron concentration throughout the entire element, from the hot junction temperature T_H to the cold junction temperature T_C should be so adjusted that the thermal emf at the temperatures prevailing in the various portion of the element. To satisfy this requirement it is necessary to make the thermoelement arms with a variable impurity content or construct them from several parts. In the low temperature zone the impurity concentration should be lower than in the high temperature zone since the value of thermal emf at a given effective mass of current carriers is governed by the ratio $T^{3/2}/n$. Where n is the number of thermo-element arms.

(5) An important requirement for practical applications of the material is its stability to chemical influences, the particular to oxidation, and good mechanical strength and elasticity are necessary to prevent the thermoelement from cracking under the effect of thermal stresses.

(6) In addition to materials suitable for the arms of the thermoelement it is also necessary to find a material for in metallic bridge connecting the arms which

should create neither additional electrical resistances at the boundary with the semiconductors nor the additional thermal stresses.

The last two conditions become particularly important when instead of a single thermoelement one is considering a pile consisting of a large number of thermoelements.

REVIEW QUESTIONS

1. What is Seebeck Thermo electric effect?
2. How Seebeck coefficients vary with temperature?
3. Describe briefly a thermoelectric power generator.
4. Define figure of merit for a thermoelectric generator.
5. Derive the expression for efficiency of a thermoelectric generator.
6. The absolute thermoelectric power of two materials (*A* and *B*) varies with the temperature as $\alpha_{SA} = 6 \times 10^{-9}\, T$ and $\alpha_{SB}\ 4 \times 10^{-9}\, T$ where, *T* is Kelvins and α_{SA} and α_{SB} are in volts/Kelvin. The materials are joined to form thermocouple whose hot and cold junctions are maintained at 300 and 10°C, respectively. The materials have a negligible electrical resistivity and thermal conductivity.
   (*a*) Find the e.m.f. produced by the thermocouple.
   (*b*) If a current of $I = 0.5$ A is supplied to an external electrical device, find the power produced.
   (*c*) Find the Peltier heat rates absorbed and rejected at the junctions.
   (*d*) Find the Thomson heat rates absorbed and rejected in materials *A* and *B*.
   (*e*) Verify that the difference between the heat rates absorbed and those rejected is equal to the power produced in (*b*).
   (*f*) Find the efficiency of the system and compare it with a Carnot system operating between the same two temperatures.
7. A thermocouple is composed of two materials The thermoelectric power of each material is independent of temperature. The hot and cold junctions are at $T_H = 500$ K and $T_C = 300$ K, respectively. Using the values $\alpha_{SA} = 2 \times 10^{-4}$ V/°K, $\alpha_{SB} = 4 \times 10^{4}$ V/°K, $R = 100\ \Omega$, and *K* (thermal conductance) $= 3.33 \times 10^{-7}$ W/°K.
   (*a*) Find the open circuit voltage (*i.e.*, the e.m.f.) of the thermocouple.
   (*b*) Find the short-circuit current of the device.
   (*c*) Find the figure of merit of the system.
   (*d*) Determine the optimum load resistance required to achieve maximum efficiency and find this efficiency.
8. Compute the figure of merit of the materials of a thermoelectric system with the following properties.

	Material 1	*Material 2*
Seebeck Coefficient α_S (volt/°K)	-196×10^{-6}	-224×10^{-6}
Specific resistivity ρ, (ohm-cm)	1.07×10^{-3}	1.05×10^{-3}
Conductivity K (W/cm°C)	14.3×10^{-3}	16×10^{-3}

[**Ans.** $Z_1 = 2.5 \times 10^{-3}$ °K^{-1}
$Z_2 = 2.98 \times 10^{-3}$ °K^{-1}]

9. (*a*) Describe the operation of thermionic converter.
   (*b*) Derive the expression for thermal efficiency of thermoelectric generator and prove that the optimum value of *M* that gives the maximum thermal efficiency is

$$M_{opt} = \sqrt{1 + Z_{max}\,T_{av}}$$

where, $M = \dfrac{R_{load}}{R_{generator}}$, Z = figure of merit

and, $T_{av} = \dfrac{T_{high} + T_{low}}{2}$

10. A thermoelectric generator that will operate between 30 and 500°C is to be constructed of *n-p* semi-conductor with the following properties:

Material	*n type*	*p type*
α (Seebeck Coefficient Volt/°K	-170×10^{-6}	210×10^{-6}
ρ (resistivity) Ω-m	14×10^{-6}	18×10^{-6}
k (thermal conductivity) W/m°K	1.5	1.1

For this system that is designed to produce 500 W at highest possible thermal efficiency, find the number of element pairs, the maximum possible thermal efficiency, and the maximum possible power output. Assume that the cross-sectional area and length of *n*-leg are 1 cm² and 1 cm respectively and that of *p* leg is 1 cm.

11. A thermoelectric generator operates between the temperature limits of 100 and 300°C. The cross-sectional area of *n*-type element is 1 cm² and the length is 1 cm. Using the optimum value for the product of internal resistance and overall thermal conductance (*i.e.*, *RK*). Calculate the maximum generator efficiency and the efficiency for maximum power. Also calculate the power output for both the cases. The following properties may be used.

Material	*n type*	*p type*
Seebeck Coefficient α_S (volt/°C)	-190×10^{-6}	190×10^{-6}
Specific resistivity ρ (ohm-cm)	1.45×10^{-3}	1.8×10^{-3}
Figure of merit Z (°K⁻¹)	2.0×10^{-3}	1.7×10^{-3}

[Ans. $\eta_{opt} = 6.39\%$, $\eta_m = 6.24\%$
$P_{opt} = 0.247$ watts,
$P_{max} = 0.477$ watts]

[Hint. For the optimum value *of RK* product the area-length relation of the *n* type and *p* type materials are calculated from

$$\left[\frac{(l/A)_n}{(l/A)_p}\right]_{opt} = \left(\frac{k_n/\rho}{k_p/\rho_p}\right)^{1/2} = \left(\frac{k_n\,\rho_p}{k_p\,\rho_n}\right)^{1/2}$$

We have $l_n = l_p = 1$, $A_n = 1$ cm², A_p can be calculated].

12. A thermoelectric generator operates between 250 and 550°C. The average value of the Seebeck coefficient is 400×10^{-6} V/K, the generator resistance is 0.004 Ω and the thermal conductance is 0.035 W/k-m. Find the open circuit voltage, the maximum power output, and thermal efficiency for maximum power output.

13. Write short notes on:
(*i*) Thermoelectric effects,
(*ii*) Seebeck, Peltier and Thomson effects,
(*iii*) Multistage thermoelectric generators,
(*iv*) Selection of Thermoelectric materials,
(*v*) Figure of merit of a thermoelectric generator.

CHAPTER 14

Thermionic Generation

14.1. Introduction

Another form of *direct conversion* of heat energy to electrical energy has been achieved in the thermionic converter. It utilizes the *thermionic emission effect,* that is, the emission of electrons from heated metal (and some oxide) surfaces. The energy required to extract an electron from the metal is an important parameter, known as the *work function* of the metal. Typical values of the order of a few electron volts. The value of the work function varies with the nature of the metal and its surface condition. In principle, a thermionic consists of two metals (or electrodes) with different work functions sealed into an evacuated vessel. The electrode with the large work function is maintained at a higher temperature than one with the smaller work function.

System consists of two electrodes held in a container filled with ionized cesium vapour. Heating one electrode, electrons are emitted, that travel to the opposite, colder electrode. The hottor electrode (or emitter) emits electrons (*i.e.,* negative charges) and so acquires a positive charge, whereas the colder electrode (or collector) collects electrons and becomes negatively charged. A voltage (or electromotive force) thus develops, between the two electrodes and a direct electric current will flow in an external circuit (or load) connecting them. The voltage, which may be 1 volt (or so), is determined primarily by the difference in the work function of the electrode materials.

Because electrons can not travel far in air, thermionic converters require that the electrodes be in a vacuum. This limit the size of the converter so that only small scale power production is feasible.

The emissions of electrons from the hotter electrode is inhibited by a *space charge* resulting from the accumulation of electrons in its vicinity. To reduce the space charge, small quantity of cesium metal is introduced into the evacuated vessel containing the electrodes. At the existing high tempera-ture, the cesium vaporizes and ionizes ; that is to say, it forms a mixture of positive

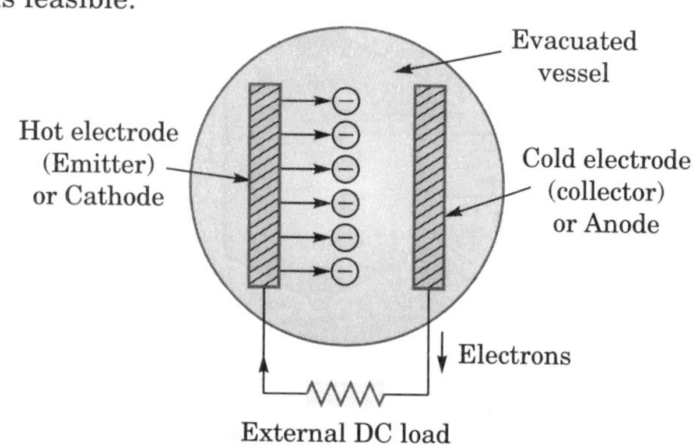

Fig. 14.1. Thermionic converter.

ions (*i.e.,* positively charged atomic particles) and electrons. By combining with the electrons in the vicinity of the emitter electrode, the possible ions help to decrease the space charge.

Deposition of solid cesium on the colder electrode (collector) is also beneficial in decreasing the work function and thus increasing the output voltage. A common thermionic electrode combination is a tungsten emitter and a cesium-coated tungsten collector.

A thermionic converter is a form of *heat engine,* in principle, heat is taken in at the upper (emitter) temperature, part is converted into electrical energy, and the remainder is discharged (or removed) at the lower (collector) temperature. As with other engines, the thermal efficiency (*i.e.,* the fraction of the heat taken up that is converted into electrical energy) is favoured by a high upper temperature and a lower temperature. The thermionic converter will continue to generate electric power as long as heat is supplied to the emitter and a temperature difference is maintained between it and the collector.

To achieve a substantial electron emission rate (per unit area of emitter) and hence a significant current output as well as a high efficiency, the emitter temperature in a thermionic converter containing cesium should be atleast 1000°C. The efficiency is then about 10%. Higher efficiencies, possibly upto 40% can be obtained by operating at still higher temperatures. Although temperature has little effect on the voltage generated, the increase in current (per unit emitter area) associated with a temperature increase results in an increase in power. Electric power is the product of voltage and current ($P = EI$).

In principle, any heat source (*e.g.,* a fossil or nuclear fuel, a radioactive material or solar energy) can be used in a thermionic generator, provided a sufficiently high emitter temperature can be obtained. Thus, thermionic conversion can be utilised in many different situations; several applications have been suggested for remote locations on the earth and in space when *solar energy* or high temperature fossil fuel combustion gases as heat source is used, the thermionic conversion system would be used in a *topping cycle* as part of a combined-cycle generation system. The hot combustion gases would supply heat to the emitter at about 1650°C, and the heat discharged from the collector at around 550°C would produce superheated steam to operate a turbine-generator in the conventional manner. The overall efficiency for conversion of heat into electricity might then be about 50%. A major problem is to design a thermionic converter that would withstand the hot, corrosive combustion gases.

14.2. Thermionic Emission and Work Function

If a metal is heated, electrons are emitted from it; this is *thermionic emission.* The thermal energy supplied to the solid results in an increase of the amplitude of the lattice vibrations. The resulting collisions with free electrons within the metal are then more energetic, so that it is possible for an electron to receive sufficient energy to expel it from the metal altogether. The energy required to extract an electron from the metal is an important parameter, known as the *work function* of the metal. As already stated typical values are of the order of a few electron volts. Of course, any lattice electron collission process may give energy to the electron in excess of that required to reject it from the metal. Any excess manifests itself as *KE* of the ejected electron. Similarly if an electron (of zero *KE*) is absorbed into the metal from the outside, it gives upto the lattice an energy equal to the work function.

Thermionic emission is the process that is common to all materials and occurs at all temperature. However the currents that can be produced are extremely small except in the special case of metals at high temperatures. The following discussion will be concerned only with temperatures in excess of 1000°C and with metals as the basic materials.

As we know in a matter, the atoms are closely spaced in a crystal lattice and the valence electron may move from atom to atom. Fig. (14.2 *a*, *b*, *c*) shows the variation of potential energy about nucleus for a single atom or 2 atoms. As in Figure (*c*), the maximum resultant potential energy is less than the potential energy at the same distance from an isolated atom.

In a metal the atoms are closed spaced in a 3 dimensional crystal lattice and the potential energy is determined by the charges of all the electrons and the nuclei. *A*, 2 *D* diagram hence cannot show the potential energy of an electron at every point in the crystal. The potential energy along a line through a row of nuclei of these atoms can however be represented as shown in Fig. 14.2 (*d*). Within the metal the potential energy distribution is a series of humbs between the nuclei and at the boundary. There is a potential energy rise, if the net charge on the metal is zero, there is no electric field outside the metal and potential energy rises a plateau flatens to ∞, that means a potential energy hill or barrier exists at the surface of the metal.

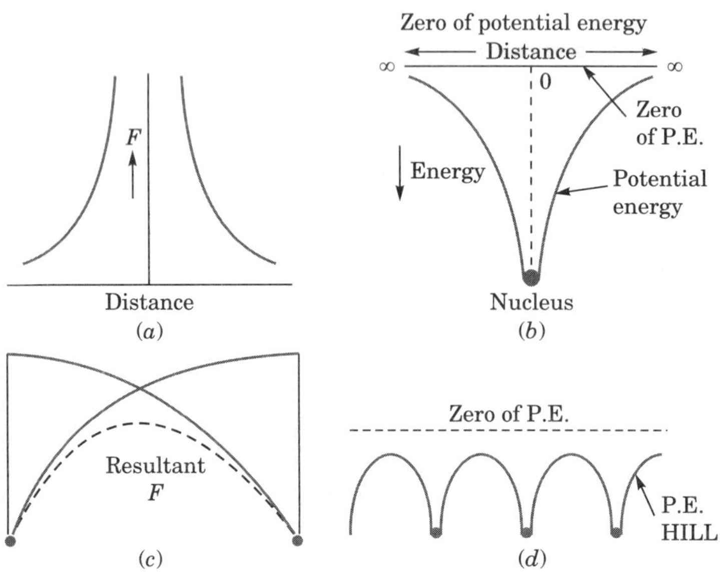

Fig. 14.2. Variation of P.E. about nucleus.

We know that there is a *K* shell and *L* shell of an atom, which consists of 2 and 8 electrons respectively and so on. The electrons in the outermost shell are the balance electrons. Normally the electrons in a atom can not leave as they do not have enough energy to cross the potential energy barrier. In a metal however the potential energy barrier between atoms are lowered owing to their close proximity.

The reason for the failure of the electrons to escape the surface of the metal is the potential energy barrier at the surface. Practically none of the electrons have total energy sufficient to enable to cross the barrier by changing their *KE* to *PE*. For an electron to have a total *KE* greater than *PE* barrier, the kinetic energy associated with its component of velocity perpendicular to the barrier must be greater than the PE of the barrier to make escape possible.

Following two points must be noted carefully :

(*i*) Within the metal, the potential energy distribution is in the form of a series of humps or "hills and valleys" between different nuclei whereas according to Classical Electrostatics, the interior of a metal is an equipotential surface.

(*ii*) Next, consider the conditions that exist near the metal surface. Broken line *MN* represents the metal surface in Fig. 14.3. Since no nuclei exist to the right of nucleus *D*, there can be no lowering and flattening of potential energy curve at it takes place in the regions between the nuclei. This leads to a very important conclusion; *a potential energy hill or barrier exists at the metal surface.*

The energy difference between the zero level of the *KE* and *PE* plateau outside the metal is called W_a and the band of energy between the zero and the *Fermy energy* level W_e is known as Fermi energy band. Fig. 14.4. Fermi level of energy W_i, it is the energy of an electron at about 0°K. The zero level is placed a little below the

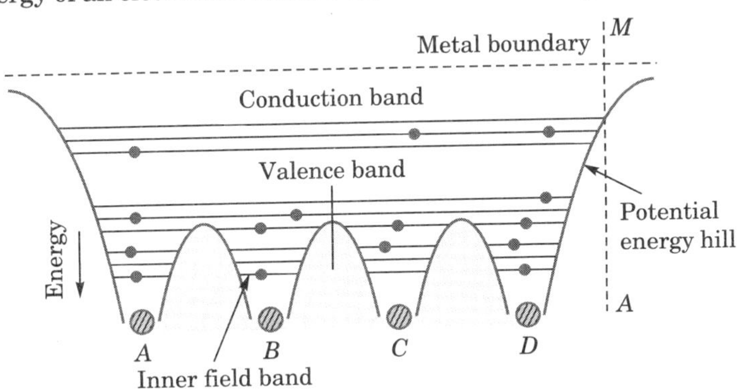

Fig. 14.3

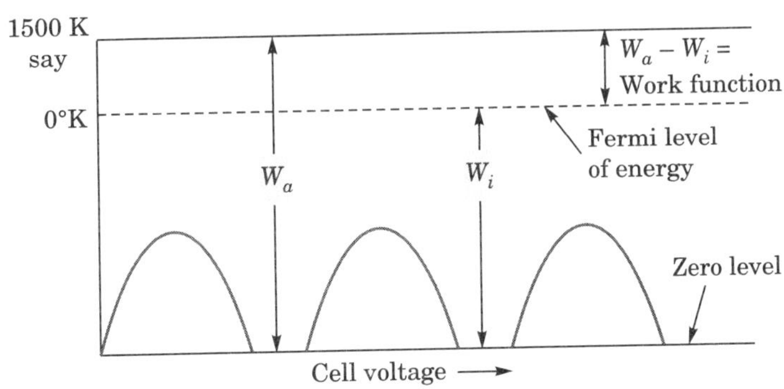

Fig. 14.4. Energy level diagram.

PE humbs in order that the band of occupied level above it may correspond to the levels actually occupied by the valence electrons.

The phenomenon of the electron emission depends on the energy difference between W_a and W_i. At 0°K the maximum associated energy of the electrons escape at this temperature. The amount of energy ($W_a - W_i$) is called the work function of the metal. It may be a function of temperature, although experiments indicate that its dependent on temperature is relatively small. If the voltage equivalent of the work function is denoted by ϕ, then $\phi e = W_a - W_i$

where, ϕ is in volts

e is charge, $\phi . e$ = work (volt × charge).

The difference $W_a - W_i$ or work function is expressed in electron volts.

It is seen that at 0°K, it is impossible for a free electron to escape from metal because maximum energy it possesses is W_i which is less than potential energy barrier W_a. In order to enable it to escape, the energy difference between W_a and W_i has to be supplied from outside. This energy difference written as work function ϕ, of the metal.

Hence, work function of a metal may be defined as the *minimum amount of energy that must be given to its fastest-moving electron at 0°K in order to enable it to escape from the metal.*

Table 14.1

Element	*Work function eV*	*Element*	*Work function eV*
Ag	4.65	Mo	4.25
Al	3.6	Na	2.0
Ba	2.52	Ni	5.0
Ca	3.0	Pt	6.10
Cs	1.75	Rb	1.82
Cu	4.3	Th	3.5
K	1.9	T	4.55
Li	2.21	Zn	3.44

Several different means of giving energy to electrons are available. They are

(1) By sufficient increase of temperature may give electrons enough energy to enable them to cross the energy barrier. The liberation of the electrons by this process is called *thermionic emission.*

(2) The electrons in a metal can also secure the additional energy required for liberation from *photons of light* impinging on the surface. This is called *photo-electric emission.*

(3) Also the electrons can be emitted by when the ions or electrons strike the metal surface from outside (*Secondary emission*).

(4) An intense surface electric field produced by an adjacent positively charged electrodes is used to modify confining potential energy barrier to such an extent that the electrons with energy near the Fermi level (W_i) penetrated and escaped. The resulting liberation of the electrons is known as field emission.

The process of thermionic emission can be studied in the light of the Fermi-Dirac free electron theory of metals. Analysis results in the following expression for the emitted current density (J).

$$J = \frac{4\pi\, m\, e\, k^2}{h^3} T^2 e^{-\phi}/KT$$

$$= AT^2 e^{-\phi}/KT \text{ amperes/m}^2 \qquad ...(14.1)$$

where, m = mass of an electron, kg.

k = Boltzmann's constant J/°K.

T = Temperature, °K.

ϕ = Work function in *ev* (electron volt).

e = charge.

h = Plank constant, Js. (Joule second)

A = a constant defined as equal to

$$\frac{4\pi\, m\, e\, k^2}{h^3}$$

$$= 120 \times 10^4 \text{ A/m}^2\,°\text{K}^2 = 1.20 \times 10^6 \text{ A/m}^2\,°\text{K}^2$$

The equation (14.1) is known as the *Richardson Dushman* equation. Although the constant A has the theoretical value of 120 A/cm²/deg K², in practice it is usually found to be considerably lower. There are a number of factors that contribute to this discrepancy. The work function of the varies with temperature owing to thermal expansion of the lattice, and varies for different crystallographic planes in the lattice. Also the presence of absorbed gases on the surface of the emitting material will affect, the apparent value of ϕ. Finally a quantum mechanism shows that a reflection will occur to 5% of the electron flow. All these factors reduce the value of A. However, there is evidence that for pure metals the theoretical value of A is attained.

Equation (14.1) shows the strong dependence of current density on work function and temperature. As an example, consider a tungsten filament emitting electrons at 2500°K. It can be readily shown from this equation that

$$\frac{\Delta J}{J} = \left(2 + \frac{KT}{\phi}\right)\frac{\Delta T}{T} \qquad ...(14.2)$$

With $\phi = 4.5\, ev$, for a typical tungsten filament, a 1% increase in temperature produces approximately a 23% increase in current. Similarly it can be shown that if at 2500°K the work function were decreased by 1%, the current would increase by approximately 21%. Further analysis of thermionic emission phenomena shows that the total average energy of an emitted electron is 2 kT. This means that every electron that leaves the cathode material extracts from it an energy equal to ϕ + 2 kT, so that the power per electron, consumed by the emitter, is

$$P_C = \frac{J}{e}(\phi + 2\,kT) \qquad ...(14.3)$$

The form of this equations show that ϕ may be regarded as the latent heat of evaporation per electron.

Equation (14.1) is derived with the assumption that emitter is surrounded by a vacuum.

14.3. Basic Thermionic Generator (the Basic Diode)

Consider Fig. 14.5 representing the basic thermionic diode generator. The cathode is heated by external means so that the temperature is raised and copious thermionic emission takes place. The emitted electrons will drift across the inter-electrode region towards the anode (the whole is in vacuum), building up a charge in this region as they do so. The electron 'Cloud' negatively charged, will inhibit the emission of further electrons. The electron emission will therefore diminish and the current reaching the anode will be limited. The current is said to be space charge limited ; and its value will be very much lower than that expected from Richardson equation (*i.e.* Eq. 14.1). The presence of the charge just outside the metal surface can be considered as effectively increasing the work function of the metal *i.e.,* making it more difficult for the electrons to get out.

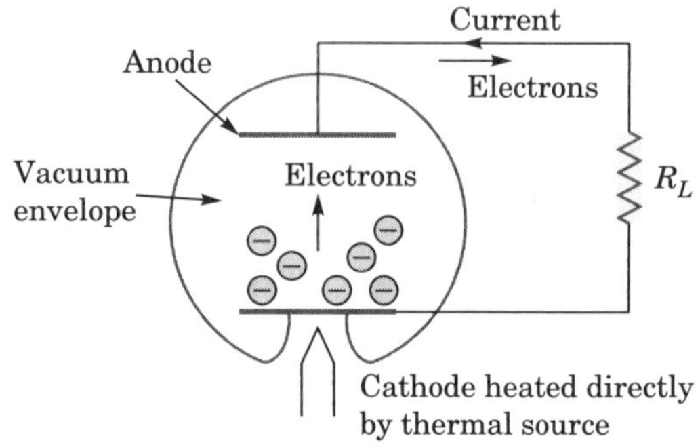

Fig. 14.5. Schematic of a basic thermionic diode.

The operation of the generator can be illustrated with reference to the potential diagram for the electron as it flows down the generator circuit Fig. 14.6. Actually as we are concerned with electrons, the potential is negative, but such diagrams are conventionally drawn positively as shown. The potential V_b, represents the potential barrier presented by the space charge in the inter-electrode region.

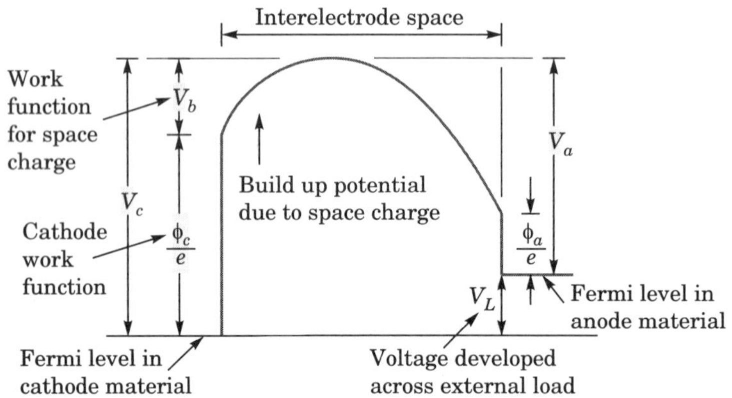

Fig. 14.6. Potential diagram for thermionic generator.

Consider an electron initially at the Fermi level in the cathode material. To leave the cathode it must acquire an energy greater than or equal to ϕc, the cathode work function, it does this by interaction with the lattice to which the thermal energy has been supplied from an external source. This energy will be sufficient to eject the electrons from the cathode, but to reach the anode a total energy of $\phi c + eV_b$ is required *i.e.,* sufficient to overcome the space charge barrier.

Having surmounted the barrier, the electron crosses the interelectrode region and enter the anode. It gives up an energy ϕ_a to the anode, ϕ_a being the work function of the anode material. As shown in the diagram this still leaves the electron with potential energy relative to the Fermi level in the cathode material since the anode is connected to the cathode through an external circuit, the electron travel through the load R_L, giving up their energy and performing useful electrical work. The thermal energy from the heat source increases the potential energy of the electron, which is converted to electrical power as the electrons return to their original energy state. Thus a direct conversion of thermal to electrical energy is achieved in which the electron current is effectively the working fluid of the conversion system.

Note that no external voltage is applied to the anode as is the case with the ordinary thermionic valve. As it is clear from Fig. 14.2 that the conventional cathode *i.e.,* the supplier of the electron, becomes the positive potential terminal in the thermionic generator. The generator works as a current source so that the voltage developed across the load depends upon the magnitude of the load resistance.

14.4. Analysis of Thermionic Generator

The emitted current density as given by Fermi Dirac is

$$J = \frac{4\pi\, m\, e\, k^2}{h^3} T^2 e^{-\phi/KT}$$

$$= AT^2 e^{-\phi/KT} \text{ amp/m}^2 \qquad ...(14.4)$$

Where A is constant and has the value equal to 120×10^4 Amp/m^2 as described in the previous section. Analysis of thermionic emission phenomenon shows that the average energy of an emitted electron is 2 KT. This means that every electron that leaves the cathode material extracts from it an energy equal to $\phi + 2\,KT$ so that the power per electron consumed by the emitter is

$$P_C = \frac{J}{e}(\phi + 2\,KT) \qquad ...(14.5)$$

This equation shows that ϕ may be regarded as the latent heat of evaporation per electron as already mentioned.

The distribution of potential work function is already shown in Fig. 14.6. The analysis is now made of a simple generator of Fig. 14.5, to ascertain the parameter that are of importance of its efficient operation. In the literature the cathode is often called the *emitter* and the anode the *collector,* as in transistor technology.

In the practical generator, some means must be provided to overcome the effect of space charge, which drastically limits the current that can be drawn. This quest is one of the major undertaking of thermionic generation development work, and various techniques have been adopted. To make the analysis reasonably general in the first instance, the presence of space charge is assumed in the following discussion. This is represented by the potential barrier V_b. The potential difference between the Fermi level of the cathode and the top of the space charge is denoted by V_C and the corresponding value of the anode, V_a. The analysis of this case can be modified for various special cases. For example, if the space charge is to be totally neglected, then putting $V_b = 0$, will yield the required modification.

With the conditions relevant to Fig. 14.6 it will be seen that the minimum energy required by an electron, to reach the anode is $\phi_c + e\,V_b$.

where, φ_c is the voltage equivalent of work-function of cathode

If e is the charge, then work = voltage × charge = $\phi_c \times e$

If W_i is the Fermi level of the energy *i.e.,* it is the energy of an electron at about 0°K, and,

W_a = the energy difference between the zero level of the *KE* and *PE* plateau outside the metal.

then, $\phi\, e = W_a - W_i$ in electron volts. (refer Fig. 14.4)

The current leaving the cathode is thus limited to, (From Richardson Dushman Eqn.)

$$J_C = A\,T_c^2\, e^{-(\phi_c + eV_b)/KY_c} \qquad ...(14.6)$$

$$= A\,T_c^2\, e^{-eV_c/KT_c} \qquad ...(14.7)$$

Where the subscript C refers to the cathode. Note that in these expressions the presence of space charge has been taken into account solely by the use of the parameter V_b. The form of the space charge barrier or its position within the interelectrode space is immaterial.

Similarly the electron current leaving the anode is :

$$J_a = A\,T^2\, e^{-eV_a/KT_a} \qquad ...(14.8)$$

Where the subscript refers to the anode. In most practical cases the design will be such that J_a is negligible compared with J_c. However it is possible for J_a to become significant and so it will be included in the analysis for the present obviously. J_a represents a loss mechanism *i.e.,* it leads to a decrease in useful load current. Therefore, the *net current density flow* is given by

$$J = J_c - J_a$$

$$= A\,T_c^2\, e^{eV_c/KT_c} - A\,T_a^2\, e^{-eV_a/KT_a} \qquad ...(14.9)$$

The equation (14.9) gives that the net current flow in the thermionic diode is the difference between the emission current from the cathode and the back emission current from the anode.

There are several factors which prevent the thermionic from achieving the theoretical Carnot efficiency ;

(1) Impedance to electrons flow resulting from the space charge between the electrodes.

(2) Thermal radiation transfer between the electrodes.

(3) Heat losses to the surroundings.

The space charge effect may be reduced by placing the electrodes very close together or by introducing positive ions into the space to neutralize the negative space charge of the electrons. Cesium vapour is commonly employed for the neutralization process because it has a low ionization potential (about 3.9 volt). When a vapour is introduced the device is called a *plasma diode.* In the vacuum diode electrode spacing as low as 0.01 mm may be used to reduce the space charge effect. Such small spacing produce very difficult fabrication problems and for this reason plasma diodes, with electrode spacings of the order of 1.0 mm have received considerable development attention.

The voltage output of a plasma diode may be written as

$$V = \phi_c - \phi_a - \phi_p \qquad ...(14.10)$$

Where, ϕ_c and ϕ_a are the cathode and anode work functions respectively and ϕ_p represents the plasma potential drop. The thermal efficiency is then given as :

$$\eta_{th} = \frac{\text{Work output}}{\text{Heat input}} = \frac{JV}{Q_c/A_c} \qquad ...(14.11)$$

where, Q_c/A_c = the cathode heat flux.

The total heat added to the cathode is equal to the sum of the electron-energy flux out of the cathode and the thermal radiation heat exchange.

$$\frac{Q_c}{A_c} = J\left(\phi_c + \frac{2\,KT}{e}\right) + \frac{Q\,\text{rad}}{A_c} \qquad ...(14.12)$$

The term $\frac{2KT}{e}$ represents the voltage equivalent to the thermal energy of the emitted electrons, the radiation heat transfer from cathode which is maintained at a very high temp., is calculated from

$$H_L = \frac{Q \text{ rad}}{A_c} = \varepsilon\, \sigma(T_C^4 - T_S^4) \qquad ...(14.13)$$

Where, ε, is an appropriate constant which is a function of the radiation properties of the electrode surfaces,

σ is the Stefan Boltzmann constant and has the value 5.668×10^{-12} Joule/sec cm^2 k^4.

T_s is the temperature of the surroundings. (In most cases $T_a \simeq T_a << T_C$) T_a is the ambient temperature.

If the cathode and anode are flat plates it can be shown that

$$\frac{1}{\varepsilon} = \frac{1}{\varepsilon_c} + \frac{1}{\varepsilon_a} - 1$$

where, ε_c and ε_a are the emissivities of the two materials.

It was noted previously that owing to the Fermi Dirac distribution of the free electrons within the cathode, they are thermionically emitted with a mean energy of $2\,kT_c$. In accordance with equation (14.6), this means that each electron emitted from the cathode carries away an energy of $(eV_c + 2\,kT_c)$. Similarly each electron emitted from the anode carries away an energy $(eV_a + 2\,kT_a)$.

Two minor loss mechanisms may also be noted. Firstly various leads are necessary in the construction of the generator, *e.g.* the high temperature cathode has to be connected to the load, which in most cases will be at very much lower temperature. These leads will thus cause an *IR* drop (V_L) and a heat loss from the cathode by thermal conduction H_L.

In the above discussion, all the heat losses and the energy considerations have been itemized. It is now possible to calculate the conversion efficiency considering the losses. The electrical power output from the generator per unit area of cathode, is the product of the net current and the voltage developed across the load, as already stated, *i.e.*

$$\text{Power output per unit area} = J \times V$$

$$= (J_c - J_a)(V_c - V_a - V_L) \qquad ...(14.14)$$

The heat loss from the cathode is due to emission cooling (*i.e.* thermionically emitted electrons carry off energy), radiation and lead conductors. But the cathode receives energy from the back current J_a. Equation (14.14) gives the power output per unit area of a vacuum thermionic generator.

The *efficiency of the generator* η_g is defined as the ratio of the electrical power output per unit area of the cathode. This can be expressed as

$$\eta_g = \frac{(J_c - J_a)(V_c - V_a - V_L)}{J_c\left(V_c + \frac{2kT_c}{e}\right) - J_a\left(V_a + \frac{2kT_a}{e}\right) + H_L + \varepsilon\sigma(T_c^4 - T_s^4)} \qquad ...(14.15)$$

where J_c and J_a are as given by equations (14.7) and (14.8) respectively.

The equation (14.15) is the general expression. To ascertain the main dependencies the small terms will be neglected. In most devices J_0 will be very much less then J_c, and T_c will be very much higher than T_s. All the effects of the leads *i.e.* H_L, and V_L, can usually be neglected, with these approximations equation (14.15) reduces to;

$$\eta_g = \frac{J_c (V_c - V_a)}{J_c \left(V_c + \frac{2KT_c}{e} \right) + \varepsilon\sigma T_c^4} \quad ...(14.16)$$

In a practical generator the term $\frac{2kT_c}{e}$ would be an order of magnitude lower than V_c. With this further approximation and substituting H_T for $\varepsilon\sigma T_c^4$ we have

$$\eta_g = \frac{V_c - V_a}{V_c + \frac{H_T}{J_C}} \quad ...(14.17)$$

Example 14.1. *Following data are given for a thermionic generator.*

Cathode work function ϕ_c = *2.5 volts.*
Anode work function ϕ_a = *2.0 volts.*
Temperature of cathode T_c = *2000°K.*
Temperature of surroundings T_s = *1000°K.*
Plasma potential drop ϕ_p = *0.1 volt.*

Assuming the value of emissivity as 0.2 for the electrode materials used, calculate the efficiency of the generator and also compare with the Carnot efficiency.

Solution. Net output voltage

$$V = \phi_c - \phi_A - \phi_p = 2.5 - 2.0 - 0.1 = 0.4 \text{ volt.}$$

Now the net current in the generator $J = J_{cathode} - J_{anode}$

but, $J_c = A\,T^2\, e^{-\phi_c/kT}$

$$= (1.20 \times 10^6)\,(2000)^2 \exp.\left[\frac{-(1.6 \times 10^{-19})\,2.5}{(1.38 \times 10^{-23})\,2000}\right]$$

Charge of an electron $e = 1.6 \times 10^{-19}$ Coulomb
Boltzmann constant $k = 1.38 \times 10^{-23}$ Joule/°K
One electron volt $= 1.6 \times 10^{-19}$ Joule

$$= 2.4 \times 10^6 \text{ amp/m}^2$$

$$J_{anode} = A\,T^2\, e^{-\phi_a/kT} = (1.20 \times 10^6)(1000)^2$$

$$\exp.\left[\frac{-(1.6 \times 10^{-19})\,2.0}{(1.38 \times 10^{-23})\,1000}\right]$$

$$= 1.2 \times 10^2 \text{ amp/m}^2.$$

The net current J can be taken $= J_c$, as J_a can be neglected in comparison with J_c.

i.e., $J = J_c = 2.4 \times 10^6$ amp/m²

The heat supplied to the cathode Q_c / A_c

$$= J\left(\phi_c + \frac{2kT_c}{e}\right) + \varepsilon\sigma\,(T_c^4 - T_s^4)$$

$$= (2.4 \times 10^6) \left[\frac{2.5 + 2(1.38 \times 10^{-23})\,2000}{1.0 \times 10^{-19}} \right]$$

$$+ 0.2,\ 5.67 \times 10^{-12},\ 10^{-4}(2000^4 - 1000^4)$$

$$= 6 \times 10^6 + 8.26 \times 10^5 \times 2 \times 10^5$$

$$= 7.026 \times 10^6 \text{ watt/m}^2$$

Efficiency of the generator

$$\eta_g = \frac{JV}{Q_c/A_c} = \frac{(2.4 \times 10^6)(0.4)}{7.026 \times 10^6} = 0.137 = 13.7\%. \quad \textbf{Ans.}$$

Carnot efficiency this device

$$\eta_c = \frac{T_1 - T_2}{T} = \frac{2000 - 1000}{2000} = 0.50 = 50\%. \quad \textbf{Ans.}$$

Hence for this plasma diode Carnot efficiency is 50%, but actual generator efficiency will not be greater than 13.7%.

REVIEW QUESTIONS

1. What is thermionic emission effect ? How space charge effect is minimized ?
2. What you mean by work function of a metal ?
3. What are the usual means of supplying energy to electron emission ?
4. Describe the principle of working and constructional details of a basic thermionic generator.
5. Explain in brief the following :
   (*i*) Potential barrier (*ii*) Thermionic emission
   (*iii*) Work function.
6. Derive the expression for power and efficiency for a thermionic generator.
7. A vacuum thermionic generator has
   Cathode work function ϕ_c = 2.5 volt.
   Anode work function ϕ_a = 2.0 volt.
   Cathode temperature = 2250°K
   Anode temperature = 1125°K
   At these conditions the value of emissivity for electrodes material may be assumed as 0.2.
   Determine :
   (*i*) Heat supplied to the cathode (*ii*) The efficiency of the device
   (*iii*) Theoretical Carnot efficiency.
   [**Ans.** (*i*) 7.149×10^6 watt/m^2; (*ii*) 17.48% ; (*iii*) 50%]

CHAPTER 15

Thermonuclear Fusion Energy

15.1. Introduction

A form of nuclear energy released by the fusion (or combination) of two light nuclei (*i.e.,* nuclei of low mass number) to produce a heavier nucleus. For two nuclei to fuse, they must come close enough to interact. However, similar electric charges repel one another, and since all nuclei carry a positive charge, an increasing force of repulsion develops as the two nuclei are brought closer together. Consequently, for the nuclei to fuse, that must have enough kinetic (or motion) energy to overcome the force of electric repulsion that keep them apart. For the force of repulsion to be small, and thus make fusion easier to achieve, the interacting nuclei should have small positive charges (*i.e.,* low atomic numbers). The element hydrogen has the lowest atomic number, since the nuclei of its three isotopes all carry a single positive charge. Hence, hydrogen isotopes, which also have the lightest nuclei, should be particularly suitable for the production of energy by the fusion of two light nuclei (similar or dissimilar) to form a heavier nucleus. The three hydrogen isotopes are ordinary or light hydrogen (H), with a mass number of 1; heavy hydrogen or deuterium (D), mass number 2; and tritium (T), mass number 3. Of the several possible reactions between pairs of hydrogen isotope nuclei, the fusion of a deuterium nucleus (or deuteron) and a tritium nucleus (or triton) has been found to take place most readily. This is referred to as the D-T reaction and is expressed as

$$D + T \longrightarrow 4\,He + n + 17.6 \text{ MeV}$$

Where $4He$ represents a common helium-4 nucleus and n is a neutron. The fusion of one deuteron and one triton releases 17.6 million electrons volts (MeV) of energy.

Nuclear fusion could be called nuclear combustion; that is, it is the release of energy that occurs when the nuclei of certain isotopes of the light elements *(viz.,* H_2, He, Li, etc.) collide and fuse together at ultra high temperatures ($>> 10^8$ °K). This is the source of energy of the sun and other stable stars which burn ordinary hydrogen as their fuel. The main isotopes being considered are the heavier isotopes of hydrogen and some isotopes of helium and lithium.

The primary fuel for fusion is deuterium, sometimes called heavy hydrogen (H_2). Deuterium is a stable isotope of hydrogen that exists together with ordinary hydrogen. At the time of creation of solar system, deuterium was formed and was retained on the earth in the waters of the ocean. There is one atom of deuterium for every 6500 atoms of ordinary hydrogen in water. This energy heritage has laid in store in the oceans for countless million of years, waiting for mankind to learn how to release it.

The total energy content of deuterium as fossil fuel is about 100,000 kW hr of energy per gram of deuterium. This is 4 times the energy released per gram in the fission of uranium, and is about 10 million times that released per gram in the combustion of fossil fuels.

Today, the total world energy consumption (all forms) is estimated to be about 6×10^{13} kW hr per year. This entire amount of energy could be supplied by the fusion of only 600 metric tonnes of deuterium. To obtain this amount of Deuterium, a deuterium isotope separation facility would require as its feed stock the amount of water that would flow at normal water pressures through a water main only 50 cm in diameter. The amount of petroleum that would be required to satisfy the same world energy need would be 120 million barrels a day—a rate sufficient to deplete the world reserves in a few years. We can deduce that the amount of deuterium D in the world oceans would be enough to sustain the present total world energy output for 100 billion years, that, is ten times the estimated age of our universe. Hence it can be said without any doubt that fusion would indeed represent an ultimate solution of mankind's energy needs for all the years to come.

Fusion by now has become a substantial international research effort. By 1971, Soviets have increased their research efforts in this field, to the extent of about 38% of the world effort. U.S. effort is 16% of the total world effort. U.S. at present spending a large sum for fusion research. The Japanese programme, although starting from a smaller base, is now growing even more rapidly; in 1973 it represented 10% of the world effort. Fusion research is indeed being taken seriously by the major industrial nations. The intellectual possibility of fusion power has been established, but many scientific and technological problems still remain to be solved. *Scientists of Prinston University America have achieved success in controlled fusion reaction using mixture of Deuterium and Tritium.*

Before we discuss the basics of nuclear fusion, let us define and differentiate the two nuclear reactions, fission and fusion. The nuclear reactions that are likely to prove useful as energy sources fall into two general categories: *Fission* and *Fusion.* In *Fission,* a heavy nucleus is split into two or more lighter nuclei. In *Fusion,* two or more light nuclei fuse (combine) to form a heavier nucleus. In both, there is decrease in mass resulting in *exothermic* energy ($\Delta E = \Delta mc^2$). Fusion reactions occur at very high temperature.

Fission. The fission process is accompanied not only by the release of energy but also by the emission of neutrons, called *fission neutrons*; two to three neutrons are emitted on the average per nucleus undergoing fission. The emitted neutrons can cause fissions in additional nuclei, thus liberating more neutrons which can cause further fissions, and so on. However, for a self sustaining fission chain to be possible with the continuous release of energy, certain requirement must be met.

$$\text{Neutron} + \underset{\text{nucleus}}{\text{Heavy}} \rightarrow \underset{\text{fragments}}{\text{Fission}} + \underset{\text{2 to 3}}{\text{Neutrons}} + \text{Energy}$$

One requirement is that the heavy nuclei must be such that they can be fissioned by neutrons of any energy; these substances are referred to as *fissile species.* The fission neutrons initially have a high kinetic (or motion) energy, but this decreases rapidly as a result of collisions with various nuclei. For a fission chain to be sustained, the neutrons of lower energy must still be capable of causing fission, this is possible only with a fissile species. All fissile substances are radioactive, and to be useful as practical energy sources they should have reasonably long half-lives so that they do not decay too quickly. Furthermore, they should exist in nature in moderate quantities, at least, or be capable of fairly easy production from available materials. Only three fissile species satisfy those conditions; uranium-233, uranium-235, and plutonium-239. Of these three species, only uranium 235 occurs in nature, the other two are made artificially. An important aspect of fission as an energy source is that it can be self-sustaining process.

A neutron can cause fission which is itself accompanied by the liberation of neutrons; these neutrons can cause further fissions, and so on. A fission chain reaction, with the continuous release of energy, is then possible. Fission can be caused by other particles, but neutrons are the only practical ones that result in a sustained reaction because two or three neutrons are usually released for each one absorbed by fission. These keep the reactions going on.

When a neutron is captured by a nucleus of an atom of U-235, the U-236 isotope is formed. This isotopic, highly unstable type of uranium exists for perhaps one millionth of a second. It splits roughly into two equal parts and has a total energy release of 200 MeV. The products formed include the fission fragments, neutrons and electromagnetic radiation or gamma rays. The fission products absorb most of the total energy released by the fission as kinetic energy, which is converted into heat as the fragment colloide. The particles decay mainly by beta and gamma radiation with some neutron particle emission and in rare causes alpha radiation. Most of the energy released as heat close to the site of fission and is released into surroundings. On the average two or three neutrons are released with each fission. When a neutron causes fission of uranium-235, a typical reaction produces barium, krypton, two or three neutrons and release of energy due to loss of mass. An example shown schematically in Fig. (15.1) is,

$$_{92}U^{235} + {}_0N^1 \rightarrow {}_{56}Ba^{137} + {}_{56}Kr^{97} + 2{}_0N^1 + \gamma \quad ...(15.1)$$

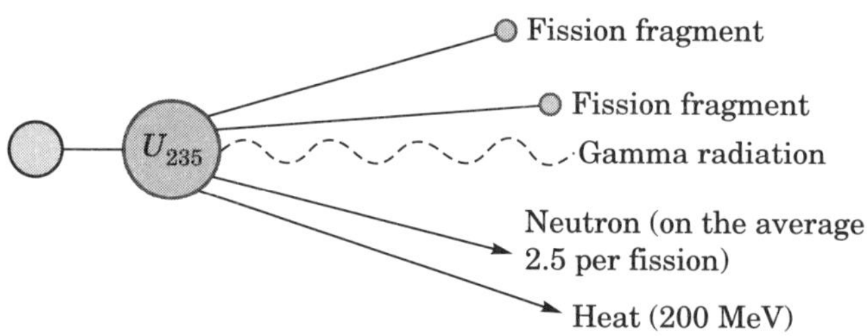

Fig. 15.1. The fission process.

The immediate (prompt) products of a fission reaction, such as Ba^{137} and Kr^{97} above, are called fission fragments. They, and their products, are called fission products.

Fusion. Fusion energy is a form of nuclear energy released by the fusion (combination) of two light nuclei (*i.e.,* nuclei of low mass) to produce heavier nucleus.

Energy is produced in the sun and stars by continuous fusion reactions in which four nuclei of hydrogen fuse in a series of reactions involving other particles that continually appear and disappear in the course of the reactions, such as He^3, nitrogen, carbon and other nuclei, but giving one nucleus of helium and two positrons,

$$4{}_1H^1 \longrightarrow 2He^4 + 2{}_{+1}e^0 \quad ...(15.2)$$

resulting in a decrease in mass of about 0.0276 amu, corresponding to 25.7 MeV. The heat produced in these reactions maintains temperatures of the order of several million degrees in their cores and serves to trigger and sustain succeeding reactions.

Artificially produced fusion may be accomplished when two light atoms fuse into a larger one as there is a much greater probability of two particles colliding than of four. The 4-hydrogen reaction requires, on an average, billions of years for completion, whereas the deutrium-deuterium reaction requires a fraction of second.

To cause fusion, it is necessary to accelerate the positively charged nuclei to high kinetic energies, in order to overcome electrical repulsive forces, by raising their temperature to hundreds of millions of degrees resulting in a plasma. The plasma must be prevented from contacting the walls of the container, and must be confined for a period of time (of the order of a second) at a minimum density. Fusion reactions are called *thermo nuclear* because very high temperatures are required to trigger and sustain them.

Of the several possible reactions between pairs of hydrogen isotope nuclei, the fusion of a deuterium nucleus (or deutron) and a tritium nucleus (or triton) has been found to take place most readily. This is referred to as the D—T reaction and is expressed by

$$\underset{\substack{\text{light nuclei}\\ \text{(nuclei of low mass}\\ \text{numbers 2 and 3)}}}{D + T} \longrightarrow \underset{\substack{\text{Heavier}\\ \text{nucleus}}}{4He} + n + 17.6 \text{ MeV.} \quad ...(15.3)$$

Whereas, 4 He represents a common helium—4 nucleus and n is a neutron. The fusion of one deutron and one triton releases 17.6 MeV of energy.

15.2. The Basic : Nuclear Fusion and Reactions

The Fusion Process. Each isotope of hydrogen has a single proton in its nucleus, which has a single electron in orbit around it. The hydrogen nucleus contains only one proton ; the deuterium nucleus has one additional neutron, and tritium has two additional neutrons.

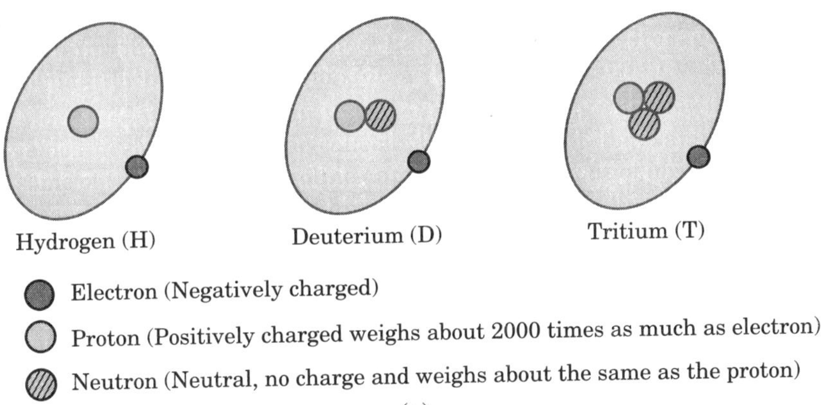

(*a*)

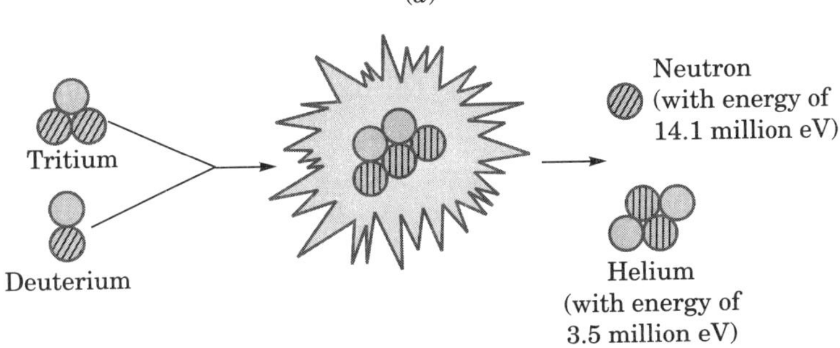

(*b*)

Fig. 15.2. The fusion process.

Therefore, deuterium weighs twice as much as hydrogen and tritium weighs three times as much.

Deuterium also known as heavy water, is abundant in ordinary water : 2 out of about every 6,500 water molecules are deuterium, rather than ordinary hydrogen. The tritium nucleus can be made in the laboratory by bombarding the lithium with neutrons. Lithium is abundant in sea water.

In this typical fusion reaction, the deuterium and tritium collide and recombine to form a helium nucleus and one neutron, releasing energy, in the process. Most of the energy is carried in the neutron, releasing energy is carried in the neutron. There is a net energy gain because the end product nuclei weigh less than the nuclei of the input fuel.

At high temperatures ($T \geq 10^8$ °K) only a very small number of particles in the entire plasma region remain in the neutral state and the gas is said to fully ionized. In such a gas the motion of individual particles is governed by much simpler laws because there are only two kinds of particles, positively and negatively charged and the complicated reactions between electrons and atomic structures, usually described by wave mechanics equations, are absent. Short range atomic forces are now replaced by long range electrostatic repulsion and attraction conditions can therefore be relatively simple, if the plasma consists essentially of the nuclei and the electrons of the gas. In plasma of gases of lightest elements (H, He etc.), it is possible that all electrons are removed from the atoms so that a gas mixture consisting of electrons and bare nuclei only is obtained. *A nuclear fusion* reaction can occur when two atomic nuclei approach very close to each other at velocities atleast large enough to overcome their mutual electrostatic repulsion, that is, the "Coulomb Barrier". With the use of primitive particle accelerators, beams of deutrons (heavy hydrogen nuclei *D)* were directed at targets containing heavy hydrogen. Nuclear rearrangement reactions, resulting from collisions between the impinging and the target electrons, were observed in which the total kinetic energy carried by the reaction products was much higher than the kinetic energy carried by impinging beam of deutrons.

Consider a *fully ionized gas* containing n_A number of nuclei of kind A per cm^3 and n_B number of nuclei of kind B. There will of course be a concentration of electrons equal to $z_A n_A + z_B n_A$, if z_A and z_B are the atomic numbers of the elements, but apart from their function of preserving charge neutrality their presence in plasma is generally ignored. It is assumed that electrons play no part in the nuclear fusion reactions resulting from impacts between the nuclei. If the nuclei of type A move through the concentration n_B of type B with velocity v, impinging particle produces, on the average $v/\lambda = v \, . \, n_B \, \sigma$ nuclear reactions per sec. Here λ is the appropriate mean free path and σ is the cross-section for a particular reaction. Thus if n_A nuclei of type A are moving about of random among that of B also moving about at random, the total number of nuclear reactions per cm^3 per sec (the reaction rate r) will be given by

$$r = n_A \, . \, n_B < \sigma v > \qquad ...(15.4)$$

where, v is the relative velocity of the two colliding particles and $< \sigma v >$ denotes the value of σv average over the Maxwellian velocity distribution. In a gas containing only one type of nuclei of total density n/cm^3 ($n_A = n_B = n$), the reaction rate will be

$$r = n_A \, n_B < \sigma v > = \frac{1}{2} \, n^2 < \sigma v > \qquad ...(15.5)$$

It must be noted that all collisions between like particles contribute to the reaction rate.

For fusion to occur, two nuclei must obviously be able to approach each other within distances of the order of nuclear dimensions *i.e.,* 10^{-12} to 10^{-13} cm or cross-section of $\sigma = 10^{-24}$ to 10^{-26} cm^2. However, being positively charged nuclei repel each other more strongly, the higher their nuclear charge. Thus fuels composed of highly charged nuclei (those having a higher atomic number) require higher and higher kinetic energy (higher temperatures of the order of 10^8 to 10^9°K) to initiate fusion reactions. Beyond lithium (at No. 3) these temperatures become so high that considering the limited energy releases involved, it would seen that no method thus for proposed would obtain a net positive release of fusion energy.

There are 4 fusion reactions, listed below, the combinations of which are most often considered for use in fusion reactors. The energy releases are given in two different units:

(*a*) *MeV* (*i.e.,* in million of eV equivalent) the kinetic energy equivalent of energy that would be carried by a singly charged particle that had been electrically accelerated by that value of electric potential; and

(*b*) In kilowatt-hours of energy released per gram of fuel ions heated (where 1 kW-hr = 3.6×10^6 W. sec. of energy).

	Reactions	σ_{max} cm^2	$\langle E \rangle$ *max* (*Energy variation*) *MeV*
(1)	$D + T \rightarrow 4He_2 + n + 17.6$ MeV (94,000 kW-hr/gram)	5.0×10^{-24}	0.10
(2)	$D + D \rightarrow 3He_2 + n + 4$ MeV (27,000 kW-hr/g)	0.09×10^{-24}	1.0
(3)	$D + D \rightarrow T + p + 3.25$ MeV (22,000 kW-hr/g)	0.08×10^{-24}	1.0
(4)	$D + 3He_2 \rightarrow 4He + p + 18.3$ MeV (94,000 kW-hr/g)	0.8×10^{-24}	0.45

Other reactions which may also take place are

(5) $T + T \rightarrow 4He_2 + 2n + 11.3$ MeV

(6) $T + 3He_2 \rightarrow 4He_2 + D + 14.31$ MeV

(7) $T + 3He_2 \rightarrow 4He_2 + p + n + 12.08$ MeV.

In the listed reactions, n is the neutron, p is the proton $1H_1$, D is the deuterium $2H_1$, T is tritium $3H_1$.

The comparative large cross-section for the DT reaction (reaction 1) at a relatively low energy suggests that it is the most favourable reaction. If the tritium produced in the reaction-2 can be prevented from escaping from the ionized gas, the D-T reaction is then possible. If $3He_2$ produced in reaction 3 can also be confined the total energy available from the D-D reaction and its secondary product is about 21 MeV. However the total energy release is determined by the small cross-section of D-D collisions and then total energy released in one reaction is distributed between charged and uncharged particles (neutrons). *In principle the heavy reaction products carry small fraction of energy.* For example in the D-D reaction the charged particles receive 60% of the released energy in the D-T reaction about 20%. Thus although the latter process gives a net gain of energy in charged particles of about

4 MeV per reaction, main interest in the reaction arises from its high probability at low impact energies (100 KeV).

Values of σ can be measured by firing beams of single energy particles (deutrons of known energy in the present case) at solid targets. The value of $< \sigma v >$ is calculated from the energy distribution of particles (usually Maxwellian) and the energy variation of $< E >$. The values of $< \sigma v >$ known as *reaction probability* are shown in Fig. 15.3 as functions of temperature and energy. It can be seen that at mean particle energies below 100 KeV the number of nuclear reactions per second in a *D-T* reaction is several times greater than in any other gas mixtures. The D-T reaction is the fastest thermonuclear process known and produces 80% of its energy in the form of 14.1 MeV neutrons which simply escape from the plasma region. These must be slowed down and captured in surrounding lithium or lithium alloy blankets in order to convert their kinetic and absorption energy into heat and to generate the much needed tritium *(T)* as per the following reactions:

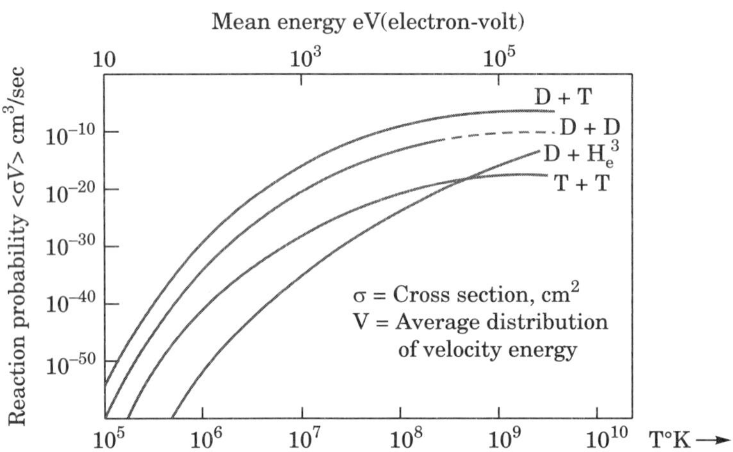

Fig. 15.3. Reaction probability of thermo nuclear fusion reactors.

7% $6Li_3$ + 93% $7Li_3$, leading to the generation of T by the following reaction

$$n + 6Li_3 \longrightarrow T + 4He_2 + 4.78 \text{ MeV}$$

or

$$n + 7Li_3 \longrightarrow T + 4He_2 + n^1 - 2.47 \text{ MeV}$$

Tritium is radioactive and decays with a half life of 12.26 years to $3He_2$ and is not naturally available. But the world supply of Li appears to be sufficient to last for many years (hundreds) and during this time it is hoped that the D-D scheme will become practicable.

It is thus possible, at such time that fusion technology permits, to conceive of a fusion fuel cycle that would use *D* as primary fuel, with the recovery, reinjection and combustion of its reaction products (T and $3He_2$). In a steady state system the D—D—T—$3He_2$ cycle could be preferentially weighted toward the D—$3He_2$ branch by storing the T for some portion of its half life, if this proved to be economically and environmentally advantageous. In fact the D—$3He_2$ reaction has also got a high reaction probability and a large energy release, with all of the latter imparted to charged particles of the reaction products.

While the pure D—D—T—$3He_2$ cycle would in many senses represents a long term goal for an economy based on the fusion reactor reaction 1, the D-T reaction is presently being contemplated for first generation fusion reactors. Fission invariably involves radioactive fission products and fissionable material with radiation hazards. By contrast, with the progress and perfection of the fusion reactor technology, fusion power should evolve towards systems in which hazards would be reduced to negligible levels and possibility of high energy conversion efficiency (through direct conversion) could be realized.

If nuclear fusion becomes a practical source of energy, it will undoubtedly be first by way of the *D-T* reaction. At a later stage, however, fusion involving deuterium nuclei only, which is more difficult to attain, may become important. The reactions (2) and (3) between deutrons, which take place at about the same rate, the tritium nucleus (T) formed in the reaction (3) immediately fuses with a deuterium nucleus by the D-T reaction. The net result is the liberation of 66 million kW-hr of energy per kilogram of deuterium.

The great advantage of D-D fusion would be that it would not be necessary to supply tritium from an outside source. Such tritium as is involved is produced in one of the D-D reactions. However, it is unlikely that the conditions for D-D fusion on a practical scale will be realized for many years. *Consequently, the following discussion will refer primarily to* D-T *fusion.*

The basic raw materials for the D-T reaction are water for deuterium and lithium minerals (for tritium). All natural waters contain a small proportion of HDO molecules in which deuterium (D) has replaced one hydrogen (H) atom in ordinary water (H_2O). Although there is only one deuterium atom for every 6500 ordinary hydrogen atoms, the amount of water is so large that more than 10 trillion (*i.e.,* 10^{13}) tonnes of deuterium are present on earth. Deuterium in the form of heavy water (D_2O), is extracted from water on a large scale at a moderate cost in U.S. and Canada. Its main use at present is in certain fission reactors.

Tritium is made by reaction of neutrons with nuclei of the element lithium, especially of its lighter isotope lithium-6. The required neutrons are produced in the D-T reaction, as seen above and lithium can be obtained from certain minerals and natural brines. If it should become necessary, this element could be extracted from sea water although at a lower cost. Because tritium is radioactive, decaying with a half life of 12.3 years, it cannot be stockpiled for any length of time.

15.3. Requirements for Nuclear Fusion

As already seen, an essential requirement for fusion is that the interacting nuclei have sufficient energy to overcome the force of repulsion. In Fusion studies, the energy of the nuclei is increased by increasing the temperature. Nuclear fusion reactions that occur at high temperature are often called *thermonuclear reactions.* In the so-called hydrogen (or fusion), bomb, energy is released in an uncontrolled manner by the fusion of hydrogen isotope nuclei at very high temperatures. The problem is to bring about these reactions under conditions that permit the controlled release of fusion energy.

In a gas at high temperature, the oppositely charged nuclei and electrons, which are normally held together in atoms by electrical attraction, become separated from each other. Such a gas, containing free positively charged nuclei (or ions) and free negatively charged electrons, is called a *plasma.* A fusion temperatures, and even well below, the hydrogen isotopes gases are actually plasmas. The unusual properties of plasmas have an important bearing on the efforts to realize controlled unclear fusion.

Ignition Temperature. As a result of the electrical interactions between the nuclei and electrons, a plasma emits energy as radiation in a fusion (D-T or D-D) plasma at moderately high temperatures, more energy is radiated than is produced by fusion. As the temperature is increased, however, fusion energy production increases more rapidly than radiation emission. At a sufficiently high temperature "ignition" can occur; that is to say, the fusion can be sustained in the plasma once it has been initiated. (The term *ignition* arises from the similarity to a fire; once it is started, the fire will continue as long as fuel is available). In order to achieve ignition, the fusion energy remaining in the plasma must exceed the radiated energy.

The products of the *D-T* reaction are a helium-4 nucleus with a positively charge and an electrically neutral neutron. Under the conditions existing in controlled fusion reactions, only the charged particle energy is deposited in the plasma, this amount to 20% of the total *D-T* fusion energy. (The remaining 80% carried by the neutrons can be absorbed and utilized outside the plasma). Hence, the ideal ignition temperature for a *D-T* fusion plasma is the temperature at which the energy emitted as radiation is less than 20% of the fusion energy. Some energy will then remain in the system to sustain the reaction. As will be seen shortly, the actual ignition temperature is higher than the ideal value.

If the radiation energy arises solely from interactions of the electrons with singly charged ions *(e.g.,* deuterium and tritium nuclei), the ideal ignition temperature for *D-T* fusion is calculated to be about 50 million °C. (In fusion studies, temperatures are expressed in terms of kilo-electron volts or KeV, where 1 KeV is equivalent to 11.6 million *K*). In practice, there are additional radiation losses because of the presence of ions with more than a single charge (*e.g.,* helium and impurity nuclei); further more, not all the charged particle energy remains in the plasma. The practical fusion ignition temperature for a D-T plasma is thus taken to be roughly 100 million °C. (For the *D-D* reactions, the corresponding ignition temperature is estimated to be 1000 million °C).

Driven Systems. In one scheme (magnetic mirrors) for producing fusion energy the loss of some charged particles is inevitable (refer *magnetic-confinement fusion).* The ignition temperature for D-T plasmas would then be much higher than 100 million °C. It is improbable that ignition could be realized in such systems, but a net production of fusion energy should nevertheless be possible. This is achieved by injecting energy continuously into the plasma and multiplying this energy by fusion reactions. Systems of this kind are said to be 'driven' as distinct from ignition systems which are self-sustaining after ignition. In all cases, however, whether ignition occurs or not, the minimum practical temperature for D-T fusion is considered to be roughly 100 million °C.

Energy Break Even Condition. (The Lawson criterion). As essential requirement for the net production of nuclear fusion energy is that the *break-even* condition be exceeded. This condition is that the plasma, be confined for sufficient time to permit the total recoverable fusion energy to balance the energy required to heat the plasma and to compensate for radiation loss. In other words, to produce net fusion power a fusion fuel charge must be heated to the combustion temperatures for fusion and must be confined in this heated condition long enough for the recovered output of the fusion energy to exceed, on the average, the energy input to the system. By making a number of simplifying assumptions, J.D. Lawson in England showed in 1957 that the energy break-even condition could be expressed by the product $n \cdot \tau$, where n is the particle density and τ is the confinement time in seconds. The particle density is usually stated as the number of deuterium or tritium nuclei per cubic centimetre (cm^3) and the confinement time in seconds; hence, $n\tau$ is commonly given in sec/cm^3 units. It follows from the fact that when the fuel density is high, the rate of fuel burning is correspondingly more rapid, leading to a shortened required confinement time before the break-even energy release is reached and *vice-versa.*

The quantity $n.\tau$ is called the Lawson number (or *confinement parameter)* and the minimum value for fusion energy break-even is referred to as the *Lawson criterion.* The Lawson criterion depends on the nature of the fusion reaction and the plasma temperature; for a D-T plasma at 100 million °C, the break-even condition is calculated to be around 6×10^{13} sec/cm^3. The attainment of this value of $n\tau$ does not imply ignition. For ignition to occur, the fusion energy production must exceed the amount required to break-even. The $n.\tau$ product for D-T ignition at 100 million °C is thus higher at about 3×10^{14}. (For D-D fusion, the break-even $n\tau$ value would be larger).

The calculation of the Lawson criterion is based on several approximations and assumptions which would not necessarily be valid in practice. Nevertheless, the confinement parameter $n\tau$ is commonly used in fusion studies as a general indication of the approach to break-even conditions. One of the objectives of current research is to attain a value close to 10^{14} sec/cm^3 at a temperature near 10^8°C.

15.4. Plasma Confinement

In a high temperature plasma, the nuclei and electrons move randomly in all directions at average speeds of several kilometres per second. Hence, unless confined in some way, all the particles would soon strike the walls of a containing vessel, thereby imparting some of their energy to the walls. The particles would then return to the plasma with less energy that is, at a lower temperature. Thus, the plasma would be cooled, and the fusion temperature could not be realized or maintained. Two quite different confinement methods are currently being investigated:

(*i*) Magnetic confinement and (*ii*) Inertial confinement.

Note that a distinction is made between containment and confinement. *Containment* applies to the vessel or chamber which contains (or holds) the plasma, whereas *confinement* refers to the means for restraining the plasma from coming into contact with the walls of the containing vessel.

Magnetic confinement is based on the fact that electrically charged particles have difficult escaping from a magnetic field. Since a fusion plasma consists almost entirely of charged particles, it should be possible to confine a plasma by a suitable magnetic field. Several different magnetic field arrangements have been (or are being) studied in the effort to confine a high temperature plasma for a sufficient time to satisfy the Lawson criterion for energy break-even. As stated the energy leakage involved in a particle escape is extermely high. In the absence of an escape barrier the loss of energy by particle emission would be of the order of 100 MW/cm^2. To bring this loss down to a value of 100 W/cm^2 *i.e.,* 25 cal/sec. would require an energy barrier against escape (evaporation) of the order of 140,000 volts and is generally created by the help of a moderately high magnetic fields.

The *inertial confinement,* a small pellet or sphere of a deuterium tritium mixture (solid or liquid) is heated by a very short burst of energy from either laser beams or beams of high energy charged particles. Note only is the temperature increased but as a result of various complex processes, the bellet material is compressed to a high density. The fusion reaction should then occur so rapidly, in about one-trillionth (10^{-12}) of a second, that inertia (*i.e.*, reluctance to move from rest) would prevent the pellet from flying apart while fusion is in progress. No other confinement is required.

Magnetic and inertial confinement represent two extremes of plasma conditions. In magnetic confinement, the plasma particle density, *n*, nuclei/cm^3, would be in the vicinity of 10^{15}, a confinement time of 0.1 sec would then be required to satisfy the Lawson criterion for energy break-even. In inertial confinement, however, the particle density would be roughly a hundred billion (10^{11}) times greater at 10^{26} nuclei/cm^3; the break-even confinement time would then be as short as 10^{12} sec.

The further details of the studies and problems of the two approaches to fusion plasma confinement are given in the following sections. The attainment of both a temperature of 100 million °C (10^8 K) and a confinement parameter $n\ \tau$ of 10^{14} sec/cm^3 in a D-T plasma has proved to be difficult. Nevertheless, the steady approach to these conditions over the years has been encouraging. World-wide efforts are thus being continued for the purpose of realizing a new, clean and almost inexhaustible energy source. It does not seem probable, however, that a commercial fusion power plant will be operative before the early years of the 21st century.

15.5. Magnetic Confinement Fusion

Introduction. The use of various magnetic field arrangements for confining high temperature plasmas to achieve controlled nuclear fusion. In a uniform magnetic field, electrically charged particles are forced to follow helical (*i.e.*, cork screw) paths along the field lines. The methods used for generating the magnetic field is described below). In a sense, charged particles are "tied" to the magnetic field lines. The magnetic field line thereby assumes the role of a guiding centre for the motion of a charged particle. The particle orbits must then not be allowed to intersect the walls of the container until the reaction is sufficiently complete. Since a high temperature plasma consists almost entirely of charged particles (atomic nuclei and electrons), it should be confined by a magnetic field.

The maximum density of a plasma that can be confined at a given temperature depends on the magnetic field strength. There is a practical limit to the field strength and hence a corresponding maximum for the confined plasma density. At a temperature of 100 million °C, required for a significant fusion of deuterium and tritium (D-T) nuclei, the maximum, *n*, would be in the range of 10^{13} to 10^{15} nuclei/cm^3. Hence, in order to satisfy the Lawson criterion for energy break even (*i.e.*, $n\tau$ greater than about 10^{14} sec/cm^3), the confinement time τ must be atleast 1 to 0.1 sec, respectively.

Although the confinement time appears short, it has been difficult to attain because several factors contribute to the escape of plasmas from magnetic fields. For example, collisions between charged particles moving along the field lines can cause the particles to move across the lines and escape from confinement. Furthermore, a plasma in a magnetic field has a tendency to become unstable and leakout of the confining field. Plasma instabilities are generally classified as large scale (or macro) instabilities and small scale (or micro) instabilities.

Macro instabilities can cause the plasma as a whole to break up and escape from its confining magnetic field. *Micro instabilities*, the other hand, lead to gradual leakage of the plasma across the field lines. Research on the magnetic confinement of high temperature plasmas was initiated independently in the United States, the United Kingdom, and the U.S.S.R. around 1951, and much of the workdone since has been concerned with understanding and overcoming plasma instabilities.

Thermonuclear Function Reactors (Plasma-details). Several magnetic field arrangements proposed for plasma confinement, but a few that appear promising will be described here. There are no assurance, however, that any will provide the basis for a practical nuclear fusion reactor. (The term "reactor" is commonly applied to a device capable of generating significant amounts of power by nuclear fission or fusion reactions).

Because tritium is radioactive and requires special handling facilities, experiments with plasmas are made with non radioactive hydrogen isotopes, either ordinary (light) hydrogen or not at all, the general behaviour is expected to be the same as when tritium is present. At a later stage of development, tritium will be introduced in order to study deuterium tritium (D-T) fusion. This process will be used in the first controlled fusion reactors.

Very efficient particle confinement is therefore required. The problem is currently to get n, and τ to the proper values so that the power output may equal the power input, *i.e.*, break-even point is reached. Some devices currently operate at high plasma densities and temperatures and have relatively short confinement times; other devices work at long confinement times but at insufficient plasma densities or low temperatures and so on and so forth.

Fig. 15.4 shows the magnetically confined Deuterium tritium plasma. *First*: one starts with a reacting Plasma which is emitting energy in the form of neutrons, charged particles and various forms of photons.

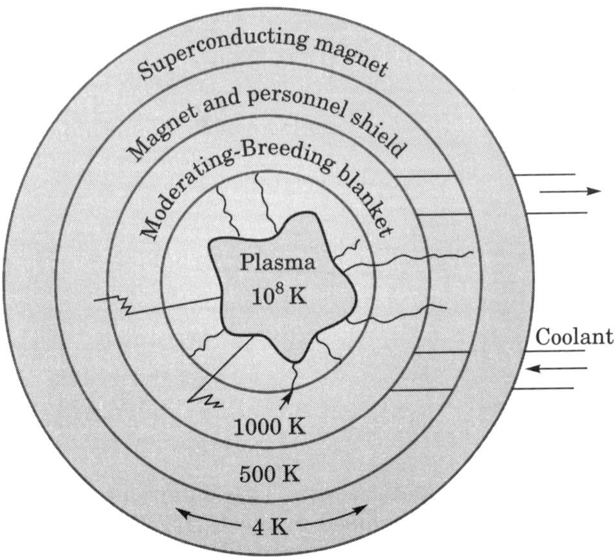

Fig. 15.4. Schematic representation of controller Thermonuclear fusion reactor based on the Deuterium tritium (D-T) cycle.

Second step is to surround the plasma with a solid wall which absorbs the charged particles and photons as well as providing a vacuum for the plasma to ignite in a magnetically confined system. This wall will absorb about 20% of the energy from the plasma and must be cooled. Typical diameters for this first wall will be 5 to 10 metres for the *toroidal* reactors such as the *Tokamak*.

Third step is to surround the vacuum wall with a moderator to slow down the neutrons, a reflector to reduce the leakage of neutrons and a coolant to carry the heat away. The region should also contain the deuterium tritium mixture so that the reaction can be continued. Approximately 1 metre of the lithium blank and first wall is required to absorb about 97% of the heat produced from the plasma. Unfortunately, some neutrons and gamma rays will escape, and the magnets (or lasers in case of pellet fusion) must be protected from these sources of irradiation. This protection is accomplished by surrounding the blanket with a shield, that completes the moderation of these neutrons that escape and absorbs the gamma rays emitted from the blanket. This shield also serves as final radiation protection for personnel in the plant.

Outside the shield will be located the magnets (or lasers), fueling equipment, heat exchangers, tritium removal devices and other equipment associated with the plant.

The overall structure of a *Tokamak reactor* is indicated schematically in Fig. 15.5.

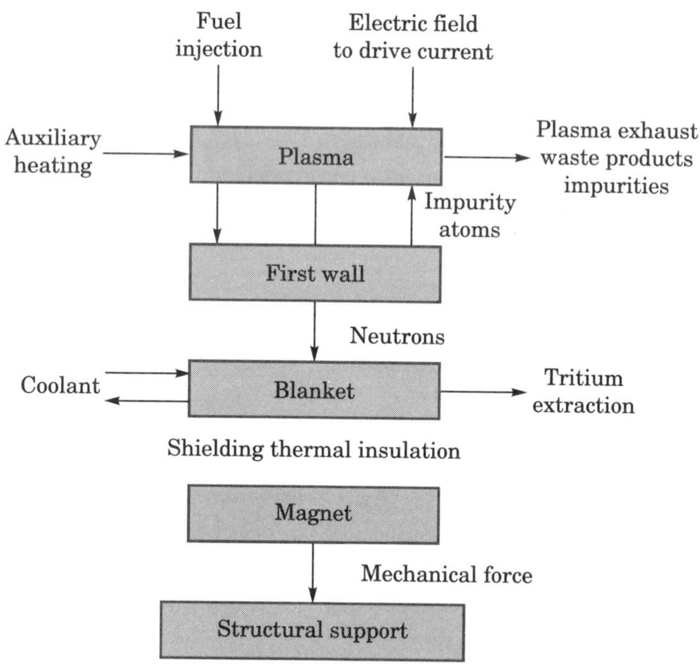

Fig. 15.5. Schematic representation of main components of a Tokamak Reactor.

Details of Tokamak Systems. The Tokamak principle of magnetic confinement in a torus was developed in the U.S.S.R. in the 1960s. (The name "tokamak" is derived from the initial letters of the Russian words meaning "toroidal" "chamber", and "magnetic" respectively). After 1968, when convincing evidence of the tokamak confinement was reported many experimental devices were built in the United States and elsewhere. As experience has been gained with tokamak operation, larger and larger machines have been constructed to increase the plasma confinement time. In particular, plasma stability is improved if the height of the torus is large relative to its horizontal diameter. In other words the torus has the shape of a "fat" doughnut with a small central hole.

These devices under different names are developed in U.S., U.K., Japan and U.S.S.R. The stainless steel toroidal containment chambers are roughly 7.5 to 10 m across and 2 to 2.5 m high. A tokamak reactor for the production of useful amounts of energy is expected to be much larger, at least twice as large in each dimension.

Basically, the tokamak magnetic field is a combination of a strong steady (for a time) toroidal, field and a much weaker, pulsed poloidal field. In a purely toroidal magnetic field, as in Fig. 15.6, the field lines are closed circles with different radii; as a consequence, the plasma drifts outward in the torous. By superimposing a weak poloidal field, the toroidal field lines acquire a slight twist so that they do not close on themselves. The twist in the field largely prevents the drift, and there is a marked increase in the plasma stability and confinement time.

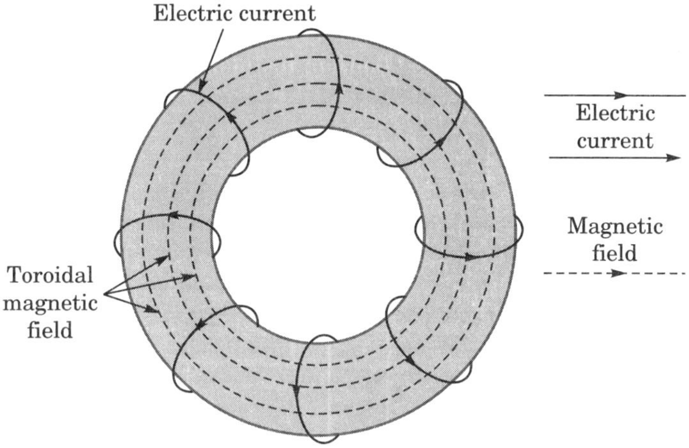

Fig. 15.6. Toroidal magnetic field.

The toroidal field component is produced by an electric current passing through coils which encircle the torus in a vertical direction. Eventually a constant direct current will probably be used to maintain a steady toroidal field, but in experimental studies a short duration field is adequate.

There are various ways of inducing the plasma current for generating the poloidal field. One procedure is to apply a pulse of primary current to horizontal coils located within the central hole of the "doughnus" (torus) and around it. The increase and decrease in the primary currents induce secondary currents within the torus, and these generate the desired poloidal magnetic field (see below magnetic fields).

In addition to produce the poloidal field, the induce currents initiate plasma formation from the gases in the torus and increase the temperature by ohmic heating. The attainable temperature is limited by the development of large-scale instabilities to about 20 million °C. Neutral-beam injection is favoured for further heating, but radio frequency heating is also being considered.

The essential requirements are the attainment of a temperature of about 100 million °C in a D-T plasma and a confinement parameter $n\,.\tau$ exceeding the 3×10^{14} sec/cm^3 necessary for energy break even and ignition. If these conditions can be realized, a tokamak fusion reactor may operate as follows.

A *D-T* plasma, formed and heated by a pulse of induced current, is confined in a torus by a combination of toroidal and poloidal fields. High energy neutral deuterium beams are then injected until *D-T* ignition is initiated. The 20% of the fusion energy carried by the helium nuclei remains in the plasma and serves to maintain the fusion reaction temperature without further neutral beam heating. During this stage, when fusion energy is being produced, the deuterium and tritium are replenished by injecting solid pellets of these substances into the plasma.

After some 10 to 20 minutes of operation, the pulse of induced plasma current will have dropped to zero; the poloidal field will then cease to exist, and stable plasma confinement is no longer possible. Residual helium gas and impurities are removed from the tokamak chamber, and the cycle is repeated, starting with the formation and heating of the D-T plasma by an induced current pulse followed by neutral beam injection.

The neutrons produced in the D-T reaction are electrically neutral and are not confined by the magnetic field; they consequently escape, carrying 80% of the fusion energy. The toroidal plasma chamber is surrounded by a "blanket containing lithium, either as liquid metal or a molten salt, and the escaping high energy neutrons enter the blanket. Here they react with lithium nuclei to produce the tritium required for the fusion process.

At the same time, the energy of the neutrons, together with the energy released in their reaction with lithium, is deposited in the blanket as heat. This can then be used to produce high temperature steam (or other vapour) to drive a turbine for generating electric power. Part of this power is required to produce the tokamak magnetic fields and to operate the neutral beam injectors and other equipment; the remainder is then available for sale.

A practical fusion reactor will undoubtedly use superconducting coils to generate the magnetic fields. When certain metals and alloys are closed to extremely to temperatures, usually below (– 263°C), they become superconductors. Once a flow of current has been started in a superconductor, it will continue indefinitely in a closed circuit after the current source is removed. Although superconducting magnets, closed by liquid helium, have been built, they are smaller than those that will be required for fusion reactors.

Methods of Plasma Heating (In magnetic confinement fusion)

(1) *Ohmic Heating.* Four main methods are available for heating a magnetically confined plasma to the high temperatures required for nuclear fusion. The simplest, called ohmic (or resistance) heating, is to pass an electric current through the plasma, the same principle is used in an ordinary electric heater. To heat a plasma, the current is generally induced from outside, to avoid the need for inserting electrical connections in the containing vessel. The heating rate depends on the resistance of the plasma and the square of the current strength. Since the resistance decreases with increasing temperature, and because instabilities may develop, the maximum plasma temperature that can be attained by ohmic heating is commonly around 20 million °C.

(2) *Neutral Beam Injection.* Further heating of a magnetically confined plasma can be achieved by *neutral-beam injection.* This depends on two factors (1) positively charged atomic nuclear (or ions) can be accelerated to high velocities (*i.e.,* high energies) by means of a high voltage, but they cannot easily penetrate a magnetic field and (2) neutral (*i.e.,* uncharged) atoms, on the other hand; can readily penetrate a magnetic field but they can not be accelerated directly. Plasma heating by neutral beam injection, used specifically for deuterium is conducted in the following manner.

Deuterium ions are generated from deuterium gas by an electrical discharge in an ion source; the ions are then accelerated by a high voltage (*e.g.*, more than 100,000 volts). The accelerated, high energy deuterium ions pass through a chamber containing neutral deuterium gas; here electrons transfer from low energy deuterium (D°) atoms to high energy ions (D^+) with the resulting formation of low-energy deuterium ions and high energy atoms; thus,

$$D^\circ \text{ (low energy)} + D^+ \text{ (high energy)} \longrightarrow D^+ \text{ (low energy)} + D^\circ \text{ (high energy)}$$

Electron

The low-energy D^+ ions are diverted and removed by a subsidiary magnetic field, while the high-energy D° atoms are injected into a magnetically confined plasma.

In the plasma, electrons are rapidly removed from the atoms to form high energy ions which are trapped by the magnetic field and can not escape. The high energy ions then transfer part of their energy to the plasma particles in repeated collisions, thereby increasing the plasma temperature. An ion temperature of 75 million°C was reached in a deuterium plasma in 1978 by neutral beam injection, and it is expected that temperatures of 100 million°C will be attainable.

(3) *Radio-frequency heating.* Radio frequency waves in the microwave and radar ranges generated outside a magnetically confined plasma, can heat the plasma to very high temperatures. If the waves have particular frequencies (or wavelengths), part of their energy is transferred to the ions or electrons or both in the plasma. These high energy particles then collide with other particles and thus increase the plasma temperature.

(4) *Compression heating.* A plasma, like a gas can be heated by compression. If the plasma is confined by a magnetic field, the compression can be achieved readily by a sudden increase in the confining field strength. In addition to increasing the temperature, compression increases the plasma density and thus contributes to the confinement parameter $n.\tau$.

Magnetic Fields. An electric current is always associated with a magnetic field such that the field lines (or lines of force) are perpendicular to the direction of current flow. The gases (*e.g.*, deuterium and tritium) required for fusion are often contained in a hollow, doughnut-shaped chamber called a torus. Magnetic fields can be generated within a torus in two general ways.

In one case, an electric current is passed through a number of rings (or coils) surrounding the torus, as in Fig. 15.6. The associated magnetic field, called a toroidal field, has its lines of force running around the long way of the torus. Alternatively, the current may be made to flow through the plasma, as in Fig. 15.7; the lines of the resulting magnetic field, called a *poloidal field,* encircle the plasma in the polar (shorter) direction. (Incidently, whenever a plasma is heated by passing a current through it, as described above, a poloidal field is produced).

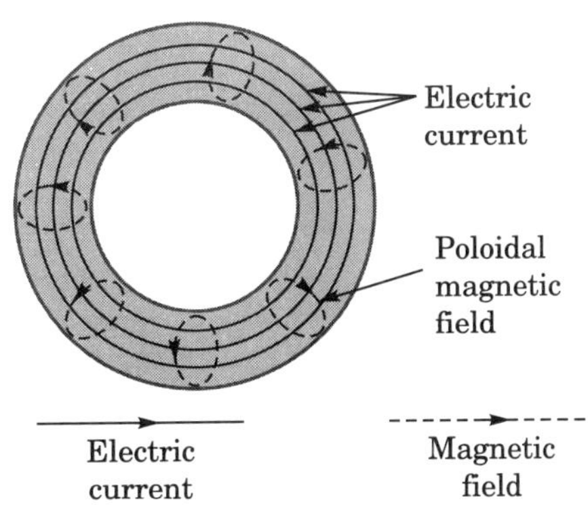

Fig. 15.7. Poloidal magnetic field.

To generate a toroidal magnetic field, the rings, which are outside the toroidal chamber are actually coils of many turns of insulated wire. Direct current of the same strength is then passed through all the coils. In order to avoid inserting electrical connections into the plasma, current within a plasma, as is required for a poloidal field (and ohmic heating), is produced by induction. In simple terms, an electric current *which varies with time* is passed through a conductor outside and parallel to the torus; this is referred to as the *primary current.* A similar *secondary current* is then induced in the plasma.

The time varying primary current is usually a pulse which increases rapidly to a maximum and then decreases slowly to zero: the secondary current changes correspondingly. An important consequence is that a poloidal field consists of a succession of pulses. A toroidal field, on the other hand, can be a steady field.

Open Ended Magnetic Mirror Systems. Plasma confinement by magnetic mirrors is based on the principle that charged particles spiraling along magnetic field lines tend to be repelled, and their direction of motion reversed, when they enter a stronger field. The reversal of direction (or reflection) of the particles leads to the name *magnetic mirror* for the region where the field is stronger.

A simple magnetic mirror system for plasma confinement would consist of closely spaced coils at each end of a straight open-ended tube Fig. 15.8 (*a*). Passage of an electric current through the coils results in a magnetic field which is stronger at the ends. The ends where the lines of force are close together, then constitute the mirror regions Fig. 15.8 (*b*). If a plasma is present between the mirrors, the charged particles may be reflected back and forth many times from one mirror to the other. If this occurs, the plasma is confined.

However, some loss of plasma cannot be avoided. If a particle has a sufficient component of motion parallel to the field lines, it will pass through the mirrors and escape from confinement. Since the directions of motion of the particles are continually changing as a result of collisions, the escape is continuous. Nevertheless, confinement of plasma between two magnetic mirrors for short periods has been demonstrated. Means for improving confinement are described below.

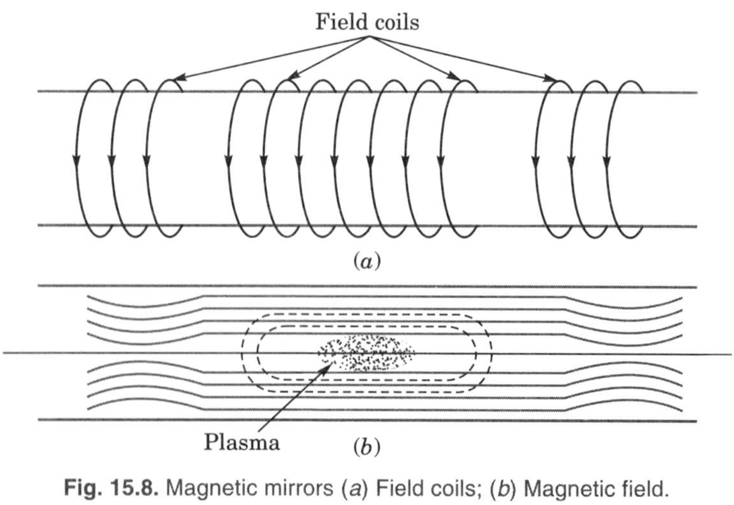

Fig. 15.8. Magnetic mirrors (*a*) Field coils; (*b*) Magnetic field.

Plasma Heating. Ohmic (or resistance) heating of a plasma in a mirror field does not seem to be practical, and so other means are used. One promising method is to inject or form a low-density, low temperature 'seed' plasma between the mirrors and then to increase the density and temperature by the injection of first intermediate energy and then high energy neutral beams.

The seed plasma can be made outside the confinement region by means of an electrical discharge in a device called a plasma "gun". The low temperature plasma is then introduced through the mirror ends by injection along the direction of the field lines. Alternatively, a seed plasma may be formed between the mirrors by heating a pellet of solid deuterium by means of a laser beam.

Magnetic Mirror Problems. Two major problems in magnetic mirror confinement are plasma in stabilities and loss of plasma by escape through the ends. Theoretical studies have indicated that large scale instabilities can be controlled if the confinement field lines always bulge into the plasma. Examination of Fig. 15.8 (*b*) shows that in the region between the mirrors the field bulges outward; the confined plasma is then not stable. For stable confinement, the magnetic field must be a minimum in the inner region where the plasma is confined and increase outward in all directions. This is often referred to as a *magnetic-well configuration.*

One way to produce such a configuration is by the so called yin-yang system which consists of two almost semicircular coils at right angles. An electric current passing through the coils produces a magnetic field having fan-shaped ends which, like the coils, are at right angles. These ends constitute the two mirror regions. There is a central region between the mirrors where the field is a minimum and where stable plasma confinement is possible. The magnetic field strength increases outward in all directions from this region, as required for a stable (magnetic-well) configuration.

Although a plasma confined between the magnetic mirrors may be stabilized in this manner, escape through the ends in unaffected. The losses may be decreased by increasing the field strength in the mirror region relative to that between the mirrors. There is, however, a practical limit to this increase. Losses may also be reduced by an increase in the plasma temperature so that collisions which can lead to the escape of particles along the field lines are decreased. This is one reason why operating temperatures in a nuclear fusion reactor using open-ended magnetic mirror confinement would probably be higher than in a toroidal system (*e.g.*, tokamak).

Tandem Mirror System. The purpose of the tandem mirror system is to reduce the end losses from a confined plasma. In a simple, open-ended arrangement, there is a single mirror at each end with plasma in between. In the tandem mirror scheme, however, each single mirror is replaced by a fairly closed spaced mirror pair (or tandem mirror), such as could be generated by yin-yang (or similar) coils or in other ways.

15.6. Inertial-Confinement Fusion

Inertial confinement fusion or pellet fusion, now being most actively pursued through the route of laser irradiation of the pellet, depends on its plasma confinement entirely on internal effects. Thus this approach would appear free from instability problem that have be set magnetic confinement. Controlled nuclear fusion in a small sphere of a deuterium-tritium (*D-T*) fuel mixture under such conditions that the disassembly (or confinement) time of the sphere at the high fusion temperature (about 100 million °C) is longer than the fusion reaction time. That is to say, as a result of inertia (*i.e.*, resistance to motion), the fuel remains confined (or holds together) long enough to permit nuclear fusion to occur to a substantial extent.

Pellet fusion to be successful at a technically feasible level of energy release per microexplosion, the particle density at which the fusion must take place must be raised to around 10,000 times the particle densities of ordinary liquids or solids. The Lawson criterion ($n.\tau$) increases as the square of the radial compression factor. To keep the energy release per microexplosion to values that are compatible with engineering limitations, such as heat transfer and apparatus size limits, only a small quantity of fuel should be reacted per microexplosion. Thus tiny pellets must be used, and this fact, taken together with Lawson criterion, leads to the necessity for very large (of the order of 10^{12} atm) density compression factors. The surface of the pellet must be irradiated at such a high power level (of order of 10^{15} $W\phi$ cm^2) and over such a short time that an almost instantaneous ablation (blowing away) of the outer surface of the pellet occurs before appreciable heat flow would be expected to occur into the interior. In principle, with the use of a sufficient number of symmetrically located converging laser beams derived from ultra-high power lasers, the required compression forces could be generated. The pellets are in the form of glass pellets filled with *D-T* gas and for economically available reactor, must be made at the cost of a penny each.

Details of inertial-confinement fusion. For a sphere with a diameter of 1 mm, the confinement time at the accepted *D-T* fusion temperature of 100 million °C is estimated to be about 10^{-10} sec (*i.e.* one-billionth of a second). If the sphere were smaller, the disassembly time would be decreased roughly in proportion to the diameter. For significant *D-T* fusion, the reaction time would have to be even shorter than this short confinement time.

As stated in a small sphere (or pellet) of a liquid (or solid) *D-T* mixture at normal density, the reaction time would be much longer than the confinement time. By compressing the sphere, the diameter would be decreased and the density of the *D-T* mixture would be increased. The decrease in diameter would result in a decrease in the confinement time, but the increase in density would produce a much greater decrease in the fusion reaction time. Therefore, at a sufficiently high compression, the reaction time could become less than the confinement (or disassembly) time and inertial-confinement fusion should be possible.

For example, in a *D-T* mixture compressed to 10,000 times normal density, the confinement time at 100 million °C of a sphere initially 1 mm in diameter would be roughly 10^{-11} sec. The reaction time, however, might be about 10^{-12} sec, thus complying with the condition for inertial confinement. Furthermore, at 10,000 times normal density, the particle density n would be close to 4×10^{20} nuclei/cubic centimeter (cm^3). The energy confinement time τ is approximately the same as the fusion reaction time 10^{-12} sec. The product $n\tau$ would then be 4×10^{14} sec/cm^3 so that, by the Lawson criterion for *D-T,* ignition and energy break-even should be realized.

Ablation-Implosion Compression. The enormous pressure, estimated to be 10^{12} times atmospheric pressure, required to compress a liquid (or solid) *D-T* mixture to 10,000 times normal density should be attainable by using the ablation-implosion principle. A thin spherical shell containing the fusion fuel is heated uniformly in an extremely short time by means of a high energy pulse. The shell surface is vaporized and ablated (or blown off) at a high speed in the outward direction. Then, in accordance with Newton's third law of motion (*i.e.* every action has an equal and opposite reaction), the remainder of the material is imploded that is, it is forced inward. In other words, the *D-T* fuel material is compressed. Under appropriate conditions, the compression necessary for inertial confinement fusion should be realized.

As the sphere of *D-T* fusion fuel is imploded, the increase in the interior pressure gradually slows down the inward motion until a brief stationary state is reached. A strong shock (or pressure) wave then moves into the compressed fuel and increases the temperature at the centre to atleast 100 million °C, at which *D-T* fusion occurs at a substantial rate.

The neutrons produced in the *D-T* reaction escape, but as a result of the high density of the compressed fuel mixture, the accompanying helium nuclei do not. The energy carried by these nuclei, representing 20% of the total fusion energy, is deposited as heat in the bulk of the fuel. With the additional heat energy available, the fusion reaction, initiated at the centre by shock wave heating, spreads through the fuel sphere in a time that is shorter than the confinement (disassembly) time.

Two general ways have been proposed for supplying the initial energy pulse required to cause surface ablation followed by compression and heating of a fusion fuel sphere (or pellet). One is by means of *intense laser beams,* and the other is to use *beams of high energy,* electrically charged particles, such as negative electrons and positive ions. The energy available for compression can be optimized by pellet design. One concept is based on the use of two shells; an outer (ablator) shell of lighter elements (*e.g.*, plastic) and an inner (pusher or compressor) shell of a heavier element *(e.g.,* gold).

Laser-Beam Fusion. The initiation of nuclear fusion by laser beams was reported first from the USSR in 1968 and was subsequently confirmed by in the United States. However, the extent of fusion, as indicated the neutrons produced, was very small. Laser beams of very much greater energy would be required to realize a practical laser fusion reactor for generating useful power.

Theory had indicated that laser beams of short-wave length radiation would provide the most efficient compression of a D-T fuel pellet. Consequently, neodymium (Nd) glass (*i.e.*, glass doped with a small amount of neodymium) lasers have been (and are being) used in basic studies of laser-beam fusion. Because these lasers have a low efficiency (*i.e.*, a small fraction of the 'pumping' energy is converted into radiation energy), a large proportion of the energy supplied is ineffective. This energy is converted into heat, and the need for cooling the solid lasing material may take Nd-glass lasers impractical for reactor use.

An alternative is the carbon dioxide (gas) laser; this has a higher efficiency than the *Nd*-glass laser and, furthermore, heat can be removed from the gas by circulation through an external cooling system. The radiation from the carbon dioxide laser has a longer wave length than from a *Nd*-glass laser, but it appears that this may not be such a drawback as had been thought.

Existing lasers, no matter what the lasing material, are not powerful enough and do not have the other requirements for a fusion reactor. Efforts are being made to develop such lasers, possibly with novel lasing materials *(e.g.,* Krypton fluoride). According to present estimates, a fusion power reactor would require a laser system which an efficiency of at least 40%, capable of delivering 1 million joules (0.28 kW-hr) of energy per pulse in pulses of less than 10^{-9} sec duration at a rate of one pulse per second.

Laser Fusion Reactor. A laser-beam fusion reactor is conceived as a strong, spherical chamber (or cavity) of some 2 m radius. A small (1 mm) diam pellet (or sphere) of solid or liquid D-T fuel is injected into the cavity when it reaches the centre, several symmetrically located high-energy laser beams are focused simultaneously on the fuel for about 10^{-9} sec. An ablation-implosion occurs, and fusion energy is released. After an interval of a second or so, another fuel pellet is injected, and the process is repeated.

Fig. 15.9 shows a schematic arrangement of a Laser-fusion reactor, in which the pellets are heated by laser light. The pellets act as miniature hydrogen bombs, exploding and delivering their energy to a wall and a cooling medium. To cause the thermonuclear reaction a beam of laser light is divided into many components and directed simultaneously as a pulse of nanosecond (10^{-9} sec.) duration to bombard a pellet from several directions. The laser energy is absorbed in the outer layer of the sphere, causing some material to be evaporated. The particles escaping from the surface impart a reaction momentum to the rest of the sphere. The resulting impulsion force compresses the material to an extremely high density, and the heating ignites the centre region of the pellet. The D-T nuclear reaction proceeds outward to burn the rest of the fuel and to cause the microexplosion that lasts about 20 picoseconds (20×10^{-12} sec). The kinetic energy of the products neutrons and helium ions is to be deposited over the surface of the container and out to a heat exchanger. This isolation of the reaction form walls is expected to prevent material damage.

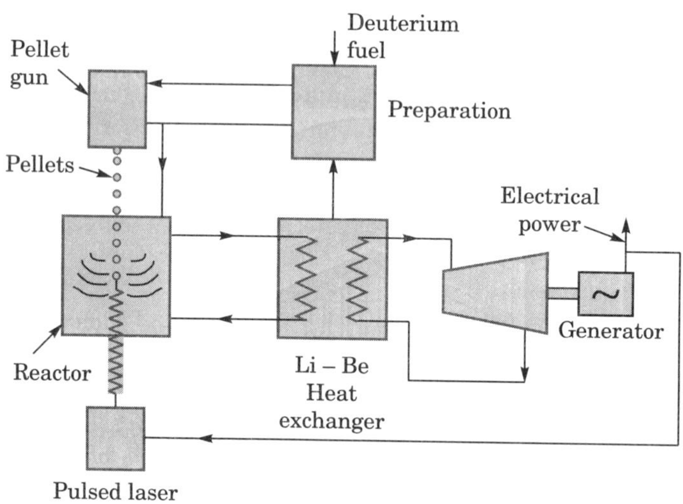

Fig. 15.9. Schematic arrangement of a laser-fusion reactor.

The helium nuclei formed in the fusion reaction are absorbed in the fuel, but the neutrons, carrying 80% of the fusion energy, will escape. They are captured in a lithium blanket where they deposit their energy as heat and also produce the tritium required as fuel material. The heat could be utilized to generate electricity in a more or less conventional manner by producing steam to drive a turbine generator. Apart from the actual fusion process, a laser beam cavity reactor would be similar in principle to (but smaller than) other nuclear fusion reactors. A problem is expected to arise from deterioration of the inner wall of a laser fusion cavity as a result of the neutron bombardment. Proposed ways for overcoming this problem are to keep the inner wall continuously covered with a thin layer of liquid lithium or to maintain a swirling vortex of this liquid within the cavity.

Although the fusion process initiated by laser beams would be self-sustaining in a single *D-T* fuel pellet, a continuous succession of fresh pellets and laser beam pulses would be required to maintain operation of the reactor. A laser fusion reactor is thus a driven energy (or power) amplifier. The energy supplied in the short pulse of laser radiation would be amplified by a factor of 100 or more in the resulting fusion reaction.

If the laser energy is 1 million joules per pulse, as suggested earlier, the fusion energy produced would be about 100 million joules (or 28 kW-hr) per pulse. For a repetition rate of one pellet (or pulse) per second, the fusion power output per cavity would be 100 megawatts (MW). Because only 80% of this energy, which is carried by the neutrons, is available outside the cavity and there are inevitable losses in the laser system and elsewhere, it is expected (optimistically) that a single cavity could generate useful power at a rate of some 20 MW. The fusion process is expected to have a minor impact on the cavity because of the small mass of D-T fuel. The proposed cavity radius of 2 m, should thus provide ample protection against the blast effect of the repeated fusion energy pulses.

Charged-Particle Beam Fusion. A pulse of energy supplied by a beam of high energy, electrically charged particles should, like a laser beam, be able to ablate and compress a fusion fuel pellet (or sphere). This capability was demonstrated with electron beams and deuterium pellets in the USSR in 1976 and in the United States in 1977. The production of neutrons showed that nuclear (D-D) fusion had occurred. There is no reason to doubt that fusion could also be realized with beams of heavier charged particles.

Proponents of charged-particle beam fusion claim that it has advantages over laser-beam fusion. Pulses of the required energy should be easier to generate, and the generation efficiency is expected to be higher than for laser beams. The coupling of the beam energy to the fuel pellet should also be better. The pellets may be larger and less spherical and, consequently, cheaper to fabricate.

Mostly studies of charged-particle beam fusion had been made with electron beams. High energy electron beams can be readily obtained with accelerators; but for fusion, pulses of unusually high current (in amperes) and short duration (less than 10^{-8} sec) are required. Furthermore, the beams must be transported from the accelerator without spreading and focused on to the small fuel pellet.

Beams of positively charged light ions, especially protons (*i.e.,* ordinary hydrogen nuclei) are also being considered for inertial-confinement fusion with such beams, fusion may be achieved with beams of higher voltage but smaller currents than for electron beams. The required pulsed proton accelerators may then be easier to build than the electron accelerators. For this and other reasons, proton beams are now preferred to electron beams for charged-particle fusion experiments.

Since last one decade interest has grown in the use of beams of accelerated heavy ions, with mass numbers from 124 to 238, to produce inertial-confinement fusion. It is possible that these charged-particle beams may prove more effective than electron, proton, or laser, beams. Because the heavy ions can travel only a very short distance in matter, essentially all their energy would be absorbed in the outer layers of the pellet. Ablation and compression of the fusion fuel pellet should then be more effective than in the other proposed methods for achieving inertial-confinement fusion.

As far as can be seen at present, a nuclear fusion reactor utilizing high energy particle beams would be similar in principle to the laser-beam reactor described earlier.

15.7. Muon Catalysed Fusion

Introduction. Muon catalysed fusion is the name given to a series of reactions resulting in the fusion of two hydrogen isotopes nuclei which are kept closed by a negative muon. This process firstly observed by Alvarez *et al* (1957). The possibility of realizing fusion-chambers based on muon catalysis is especially attractive if one considers the advantage of work with a fusion chamber that is operated relatively at low temperatures. There is a wealth of experience accumulated by nuclear community of working at these temperatures. Of course, it is almost futile to take of pure fusion reactors based on muon catalysis alone because of the unfavourable energetics associated with the production of negative mesons. However, the preliminary estimates of energy balance for muon catalysed fusion chambers combined with the fissionable nuclide-blanket show a net positive gain.

There is another advantage in selecting muon catalysed fusion as driver for fissionable blankets. The muon catalysed fusion chambers are very useful in themselves as one can investigate experimentally the validity effusion cross-section for advanced fusion fuels like (D, D), (D, He^3) (D, Li^8) without bothering about the problems connected with excessive plasma temperature one requires while trying to realize thermonuclear fusion environment. Also it would be extremely useful to simulate radiation environment with muon catalysed advanced fusion fuel for physics and engineering investigations of the likely energy extraction systems of later-day thermonuclear reactor based on advanced fusion fuels.

There is also an interesting possibility of utilizing muon catalysed fusion to ignite usual thermonuclear reactions in an inertially confined deuterium tritium pellet.

Physics of Muonic Catalysis. The phenomenon of catalysis of nuclear reaction in cold hydrogen by μ mesons is brought about by a whole lot of atomic and nuclear processes which are set into motion before hydrogen isotopes are made to fuse while keeping close company within a muonic hydrogen molecular ion by a single μ-meson. The fusion reactions of our interest among hydrogen isotopes are given earlier. Under ordinary conditions these reactions can take place only when the reacting nuclei are imparted enough kinetic energy to tunnel significantly through the Coulomb barrier. For example, the nuclei are heated to high temperatures in thermonuclear fusion. It is to be noted that the kinetic energy required to be supplied to the nuclei is usually much smaller than the height of the Coulomb-barrier. Also the probability of barrier penetration is more critically dependent on barrier width. These bring the nuclei close enough, say by chemical forces. In an ordinary hydrogen molecule, a nuclear reaction is impossible because of large distance between the two nuclei. In fact, it is easy to see this from the following relation for barrier penetration

$$B = \exp\left(\frac{-Km}{\sqrt{m_e}}\right) \quad \text{...(15.6)}$$

where, K is a numerical factor around 3, m is the reduced mass of the hydrogen nuclei. m_e is the mass of electron, if electron could be replaced by some "electron like" heavier particle to produce chemical binding in the hydrogen molecule, B would increase. Only suitable "electron- like" particle turns out to be negative meson as it does not react with nucleus. The subnuclear μ meson is 207 times as heavy as electron with the same electric charge. This leads to very small value of

Bohr radius ($a\mu$) for muon. Muon has got life time of 2.2×10^6 sec and it decays into an electron, neutrino and antinutrino. This short lifetime of muon necessitates its utilization through fast processes. When muons are injected into some dense matter, they slow down rather quickly and at low energies are captured to form muonic atoms. For example, slowing down in a 50% D_2 + 50% T_2 medium will form muonic atoms D_μ and T_μ. These muonic atoms move around in the host medium and undergo one of the two main processes:

(*i*) muon transfer from lighter to heavier nucleus, *i.e.*, colliding with *t* in T_2 molecule will yield *T* as follows:

$$D\mu + T \longrightarrow T\mu + D \qquad ...(15.7)$$

(*ii*) the muonic molecular ion formation through interaction heavier muonic atom with lighter or equally heavy nuclei present like;

$$T\mu + D \longrightarrow DT\mu \qquad ...(15.8)$$

It is to be added here that the probability of formation of *DT* molecular ion through the collision of *D* and *T* is rather small as compared to the probability of muon transfer as shown in Eqn. (15.7). It can be explained from the fact that while the presence of a third particle (electron or deutron) is necessary to transfer the binding energy released in the formation of $DT\mu$ molecular ion, the binding energy excess released in the formation of heavier muonic atom $T\mu$ is shared by the product particles, $t\mu$ and *D*. After the formation of muonic molecular ion like $DT\mu$, the heavy hydrogen nuclei present within it, namely *D* and *T* in the present case, undergo fusion almost instantaneously. It happens because of the extreme closeness of these nuclei within the muonic molecular ion the width of the *Coulumb barrier is reduced drastically* by the binding particle muon. The fusion of *D* and *T* within $DT\mu$ can be represented as:

$$DT\mu \longrightarrow He^4 + n + \mu + 17.6 \text{ MeV}. \qquad ...(15.9)$$

The energy released in the above reaction is distributed essentially between He^4 and *n*. The muon freed in reaction (15.9) can repeatedly catalyze such nuclear reaction till either it decays or is captured. It is possible to give 100 (*D, T*) fusion events per muon.

It is obvious from the foregoing discussion that the number of (*D, T*) fusion catalyzed by a single muon will be governed by the 'reaction rates', associated with various processes leading to formation of a "meso-molecule".

In the mesomolecule a virtual photon can be absorbed by the muon thus giving (refer Fig. 15.10).

$$pD\mu \longrightarrow He^3 + \mu$$

The energy of the muon is

$$E\mu = 5.4 \text{ MeV}.$$

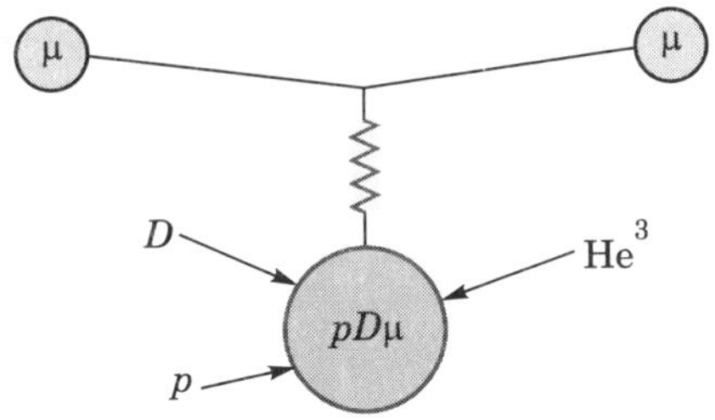

Fig. 15.10. Concept of μ catalysed fusion by photon absorption.

Energy production from Muon catalysed fusion. The large value of the $DT\mu$, mesomolecule formation rate has revived the idea of using muon catalysed fusion (m.c.f.) for energy generation.

In this regard, Petrov has proposed an indigeneous scheme, so called Mesocatalytic Reactor (MR) or Hybreader, where m.c.f. is used together with the electronuclear breeding in order to produce a positive energy outline. The logical scheme of the Mesocatalytic reactor is shown in Fig. 15.11.

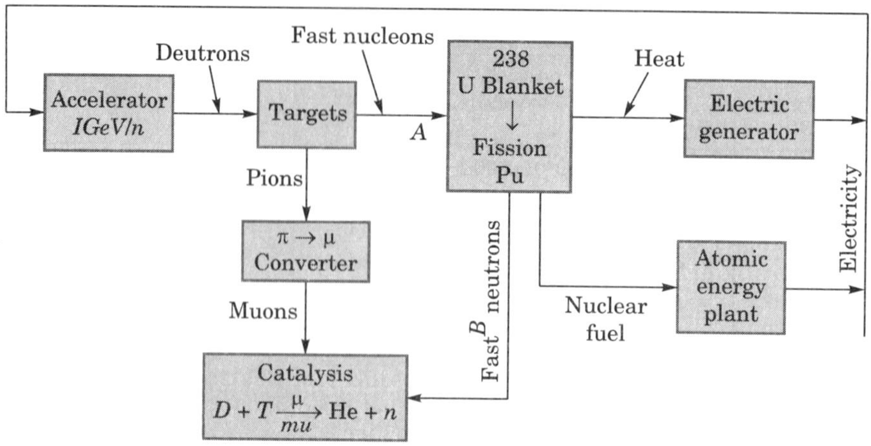

Fig. 15.11. Schematic diagram of meso-catalytic reactor.

Light nuclei, such as *D* or *T*, are accelerated up to an energy of about 1 GeV/nucleon and hit a target where fast nucleons and π mesons (a meson is a particle equal in charge, but having greater mass than an electron or positron, and less than a neutron or proton) are produced. As π are most likely produced in *n-n* collisions, it is convenient to have both beam and target rich of neutrons.

The fast neutrons impinge onto a U-238 blanket, where they cause the fission of uranium and produce fissile isotope (Pu). In this way one gets out heat, which can be converted into electric energy in the electric generator as well as nuclear fuel, which can be used in an atomic energy plant.

Essentially, this is the scheme of electronuclear energy production.

The pions, which have been trapped in a magnetic device around the target, will decay into muons and neutrinos in the converter. In principle, this system provides a source of muons which is much more economical than the standard way of producing a muon beam. Petrov and Shabelski (1981) estimate that the beam energy necessary for the production of one negative point in a cylindrical berillium target is about 4.5 GeV.

In a suitable deutrium and tritium mixture inside the so-called synthesizer the muon will catalyse nuclear fusions. Petrov observes that, by using for electronuclear breeding the 14 MeV neutron originated from the *DT* fusion, it is possible to recover much more energy than the fusion energy itself.

Petrov's schemes brings following two ideas into the subject of m.c.f.:

(*i*) It presents an 'economic way' of getting a source of muons, in that all the energy of the primary beam is eventually used for energy production—through channels *A* and *B* of Fig. 15.11, and the particle losses of the muon beam are appreciably reduced.

(*ii*) It shows that, by a suitable use of the fusion products, it is possible to recover much more energy than from the fusion itself.

The theoretical predictions (later confirmed by experimental results) of a particularly high value of the ($DT\mu$) formation rate have revived the idea of using the m.c.f. for energy production. If this respect several points have to be further investigated. First, one has to clarify the formation mechanism of the ($DT\mu$) meso-molecule and provide definite estimates on the number of possible fusions.

15.8. Characteristics of D-T Reaction

1. The products of the *D-T* reaction are not radioactive and therefore, present no long term disposal problems.

2. The tritium is radioactive and does not occur in nature, it must be bred in the reactor by the reaction of neutrons with lithium. The real fuel for *D-T* system is deuterium and lithium.

3. The absorption of neutrons by structural components will produce radioisotopes which will have to be disposed of after the plant is closed down. The nature of these isotopes is somewhat different from that of fission reactors and represents a smaller, but not completely negligible problem.

4. Most of the energy (80%) in a D-T reaction is carried away by neutrons. This kinetic energy must be converted to thermal energy by absorption etc. and then subsequently to electric energy.

5. The energy of fusion neutrons is about 14×10^6 eV compared to about 2×10^6 eV from the fission neutrons. The higher energy neutrons cause more and somewhat different types of damage in structural materials than is found in fission reactors. They also require more massive structures for complete thermalization and removal of the neutrons.

6. Approximately four times as many neutrons must be produced per unit of energy in fusion reactors than in fission reactors. Roughly speaking, 20 MeV are released per fusion neutron, whereas 80 MeV are released per fission neutron.

7. Unlike plutonium, tritium cannot be made into bombs, so it may be acceptable. There is no possibility of a nuclear explosion although there is some possibility of fire hazards.

15.9. Advantages of Nuclear Fusion

Some of the expected advantages of nuclear fusion as an energy source are:

1. An ample supply of raw materials for fuel (deuterium and tritium) at a moderate cost.

2. The amount of fuel in the fusion system at any time would be so small that a large accidental release of energy would be impossible.

3. Inherent safety with no danger of overheating if the cooling (heat removal) system should fail.

4. Minimum problems of radioactive release to the environment and of waste disposal.

5. High *thermal efficiency* for the conversion of fusion energy into electrical energy because of the high temperatures which should be attainable.

It is because of the above potential advantages that much effort is being expended in many countries in order to make fusion power a practical reality. Although the general requirements are well understood; many scientific and technological problems still remain to be solved.

15.10. Fusion Hybrid

A concept for utilizing the neutrons produced by nuclear fusion especially of deuterium and tritium, to convert a fertile nuclear species (*e.g.*, uranium-238 or thorium-232) into a fissile species (*i.e.*, plutonium-239 or uranium-233 respectively). The fissile material would then be used to generate energy in a conventional nuclear fission reactor. By making use of fertile material in this manner, a fusion hydrid provides an alternative to the *breeder reactor*. Because of the additional energy that would be available from fission, the requirements of a fusion hydrid would be less severe than for a pure fusion system.

The earliest nuclear fusion reactors would be based on the deuterium tritium (D-T) reaction in which 80% of the energy released is carried by the neutrons produced. In a pure fusion reactor, these neutrons would enter a lithium containing blanket in which they would deposit their energy. Interaction of the neutrons with lithium nuclei would generate the tritium required for the D-T fusion fuel.

The main difference in a fusion-hydrid reactor would be that the blanket contains fertile material, say uranium-238, in addition lithium. The high energy fusion neutrons can interact with uranium 238 nuclei in two ways that result in an increase in the number of neutrons. First, fission of uranium-238 by the fusion neutron is accompanied by the liberation of three or four neutrons to replace the one causing fission and second, a high energy neutron can reject one or two neutrons of lower energy from a uranium-238 nucleus. Some of the neutrons are captured by uranium-238 and converted into fissile plutonium-239, whereas other interact with lithium to produce tritium.

After a period of operation, the blanket would have to be removed and the fissile material extracted for use as a reactor fuel. The net result would be the release of several time more energy than could be obtained by fusion alone. The only additional material required would be either normal *(i.e.,* neutral) uranium or the depleted uranium residues of the isotope enrichment process. The foregoing description has referred in particular to uranium- 238, but it is equally applicable to thorium-232 as the fertile species.

15.11. Cold Fusion (Current Status)

Introduction. On March 31, 1989, Prof. Stanley Pons, University of Utah, USA, presented for the first time, the scientific details of his experiment demonstration of the generation of net energy from a simple laboratory experiment. Many laboratories have confirmed Prof. Pon's findings and are continuing with more elaborate experiments.

The experimentors conclude that the *deuterium* pressed within the *palladium crystal lattices* and aided by a small electric current has undergone nuclear fusion, overcoming the strong coulomb barrier, and that it is this process that accounts for both the generation of heat and the presence of a small number of neutrons.

The Experiment. The Pons-Fleischmann experiment is a simple *electrolytic process.* It consists of a bottle of water and two metal electrodes. The electrodes are immersed in water and a small amount of electricity is passed between them. Small amounts of various salts are dissolved in the water to increase the electrical conductivity of the medium water. They hold the electro chemical cells which they have created a sustained nuclear fusion reaction at room temperature.

Under ordinary circumstances, with proper metal electrodes and adequate amount of electrical current hydrogen gas will accumulate at the cathode-the negatively charged electrode. In the Pons-Fleischmann experiment, a palladium cathode and a platinum anode is placed in water which contained 99.5% deuterium and a small electric current is passed through the heavy water. The heavy hydrogen (deuterium) that is generated is slowly absorbed by the palladium cathode (in about 3 weeks). At the end of this period, the experimentors noticed a significant generation of heat in the palladium cathode. The amount of heat generated was extra-ordinary. More than 10 watts per cubic centimeter of the palladium cathode of energy was generated— about 10 times the energy input used to keep the cell in operation. The experiment generated 10 times more energy than it consumed.

The particular type of fusion is distinctly different from *"hot fusion"* in which the plasma reaches a very high temperature sometimes as high as 100 million degrees. An analogy has been drawn to *muon catalysed fusion* which also occurs at low temperature.

The muon is a short-lived elementary particle generated by high energy particle accelerators. The short-lived muon acts like a heavy electron, because it has the same charge as the electron and about 200 times the mass. It is found that if the muon is introduced into molecular hydrogen which contains deuterium and tritium (a yet heavier isotope of hydrogen which does not occur naturally and consists of two neutrons and one proton in its nucleus), the muon can displace one of the ordinary electrons within the hydrogen molecule. When the muon does this, it reduces the intermolecular distances, a major obstackle to fusion, by a factor of 200. This reduction from the normal separation of 0.74. Angstroms (a unit of length—10 billionth of a meter—used primarily to express wavelengths of optical spectra) is sufficient to increase the rate of fusion by 80 orders of magnitude. During its short lifetime, the muon can catalyse upwards of 150 fusion reactions.

Indian Research on Cold Fusion. Keeping aside the controversy as to whether or not the Utah experiment was *"cold fusion"*, there are two immediate issues with which the scientists must deal now. First is the potential of the exhibited process— Whatever it is—for commercial generation of power. Here the most noteworthy efforts have been undertaken in India at the Bhabha Atomic Research Centre (BARC). Following verification of the Utah experiment for a complete one month, a decision was taken at BARC to make electrolytic cells which may become the forerunners of future commercial reactors. It has been noted that at this point in time, the temperature recorded at cathode palladium or titanium is about 70°C. This temperature is too low for economic generation of steam to run a turbine to generate electricity. Higher temperatures are needed. But if researchers can increase the temperature in the cathode significantly, two practical problems will arise. First, palladium and titanium both tend to burn and corrode at higher temperatures. Second, heavy water (D_2O) boils away at 100°C.

To deal with the first problem the BARC scientists are working on the design of an annular electrode, through the centre of which flowing cold water will carry off the heat generated. An electrode shaped like a tube will not only be more uniform in its physical properties in that the centre of the metal portion will be subject to similar conditions as outside, but it can also carry the cold water that takes the heat out. Also it keep the currents passing through the cell low so that temperatures do not rise high.

If this design of electrode turns out to be success, the second problem boiling of heavy water can be easily tackled by putting the whole assembly in a pressure vessel. If the scaled-up reactor works, the potential for generating electrical power in abundance, at a fairly low cost and with total safety is assured. It can also be expected that the operating costs of running such reactors would be much less than that of a fission reactor or a coal-fired power plant because the conception of heavy water, the most expensive of the consumed items in such a reactor, will be negligible than for a reactor which can produce 1000 Megawatts of power.

All in all BARC attempt to capitalize on the Utah experiment is by for the most impressive response and one that exemplifies the responsibility of scientists for their nation's well being.

On the cold fusion, Prof. Stanley Pons and Martin Fleischmann experiment, was followed by hundreds of experiments in the world, and mostly agreed principally to their findings.

REVIEW QUESTIONS

1. What is fusion reaction? Differentiate between Fission and Fusion.
2. What are the main four fusion reactions, which are considered for use in fusion reactors? Which one is most favourable reaction?
3. What is the main advantage of D-D reaction?
4. What are the main conditions or requirements for thermonuclear fusion? Explain in brief.
5. State the Lawson criterion on what factors does it depend?
6. What are the main methods of plasma confinement? Describe briefly.
7. Describe the Magnetic-Confinement fusion, considering *D-T* reaction.
8. Describe Tokamak reactor systems for power generation.
9. Describe the different methods of plasma heating.
10. Describe the working of internal-confinement fusion or pellet fusion reactor.
11. Write short notes on:
    (*i*) Magnetic heating
    (*ii*) Advantages of nuclear fusion
    (*iii*) Pellet fusion reactor
    (*iv*) Plasma heating
    (*v*) Cold fusion.

CHAPTER 16

Energy Storage and Distribution

16.1. Introduction

Energy is useful only if available when and where it is wanted. Carrying energy to *where* it is wanted is called *distribution*; keeping it available until *when* it is wanted is called *storage*.

Means for storing energy in a readily recoverable form when the supply exceeds the demand for use at other times. Storage of primary fuels (*e.g.*, coal, oil and gas) is a form of energy storage, but the term generally applies to actual energy and to secondary fuels (*e.g.*, hydrogen) rather than to primary fuels.

The effective utilization of intermittent and variable energy sources such as solar energy and wind energy, often requires energy storage. If the intermittent energy is converted into electricity, as it is with solar photovoltaic cells and in most cases of wind energy utilization, electrical energy in excess of the demand might be fed directly into a utility grid. If this is not possible, some form of energy storage would be required. Furthermore, where solar energy is used to produce steam in generating electricity, thermal (heat) energy storage is necessary for continuous operation. Storage of thermal energy is also desirable when solar energy is used for space and water heating.

In some circumstances, electricity may be generated, either on land or at sea, at a location that is too distant from a load (or consumption) centre for conventional transmission lines to be used, for example ocean thermal energy conversion. Means must then be found for both storing the energy and transporting it economically to a load centre.

Electrically propelled vehicles, which are expected to come into increasing use, require some form of energy storage. Since the vehicle must carry its energy supply, equivalent to the gasoline in a conventional automobile, the storage system should be readily transportable.

Electric utilities generally useless efficient (intermediate and peaking) units, in addition to the more efficient base load equipment, to meet the additional demand for power during the daytime. With thc availability of energy storage facilities, however, the less efficient units can be eliminated. The most efficient plants are operated continuously at the optimum (or rated) power level, excess electrical energy generated at night and during weekends is stored for use when the demand exceeds the base load. This procedure, called *load levelling,* can reduce the overall cost of generating electrical power.

Renewable energy supplies have different requirements for storage and distribution than do fossil or nuclear energy supplies. In particular the low intensity and wide distribution of renewable sources favour decentralized end use. Therefore energy from renewable sources will often not require much further distribution since the sources are already distributed. However, some renewable sources can fruitfully be harnessed in a moderately centralized manner (*e.g.*, large hydroelectric scheme).

Since use of renewable energy supplies constitutes a diversion of a continuing natural flow of energy, there are problems in matching supply and demand in the time domain, *i.e.*, in matching the rate at which energy is used. This varies with time on scales of months (*e.g.*, house heating in temperature climates), days (*e.g.*, artificial lighting) and even seconds (*e.g.*, starting motors). In contrast to fossil fuels and nuclear power, the initial input power of renewable energy sources is outside our control. We have the choice of either matching the load to the availability of renewable energy supply, or storing the energy for future use.

16.2. Energy Storage Systems

Methods for energy storage may be classified according to the form in which energy is stored; the following categories appear to be the most important:

1. Mechanical energy storage
   (*i*) Pumped hydroelectric storage
   (*ii*) Compressed air (*iii*) Flywheel.
2. Electrical storage: the lead acid battery
3. Chemical energy storage
   (*i*) Hydrogen (*ii*) Ammonia
   (*iii*) Reversible chemical reactions.
4. Electromagnetic energy storage
5. Thermal (heat) energy storage
   (*i*) Sensible heat (*ii*) Latent heat
   (*iii*) Chemical reactions
6. Biological storage

The primary electric-generating plant is continuously operated in a base load mode, which results in excess electricity production during the off peak periods. Electrical or electrical-mechanical energy storage is then used to hold this excess electricity for use during peak demand.

In thermal storage, all schemes deal with storing energy in a thermal form in a material during periods of low power demand and releasing it back during periods of high demand. The primary electric-generating plant is operated to meet the real-time electrical demands during off-peak hours. The available thermal energy input to the plant may be essentially constant as is that from fossil or nuclear fuel, or varying as from solar incidence. The excess thermal energy is stored as such and withdrawn to be converted to meet peak electrical demands.

Energy storage systems like pumped hydro, compressed air and super conductive magnets are, or will be, suitable for large utility energy storage. Others, like flywheel and batteries, are in the development stages and will probably be suitable for medium utility energy storage.

This chapter discusses these different storage forms, in sufficient detail for users to choose the most suitable system for a particular application.

16.3. Mechanical Energy Storage

16.3.1. Pumped Hydroelectric Storage

Electric power in excess of the immediate demand is need to pump water from a supply (*e.g.*, lake, river or reservoir) at a lower level to a reservoir at a higher level. When the power demand exceeds the supply, the water is allowed to flow back down through a hydraulic turbine which drives an electric generator. The overall efficiency of *pumped storage,* that is, the percentage of the electrical energy used to pump the water that is recovered as electrical energy, is about 70%.

Pumped hydroelectric storage is the most economical means presently available to electrical utilities. It could also be used for storing electrical energy produced from *solar* and *wind energy.*

There are relatively few suitable sites where there is a water supply at a lower level and reservoir can he constructed at a higher level, but the use of natural or excavated underground caverns as lower reservoirs, now being developed, should greatly increase the number of possible sites.

Pumped storage is an indirect method for temporarily storing substantial amounts of electrical energy by pumping water from a lower to a higher level. Pumped storage can be used in conjunction with electrical generating plants of all types, regardless of the energy source. In a pumped-storage facility, the power generated in excess of the demand is used to pump water from a lower reservoir (*e.g.*, a lake, river or underground cavern) to an upper reservoir. During periods of peak demand, when the power demand exceeds the normal generating plant *capacity,* water from the upper level is allowed to flow through a hydraulic turbine at the lower level. Fig. 16.1 the turbine then drives a generator to produce electricity in the usual way.

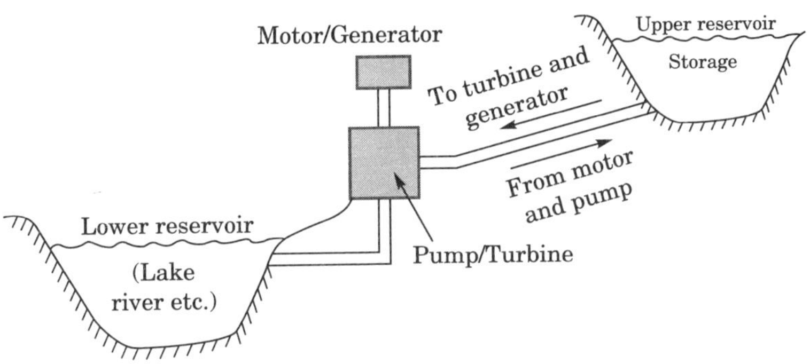

Fig. 16.1

In most pumped-storage plants, the turbine generator system is reversible and can serve to pump water from the lower to the upper level, as well as to generate electricity. In the pumping mode, the generator becomes a motor, driven by electricity produced by a generator in the main plant, and the turbine then operates as a pump. Start up of the turbine/generator or reversal from motor/pump to turbine/generator requires only a few minutes, so that power could be stored after a short delay in the event of a failure in the main plant.

The altitude difference between upper and lower levels (*i.e.*, the water head) in pumped storage facilities ranges from less than 30 m to some what more than 300 m. As a general rule, Francis type reversible turbines are used, but for low heads propeller type turbines, are preferred.

In the pumped-hydro system the high heads are desirable, some topographies do not allow them, and thus pumped-hydro systems are often classified as *above-ground,* which includes the preferred high head and medium head, and *underground.*

In *underground pumped hydro system,* the upper reservoir may be at or near ground level. The lower reservoir is placed underground in natural caverns, old mines, or other underground cavities. This type of system overcome the requirement of a suitable topography. Such a system is shown in Fig. 16.2. In all systems, a reversible pump-turbine or motor-generator set is a principal piece of equipment.

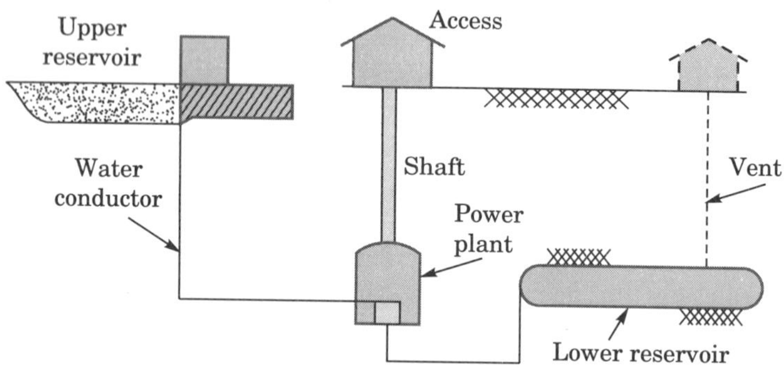

Fig. 16.2. Schematic of an underground pumped hydro storage system.

Pumped hydro, like compressed air, is a potential-energy storage system suitable for large utility energy storage. It is the most developed and used of all storage systems. The principle behind pumped hydro is simple and follows the law of potential energy (PE) that is, the raising of mass to an elevation, height or head H. It is given by

$$PE = \rho g_0 Q_0 H, \quad ...(16.1)$$

where, Q_0 is the natural flow rate of water at the site,

PE = potential energy (Joule)

g = gravitational acceleration 9.81 m/sec^2

H = the vertical distance through which water falls, m.

Also $$PE = g\, m\, H \quad ...(16.2)$$

where, m = mass (kg/sec)

$= \rho\, Q_0$

The operating heads on the pump turbine, in the pumping mode H_p and in the turbine-generating mode H_T are different and are made up of two components each,

$$H_p = H + H_l \quad ...(16.3)$$

$$H_T = H - H_l \quad ...(16.4)$$

where, H is the static head or height and H_l represents the losses during flow conditions (which are different because of different flow rates).

The pumping and generating powers are given by replacing the mass in equation (16.2) with the mass flow rate kg/s and using the proper head, or with

$$P_p = g\, \rho\, Q_P' H_p \quad ...(16.5)$$

and $$P_T = g\, \rho\, Q_T' H_T \quad ...(16.6)$$

where, P_p and P_T = pumping and turbine powers respectively (W)

ρ = density of water (kg/m^3)

Q_p' and Q_T' = volumetric flow rates in pumping and generation, respectively, m^3/s.

Equation (16.2) shows that 1000 kg raised 100 m will store 9.81×10^5 J or 0.2725 kWh. Thus large masses must be elevated to sufficiently large heights to store large quantities of energy. Fortunately large masses are available in pumped hydro systems by the elevation of large quantities of water from a lower to an upper reservoir. One or both of these reservoirs may be artificially excavated or may be a natural river or lake.

The efficiency of a pumped-storage plant is commonly from 65 to 75%. In spite of the loss of 25 to 35%, pumped storage can be economical. The stored water generates power at times of peak demand when it would otherwise be supplied by a gas turbine or diesel engine that is expensive to operate. On the otherhand, apart from capital and maintenance charges, the costs of operating a pumped storage plants are small.

Advantages of Pumped Storage Plants. Pumped storage plants have the following advantages:

(*i*) As compared to other peaking units, pumped storage plants have relatively low capital cost and thus an economical source of peaking capacity.

(*ii*) The pumped storage plant is as rugged and dependable as conventional hydel power station and can pick up load rapidly in a matter of few minutes.

(*iii*) Such plants are readily adoptable to automation as well as remote control.

(*iv*) It is free from effects of environmental pollution.

(*v*) These types of plants allow a great deal of flexibility in the operational schedules of the system.

(*vi*) The power required for pumping is available at cheaper rate.

(*vii*) The pumped storage plants allow the entire thermal or nuclear power generation to take up the base load. Thus the load factor of these units improve giving rise to overall greater system efficiency.

(*viii*) Standby capacity is available on short notice. Power engineers in utilities having pumped-storage installations have long realised the benefit of quick switching on and off capability of these installations. Pumped storage plants need a starting time of only 2 to 3 seconds and can be loaded fully in about 15 seconds. In the event of an outage on a unit, a pumped storage plant can be called upon to meet the generation deficiency, (occurring due to outage) thus ensuring reliable supply and avoiding the necessity of load shedding.

(*ix*) Since the base load plants need not be used to supply peak loads, the forced and maintenance outages of these plants are likely to be reduced.

(*x*) Pumped storage plants can he used for load frequency control.

16.3.2. Compressed Air Storage

This type of storage is analoguous to pumped hydro storage. Whereas in pumped hydro system excess energy generated by a base-load plant during periods of low demand is used to increase the potential energy of hydrostatic pressure of water, compressed air energy storage compresses and stores air in reservoirs, aquifiers, or caverns. The stored energy is then released during periods of peak demand by expansion of the air through an air turbine. In general, the efficiency of compressed air storage is comparable to that of pumped-hydro storage.

In a gas-turbine, roughly 60% of the power output is consumed in compressing air for combustion of the gas. In the com-pressed air storage system, electrical energy in excess of the demand is used to compress air which is stored in a reservoir for later use in a gas turbine to generate electricity. Compressed air storage could serve for electric utility load levelling or for storing electrical energy generated from *solar* or *wind energy*. The overall recovery efficiency is estimated to be about 65 to 75%. A wind turbine, for example, could be created which would directly pump air into a suitable pressurized storage tank. Then later when the wind is not blowing, the energy stored in the air could be utilized to drive an air turbine whose shaft would then drive a generator, this supplying the needed electrical power when the wind is not blowing a compressed air storage may also be applicable to other *solar-electric* conversion systems.

In a conventional gas turbine, the compressor and turbine are connected. In a compressed-air energy storage system, however, the turbine and compressor are uncoupled so that they can operate separately. Furthermore, the electric generator, normally connected to the turbine, must also be capable of functioning as a motor when electricity is supplied. (refer Fig. 16.3).

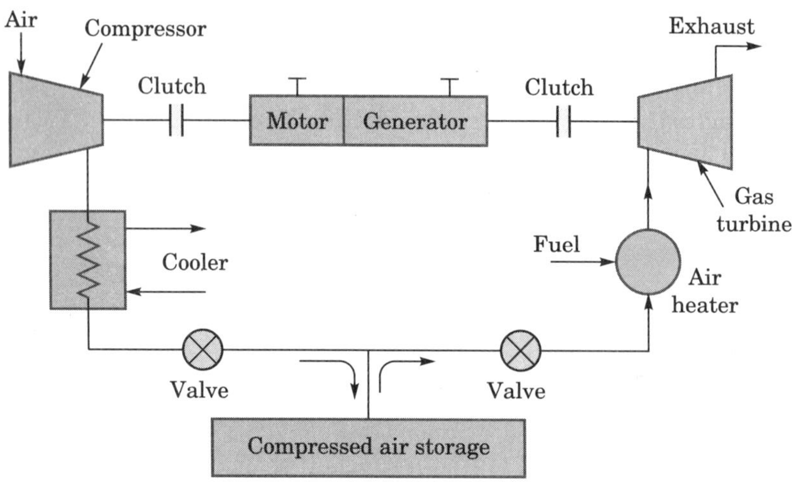

Fig. 16.3. Compressed-air-energy storage.

Electric power in excess of the immediate demand is supplied to the motor/generator which drives the compressor; the compressed air, at about 70 atm (7 MP_a), is stored in a suitable reservoir. The air is heated during compression and may have to be cooled prior to storage to prevent damage to the reservoir walls. When additional power is needed to meet the demand, the compressed air is released and heated using gas or oil fuel. The hot compressed air is then expanded in a gas turbine connected to the motor/generator unit which now acts as a generator.

Compressed-air storage reservoir would probably be too large and too expensive for above ground construction; hence, underground reservoirs, preferably existing ones, are being considered. Among the possibilities are natural caverns, deep aquifiers, depleted gas or oil reservoirs, mined-out rock or salt caverns, and abandoned mines.

Salt Caverns have been used in the past to store petroleum products. They are stable under compressed air storage loading for the duration of plant life. The major concerns are cavern geometry, size and spacing, long-term creep and creep-rupture of rock salt, and air leakage. *Aquifiers* are naturally occuring porous rock formation. They have been used for natural gas storage. Hard-rock caverns type of reservoirs require water- compensating surface, due to their size. It is to maintain air pressure. Therefore these types are mostly compare the two types mentioned above, but they are stable.

When the air is compressed for storage, its temperature will rise. The heat of compression may be retained in the compressed air or in another heat-storage medium and then restored to the air before expanding through the turbine. This is called *adiabatic storage* and results in high storage efficiency. Restoring the heat to the air also prevents the turbine parts from freezing if low temperature air is allowed to dissipate, additional heat could be added by fuel combustion to retain the high storage efficiency, but the results would be extra expense and maintenance problems. This is called a *hybrid* system.

16.3.3. Energy Storage via Flywheels

The basic idea of flywheel energy-storage, sometimes referred to as a "super flywheel", is to accelerate a suitably designed physical rotor to a very high speed in a vacuum, as via an electric motor, at which state high energy storage densities are achieved. The energy is stored as kinetic energy most of which can be electrically retrived when the flywheel is run as a generator, (armature is rotated by the flywheel). Flywheel could conceivably be used for electric utility peaking units, for storage of *solar* and *wind* energy, and for *vehicle propulsion*. The energy recovery efficiency is estimated to be upto 90%. The proponents of super flywheel energy storage claim this storage method has higher energy storage per kg than conventional flywheel or lead acid storage batteries.

More recently interest in flywheel energy storage has been generated by motor vehicle designers. In the so-called hybrid automobile for example, the flywheel stores some of the energy of gasoline engine during periods of low vehicle demands and releases it during periods of high demands, such as during acceleration, hill climbing, etc, and thus operates the engine at a more steady and hence more efficient output.

Flywheels have been used extensively to smooth out power pulses from reciprocating engines. They store off-peak energy as kinetic energy. The fluctuations in speed caused by torque variations are reduced to a minimum by the use of flywheels. As kinetic energy is proportional to the mass times velocity squared, the changes in velocity from the addition or subtraction of kinetic energy are reduced by the use of a large mass. Conversely the energy stored in a flywheel can be increased by increasing the velocity. Energy stored in a flywheel (E) is equal to the kinetic energy, given by

$$E = 1/2\, mv^2 = 1/2\, m(2\pi Rn)^2$$
$$= 2\pi^2\, mR^2\, n^2 \quad \text{...(16.7)}$$

where, velocity of the flywheel = $2\pi Rn$

E = energy (Joule)

m = mass of the flywheel (kg)

R = radius of gyration (m) = 0.7071 R_0

R_0 = outer radius

$$n = \text{revolutions per second} = \frac{\text{revolutions/min.}}{60}$$

The energy E absorbed or released by a flywheel between speed of rotation n_1 and n_2 is thus given by

$$\Delta E = 2\pi^2 mR^2 (n_2^2 - n_1^2) \qquad ...(16.8)$$

The ratio of the variation in rotational speed to the mean speed n is called the coefficient of speed fluctuation k_S, given by

$$k_S = \frac{n_2 - n_1}{n} = \frac{2(n_2 - n_1)}{n_1 + n_2} \qquad ...(16.9)$$

$$\left(\text{where, } n = \frac{n_1 + n_2}{2}\right)$$

Now $\Delta E = 2\pi^2 mR^2 (n_2 + n_1)(n_2 - n_1)$

$= 2\pi^2 mR^2\, 2n\, k_S\, n$ (from Eqs. 16.8 and 16.9)

$= 4\pi^2 k_S\, mR^2 n^2$ ...(16.10)

The value of the coefficient k_S depends upon the desired closeness of speed regulation. For example it varies from 0.005 for fine to 0.2 for coarse regulation. Thus for a given energy absorption ΔE, m and or R^2 must be high for close speed regulation. *Maximum energy density* of a rotating steel disk is

$$W_m = \frac{E}{m} = \frac{1/2\, I\, \omega^2}{m} \qquad ...(16.11)$$

The kinetic energy of rotation of an object is E which is equal to $1/2\, I\, \omega^2$, where, I is the moment of inertia of the object about its axis, and ω is its angular velocity (rad/sec). In the simplest case, where the mass m is concentrated in a rim of radius $r = a$, then $I = ma^2$. For a uniform disk of the same mass, I is lower $(ma^2/2)$ because the mass nearer the shaft contributes less to I.

Now, $$W_m = \frac{1}{2} ma^2 \omega^2 \frac{1}{m}$$

$$= \frac{1}{2} a^2 \omega^2 \qquad ...(16.12)$$

For a flywheel to be a useful store of energy (and not just a smoothing device), it follows from Eqn. (16.12), that it must rotate as fast as possible. However, its angular velocity is limited by the strength of the material which has to resist the centrifugal forces tending to fling it apart.

For a uniform wheel of density ρ, the maximum tensile stress is

$$\sigma_{max} = \rho\, \omega^2 a^2 \qquad ...(16.13)$$

In general, $I = k\dfrac{ma^2}{2}$ for a particular shape, where k is a constant ~ 1, so

$$W_m = k\frac{a^2 \omega^2}{2} \qquad ...(16.14)$$

and $$W_{m_{max}} = \frac{k\, \sigma_{max}}{2\rho} \qquad ...(16.15)$$

Conventional materials such as steel give rather low energy densities. For a rotating steel disk, maximum energy density with $k = 1$,

$$W_{m_{max}} = \frac{1 \times (10^6 \text{ N/m}^2)}{2 \times 7800 \text{ kg/m}^3} \quad \text{(from Eqn. 16.15)}$$

$$= 0.06 \text{ MJ/m}^3$$

Much higher energy densities can be obtained by using light weight fibre composite materials, such as fibreglass in epaxy resin, which have higher tensile strength σ_{max} and lower density ρ. To make the best use of these materials, flywheel should be made in unconventional shapes with the strong fibres aligned in the direction of maximum stress. Such systems can have energy densities of 0.5 MJ/kg (better than lead acid batteries) or even higher.

Materials for energy storage flywheels must have high tensile strengths, high strength density ratio, high resistance to cyclic crack growth, and high strength density to cost ratios. Those under consideration include some alloys, such as so-called maraging steels, and more promising, composites such as fibre reinforced plastics. One composite that shows particular promise is a 62 volume % *S*-glass in epoxy composite. Other attractive composites are graphite-epoxy and kelvar epoxy.

Flywheels for energy storage are systems that include, besides the flywheel itself, a number of subsystems. These are a housing; bearings with ball bearings believed the most suitable; a vacuum pump to minimize windage losses inside the housing; seals to minimize oil and air leakage into the vacuum chamber; and sometimes a containment ring to protect nearby personnel and equipment from flying fragments in case of flywheel rotor fracture.

Losses in a flywheel energy-storage system include windage, bearing and real friction, vacuum pump input power, and eddy current (hysteresis) and other inefficiencies in the motor generator (or in transmission systems). In early designs these losses were prohibitively large, and much development work still needs to be done to arrive at a technically attractive system.

For use in smoothing demand in large electricity networks, flywheel have the advantage over pumped hydro systems that they can be installed anywhere and take up little land area. Units with a 100 tonne flywheel would have a storage capacity of about 10 MWh. Larger storage demands would probably best be met by cascading many such 'small' units.

Flywheel also offer an interesting alternative to storage batteries for use in electrically powered vehicles especially since the energy in a flywheel can be replenished more quickly than in a battery.

16.4. Electrical Storage: The Lead Acid Battery

Storage batteries

Electricity is a high grade form of energy, and therefore great effort is made to find cheap and efficient means for storing it. A device that has electricity both as input and output is called an (electrical) storage battery or (electrical) accumulator. Usually the combination of electrolyzer and fuel cell is not included as 'electrical storage', however. Batteries form an essential component of almost all *photovoltaic* and *small wind electric system,* and there is steady development of battery powered vehicles.

Although many electrochemical reactions are reversible in theory, few are suitable for a practical storage battery, which will be required to cycle hundreds of times between charging and discharging currents of 1-100 A. The most widely used storage battery is the lead acid battery, invented by Plante in France in 1860.

When a storage battery is charged, by connecting it to a source of direct electric current, chemical changes occur in the battery materials. As a result, electrical energy is converted into stored chemical energy. When the battery is discharged, by connecting a load (*e.g.*, a motor) between the terminals, the chemical reactions are reversed, and the stored chemical energy is reconverted into electrical energy. The energy recovery efficiency of a storage battery varies with the type of battery and the rate of discharge, but 75% should be attainable. However the efficiencies are often lower.

Batteries are modular in nature; that is to say, they are built up of individual units of moderate size. This means that the energy storage capacity can be varied over a wide range simply by varying the number of units that are connected together. Consequently, batteries are adaptable to any type of storage in which the energy input is in electric form. Potential applications are utility *peak shaving* (and possibly load levelling), *vehicle propulsion,* and storage of electrical energy generated from *wind energy* or *solar cells.* The capability of rapid operation reversal, from charge to discharge, makes batteries especially convenient for electric utility applications. Moreover, they permit dispersed distribution by locating storage facilities near load centres.

By far the most common type of storage battery is the lead acid battery used in the SLI (Starting, Lighting and Ignition) systems of essentially all automobiles and other road vehicles. These batteries are usually maintained in a fully charged condition by a generator which produces direct current when the engine is running. For utility applications or vehicle propulsion, the operating conditions are quite different. In such uses, the battery is subjected to a deep (*i.e.*, almost complete) discharge and then recharged roughly once a day. When subjected to such repeated cycling, the lead acid batteries used for automobile SLI systems have a short life time.

Studies in the progress will undoubtedly result in the production of lead-acid batteries/more-suited to utility and vehicle applications. However, there are some basic limitations to lead-acid batteries (*e.g.*, their heavy mass per unit of stored energy) that can not be overcome. Consequently, several other kinds of storage batteries, usually, entirely different materials and some operating at high temperatures, are under development in the hope of finding a type (or types) that has a low (or moderate) cost, a long life time, and is lighter than the lead-acid battery.

The battery requirements for stationary (utility) applications are low (or moderate) cost and a long life time; the mass is less important than for vehicle propulsion. For stationary storage, a battery should be capable of atleast 3000 deep discharges over a life time of 10 to 15 years. The discharge time for peak power supply would be 8 to 10 hours and the charge time roughly 10 hours. No existing storage battery can meet these requirements, but there is hope that one or more of the new types will eventually do so. Central-station power plants usually generate and transmit alternating current (a.c.), but direct current (d.c.) is required for battery charging. Furthermore, upon discharge, the batteries produce d.c. which must be converted into a.c. for feeding into transmission lines. Hence, the battery test facilities will involve the testing of equipment for converting a.c. into d.c., and d.c. back to a.c. in an efficient manner.

Basic Battery Theory. A battery is a combination of individual cells. A cell is the elemental combination of the materials and electrolyte constituting the basic electrochemical energy storer. A battery can also be thought of as a black box into which electrical energy is put, stored electrochemically, and later regained as electrical energy.

A generalized cell consisting of two electrodes called the anode and cathode immersed in a suitable electrolyte. When an electrical load is connected between the electrodes charge separation occurs at the interface between one electrode and the electrolyte, freeing both an electron and an ion. The electron flows through the external load and the ion through the electrolyte, recombining at the other electrode.

The polarity and magnitude of the cell terminal voltage is, in general, a function of the electrode materials, electrolyte, cell temperature, and other factors. By combining appropriate numbers of cells in both series and parallel, the battery can deliver the desired voltage and current. An appropriate combinations of cells can provide the desired power output. Fig. 16.4.

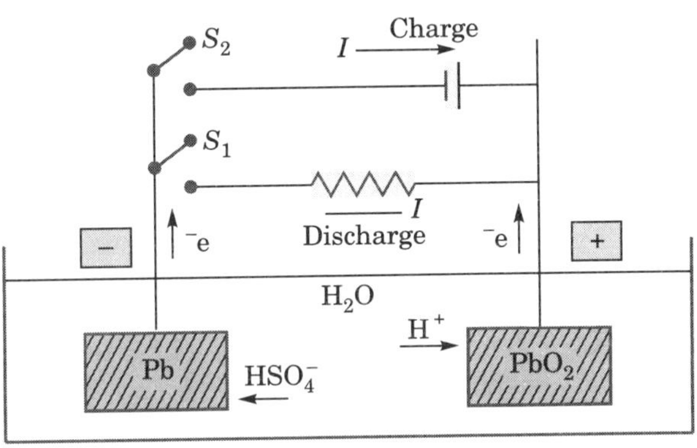

Fig. 16.4. Schematic diagram of *lead acid cell.*
The charge carriers move in the direction shown during the discharge reaction.
The reaction and carrier movements are reversed during charging.
(Switch S_1 open and S_2 closed.)

The *lead acid battery* is built up from cells, one of which is shown schematically in Fig. 16.4. As in all electrochemical cells, there are two solid plates immersed in a conducting solution (electrolyte). In this case the plates are in the form of grids holding pastes of lead and lead dioxide respectively. The electrolyte is sulfuric acid, which ionizes as follows.

$$H_2SO_4 \longrightarrow H^+ + HSO_4^- \quad ...(16.16)$$

During discharge, the reaction at the negative electrode is

$$Pb + HSO_4^- \longrightarrow PbSO_4 + H^+ + 2e^- \quad ...(16.17)$$

Lead (Pb) is oxidized to Pb^{2+} which is deposited as $PbSO_4$. The sulphate takes the place of the Pb paste in the plate. The electrons so liberated travel through the external circuit to the positive electrode, where they contribute to the reaction:

$$PbO_2 + HSO_4^- + 3H^+ + 2e^- \longrightarrow PbSO_4 + 2H_2O \quad ...(16.18)$$

This $PbSO_4$ likewise replaces the PbO_2 in that plate. The electrical current through the solution is carried by H^+ and HSO_4^- ions from the sulfuric acid, which themselves take part in the plate reactions.

Knowing the reactions involved and the corresponding standard electrode potentials (given in the chemical tables), the theoretical energy density of any proposed battery can be calculated. Theoretical energy density of lead acid battery (W_m°) is calculated and its value comes out to be 0.60 MJ/kg of active material as shown below:

The reactions (16.16) and (16.17) show that to transfer 2 mol of electrons requires:

1 mol Pb = 207 gram

1 mol PbO_2 = 239 gram

2 mol H_2SO_4 = 196 gram

Total active material = 642 gram.

But 2 mol of electrons represents a charge:

$$(2 \text{ mol}) (-1.60 \times 10^{-19} \text{ C})(6.02 \times 10^{23} \text{ mol}^{-1})$$
$$= -(2)(9.6)(10^4) \text{ C} = -1.93 \times 10^5 \text{ C}$$

The standard electrode potential for $Pb/PbSO_4$ is 0.30 V and for ($PbSO_4/Pb^{4+}$) is – 1.62 V.

So the theoretical cell emf for ($Pb/PbSO_4/H_2SO_4/PbSO_4/PbO_2$) is E_{cell} = + 1.92 V.

With the PbO_2 plate positive, according to the IUPAC sign convention.

The actual cell EMF depends on the concentration of reagents, and can be calculated by standard electrochemical methods. In general, the voltage of a cell operating at low currents differs by only a few per cent from the theoretical cell voltage. In particular, lead acid batteries are usually set to give 2.0 V per cell.

Therefore the work done in moving 2 mol of electrons is (1.93×10^5 C) (2.0 V) = 0.386×10^6 J. Thus the energy stored in 1 kg of active ingredients is, in theory,

$$W_m° = (0.386 \times 10^6 \text{ J})/(0.642 \text{ kg}) = 0.60 \text{ MJ/kg}.$$

Unfortunately, the energy density W_m of any practical battery is always much less than the theoretical value W_m°, if the total mass of the whole battery is considered. Most commercial batteries have $W_m \sim 0.15\, W_m$°, although more careful (and more experience) designs can reasonably be expected to achieve energy densities up to 25% of the theoretical values.

In the specific case of the lead acid battery, the main reasons for this 'under achievement' are:

(1) A working battery necessarily contains non active meterials, *e.g.*, the case, the separators (which prevent the electrodes short circuiting) and the water in which the acid is dissolved. (The acid concentration must not be too high or the battery will discharge itself). Since the mass of an actual battery exceeds the mass of the active ingredients, the energy density is less than the theoretical value calculated from the active mass alone.

(2) The reactions cannot be allowed to go to completion. If all the lead was consumed by reaction (16.17), there would be no electrode left for the reverse reaction to operate at *i.e.*, the battery could not be cycled. Similarly, if the concentration of H_2SO_4 is allowed to fall too low, the electrolyte ceases to be an adequate conductor. In practice, the battery cannot be allowed to discharge more than about 50% of its stored energy, or it will be ruined. Such a discharge is called a 'deep discharge'.

A further limitation of real batteries is similar to all car owners: they do not last forever. Solid Pb is almost twice as dense as the $PbSO_4$ found in the discharge reaction (16.17). Therefore it is difficult to fit the $PbSO_4$ crystals into the space originally occupied by the Pb paste in the negative electrode. In practice, some $PbSO_4$ falls to the bottom of the cell in every discharge. This constitutes an irreversible loss of active material. This loss is worse if the battery is allowed to fully discharge; indeed, it may rapidly become impossible to recharge the battery.

The other main factor limiting the life of even a well maintained battery is self-discharge of the positive electrode. This is particularly actute in vehicle (SLI) batteries in which the grid is not pure Pb but a lead-antimony alloy, which is stronger and better able to stand the mechanical stresses during motion. Unfortunately antimony promotes the reaction.

$$5PbO_2 + 2Sb + 6H_2SO_4 \longrightarrow (SbO_2)_2\, SO_4 + 5PbSO_4 + 6H_2O \qquad ...(16.19)$$

which also slowly, but irreversibly, removes active material from the battery.

Batteries for stationary applications (*e.g.*, photovoltaic lighting systems) can use Sb-free plates and have longer life (upto 7 years) if not excessively discharged.

The performance of a battery depends on the current at which it is charged and discharged, and the depth to which it is regularly discharged.

16.5. Chemical Storage

16.5.1. Introduction

Energy can be held in the bonds of many chemical compounds and released by exothermic reactions, notably combustion. Sometimes it is necessary to apply heat or other catalysts (*e.g.*, enzymes) to promote the desired reaction. Biological compounds are a special case. Here we discuss the most important inorganic compounds which have been suggested as practical energy storers by means of their combustion in air.

16.5.2. Energy Storage via Hydrogen

Energy can be both stored and transported or hydrogen, which serves as a secondary fuel. The input energy, usually electrical but possibly thermal, serves to decompose water (H_2O) by electrochemical (or chemical) reaction into its constituents elements hydrogen and oxygen. These substances can then be recombined to release the stored energy as required instead of using the oxygen produced from water in this energy recovery process, oxygen from the air is commonly employed. The pure oxygen from water can then be sold for industrial applications (*e.g.*, in iron and steel fabrication). Hydrogen can be transported either as compressed hydrogen gas, as liquid hydrogen (at low temperature), or in the form of a solid compound with certain metals or alloys. Consequently, hydrogen may be useful as a means of storing and transporting energy generated in remote locations far from load centres.

The most convenient means for producing hydrogen and oxygen from water is by electrolysis, that is, by passing a direct electric current through water containing an acid or alkali to make it an electrical conductor. The input energy is then in the form of electrical energy. It may also be possible to decompose water by heat (*i.e.*, with thermal energy input) as a result of a series of chemical reactions. The chemical energy in hydrogen (and oxygen) can be converted into thermal, mechanical, or electrical energy. One possibility is to burn hydrogen in air, in a manner similar to natural gas, to produce heat (thermal energy) for use in the home or in industry. Hydrogen can also serve as the fuel, in place of gasoline in automobile truck, and even air craft engines.

Electrical energy can be obtained from hydrogen in several ways. For example, steam from a water boiler heated by burning hydrogen could be used to drive a conventional steam turbine with attached electric generator. Alternatively, hydrogen can provide the fuel for a gas (combustion) turbine which in turn drives a generator. The maximum overall efficiency (possibly 55 to 60%) for recovery of the input energy, however, would be obtained by means of a *fuel cell*; in such a cell, electrical energy is generated directly from hydrogen and oxygen. Fuel cells could also be used in homes and apartments houses, by industry, for peak saving by utilities, and in electric vehicles.

Energy from various solar electric systems can be stored in hydrogen. A wind-electric or photovoltaic system, for example with a d.c. output, the power can be fed directly into an electrolyzer tank which produces hydrogen and oxygen from ordinary water as already stated. The gases may be produced either under pressure or near atmospheric pressure and then *via* external pumps compressed to the desired pressure. The latter approach, however requires auxiliary energy. The hydrogen and oxygen gas produced can be stored either in gas or liquid forms for a long time. It can when needed, be quickly and easily converted again directly into electrical energy *via* the well known fuel cell (see in previous chapter 10). The system thus effectively stores the sun's energy as hydrogen and oxygen, and from this storage a smooth reliable power output may be taken for a limited time set by the hydrogen storage capacity.

To store hydrogen in large quantities is not common. Most promising is the use of underground caverns, such as those from which natural gas is now extracted. But storage of gas-even if compressed is bulky. Hydrogen can be liquefied, but since its boiling point is 20 K these stores are awkward to maintain. Chemical storage as metal hydrides, from which the hydrogen can be released by heating, is more manageable and allows large volumes of H_2 to be stored. For example:

$$FeTiH_{1.7} \xrightarrow{T \sim 50°C} FeTiH_{0.1} + 0.8\ H_2 \qquad ...(16.20)$$

This reaction is reversible, so that a portable hydride store can be replenished with hydrogen as a central 'filling station'. The heat released in this process can be used for distric heating, and the portable hydride store can be used as the 'fuel tank' of a vehicle. The main difficulty is the weight and cost of the metals used. Hydrogen can also be distributed through the extensive pipeline networks used for natural gas in many countries. It is also possible to convert H_2 very efficiently to electricity, by means of fuel cells as already described.

16.5.3. Ammonia

Unlike water, ammonia can be dissociated at realizable temperatures:

$$N_2 + 3H_2 \rightleftharpoons 2NH_3 \qquad ...(16.21)$$

In conjunction with a heat engine, these reactions form the basis of what may be the most efficient way to generate continuous electrical power from solar heat. Collectors with concentrators can achieve temperatures high enough to operate a heat engine at reasonable efficiency, which can be used to generate electricity.

16.5.4. Reversible Chemical Reactions

In general sense, a reversible chemical reaction is one that proceeds simultaneously in both directions. Of interest here are reversible reactions that occur predominately in one (forward) direction with the absorption of heat at a higher temperature, and predominately in the opposite (reverse) direction with the emission of heat, at a lower temperature. The products of the forward reaction store thermal energy (heat) as chemical energy which can be recovered as thermal energy when the conditions are changed to permit the reverse reaction to occur. This kind of energy storage may be useful for storing high temperature heat obtained by *concentrating solar energy.*

To be suitable for heat storage, the reaction system should involve materials that are inexpensive and not too difficult to handle. In addition, the forward and reverse reactions, as required, must be able to proceed at reasonable temperatures. A catalyst (or catalysts) may be needed to speed up the desired reactions, especially at the lower temperature.

Thermo-chemical storage systems are suitable for medium or high temperature applications only. For storage of high temperature heat, some reversible chemical reactions appear to be very attractive. Advantages of thermo-chemical storage include high energy density storage include high energy density at ambient temperatures for long periods without thermal losses and potential for heat pumping and energy transport over long distances. This type of storage is illustrated by a hypothetical reactions:

$$A + B \rightleftharpoons AB \quad \text{...(16.22)}$$

The forward reaction takes place with absorption of heat from heat source (may be solar energy), and the heat is stored in the form of products. When heat is desired, the products are to be remixed to allow the reversible reaction to take place with liberation of heat. The original reactant is regenerated completing the cycle. Both forward and reverse reactions take place at constant but different temperatures, the forward reaction occurring at a higher temperature than the reverse reaction.

The basic measures of merit to be followed in the selection of thermochemical reactions are the following:

(*i*) The reaction should be nearly complete within the temperature range of available heat source or solar collectors.

(*ii*) The reverse reaction should be nearly complete at the temperature at which useful energy is to be extracted.

(*iii*) The reaction should be able to release energy for the use at temperatures close to those supplied by the collector (in case of solar energy).

(*iv*) The energy absorbed per unit volume of products stored should be as large as possible to minimize the volume of storage, and the products should be in liquid form.

(*v*) The reaction should be completely reversible without any side reactions.

(*vi*) Reactions should be rapid.

(*vii*) Large enthalpy changes of reaction are desired.

(*viii*) For an uncatalyst reaction, the products must be readily separable prior to storage. For a catalyzed reaction the products must be stable during the storage period and the catalyst should be easily separable from the products.

In addition, it is obviously preferable that the chemicals required should be readily available at low cost and should be easy to handle.

Following few chemical reactions are illustrated here from the stand-point of their suitability based on the above criteria and are found suitable for use in solar applications. Table (16.1) gives the details of three reactions.

Table 16.1

Reaction	*Temperature of forward reaction* °C	*Temperature of reverse reaction* °C	*Energy stored per unit vol. of storage material* kJ/m³
$CH_4 + H_2O \rightleftharpoons CO + 3H_2$	780	610	209.4×10^3
$SO_3 \rightleftharpoons SO_2 + \frac{1}{2} O_2$	1025	590	460.6×10^3
$NH_4HSO_4 \rightleftharpoons NH_3 + H_2O + SO_3$	498	435	2143.7×10^3

In the above table first reaction is methane-syngas reaction, which is occurred in the presence of a nickel catalyst.

Suitably of the reaction is that the technology required is available. Moreover the reverse reaction (the methanation process) is being commercially used for the manufacture of methane on a large scale. The main problem is that the products CO and H_2 are gases and have to be stored at a pressure of about 100 bars (10^7 N/m^2).

Second reaction is of decomposition of sulphur trioxide to sulphur dioxide and oxygen in the forward direction and recombination of sulphur dioxide and oxygen to form sulphur trioxide in the reverse direction, in the presence of a catalyst. This storage system has been suggested for use in 100 MW central tower solar power plant operating on the Brayton cycle with helium as the working fluid. A disadvantage associate with this system is that the although SO_2 can be stored as a liquid, O_2 has to be stored as a gas under a pressure of about 100 bars (10^7 N/m^2). Problems of corrosion and safety may also require careful attention.

In the third reaction catalyst is not needed, and it satisfies most of the criteria mentioned earlier.

The double conversion of reciprocal salt pairs with favourable transition temperatures and temperature and high enthalpy of transition are suitable for low temperature heat storage. A system of two salts with different cations and anions is called a reciprocal salt pair. In this system of reaction

$$AB + CD \rightleftharpoons AD + CB + \Delta H \qquad ...(16.23)$$

is possible. This type of reaction is called the double conversion of reciprocal salt pair. The approximate transformation temperatures of more than 500 salt pairs were determined by Reiter of which about 30 pairs seen to be suitable for solar heat storage.

Two of the most promising reciprocal salt pairs involving the salt hydrate $Ba(OH)_2 8H_2O$ have been investigated by Reiter and Rota (1984) experimentally. These salt pairs are:

$$2KNO_3 + Ba(OH)_2.8H_2O = 2KOH + Ba(NO_3)_2 + 8H_2O \qquad ...(16.24)$$

(Transition temperature 65°C and H_R = 670 J/cm^3)

$$2NaNO_3 + Ba(OH)_2.8H_2O = 2NaOH + Ba(NO_3)_2 + 8H_2O \qquad ...(16.25)$$

(Transition temperature 72°C and H_R = 670 J/cm^3)

The investigated heat storage materials exhibited favourable temperatures of transition, high enthalpies of transition and insignificant supercooling. A type of thermo energy storage concept involving an oxide/hydroxide reversible chemical process has been under investigation at the Energy System Group of Rockwell international for several years. The basic reactions are of the following types:

$$M(OH)_2 + \text{heat} = MO + H_2O \qquad ...(16.26)$$

where, M refers to various alkaline earth metals such as magnesium calcium and barium. Forward reaction where the hydrooxide is heated to the oxide and dehydration takes place in the energy storage phase. The reverse reaction where the water is added to the oxide and evaluation of considerable amount of heat takes place, in the energy recovery phase. Most of the inorganic oxides yield heat energy at high temperatures during the hydration cycle.

Several methods of getting heat energy into and out of a reaction vessel have been identified. For small systems such as might be considered for medium heating and cooling systems, a basic concept depicted in Fig. 16.5 will be idle. Incidently the disadvantage of limited storage time before the energy is lost as a result of natural heat transfer processes of the sensible heat and latent heat methods is eliminated in the thermochemical method.

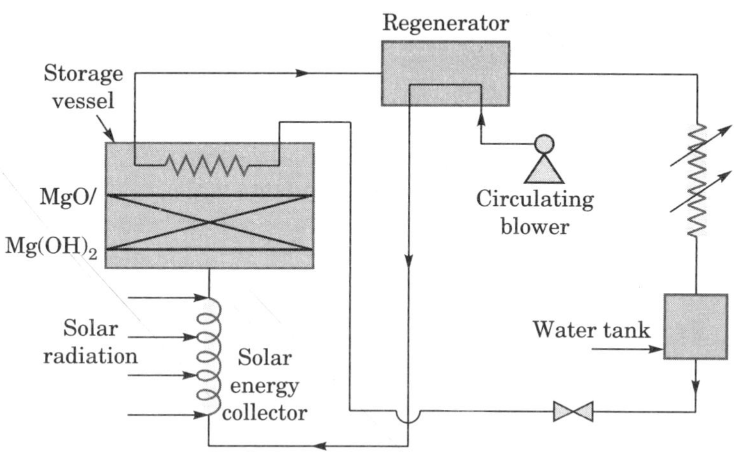

Fig. 16.5. Basic oxide/hydroxide storage concept.

16.6. Electromagnetic Energy Storage

Electromagnetic energy storage requires the use of superconducting materials; these materials (metals and alloys) suddenly lose essentially all resistance to the flow of electricity when cooled below a certain very low temperature. The temperature below which they become superconductive is called the *transition* or the *critical temperature*. All superconducting metals have transition temperatures in the *cryogenic* (low temperature of the order of – 273°C to – 150°C or 0 K to 123 K) range. If maintained below this critical temperature a superconducting metal (or alloy) can carry strong electric current with little or no loss. Useful superconducting materials available commercially are a niobium titanium (Nb-Ti) alloy at temperatures below – 263°C and a compound of niobium and tin (Nb_3-Sn) below – 255°C.

Superconducting magnetic energy storage is a concept that initially received attention for pulsed energy storage in which the charge and discharge times have been short. It subsequently became apparent that such a concept is suitable for large-scale energy storage by an electric utility. The concept is based on the principle that energy can be stored in the magnetic field associated with a coil. If the coil is made of a material in a superconducting state, *i.e.,* maintained at a temperature below its critical temperature, then once it is charged, the current will not decay and the magnetic energy can be stored indefinitely. The stored energy can be released back to the network by discharging the coil. An electromagnetic field, such as is produced by an electric current flowing through a wire (or wire coil), can store energy. Under ordinary conditions, losses result from the resistance of the wire, and energy must be supplied continuously to maintain the current. But if the wire coil were made of superconducting material and kept at the required low temperature, resistance losses would be very small and, once initiated an electric current would remain almost constant. Electrical energy supplied as direct current to the wire coil would then be stored in the electromagnetic field. By attaching the coil to a load, the stored energy could be recovered as electrical energy (direct current). This is the basis of the Superconducting Magnetic Energy Storage (SMES) system under study for possible eventual use in electrical utility load levelling and/or peaking applications.

The energy E stored in a coil in which a current I circulates is given by

$$E = \frac{1}{2} LI^2 \quad ...(16.27)$$

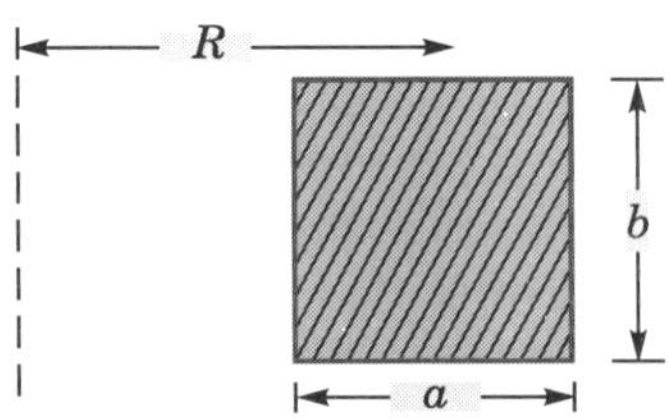

Fig. 16.6. Dimensions of a cylindrical coil with rectangular conductors.

where, E = energy, (Joules)

(Joule = Watt × Second)

L = Inductance H

$$\left(H = \text{Henry} = \frac{\text{Volt} \times \text{Second}}{\text{Ampere}} \right)$$

I = Current, A (Ampere)

The inductance L of a coil is a function of its dimensions, which are characterized for a coil with conductors of a rectangular cross-section (see Fig. 16.6) by

$$\xi = \frac{2R}{\sqrt{ab}} \quad ...(16.28)$$

$$\delta = a/b \quad ...(16.29)$$

and $V = 2\pi Rab$

$$= \frac{8\pi R^3}{\xi^2} \quad ...(16.30)$$

where, R = mean radius of coil, m.

a and b = width and depth of conductor, m.

V = volume of conductor in one coil turn m^3 and L is given by

$$L = f(\xi, \delta)\, RN^2 \quad ...(16.31)$$

where, $f(\xi, \delta)$ = form function having the units (V.s)/(Am)

N = number of turns of coil.

The energy stored in the coil, Eqn. (16.27), can now be written as

$$E = 1/2\, f(\xi, \delta)\, RN^2I^2 \quad ...(16.32)$$

Using a current density j, given by

$$j = NI/ab \quad ...(16.33)$$

$$\therefore \qquad E = 1/4\pi^{-5/3} f(\xi, \delta)\, x^{-2/3}\, V^{5/3} j^2 \qquad ...(16.34)$$

A coil that gives the maximum value of inductance to volume L/V is called a Brooks coil, it has the dimensions

$$a = b \text{ and } R = 3/2b$$

so that for a Brooks coil, $\delta = 1$, $\xi = 3$ and the energy stored in it E_B is given by

$$E_B = 3.028 \times 10^{-8}\, V^{5/3} j^2 \qquad ...(16.35)$$

For a cylindrical coil, other than Brooks, the energy is given as a function E_B (Eq. 16.35) by a factor F less than 1.0.

$$F = E/E_B \qquad ...(16.36)$$

F is a function of ξ and δ of the particular coil.

A number of problems are associated with superconductivity storage. They include the operation and maintenance of a cryogenic (*i.e.,* Refrigeration) plant for producing the liquid helium required for the very low superconductivity temperatures.

Furthermore, special structures would be needed to withstand the strong magnetic field of an SMES unit.

The main mechanical design problem in magnetic energy storage arises because of the need of a very large structural mass to contain the magnetic field energy. This causes a large radial outward force from the solenoid. The mass is proportional to the material density and the stored energy, and is inversely proportional to the stresses. Such mass, if made of stainless steel, would amount to about 160 kg/kWh and result in unacceptable costs. This consideration led to the selection of bed rock as the structural material with an excavated circular tunnel that would bear the radial outward force and transmit to the surrounding bed rock.

The energy transfer between the three phase a.c. from the grid to the d.c. magnet is accomplished by an a.c.—d.c. power converter and inductor using a Graetz thyristor bridge circuitry. The turn around efficiency (charge-discharge) of this circuitry is said to be greater than 95%.

It appears at present that superconductivity storage would, at best, be economical only in large installations on the whole SMES is rated as being a promising but long term prospect for energy storage.

Superconducting materials are classified as Type I and Type II. Those of type I are mostly pure metals; they have relatively low critical currents and magnetic fields and are of little practical value. Type II superconductors, on the other hand, which are either compounds of solid-solution alloys of two or more metals, have certain characteristics that make them potentially useful materials. As a general rule, Type II superconductors retain their superconductivity at high magnetic fields and high currents. However, the critical temperature above which superconductivity ceases is decreased as the magnetic field and the current are increased. Of the high current Type II superconductors generally known two of which are available commercially, are of special interest. One is a solid solution alloy of 30 to 40 atom % of niobium (Nb) with titanium (Ti), represented by Nb-Ti; its transition temperature is about 10 K. The other is a compound of niobium and tin (Sn) formula Nb_3Sn, with a transition temperature of 18 K. Not only temperature, which is advantageous from the cooling stand-point, but the critical current and magnetic field are higher for Nb_3Sn than for Nb-Ti. On the other hand, Nb_3Sn is a brittle material that have proved difficult to fabricate, because direct wire drawing is impracticable. The problem, has been solved by producing filaments of niobium, which is ductile material, and converting them into Nb_3Sn. For example, rods of niobium are embedded in a copper tin alloy and drawn into wires; upon heating, tin diffuses into the niobium filaments to form thin surface layers of Nb_3Sn. Alternatively, filaments of niobium in copper are coated with tin and heated; the tin diffuses through the copper and converts the niobium into Nb_3Sn. The outer layer of tin, which may be undesirable, is then removed by acid. Superconducting taps (ribbons) required to certain applications are fabricated in a somewhat analogous manner by rolling sheets rather than drawing wires.

Much of the activity of magnetic energy storage for utility use has taken place in U.S.A. and Japan. A 10 MWh unit is planned to be built in Japan.

16.7. Thermal Energy Storage

The need for storing thermal (or heat) energy arises in several situations, for example, in connection with the applications of solar and wind energy. The storage methods in common use are described in the sections dealing specifically with these applications. The discussion here will deal, therefore, with general principles of heat storage as either *sensible heat* or *latent heat.*

Energy can be stored by heating, melting or vaporization of materials, and the energy becomes available as heat when the process is revered. Storage by causing a material to rise in temperature is called *sensible heat storage.* Storage by phase change, the transition from solid to liquid or from liquid to vapour is another mode of thermal storage, known as *latent heat storage,* in which no temperature change is involved. It is possible for both sensible and latent heat storage to occur in the same material, as when solid is heated, then melted, then raised further in temperature.

Thermal energy storage is essential for both domestic water and space heating applications and for the high temperature storage systems needed for thermal power applications. Storage is also required for process industries and horticultural. The choice of the storage material depends on the particular application and for many domestic applications, water and/or rock storage systems have been developed. Water and rock are typical examples of material which store energy as specific heat (sensible heat), but their use is limited by their comparatively low specific heats. The heat of fusion (latent heat) which is involved when a substance changes state from a solid to a liquid provides an attractive method of storing a given amount of heat within a much smaller volume. Glauber's salt ($Na_2SO_4.10H_2O$) is the least expensive and most readily available salt hydrate. For high temperature storage in the order of 200 to 300°C, other salts have been used and the heat of hydration of inorganic oxides principally MgO and CaO.

16.7.1. Sensible Heat Storage

Sensible energy storage is the first of the thermal energy storage systems to be discussed. In general, thermal energy storage systems can operate at many desired temperature levels depending upon the use and choice of the system and material, ranging from refrigeration temperatures to 1250°C. They have found wide use in many industrial applications, such as in the manufacture of cement, iron and steel, glass, aluminium, paper, plastics and rubber, and in food processing.

Sensible heat refers to thermal energy that results in an increase of temperature when added to a material (or decrease of temperature when taken from it). For example, water at temperatures above the freezing point and below its boiling point can store energy as sensible heat. So also can other liquids, air and solids provided they do not change their form (*e.g.,* by freezing, melting or boiling). Water or ceramic bricks are used for heat storage on a large scale in industry, and either water or a combination of air and rock provide storage for heat derived from solar energy. Thermal energy storage as sensible heat in high temperature steam can provide utilities with a means for generating electric power to satisfy peak demands.

If the temperature is too high for the use of water without pressurization, special hydrocarbon oils with high boiling points, either alone or mixed with rock, can serve to store sensible heat. Molten salt mixtures can be used for heat storage at very high temperatures. One such mixture is Du Pont's HITEC, containing sodium nitrite and sodium and potassium nitrates, it melts at 142°C and can be used to store sensible heat upto about 550°C. Mixtures of sodium and potassium nitrates have been proposed for somewhat higher temperatures. For temperatures upto about 815°C, molten fluorides (*e.g.*, lithium fluorides and mixture of sodium and magnesium fluorides and of sodium and zinc fluorides) are possibilities.

The ability of store thermal energy (or the storage density), in a given container of volume V is,

$$\frac{Q_s}{V} = \rho\, C_P\, \Delta T \qquad ...(16.37)$$

where, Q_S = Energy stored or thermal energy

ρ = Density of the storage medium

C_p = Specific heat of the material

ΔT = Difference in temperature or temperature rise.

Thus the ability to store heat depends upon the product ρC_p, water has the highest value, but other materials have low values. Materials which are generally used for this type of storage are: (*i*) water, (*ii*) rock, gravel or crushed stone, (*iii*) iron shot, (*iv*) concrete (*v*) refractory materials like magnesium oxide, aluminium oxide (alumina) and silicon oxides.

Thus any of the following devices could be employed as sensible energy storage:

- Pressurized-water storage
- Packed solid beds
- Refractory materials
- Organic solid beds.

Properties of some sensible heat storage materials are given in Table 16.2.

Table 16.2. Properties of some sensible heat storage materials

Property	*Water*	*Rocks*	*Material Pebbles*	*MgO*	*Al_2O_3*	*SiO_2*
Density, ρ (kg/m^3)	958.4	2245	1350	3575	4000	2600
Specific heat, C_p (kJ/kg. K)	4.22	0.81	0.90	1.06	1.02	1.26
ρCp (kJ/m^3 . K)	4044	1818	1215	3790	4080	3276
Thermal conductivity (W/m°K)	0.683	0.13	0.85	10.5	6.3	2.3

Water storage. The most common heat transfer fluid for a solar system is water, and the easiest way to store thermal energy is by storing the water directly in a well insulated tank. The optimum tank size for flat-plate collector system is usually about 2 gal/ft^2 (70 kg/m^2). Water has the following characteristics for storage medium:

(*i*) It is an inexpensive, readily available and useful material to store sensible heat.

(*ii*) It has high thermal storage capacity.

(*iii*) Energy addition and removal from this type of storage is done by medium itself, thus eliminating the temperature drop between transport fluid and storage medium.

(*iv*) Pumping cost is small.

(*v*) In case of solar energy storage, collectors and storage units may be operated by natural circulation in domestic solar water heater application, or forced circulation may be used. A schematic diagram of a power plant with pressurized-water sensible energy storage is given in Fig. 16.7. The system can have a nuclear reactor or a fossil fueled furnace as a primary heat source. The base-load portion of the plant is capable of supplying more steam than needed during periods of low demand. The excess steam is extracted at high pressure *via* turbine extraction (as in feed water heating) during these periods of low demand. The extracted steam is fed to steam accumulators and mixed with water, thus producing saturated pressurized water. The accumulators are later discharged through a small peaking turbine during periods of high demand. Discharge continues until a low specified pressure is reached in the accumulators. It is observed that there is low and varying steam temperature entering the peaking turbine.

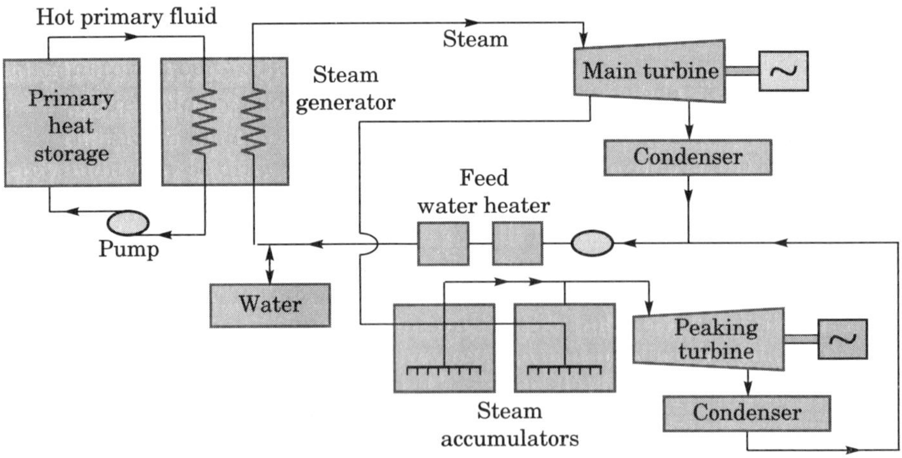

Fig. 16.7. Schematic flow diagram of a power-plant with a pressurized-water sensible energy storage system.

Typical values of accumulator high and low pressures are 20 bar, corresponding to a saturation temperature of about 212°C, and 2 bar, corresponding to a saturation temperature of about 120°C. It is to be noted that while this steam involves steam condensing in water during accumulator charge and re-evaporating during discharge, the storage medium is the pressurized water in the accumulator and operates over a relatively wide temperature range. Hence it is an example of sensible rather than latent heat storage.

The storage density of the thermal energy utilized in the peaking turbine per unit volume of the higher pressure saturated water is given by

$$\text{Storage density} = \frac{h_{f_1} - h_{f_2}}{v_{f_1}} \qquad \text{...(16.38)}$$

where, v_f = specific volume of saturated water

h_f = enthalpy of saturated water

Subscripts 1 and 2 refer to the stored and emptied pressures respectively.

Pebble bed storages

Pebble bed storages have been used in the solar houses or with hot air collector system. For sensible heat storage with air as the energy transport mechanism, rock, gravel, or crushed stone in a bin has the advantage of providing a large cheap heat transfer surface. Its thermal capacity, however is only about half that of water, and the bin volume will be about 3 times the volume of a water tank that is heated over the same temperature interval. Water is superior because of its lower material cost and lower volume required per unit of energy stored. (See Art. 4.2 for further details).

Energy storage is required in *solar-thermal electric power plants* (and solar process heat application) because of normally variable solar insolation, non-solar periods, and abrupt insolation changes in inclement weather. Conventional (fossil-fired) backup system may, of course, be used during such periods, as in hybrid systems but thermal storage is another option that must be used in pure solar systems. There are two types of thermal storage that are considered with solar systems. These are:

(*a*) Single-tank, or thermocline and

(*b*) Dual tank, or hot cold systems.

The single-tank or thermocline, storage system is already described (see Art. 4.2). The complete schematic of power plant storage system is shown in Fig. 16.8 in which storage takes place by circulating some of the hot primary coolant through the storage medium and returning the cooled primary coolant from the bottom of the storage tank back to the receiver for reheating.

Heat extraction during times of need is accomplished by reversing the process. Cold primary coolant from the power plant is heated by the storage medium, drawn from the top of the tank to the power system, and then returned to the bottom of the tank. The thermal gradient of hot at top, cold at bottom maintains stratification, allowing the hot fluid to remain afloat on the top, and gives the tank thermal stability. In addition, a solid storage medium of low thermal conductivity and high volumetric heat capacity, such as rock, is used to help impede mixing of hot and cold fluids. A solid storage medium is a necessity in case the primary coolant is a gas (such as air or helium) because of the low heat capacity of gases. A porous solid makes a good storage medium for gases. A variation of the single tank system is one in which a separate storage fluid, other than the primary coolant, may be used. This would be a liquid of low thermal conductivity and high volumetric heat capacity, usually an oil. It may be used alone as a single medium or be assisted by a solid storage medium, such as rock in a *dual medium* system as above, to reduce the volume of costly storage liquid and impede its mixing. Heat would be added to and extracted from the primary coolant *via* intermediate heat exchangers.

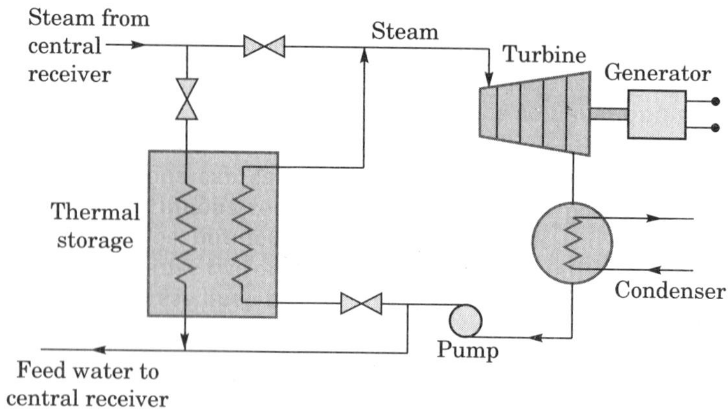

Fig. 16.8. Electric power generation using thermal storage.

The *Solar one power plant* uses a dual-medium thermocline storage system, shown in Fig. 16.9. The liquid is oil. The solid is composed of 2.5 cm diameter gravel plus sand. The oil is uniformly distributed over the rock bed and extracted uniformly from it by diffuser manifolds. The tank is large, being about 15 m high and more than that in diameter.

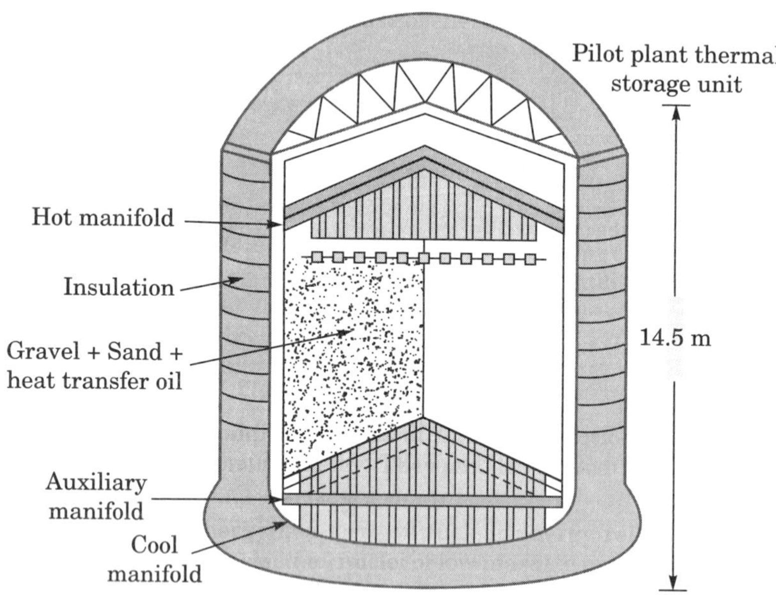

Fig. 16.9. Thermocline storage system.

The *dual tank,* or hot-cold, system uses two tanks. The fluid, a liquid, is, unlike the stratified single tank or thermocline system, at one temperature, hot in one tank, cold in the other. The amount of sensible energy stored varies by varying the level of the fluids in a well insulated tanks Fig. 16.10. Thus during storage, cold is drawn from the cold tank, heated, and added to the hot tank. During extraction the operation is reversed. This dual-tank technology is borrowed from refinery work. It is suitable only for liquid coolants such as sodium molten salt, or oil. The hot tanks used with sodium and molten salt need be made of high alloy material such as austenitic stainless steel.

Thermal energy storage densities vary widely among the different storage media. Some typical values for the temperature range 290 to 565°C are given below.

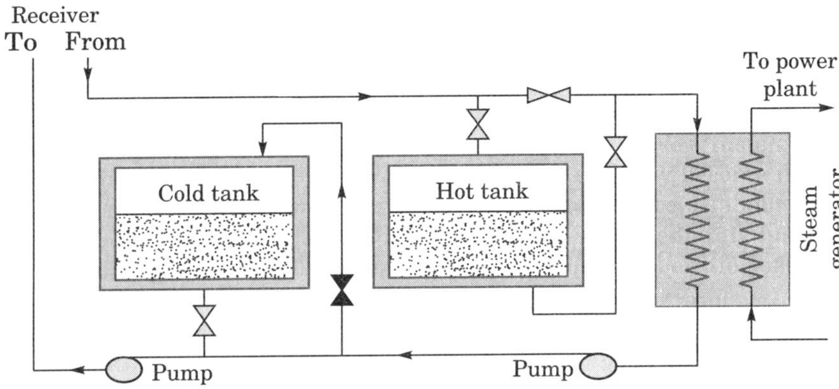

Fig. 16.10. Schematic of a dual tank, or hot-cold thermal storage system.

Table 16.3

Medium	*Energy Storage density (MWh/m^3)*
Sodium	0.08
Rock (25% void fraction)	0.15
Molten salt	0.22

The solar one pilot plant has an oil-rock storage medium that operates at the much lower temperature range of 220 to 300°C and has an energy storage density of 0.05 MWh/m^3.

16.7.2. Latent Heat Energy Storage or phase change energy (storage)

Introduction. Latent heat is thermal energy that is stored in (and can be removed from) a substance or mixture when it goes a change of phase (*e.g.,* of physical form) while the temperature remains unchanged. The heat that can be stored per unit mass (or volume) in this manner is usually several times greater than for sensible heat storage.

The phase change from solid to liquid, taking place at the melting point of the solid, is accompanied by the absorption of latent heat without a change in the temperature. The heat is recovered when the process is reversed (*i.e.,* the liquid is converted to solid) at the same temperature. Hence solid and liquid phases of the same material when present together can store thermal energy as latent heat. Saturated steam (*i.e.,* steam and moisture) and freezing water (*i.e.,* water and ice) can be used for latent heat storage.

Another type of phase change, involving a change in chemical composition, that has been used for latent heat storage, is the transition between solid sodium sulphate decahydrate ($Na_2SO_4.10H_2O$) on the one hand and solid anhydrous sodium sulfate (Na_2SO_4) and an aqueous solution, on the other hand. At a temperature of 32°C, heat is stored when the $Na_2SO_4.10H_2O$ phase changes to Na_2SO_4 phase (plus solution) and can be recovered when the process is reversed. A somewhat similar phase transition between sodium thiosulfate pentahydrate ($Na_2S_2O_3 . 5H_2O$) and the anhydrous salt ($Na_2S_2O_3$) plus solution occurs at 48°C. It has been considered for latent heat storage of solar energy for space heating.

Advantages of phase change energy storage

(*i*) The storage density here is equal to the product of the latent heat of fusion (or vaporization) times the density of the storage material (L. ρ). It is greater than that in sensible heat storage because the latent heats are much larger than the specific heats of the single phases of the materials.

(*ii*) The system has the additional advantage of operating at essentially constant temperature with low volume changes during phase changes.

(*iii*) It also has the advantage of a wide choice of materials with different fusion and evaporation temperatures, which allows a choice of operating temperatures and the ability to generate high temperature steam for the peaking unit. Some sensible heat storage may be added to latent heat storage by further raising the temperature of the resulting molten solid or vapour.

Latent heat energy storage is not till yet, considered a simple, operationally reliable solution to the problem of energy storage in electric-generating power plants. It is, however, included here as a potential solution along with the problems that must be overcome if such a solution is to become a viable one. Although little has been done on the application of latent heat energy storage to large power plants, much work has been done on its we for residential and solar heating applications using fused salts that are available for high and low-temperature operating ranges.

Materials for phase change energy storage. Materials that undergo a change of phase in a suitable temperature range may be useful for energy storage if the following criteria can be satisfied:

(*i*) The phase change must be accompanied by a high latent heat effect.

(*ii*) The phase change must be reversible over a many of cycles without serious degradation.

(*iii*) The phase changes must occur with limited supercooling.

(*iv*) Means must be available to contain the material and transfer heat into it and out of it.

(*v*) The cost of materials and its containers must be reasonable.

(*vi*) Its phase change must occur close to its actual melting temperature.

(*vii*) The phase change must have a high latent heat effect, that is, it must store large quantities of heat.

(*viii*) The material must be available in large quantities.

(*ix*) The preparation of phase changing material for use must be relatively simple.

(*x*) The material must be harmless (non toxic, non inflammable, non combustible, non corrosive).

(*xi*) It should have low cost.

If these criteria can be met, phase change energy storage systems can have high capacities (relative to energy storage in sensible heat type systems) when operated over small temperature ranges, with substantial reductions in volume and weight.

No material meets all these requirements but some fluoride salts meet some of them. One of the salts considered most suitable for latent heat storage is the

70% NaF—30% FeF_2 eutectic salt, which has a fusion temperature of about 680°C and potentially possesses the high storage energy density of any thermal energy storage material about 1500 MJ/m^3. $ZnCl_2$ is another material with a fusion temperature of about 370°C and a potential storage energy density of about 400 MJ/m^3.

There are several materials that a undergo a change of phase. Glauber's salt (Na_2SO_4. $10H_2O$), water, $Fe(NO_3)_2 6H_2O$ and salt eutectics are mostly used.

***(i)* Glauber's salt ($Na_2SO_4.10H_2O$)** or sodium sulphate decahydrate. Phase changes from solid to liquid requires less energy than those from liquid to gas, but some solid to liquid changes still provide useful amounts of storage potential. Phase change materials involving water of dehydration have long been experimented with *solar energy*. One of these is Glauber's salt, which decomposed at about 32°C to give a solution plus Na_2SO_4, with a heat of fusion of 243 kJ per kg (56 kcal/kg) and has been proposed mainly for storing domestic heat. The reaction is

$$Na_2SO_4.10H_2O + \text{Energy} \rightleftharpoons Na_2SO_4 + 10H_2O \quad ...(16.39)$$

Energy storage is accomplished by the reaction proceeding from left to right on addition of heat. Energy extraction from storage is the reverse procedure, with the reaction proceeding from right to left and the thermal effects reversed. It has been found that the performance decades on repeated cycling, with the thermal capacity of the system reduced, Na_2SO_4. $10H_2O$ has an incongruent melting point, and as its temperature increases beyond the melting point, it separates into liquid (solution) phase and solid Na_2SO_4. Since the density of salt is higher than the density of the solution, a phase separation occurs. Attempts have been made to use gels or other agents to avoid phase separation.

***(ii)* Water.** The largest quantity is for change of phase from water to steam storing 548 kcal/kg or less depending on the temperature at which the phase changes occur. Steam, however is a difficult medium to contain and attempts to store steam in gigantic pressure vessels have been made and abandoned as impractical and dangerous. In general, it is preferable to work with a phase change from solid to liquid, rather than from liquid to gas. The transition from ice to water is an excellent historical example that has been used for energy storage for centuries.

***(iii)* $Fe(NO_3)_2$.$6H_2O$.** This compound has congruent melting point and melts in the same manner as pure compounds.

***(iv)* Organic Compounds.** Organic substances serve as heat storage materials due to the desirable properties they possess in comparison with inorganic compounds. Some of these advantages include their ability to melt congruently, their selfnucleating properties, their capability to freeze without super cooling and their compatibility with conventional materials of construction. *Paraffin* and fatty *acids,* are the two candidates organic substances qualified as heat of fusion materials. Paraffins as we normally understand them are substances, having a waxy consistency at room temperature. Paraffins are, in a chemical sense, a complicated mixture of several compounds. They possess generally high heats of fusion and are available in large temperature ranges. They are known to freeze without supercooling. Fatty acids are characterized by $CH_3(CH_2)_{2n}$ COOH with heat of fusion values comparable to that of paraffins. Some fatty acids of interest to *low temperature latent* heat storage are listed in Table (16.4). Fatty acids are known to possess a reproducible melting and freezing behaviour and freeze with little or no supercooling. Hence they qualify as good phase change thermal energy storage materials with a major drawback of higher cost (about 2 to 2.5 times greater than that of paraffins).

Table 16.4. Physical Properties of Fatty Acids

Material	*Melting point °C*	*Heat of Fusion kJ/kg*	*Density gm/cc*	*Specific heat kJ/kgK*	*Thermal conductivity W/mK*
Caprillic acid	16.3	149	1.033 (10°C)	—	0.148 (20°C)
Capric acid	31.5	136	0.886 (40°C)	—	0.149 (40°C)
Lauric acid	42.44	178	0.870 (50°C)	1.6	0.147 (50°C)
Myristic acid	54	187	0.884 (80°C)	1.6	—
Palmitic acid	63	187	0.847 (80°C)	—	0.165 (70°C)
Stearic acid	70	203	—	—	0.172 (70°C)

(*v*) **Salt Eutectics.** Salt Eutectics offer the possibility of lowering the temperature of a phase change to below the normal melting points of any of the compounds forming the mixture. The term eutectic is generally used to describe a mixture of chemicals having a desired temperature behaviour, strictly, however the term refers to that particular mixture where the melting point is lowest. Some of the lowest melting point eutectics are useful as heat transfer media. Possible eutectics are:

(*i*) $CaCl_2$—$MgCl_2$—H_2O

(*ii*) $Mg(NO_3)_2$ $6H_2O$—$Al(NO_3)_3$, $9H_2O$

(*iii*) Acetamide—stearic acid

(*iv*) Urea NH_4NO_3.

The Table (16.5) provides information about eutectic mixtures of possible interest in thermal energy storage (mainly in solar energy storage system).

Table 16.5

Composition wt %	*Melting point °C*	*Latent heat kcal/kg/(kJ/kg)*
$CaCl_2$—$MgCl_2$. H_2O (41.10.49)	25	41.7 (175)
Mg $(NO_3)_2$ $6H_2O$. Al $(NO_3)_3$.$9H_2O$ (53.47)	61	35.4 (148)
Acetamide—stearic acid (17.83)	65	52 (218)
Urea NH_4NO_3 (45.3—54.7)	49	41 (172)

Heat transfer to and from a phase change material must be given careful consideration. The material must be contained in such a way that the heat can be transferred to and from the material with a minimum temperature drop. This has been done experimentally by placing the material in small containers (cylindrical canes, tubes or trays) with the containers placed in bins or ducts. The heat transfer fluid usually air is circulated over the container as in a packed bed.

The heat transfer problem external to the containers is similar to a packed bed internally, two additional phenomenon must be considered. First the latent heat must be considered, which effectively adds a high specific heat over a very small temperature range; and second, the thermal resistance to heat transfer within the material is variable with the degree of solidification and weather heating or cooling of the material is occurring. As the heat is extracted from a phase change material, crystallization will occur at the walls and then progressively inward into the material; at the end of the crystallization, heat must be transferred across layers of solid to the container walls. As a solidification material is heated, melting occurs first at the walls and then inward towards the centre of the container. Other practical factors to be considered with phase change storage are corrosion, side reaction, vapour pressure, toxicity and cost.

High temperature latent heat storage. We have discussed so far phase change material used in *low temperature storage,* such as organic compounds, inorganic salts and compounds. Table 16.6 gives a list of thermal storage materials for the temperature range of 200°C– 450°C.

Table 16.6

Material	*Melting point °C*	*Density gm / cc*	*Latent heat kJ / kg*
$ZnCl_2$	283	2.91	168.8
$Na(OH)_3$	307	2.26	171.8
Na(OH)	319	2.15	159.0
B_2O_3	460	1.84	330.4
KOH—$ZnCl_2$ (64 : 36)	190	2.06	230.0
KCl—$MgCl_2$—NaCl (20 : 50 : 30)	396	2.25	412
$MgCl_2$ NaCl (38 : 62)	435	2.16	328.0
KCl—$ZnCl_2$	235	2.48	198.0

Heat of fusion system using inorganic salt mixtures have a distinct advantage for storing energy at high temperatures, since they can reduce storage volumes and produce more energy at constant temperatures. A system was developed at the Martin Marietta Denver Aerospace Center, Colorado in 1982, to provide low cost thermal energy storage using molten salts. The system used molten nitrate salt (60% $NaNO_3$.40% KNO_3 by weight) and a dual tank approach with one to store the 'hot salt' at 566°C and the second to store the 'cold' at 560°C. The hot salt is heated to 566°C in the solar receiver and pumped to the hot storage tank. This energy is used to produce steam at 538°C, which can be used to drive a turbine or used as industrial process heat. The salt exists in the heat exchanger at a temperature of 289°C, which is sufficiently above the salt freezing point to prevent any solidification. The cold salt returns to the cold tank where it is stored for recycling. Preliminary experimental results indicate that externally insulated tanks are practical for this type of energy storage and the overall concept itself is appropriate for solar thermal central receiver system.

Other materials for energy storage being suggested are silicon germanium and sulfides of germanium. They have high heats of fusion, and like water, they expand upon freezing, so that they tend to float upon solidification, which has advantageous in heat transfer. Silicon and germanium, however, have fusion temperature that are too high for power plant operation and are very reactive. Germanium sulfides have usable fusion temperatures but tend to solidify to a glassy consistency rather than crystallizing, thus posing an undesirable heat-transfer barrier.

16.8. Biological Storage

The growth of plants by photosynthesis, and the storage of the solar energy in oxygen, is discussed in chapter 7. This energy is released in the combustion of biological and fossil fuel material. We may therefore consider the energy as stored in the material itself, although strictly this is not correct. The heats of combustion of many of these materials are listed in *Appendix E Table E-5.* The many chemical paths for producing bio fuels are described in chapter 7.

Some of the bio-fuels are liquids and gases that may be used in internal combustion engines, and are therefore important for transport in place of conventional petroleum fuels. The generation of electricity by such fuels in small diesel engines is also important.

16.9. Distribution of Energy

16.9.1. Introduction

Table 16.7 sets out the major methods by which commercial energy is distributed to its point of use.

The methods are categorized by whether they involve continuous flow (*e.g.*, pipelines) or batch movement (*e.g.*, lorries), and by their suitability for use over long, medium or short distances. Also shown are numerical values representing the energy flows in a typical unit of the type in question (*e.g.*, an individual gas pipe line). Although these flows vary greatly in scale, the energy flow through each system *per end-user day* is in practice remarkably similar, *i.e.* ~ 10 MJ/user/day. ≈ 100 W/user.

For renewable energy supplies short distance distribution is more important than long distance, because the sources themselves are usually wide spread and of low intensity. In particular, short haul carriage of biomass, and movement of heat (*e.g.* through building), are more significant. The renewable energy supplies that are mechanical in origin, *e.g.*, hydro, wave and wind, are usually best distributed by electricity. In this way electricity is a carrier or vector of energy, and not necessarily the main end use requirement. Movement of gas, perhaps on a large scale, would be required if hydrogen becomes a common store of energy.

16.9.2. Gas Pipelines

In the pipelines usually used for carrying fuel gases, the flow is turbulent but not supersonic. In this case the pressure gradient along a small length of pipe of diameter D is given as (neglecting compressibility of the gas):

$$\frac{dp}{dx} = -2f\frac{\rho u^2}{D} \quad \text{...(16.40)}$$

where, f = frict of the fluid = 16 $\gamma/(uD)$ (laminar)

γ = kinematic viscosity

ρ = density of the fluid

u = mean speed of the fluid

In a steady flow of gas, both ρ and u vary along the length of a long pipe, but the mass flow rate

$$\dot{m} = \rho Au \quad \text{...(16.41)}$$

is constant. $A = \pi/4\, D^2$ is the cross-sectional area.

Also the density ρ varies with the pressure p:

$$\rho = (RT/M)\, \rho \equiv K.\, \rho \quad \text{...(16.42)}$$

with K more or less constant for a given gas. R *is* the universal gas constant, T is absolute temperature, and M is the gram molecular mass/1000, so as to have units of kg/mol. If the Reynolds number $\left(R_e = \frac{uD}{4}\right)$ is large f will not vary appreciably along the pipe, and we can integrate eq. (16.40) between stations x_1 and x_2 to

$$p_1^2 - p_2^2 = \frac{64fRT\dot{m}^2\,(x_2 - x_1)}{\pi^2\, MD^5} \quad \text{...(16.43)}$$

Table 16.7. Summary of Major Means And Flows For Distributing Energy

	Long distance (> 1000 km)	*Flow*		*Medium distance (1–1000 km)*	*Flow*		*Short distance 10 m–1 km*	*Flow*	
		(MW per unit)	*(MJ per user per day)*		*(MW per unit)*	*(MJ per user per day)*		*(MW per unit)*	*(MJ per user per day)*
Continuous	Oil pipe line	15000	60	Oil pipe line	10,000	60			
	Gas pipe line (high pressure)	500	20	Gas pipe line (high pressure)	500	20	Gas pipe line (low pressure)	–	7
				Electricity (high voltage line)	100	20	Electricity (low voltage)	–	10
							Heat in gas, vapour or liquid		
Batch	Oil tanker	1200		Oil (or substitute *e.g.* ethanol)	–	–			
				In vehicle					
				As cargo	200	–			
				As fuel in tank	–	28			
	Coal in ships			Coal in trains			Biomass on lorry	15	–
				Biomass on lorry	15	–			
							Wood 'by hand'	0.03	15

Thus the pressure falls off rapidly along the length of the pipe, and frequent pumping (recompression) stations are needed to maintain the flow.

According to Eqn. (16.43), larger pipes (bigger D) will require much less pumping. The most economical balance between pipe size (capital cost) and pump separation (running cost) depends largely on the accessibility of the pipe. Construction costs are very variable.

The compressibility of the gas offers another benefit. The pipe itself can be used as a temporary store, simply by pumping gas in faster than it is taken out so that compressed gas accumulates in the pipe, such storage is very substantial.

16.9.3. Electricity transmission

Consider two alternative systems transmitting the same useful power p to a load R_L at different voltage V_1, V_2 in wire of the same resistance per unit length Fig. 16.11. The corresponding currents are $I_1 = p/v_1$, $I_2 = \frac{p}{v_2}$, and therefore the ratio of power lost in the two systems is

$$\frac{p_1^1}{p_2^1} = \frac{I_1^2 R_w}{I_2^2 R_w} = \left(\frac{P}{V_1}\right)^2 \left(\frac{V_2}{P}\right)^2 \qquad ...(16.44)$$

$$= \frac{V_2^2}{V_1^2} \qquad ...(16.45)$$

Significantly less power is dissipated in the system working at high voltage. The low voltage system can have the same loss as the high voltage system only with thick, and therefore expensive, cable. For instance if electricity is to be transmitted at domestic main voltage (~ 110 V or ~ 220 V), the cost of cable becomes prohibitive for distance greater than about 200 m. The difficulty becomes even greater at very low voltage ~ 12 V. These factors govern the design of all electrical power networks. Normal rotating generators work best at voltages ≤ 10 kV. The case with which A.C. can be transformed up and down in voltage explain why A.C. transmission systems are now standard for all but the smallest networks. As indicated in Fig. 16.11 (*b*), power is generated at low voltage, stepped up for transmission, and then down again to a safer level for consumption. The transmission voltage is limited by dielectric breakdown of the air around the overhead cables and by the insulation of the cables from the metal towers (at earth potential). Improvements in insulation have allowed transmission voltages for long lines to increase from 6000 V in the year 1900 to over 200000 V today.

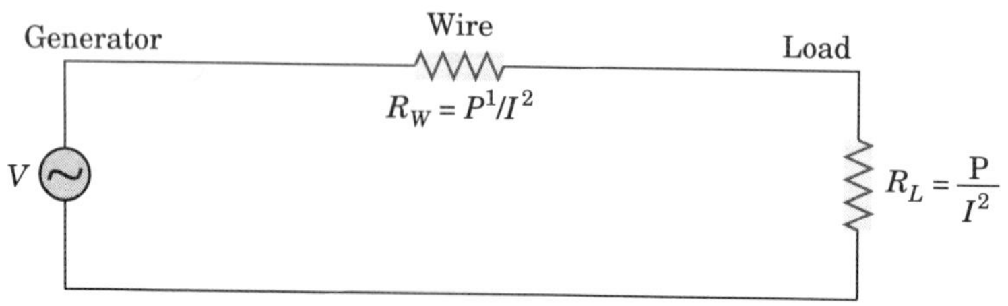

(*a*)

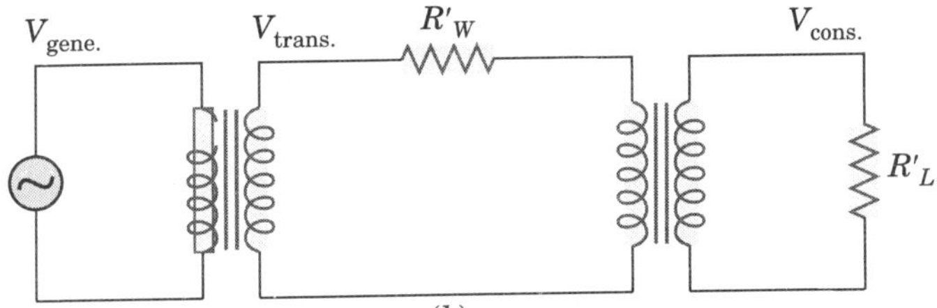

(b)

Fig. 16.11. Electrical transmission.
(a) Power transmission to a load of resistance R_L, through a wire of finite resistance R_W.
(b) More likely realization, with generated voltage transformed up for transmission and then down for consumption.

Grids using even higher voltages are now being constructed, but will probably make only a marginal improvement in costs. The same is true for very high voltage D.C. systems, which have certain advantages, in transmission but require more expensive step-up equipment. Superconducting lines of zero resistance are attractive but can operate only at very low temperatures (≤ – 260°C). Maintaining such low temperatures is difficult over large distances and such lines are not yet a commercial proposition.

Electrical power companies usually link power stations into a common national network (grid). Thus if one station shuts down for maintenance, repair or economy, its demand can be met by the others in the network. Power from variable renewable sources can be fed directly into such a grid. Since all major grid systems can cater for rapid demand fluctuations of ~ 20%, it is also possible for the same systems to cater for rapid supply fluctuations of ~ 20%. Thus the inclusion of unpredictable supply *e.g.* wind power is possible. If hydro-power is available, then rapid control of the grid is easy whatever the supply.

16.9.4. Batch Transport

Biomass and coal can be carried in suitable containers by hand, road, rail or boat. However, the low density and bulky nature of most biomass means that it is rarely economic to distribute it over large distances (> 1000 km). Even over medium distances (1 – 1000 km), it is unlikely to be economic to distribute biomass for its energy value alone. Economic, and ecological, use of biomass energy sources therefore depends on harnessing energy from a flow of biomass which is already occurring for some other purpose, *e.g.* the extraction of sugar from sugarcane often which the spent cane is used to fuel the factory. In this case the transport of fuel may be regarded as 'free' or nearly so. Biofuels can however be transported over medium to long distances after conversion from raw biomass (*e.g.*, by pyrolysis).

Alternatively, the biomass can be used very close to its source. This is usually the case with firewood, which remains a fuel of major importance in most developing countries.

16.9.5. Heat

The movement of heat within a building through hot air 'ducts' (whether specially built as such or not) and through steam pipes is a major means of distributing energy over short distances. This is especially true in cold climates, where space heating dominates energy use. Heat transport by steam is also used in many industrial processes. Heat transport is limited to short distances by heat losses from the sides of the duct. The major resistance to heat loss is by conduction through the insulation, the heat loss is given by relation

$$Q_{loss} = -kA\ \Delta t/\Delta x \qquad ...(16.46)$$

where, k is conductivity of insulation material normally mineral wool

A = area of cross-section of pipe which delivers heat

ΔT = temp. difference of steam and ambient

Δx = insulation thickness.

The loss is independent of the flow rate in the pipe. Obviously very large heat flows ($\geq$ 10 MW) are needed if the losses are to be proportionately small. *District heating* of this kind operating successfully in many cities.

The *heat pipe* offers another way to move large quantities of heat over very short distances. It is a tube, containing vapour with the condensate recycled by a wick, and offers an effective conductivity much greater than that of copper. Usually, heat pipe is a means for transporting heat efficiently from a source to a cooler receiver by utilizing the latent heat of vaporization of a liquid. A different kind of heat pipe is described at the end of this section. In essence, a heat pipe is a closed space containing a suitable working liquid and its vapour. One part of the space is in contact with the heat source and another with the cooler material to which the heat is to be transported. The interior wall of the space is lined with a porous material called a *wick.* For high temperature operation, the wick can be made of several layers of a woven metal wire mesh. A simple form of heat pipe is shown in Fig. 16.12, but more complex arrangements which satisfy the basic conditions outlined above are possible.

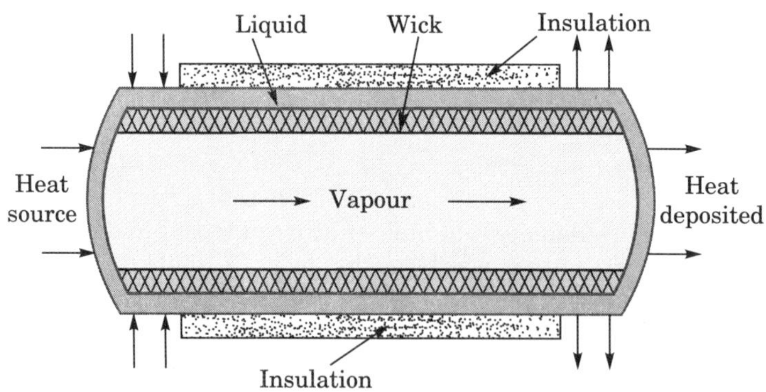

Fig. 16.12. Simple heat pipe.

In the hotter (*i.e.,* heat source) part of the heat pipe, the working fluid is vaporized, thereby taking up the latent heat of vaporization. The vapour diffuses toward the cooler region because the pressure is lower there and condenses to liquid. In doing so, it deposits the heat of vaporization taken up from the source. The liquid is returned to the heat source region by capillary action of the wick. There is thus a continuous movement of vapor from the heat source to the receiver and of condensed liquid back to the source, accompanied by the transfer of heat.

Since the working liquid is returned to the heated region by capillary action, the heat source maybe located above the heat receiver. However, if the receiver is at a higher level than the heat source, the return of the liquid is assisted by gravity. The vapour always moves from the higher pressure (*i.e.,* the heat source) region to the lower pressure (*i.e.,* heat receiver) region regardless of their relative positions.

A special feature of the heat pipe is its ability to transport heat at a high rate even when there is only a small temperature difference between the source and receiver. For other, methods of heat transport (*i.e.*, conduction, convection, and radiation), the rate is highly dependent on the temperature difference, but in a heat pipe, the important consideration is the heat of vaporization of the liquid. Another advantage is that a heat pipe can serve as a concentrator, heat can be taken up by vaporization from a large area and deposited during condensation on a small area.

For operation of a heat pipe at high temperatures, sodium, melting at 97°C, has been used as the working fluid/liquid. Its large heat of vaporization and good thermal conductivity make it particularly suitable for this purpose. A sodium heat pipe might be used to transfer heat from a source possibly stored heat, to a *Stirling engine*.

16.9.6. Chemical Heat Pipe

The purpose of the chemical heat pipe concept under development is to transfer heat from a source to a distant receiver, up to 160 km (100 miles) away. The operating principle is quite different from that of the physical heat pipe described above.

The chemical heat pipe is based on a reversible chemical reaction, that is, a reaction that proceeds predominantly in one direction or in the opposite direction according to the conditions (*e.g.*, temperature). Heat is supplied from a source (fossil fuel, nuclear, or solar) to substances that absorb the heat and undergo a chemical reaction. The products are then pumped through a pipe line to a distant receiver where the conditions are changed so that the chemical reaction is reversed and heat is released. The original interacting materials are recovered and returned by another pipeline to the heat source for reuse. The principles are the same as described under Reversible Chemical Reactions in the section on *Energy Storage*.

Energy Conservation

17.1. An Economic Concept of Energy

It is impossible to forecast future economic trends accurately except in one respect, if a non-renewable asset is being used up, its price is bound to go up relative to everything else.

The vast bulk of the energy used in the world today is in the form of non-renewable oil, natural gas and coal. These resources were laid down many millions of years ago and are at present being consumed at the rate of almost 9 billion tonnes of coal equivalent or 0.32 million petajoules (1 petajoule = 10^{15} joules) per annum. This compares with proved recoverable world resources of 20.3 million petajoules of coal and 11.2 million petajoules of oil and natural gas. However, this is not the whole story. Advanced capitalist industrial countries, which make upto 15.75% of the world's population, consume 54.5% of these 0.32 million petajoules each year. It is unlikely that the world's under developed countries are going to achieve the living standard of the opulent west overnight. If they did, the annual world's energy demand would rise to 1.11 million petajoules each year even if the living standards in all western countries remained constant. Obviously, looked at from the point of view of western society, things are not particularly desperate as yet. There may be more coal and oil to be found than we know of at present. The snail's pace development of fast-breeder nuclear reactors and fusion reactors may yet to be accelerated. Also the poor, with their very modest demands for the earth's stored energy resources, will be with us for many years yet.

On the other hand, although actual energy shortages are not likely to be experienced for many decades, it is certain that since fossil fuels are a wasting asset (like land, coal and oil are not being made any longer), the costs are bound to go up. For this reason we must abandon many wasteful practices in energy utilization which have grown up over the years, particularly during periods when fossil fuel supplies seemed to be inexhaustible.

We must conserve energy by all possible means as a bounden duty to those who come after us, our as yet unborn descendants. We must also strive after good energy economy out of quite hard-headed and commercial considerations, to reduce manufacturing costs and stay complete *vis-a-vis* our commercial rivals.

Although all forms of energy are expressed in the same units (joules, megajoules, gigajoules etc.), the financial value of energy varies enormously with its grading. This means that electrical and mechanical energy are the most costly, followed by high grade heat energy. At the other extreme, thermal energy which is only a few degrees above ambient has virtually no commercial value at all. This is, of course, the weakness of all these bright ideas about trying to equate the energy contained in solar radiation reaching the earth, or the heat content of geothermal fluids, with high grade energy obtainable from fossil fuel or nuclear power. For economic operation, one should never use energy at an appreciably higher grading rate than needed. For example, electrical energy which has the highest energy grading of all, should be used for such purposes as mechanical energy generation, production of light, sound and very high temperatures in electrical furnaces. It is also the only suitable source of energy for electronic and similar purposes. Electric fires, on the other hand, where electricity is basically used only for raising the ambient air temperature to about 20°C, are an extremely wasteful use for electricity. This applies both in the domestic and the industrial context. At present there is over-production of electricity at night and therefore some of the night electricity is sold off at out price for space heating purposes. This is an inherent waste. It is far better to find more efficient, methods of energy storage such as flywheels, compressed air or pumped water, in which there is far less thermodynamic irreversibility.

Under industrial conditions there are often several tasks which require energy, hut which lie at different grading levels. It makes sense to use the waste heat from one process to serve the needs of another. For example, glass works produce waste heat at between 400 and 500°C. This is quite sufficient to raise intermediate pressure steam for running back pressure turbines to produce electricity and heat in the form of low pressure steam at, say 120°C. This, in its turn, could be used to evaporate moisture from agricultural products. The water vapour obtainable from such processes would probably be condensed to provide warm water at about 60°C. This could be employed for space heating or for the supply of heat at fish farms or greenhouses. If this is done, the original supply of high grade energy obtained by burning coal, oil or natural gas performs no less than four separate tasks.

The various glass constituents are melted and held above the solidification temperature (about 1500°C).

Medium pressure steam is used to produce electricity (500°C).

The exhaust steam from the back pressure turbine is used for crop drying (120°C).

The condensed water vapour is able to heat up water to be used for space heating, fish farm or greenhouses (60°C).

The glass industry represents the highest heat rejection grading with a temperature of upto 500°C.

Pottery Kilns provide waste heat at around 350°C. Most other industries reject heat at temperatures of 200°C and lower. Yet waste heat of this type is ideally suited for evaporating water from materials used in the paper and board industries, agricultural food industries, textile industries etc. It has been calculated that the energy needed for industrial evaporation processes in which a total of 20.45 million tonnes of water is removed from wet products in Great Britain alone, amounts to the staggering quantity of 2.37 million tonnes of coal equivalent or 85.4 petajoules worth millions of currency.

For all industrial evaporation process, heat exchange media at a temperature of below 200°C are quite adequate. This means that waste heat rather than expensive prime fuel heat could be used with complete success.

Agriculture and *horticulture,* these two industries employ huge quantity of low grade heat for which waste heat produced by other industries would be extremely adequate.

In producing paper, pulp and board some 30% of energy used is for the drying process, $1\frac{1}{2}$ tonnes of water have to be evaporated for each tonne of paper produced; drying is normally done by drums running at 200°C, providing exhaust air at between 70 and 120°C, which could be used to provide further waste heat.

In *petrothermal* and *general chemical industries* much of the heat rejected is at temperatures as high as 200°C. At present the petrochemical industry tends to make little use of the vast quantities of waste heat generated, which can amount to much as 60 MW for a larger refinery.

17.2. Principles of Energy Conservation and Energy Audit

Energy conservation means reduction in energy consumption but without making any sacrifice of quantity and quality of production. In other words, for the same energy consumption, higher production. It does not prevent you, use of energy by fixing some limit quantitatively within the agreement but insists for use efficiently thus decreasing the cost of your production to some extent by way of reduction in the energy bill. Thus energy saved is the money earned which would be used in other productive means. It is therefore imperative that electricity which is in shortage, be utilised efficiently and the areas, of where the energy is wastefully used are to be identified and corrective measures are searched for adoption. This could be done by *"Energy Audit"*. Energy audit is a technical survey of a plant in which the machinewise/section-wise/departmentwise pattern of energy consumption is studies and attempts to balance the total energy input correlating with production. As a result of the study, the areas where the energy is wastefully used and the improvements are felt are identified and corrective measures are recommended for adoption on short term/long term basis giving priorities so that the overall plant efficiency could be improved. While doing so, a clear picture of financial involvement is given for each type of recommended measure supported by payback period to facilitate the easy adoption.

Like capital, labour and material, energy is one of the production factors used to produce final goods and services. In economic terms, energy is demand-derived goods. That is energy is an intermediate good whose demand depends on the demand of final (end use) goods and services it produces.

Energy conservation can be defined as the substitution of energy with capital, labour, material and time. This definition also covers the substitution of scarce type of energy (*e.g.*, oil) with abundant types of energy (*e.g.*, coal) or the substitution of energy with convenience. For example, people will turn lights off when they are not in their offices.

Principles. The two principles governing energy conservation policies are maximum thermodynamic efficiency and maximum cost effectiveness in energy use. The first and second laws of thermodynamics measure the efficiency of energy use, allocation of available production factors determines the cost effectiveness of conservation.

Maximum thermodynamic efficiency in energy use is defined as maximum workdone production by using a given amount of primary energy input, as defined in the following form:

Maximum work = (Energy input) — (Energy loss in transfer)
— (Energy discharge).

Work. Work is an intercreation between a system (*e.g.*, boiler) and its surroundings (*e.g.*, turbines). It is done by a system on its surrounding, if the sole external effect of the interaction could be movement of a desired mass in the surroundings. In most cases, work is done because of force on a moving boundary of a system. The magnitude of workdone is measured by the number and force of turn made by the turbine.

Energy input. Energy inters most process equipment (energy systems) as chemical energy in the form of fossil fuels, as sensible enthalpy in the fluid streams, as latent heat in vapour steams, or as electrical energy.

Energy loss. Energy lost to the ambient environment from the process equipment through radiative and convective heat transfer, thus loss of energy occurs.

Energy conservation is the energy demand management that aim at increasing the efficiency of use. An energy audit helps to understand more about the ways different energy sources are used in the industry and helps to identify areas where waste can occur and where scope for improvement may be possible. The energy audit would give a positive orientation to the energy cost reduction, preventing maintenance and quality control programmes which are vital for production and utilies activities. Energy audit attempts to balance total input of energy with its use. The scope of an energy audit varies according to the facilities being audited like refineries, petro-chemical plants and thermal power plants where cascading of energy and complex energy balance are involved. Energy audit is one of the concepts used in the energy management industries and it involves methodological examination and comprehensive review of energy use in industries. Energy audit is the starting point of an energy management plan. A typical energy audit of a factory would be as shown in Fig. 17.1.

Energy audit broadly covers the following questions:

(*i*) How much energy are we consuming ?

(*ii*) Where is the energy consumed ?

(*iii*) How efficiently is the energy consumed ?

(*iv*) Can there be improvements in energy use ?

The energy audits result in energy conservation proposals or projects. Initially, proposals with minimum investments are identified. The criteria used for ranking proposals are saving obtained from a project compared to the investments needed for achieving that savings. For example, a fertilizer unit by changing the burners in catalytic cracking systems can obtain a saving of ₹ 30 lakh per annum and this will involve an investment of ₹ 10 lakh for changing the existing inefficient burners. The savings to investment ratio (SIR) in simple terms is 3.0.

The energy audit thus can give a positive orientation to the energy cost reduction, preventive maintenance and quality control programmes which are vital for production and utility activities. Such an audit programme will help keep alive variations which occur in the energy costs, availability and reliability of supply of energy, decide on appropriate energy mix, indentifying energy conservation technologies.

The energy measurement and energy monitoring are the basic necessary steps for energy audit. The measuring of energy flow in the main and sub-circuit particularly for the major energy consuming and equipment will give clear clue for energy conservation measures.

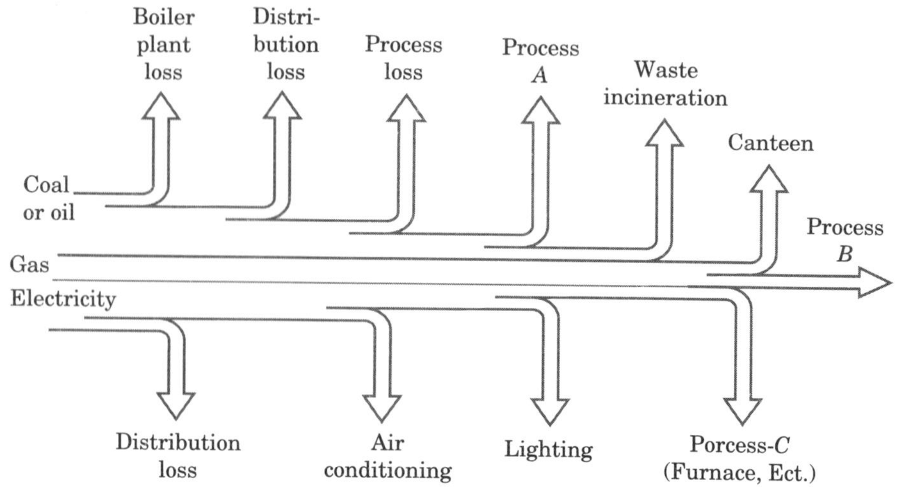

Fig. 17.1. Sankey diagram for energy audit.

Alongwith the measuring of energy flow it would be beneficial to the users to be able to monitor the energy circuit from time to time. To detect any leakage or abnormalities (*e.g.* efficiency of combustion, effectiveness of thermal insulation, leakage in stream traps, etc.), the use of combustion efficiency monitor meter, radiation loss indicator and ultrasonic leakage detector will be effective for energy monitoring.

Types of energy audit. The primary objective of energy audit is to determine ways to reduce energy consumption per unit of product output or to lower operating costs. The extent and the type of energy audit should result in gains commensurate with the efforts.

There can be two types of energy audit:

(*i*) Preliminary audit, (*ii*) Detailed audit.

The action plan towards the achievement of energy conservation through energy audit may be drawn up into three phases: (*i*) short term, (*ii*) medium term, (*iii*) long term.

Preliminary audit. Preliminary audit is carried out in the limited time *i.e.,* from 1 to 10 days and it highlights the energy cost and wastages in the major equipments and processes. It also gives the major energy supplies and demanding accounting. The questionnaire containing the industrial details of energy consumption process carried out, energy need to unit product, load data, etc. must be completed before the pre-audit visit. The financial report regarding the industrial audit will be prepared within two weeks of time.

Detailed audit. Detailed audit includes engineering recommendations and well defined projects with priorities. It accounts for the total energy utilised in plants. It involves detailed engineering for options to reduce cost/consumption. The duration for the visit would be 1 to 10 weeks. It requires advance notice to the departmental head arranging for office and secretarial support. The final report would be prepared within three months of audit.

The short term action

This plan requires no capital investment or least investment to avoid energy wastages and minimising non-essential energy users and improving the system efficiency through improved maintenance programme.

The medium term action

Plan requires a little investment to achieve efficiency improvement through modifications of existing equipments and other operations.

The long term action

Plan is aimed to achieve economy through latest energy saving techniques and innovations. The capital investments are required to be studied thoroughly while finalising the long term action-plan.

17.3. Energy Conservation Approach/Technologies

Conservation of energy is using energy more efficiently by substituting time, convenience, labour and capital, for effective optimisation of costs. Energy today has become an indispensable component of industrial production, agriculture, employment and economic growth. Natural gas and oil are dominant in energy use and their shortages have caused economic chaos in the past and the recent past and the recent Gulf crisis have paralysed economies. Energy available for future focuses a lot of attention on use of energy efficient equipment, adoption of energy conservation techniques and switching over to new energy technologies where possible.

The governing principles for energy conservation centres around maximum conservation of efficiency and maximum cost effectiveness in energy usage. The energy efficiency includes obtaining maximum quantitative energy for a given input and achieve energy quality of output over availability of energy input. The cost effectiveness is achieving maximum benefits at minimal costs.

Steps towards energy conservation requires 3 level of efforts:

(*i*) Formulation of administrative and information programme relatively easy and in expensive to implement like tune ups, light turn offs, small adjustment in production process etc., effective in reducing current energy consumption to the extent of 3 to 7%. The responsibility for implementation of this programme falls on the operators and low level managers.

(*ii*) Re-equipping, retrofitting and re-cycling through small incremental investments for gaining 5 to 10% savings. Middle level managers will make these decisions.

(*iii*) Major production process changes through large scale capital expenditure for obtaining savings from 20 to 90% gain, depending on the nature of operations and facilities involved. Facilities and process that waste are candidates for replacement to make them energy efficient. Synthesising measures include 3 criteria:

1. **Method of installation:**

   (*a*) Re-cycling. (*b*) Retro fitting,

   (*c*) New process

2. **Method of energy use:**

   (*a*) Waste energy recovery, (*b*) Waste material usage,

   (*c*) Waste energy utilization,

   (*d*) Process efficiency improvement/co-generation.

3. **Size of investments:**

   (*a*) Administrative and information process to create awareness and reduce individually controlled energy use.

(*b*) Small incremental investments to recover wastage energy, alter process flows and retrofit facilities for better utilization.

(*c*) Major capital expenditure to re-design production process overtime.

Energy Saving Devices Eligible for Higher Depreciation

(*a*) Special boilers and furnaces:

(*i*) Igni-fluid/fluidised bed boilers.

(*ii*) Flameless furnaces.

(*iii*) Fluidised bed type heat treatment furnaces.

(*iv*) High efficiency boilers (thermal efficiency higher than 75% in case of coal fired and 80% in case of oil/gas fired boilers).

(*v*) Waste heat boiler design for gas turbine combined cycle station.

(*b*) Instrumentation and monitoring systems for monitoring energy flows:

(*i*) Automative electrical load-monitoring systems.

(*ii*) Digital heat loss meters.

(*iii*) Micro-processor-based control systems.

(*c*) Waste heat recovery equipment and generation system:

(*i*) Economisers and feed water heaters.

(*ii*) Recuperators and air preheaters.

(*iii*) Back pressure turbines for co-generation.

(*iv*) Heat pumps.

(*v*) Vapour absorption refrigeration system.

(*vi*) Organic Rankine cycle power systems.

(*d*) Power factor-correcting devices:

Shunt capacitors and synchronous condenser system.

Renewable energy devices eligible for higher depreciation:

(*i*) Flate plate solar collectors.

(*ii*) Concentration and pipe-type solar collectors.

(*iii*) Solar cookers.

(*iv*) Solar water heaters and systems.

(*v*) Air/gas/fluid heating systems.

(*vi*) Solar crop driers and systems.

(*vii*) Solar refrigeration, cold storage and air-conditioning systems.

(*viii*) Solar stills and desalination systems.

(*ix*) Solar-power generating systems.

(*x*) Solar pumps based on solar thermal and solar photo-voltaic conversion.

(*xi*) Solar photo-voltaic modules and panes for water pumping and other applications.

(*xii*) Wind mills and any specially designed devices which run on wind mills.

(*xiii*) Any special devices including electric generators and pumps running on wind energy.

(*xiv*) Biogas plants and biogas engines.

(*xv*) Electrically operated vehicles including battery powered or fuel cell powered vehicles.

(*xvi*) Agricultural and municipal waste conversion devices producing energy.

(*xvii*) Equipment for utilising ocean waves and thermal energy.

(*xviii*) Machinery and plants used in the manufacture of any of the above items.

Some of the above important energy conservation technologies shall be discussed in the following articles.

17.4. Co-generation

A procedure for generating electric power and useful heat in a single installation, the useful heat may be in the form of steam, hot water, or hot air. In a *heat engine* (*i.e.*, a device for converting heat into mechanical work) part of the heat taken up from a "source" (*e.g.*, a burning fuel) at a higher temperature must be discharged to a "sink" at a lower temperature. In a co-generation system, the mechanical work is converted into electrical energy in an *electric generator,* and the discharged heat, which would otherwise be dispersed to the environment, is utilised in an *industrial process* or in other ways. The *net result is an overall increase in the efficiency of fuel utilization.*

The general type of co-generation systems may be distinguished.

1. The major purpose of the system is to generate electricity (*i.e.*, by a utility in a central power station) with heated water as a by-product. This type of co-generation is referred to in the section on *Waste heat utilisation.*

2. The system is designed to supply both the electrical and heat requirements of a large building complex or a community. This type is described in the section on *total* (integrated) *energy system.*

3. In most existing co-generating systems, the objective is to produce both electricity and industrial process heat in the form of steam, hot water, and/or hot air. Many industries (*e.g.*, petroleum refineries, chemical plants, paper and pulp manufactures and others) use large amounts of process heat as well as electricity to operate pumps, etc. In these cases co-generation can result in a 30% lower fuel consump-tion than for the separate generation of electricity and process heat.

Co-generation is thus the simultaneous generation of electricity and steam (or heat) in a single power plant. It has long been used by industries and municipalities that need process steam (or heat) as well as electricity. Examples are chemical industries, paper mills, and places that use district heating. Co-generation is not usually used by large utilities which tend to produce electricity only. Co-generation is advisable for industries and municipalities if they can produce electricity cheaper, or more conveniently, than that brought from a utility.

From an energy resource point of view, co-generation is beneficial only if it saves primary energy when compared with separate generation of electricity and steam (or heat). The *co-generation plant efficiency* η_{CO} is given by

$$\eta_{CO} = \frac{E + \Delta H_S}{Q_A} \quad \text{...(17.1)}$$

where, E = electric energy generated

ΔH_s = heat energy, or heat energy in process steam

= (enthalpy of steam entering the process) – (enthalpy or process condensate returning to plant)

Q_A = heat added to plant (in coal, nuclear fuel etc.)

For separate generation of electricity and steam, the heat added per unit *total* energy output is:

$$\frac{e}{\eta_e} + \frac{(1-e)}{\eta_h} \quad \text{...(17.2)}$$

where, e = electrical fraction of total energy output

$$= \frac{E}{E + \Delta H_S}$$

η_e = electric plant efficiency

η_h = steam (or heat) generator efficiency

The *combined efficiency* η_c for separate generation is therefore given by

$$\eta_c = \frac{1}{\left(\frac{e}{\eta_c}\right) + \left[\frac{(1-e)}{\eta h}\right]} \quad ...(17.3)$$

and co-generation is beneficial if the efficiency of the co-generation plant (Eqn. 17.1) exceeds that of separate generation (Eqn. 17.3).

Types of Co-generation

There are two broad categories of co-generation:

1. *The topping cycle,* in which primary heat at the higher temperature end of the Rankine cycle is used to generate high-pressure and temperature steam and electricity in the usual manner. Depending on process requirements, process steam at low pressure and temperature is either (*a*) extracted from turbine at an intermediate stage, such as for feed water heating, or (*b*) taken at the turbine exhaust, in which case it is called a *back pressure turbine.* Process steam pressure requirements vary widely, between 0.5 and 40 bar.

2. *The bottoming cycle,* in which primary heat is used at high temperature directly for process requirements. An example is the high-temperature cement kiln. The process low-grade (low temperature and availability) waste heat is then used to generate electricity, obviously at low efficiency. The bottoming cycle thus has a combined efficiency that most certainty lies below that given by Eqn. (17.3), and therefore is of little thermodynamic or economic interest.

Only the topping cycle, therefore can provide true savings in primary energy. In addition most process applications require low grade (temperature availability) steam. Such steam is conveniently produced in a topping cycle. Some are:

(*a*) Steam-electric power plant with a back pressure turbine.

(*b*) Steam-electric power plant with steam extraction from a condensing turbine Fig. 17.2.

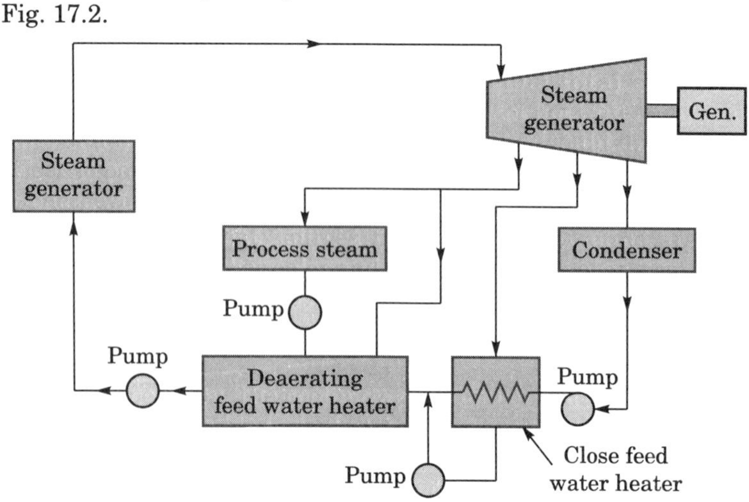

Fig. 17.2. Schematic of basic cogeneration plant with extraction-condensing turbine.

(*c*) Gas turbine power plant with a heat recovery boiler (using the gas turbine exhaust to generate steam).

(*d*) Combined steam-gas turbine power plant. The steam turbine is either of the back-pressure type:

(*i*) or of the extraction condensing type, (*ii*) above.

The most suitable electric to heat generation ratios vary from type to type. The back pressure steam turbine plant (*a*) is most suitable only when the electric demand is low compared with the heat demand. The combined-cycle plant (*d*) is most suitable only, when the electric demand is high, about comparable to the heat demand or higher, though its range is wider with an extraction condensing steam turbine than with a back-pressure turbine. The gas turbine (*c*) lies in between. Only the extraction condensing plant (*b*) is suitable over a wide range of ratios.

Gas turbines and Diesel engines

If natural gas or a suitable petroleum product is available as fuel, it may be used either in a gas turbine or a diesel engine, to drive an electric generator. These prime movers have a fairly high thermal efficiency (*i.e.*, proportion of heat supplied that is converted into useful work or electrical energy) because they take advantage of the high temperatures of the combustion gases that are the heat source. Instead of being discharged to the atmosphere, the hot exhaust gases, at temperatures above 315°C (600°F) pass to a *waste-heat* (heat exchanger) boiler. Water under pressure, flowing through finned tubes surrounded by the hot gases is converted into steam at the desired temperature (and pressure). Process steam is usually in the temperature range of 150—220°C. If the temperature of the steam from waste heat boiler is higher than needed, its temperature may be decreased and useful energy recovered, by passage through a steam-turbine generator (see combined cycle generation). The exhaust steam from the turbine is used as process steam or to produce hot (near boiling) water. To conserve natural gas and petroleum products, it would be desirable to substitute coal for these fuels in co-generation systems. If fly-ash could be removed from the combustion process, finely powdered coal might be used directly to operate gas turbines. Coal has been considered as fuel for diesel engines. The techniques for using coal require development, but other less direct methods for utilising this fuel in co-generation systems, as outlined below may be closer to practical realization.

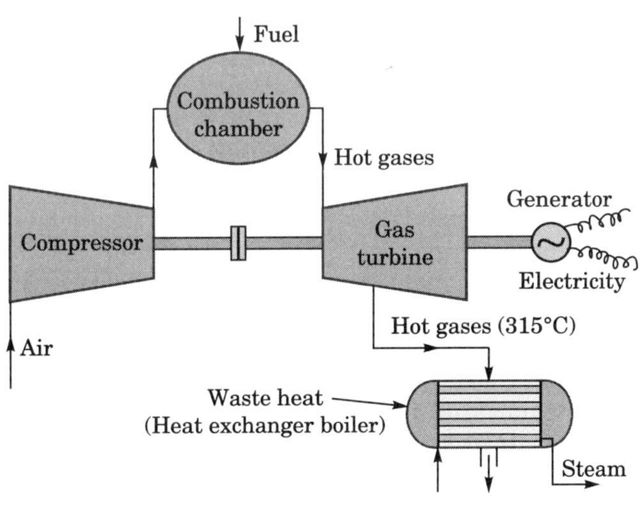

Fig. 17.3. Gas turbine with waste heat boiler.

Air is passed through pipes immersed *in a fluidised-bed combustion furnace* (or combustor). The emerging air at high temperature (and pressure) operates an open-cycle gas turbine-generator system. The exhaust air can be used directly to provide industrial heat or it can generate steam in a waste-heat boiler. The flue gas from the combustor is used to preheat the combustion air.

Another method of co-generation is to burn the coal in air in a combustor (fluidised bed or conventional) which serves as the heat source for a closed-cycle gas turbine generator. The heat source and the working gas (*e.g.*, air, carbondioxide, or helium) are then independent. The hot flue gases from the combustor generate process steam in a waste-heat boiler.

A third possibility is to convert the coal into a low heat fuel gas and to use this as fuel for a gas-turbine (open or closed cycle) or diesel engine. The exhaust gases from the turbine or engine are then utilised in a waste heat boiler to produce steam or hot water.

Steam turbines

In using a steam turbine for co-generation, the exhaust steam is not condensed, as it is in conventional steam-turbine generator, but is withdrawn at the desired pressure for process use. The turbines are thus of the non-condensing (or back pressure) type in which the steam leaves the turbine at a significant pressure rather than at the very low pressure in a condensing turbine. Steam is discharged from a back pressure turbine at a higher temperature than from a condensing turbine, consequently, the former has a lower thermal efficiency for electric power generation. However this is not important because the discharged steam is required for an industrial process. Further more, elimination of condenser and cooling water requirements reduces the capital and operating costs of power generation.

17.5. Waste Heat Utilization

As used here, refers to the beneficial use of the heat discharged in the steam turbine condenser cooling water from an electric power generation plant. From about 50 to 65% of the heat supplied either by combustion of a fossil fuel (coal, fuel oil, or natural gas) or by fission of a nuclear fuel in existing steam electric plants is removed by the cooling water. This heat is dissipated to the environment by direct discharge of the warmed water to the ocean or, a river or by passage through a cooling lake or cooling tower. Various possibilities are being considered for making use of the large amount of heat that is lost in this way.

Fossil fuels can be utilized more efficiently, with less heat discharged for a given total electric power output, if the steam turbine is preceded by another heat engine that operates at a higher temperature. The heat discharged from the high temperature cycle would then be used to generate steam for a conventional turbine (*Topping Cycle*) in general, the first (or top) heat engine cycle of a combined cycle generation system. In such a system, the heat discharged from topping cycle is used to generate steam for operating a steam turbine. The temperature at which heat is taken up in a steam turbine is limited to about 540°C by the properties of materials, but much higher temperatures are attained by the combustion of fossil fuels. The purpose of the topping cycle is to utilize this high temperature heat and thus increase the overall thermal efficiency (*i.e.*, proportion of heat taken up that is converted into useful work or electrical energy). There is a corresponding decrease in the condenser cooling water requirement for a given power output.

The most highly developed topping cycle is the *gas turbine,* but the *diesel engine* could be used in smaller installations. The *potassium vapour cycle* and *thermionic* conversion are possibilities, but the major interest at present is in *magneto-hydro-dynamic conversion,* largely because it has the potential for a high thermal efficiency using coal as the fuel.

Combined cycle with heat-recovery boiler

Fig. 17.4 shows a schematic flow diagram of such a combined cycle. A simple gas turbine cycle, consisting of air compressor (AC), combustion chamber (CC) and gas turbine (GT) is used with the turbine exhaust gas going to heat a heat-recovery boiler (HRB) to generate superheated steam. That steam is used in a standard steam cycle, which consists of turbine (ST), condenser (C), pump (CP), closed feed water heaters (FWH), and deaerating heater (DA). The HRB consists of an economiser (EC), boiler (B), steam drum (SD), and superheater (SH). The gas leaves the HRB to the stack. Both gas and steam turbines drive electric generators (G).

The gas turbine is usually operated with a high air-fuel ratio, approximately 400% theoretical air, to make sufficient air available in the gas turbine exhaust for further combustion.

For low-powered combined cycles the steam turbine output is less than the gas turbine output, by as much as 50%, and the number of feed water heaters is small, often one deaerator and one closed-type. To increase the output for short periods during load peaks, supplementary fuel burners may be fitted to the HRB to increase the steam mass-flow rate. This can also be done on a continuous basis, but limits on the amount of fuel added as posed by the design of the HRB, which usually lacks refractory lining, water cooled walls, etc. A limit of about 760°C on the gas temperatures in the HRB is common. This however is usually sufficient to increase the steam-turbine output by perhaps 100% and the total cycle output by 30%.

In large combined-cycle plants used for base-load operation, where efficiency is of prime importance, separate supplementary firing equipment (SF) is interposed between the gas turbine and the HRB. The steam turbine output is usually greater than the gas turbine output by up to 8 : 1. The steam cycle is therefore designed for high efficiency with reheat and a full complement of feed water heaters. A forced-draft fan may be installed ahead of the SF to operate the steam cycle on its own when the gas turbine is cut off.

The fuel used in supplementary firing may be the same high-grade fuel used in the gas turbine. This is the simplest solution as such fuel causes few problems in the SF and HRB. However, cheaper lower grade fuels such as heavy oil or coal, can also be used in the SF.

Legend

AC — Air Compressor
CC — Combustion Chamber
GT — Gas Turbine
G — Generator
SF — Supplementary furnace
ST — Steam turbine
C — Condenser
P — Pump
EC — Economiser
B — Boiler
SD — Steam drum
SU — Superheater
FWH — Feed Water Heater
BFP — Boiler Feed Pump

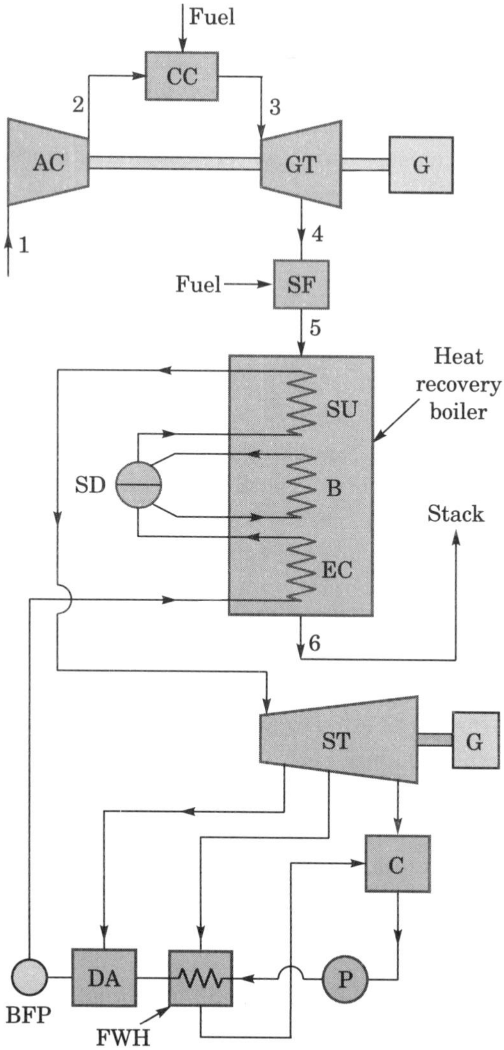

Fig. 17.4. Combined cycle with heat recovery boiler.

Variations of the cycle shown in Fig. 17.4 are used to extract the maximum amount the energy from the gas leaving the heat-recovery boiler before exhausting it to the stack. Depending upon the temperature of that gas, it may be used for (*i*) partial heating (regeneration) of the compressed air leaving the compressor, (*ii*) feed water heating of the steam cycle in a closed-type feed water heater or (*iii*) generating steam in a dual-multi-pressure steam cycle. The last variations is described in the following section.

Another approach is to add a *bottoming cycle* to the steam turbine. The heat discharged from the latter would be used to operate a low temperature turbine. Instead of converting the waste heat into mechanical work in a heat engine, it may be more practical to utilise the heat directly. The more promising uses fall into three categories: (1) space heating and cooling, (2) agriculture, including raising livestock, and (3) aquaculture.

Space heating and cooling

Warm water can be distributed to homes, offices, stores, and industry, as it is in several European countries and to some extent in the United States, for space heating and domestic use. This is referred to as district heating. Cooling could also be achieved with hot water (nearly 100°C) using absorption cooling refrigerating system. However, to be useful for space heating and cooling, the temperature of the water should be (50 to 100°C), whereas the average temperature of the steam-turbine condenser discharge water is only about 40°C.

It is possible to operate a turbine at a higher temperature, but this would result in a lower thermal efficiency (*i.e.*, proportion of heat converted into mechanical work or electrical energy). In addition, the accompanying increase in the exhaust steam pressure would result in a smaller power output for a given turbine. An alternative approach, which might be preferred because it could be more readily adopted to a changing heat demand, would be draw off steam at an intermediate stage of the turbine and use it to increase the temperature of the condenser discharge.

It is unlikely that the heated water from an electric power plant will provide district heating and cooling for an existing community because of the high cost of installing the required underground distribution system. But in developing a new residential and commercial complex, the hot water pipes could be laid at the same time as the service-water lines. It is improvable that such a complex would be built close to a steam electric plant (or vice versa), but it has been estimated that water at a temperature approaching 100°C could be transmitted in wide pipes for distances upto 65 km without serious heat losses.

Warm water in Agriculture

Water at a temperature of 24 to 46°C, which is within the range of condenser discharges, could be used to heat greenhouses in winter. Flowers, tomatoes, and cucumbers have been grown in this way throughout the year. Other agricultural uses of warm water that have been tested are for protecting fruit trees from frost by spraying, for irrigation, and for soil warming to extend the growing seasons of trees and vegetables.

Warm water in aquaculture

Fish farming is already being practiced to some extent in the United States, but it could be greatly extended by using warm water from power plant discharges. Catfish and tilapia an excellent food fish from Africa, exhibit optimum growth, relative to nutrients supplied, at a temperature of about 32°C. If this temperature could be maintained throughout the year, the fish yield from a given pond could be increased. Lower temperatures are required for spawning and egg development; conseqently, fingerlings would have to be raised in a hatchery and transferred to the warmer water at the appropriate stage of development.

An essential requirement for economic aquaculture is a supply of low-cost nutrients. It has been suggested that various small plants and animal organisms, upon which fish normally feed, could be produced in warm water ponds supplied with various animal and food processing wastes.

Several edible marine shellfish (*e.g.*, shrimp, oyster, and lobster) also exhibit maximum growth in water at temperatures some what above normal. Experiments conducted in the United States; the United Kingdom, and Japan have shown that shellfish can be cultivated successfully under controlled conditions in sea water that has been warmed by condenser discharges.

17.6. Combined Cycle Power Generation

The production of electricity using two or more *heat engines* in tandem as primemovers to operate electric generators. In a heat engine (*i.e.,* a device for converting heat into work) part of the heat taken up from a 'source' (*e.g.,* a burning fuel) at a higher temperature is discharged to a heat "sink" at a lower temperature. In a combined cycle system, the heat discharged from one heat engine serves as the source for the next engine. The net result is a greater overall operating temperature range (*i.e.,* between the initial heat source and the final sink) than is possible with a single heat engine. The *thermal efficiency* of the combined system (*i.e.,* proportion of heat converted into useful work or electrical energy) is thus greater than for the two heat engines operating independently. Less fuel is then required to generate a given amount of electrical energy.

As a general rule, combined cycle systems have two heat-engine stages, the second stage being a conventional *condensing steam turbine.* The highest inlet steam temperature for a turbine is limited by properties of materials to about 540°C. However the flame temperatures in a boiler burning fossil fuels may be more than 1650°C. Consequently in a combined cycle system, the steam turbine is preceded by a *topping cycle* heat engine which can utilise heat at high temperatures. The working fluid leaves the topping cycle at a sufficient high temperature to generate steam for the steam turbine.

Because the technology is well developed, the *gas turbine* is most commonly used as the topping cycle engine. The turbine exhaust gases, at a temperature of 600°C or more, pass to a waste heat (heat exchanger) boiler where steam is produced from water under pressure flowing through finned tubes surrounded by the hot gases. If it is desirable, the steam temperature may be increased in a subsidiary heater before it enters the steam turbine. In addition to using fuel more efficiency, a combined gas and steam turbine generating system require less condenser cooling water for a given electrical output than a steam-turbine generator alone.

Gas turbines generating use liquid petroleum products or natural gas as the fuel, but coal could be preferred primary fuel for combined cycle generation because of its availability. One possibility is to convert the coal into a low heat value fuel gas, which is a suitable gas turbine fuel. Another approach is to burn coal with compressed air in a *pressurized fluidized bed combustion* furnace. After removing particulate matter, the hot combustion gases are mixed with compressed air that has been heated by passage through pipes immersed in the fluidized bed. The mixture of hot pressurized gases then drives the gas turbines.

Other topping cycles are being studies as alternatives to the gas turbine in combined-cycle systems. The most important are *magneto hydrodynamic* (MHD) *conversion* and the *potassium-vapor cycle.* In both cases, combustion of coal would provide the heat for conversion into electrical energy.

At least 50% of the heat energy supplied to a steam turbine is discharged at a fairly low temperature, averaging about 40°C, in the condenser cooling water. A possible way of converting part of this heat into useful work (*e.g.,* electrical energy) is by means of a *bottoming cycle* in which the heat discharged by the steam turbine is used in a *vapour turbine* with a working fluid of low boiling point. However, the economics of this form of co-generation is uncertain.

Hence the combined cycle concept combines two thermodynamic cycles *viz.* (*a*) Brayton cycle, (*b*) Rankine cycle. The Brayton cycle has the capability to operate at very high combustion temperature. The Rankine cycle which is limited to steam temperature of approximately 540°C is ideally suited for recovering waste heat from the gas turbine exhaust. When operating together, the combined cycle recovers more energy from the fuel than either cycle alone can do.

The combined cycle power generation system has the following advantages over combined system;

(*i*) Higher efficiency, (*ii*) Low specific cost of the gas turbine, (*iii*) Smaller space requirement, (*iv*) Less cooling water demand.

Different systems for combined cycle power generation

The combined cycle power generation from coal *two* different routes may be considered. These are;

(1) Combined cycle power generation through pressurized fluidised bed combustion of coal (PFBC).

(2) Integrated gasification combined cycle (IGCC) power generation.

1. Combined cycle power generation through pressurised fluidised bed combustion of coal (PFBC)

The major advantages of pressurized fluidised bed combustion of coal are:

(*i*) Ability to use a wide range of fuels, specially high sulphur, high ash coals.

(*ii*) Elimination of separate fuel desulphurization unit.

(*iii*) Low combustion temperature, restricting NO_x formation.

(*iv*) Increased heat transfer coefficient.

(*v*) Reduce combustor size and number of fuel feed points compared to atmospheric FBC with similar power levels.

(*vi*) Improved volumetric heat releases relative to atmospheric FBC.

(*vii*) Increased thermal efficiency in combined cycle operation.

The two basic systems used with PFBC are:

(*a*) Pressurised steam-cooled combustor.

(*b*) Pressurised air-cooled combustor.

Steam boiler concept. In this coal is burned in a PFBC (refer Fig. 17.5) and the combustion gases from PFBC would drive the gas turbines, white steam produced in the boiler in the PFBC drives the steam turbines. The 70% of the total power is generated by steam turbine and 30% by gas turbine. The main elements of steam cycles are:

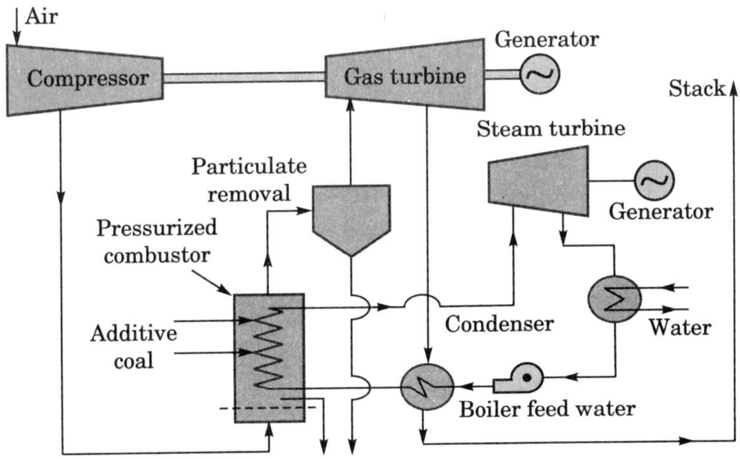

Fig. 17.5. Steam boiler concept.

(*i*) A pressurised fluid bed boiler from which superheated steam and reheated steam are fed to a power turbine.

(*ii*) A gas turbine fed by the combustion gases from the PFB combustor.

(*iii*) An economiser on the gas turbine exhaust which transfers heat to the low temperature side of the steam system.

(*b*) **Air heater concept.** In this system (refer Fig. 17.6), coal is burned directly in a PFBC and the air passing through the tubes in the bed at high pressure is heated to high temperature (1000°C). The air stream from the compressor splits into two parts with 65% going through the tubes and 35% through the bed for combustion purposes. Only the combustion gases have to pass through hot gas clean-up process, so that the hot air that went through the tubes by passes this process. The recombined flow of hot air and hot gases then passes through the gas turbine and then through the waste heat boiler. In this cycle about 70% of the power output is provided by the gas turbine and 30% by the steam turbine generator. The overall efficiency of this system is about 40-41%. The main elements of the air cycle includes:

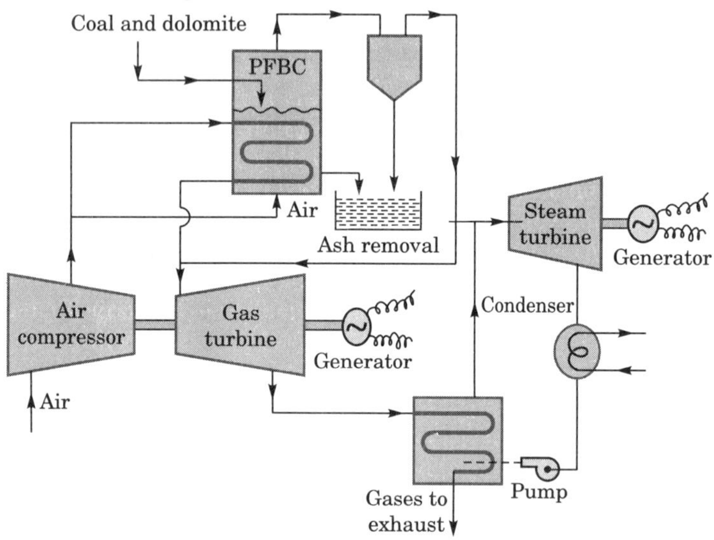

Fig. 17.6. Air-cycle system.

(*i*) A pressurized fluidized bed air heater.

(*ii*) A mixing chamber in which air heated in bed tubes is mixed with hot combustion gases and delivered to a gas turbine.

(*iii*) A waste heat boiler in which heat recovered from gas turbine exhaust is used to supply steam to a power turbine.

An outstanding feature of PFBC system is that pressurised operation reduces the space required for combustion vessel since volume is reduced in proportion to the operating pressure. A bed area for atmospheric unit is 2 m^2 per MW at fluidizing velocity of 2 m/s, whereas for pressurised unit, it is 0.2 m^2 per MW at 10 bar.

(2) Power generation through integrated gasification combined cycle (IGCC). The second alternative for combined cycle power generation from coals which has made significant headway on its development is integrated gasification of coal and combined cycle thermal power generation. In this system, coal is gasified and the gasifier may be coupled with the power generation system in *three modes.* These are:

(*a*) **Combined cycle with waste heat boiler.** This is one of the most popular scheme in the combination of the gas turbine cycle and steam turbine cycle. A flow diagram of the scheme is presented in Fig. 17.7. Hot gas from the turbine is utilised to generate steam that in turn is used to drive the steam turbine. In this system, the combustion of the fuel gas is effected only at one point in the cycle *i.e.,* the combustion chamber of the gas turbine and the power is generated in two stages. The waste heat steam generator couples the gas turbine part of the system with the steam turbine part.

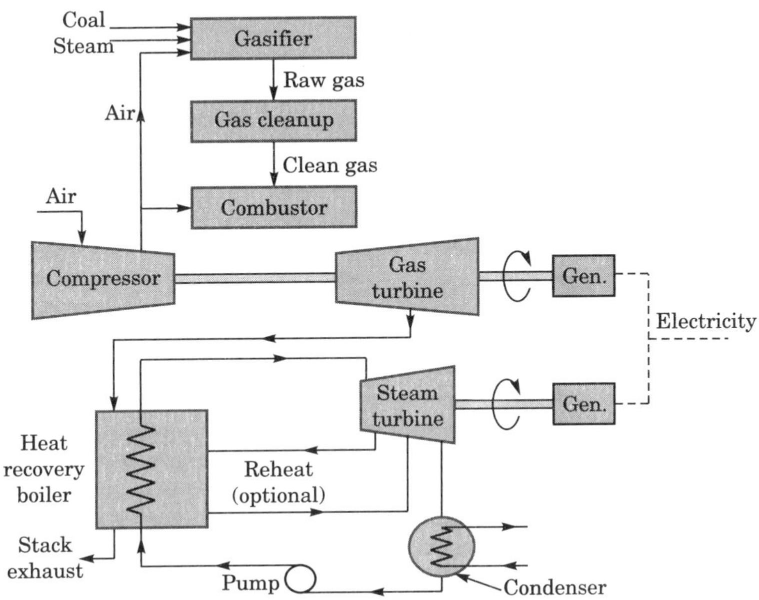

Fig. 17.7. Combined cycle with waste heat boiler.

(*b*) **Combined cycle with exhaust fired boiler.** In recent years, another arrangement of waste heat steam generator have been developed. In addition to the unfired boiler just described; the exhaust gas fired boiler is also available for combined cycle selection. The exhaust gas fired boiler is similar to the conventional steam generator equipped with a complete set of combustion equipment. Additional fuel is supplied through the gas turbine compressor. In general, the combined cycle system with exhaust gas fired boiler has higher efficiency, but the initial investment also is high.

(*c*) **Combined cycle with supercharged boiler.** Another system is the combined cycle with supercharged boiler. The fuel is fired in the supercharged boiler which operates the high pressure. The steam generated goes through the steam cycle and the exhaust gas from the boiler is expanded through the gas turbine. The system has been used by STEAG, FRG in combination with Lurgi Dry Ash process. The system is shown in Fig. 17.8.

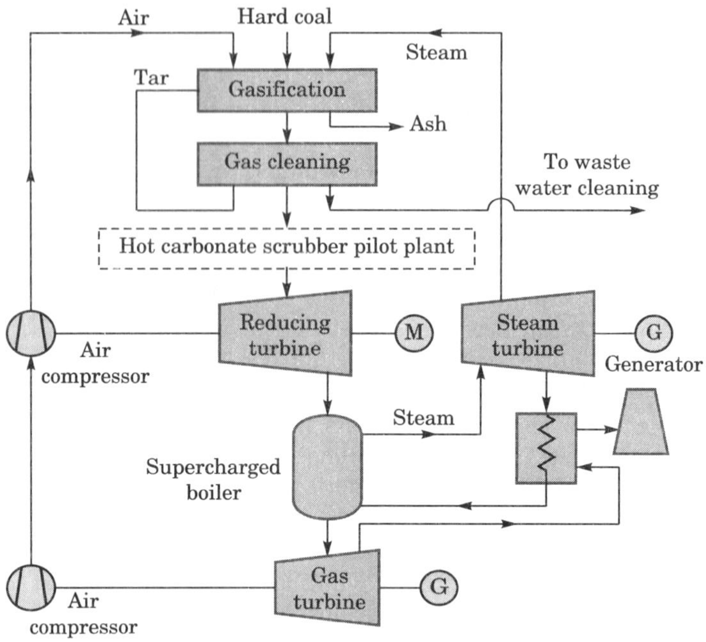

Fig. 17.8. Combined cycle plant with superchanged boiler.

Any of these three concepts may be utilized for power generation through combined cycle modes provided suitable mode can be made available through any efficient gasification process. Integration of the gasification system into combined gas and steam turbine power plants is technically feasible and provides a high potential for improvement in the efficiency as against conventional coal fired plants. In addition, IGCC power plants are economically and ecologically superior to conventional power generation technologies.

Although further development work is necessary for certain components such as gasifier, heat recovery boiler and high temperature gas turbine, the results obtained todate give confidence that this technology will be ready for commercial applications in not too distant a future. This IGCC technology allows installation of phased construction and the avoidance of excess capacities which is of particular interest from the economic point of view. In addition gasification is suitable for retrofitting existing steam or combined gas and steam turbine power plants to IGCC systems.

Advantages of combined cycle power plants

The following are some of the salients merits of a combined cycle:

(1) The efficiency of the combined cycle plant is better than a simple gas turbine or steam cycle.

(2) The capital cost of combined plant with supplementary firing is slightly higher than a simple gas turbine plant and much below those of a classical steam plant of the same power capacity.

(3) The combined plant is more suitable for rapid start and shutdown than a steam plant.

(4) The cooling water requirement of a combined cycle is much lower than a pure steam plant having the same output.

(5) The combined steam offers self-sustaining features if unfortunately, power station is shut down due to some fault, the gas turbine offers a way to start the station from the cold shut conditions. No outside power source is required. Gas turbine is always equipped with a diesel engine to start from cold.

(6) Many utilities are planning and installing simple gas turbine units which will later to be converted into combined cycle operation. This two phase development requires short installation time for peaking power plus the future capability for efficient operation for base load generation.

(7) The present trend to increase the thermal efficiency of gas turbine plant is to increase the turbine inlet temperature. Higher turbine inlet temperature reduces the heat rate, fuel cost and generation cost. The present combined cycle efficiency may reach 50% soon then better turbine material would be available.

(8) The environmental standards of many old fossil fuel plants are not acceptable and they are likely to be closed. These can be renovated by replacing the old boiler with a gas turbine unit and heat recovery boiler. With these modification exhaust emission can be reduced and thermal efficiency and generating capacity can be increased.

17.7. Heat Recuperators

Heat recuperators, or *heat exchangers,* as they are also called, are pieces of equipment which can abstract sensible heat from one stream of flowing fluid and supply it to another stream. They are essential feature of all production processes in the chemical industry. Because of the importance of improving heat recovery, consequent on the very appreciable rise in prime energy costs, heat exchangers are becoming increasingly important in the heating and ventilating field as well.

Main Uses of Heat Recuperators

1. To extract useful heat from waste hot liquids and gases. The heat is transferred to secondary fluids, which can then be used for either space heating or for the supply of hot water to kitchens and bathrooms.

2. To operate calorifiers, which are particularly widely used in the district heating field. Thermal energy is transferred from the circulating fluid, which has had to be closed with poisonous substances such as hydrazine, morpholine and caustic soda in order to protect mild steel pipes from corrosion. The heat is given off via heat exchangers to highly purified town water to enable it to be used for cooking and washing purposes.

3. In district and group heating practice, heat exchangers are used to provide indirect hot water supply to, for example, high building. This supply hot water may be at a pressure insufficient to enable it to service either the top floors of a tall building or one sited on top of a hill. In such cases it is advantageous to use water/water heat exchangers to transfer the heat to the secondary medium, which can then be pumped to the top by a separate system.

4. For normal heat transfer from steam heaters or flues to circulating air, in order to raise this air to the required working temperature.

5. For the operating of air-conditioning equipment, in which heat is being abstracted from room air by the refrigeration fluid or by chilled air.

6. For the supply of heating to swimming pools, where heat generated by either conventional heat sources or by solar batteries is transferred to the large volume of swimming pool water.

7. For heat recovery from exhaust air, flue gases and other sensible heat sources.

Classification of heat exchangers

Heat exchangers can be subdivided conveniently into three categories:

1. Liquid/liquid heat exchangers.
2. Liquid/gas heat exchangers or gas/liquid heat exchangers.
3. Gas/gas heat exchangers.

Liquid/liquid heat exchangers. The most common system in use is one in which one fluid flows through a series of parallel pipes, while the other passes through the space between the tube bundle and the casing which contains the tube bundle. The heat transfer surface is the total circumferential area of all the tubes. The hot medium can flow either in the tube bundle or in the casing which surrounds it. The real criterion of choice is the liability of either of the two fluids to cause scaling or other deposition of impurities. Cleaning out the tube bundle (calandaria) is a reasonably easy procedure for most heat exchangers of this type, and for this reason the dirtier fluid or the one more liable to scaling is always allowed to flow through the calandaria. The cleaner fluid goes across the tube bundle.

An alternative design to this shell and calandria system is one in which there are a number of parallel plates with alternate spaces being occupied by the fluid which gives up its heat, and the one which receives it. Heat exchange between the two fluids follows the basic equation, *i.e.*,

$$dQ = dA.U.(t_h - t_c) \text{ J/s (W)} \qquad ...(17.4)$$

where, dQ is the heat transferred through an area of dA m^2. The U value between the two flowing fluids, including the interface (metal plus encrustations), is U W/m^2 K, and the temperature of the hot and cold flowing fluids are t_h and t_c °C respectively.

Equation (17.4) shows clearly that the heat transferred is a product of the following three variables:

(*i*) The area of interface between the two flowing liquids,

(*ii*) The U-value of the interface, and

(*iii*) The temperature difference between the two fluids.

Co-current and counter-current heat exchangers

It is possible to operate heat exchangers in two ways, co-current and counter current. In the co-current mode the hottest part of the heat donating fluid give off its heat to the coldest part of the heat receiving fluid, so that at the beginning of the heat exchange process $(t_h - t_c)$ is a maximum. As the heat exchange continues, so the value of $(t_h - t_c)$ is reduced, being a minimum at the outlet from the heat exchanger. Such a system is best suited to small and cheap heat exchangers as there is a rapid heat exchange at the initial position, which drops off rapidly as the $(t_h - t_c)$ values falls. It is obvious that co-current heat exchangers are not economical to use if a very high efficiency of heat exchange is needed. They are also unsuitable for long flow paths, as the cost of a longer heat exchange surface becomes unviable.

Co-current heat exchangers should be squat in design, with the maximum heat exchanger area, at the position where the fluids enter. At this position too, one should seek to obtain a maximum U-value by the induction of adequate turbulence. Co-current heat exchangers are most suitable under the following circumstances:

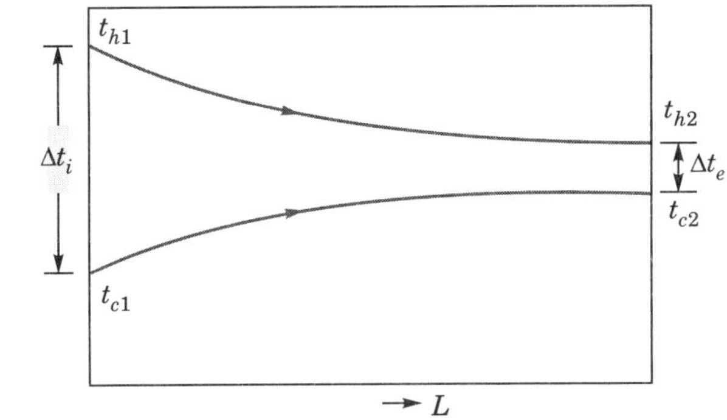

(*a*) Cocurrent or parallel flow heat exchanger.

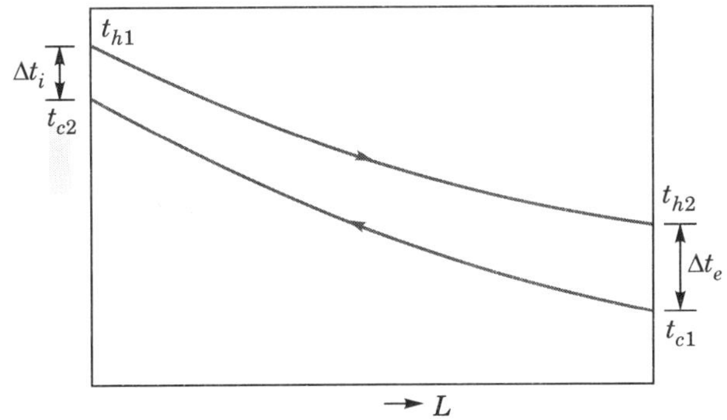

(*b*) Counter-current heat exchanger.

Fig. 17.9. Temperature-distance curve for current and counter current heat exchangers.

(1) When it is desired that as much heat as possible should be transferred from one fluid to the other, using equipment which is as cheap as possible to buy and maintain, but where it is not vital that all the heat or a very high proportion of it is used.

(2) Where high viscosity of the fluids makes it impractical to induce high turbulence over a considerable length of the heat exchanger, because this would raise pumping costs to intolerable levels.

In the counter-current mode of operation the hottest inflow faces the warmest outflow. This means that throughout the heat exchange operation, the value of $(t_h - t_c)$ remains essentially constant. By and large the efficiency of such heat exchangers is directly proportional to their length and the surface area of the calandria.

Heat exchangers operating on the counter-current principle are much easier to design than those on the co-current principle and are mainly used for the following purposes:

(*i*) When it is necessary to transfer as much heat as possible from the heat donating fluid to the heat receiving fluid, using a costly and complex heat exchanger plant.

(*ii*) Where the difference in temperature between the fluids is low.

(*iii*) When the temperature of the heat donating fluid leaving the heat exchanger is lower than the temperature of the heat receiving fluid leaving the heat exchanger.

For example, it is quite feasible with counter-current heat exchangers to have a heat donating fluid entering the heat exchanger at say, 150°C and leaving the exchanger at 80°C, while the heat receiving fluid is heated up from 40°C to 120°C or even more. This would, of course, be impossible to achieve with co-current operation.

In stream line flow, liquid molecules flow along in a parallel fashion and in consequence heat transfer from the centre of the fluid to the walls of the heat exchanger tube proceeds by conduction only. As Table 17.1 shows, thermal conductivities of the fluids are remarkably poor compared with those of metals.

Table 17.1. Thermal Conductivities of Various Solids and Liquids

Material	*Thermal conductivity W/M °K at 20°C*	*Material*	*Thermal conductivity W/M °K at 20°C*
Aluminium	237	Water	1.964
Copper	166	Toluene	0.44
Iron	147	Petrol	0.47
Magnesium	159	Glycerol	0.97
Silver	427	Oil	0.75
Zinc	115	Air (no convection)	0.025

It is therefore always necessary to ensure that fluids in heat exchangers move turbulently; *i.e.*, in such a fashion that constant mixing occurs.

When turbulent motion occurs one can accept that the entire body of the fluid has the same temperature because of the turbulence. The only conduction heat transfer which is needed is across the boundary layer. This is the layer of stagnant fluid between the turbulent fluid and the wall of the tube or other interface which contains it. Turbulence can be inducted in a fluid if Reynold's number exceeds about 2000. Reynolds number is commonly quoted as:

$$NR_e = \frac{VD\rho}{\mu} \quad ...(17.5)$$

where, V is the fluid velocity in m/s,

d is the diameter of the pipe containing it or other linear dimensions for different systems in m.

ρ is density of the fluid in kg/m^3

μ is viscosity of fluid in kg/m.s.

It can be seen that if the various units are entered as given, Reynold's number is a dimensionless function; it is the same when imperial units are used as it is when the newer SI units are employed.

Liquid/gas and gas/liquid heat exchangers

We have already dealt with the flow of liquids in heat exchangers. Gases can be either non-condensable, such as air or condensable such as steam.

Non-condensable gases have a far poorer or heat transfer than equivalent fluids and, in addition, a very small specific heat per unit volume. This means therefore that air and similar gases have to be blown at a high rate through heat exchangers with a very large surface area, to be at all effective. On the other hand, there is never any problem with regard to stream line motion at low speeds of flow, as would be case with liquids. Heat exchangers in which the primary flow is a liquid such as hot water and the secondary flow is air include air heaters, air conditioning equipment and heat reclamation plants.

Heat exchangers in which the primary flow is a non-condensible gas and the secondary heat medium is a liquid, would be the economisers, in which hot flue gases heat up water for turbine feeds or for other purposes. Here again we find that heat reclamation units. In these the sensible heat of exhaust air is removed by a fluid, and the heat resupplied to incoming cold air.

Condensible gases like steam are far more effective as heat exchange media because of the considerable heat evolved when the gas exchanges into a liquid.

Condensible steam is a very important heating medium and is used in steam heaters. It is always necessary to provide means of run-off and drainage. The film heat transfer coefficient for vertically heated steam calandria is as high as 10,000 W/m^2 K with some designs, so that extremely compact heat exchangers can be built. In fact, in such exchangers nearly all the resistance to heat exchange is on the tube/water side. This means, in effect, that a steam-heated exchanger of the same capacity as its equivalent water heated exchanger need only half the interface area of the latter, if we assume that all other conditions such as temperature difference and secondary water speed are kept the same.

Uses of liquid/gas and gas/liquid heat exchangers

These types of heat exchangers are used for the following main purposes:

(*i*) For blowing hot-air into environmental areas using hot water or other fluids as the source of heat.

(*ii*) For operation of an air conditioning system in which air is cooled by chilled circulating water.

(*iii*) For the extraction of sensible heat from exhaust air for heat recovery.

The main problem involved with liquid/air heat exchangers is the very poor thermal conductivity of air when stagnant. It is only 0.02423 W/m K at 0°C rising linearly to 0.03185 W/m K at 100°C. The corresponding figures for stagnant water are 0.567 W/m K and 0.672 W/m K, respectively. This means that heat exchangers which employ air as a heating medium must have very much larger contact surfaces than plain liquid/liquid heat exchangers. In addition, every effort must be made to reduce the thickness of the boundary layer of stagnant air, which adheres to the surface of the heat exchanger and constitutes the dominant resistance to heat flow. In other words, one must induce maximum convection currents.

Gas and gas heat exchangers

All gases including air have a very low specific heat and are also poor thermal conductors. In consequence air/air heat exchangers require very considerable interface area and appreciable turbulence to restrict the thickness of the dead laminar flow. Virtually the only type of air/air heat exchanger used commercially is a multiple plate system. In this system the supply air and exhaust air are completely separated from each other, being pumped at considerable speed past this metal plates.

Air/air heat exchangers complete with thermal wheels when applied to the recovery of heat from exhaust air streams. There are two main fields in which they are superior to thermal wheels.

1. When handling very damp high temperature exhausts to warm up incoming dry air. Thermal wheels would get water logged under such circumstances and in consequence soon become quite use less. In the case of air/air heat exchang-

ers, there is always provision for disposed of condensed water. If condensation takes place in an air/air heat exchanger the efficiency actually increases, because heat transfer is improved due to the utilization of latent heat released as water vapour in the air condenser.

2. When handling exhausts which contain extremely obnoxious gases, solvents and odours. Although thermal wheels have purge sections, they may not be fully effective.

The interface leaves from which the air/air heat exchangers are constructed are usually of aluminium. This is satisfactory for most designs except those where the exhaust air is particularly corrosive. Under such circumstances the surface in contact with the exhaust gases are coated with a plastic film. In all air/air heat exchangers provision is made for easy dismantling to fascilate cleaning.

17.8. Heat Regenerators

Waste heat from exhaust gases can be recovered by using either heat recuperators or heat regenerators. The former term encompasses the whole range of heat exchangers and the like, while the term heat regenerators implies the presence of an intermediate phase.

Basically what happens in a heat in a heat regenerators is that exhaust heat is absorbed by a soled thermal storage material. This heat is then given off, to the incoming fresh air supply. The classical method of using heat regeneration is used in the gas industry to make hydrogen, hot producer gas is passed through chequer brickwork. This is raised to incandescence by the exothermic reaction which takes place when carbon monoxide burns to carbon dioxide. After a certain time the flow of producer gas and air is switched to another chequer brick chamber and a mix of steam and carbon monoxide is passed through the incandescent chamber. The following reaction, which is endothermic (absorbs heat) takes place:

$$CO + H_2O \longrightarrow H_2 + CO_2. \qquad ...(17.6)$$

The heat required to complete the reaction is taken from incandescent bricks. The reaction proceeds until the brickwork has virtually cooled down. The flow of gases is then switched once more, with the exothermic reaction:

$$CO + \frac{1}{2} O_2 \longrightarrow CO_2 \qquad ...(17.7)$$

heating up the cooled chequer brick chamber, while the heated chamber enables the endothermic reaction to take place. Many other industries also make use of such regenerative processes.

Today heat regeneration also plays an important part in simple energy conservation practice. Waste heat is taken off exhaust gas streams. It is then transferred to clean and fresh external air, thus reducing the load on the heating system and saving an appreciable amount of precious fuel.

The thermal wheel

The thermal wheel heat generator was invented and patented originally by the Swedish engineer F. Ljunstron in 1922. The thermal wheel consists of a thick disc-shaped matrix, made from metal or other material, which permits gases to pass through its thickness. It revolves slowly around its axis, driven by a low-power electric motor. In the heat absorption part hot waste gases are pumped through the matrix of the wheel, heating up this section. As the wheel revolves, the heated sector passes through the purge section. In this, fresh air is blown through, to remove the last traces of exhaust air which may remain in the passage of the wheel matrix. The heated sector next passes through the supply air section. Where cold external air is blown through the hot wheel matrix. The heat stored in the matrix is given off to the incoming air. The transfer of heat from the hot but impure outlet gases to the cold but clean supply air is a continuous one. The principle of the heat wheel can be used for numerous different domestic, commercial and industrial purposes.

Different companies use different methods of constructing the slowly rotating thermal wheel. The main criteria of construction must be:

(*i*) Strength and durability.

(*ii*) High thermal storage capacity.

(*iii*) Ease of heat transfer with a minimum of pressure drop of exhaust gases and supply air.

(*iv*) Correct thermal resistance design depending upon temperature of heat supply gas used.

(*v*) Corrosion resistance.

17.9. Heat Pipes

Heat pipe principle. Usually, a means for transporting heat efficiently from a source to a cooler receiver by utilizing the latent heat of vaporization of a liquid. In essence, a heat pipe is a closed space containing a suitable working liquid and its vapour. One part of the space is in contact with the heat source and another with the cooler material to which the heat is to be transported. The interior wall of the space is lined with a porous material called a *Wick*. For high temperature operation, the wick can be made of several layers of a wooven metal wire mesh. A simple form of heat pipe is shown in Fig. 17.10, but more complex arrangements which satisfy the basic conditions outlined above are possible.

In the hotter (*i.e.,* heat source) part of the heat pipe, the working liquid is vaporized, thereby taking up the latent heat of vaporization. The vapour diffuses towards the cooler region because the pressure is lower there and condenses to liquid. In doing so, it deposits the heat of vaporization taken up from the source. The liquid is returned to the heat source region by capillary action of the wick. There is thus a continuous movement of vapour from the heat source to the receiver and of condensed liquid back to the source, accompanied by the transfer of heat.

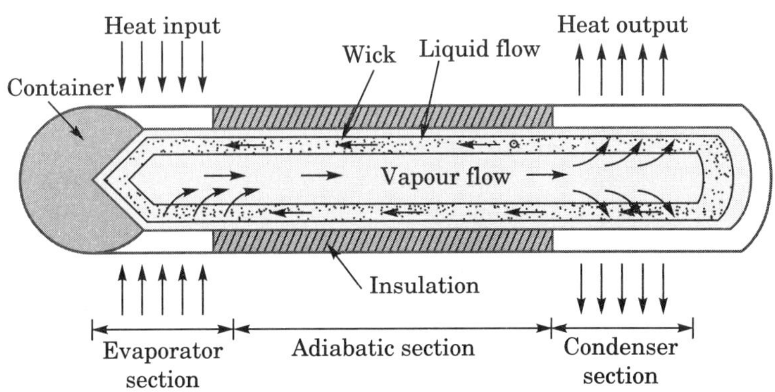

Fig. 17.10. A simple heat pipe.

Since the working liquid is returned to the heated region by capillary action, the heat source may be located above the heat receiver. However, if the receiver is at a higher level than the heat source, the return of the liquid is assisted by gravity. The vapour always moves from the higher pressure (*i.e.,* the heat source) region to the lower pressure (*i.e.,* heat receiver) region regardless of their relative positions.

A special feature of the heat pipe is its ability to transport heat at a high rate even when there is only a small temperature difference between the source and receiver. For other methods of heat transport (*i.e.,* conduction, convection, and radiation), the rate is highly dependent on the temperature difference, but in a heat pipe, the important consideration is the heat of vaporization of the liquid. Another advantage is that a heat pipe can serve as a concentrator; heat can be taken up during vaporization from a large area and deposited during condensation on a small area.

For operation of a heat pipe at high temperatures, sodium, melting at 97°C, has been used as the working liquid. Its large heat of vaporization and good thermal conductivity make it particularly suitable for this purpose. A sodium heat pipe might be used to transfer heat from a source, possibly stored heat, to a *Stirling engine.*

Thus in the heat pipe heat is transferred by convection and change of state. A liquid is evaporated at the absorber or evaporation end, the heat being abstracted in the form of latent heat. This is enormously larger in amount than sensible heat.

In any physical heat pipe can equilibrium has to be set up so that the rate of evaporation and condensation equals the rate of travel of condensed fluid back to the evaporation section. It is the rate of fluid travel which governs the whole efficiency of the heat pipe. One can express the heat transfer rate as being equal to

$$= L \times G \quad \text{J/s}$$

where, L is the latent heat of the fluid in J/kg, and G is the mass flow rate of the fluid from the condenser section to the evaporator section in kg/s.

The rate of fluid flow is determined by the following factors:

(*i*) *The pressure drop caused by differences in pressure between the evaporator section and the condensing section.* As the pressure in the condensing section is lower than that of the evaporator section this causes the vapour to flow along the pipe. At the same time this pressure difference also impeds the liquid flow, although the effect is not very large.

(*ii*) *Capillarity* is the main effect causing the fluid to flow from the condensing end to the evaporator end of the heat pipe.

(*iii*) *The gravitational head* depends upon whether the evaporator is at the same level as the condenser, in which case this effect is zero. If it is above the level of the condenser, the flow rate of fluid is impeded, and therefore the capacity of the heat pipe is lowered. Equally if the evaporator section is below the level of the condensing section, gravity helps the fluid flow and therefore increases the heat transfer capacity of the heat pipe.

There are certain other methods of increasing the fluid flow rate and therefore the heat transfer capability of the heat pipe, as for example, the use of centrifugal force in a rotating heat pipe system, or the incorporation of a pumping device.

For most simple heat pipe systems, a wick is all, that is provided to enable the fluid to return from the condenser to the evaporator. A wick is basically a porous layer which can be considered to constitute a large number of tiny capillary tubes, side by side.

Elementary physics explains that a liquid rises in a capillary due to the wetting effect of the fluid upon the solid material; the following equation gives a capillary rise in such tubes:

$$h = \frac{2\sigma \cos \theta}{\rho . g . r} \qquad ...(17.8)$$

where, h is the height of capillary rise in metres

σ is the surface tension of the liquid in N/m

θ is the angle of contact made by the fluid with the surface of the solid material, in degrees

ρ is the density of fluid in kg/m^3

r is the radius of the capillary in m

γ is the acceleration due to gravity in m/s^2.

When a liquid wets a surface completely, as is the case when one uses hydrophilic fluids, such as water, alcohol or ether against ceramic or similar surfaces, θ becomes 180°C so that cos θ approaches 1.

In a wick the capillaries are usually considered to be arranged horizontally, so that gravitational effects do not come into play, except in the circumstances already mentioned. The fluids flows through these capillaries in strictly streamline motion because Reynold's number (R_e) for the flow, expressed as:

$$R_e = \frac{V . d . \rho}{\mu} \qquad ...(17.9)$$

where, V is the velocity of flow in m/s

d is the effective diameter of each capillary in m

ρ is the density of fluid in kg/m^3

μ is the viscosity of the fluid in kg/m

It is always far below 2000, which is the transition point between streamline and turbulent motion. The driving the force for the flow of the fluid in the wick is surface tension.

It is almost impossible to work out the action of a specific wick because of the way the channels are twisted and the uncertainties of measuring capillary pores. A useful practical equation to evaluate the travel of different fluids in wicks of different dimensions is Darcy's equation, which applies to horizontal systems.

This can be simply stated as:

$$G = \frac{K \sigma \mu A}{\rho . l} \qquad ...(17.10)$$

where, G is the mass flow rate of the fluid along the wick in kg/s.

σ is the surface tension of the fluid against the wick material in N/m.

μ is the viscosity of the fluid at the temperature in equation in kg/m.s.

A is the cross-sectional area of the wick in m^2.

ρ is the density of the fluid in kg/m^3 at the temperature concerned.

l is the length of the wick in m.

K is the constant applying to the wick material (dimensionless).

It can be seen that the rate of flow of fluid along a heat pipe is directly proportional to the cross-sectional area of the wick and inversely proportional to its length. As the rate of heat transfer in a given heat pipe is limited by $G \times L$

where L is the latent heat of evaporation of the fluid in terms of J/kg, it can be seen that the maximum capacity of any given horizontal heat pipe is expressed as:

$$\text{Maximum heat transfer} = \frac{K\sigma\mu AL}{\rho . l} \text{ J/s.} \qquad ...(17.11)$$

Numerous fluids such as Ammonia, Methanol, Ethanol, Diethyl ether, Water, sodium, potassium, entire range of fluoro-hydrocarbons, toluene, and diphenyl are commonly used in heat pipes, as well as helium, molten lithium and even molten silver for extreme use purposes. Materials of construction depend upon the fluids used, but normally include aluminium, copper and stainless steel.

Wick Design. The wick for the return flow of the fluid from the condenser end to the evaporator end of each heat pipe is chosen primarily for adequate wettability and compatability with the fluid to be transported. The smaller is pore size, the better the capillary head exerted, but the lower the rate of fluid flow will be. For this reason low performance wicks in horizontal and gravity-assisted heat pipes should have fairly large pipes, of the orders of 50-100 μm in radius.

When the fluid has to be pumped against gravity, one needs very small pores to obtain very good capillary pressure, but the flow of fluid through such a wick is very slow and the capacity of these heat pipes is small. Thick wicks can transport more fluids and therefore increase the capacity of a given heat pipe unless the thickness of the wick initiates against the easy evaporation of the fluid at the evaporator end.

Types of wicks used

Nickel is used in the form of powder, felt and foam. In the form of powder the capillary height capable of being overcome may exceed 400 mm, but for the other materials capillary pressures are low. Nickel felts are used mainly with heat pipes filled with water, methanol and other organic fluids.

Copper is used as a powder, where pore sizes can be as small as 10 μm in radius, with a resulting capillary height exceeding 1.5 m. The thermal capacity is poor because of slow fluid flow. Copper can also be used as a foam for short, high capacity heat pipes. Like nickel, it is used for water and organic fluids.

Stainless steel in the form of mesh or twill can be made in pore sizes between 30 and 130 μm. It is used in contact with gases such as helium, nitrogen and ammonia for cryogenic heat pipes, as well as liquid metals such as mercury, sodium and potassium. In many cases there is a fluid flow area between the wick and the tube wall, which can consist of grooves or ducts.

Nature of fluids to be used in a heat pipe:

Fluids used in a heat pipe must:

(*i*) Be chemically stable over long periods of time.

(*ii*) Be easy to purify and degassify.

(*iii*) Be reasonably cheap.

(*iv*) Not react with the materials of construction of the heat pipe wick.

(*v*) Boil at the approximate temperature of heat input and delivery without requiring the heat pipe to be pressurized excessively as this would involve a considerable and disadvantageous increase in the thickness of tube walls.

Classification of Heat Pipes

Heat pipes do not have to have the circular cross-section the name implies; various other shapes are also used. In addition, several methods are employed to speed the flow of the liquid from the condensing section back to the evaporator section. The speed of this flow is the capacity determining factor for any given system. The basic types of heat pipe are the following:

(*a*) **Multiple tube type capillary heat pipe.** In these a number of tubes lead from the evaporator section to the condenser take-off section, capillarity being the only means of obtaining a return of the fluid. The heat transfer coefficient between an external forced air stream and a fluid inside a heat exchanger is about 50 W/m^2 K, while internal heat exchange values of upto 10,000 W/m^2 K are easily achieved with horizontal heat pipes. For this reason the extended surface area of the evaporator and condenser surfaces must be between 20 and 30 times the cross-sectional area of the heat pipes joining them. Circumferentially grooved wick types of heat pipe, using water as working fluid are claimed to have heat transfer coefficients as high as 80,000 W/m^2 K.

(*b*) **Gravity-induced fluid flow heat pipe.** In these the wick is comparatively small being mainly present to provide thermal insulation between liquid and vapour. The liquid flows from the condensing section to the evaporator section by gravity.

(*c*) **Osmotic flow heat pipe.** This type was developed by the U.S. energy conversion systems Inc, and works as follows: a solution of sugar in water used as the working fluid. On the evaporator side heat acts upon this sugar solution, so that pure water vapour is driven off. This water vapour condenses on the condensing side, to give off its latent heat. The vapour section is separated from the liquid flow section by a semi-permeable membrane of cellulose. Osmotic pressure therefore develops between the sugar solution and the water produced during condensation. In consequence the water is fed into sugar solution, being thus driven to the evaporator section where water is evaporated once more. The advantage of this system is that osmotic pressure is much greater than the simple capillary pressure which one relies on in normal heat pipes. With osmotic heat pipes it becomes feasible to place the condenser well below the evaporator, which is impossible with capillary heat pipes. On the other hand, flow rates of water through the membrane are low and therefore such heat pipes have a much poorer capacity.

(*d*) **Electro-Osmotic heat pipes.** If one applies an electric field along the level of the heat pipe, the speed of the fluid along the wick is enhanced by the effect of electro-osmosis. It is necessary to use fluids with high dielectric properties. With some designs it has been found possible to pump fluids for distances of upto 500 mm against gravity, which is difficult to achieve with conventional heat pipes relying only on capillarity.

(*e*) **Inverse Thermosiphon.** In these heat pipes the evaporator section is above the condensing section. There is also a tube, which dips into the condensing sump into which a small auxiliary heater is fitted. As the vapour condenses in the lower section, it enters the return tube. The auxiliary heater induces the formation of vapour bubbles in this liquid, so that the density of the liquid/vapour column is now less than the pure liquid which surrounds it. This difference in density helps return the liquid into the evaporator section.

(*f*) **Heat Plates.** These are not heat pipes at all, but plates, in which the meshed wicks alternate with vapour passages. As with normal heat pipes the evaporator section is on one side and the condenser section on the other. Flat heat plate are normally only used horizontally.

(*g*) **Flexibility heat pipes.** In applications where vibration of either the evaporator section or the condenser section is encountered it may be difficult to fit a rigid heat pipe. Heat pipes are developed with flexible bellows to make it possible to transfer heat from one section to the other. The design of a wick able to remain operational under such circumstances presented difficulties, but the problem has now been solved.

(*h*) **The rotating heat pipe.** As has already been pointed out, the governing factor in the design of any heat pipe is the speed with which fluid passes from the condensing section to the evaporator section. The rotating heat pipe is an ingenious way of using centrifugal force to augment gravitational and capillary forces. The rotating heat pipe has the shape of a truncated cone, revolving on its axis, with the evaporator section being wider than the condensing section. As soon as the fluid condenses in the condensing section it becomes subject to centrifugal force and flows back to the evaporator section at a rate proportional to the angular momentum exerted.

It is possible to overcome gravity with such heat pipe designs, and the condensing section can be much lower than the evaporator section. Systems of this type are widely used for cooling rotors in electric motors and generators.

Applications of Heat Pipes

Heat pipes are used as extremely effective heat exchangers for heat energy transfer at thermal levels which vary from the temperature of liquid helium (boiling point 4.55 K) upto more than 2200°C, using liquid silver as a heat transfer medium. Because of the very remarkable efficiency of heat transfer from one end of the heat pipe to the other, and the fact that there is very little difference between the heat supply temperature and the heat take-off temperature, heat pipes are often employed for purposes where normal heat exchangers cannot be used.

An important factor in the management of heat pipe units is tilt control. The system is designed so that units can be rotated in order to vary their inclination. If the evaporator or hot side is tilled so as to lie above the condenser or cold side the heat pipe capacity is modulated or reduced eventually to zero, thereby shutting down the heat exchanger. Till is used as an accurate control of the heat to be transferred.

Reclamation or Recovery of flue gas heat. Very large quantities of sensible heat are normally lost to the environment in the form of exhaust gases from furnaces of all kinds. Waste heat boilers can be heated by gas/liquid heat exchangers in which circulating coils of water are passed through these exhaust gases. The problem with such heat exchangers is the very marked drop in temperature encountered. Because of the need to restrict the heat transfer area to a reasonable fig., it is necessary to circulate the water at a temperature much lower than that of the flue gases. Comparatively low pressure steam is therefore produced in the waste heat boilers. If one employs heat pipes, one can choose a fluid which boils at a temperature only a few degrees below that of the flue gases, and the waste heat can then be given off at almost the same temperature and be reused for such purposes as supply air preheating. As the flue gas has been cooled appreciably, it is now possible to introduce further heat pipe systems with fluids boiling at lower temperatures to recover the lower grade sensible heat as well.

Heat pipe heat exchangers are used in many cases where heat has to be transferred between hot exit gases and cold inlet gases when there is need to avoid mixing the gases, as may happen with rotary heat exchangers, even when purge conditions are good. As in normal heat exchanger practice, the outflow and inlet gases are separated physically.

Heat pipes compete with heat recuperators in this field. They offer the advantages of reduced flow resistance because heat absorber surfaces are far fewer, due to the much improved heat transfer characteristics of the heat pipe system. In general one designs heat pipe heat exchangers systems with an air flow of about 2-4 m/s past the heat exchanger surfaces in order to keep pressure losses past tube bundles fairly low.

Thermal recovery units in an air-conditioning system

These are counter flow air-to-air heat exchanger units with the outside appearance of a normal plate fin chilled water or steam coil. Such a unit, however, differs from a normal heat exchanger in that each tube which connects the two is a heat pipe.

Hot air is passed through one side of the exchanger and cold air through the other in the opposite direction. A sealed partition separates the two air flows to prevent any cross-contamination. It is possible to recover between 60 and 80% of the sensible heat difference between the two air streams. The same system can also be used to save cooling costs in air conditioning systems. It is claimed that in such a case, which gives almost the best return on expenditure, the cost of the heat pipe heat exchanger system can be recovered by cost savings in cooling equipment only. Units of this type are claimed to have the advantages of being free from cross-contamination, more compact per unit of performance, and needing negligible maintenance due to having no moving parts. They can be applied for numerous purposes. The following are suggested:

(*a*) **Use in industrial plant.** Wherever heated gases are exhausted to the outside it is possible to use the sensible heat for preheating inlet air. Due to the absolute separation of the two streams, incoming air is completely clean and pollution free.

(*b*) **Use in public buildings.** Heat recovery units are suggested for hospitals, schools, factories, offices, hotels, departmental stores, auditoria and all other kinds of similar buildings. The exhaust air, which may be laden with cigarette smoke, food odours, germs and viruses is passed through the evaporator section of the heat recovery unit, where its sensible heat is employed in the evaporation of the heat pipe fluid, usually one of the Freon groups of refrigerants. Incoming cold air is blown through a completely separate channel and is heated up by the condensation of this evaporated fluid. In buildings with air conditioning, cold air is recovered in the same way by pre-cooling incoming warm external air in summer.

Considerable savings can be made in both cases in both installation costs and running costs of heating and cooling equipment.

(*c*) **Indoor swimming pools.** Considerable ventilation is required to reduce the moisture content of air. Normally this would be accompanied by appreciable sensible heat losses. Heat pipe heat recovery systems prevent these heat losses.

Industrial heat recovery

Because flow of the source heat medium is completely separate from the supply of air flow, heat pipes can be used to recover heat from highly corrosive, poisonous and otherwise objectionable media in industry. Care must be taken in all cases to reduce corrosion and scaling of the evaporator side, and to see to it that adequate means exist to drain off harmful condensate liquids.

As there is a large range of heat pipe fluids suitable for use in industrial equipment, it is possible to operate within any desired temperature range. Equally, commercial equipment exists to deal with volume flows ranging from 250 l/s upto over 40 m^3/s. The only maintenance normally required is cleaning, and even this can be reduced by adequate filtering of the inlet and exhaust air flows.

List of industrial processes from which heat can be abstracted in order to preheat air used for space heating in other areas:

Paint drying ovens,	Curing ovens
Spray dryers	Forging areas
Boilers	Rubber vulcanizing units
Textile ovens	Plating processes
Desiccant dehumidifiers	Paint spray booths
Bleaching ovens	Foundaries
Brick kilns	Baking ovens
Paper dryers	Timber dryers
Heat treatment areas	Waste steam exhaust
Reverbatory furnaces	Grinding areas
Vinyl ovens	Casting Plant

Process to process heat transfer

Apart from thermal recovery units, heat pipes are also used in industry for more direct purposes.

Indirect heating and cooling systems

In many industrial processes it is necessary to provide heating or cooling in precise positions, at which it is not feasible to place any other heating or cooling devices except heat pipes. As heat pipes can use a variety of fluids, they are capable of universal application. The following are some examples.

(*a*) **Cooling electronic components.** In complex electronic devices heat often has to be liberated which could cause considerable damage if not dissipated. Heat pipe systems, usually of the plate variety, carry this excess heat from the individual components to a heat sink which is then cooled down by a refrigerant circuit. Such systems are very common indeed and constitute possibly one of the most promising future uses for heat pipes. The main components cooled in this way are transistors, integrated circuit packages and transformers.

(*b*) **Supply of heat or cold to moulding machines.** In many metals or plasting casting processes (*e.g.*, injection moulding) it is necessary to supply heat to fairly in accessible parts of the system in order to keep the material fluid. Heat pipes are convenient systems to do this. Equally, in other kinds of casting processes, the liquid metal or plastic is being shaped and must be cooled down rapidly while in the mould. In such cases heat pipes are the most convenient system to carry out such a task.

(*c*) **Improvement in running of engines.** The Stirling engine, an engine in which the gas working fluid is heated externally by coal, refuse and numerous other fairly unpromising fuels, is once more being studied after being first invented back in 1816 by a Scottish Clergyman. It is necessary to transport heat from a single heat source to the various cylinders of the unit and this has now been found to be achieved most easily by a heat pipe. In the case of normal petrol engines experiments are now being made using heat pipes to transfer exhaust heat to the inlet manifold, to obtain better running.

Numerous other fields of use for heat pipes have been suggested by workers in the field, P. Dunn and D.A. Reay, who list the following additional uses:

Solar collectors	Gas turbine regenerators
Air craft brake cooling	Calorimeters
Furnace muffle tubes	Motor cycle oil cooling
Helium target cooling,	
Cryoprobes for surgery	Transformer cooling.

17.10. Stirling Engine

Introduction. A Stirling Engine (SE) is a mechanical device which operates on a closed regenerative thermodynamic cycle, with cycle compression and expansion of the working fluid at different temperature levels and where the flow is controlled by volume changes, so that there is a net conversion of heat to work or *vice-versa.*

A *heat engine* of the reciprocating (piston) type that differs from others of this type (*e.g., Internal-combustion engines*) in having a working gas and a heat source that are independent. Hence in principle, the Stirling engine can be designed to utilize any heat source (*e.g.*, fossil, or nuclear fuel or solar energy) and any convenient working gas (*e.g.,* air).

In internal-combustion (gasoline or diesel) engines, compression and expansion of the working fluid occur in the same chamber (*i.e.,* the cylinder). But in the Stirling engine the working gas is compressed in one region of the engine and is transferred to another region where it is expanded. The expanded gas is then returned to the first region for recompression. The working gas thus moves back and forth in a closed cycle.

A repeated succession of operations (or cycle) representing the idealized behaviour of the working fluid in the *Stirling Engine.* The *Stirling cycle* is illustrated and described in Fig. 17.11. In the description, each stage is assumed to have been completed before the next stage is initiated. However, in an actual engine there is a gradual rather than a sharp transition from one stage to the next; hence, the sharp points in the figure would actually be rounded off. The cycle is composed of 4 heat transfer processes:

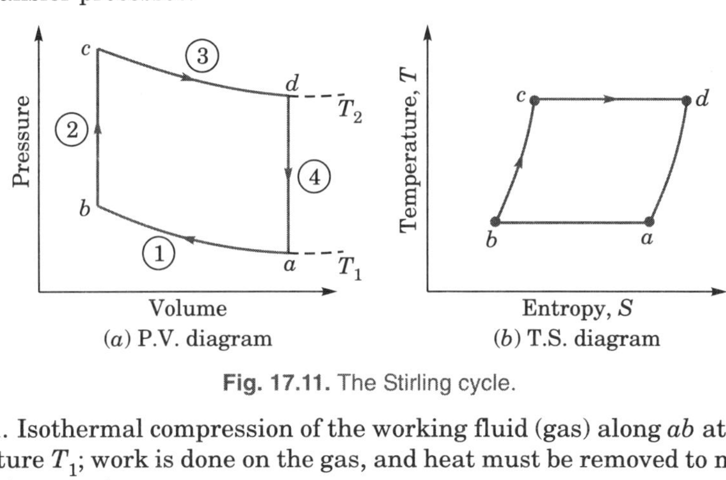

(*a*) P.V. diagram (*b*) T.S. diagram

Fig. 17.11. The Stirling cycle.

1. Isothermal compression of the working fluid (gas) along *ab* at the lower temperature T_1; work is done on the gas, and heat must be removed to maintain a constant temperature.

2. Heat addition to the gas at constant volume along *bc*; the pressure rises, and the temperature increases to the upper temperature T_2.

3. Isothermal expansion of the gas along *cd* at the temperature T_2; mechanical work is done by the gas, and heat must be supplied to maintain the constant temperature.

4. Heat removal (rejection) from the gas at constant volume along *da*; the temperature falls to T_1, and the gas returns to its initial state at *a*. Work is done by the gas when it expands in stage 3 and on the gas when it compressed in stage 1. The thermal efficiency (*i.e.,* the fraction of the heat supplied in stage 3 that is converted into net mechanical work) is increased by increasing the upper working temperature T_2 and/or by decreasing the lower temperature T_1.

The ideal cycle assumes all processes to be thermodynamically reversible. Expansion and compression of fluid are assumed to be isothermal, but this means infinite rates of heat transfer between the cylinder walls and the working fluid. The mass of the working fluid is assumed to be at any particular time in the compression or expansion space. The effects of voids in the regenerator matrix and the cylinder clearance space were neglected. The two pistons were caused to move in some idealized discontinuous way to achieve the prescribed working fluid distribution. Besides this, regenerating was assumed to be perfect.

The basic principle of the Stirling Engine was patented by Robert Stirling, a Scottish minister, in 1816. The air engine as the Stirling engine was called because air was the working gas, was improved and used, especially for pumping water, in the first half of the 18th century. Heat was supplied by burning wood or coal. With the development of steam and particularly internal combustion engines interest in the Stirling engine waned, but it was revived around 1940 at the Philips Research Laboratories in the Netherlands. The original purpose was to develop a reliable engine for generating electricity in remote locations. Subsequently, it was realized that the Stirling engine had the potential for use in road vehicles as well as in many other ways.

General Principles

Several configurations of the Stirling engine have been proposed, but they all have the same basic components: a compression space and an expansion space, with a heater, regenerator, and cooler in between. Heat is supplied to the working gas at a higher temperature by the heater and is rejected at a lower temperature in the cooler. The regenerator provides a means for storing heat deposited by the hot gas during one stage of the cycle and releasing it to heat the cool gas in a subsequent stage.

As in all heat engines, the thermal efficiency (*i.e.,* the proportion of the heat supplied that is converted into useful mechanical work) is increased by supplying heat at the highest possible temperature and rejecting it at the lowest possible temperature.

Stirling engine configurations fall into two general categories: single acting and double acting. In single acting systems there are two piston like moving components which may be in one cylinder or in two connected cylinders. In the simplest single-acting configuration, one of the moving components, called the displacer, serves to transfer the working gas from the expansion space at one end of the cylinder to the compression space at the other end and back again. Only a small amount of work is required to operate the displacer. The other moving component is a conventional working piston. The piston does work on the gas to compress it, and work is done on the picton by the expanding gas. Both displacer and piston may be connected to the same crankshaft (or in another way) so as to cause the displacer to move somewhat ahead of the piston.

The positions of the displacer and piston in the four states (a, b, c, d) of Stirling cycle in Fig. 17.11 are represented in Fig. 17.12. The letters H stand for heater, R for regenerator, and C for cooler. The conditions in these states are as follows:

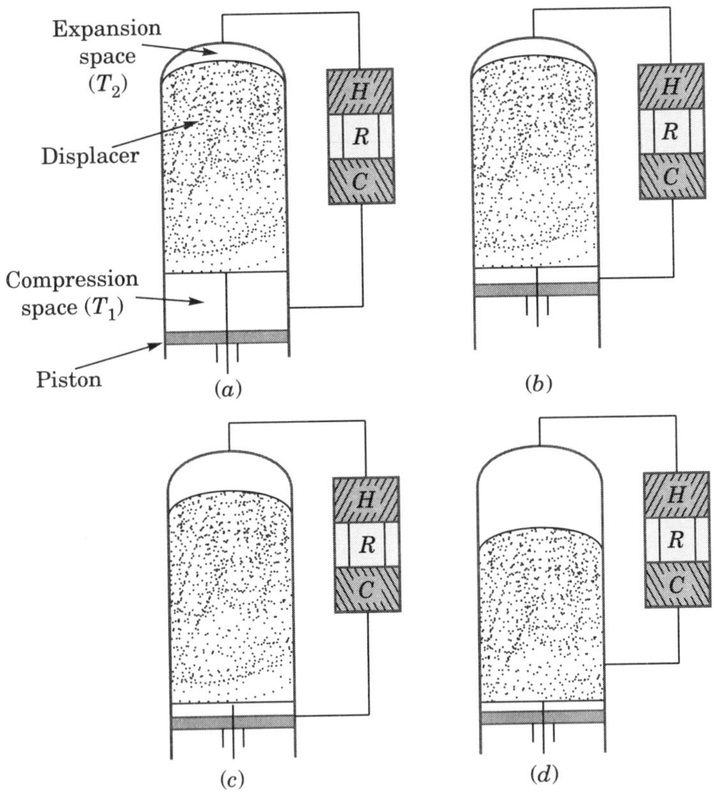

Fig. 17.12. Single acting Stirling engine with displacer and piston.

(a) The working gas is at its maximum volume in the compression space at the lower temperature T_1.

(b) The gas in the compression space has been compressed by movement of the piston, work has been done on the gas. The heat generated by compression is removed by the cooler so that the gas temperature remains essentially constant at T_1.

(c) The compressed gas has been transferred to the expansion space, through the regenerator and heater, by movement of the displacer. Heat is taken up from both regenerator (see below) and heater, and the gas temperature increases to T_2.

(d) The hot gas in the expansion space is expanded at temperature T_2, and work is done by pushing back the piston. Heat supplied by the heater maintains the temperature at T_2.

Subsequently, movement of the displacer transfers the gas back to the compreaaion space, through the regenerator and cooler, to restore the initial state a and thus complete the cycle. Heat from the working gas is deposited in the regenerator and is later recovered when the gas passes from state b to state c.

Double-acting systems have two or more cylinders, each containing a piston which acts as both displacer (at one end) and working piston (at the other end). In a configuration proposed by H. Rinia in the Netherlands in 1946, the compression space in each cylinder is connected, through a cooler, regenerator and heater, to the expansion space in an adjacent cylinder. A simple form of the Rinia arrangement with two cylinders is shown in Fig; (17.13); the in and out motions of the two pistons are half a stroke apart. (A stroke is a complete in or out motion of a piston in a cylinder).

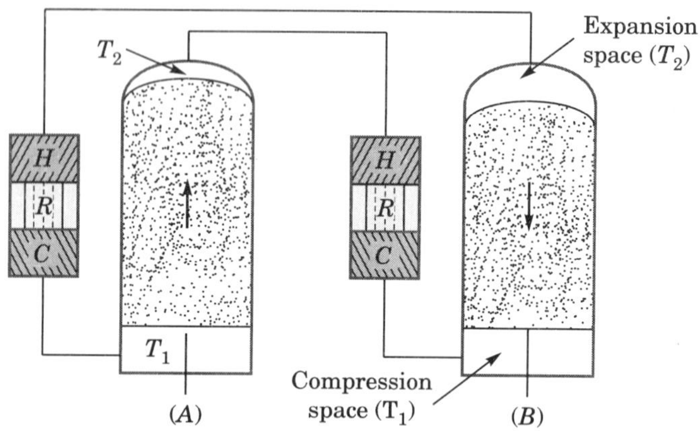

Fig. 17.13. Simple Rinia arrangement of Stirling engine.

There are two separate quantities of working gas; one is contained in the upper part of cylinder *A* and the lower part of *B,* whereas the other is in the lower part of *A* and the upper part of *B.* With this system, two compression and two expansion (or working) strokes occur in each cycle.

Main components of the Engine

There are five main components:

(1) *Heater.* Main function of heater is to transfer heat from outside source to inside working medium during expansion process. Heater may consist of a bank of, closely packed tube, annular type or any other configuration with efficient heat transfer capacity.

(2) *Regenerator.* Heat of the working fluid is stored in one process and given back to the fluid in another process, both processes being at constant volume. Efficiency of regenerator is a critical criteria and has a predominant effect on the engine generally, closed packed fine wiremesh type regenerator are used.

(3) *Coolers.* Rejects heat to outside environment (norrnally cooling water) during the compression process. The cooler is formed from bundles of small tubes surrounded by annulus chamber through which water is circulated.

(4) *Displacer.* The displacer piston is simply a hollow steel shell which fits loosely in the cylinder and its motion is controlled through the piston rod. There is very little pressure difference between the top to bottom of displacer. One end of the displacer operated approximately at the temperature of the heater while the other end at the temperature of the cooler. Therefore the displacer piston walls are made as thin as possible to reduce conduction losses. Radiation shields are introduced to reduce radiant heat transfer from displacer. The movement of the fluid is due to the displacer.

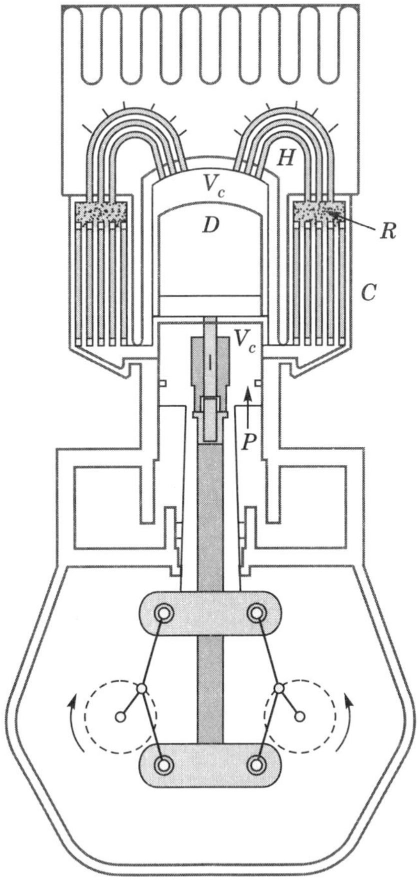

Fig. 17.14. Rhombic drive stirling engine.

(5) *Power Piston.* It provides work of compression to the gas and delivers the work of expansion so it experiences a considerable pressure difference between its upper and lower extremities. It fits more closely into the cylinder and is equipped with suitable rings or other sealing means to prevent gas leakage between piston cylinder walls. The piston operates entirely in the cold zone. The power is delivered by the power piston.

The Practical cycle

In the operation of the engine a significant departure from reality arises as a consequence of the continuous, rather than discontinuous motion of the pistons. This results in a P.V. diagram which is a smooth continuous envelope. Fig. 17.15. Aerodynamic flow losses in the regenerator and other heat exchangers cause differences in the pressure of the working fluid in the compression or expansion spaces. Flow losses are important because they cause a decrease in the area of expansion space in P-V diagrams resulting in a decrease of net cycle operating pressure. The ideal cycle assumption of reversibility of processes, isothermal expansion and compression and perfect regeneration are practically not possible. The effect of void is to decrease engine efficiency. Thus the actual effciency is fraction of Carnot efficiency

$$\left(\eta_{\text{Carnot}} = \frac{T_{\max} - T_{\min}}{T_{\max}}\right)$$

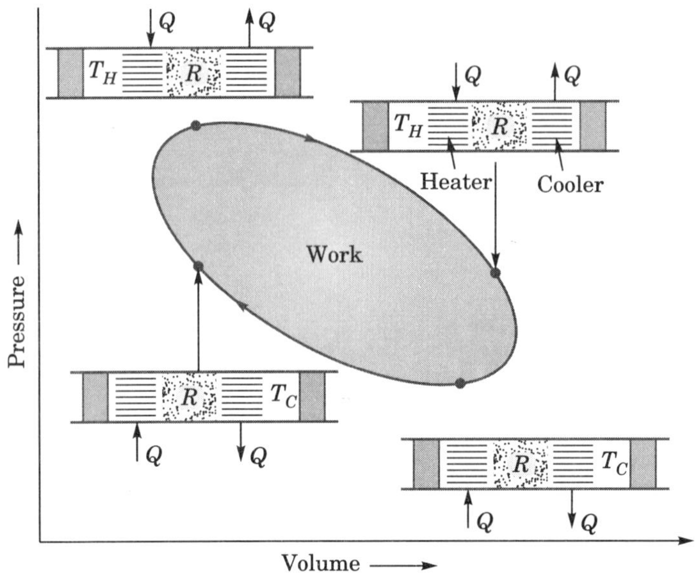

Fig. 17.15. Essential character of Stirling engine.

Working Fluid

The design flexibility of Stirling cycle devices is evident from the wide variety of mechanical configurations that have been developed as well as many differing applications that have been shown to be technically feasible. The choice of working fluid is one option that strongly influences engine design. Hydrogen permits that most compact engine (for a given power output and efficiency) of any gaseous working fluid investigated and therefore has been choice in Stirling development programs directed at automative application where engine size is a major concern. Systems using Helium or Air are presently under development for applications where size is not as important a consideration because the use of such gases leads to substantially larger engines.

Uses of Stirling Engine

1. Automobile Engine. The Stirling engine is attractive for automobiles for the following reasons:

(1) Low air pollution levels that are possible,

(2) Low noise levels because there are no explosions as in internal-combustion engines, and

(3) Ability to use a variety of fuels (*i.e.,* heat sources), such as natural or synthetic gaseous or liquid hydrocarbons, stored solar energy, or even possibly powdered coal. Some of the special features of automative Stirling engines are outlined below.

Heater. Stirling engines of the double acting type are preferred because of their high specific power (*i.e.*, power per unit mass). Four cylinders can be grouped together in a closely spaced arrangement that permits use of a single fuel burner to heat the working gas for all the cylinders. In the heater the gas passes through a number of tubes surrounded by the hot combustion gases produced by the burning fuel. Because the heat source is independent of the working gas, the burner and air supply are designed to ensure essentially complete combustion of the fuel. The residual hydrocarbons, carbon monoxide, and particulates in the exhaust gases can thus be kept at lower levels than with internal combustion engines. Moreover, the flame temperatures are lower in the Stirling engine; hence, the **nitrogen oxides** content of the exhaust is also lower.

The heater tubes, which are in direct contact with the burning fuel, may determine the life of a Stirling engine. In the past, these tubes have been made from an expensive cobalt alloy, but a cheaper material would be necessary for a practical automobile engine. Efforts are under way to develop less costly alloys and refractory materials (*e.g.*, silicon carbide or nitride) for heater tubes as well as for the hot ends of the cylinders.

Cooler. In the cooler, heat is removed by cold water flowing around a bundle of parallel tubes through which the working gas passes. The heat taken up by the water is dissipated to the atmosphere by circulation through a radiator. In order to maintain the thermal efficiency, the water must return to the cooler at the lowest possible temperature.

The radiator in a Stirling engine would be similar in design to a conventional, automobile radiator, but it must be much more efficient in transferring heat to the air. In an internal combustion engine, the cooling water plays no part in the heat engine cycle; heat is rejected with the exhaust gases, and the purpose of the cooling water is to prevent overheating and weakening of the cylinder walls. But in the Stirling engine, heat rejection to the cooling water and its dissipation in the radiator are essential to the operation of the cycle.

The radiator in a Stirling engine would be similar in design to a conventional automobile radiator, but it must be much more efficient in transferring heat to the air. In an internal combustion engine, the cooling water plays no part in the heat engine cycle; heat is rejected with the exhaust gases, and the purpose of the cooling water is to prevent overheating and weakening of the cylinder walls. But in the Stirling engine, heat rejection to the cooling water and its dissipation in the radiator are essential to the operation of the cycle.

Regenerator. The regenerator is usually smaller than either the heating or cooling system. It may consist of a stack of fine-mesh wire screens or a network of metallic strips or wires.

Working gas. Although the nature of the working gas is not material to the Stirling cycle, certain physical properties (*i.e.*, heat capacity, thermal conductivity, density, and viscosity) affect the specific power. Hydrogen gas can best meet the requirements for an automobile engine, but it has some drawbacks. Hydrogen can diffuse through hot metals and can cause embrittlement; furthermore, the gas is flammable and can form an explosive mixture with air. It is thought, however, that these difficulties can probably be overcome. If this not possible, helium will be preferred working gas.

Speed Control. In an internal combustion engine, adjustment of the quantity of fuel admitted to the cylinder results in an intermediate response in the operating speed. Control by fuel adjustment is not practical in a Stirling engine for an automobile, especially because the response would be slow. The method adopted in most experimental engines is based on changes in the pressure of the working gas or in the "dead" volume (*i.e.,* the volume of gas outside the expansion and compression space) or in a combination of both. An increase in pressure or a decrease in the dead volume produces an increase in power output. Another approach to speed control is based on a design which permits variation in the length of the piston stroke.

Start up. Cold startup of a Stirling engine is slower than a gasolme engine. The fuel is ignited by means of an electric spark or a hot wire, and the burner is allowed to operate for several seconds. When the working gas has become hot enough to produce power, the engine is started with a battery-operated electric motor in the same way as in a conventional *automobile.*

Heat Storage. The use of stored heat energy, rather than energy generated by burning fuel, may have advantages for operating for operating Stirling engines in automobiles. The heat would be stored in a molten salt (or salt mixture) contained in a well insulated vessel on the vehicle. Lithium fluoride, which melts at 849°C and has a high heat storage capacity per unit mass, has been proposed for use with a Stirling engine. A mixture of sodium and magnesium fluorides, melting at 832°C, is a cheaper alternative, but it has a somewhat lower storage capacity. Transport of heat from the storage unit to the engine could be achieved by using the *heat pipe* principle with liquid sodium as the transfer medium.

The heat storage unit in a Stirling engine would have to be replaced periodically, possibly at intervals of about 250 km. The heat could be provided by off-peak electric power, by burning coal, or possible solar energy.

2. Stationary Engines. The requirement of a relatively high power output per unit mass for an automobile engine does not apply to a stationary engine; hence, there are fewer constraints on the design of a stationary Stirling engine. For example, such an engine can use helium rather than hydrogen as the working gas; problems of safely and of permeability and embrittlement of metals at high temperature are then avoided. The pressure of the working gas can be lower in a stationary engine, thus permitting a greater choice of materials for the cylinders. There is also more flexibility in the use of fuels and heating methods, especially in conjunction with the heat-pipe principle for heat transport. Finally several mechanical features are simplified in a stationary engine because a compact configuration is not essential, as it is in an automobile.

There are many potential applications for a Stirling engine capable of making efficient use of non-scarce fuels (*e.g.,* coal) or possibly solar energy as the heat source. Stirling engines have been proposed for generating electric power in remote locations; the fuel would then presumably be a hydrocarbon oil. The Stirling engine is also being considered for driving the compressor of a **heat pump** for space heating and cooling. The overall efficiency is expected to be better than when an electric motor is used, as is generally the case.

The Stirling engine can be incorporated into a **total energy system** to meet both the electrical and thermal demands (*i.e.,* for hot water and space heating and cooling) of a building or community. The engine would drive a generator to produce electric power in the usual manner, and the heat normally dissipated to the atmosphere would satisfy the thermal demands.

The major waste heat sources from a Stirling engine are (1) the heat discharged to the water in the cooler and (2) the exhaust gases from combustion of a hydrocarbon fuel in the heater. For simplicity, these waste heat sources are described separately, although various combinations are possible to meet local requirements.

In normal operation, the cooling water temperature of a Stirling engine is maintained as low as possible (*e.g.,* 32°C) to improve the thermal efficiency of the heat-engine cycle. By allowing the temperature to increase to about 60°C, there would be some loss of efficiency, but the water leaving the cooler would be suitable for domestic (and similar) purposes.

Even with an air preheater, the temperature of the combustion exhaust gas leaving a Stirling engine heater would be at least 260°C. By means of a *heat exchanger,* these gases could be used to heat air or water sufficiently for direct space heating. Water near its normal boiling point or low-temperature steam could provide cooling by an absorption system. Process steam might also be generated for industrial use.

Heat Pipe and Stirling Engine Combination

Introduction: Compact power generating systems capable of using the combination of a variety of solid fuels are under evaluation as local power sources. One such system is a *heat pipe* based, Biomass energy driven Stirling Engine. This system is unique in that it integrates three previously unrelated technologies into one system. Until now, the three technologies have been commercially pursued for separate and distinctly different products. By proper integration, an effective general purpose power system has been created.

This system consists of a biomass energy system which includes a cyclone furnace and uses rice husk as fuel, heat pipes and Stirling engine. The heat pipes collect heat from the cyclone furnace and transfer it to the engine's heat exchanger for conversion to shaft power. The Stirling engine as a closed cycle, regenerative machine converts the heat received from the heat pipes into mechanical power by alternatively heating, expanding, cooling and compressing a fixed charge of working gas. The main emphasis is to understand the significance of decentralised power systems in rural areas coupled with the abundance of biomass, agricultural and animal waste, solar energy thus generating technology for economically exploiting the non-fossil sources of energy for agricultural and industrial developments.

Heat pipes. The use of heat pipes enables a number of unconventional energy sources to be utilized by a Stirling engine. Heat pipes can be used with any thermal energy source and if coupled to a Stirling engine would then ensure total multifuel capability. In the future, therefore, with the present fuel problems and the need to be utilize all energy sources, the use of the heat pipe based Stirling engine should become more prevalent and the designers and users of the Stirling engine will need to be familiar with heat pipe technology.

Although there are many forms of heat pipe, the basic principle of operation is the same for all *i.e.*, the use of an entrapped phase change material to transport heat. The rudimentary structure of a heat pipe in conventional form is already shown in Fig. 17.16. It consists essentially of a closed metal tube whose linear surfaces are lined with a porous material such as a fine wire gauge, which promotes capillary action during operation. This porous material is usually referred to a *"WICK"*. A liquid is contained within the tube by saturating the *WICK* and the remaining internal volume is filled with the vapour phase of the liquid. One end of the tube is called the evaporator and the other end, the condenser. Heat is applied at the evaporator end, thus evaporating the liquid; the resulting vapour is transported because of the resulting pressure differential to the condenser end of the heat pipe where it 'condenses' and releases its enthalpy of vaporization. The vapour thus returns to its liquid form and is transported back to the evaporator end by capillary action along the wick. In same instances this flow back is aided by the altitude of the heat pipe, *i.e.,* gravity assisted.

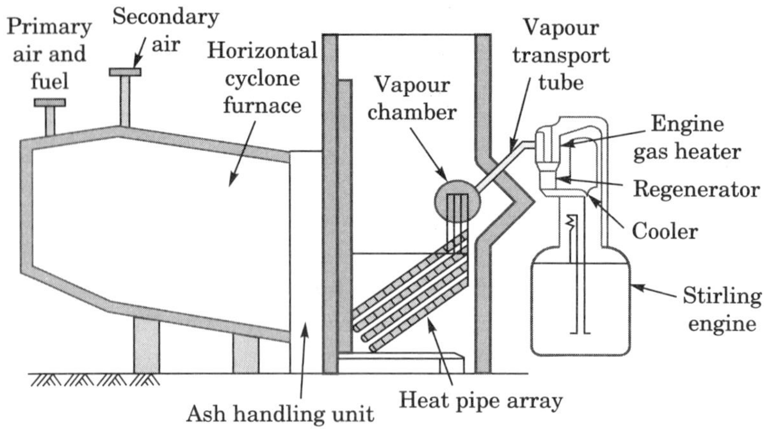

Fig. 17.16. Cyclone furnace/Stirling engine/Heat pipe power system.

Design of the heat pipe heat transfer system is complicated by the need for a concentrated heat transfer system at the engine heater head and a distributed heat collection system speed throughout the cyclone furnace. In order to achieve this area change and at the same time insure an even distribution of the working fluid through repeated start-up and shut-down, a two stage heating system may be deviced consisting of an array of 20 heat pipes extending diagonally through the furnace and feeding into a receiver. This receiver located within the main furnace assembly serves as the condenser region for the heat pipe array and as the evaporator of a single heat pipe which connected to the engine heater head. The upper end of the array heat pipes are wicked on both the interior and exterior to provide this dual function. The heat pipes may be designed to operate in a gravity assist mode with heat throughput limited by entrainment of the sodium liquid in the counter flowing vapour stream.

Fuel Handling System

A fuel handling system supplies thermal energy input to the Stirling engine. Fuel is burnt in the system and heat is supplied to the cylinder externally. The system consists of following components and units:

1. Fuel storage unit
2. Fuel conveyor unit
3. Air handling unit
4. Combustion unit
5. Ash handling unit
6. Air preheating unit.

1. Fuel storage unit. Fuel is stored in a hopper. Once it is charged it should supply fuel to the furnace for several hours. The size of the hopper should be sufficiently big.

2. Fuel conveyor unit. Unit has to be designed to convey fuel from hopper to the furnace. This is essential because sufficient distant is kept between the furnace and the hopper. Pneumatic conveyance ensures proper mixture of air and fuel. High inlet velocity at the duct creats turbulence and facilitates the combustion process. This is important because silica in husk prevents good combustion. So husk requires higher quantities of excess air.

3. Air Handling Unit. Correct amount of air has to be supplied for proper combustion as there exists optimum air-fuel combination for maximum efficiency. Air handling unit is responsible for supplying required quantity of air and forcing it through various components of the system.

4. Combustion Unit. This is the heart of the system. The husk-air mixture is admitted in the furnace and burnt. Heat is generated. Since rice husk is a low quality fuel, the efficiency of the furnace should be high enough to extract maximum possible heat out of it. Burning of husk in fluidized or suspension state is reported to give more complete combustion.

5. Ash Handling Unit. Fly ash can be removed, by putting obstacles in the path of combustion gases. Fly ash strikes against these and falls in the ashpit below from where it can be removed manually or mechanically.

6. Air Preheating Unit. Heat from exhaust gases can be used by incorporating air preheater. The temperature of inlet air can be increased to about 300°C. This facilitates the combustion process and helps in attaining higher temperature.

Stirling engines have a bright future in India with viable, solar and biomass energy propositions. The agricultural waste can be used to increase the quality of life in the rural sector. This will also reduce the outflow of foreign exchange for petroleum products. The system discussed above is an example of a decentralised power system using waste rice husk to drive a rice mill. This example can be extended to suit various other applications.

17.11. Heat Pumps

In general, a device operating in a manner opposite to a *heat engine*; it takes up heat at a lower temperature and by utilizing mechanical work, discharges heat at a higher temperature. More specifically, the term heat pump is applied to a machine that can be used for space heating in the winter and for cooling (air-conditioning) in the summer. The most common type of heat pump for domestic use is the air to air system in which heat is both taken up from and transferred to air. Water-to-air and water-to-water heat pumps are also available, and air-to-water heat pumps has been designed for domestic hot-water heating.

Operating Principles

In essence, a heat pump is a vapour-compression type refrigerator, consisting of an evaporator unit (where latent heat is taken up) and a condenser unit (where heat is discharged), with a mechanical compressor between them. One of the units is installed in the interior of the building and is connected to the air-distribution system; the other unit is outdoors. In the winter, the interior unit becomes the condenser and discharges heat into the building; heat is taken up from the outside air by the evaporator. In the summer, the situation is reversed. The indoor unit becomes the evaporator and removes heat from the interior air, thereby cooling it, heat is discharged to the surroundings from the outdoor (condenser) unit.

The operation of a heat pump for space heating and cooling may be understood from Figs. 17.17, which show heating and cooling modes respectively. The components are the same in both cases, but the refrigerant (working fluid) flows in opposite directions. The change in operation mode, from winter to summer (and *vice versa),* is achieved by a reversing valve and a combination of expansion and one-way (check) valves. A blower (or fan) draws room air across the coils in the inside (right-hand) unit to provide space heating in winter or cooling in summer.

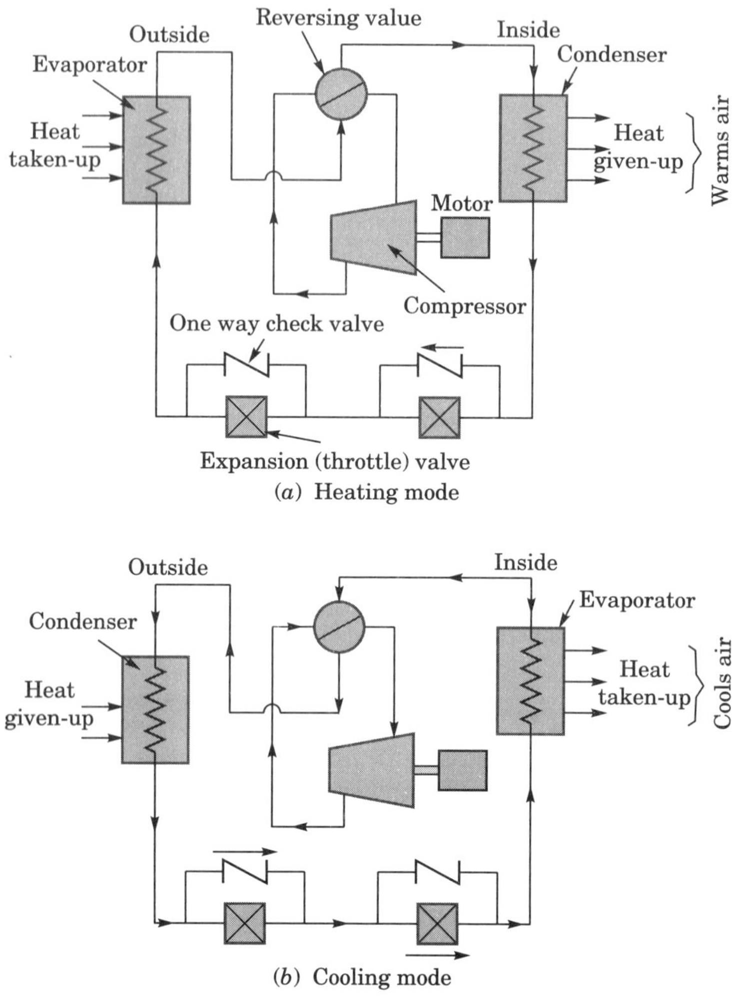

Fig. 17.17. Heat pump.

Hence the heat pump uses the refrigeration cycle to extract heat from air or water by using a fluid refrigerant which is at a lower temperature. A compressor is used to pump the refrigerant round a circuit, as shown in the Fig. 17.17. Heat is removed from air or water by the refrigerant as it flows through the evaporator coil. The absorption of heat by the refrigerant changes its state from a cold liquid/ vapour mixture to a cool gas which is then drawn into a compressor in which its pressure and temperature are increased. This increase in temperatures allows heat to be transferred from a condenser coil direct to air or water. The cooling effect of the condenser converts the gas to a hot liquid still under high pressure and by means of a thermal expansion valve into a low pressure mixture of cold liquid and vapour. The cycle is then ready to be repeated.

The heart of the heat pump is the compressor. This can be driven either by an electric motor, by wind or water power, or by a petrol, gas or diesel engine. A refrigerant is in the form of a gas when it flows into the compressor. The work done in compression increases its density and it also heats up appreciably. The hot compressed gas then flows into the condenser, which is an efficient heat exchanger, where the excess heat is removed. As the gas condenses to form a liquid, its latent heat is used to heat up either water or air via the heat exchanger.

The nature of the refrigerant is of some importance. It is essential to avoid fluids where the temperature at which heat is taken off is above the fluid's critical point. The critical point is defined as the temperature above which a gas cannot be liquified by pressure alone. For example, a normal heat pump could not use either oxygen or nitrogen as a refrigerant medium because the temperature of the hot section is very much higher than the critical point of either of these two gases.

For optimum performance, it is desirable to take off as much heat as possible in the condenser/heat exchanger. The liquid which has now had its excess heat removed next flows to the expansion section, where it evaporates. As this evaporation takes place, a refrigerant effect is produced due to the need for a liquid to absorb latent heat as it vaporizes.

Heat pumps have been found particularly suitable under the following circumstances:

(*i*) when cheap electricity is available.

(*ii*) when natural gas is available.

(*iii*) where refrigeration energy is used.

(*iv*) where very cheap low grade heat is available.

(*v*) using unconventional fuels.

(*vi*) for space heating/cooling in countries with appreciable differences between external winter and summer temperatures.

Applications of Heat Pumps

1. **As a domestic heat pumps.** Normal day time electricity tends to cost up to three times as much as natural gas per unit heat output, and thus the heat pump has to have a COP of atleast 3, to compete in running costs with a simple gas fired central heating system. As a heat pump is more expensive to install than a direct gas burner, it is not viable under such circumstances.

Domestic heat pumps can however be justified on the following grounds:

(*a*) If the purpose of the heat pump is dual *i.e.,* in winter it provides space heating, using the ground or external air as a heat sink, while in summer it provides room cooling. It is best to combine the heat pump with an energy storage device to obviate the use of costly daytime electricity.

(*b*) While in urban areas gas is piped to most dwellings, this is not the case in rural areas. Electricity, is usually for cheaper to distribute than gas. Coal and oil have some disadvantages for space heating. The former is dirty and laborious to handle, while suitable heating oils are costly. In such a case heat pumps are direct competitors with normal electric heating; as a bonus one obtains a high heat output caused by the COP, as well as refrigeration capacity. Heat pumps are therefore of particular values to the farming community.

(*c*) Solar heaters have the disadvantage that heat is being collected part of the year and part of the day only and is necessarily low grade. Provided solar heating is combined with a suitable energy store, the low grade energy held in it can be used to improve the COP of a heat pump very markedly. Under such circumstances a good return can be obtained even when an appreciable fraction of the electricity used is charged at daytime rates.

(*d*) A system of domestic heating where heat pumps are particularly viable is one where heat has to be recovered from the exhaust air. One of the best ways of doing so is to use the exhaust heat exchanger to supply law grade heat to the evaporator section.

(*e*) Heat pumps are used in all types of experimental schemes. Prime mover energy comes from sources such as water wheels, windmills and the like. Such schemes tend to idealistic self sufficiency exercises rather than practical applications. There are however occasions when they are economically viable.

(*f*) Heat pumps can be made viable when right-time electricity is being sold at a considerable discount. Under such circumstances water can be heated at night using cheap electricity and a favourable COP. This is stored in well insulated tanks. The hot water can be used for space heating and consumption purposes on the following day.

Although at the present moment it is difficult to justify installing heat pumps in dwellings except under the conditions mentioned above, the long term future for domestic heat pumps is quite excellent. Full air-conditioning, which necessitates air chilling for part of the year atleast, is already the standard in the United State, parts of Canada, parts of Europe and Japan. Heat pumps are infinitely more economical than directly fired room heaters for winter and separately installed window air conditioning units for summer.

2. **Industrial and commercial applications of Heat Pump**

Industrial uses of heat pumps can be categorized with two types:

(*a*) Where it is necessary to install refrigeration equipment, as in food stores, ice cream factories, and the like. Under such circumferences the refrigerated area serves as the supplier of the low grade heat for the evaporation section of the heat pump. The condensation heat is used for space heating and process heat. The coefficient of performance of such a system can be raised by using the sources of waste heat to raise the temperature of the refrigerant after it leaves the refrigeration area. Such a system is infinitely preferable to the traditional system where waste heat is simply dissipated in a cooling tower.

(*b*) For all purposes where it is necessary to provide process heat at a higher temperature than that of the waste heat available. The heat pump in this case recovers thermal energy present in liquid and gaseous effluents and converts it into high grade heat for drying evaporation, boiling and space heating. Although it is necessary to expend some high grade electric or other energy for driving the compressor, the advantageous COP obtained makes such a process an economic proposition.

Heat pumps are widely used for drying. At the condensing side of the system air is heated sufficiently to drive moisture, out of the damp materials to be dried. The moist air is then used as a heat sink on the evaporator section of the heat pump. Water condensers and drains off. The air which has had its moisture removed is then recycled to the warm part; where it is heated once more, to absorb further quantities of water from the medium to be dried.

In the various food, chemical and pharmaceutical industries it is necessary to remove tonne of water each year by evaporation. Drying by heat pump is far more efficient than conventional drying. In conventional drying one cannot recirculate the air which has carried out the job of drying the medium, as the air is saturated with moisture.

It has to be released damp to the atmosphere. With heat pump drying, the water and sensible heat content of the damp air are both removed and the air is then recycled for further drying.

Many liquid *effluents* not only carry away with them valuable sensible heat, but constitute an environmental hazard because they are warm. Discharging warm liquids into rivers is often forbidden because raising water to temperatures by even a few degrees can kill off fish. Yet it is often inapplicable to remove the heat from such effluent flows by using simple heat exchangers as there is little need inside the works for low grade heat.

Heat pumps are most useful in this context because they can cool down the warm effluents to a temperature at which they are harmless. At the same time, with the expenditure of quite a modest amount of prime energy, the thermal level of the extracted heat can be raised sufficiently for this heat to be useful for process heating.

17.12. Energy Storage

Whenever we produce heat or power by non-traditional methods (*i.e.*, those which do not involve the combustion of fossil fuels), we face the problem of matching demands with supply. Nuclear power stations operate at a virtually steady output day and night, summer and winter and cannot easily be switched off or turned down as can be done with coal and oil fired stations. Solar power is at a maximum during midday in summer: at periods when demand for electricity is quite low and heat demand is virtually zero. Tidal power, wave power, wind power and the like also peak at varying times which do not coincide with periods of maximum energy demand.

Increasing therefore, it will become necessary to build systems of energy storage into our energy economy. These aim to store heat or electric power during periods of high supply but low demand, such as at night and during the summer, and release this energy once more at periods of maximum demand.

The main methods of energy storage used are the following:

- Sensible heat,
- Latent heat.
- Chemical,
- Electrochemical,
- Compressed air,
- Fly wheels,
- Pumped water etc.

These methods have already been discussed, in chapter "Energy Storage".

17.13. Renewable Energy Sources/Devices

Renewable energy sources could in theory provide 10-13 TW annually equal to current global energy consumption. Today they provide about 2 TW annually, about 21% of the energy consumed world-wide of which 15% is biomass and 6% hydropower. Application of renewable energy resources, *e.g.*, solar-thermal and photovoltaic, biogas, wind energy should be encouraged because it will not only save conventional energy sources but will also protect the environment from pollution.

Solar Thermal Energy

For decades to come the sun will grow in importance as one of the most attractive alternatives to fossil fuels as a measure source of energy to mankind. The solar radiation on the earth surface, though eratic due to cloud cover, discontinuous due to the night and seasons and diffuse but enough to provide energy for industrial processes, commercial building, private homes, electricity generation and many other end uses. Now solar heating and cooling systems have been commercialised. Large size solar thermal power plants (100 MW) are in the developing stages. Several different systems are being studied for both **central** and **distributed** approaches and many solar collector designs are being explored.

Solar Photo Voltaics (SPV) Systems

In SPV systems, solar energy is converted directly into electricity, which will go a long way in helping rural electrification programme in India. A serious obstacle to be overcome is the high initial capital cost, compared to conventional power. In the systems developed in the country the cost is about ₹ 100 per peak Watt. This has to be reduced by a factor of three to four through break through in technology, if SPV is to become economically viable over a wide range of applications. A wide ranging programme of R & D in the area of SPVs has been initiated with a view to reducing costs, improving production technologies. A production base for photovoltaic systems has been created in two Public Sector Undertakings, Central Electronics Limited and BHEL. Solar pumps for drinking water supply and irrigation and photovoltaic street lighting in villages were initiated for demonstration and extension programme under a National Solar Photo Voltaics Energy Development Programme (NASPEDP).

Wind Energy

Wind pumps for irrigation and drinking water have also been developed in India. Against 1385 targeted wind pumping mills about 1000 have been installed under demonstration programme by the end of the 7th plan. However, wind energy potential is site-specific. R&D in wind energy on superior pump systems suitable for low wind speed and deep well pumping and the development of indigenous wind generation system have to be geared up for reducing the dependency on the hydro thermal electric power in rural areas. Small wind battery charging have also reached a degree of technological maturity, hence considerable technological development is still required for large scale wind-electric generation in the country.

Biomass and Biogas. Biomass is a major energy source today for rural population of our country. There is a plenty of biomass available throughout the world which can be utilized for energy demand of rural people for domestic activities, if managed properly. Biomass can be converted into liquid, solid and gaseous fuels through biological, thermal and hydrolytic conversion processes. Thus alcohol, biogas and producer gas fuels can be produced by enzymatic action, anaerobic fermentation and gasification.

The solid fuel from biological waste can be prepared through *pyrolysis* and briqueting process. The biogas can be used as the fuel for illumination, cooking and shaft power. The biogas slurry is a good source of enriched manure. *Biogasification* also reduces pollution. There are 15 lakhs plants in our country producing 60 million cu m biogas per day.

Biogas plant works as a factory for the production of fuel and manure. Biogas is very good fuel for rural domestic kitchen, illumination and shaft power. Biogas stoves used for cooking give a thermal efficiency of about 40% as against 10% for the conventional chullahs. Biogas lamps are used for illumination in the vicinity of the gas plants. Biogas can also be utilized in the commercial diesel engines. It can substitute upto 75% of diesel oil requirement. On the average biogas is consumed at a rate of approximately 200 l/hr in a petromax, 400 l/hr in a biogas stove and 600 l/bhp-hr in a diesel engine working in dual fuel mode. The digested slurry discharged from the biogas plants is rich in plant nutrients and also adds humus to the soil, since human and animal faeces and other agrowaste materials are anaerobically fermented inside the biogas plants, it helps in maintaining cleanliness within the village and also in reducing environment pollution. It has been found that energy conservation in anaerobic digestion of one tonne of cattle dung as compared to FYM preparation and dung cake preparation for domestic fuel worked out to be 197445 kcal or 148126 kcal respectively.

Micro-hydel Systems. Power from micro-hydel projects which can provide renewable energy of rural needs in a decentralised manner remains largely untapped in the country so far, especially from small streams and canal droppings. During the 6th plan efforts to set-up the necessary R and D base were initiated by central Government funding for developing technology for tapping low head and low-capacity micro-hydel potential. R and D efforts should also aim at developing central systems, micro-processors and civil engineering technology for low and ultra-low head micro hydel systems.

Solar Cookers. Solar cooker can be another appliance for cooking in which there is no smoke, no soot. It saves money and fuel, preserves the nutritional value of the food. Of course, it has got limitation of taking more time to cook but an ideal appliance for rural labour and rural population who can keep the food in this cooker while working in the field and take food during lunch time.

The use of improved chulhas which may reduce deforestration, meet cooking energy needs, save fuel and reduce health hazards, while cooking. It is estimated that by using 50 lakhs of chulhas, saving of 48 lacs tonnes of firewood every year can be achieved.

A major constraint in large-scale utilization of renewable energy system is their high initial cost. The costs of solar, wind or biomass devices are considerably higher than that of installing alternative based on conventional energy sources. This becomes a deterrant to most potential users for renewable energy systems in our country where capital is relatively scarce and individual earnings are not high especially in rural areas. The low level of production, in turn inhibits cost reduction that may to some extent be achieved through a large volume of production. The problem of high cost of renewable energy systems, has to overcome through intensive efforts of technology development, materials and manufacturing innovation. However, fiscal and other economic measures like loans from bank on low interest and subsidies, would also have an important role in the creation of demand of renewable energy technologies. In this context, pricing of conventional energy to the users does not always reflect the real cost of introducing and distributing them. A comprehensive system of energy pricing thus needs to be developed which takes into account the subsidies for different sources of energy, conventional as well as non-conventional and renewable sources of energy. Again for popularising their uses in large scale, practical sized demonstrated programmes should be undertaken in a large scale.

17.14. Instrumentation and Control

Efficiency in the use of primary energy can be achieved and maintained only by the careful control of the conditions under which it is converted to produce heat or mechanical/electrical energy. To control any physical condition a primary requirement is that its value be known; to obtain this information appropriate measurements are required. These include measurements of pressure, temperature, rate of flow, density and gas composition.

Associated with these basic measurements are the transmission of these values over a distance, distance monitoring and recording instruments, and remote control of the process.

The rapid expansion of instrumentation in recent years has led to the appearance of a wide range of measuring equipment for sensiny and registering the value of virtually all physical variables encountered in industry.

The control and instrumentation of a waste incineration plant (with or without energy recovery) must be designed, executed and operated in a manner which provides complete safety and achieves efficient and smokeless combustion. An automatically stoked incinerator is usually safer and more efficient than a manually stoked plant.

Control systems should not be over-specified or be excessively elaborate. The most effective control system is usually the simplest one, provided that each plant function is adequately controlled and sequenced. Certain control malfunctions cannot be readily detected without an appropriate instrument indication or recording.

An installation for measuring and controlling the value of a physical variable can be generally be divided between (*a*) the detecting element-which senses the value of the variable under consideration, (*b*) the measuring element-which registers this value in appropriate units, and (*c*) the control system-which permits the process to be manually or automatically operated and held to prescribed conditions.

The fundamental need to obtain to good basic measurements cannot be over-emphasized and the importance of selecting the best transducer for the job is only paralleled by the necessity to install the transducer to comply with the manufacturer's design criteria.

Although a very wide range of variables may be measured on process and production plant, these can be reduced to a small number for the great a majority installations. Among the most common are:

- Temperature
- Pressure
- Density
- Displacement

These measurements are supplemented by gas composition analysers of various types:

- Carbon dioxide
- Carbon monoxide
- Oxygen
- Sulphur oxides
- Nitrogen oxides
- Smoke
- Grit and dust.

General Techniques of Measurement

- Performance specifications
- Measurement techniques
- Temperature measurement
- Flow-measuring systems
- Density measurement
- Smoke density measurement.

Automatic Control systems

- Automatic values
- Analogue control systems
- Digital system applications
- Metering pump-speed control system
- Interfacing pneumatic controller with digital computers.

Special Measurements used in fuel technology

- Temperature
- Total heat flow
- Combustion
- Gas-borne particle in flue gases
- Chimney emissions
- Fire side fouling and corrosion
- Water and steam purity.

Appendix A

Table A-1: Common Measurements

1 sq. m	= 10.76 sq. ft.,
1 m	= 3.28 feet
1 cu ft.	= 28 litres = 0.028 m^3
1 gallon	= 4.55 litres
1 litre	= 0.22 gallons
1 m^3	= 220 gallons
1,000 litres	= 1 cu m
1 cu ft	= 6023 gallons
1 gallon of water weighs 10 lbs	
1 litre of water weighs 1 kg	
1 lb	= 0.454 kg
1 kg	= 2.2 lbs
2,240 lbs	= 1 ton = 1.02 tonnes
1000 kg	= 1 tonne = 0.984 ton
1 part per million	= 1 milli gram/litre
	= 1 gm/m^3
1% = 100,000 p pm	= 10000 m gm/litre
	= 10 gm/litre
1 cu ft/lb	= 0.062 m^3/kg
1 m^3/kg	= 16.1 cu ft/lb
1 lb/cu ft	= 16.2 kg/m^3
1 kg/m^3	= 0.062 lb/cu ft
1 acre	= 4,840 sq. yards
	= 0.405 hectares
1 hectare	= 10,000 sq metres
	= 2.47 acres.

Pressure

1 in. of water	= 0.25 m bar
1 lb/sq in. (psi)	= 68.95 m bar
1 atmosphere	= 1.013 m bar
1 British thermal unit (BTU)	= 0.252 kcals
1 BTU	= 1,055 Joules
1 kcal	= 3.97 BTU
1 Joule	= 9.5 × 10^{-4} BTU
1 kcal	= 4.19 kJ
1,00,000 BTU	= 1 Therm
	= 29.3 kilowatt hours
1 BTU/cu ft	= 0.038 J/cm^3 (or mega joules/m^3)
1 J/cm^3	= 27.0 BTU/cu ft
1 BTU/lb	= 2,320 J/kg
1 J/kg	= 4.29 × 10^{-3} Btu/lb
1 BTU/hr	= 0.0011 mJ/hr

Table A-2: Conversion of Units

Energy:	1 kcal	= 4.187 kJ
	1 kWh	= 860 kcal
	1 Wh	= 0.860 kcal
	1 kcal	= 426.99 kg m
		= 1.163 × 10^{-3} kWhr
	1 h.p.	= 632.34 kcal = 0.736 kWhr.
Pressure:	1 atm (technical)	= 735.6 mm Hg
		= 1 kg/sq cm
	1 ata (physical)	= 760 mm Hg
		= 1.034 kg/sq cm
	1 kgf/cm^2	= 9.81 N/cm^2

Force: 1 kgf = 9.81 N

Specific gravity: 1 kgf/m^3 = 9.81 N/m^3

Density: 1 kgf. sec^2/cm^4 = 9.81 kg/m^3

Dynamic viscosity: 1 kgf. sec/cm^2 = 9.81 N/m^2

Heat capacity: 1 kcal/kg°C = 4.187 kJ/kg°K

Enthalpy: 1 kcal/kg = 4.187 kJ/kg

Heat flux: 1 kcal/h = 1.163 W

Rate of heat flow: 1 kcal/m^{2}h = 1.163 W/m^2

Volume rate of heat flow: 1 kcal/m^3.h = 1.163 W/m^3

Thermal conductivity: 1 kcal/m.hr.°C = 1.163 W/m°K

Heat transfer coefficient: 1 kcal/m^2.hr.°C = 1.163 W/m^2 °K

1 BTU/ft^2/°F/hr. = 20.44 kJ/m^2/°C/hr.

Radiation coefficient:

1 kcal/m^2.h.°K = 1.163 W/m^2.k^4.

Table A-3: Approximate Total Normal Emissivities of Various Surfaces at 20°C

Surface	*Emissivity*
Aluminium, heavily oxidised	0.20—0.30
Aluminium polished	0.09
Asbestos sheet	0.93—0.96
Brass polished	0.06
Brick (rough)	0.93
Carbon, lamp black	0.96
Copper, heavily oxidised	0.75
Copper, polished	0.03
Enamel, white fused on iron	0.90
Iron oxidized	0.74
Paint, aluminium	0.27—0.67
Paint, flat back locquer	0.97
Sheet polished	0.07
Sheet plate rough	0.95
Tin bright	0.06
Water	0.95

Table A-4: Absorptivity of Various Surfaces for Sun-rays

Surface	*Absorptivity*
Aluminium polished	0.26
Asphalt	0.89
Black paint	0.97—0.99
Brick red	0.70—0.77
Copper polished	0.26
Galvanised iron (new)	0.66
Galvanised iron directly	0.89
Iron oxidised, rustly	0.74
Iron polished	0.45
Paper, white	0.27
Tile, red and brown	0.65—0.74
White paint	0.12—0.26

Appendix B

Solar Radiation Data for Some Location in India (MJ m^{-2} day^{-1})

Location		Months											
		Jan.	*Feb.*	*March*	*April*	*May*	*June*	*July*	*Aug.*	*Sept.*	*Oct.*	*Nov.*	*Dec.*
1. Ahmedabad tilt = 30°	H_b	18.72	20.01	19.50	17.21	15.70	9.11	3.96	4.35	11.11	18.38	19.80	19.83
	H_d	4.85	5.58	6.18	7.35	7.32	10.15	11.38	10.53	8.48	5.50	4.09	4.31
2. Kolkata tilt = 30°	H_b	13.68	14.86	14.19	12.45	10.25	5.07	4.15	4.82	6.81	10.88	14.28	14.80
	H_d	5.87	6.50	7.59	8.80	9.82	10.20	10.10	9.58	8.80	7.19	5.76	5.30
3. Chennai tilt = 20°	H_b	14.69	18.71	18.52	15.99	11.94	8.49	8.94	8.42	10.82	10.17	9.67	10.14
	H_d	7.01	6.28	6.50	7.12	8.53	9.58	10.27	10.03	8.90	8.02	7.49	7.56
4. New Delhi tilt = 40°	H_b	16.15	17.65	17.42	14.99	12.14	7.72	6.10	7.18	12.62	17.80	19.07	16.96
	H_d	4.94	5.61	6.52	7.74	8.72	10.48	9.31	8.38	7.11	5.34	4.42	4.60
5. Poona tilt = 30°	H_b	19.93	21.52	19.40	16.47	13.89	6.99	2.94	3.29	8.16	15.74	19.26	18.74
	H_d	4.31	4.45	5.51	6.67	6.98	10.27	11.04	11.33	9.74	6.24	4.58	4.47
6. Visakha-patnam tilt = 30°	H_b	20.65	20.00	17.25	14.41	11.63	5.65	4.36	6.11	8.74	13.53	17.34	18.32
	H_d	5.46	5.43	6.36	7.18	8.10	9.30	9.80	9.82	8.65	6.62	5.79	5.34

Appendix C

Table of Thermodynamic Properties

(C—1) Common properties of some materials (in SI units)

Material	*Density* kg/m^3	*Thermal conductivity* $(W/m°C)$
Air at normal pressure 20°C	1.204	0.0260
Expanded polystyrene at 10°C	16.0	0.0346
Foam polyurethane, rigid	24.0	0.0245
Mineral, batts, at –2°C	32.0	0.0346
Foam PVC rigid at 21°C	60.0	0.0360
Expanded ebonite at 10°C	64.0	0.0303
Cork board, dry at 18°C	144.0	0.0418
Sheets of asbestos cement at 30°C	150	0.319
Pine fibre board at 24°C	256	0.519
Pine 15% wet	570	0.138
Oak 14% wet	770	0.160
Dry plaster, gypsum at 23°C	881.0	0.170
Ice at –1°C	918.0	2.26
Water at 4°C	1000.0	0.596
Ammonia, 20°C	617.0	0.570
Typical building concrete	2400.0	1.73
Standard glass	2515.0	1.05
Aluminium	2675.0	211.0
Steel	7850	47.6
Copper	8795	385.0
Brass	8600	85.429
Iron	7220	62.764
Lead	11400.0	34.869
Aluminium foil (50°C)	20.0	0.047
Asbestos fibre (50°C)	470.0	0.110
Brick masonry (20°C)	800 to 1500	0.23 to 0.29
Cellotex (20°C)	215	0.047
Chalk (50°C)	2000	0.093
Coal (20°C)	1400	0.186

Cork sheet (30°C)	190	0.042
Fire brick (100°C)	550	0.140
Leather (30°C)	1000	0.159
Marble (90°C)	2700	1.302
Mica	290	0.581
Mineral wool (50°C)	200	0.047
Plaster (20°C)	1680	0.779
Rubber (0°C)	1200	0.163
Sand dry (20°C)	1500	0.325
Sand wet (20°C)	1650	1.127
Soil dry	1500	0.138
Soil wet	1700	0.657

Specific heat (kJ/kg°C)

Water, at 20°C and normal pressure	4.19
Ice, at – 21°C to 1°C	2.10
Steam (C_p) at 100°C and normal pressure	1.95
Air (C_p) at 20°C and normal pressure	1.012
Concrete at 18°C	0.837
Ammonia liquid (20°C)	4.731
Asbestos fibre (50°C)	0.816
Chalk (50°C)	0.879
Coal (20°C)	1.424
Concrete (20°C)	1.130
Cork sheet (20°C)	1.884
Glass (20°C)	0.670
Ice (0°C)	2.261
Marble	0.419
Mica	0.879
Mineral wool (50°C)	0.921
Rubber (0°C)	1.382
Aluminium (0°C)	0.921
Brass (0°C)	0.877
Copper (0°C)	0.381
Iron (20°C)	0.502
Lead (0°C)	0.1298

Heat of vaporization (kJ/kg)

Water, at 20°C	2454.0
Water, at 100°C	2257.0
R_{12}, at 0°C sat.	151.5
R_{22}, at at 0°C sat.	205.4
R_{11}, at 0°C sat.	188.9
R_{500}, at 0°C sat.	183.0
R_{717}, at 0°C sat.	1263.3

(C—2) Common constants

NTP is normal temperature T_0 of 273.15 K = 0°C and normal pressure P_0 of 101.325 kpa.

g_0 is standard gravity 9.80665 m/s^2.

Velocity of sound in air, 344 m/s at P_0, 20°C and 50% *RH*.

(C—3) Gas constants

R_u = 8314.4 J/mole K (universal)

R_a = 287.045 J/kg K (air)

Boltzmann's constants 1.380 × 10^{-23} J/°K

Faraday's constant 9.65 × 10^{7} C/kg mole

Planck's constant 6.626 × 10^{-34} Js

Stefan-Boltzmann constant 5.67 × 10^{-8} J/m^2 s°K^4

Volume of 1 kg mole of ideal gas at NTP 22.42 m^3.

Appendix D

D–1 Properties of Saturated Water at 298°K in SI Units

Molecular weight	M	18.02
Gas constant	R	461.4 J/kg °K
Density	ρ	999 kg/m^3
Specific heat	S	4181 J/kg °K
Dynamic viscosity	μ	9.239×10^{-4} kg/m-s
Kinematic viscosity	ν	9.248×10^{-7} m^2/s
Prandtle number	P_r	6.395
Thermal conductivity	K	0.604 W/m °K

D–2 Properties of Air at Standard Conditions in SI Units

Molecular weight	M	28.97
Gas constant	R	287 J/kg °K
Specific heat at constant pressure	C_P	1005 J/kg °K
Specific heat at constant volume	C_v	718 J/kg °K
Density	ρ	1.185 kg/m^3
Dynamic viscosity	μ	1.84×10^{-5} kg/m-s
Kinematic viscosity	ν	1.55×10^{-5} m^2/s
Thermal conductivity	K	0.0257 W/m °K
Ratio of specific heats $\gamma(C_p/C_v)$	γ	1.3997
Prandtle number	P_r	0.720

Appendix E

Table E—1: Physical properties of dry air at atmospheric pressure

Temperature T	*Density* ρ	*Specific heat* C_p	*Kinematic viscosity* $\nu = \mu/\rho$	*Thermal divisibility* K_d	*Thermal conductivity* K	*Prandtle number* $Pr = \frac{\mu C_p}{K}$
°C	kg/m^3	$10^3 J/kg\,K$	$10^{-6}\,m^2/s$	$10^{-6}\,m^2/s$	$10^{-2}\,W/m\,K$	—
0	1.30	1.01	13.3	18.4	2.41	0.72
20	1.20	1.01	15.1	20.8	2.57	
40	1.13	1.01	16.9	23.8	2.72	
60	1.06	1.01	18.8	26.9	2.88	0.70
80	1.00	1.01	20.8	29.9	3.02	
100	0.94	1.01	23.0	32.8	3.18	0.69
200	0.75	1.02	34.6	50.0	3.85	0.68
300	0.62	1.05	48.1	69	4.50	
500	0.45	1.09	78	115	5.64	
1000	0.28	1.18	174	271	7.6	0.64

Table E—2: Physical properties of water (at moderate pressure)

Temperature T	*Density* ρ	*Kinematic viscosity* $\nu = \frac{\mu}{\rho}$	*Thermal diffusivity* K_d	*Thermal conductivity* K	*Prandtl number* P_r
0°C	$10^3\,kg/m^3$	$10^{-6}\,m^2/sec$	$10^{-6}\,m^2/s$	W/mK	-
0	0.9998*	1.79	0.131	0.55	13.7
20	0.9982	1.01	0.143	0.60	7.0
40	0.9922	0.66	0.151	0.63	4.34
60	0.9832	0.48	0.155	0.65	3.07
80	0.9718	0.37	0.164	0.67	2.23
100	0.9584	0.30	0.168	0,68	1.76

*The maximum density of water occurs at 3.98°C and is 1000.0 kg/m^3.

Table E—3: Water Vapour in Air

Temperature *T* *(°C)*	*(Saturated) vapour pressure* *Pv* *(kN/m²)*	*Mass of H_2O in 1 m³ of saturated air* *X* *(g/m³)*
0	0.61	4.8
10	1.23	9.4
20	2.34	17.3
30	4.24	30.3
40	7.38	51.2
50	12.34	82.9
60	19.9	130.0
70	31.2	197.0
80	47.4	291.0
90	70.1	
100	101.3	

Note: $X = (2.17 \times 10^{-3}$ kg Km²/N) p_v / T.

Other properties of water:

Specific heat capacity $C = 4.2$ kJ/kg K (0°C < T < 100°C)

Latent heat of freezing $L_f = 334$ kJ/kg

Latent heat of vaporization $L_v = 2.45$ MJ/kg (20°C)

$= 2.26$ MJ/kg (100°C)

Surface tension (against air) = 0.073 N/m (20°C).

Table E—4: Emittance of Common Surfaces

Material	*Temperature* (°C)	*Emittance* ∈ %
Aluminium		
Polished	100	9.5
Roughly polished	100	18
Iron, roughly polished	100	17
Tungston filament	1500	33
Brick (rough, red)	0—90	93
Concrete (rough)	35	94
Glass (smooth)	25	94
Wood (oak, planed)	90	90

Table E—5: Calorific values of various fuels

Fuel	*Gross calorific value**		*Remarks*
	MJ/kg	*MJ/lit. at 15°C*	
Crops			
Wood			Varies more with moisture content than species of wood
Green	– 8	– 6	
Seasonal	– 13	– 10	
Oven dry	– 16	– 12	
Vegetation : dry	– 15		Examples : grasses, hay
Crop residues			For dry material.
Rice husk			In practice residues may be very wet.
Bagasse (sugar cane solids)	12—15		
Cow dung			
Peat			
Secondary bio fuels			
Ethanol	30	2.5	C_2H_5OH : 789 kg/m^3
Methanol	23	1.8	CH_3OH
Biogas	28	20×10^{-3}	50% methane + 50% CO_2
Produces gas	5—10	$(4 - 8)\ 10^{-3}$	
Char coal :			Depends on composition
Solid pieces	32	11	
Powder	32	20	
Coconut oil	39	36	
Cocohol	39	33	Ethyl esters of coconut oil
Fossil Fuels			
Methane	55	38×10^{-3}	Natural gas
Petrol	47	34	Motor spirit, gasoline
Kerosene	46	37	
Diesoline	46	38	Automotive distillate, derv.
Crude oil	44	35	
Coal	27		Black coking grade

*Gross Calorific value (also called heat of combustion) is the heat evolved in a reaction of the type

$$CH_2O + O_2 \longrightarrow CO_2(\text{gas}) + H_2O(\text{liquid}).$$

Some authors quote the *net calorific value* which is the heat evolved when the final H_2O is gaseous.

Appendix F

Temperature Milli—volt Relation for Various Thermocouples (Reference Junction at 0°C)

Temperature 0°C	*Copper constantan*	*Iron constantan*	*Chromel Ni/Cr*	*Platinum 10% Rhodium*
– 190	—	—	– 5.60	
– 180	– 5.211		– 5.43	
– 170	– 5.021		– 5.24	
– 160	– 4.818		– 5.03	
– 150	– 4.603		– 4.81	
– 140	– 4.374		– 4.58	
– 130	– 4.134		– 4.32	
– 120	– 3.881		– 4.06	
– 110	– 3.617		– 3.78	
– 100	– 3.341		– 3.49	
– 90	– 3.053		– 3.19	
– 80	– 2.755		– 2.87	
– 70	– 2.446		– 2.54	
– 60	– 2.126		– 2.20	
– 50	– 1.796		– 1.86	
– 40	– 1.456		– 1.52	
– 30	– 1.106		– 1.14	
– 20	– 0.746		– 0.77	
–10	– 0.378		0.39	
0	0.000	0.000	0.000	0.000
10	0.388	0.52	0.40	0.055
20	0.783	1.04	0.80	0.112
30	1.188	1.56	1.20	0.172
40	1.600	2.08	1.61	0.233
50	2.021	2.60	2.02	0.297
60	2.449	3.14	2.43	0.363
70	2.885	3.67	2.85	0.430
80	3.329	4.21	3.26	0.499
90	3.780	4.74	3.68	0.570
100	4.239	5.28	4.10	0.642
110	4.704	5.83	4.15	0.716
120	5.176	6.38	4.92	0.791
130	5.654	6.92	5.33	0.867
140	6.140	7.47	5.73	0.945
150	6.631	8.02	6.13	1.024
160	7.129	8.57	6.53	1.104
170	7.632	9.12	6.93	1.185
180	8.142	9.68	7.33	1.268
190	8.657	10.23	7.73	1.351
200	9.178	10.78	8.13	1.435

Temperature 0°C	*Copper constantan*	*Iron constantan*	*Chromel Ni/Cr*	*Platinum 10% Rhodium*
220	10.236	11.88	8.94	1.605
240	11.315	12.99	9.75	1.778
260	12.413	14.09	10.57	1.954
280	13.531	15.20	11.39	2.133
300	14.666	16.30	12.21	2.314
320	15.819	17.40	13.04	2.498
340	16.989	18.51	13.88	2.683
360	18.175	19.61	14.71	2.871
380	19.376	20.72	15.55	3.059
400	19.982	21.27	15.98	3.154
420		24.58	18.51	3.730
450		27.39	20.65	4.220
500		30.25	22.78	4.419
550		33.16	24.91	5.226
600		36.13	27.03	5.742
650		39.19	29.14	6.265
700		43.32	31.23	6.798
750		45.48	33.30	7.339
800		48.05	35.34	7.888
850		51.82	37.36	8.446
900		54.99	39.35	9.014
1000		58.16	41.31	9.591
1050		61.33	43.25	10.172
1100		64.50	45.16	10.757
1200			48.89	11.946
1300			52.46	13.155
1400			for 1370°C	14.371
1500			*mv* = 54.88	15.580
1600				16.770
1700				17.992
1760				18.592

Renewable Energy Status in India

Solar Technology		
Solar cookers	3,00,000	(MNES 1993)
Solar water heaters	19,000	(MNES 1993)
Solar PV lighting systems	40,000	(MNES 1993)
PV water pumps	741	(MNES 1992)
PV power units	316.9 kW	(MNES 1992)
Wind Technology		
Wind generators	316 (77 MW)	(MNES 1993)
Wind pumps	2,922	(MNES 1992)
Biomass Technology		
Domestic biogas plants	1.7 million	(MNES 1993)
Institutional biogas plants	1,000	(MNES 1993)
Biomass gasifiers/sterling engines	1,040 engines	(MNES 1992)

Index

A

A.C. commutation generator, 226
Abbot silver, 48
Absorber, 52
Absorption air conditioning, 126
— of solar radiation, 32
— refrigeration, 125
Actionometer, 49
Active solar heating, 124
Active space heating system, 120
Aero dynamic force, 218
Aero foils, 176, 209
Aerodynamic lift, 204
Aerogenerators, 176
AFPRO, 266
Agricultural waste, 292
Agriculture and Industrial process heat, 149
Air craft, 491
Air heater, 59
Air heating system, 5, 9
Air mass, 40
Air pollution, 297, 644
Albitade angle, 43
Algae, 278
Alternative liquid fuels, 318
Anaerobic digestion, 247, 251
Analysis of thermionic generator, 559
Anderson cycle, 395
Angle of incidence, 42
Aquifers, 533
Arrays of solar cells, 141
ASTRA, 266
Atmospheric transmission factor, 41
Attached sun space, 292
Axial flow turbine, 356
Azimuth of solar collector, 35

B

Bacteria, 252
Basic battery theory, 450
— diode, 532
— hot air system, 122
— hot water system, 121
— photovoltaic system, 145
— principle for wind machine, 177
— thermionic generator, 557
Batch transport, 625
Batoh plant, 260
Batteries for bulk energy storage, 462
Batteries, 450
Battery storage, 102, 147
Beam radiation, 54, 401
Bio chemical conversion, 238
Bio fouling in OTEC system, 384
Bio fuels, 242
Bio gasification, 303
Bio logical storage, 621
Bio-fouly, 384
Bio-gas generation, 251
Biogas utilization, 279
Bio-gas, 19, 241
Bio-gas-plants, 259
Biological energy conversion, 314
Biomass and bio-conversions, 241
Biomass conversion, 247
Biomass gasification, 301
Biomass pyrolysis, 300
Biomass, 246
Biophotolysis, 482
Bird life, 237
— anaerobic, 252
— aerobic, 252
Blade design, 208
Blades and thrust on turbines , 184,
Boltzmann's constant, 557
Box type solar oven, 159
Bulb turbine, 419

C

Capital costs and economics concentrating collector, 80
Carnot efficiency, 377
Casuarina, 310
Central receiver system, 130
— receivers collector, 84
Chemical energy sources, 229, 427
— heat-pipe, 627
— storage, 103
Chill factor, 256
Chinese digester, 275
Chlorine electrode cells, 459
Classification of batteries, 454
Claude cycle, 379
Climate variation, 365
Closed cycle for MHD, 510
— cycle for OTEC, 381
Coal gasification plants, 480
Coal, 4
Co-current-and counter-current heat exchanges, 648
Co-generation, 635
Cold fusion, 590, 603
Collective and heat transport, 123
Collector efficiency, 71
Collectors, 51
Combined cycle power plant, 646
Combined cycle with heat recovery boiler, 639
Community biogas plants, 277
Comparison of flashed steam, 353

Compound parabolic concentrator, 85
Compressed air storage, 105, 403, 597
Conductive losses, 71
Convective loop, 120
— Concentrating, 70
— Flat plate, 50
— Porous absorbers, 66
Conventional energy sources, 3
Conversion efficiency of fuel cells, 437
— process, 321
Convective losses, 71
Cross draught classifiers, 305
Cut out speed of WECS, 207
Cylindrical parabolic collector, 87

D

Dam, 392
Darrius rotor, 215
Daylength, 45
Declination, 48
Deen bandhu biogas plant, 268
Definitions of fundamental quantities, 451
Desalination, 111
Desilting tank, 417
Diffuse radiation, definition of, 47
Digester, 252
Digester design, 280
Direct (solar) radiation, 41
Direct gair, 118
Direct heat applications, 232
Distillation (solar), 150
Distribution of energy, 622
Diversion structure, 417
Dolphin machine, 409
Dome, 261
Doping of semi conductor, 148
Double basin arrangement, 397
— cycle system, 400
— flash cycle, 345
Double output Induction generator, 225
Drag forces, 210
Dry rock (geothermal), 351
D-T reaction, 589
Dual medium, 615
Duct wall, 525

E

Efficiency
— of solar cell, 150
Electrical filter press, 469
— generation application, 233
— resistivity, 367
— storage, 101, 601
— tank type, 468
Electricity transmission, 624
Electro chemical process, 102
Electro magnetic energy, 106
Electro osmotic heat pipes, 657
Electrodes, 524
Electrolyser, 367
Electrolysis, 467
Electrolyte, 170
Electromagnetic interforce, 237
End-walls, 165
Energy audit, 630
— break even condition, 572
— consumption, 1
— density, 600
— efficiency, 450
— from biomass, 241
— from the ocean, 375
— from tides, 387
— plantation advantages, 261
— plantation, 20, 295
— production, 587
— sources, 22
— storage systems, 594
— storage via fly wheels, 589
— storage, 228, 593
Enzymatic hydrologies, 252
Epply pyranometer, 48, 49
Estimation of average solar radiation, 51
Ethanol fermentation, 301
Ethanol production, 320
Eucalyptus plantation, 297
Evaporative cooling, 168
— of battery, 100
— photo synthesis, 251
Extinction coefficient
Extraction thermal energy, 109
— of glass, 69

F

Farming of energy, 296
Fermentation system, 325
Fermentation, 247
Fermi energy, 142
Fermilevel semiconductors, 142
Fertilizer from biogas digester, 295
Field coils, 524
Figure of merit, 543
Fill factor of solar cell, 152
Filter-press electrolyzer, 469
Fin efficiency, 76
Firm generating, 234
Fixed dome type plant, 262
Flame traps, 498
Flashed steam system, 343
Flat plate collector, 59
Floating drum plant, 261
— tank, 408
Fluidised bed combustion boilers, 11
Fluidised bed gasifier, 307
Fluidization, 9
Fly wheel storage, 105
Forebay, 417
Fossial and nuclear fuels, 464
Fossil fuel methods, 477
Fouling factor, 382

Fresnel lens, 82
Fuel cells, 20, 427
— Applications, 448
— Biogas plants, 262
— Classification, 431
— Hydrazin, 427
— Hydro carbon, 427
— Hydrogen, 430
— Ion exchange, 431
— Methanol, 435
— Molten carbonate, 433
— Solid electrolyte, 435
Fuel handly system, 670
Fuel properness of biogas, 290
Fusion energy, 567
Fusion hybrid, 589

G

Gas and gas heat exchangers, 651
Gas conductivity, 519
Gas holder, 286
Gas, 5
Gasification of biomass, 249
Gasifiers, 301
— applications, 307
— classification, 303
Gates and locks, 393
Generator control, 227
Generators (electrical), 224
— Induction, 225
— Double output induction, 226
— Glazing, 166
— Synchronous, 225
Geopressured resources, 337, 350
Geothermal energy applications, 361
— energy, 22, 333
— exploration, 366
— well drilling, 367
Gibbs free energy, 5
Glass, 62
— as cover on solar collectors, 66
— transmittance, 78
Glauber's salt, 100, 619
Global green house, 163
Gobar gas (bio gas), 219
Gothic arch green house, 163
Green house effect, 62
Green house environment and control, 166
Ground reflectance, 62

H

Heat distribution, 125
Heat exchangers, 383
Heat flow rate, 367
Heat for bio-mass conversion, 111
Heat losses, 77
Heat pipe, 625
Heat-plates, 657
Heat pump, 668
Heat storage, 124, 166
Heating and cooling building, 110
Heating animal housing, 111
Heat-pipe and stirling engine combustion, 609
— pipes, 653
— pump, 671
— recuperators, 647
— regenerators, 652
— transport system, 63
Helical rotor expander, 359
Heliostats, 133
Hero's Turbine, 357
Hetrojunction solar cells, 141
High temperature system, 133
— production, 451
— storage, 485
— transportation, 488
— utilization, 489
Homojunction solar cells, 141
Horizontal axis machines, 197
Hot air system, 123
Hot dry rock resources, 337, 351
Humidity, 164
Hybrid cycle, 385
— system, 354, 442
Hydraulic accumulator, 450
Hydride as portable-energy storage, 486
Hydrogen energy, 23
— fuel cells, 430
— production, 166
— storage, 104
— transportation, 488
Hydrolysis, 301
Hydrothermal sources, 338
Hyper thermal field, 336

I

Impulse machine, 361
Impulse turbine, 356
Incident angle, 44
Indian research on cold fusion, 591
Induction generator, 224
Industrial heat-recovery, 659
— process heat, 111
— waste, 364
Inertial confinement fusion, 581
Infrared radiation, 58
Insolation, 39
Instrumentation and control, 677
Interconnected system, 235
Intermittent absorption cooling, 129
Inverters, 147
Iodine sulphur cycle, 476
Irrigation water pumping, 137
Isodynes, 187
Isovents, 186

J

Janta biogas plant, 267
Jar digester, 266
Joul effect, 535
Junction p-n, 141

K

Kachra gas plant, 276
Kaplan turbine, 403
Keller rotor, 358
Kipp, 50
KVIC, 266

L

Langley, 52
Laser beam fusion, 583, 585
Laser fusion reactor, 583
Latent heat storage, 100
Latent-heat-energy storage, 617
Latitude, 42
Lead acid battery, 455, 603
Lift and drag, 186
Lift force, 217
Ligno cellulose, 247
Line focusing collectors, 80
Line packing, 485
Liquid dominated systems, 316, 343, 350
Liquid metal system, 512
Liquid storage, 485
Liquification, 249
Lithium bromide,
Load control, 228
Local solar time, 45
Longitude, 49

M

Magma resources, 338
Magnetic confinement fusion, 574
— fields, 526, 579
— mirror problems, 581
Magneto hydrodynamics, 25
Magnus effect, 187
Material for channel, 525
Materials, 169
Maximum power, 181
Mechanical energy storage, 105, 595
Medium temperature system, 133
Methanol, 325
Methods of production of hydrogen, 169
MHD
— Design and development, 513
— Generator, 25, 503
— Principle, 503
— Systems, 507, 524
Micro hydropower, 414
Mini and micro hydel project, 416
Mini hydel power, 414
Mirror, 137
Mirror-strip reflector, 82
MIS solar cell, 142
Modulus of solar cell, 143
Moulter carbonate cells, 433
Multiple disk turbine, 358
Muon catalyst fusion, 586

N

Natural circulation, 115
Neap tides, 390
Nickel iron cell, 457
Nodding duck, 410
Non conventional sources, 12
Non-porous absorber plate, 64
n-type semiconductor, 141
Nuclear fission, 565
— fusion, 565
— option, 9
— power, 7
Number of blades, 208

O

Ocean kelp, 279
— thermal energy conversion, 20
— tides, 386
— waves, 404
Oil drum digester, 414
Oil, 5
Open cycle systems, 508
Open cycle, MHD systems, 442
Operational and environmental problems, 368
OTEC, 375
Ozone, 40

P

Packed bed exchange storage, 99
Parabolic trough concentrator, 81
Parameters for plant-growth, 164
Passive heating system, 117
— Activation, 439
— Concentration, 439
— Resistance, 439
Passive solar device, 118
Pathogen reduction, 254
Pebble bed storage, 615
Peltier effect, 532
Penstock, 418
Performance of wind machines, 220
Petrothermal systems, 351
Phase change storage, 99
Photoelectric emission, 556
Photoelectrolysis, 483
Photolytic, 171
Photosynthesis, 249
Photosynthetic efficiency, 249, 251
Photovoltaic cell, 112
Photovoltaic effect, 113
Physics of muonic catalysis, 586

Plasma confinement, 573
— heating, 578
p-n junction solar cell, 142
Point focussing collector, 83
Polarization,
Poloidal field, 579
Poltery kilns, 629
Ponds, solar, 106
Power in the wind, 178
— coefficient, 180
— duration curve for wind, 180
Practical cycle, 665
Pragati biogas plant, 269
Principal of working, 154
Production of power, 111
Production requirement, 281
p-type semiconductor, 140
Pumping application, 230
Pyranometers, 48
Pyrheliometer, 47
Pyrolysis, 300

R

Radial out flow, 357
Radiation losses, 71
Radiation, 37, 38
Range of tides, 390
Rate of feeding, 258
Reaction turbine, 356
Receiver pipe, 83
Receiver sub system, 135
Reduction and oxidation process, 250
Reflection at interfaces, 68
Regenerator, 667
Renewable energy prospects, 29
— energy sources, 27
— energy, 28
Retention time, 281
Reversible chemical reactions, 606
Richardson dushman equation, 557
Richardson dushman, 557
Rim type turbines, 394
Road vehicles, 490
Rock (hot), 351
Roof storage, 119
Rotor, 204

S

Safety and management, 493
Safety systems, 236
Salt gradient pond, 107
Savonius rotor, 213
Scattering of radiation, 40
Scattering, 40
Schottky junction, 140
Seeback effect, 531
Seeback thermosetting, 332
Seeding, 257, 523
Seismic monitory, 367
Selection of materials, 548
Selective absorber coatings, 94
Surface for solar absorber, 98
Semiconductor junctions, 141
Semiconductors, 140
Sensible heat storage, 98, 612
Short term action, 632
Silicon solar cells, 140
Simplified analysis, 520
Site selection, 385
Slope, 44
Small hydel development, 413
Small scale hydroelectric, 44
Sodium sulphur cell, 461
Solar azimuth angle, 411
Solar cell principles, 139
— cell array, 146
— cell modules, 146
— collectors, 57
— concentrators, 134
— constant, 37
— cooking, 156
— declination, 23
— distillation, 15
— distribution, 150
— energy collectors, 57
— energy, 13
— storage, 96
— furnaces, 154
— green houses, 160
— heating, 117
— photovoltaic arrays, 138
— pond, 106
— power tower, 152
— production of hydrogen, 169
— pumping, 152
— radiation data, 37
— radiation geometry, 42
— radiation, 37
— still, 158
— thermal electrical conversion, 130
— thermal energy, 675
— water heating, 114
Solid oxide electrolyte cells, 434
Sorghums, 298
Space cooling, 126
Space heating systems, 118
Space heating, 117
Spill way, 418
Spring tides, 390
Stationary engines, 668
Stirling engine, 661
Storage batteries, 102, 601
— of energy, 592
— tank, 98
Summer green houses, 163
Sun at-zenith, 40
Sunshine recorder, 50
Superconducting materials, 106, 611

Superconductivity,526
Surface azimuth angle, 28
Synchronous generator, 224
Synthesis gas, 325
System with battery storage, 234

T

Tandem mirror system, 581
Taper digester, 264
Thermal analysis, 74
Thermal efficiency, 506
— electric conversion, 131
— energy storage, 612
— ionization, 522
— reactor, 8
Thermal storage wall, 119
Thermionic converter, 26
Thermionic emission effect, 553
— chemical methods, 170
— chemical reaction, 309
— electric power generator, 538
— electric power, 26, 531
— nuclear fusion reactor, 574
— nuclear fusion, 567
— siphon system, 574
— chemical cyclic processes, 473
— chemical reactions, 300
Thomson coefficient, 535
Thomson-effect, 535
Thwaits, 187
Tidal energy double basin arrangement, 396
— single basin arrangement, 394
— energy, 21
Tidal plants components, 391
Tidal power generation, 403
Tidal power, 389
Tidal range, 390
Tides, 387
Tip speed ratio, 205
To kamak systems, 576
Topping cycle, 553
Torque coefficient, 207
Total energy system, 668
— flow concept, 347
— radiation, 55
Tower concept, 127
Toxicity, 259
Transmission control, 198, 227
Transmissivity, 54
Transmittance absorptance product, 70
Tube turbine, 419
Turbine tower system, 212
Types of co-generation, 636
Types of electrodes, 441
Typical air collector, 63
Typical liquid collector, 60
Total induction generator, 227

U

Ultraviolet bana, 52
Utility grid, 223
Utilization of hydrogen gas, 489

V

Vapour absorption system, 16
Vapour dominated system, 337
Vector diagram of forces, 219
Ventilation, 166
Vertical axis wind machines, 212
Voltage factor of solar cell, 150

W

Warn water in agriculture, 641
Waste storage, 614
Water chemistry, 367
Water conductor system, 417
Water Power, 6
— heaters, 127
— hyacinth, 278
— storage, 99
Wave energy conversion by floats, 408
Wave energy, 229
Westing house electrochemical thermal sulphur cycle, 473
Wet sprocess, 247
Wet world energy futures, 2
Wick design, 656
Wind energy collector, 201
— Principles of wind energy connection, 177
Wind energy, 175, 229, 676
— data, 189
— energy collectors, 17, 201
— energy conversion systems, 185, 196, 407
— maps, 197
— power, 197
— stream variation, 210
— surveys, 191
— function, 553
— mille heat, 198
— turbines, 176
— velocity, 236
Winter green house, 163
Work Heat utilization, 638
Work out put and EMF of fuel cell, 441
Working fluid, 666
World renewable energy resources, 30

Y

Yaw control, 210
Yield of crops, 305
YSZ, 524

Z

Zenith angle, 44
Zirconia, 524